AF388888

Dietary, Lifestyle and Children Health

Dietary, Lifestyle and Children Health

Editor

Zhiyong Zou

Basel • Beijing • Wuhan • Barcelona • Belgrade • Novi Sad • Cluj • Manchester

Editor
Zhiyong Zou
Peking University
Beijing, China

Editorial Office
MDPI
St. Alban-Anlage 66
4052 Basel, Switzerland

This is a reprint of articles from the Special Issue published online in the open access journal *Nutrients* (ISSN 2072-6643) (available at: https://www.mdpi.com/journal/nutrients/special_issues/Dietary_Lifestyle_Children).

For citation purposes, cite each article independently as indicated on the article page online and as indicated below:

Lastname, A.A.; Lastname, B.B. Article Title. *Journal Name* **Year**, *Volume Number*, Page Range.

ISBN 978-3-0365-8854-4 (Hbk)
ISBN 978-3-0365-8855-1 (PDF)
doi.org/10.3390/books978-3-0365-8855-1

© 2023 by the authors. Articles in this book are Open Access and distributed under the Creative Commons Attribution (CC BY) license. The book as a whole is distributed by MDPI under the terms and conditions of the Creative Commons Attribution-NonCommercial-NoDerivs (CC BY-NC-ND) license.

Contents

Preface

Childhood is a critical period for the development of a healthy lifestyle and the prevention of chronic diseases in adulthood. However, rapid changes in lifestyle and dietary behaviors have significantly increased the risk of diet-related chronic diseases in children. In recent years, the prevalence of diseases such as obesity and hypertension has been increasing and unhealthy lifestyles are becoming an epidemic, posing a potential future burden of adult chronic disease. Therefore, it is important to assess children's health behaviors and reduce the risk of chronic non-communicable diseases in early adulthood. This book contains studies on dietary and lifestyle behavioral factors associated with healthy child development.

This book is dedicated to increasing and updating our knowledge of the association between dietary quality/diversity and childhood diseases. It is focused on food variety, such as a high fruit and vegetable intake, and the importance of healthy diet patterns in preventing chronic diseases in children. The book also explores the impact of other lifestyles, such as digital screen time, physical activity, and sedentary behavior on children's metabolic diseases. Finally, to evaluate the lifestyle of children, including nutrition, this book provided various tools and methods which could help with the assessment of children's health. It provides the foundation of scientific knowledge for the interpretation and evaluation of future advances in child nutrition and healthy development.

Readers interested in a broad perspective of dietary factors and the involvement of healthy lifestyle factors in the prevention and management of childhood non-communicable chronic diseases will find this book interesting as it explores state-of-the-art discoveries and new research directions.

We thank the authors for submitting their manuscripts and for returning their revisions in time. We also thank the reviewers for their hard work in publishing this book. We also received great support from the Editorial Office of *Nutrients*, which substantially reduced the workload of the Guest Editors. The combined efforts of everyone made this book possible.

Zhiyong Zou
Editor

Editorial

Dietary, Lifestyle, and Children Health

Xiaoran Yu and Zhiyong Zou *

Institute of Child and Adolescent Health, School of Public Health, Peking University, Beijing 100191, China; yuxr0811@outlook.com
* Correspondence: harveyzou2002@bjmu.edu.cn

Childhood is a critical period for the development of a healthy lifestyle and the prevention of chronic diseases in adulthood. However, the prevalence of childhood obesity is increasing and unhealthy lifestyles are becoming an epidemic, posing a potential future burden of adult chronic disease. Therefore, in order to assess children's health behaviors and to reduce the risk of chronic non-communicable diseases in early adulthood, this Special Issue of *Nutrients*, "Dietary, Lifestyle, and Children Health", features 28 original studies and 3 systematic reviews on dietary or other lifestyle behavioral factors about children or adolescents. These findings may help to broaden our knowledge and to open up new research directions.

Dietary quality and diversity are as essential to human health as air is to human life. The study by Callum Regan [1] indicated that higher diet diversity and increased fruit and vegetable consumption could be a strategy to improve health-related quality of life among adolescents. Three papers in this Special Issue explored the relationship between vegetable and fruit intake, and childhood illness. These studies reported that moderate fruit consumption was associated with lower odds of lipid disorders [2] and that high fruit and vegetable intake (FVI) could reduce asthma-related illness [3]. However, insufficient FVI and low potassium intake aggravate early renal damage in children [4]. Cereals are a major source of dietary energy and a good source of many micronutrients and dietary fiber. The article by Xiaotong Wang and Tongtong He [5] found that high soy food intake (>3 times/week) was significantly associated with a lower prevalence of some chronic diseases in southern Chinese children. In addition to the child's own diet behavior, maternal healthy lifestyle factors are associated with their offspring's health. Dietary patterns during pregnancy were found by Siyuan Lv and Rui Qin to possibly have cumulative effects on an offspring's neurodevelopment [6]. Regarding lactation, there is clear evidence of the importance of breastfeeding, compared with formula feeding, in the protection against chronic diseases. However, research by Jiajia Dang confirmed that prolonged breastfeeding (≥12 months) is detrimental and associated with an increased risk of offspring obesity [7].

Healthy diet patterns have a positive effect on the prevention of chronic non-communicable diseases in children. Given the malnutritional problems prevalent among children and adolescents, health education focusing on behavior intervention and nutrition education are necessary for containing nutritional problems among children [8]. An important prerequisite for better regulation of children's diets is the monitoring and assessment of children's dietary behavior. In this Special Issue, a systematic review written by Anne Krijger [9] explored many dietary and lifestyle screening tools, however, with poor prospects for the application of these screening tools. There is a need for an easy-to-administer screening tool for children in order to improve a child's lifestyle at an early age. Additionally, in this Special Issue, Sheng Ma [10] first described the most commonly consumed food items, in terms of 29 food groups, nationally and by province and developed a low-burden, food group-based diet quality questionnaire (DQQ) for China to evaluate diet quality at the population level. Immediately afterward, Huan Wang et al. [11] confirmed that DQQ could be a valid tool to assess diet quality for Chinese children and adolescents. The Mediterranean

Citation: Yu, X.; Zou, Z. Dietary, Lifestyle, and Children Health. *Nutrients* **2023**, *15*, 2242. https://doi.org/10.3390/nu15102242

Received: 21 April 2023
Accepted: 24 April 2023
Published: 9 May 2023

Copyright: © 2023 by the authors. Licensee MDPI, Basel, Switzerland. This article is an open access article distributed under the terms and conditions of the Creative Commons Attribution (CC BY) license (https://creativecommons.org/licenses/by/4.0/).

diet is one of the dietary patterns with the healthiest outcomes, which has a positive role in protecting people from chronic non-communicable diseases. A longitudinal study found that children who adhere well to the Mediterranean diet have a lower risk of transient glomerular impairment [12]. Another randomized pilot trial indicated that a nutritional intervention based on the Mediterranean diet helps to reduce high glycated hemoglobin (HbA1c) and insulin levels and, therefore, to reduce the prevalence of prediabetes [13]. Their findings further underscore the necessity of early development of a healthy diet and contribute to providing some basis for establishing a targeted dietary intervention for children with certain diseases.

In addition to healthy dietary patterns, other lifestyle factors, including a healthy body mass index, regular exercise, no smoking, and sufficient sleep duration, are associated with a lower incidence of chronic non-communicable diseases and longer life expectancy. Two papers in this Special Issue focus on the impact of sedentary behavior and digital screen time on chronic disease risk. First, the study by Ning Yin [14] indicated that higher sedentary behavior time is associated with a higher risk of metabolic syndrome components among children aged 6–14. Moreover, Agata Rockad et al. [15] surveyed the parents of preschool and school-aged children during the COVID-19 blockade in Poland and found that the majority of children were exposed to screens during meals, which is a risk factor of obesity. Additionally, it may negatively affect physical activity. This suggests an urgent need to develop effective strategies to limit excessive screen time and promote healthy eating habits and physical activity in children.

To summarize, the studies featured in this Special Issue are critical in identifying and assessing dietary factors, as well as other healthy lifestyle factors during childhood and adolescence. These findings not only will help to regulate eating behaviors and to improve the quality of children's diet but also will help healthcare professionals to manage healthy behaviors in children, thereby reducing the prevalence of childhood chronic diseases.

Author Contributions: Writing—original draft preparation, X.Y.; writing—review and editing, Z.Z. and X.Y.; supervision, Z.Z.; funding acquisition, Z.Z. All authors have read and agreed to the published version of the manuscript.

Funding: National Natural Science Foundation of China (82073573).

Conflicts of Interest: The authors declare no conflict of interest.

References

1. Regan, C.; Walltott, H.; Kjellenberg, K.; Nyberg, G.; Helgadottir, B. Investigation of the Associations between Diet Quality and Health-Related Quality of Life in a Sample of Swedish Adolescents. *Nutrients* **2022**, *14*, 2489. [CrossRef] [PubMed]
2. Liu, J.; Li, Y.; Wang, X.; Gao, D.; Chen, L.; Chen, M.; Ma, T.; Ma, Q.; Ma, Y.; Zhang, Y.; et al. Association between Fruit Consumption and Lipid Profile among Children and Adolescents: A National Cross-Sectional Study in China. *Nutrients* **2021**, *14*, 63. [CrossRef] [PubMed]
3. Hosseini, B.; Berthon, B.S.; Jensen, M.E.; McLoughlin, R.F.; Wark, P.A.B.; Nichol, K.; Williams, E.J.; Baines, K.J.; Collison, A.; Starkey, M.R.; et al. The Effects of Increasing Fruit and Vegetable Intake in Children with Asthma on the Modulation of Innate Immune Responses. *Nutrients* **2022**, *14*, 3087. [CrossRef] [PubMed]
4. Li, M.; Amaerjiang, N.; Li, Z.; Xiao, H.; Zunong, J.; Gao, L.; Vermund, S.H.; Hu, Y. Insufficient Fruit and Vegetable Intake and Low Potassium Intake Aggravate Early Renal Damage in Children: A Longitudinal Study. *Nutrients* **2022**, *14*, 1228. [CrossRef] [PubMed]
5. Wang, X.; He, T.; Xu, S.; Li, H.; Wu, M.; Lin, Z.; Huang, F.; Zhu, Y. Soy Food Intake Associated with Obesity and Hypertension in Children and Adolescents in Guangzhou, Southern China. *Nutrients* **2022**, *14*, 425. [CrossRef] [PubMed]
6. Lv, S.; Qin, R.; Jiang, Y.; Lv, H.; Lu, Q.; Tao, S.; Huang, L.; Liu, C.; Xu, X.; Wang, Q.; et al. Association of Maternal Dietary Patterns during Gestation and Offspring Neurodevelopment. *Nutrients* **2022**, *14*, 730. [CrossRef] [PubMed]
7. Dang, J.; Chen, T.; Ma, N.; Liu, Y.; Zhong, P.; Shi, D.; Dong, Y.; Zou, Z.; Ma, Y.; Song, Y.; et al. Associations between Breastfeeding Duration and Obesity Phenotypes and the Offsetting Effect of a Healthy Lifestyle. *Nutrients* **2022**, *14*, 1999. [CrossRef] [PubMed]
8. Tian, T.; Wang, Y.; Xie, W.; Zhang, J.; Ni, Y.; Peng, X.; Sun, G.; Dai, Y.; Zhou, Y. Multilevel Analysis of the Nutritional and Health Status among Children and Adolescents in Eastern China. *Nutrients* **2022**, *14*, 758. [CrossRef] [PubMed]
9. Krijger, A.; Ter Borg, S.; Elstgeest, L.; van Rossum, C.; Verkaik-Kloosterman, J.; Steenbergen, E.; Raat, H.; Joosten, K. Lifestyle Screening Tools for Children in the Community Setting: A Systematic Review. *Nutrients* **2022**, *14*, 2899. [CrossRef] [PubMed]

10. Ma, S.; Herforth, A.W.; Vogliano, C.; Zou, Z. Most Commonly-Consumed Food Items by Food Group, and by Province, in China: Implications for Diet Quality Monitoring. *Nutrients* **2022**, *14*, 1754. [CrossRef] [PubMed]
11. Wang, H.; Herforth, A.W.; Xi, B.; Zou, Z. Validation of the Diet Quality Questionnaire in Chinese Children and Adolescents and Relationship with Pediatric Overweight and Obesity. *Nutrients* **2022**, *14*, 3551. [CrossRef] [PubMed]
12. Li, M.; Xiao, H.; Shu, W.; Amaerjiang, N.; Zunong, J.; Huang, D.; Hu, Y. Good Adherence to the Mediterranean Diet Lowered Risk of Renal Glomerular Impairment in Children: A Longitudinal Study. *Nutrients* **2022**, *14*, 3343. [CrossRef] [PubMed]
13. Blancas-Sanchez, I.M.; Del Rosal Jurado, M.; Aparicio-Martinez, P.; Quintana Navarro, G.; Vaquero-Abellan, M.; Castro Jimenez, R.A.; Fonseca Pozo, F.J. A Mediterranean-Diet-Based Nutritional Intervention for Children with Prediabetes in a Rural Town: A Pilot Randomized Controlled Trial. *Nutrients* **2022**, *14*, 3614. [CrossRef] [PubMed]
14. Yin, N.; Yu, X.; Wang, F.; Yu, Y.; Wen, J.; Guo, D.; Jian, Y.; Li, H.; Huang, L.; Wang, J.; et al. Self-Reported Sedentary Behavior and Metabolic Syndrome among Children Aged 6-14 Years in Beijing, China. *Nutrients* **2022**, *14*, 1869. [CrossRef] [PubMed]
15. Rocka, A.; Jasielska, F.; Madras, D.; Krawiec, P.; Pac-Kozuchowska, E. The Impact of Digital Screen Time on Dietary Habits and Physical Activity in Children and Adolescents. *Nutrients* **2022**, *14*, 2985. [CrossRef] [PubMed]

Disclaimer/Publisher's Note: The statements, opinions and data contained in all publications are solely those of the individual author(s) and contributor(s) and not of MDPI and/or the editor(s). MDPI and/or the editor(s) disclaim responsibility for any injury to people or property resulting from any ideas, methods, instructions or products referred to in the content.

Systematic Review

Lifestyle Screening Tools for Children in the Community Setting: A Systematic Review

Anne Krijger [1,2,*], Sovianne ter Borg [3], Liset Elstgeest [2,4], Caroline van Rossum [3], Janneke Verkaik-Kloosterman [3], Elly Steenbergen [3], Hein Raat [2] and Koen Joosten [1]

[1] Department of Pediatrics and Pediatric Surgery, Erasmus MC-Sophia Children's Hospital, University Medical Center Rotterdam, 3000 CB Rotterdam, The Netherlands; k.joosten@erasmusmc.nl
[2] Department of Public Health, Erasmus MC, University Medical Center Rotterdam, 3000 CA Rotterdam, The Netherlands; l.elstgeest@erasmusmc.nl (L.E.); h.raat@erasmusmc.nl (H.R.)
[3] National Institute for Public Health and the Environment, 3720 BA Bilthoven, The Netherlands; sovianne.ter.borg@rivm.nl (S.t.B.); caroline.van.rossum@rivm.nl (C.v.R.); janneke.verkaik@rivm.nl (J.V.-K.); elly.steenbergen@rivm.nl (E.S.)
[4] Reinier Academy, Reinier de Graaf Hospital, 2600 GA Delft, The Netherlands
[*] Correspondence: j.j.a.krijger@erasmusmc.nl; Tel.: +31(0)6-2461-2722

Abstract: Screening of children's lifestyle, including nutrition, may contribute to the prevention of lifestyle-related conditions in childhood and later in life. Screening tools can evaluate a wide variety of lifestyle factors, resulting in different (risk) scores and prospects of action. This systematic review aimed to summarise the design, psychometric properties and implementation of lifestyle screening tools for children in community settings. We searched the electronic databases of Embase, Medline (PubMed) and CINAHL to identify articles published between 2004 and July 2020 addressing lifestyle screening tools for children aged 0–18 years in the community setting. Independent screening and selection by two reviewers was followed by data extraction and the qualitative analysis of findings. We identified 41 unique lifestyle screening tools, with the majority addressing dietary and/or lifestyle behaviours and habits related to overweight and obesity. The domains mostly covered were nutrition, physical activity and sedentary behaviour/screen time. Tool validation was limited, and deliberate implementation features, such as the availability of clear prospects of actions following tool outcomes, were lacking. Despite the multitude of existing lifestyle screening tools for children in the community setting, there is a need for a validated easy-to-administer tool that enables risk classification and offers specific prospects of action to prevent children from adverse health outcomes.

Keywords: screeners; nutrition; physical activity; sedentary behaviour; lifestyle risk; obesity

Citation: Krijger, A.; ter Borg, S.; Elstgeest, L.; van Rossum, C.; Verkaik-Kloosterman, J.; Steenbergen, E.; Raat, H.; Joosten, K. Lifestyle Screening Tools for Children in the Community Setting: A Systematic Review. *Nutrients* **2022**, *14*, 2899. https://doi.org/10.3390/nu14142899

Academic Editor: Zhiyong Zou

Received: 20 June 2022
Accepted: 13 July 2022
Published: 14 July 2022

Publisher's Note: MDPI stays neutral with regard to jurisdictional claims in published maps and institutional affiliations.

Copyright: © 2022 by the authors. Licensee MDPI, Basel, Switzerland. This article is an open access article distributed under the terms and conditions of the Creative Commons Attribution (CC BY) license (https://creativecommons.org/licenses/by/4.0/).

1. Introduction

A healthy lifestyle is essential for optimal growth and development as well as for later-life health of children [1,2]. The World Health Organization proposed the concept of a healthy lifestyle to be 'a way of living that lowers the risk of being seriously ill or dying early' [3]. A large number of factors can be considered as lifestyle. In children, nutrition, physical activity (PA), sedentary behaviour and sleep are lifestyle factors that were found to be associated with health outcomes [4–7]. Overweight, obesity and other cardiovascular risk factors are common consequences of an unhealthy lifestyle and may already appear during childhood [4]. The adequate evaluation of children's lifestyle can contribute to preventive actions that combat the increasing prevalence of lifestyle-related conditions.

To evaluate the lifestyle of children, including nutrition, various tools can be used. Two groups of lifestyle tools can be distinguished: lifestyle assessment tools and lifestyle screening tools [8]. Lifestyle assessment tools, such as food frequency questionnaires, 3-day food diaries and physical activity trackers, are used to examine the child's behaviour and/or characteristics in detail. To be of service to youth healthcare, which has a preventive function but limited consultation time, this paper focuses on lifestyle screening tools that identify risk (factors) on an individual level. Lifestyle screening tools usually comprise more general items than lifestyle assessment tools, are used for quick evaluation and assign a certain value to the lifestyle behaviour and/or characteristics of the child. In practice, a commonly used method for this is a short questionnaire. Outcomes of lifestyle screening tools may vary; they can, for example, result in an overall lifestyle score or highlight areas for improvement ('red flags'). Given the rapid value judgment, lifestyle screening tools can be helpful in clinical practice or community screening. Here, they can serve as a basis to enter into dialogue with the parents or provide advice for further actions, for instance, referral to a dietitian or starting an intervention. Next to the design characteristics of lifestyle screening tools (such as the number of items, covered topics and intended target group), the psychometric properties (i.e., reliability and validity) and implementation methods (such as the manner in which the outcomes or advice for further action are formulated (prospects of action)), practical application and tool format (online, on paper, etc.) are likely to affect the usability and effectiveness of such screening tools.

Reviews specifically on nutrition screening tools for children have mainly focused on tools developed for hospital settings [9–13]. A recently published systematic review by Becker et al. targeted the reliability and validity of nutrition screening tools for children up to 18 years of age, including tools for the community setting [14]. The community health care setting, represented by preventive and primary health care services, is the perfect place for the usage of lifestyle screening tools. This is because most children with a suboptimal lifestyle reside in the community setting and will not be admitted to a hospital. A thorough overview of existing lifestyle screening tools for children aged 0–18 years in the community setting, not limited to nutrition, is yet lacking.

Therefore, our systematic review aims to comprehensively describe lifestyle screening tools for children in the community setting. The present study is embedded in a Dutch governmental project that intends to develop a lifestyle screening tool for children aged 0–4 years. This screening tool will ultimately lead to timely measures to prevent children from negative lifestyle-related health outcomes. The specific questions to be addressed within our review are:

(1) What lifestyle screening tools for children in the community setting are available?
(2) What are the main features of these lifestyle screening tools regarding design, psychometric properties (i.e., reliability and validity) and implementation?

2. Materials and Methods

This systematic review is reported as indicated in the Preferred Reporting Items for Systematic reviews and Meta-Analyses (PRISMA) guideline [15]. An a priori systematic review protocol was developed (available upon request).

2.1. Search Strategy

We performed systematic searches in the electronic databases of Embase, Medline (PubMed) and CINAHL to identify articles addressing lifestyle screening tools for children in the community setting, published between January 2004 and July 2020. Based on the study objectives, the PICO model [16] was used to further specify the search strategy. The population (P) was defined as children up to 18 years of age in the community setting, the intervention/exposure (I) as lifestyle screening tools and the outcomes of interest (O) as indicators of an unhealthy lifestyle. We did not include a comparison to a control group (C) as we did not study an intervention effect. Search strings were developed with assistance from a librarian. Search terms were divided into the categories 'child', 'screening' and 'lifestyle', which were combined with 'AND'. Emtree terms and MeSH terms were used to identify relevant articles (Supplementary file S1). Search filters to restrain the results to humans and English or Dutch language were applied. The search strategies were not limited to specific lifestyle factors.

As nutrition is such an eminent part of lifestyle, we performed additional literature searches focusing on nutrition screening tools. Hence, we updated the searches by Becker et al. and an exploratory systematic search that was conducted in 2019 (unpublished research, for details, see Supplementary file S1). Similar to the broader search on lifestyle screening tools, filters to limit the results to humans and English or Dutch language were applied.

Full details on the search strings are provided in Supplementary file S1. Search results were exported to EndNote X9 reference management software and deduplicated.

2.2. Eligibility Criteria

For the inclusion of an article, the following predefined criteria had to be met:

1. The study described a screening tool to identify lifestyle risk (factors) on an individual level for
2. children up to 18 years of age in
3. the community setting.
4. The tool had to be applied by a parent/caregiver, health professional (e.g., physician, nurse) or by the child him- or herself, and
5. the study was published in English or Dutch
6. between January 2004 and July 2020.

Exclusion criteria comprised:

1. studies reporting on lifestyle questionnaires, with a purpose other than screening for lifestyle risk (factors) on an individual level (e.g., general questionnaires in national surveys);
2. studies on lifestyle assessment tools (e.g., (derivatives of) food frequency questionnaires, diet quality scores, anthropometry);
3. studies on a single specific lifestyle or nutrition factor (e.g., solely screen time or vegetable intake);
4. studies reporting prevalence rates of malnutrition or growth charts as a measure of nutrition risk;
5. tools to identify eating disorders;
6. tools developed for hospital settings or specific patient groups;
7. commentaries and conference abstracts.

2.3. Screening, Selection and Data Extraction

Applying the abovementioned inclusion and exclusion criteria, two reviewers (A.K. and S.t.B.) independently screened titles and abstracts of the obtained articles. Thereafter, they selected the relevant articles based on full texts according to the same inclusion and exclusion criteria. Additionally, articles included in the review of Becker et al. [14] and identified with the exploratory search on nutrition screening tools were checked for eligibility. Discrepancies in opinion on inclusion by the reviewers were resolved by discussion until consensus or in consultation with a third reviewer (L.E.). A.K. and S.t.B. then extracted the data from the included studies. Reported general information (reference, title), study characteristics (study objective, study year, country of origin, study design, sample size, age, outcome measures, results) and tool characteristics (tool name, tool aim, target population, person administered, administer duration, administer frequency, administer method, addressed domains, number of items, response format, tool outcome, prospect of action, strengths, limitations) were entered into a predesigned data extraction table. The usability of the data extraction table was tested beforehand by extracting data from 10% of the articles in duplicate by A.K. and S.t.B.. Articles reporting on the same tool were grouped. Articles covered in included reviews were also assessed for eligibility.

2.4. Data Analysis

By summarising the characteristics of the included studies and corresponding lifestyle screening tools, we performed an initial data synthesis. Subsequently, qualitative analysis was performed by tabulating and assorting by specific features, such as target age (toddlers, 1–3 years old; preschoolers, 3–5 years old; school age, 6–12 years old; adolescents, 13–18 years old), number of tool items and prospects of action. This enabled us to aggregate the data further and to explore similarities and differences between the identified screening tools.

3. Results

A total of 2698 articles were identified for screening (Figure 1). After the full-text review of 105 articles, 48 met the inclusion criteria and were included in the qualitative analysis. The most common reasons for exclusion were: not describing a screening tool or describing a general questionnaire instead of a screening tool. We included two systematic reviews [14,17], yielding no additional screening tools for inclusion. The other 46 articles [18–63] described 41 unique screening tools. The majority of the included articles reported on the development and validation of screening tools, whereas their implementation was rarely addressed. Studies were performed between 2001 and 2019 in sixteen different countries (both Western and non-Western), with nearly half conducted in the United States ($n = 20$).

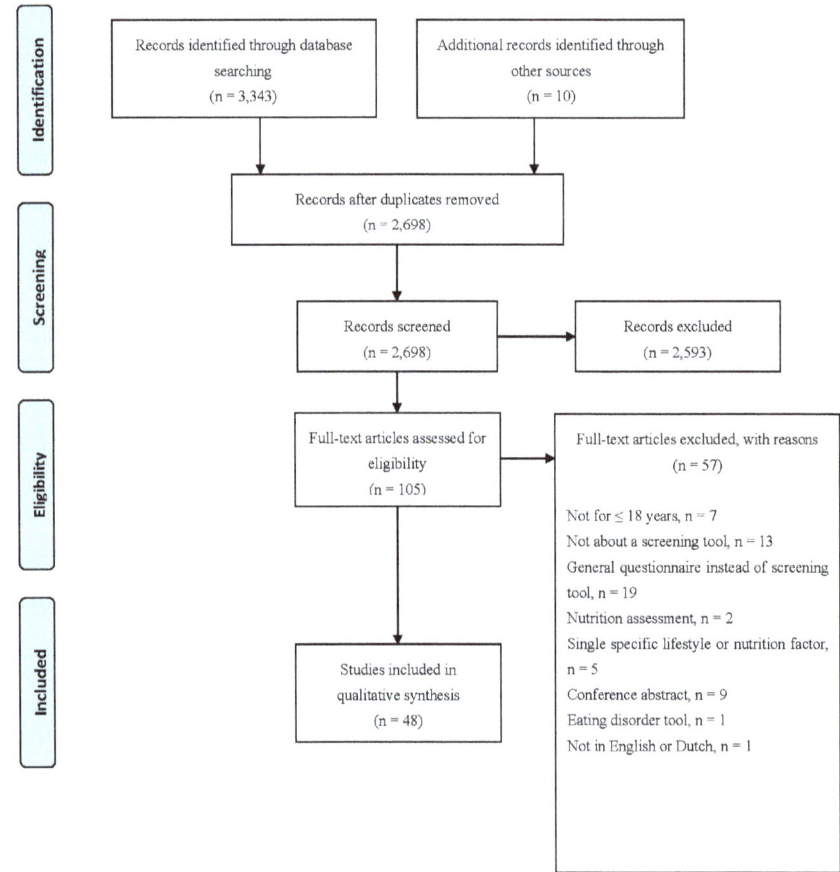

Figure 1. PRISMA flowchart of methodology.

3.1. Design of Screening Tools

Table 1 demonstrates various characteristics of the included lifestyle screening tools. The majority of tools were developed to screen lifestyle behaviour and habits. Although not always explicitly stated in the tool's aim, articles mostly described that the tool focused on factors associated with obesity risk. Ten screening tools were distinctively designed for toddlers (1–3 years old) or preschoolers (3–5 years old) [18–31] and another nine for school-aged children (6–12 y) [32–39]. Fourteen tools were described as either designed for children in general or did not specify the children's target age (0–18 y) [40–55]. Eight tools were specifically designed for adolescents (13–18 y) [56–63]. The tools aimed at toddlers and preschoolers were to be administered by parents or health care professionals. Children of school age reported themselves ($n = 6$) or their parents did ($n = 3$). One tool for children without specified age was divided into a part completed by the child and a part completed by the parents [55]. Tools for adolescents only were exclusively self-reported. Tools administered to parents could include proxy-reported items on the child but also self-reported items regarding parents themselves, such as self-efficacy for a healthy lifestyle or parental feeding practices. The number of items per tool ranged from 3 to 116, with a median of 22 items (interquartile range (IQR): 17, 34). No article described the rationale for the number of items. All tools used multiple choice questions (some combined with open questions), mainly on Likert-type scales. Two tools used visuals to increase comprehensibility [30,37]. These visuals included portion sizes and images to make the tool more appealing. The time needed to complete the tool was reported for only thirteen tools [18–20,30,31,34,37–40,47,52,60,63]. From those who reported the time, the time needed ranged from 3 [18–20] to 90 [37] minutes; six tools could be completed within 15 min [18–20,31,38,40,52].

Table 2 shows the encompassed lifestyle domains with specified items of the included screening tools. Specification of the nutrition items is demonstrated in Table 3. The domains covered most were nutrition ($n = 39$), PA ($n = 25$) and sedentary behaviour/screen time ($n = 21$) (Figure 2). The median of the number of covered domains was three. Tools for toddlers and preschoolers covered, with a median of two, fewer domains. All screening tools intended for toddlers and preschoolers covered nutrition. None of the screening tools specifically for toddlers included PA items, whereas, in other tools, PA was mainly evaluated by estimating the frequency and duration per week. Sedentary behaviour was not determined as such but evaluated with screen time as proxy. Sleep and hygiene were included in four and five tools, respectively, mainly as sleep duration ($n = 2$) and dental care ($n = 4$). Huang et al. included neighbourhood safety [55]; environmental factors in other tools were generally related to nutrition and PA (e.g., parental modelling). As for the items on nutrition, the intake of specific food groups, dietary habits and psychological factors were predominantly evaluated (Table 3). Of all the tools that evaluated the consumption of food groups ($n = 27$), most asked about vegetables ($n = 25$), fruits ($n = 25$), sugar-sweetened beverages ($n = 16$) and unhealthy snacks/fast food ($n = 16$). Commonly addressed eating habits were consuming breakfast ($n = 9$), eating at the table or while watching TV ($n = 6$) and eating with the family together ($n = 5$). Psychological factors mainly included (parental) beliefs and attitudes towards healthy eating. In addition, nutrition knowledge ($n = 4$) and food costs ($n = 2$) recurred in several tools.

Table 1. Characteristics of lifestyle screening tools for children in the community setting.

Tool Name	Tool Aim	Target Population	Administered by	Number of Items	Item Response Format	Tool Scoring	Prospect of Action
1. NutricheQ [18–20]	Assess dietary risk	Toddlers	Parent [b]	11 [c]	3-point Likert scale	Subsection score and total score; ranging from 0 to 22 Cut-offs for low, moderate and high risk are available per section	Tool identifies children who may need blood screening and nutritional intervention
2. Toddler Feeding Questionnaire (TFQ) [21]	Assess indulgent, authoritative and environmental feeding practices	Toddlers	Parent [a]	24	5-point Likert scale (never-always)	Subscale scores	NR
3. Toddler NutriSTEP [22]	Assess nutritional risk	Toddlers	Parent [b]	17	Likert-type scale	Total score; ranging from 0 to 68 Cut-offs for low, moderate and high risk	Treat impaired state and refer to needed services
4. Toddler Dietary Questionnaire (TDQ) [23]	Assess dietary risk	Toddlers	Parent [b]	19	Likert-type scale	Total score; ranging from 0 to 100 Cut-offs for low, moderate, high and very high dietary risk	Health care professionals may refer to a dietitian based on identified risk
5. Child Eating Behavior Questionnaire (CEBQ) [24]	Assess eating behaviours	Preschoolers	Parent [b]	35	5-point Likert scale (never-always)	Subscale scores	NR
6. Nutrition Screening Tool for Every Preschooler (NutriSTEP) [25–27]	Assess nutritional risk	Preschoolers	Parent [b]	17	Likert-type scale (varying)	Total score; ranging from 0 to 68 Cut-offs for low, moderate and high risk	Parents receive results, customised feedback and resources such as links to credible health information websites
7. Preschooler Dietary Questionnaire (PDQ) [28]	Assess diet and provide a dietary risk score	Preschoolers	Parent [b]	19	Likert-type scale	Total score; ranging from 0 to 100 Cut-offs for low, moderate, high and very high dietary risk	Health care professionals may refer to a dietitian based on identified risk
8. Preschoolers Diet–Lifestyle Index (PDL-index) [29]	Assess adherence to diet–lifestyle recommendations	Preschoolers	Health care professional [b]	11	Likert-type scale (varying)	Total score; ranging from 0 to 44	Tool may guide health care professionals in counselling parents and policy makers in developing interventions

Table 1. *Cont.*

Tool Name	Tool Aim	Target Population	Administered by	Number of Items	Item Response Format	Tool Scoring	Prospect of Action
9. Healthy Kids [30]	Assess diet, lifestyle and parenting domains to determine obesity risk	Children aged 2–5 y from low-income families	Parent [a,b]	19	Combination of closed and open questions	Total score; ranging from 19 to 95	Tool can be used to target counselling or nutrition education for families and to supplement physical examination
10. Tool by Das and Ghosh [31]	Assess nutrition knowledge	Children aged 3–6 y	Parent [a]	32	Closed questions	Total score; ranging 0–32	NR
11. Start the Conversation 4–12 (STC-4–12) [32]	Assess and counsel nutrition and PA barriers and behaviours	Children aged 4–12 y	Parent [a,b]	22	Likert-type scale (varying)	No score	The tool provides tips that serve as cues for action for parents and guide counselling by health care professionals
12. Healthy Families Survey [33]	Assess nutrition and PA behaviours	Elementary school children	Parent [a,b]	45	Combination of closed and open questions	Subscale scores	NR
13. Knowledge, Attitudes and Habits (KAH-) questionnaire [34]	Assess knowledge, attitudes and habits towards a healthy lifestyle	Elementary school children	Child [a]	48	3-point Likert scale	Subscale scores and total score; ranging from 0 to 96	NR
14. Parental Self-efficacy Questionnaire [35]	Assess parental self-efficacy for enacting healthy diet and PA behaviours in their children	Children aged 6–11 y	Parent [a]	34	11-point Likert scale	Subscale scores and total score; ranging from 0 to 340	NR
15. Tool by Chacko and Ganesan [36]	Assess dietary gaps	School children aged 6–17 y	Child [a]	10	2-point Likert scales (yes–no)	Total score; ranging from 0 to 10	Parents and children can receive corrective counselling on the identified gaps
16. Food, Health and Choices questionnaire (FHC-Q) [37]	Assess energy balance behaviours and related theory-based psychosocial determinants	Upper elementary school children	Child [a]	116	Likert-type scale	Subscale scores	NR
17. Healthy Eating and Physical Activity Self-Efficacy Questionnaire for Children (HEPASEQ-C) [38]	Assess self-efficacy of healthy eating and PA	Upper elementary school children	Child [a]	9	3-point Likert scale (there is no way I can do this–I believe I can do this)	Total score; ranging from 9 to 27	NR

Table 1. *Cont.*

Tool Name	Tool Aim	Target Population	Administered by	Number of Items	Item Response Format	Tool Scoring	Prospect of Action
18. Healthy Eating and Physical Activity Behavior Recall Questionnaire for Children (HEPABRQ-C) [38]	Assess recall of healthy eating and PA	Upper elementary school children	Child [a]	10	Combination of closed and open questions	Total score; ranging from 0 to 21	NR
19. Eating Behavior Questionnaire for School Children [39]	Assess eating behaviours	School children	Child [a]	23	5-point Likert scale (never-always)	Subscore per domain	NR
20. Tool by Drouin and Winickoff [40]	Assess health-related behavioural risk factors	Children aged 0–18 y	Parent [b]	3	Closed questions	No score	Parents receive a handout with information about identified risk factors Health care professionals receive the survey results and an evidence-based suggested course of action
21. Child Nutrition and Physical Activity (CNPA) Screening Tool [41]	Assess behaviours that increase the risk of obesity	Children aged 2–18 y	Parent [a,b]	22	4-point Likert scale and open questions	Subscores for generated readiness to change and perception factors only	Tool provides health care professionals means to start the conversation about a healthy lifestyle with parents
22. Electronic Kids Dietary Index (E-KINDEX) [42]	Assess food habits, dietary beliefs and practices related to obesity	Children	Child [a], parent [b] or health care professional [b]	30	Likert-type scale (varying)	Subscale scores and total score; ranging from 1 to 87	In clinical practice, the score can be used as visual educational tool, provide continuous feedback and individual items may be used as specific goals for obesity status improvement
23. Family Health Behavior Scale (FHBS) [43]	Assess family eating and PA habits related to obesity	Children	Parent [a,b]	27	5-point Likert scale (never-nearly always)	Subscale scores and total score	NR
24. Family Nutrition and Physical Activity (FNPA) screening tool [44,45]	Assess risk factors for overweight/obesity in the home environment	Children	Parent [a,b]	20	4-point Likert scale (never-always)	Subscore per domain and total score	Korean version: based on scores, interventions such as counselling and education should be developed and provided

Table 1. *Cont.*

Tool Name	Tool Aim	Target Population	Administered by	Number of Items	Item Response Format	Tool Scoring	Prospect of Action
25. HABITS questionnaire [46]	Assess weight-related behaviours and intervention targets	Children	Child [a]	19	Likert-type scale (varying)	Subscale scores	Tool can establish a dialogue about weight-related lifestyle behaviours between health care professional and families
26. Healthy Living for Kids Survey (HLKS) [47]	Assess healthy lifestyle perceptions and behaviours	Children	Child [a]	59	Likert-type scale (varying)	Subscale scores and total score	Education of parents and children to redress inaccurate perceptions of a healthy lifestyle
27. HeartSmartKids (HSK) [48] (HeartSmartKids, LLC, Boulder, US)	Assess lifestyle habits to guide behaviour change counselling	Children	Child [a]	21	Likert-type scale (varying)	NR	Patient-specific education handouts with lifestyle recommendations are generated
28. Home Self-Administered Tool for Environmental Assessment of Activity and Diet (HomeSTEAD) [49]	Assess home environment factors related to children's diet and PA	Children	Parent [a]	86	5-point Likert scale	Subscale scores	Promotion of healthy feeding practices
29. Lifestyle Behavior Checklist (LBC) [50,51]	Assess parental perceptions and self-efficacy in managing problems related child eating, activity and weight issues	Children with obesity	Parent [a]	25	Combination of closed and open questions	Subscale scores	NR
30. Pediatric Adapted Liking Survey (PALS) [52]	Assess dietary behaviours linked to caries and obesity risk	Children	Parent [b]	33	Horizontal visual 5-point Likert scale, (hates it–loves it)	Subscore per domain; ranging from -100 to $+100$	Tailored motivational diet-related messages for dental caries and obesity prevention
31. Short-Form, Multicomponent Dietary Questionnaire (SF-FFQ4PolishChildren) [53]	Assess dietary and lifestyle behaviours	Children	Child [a] or parent [b]	44	Likert-type scale (varying)	Subscore per domain Cut-offs for low, moderate and high subscores	NR
32. Tool by Hendrie et al. [54]	Assess family activity environment	Children	Parent [a]	25	5-point Likert scales (strongly disagree–strongly agree)	NR	NR

Table 1. *Cont.*

Tool Name	Tool Aim	Target Population	Administered by	Number of Items	Item Response Format	Tool Scoring	Prospect of Action
33. Tool by Huang et al. [55]	Assess correlates of PA and screen time behaviours	Children	Child [a] and parent [a,b]	46	Likert-type scale (varying)	NR	NR
34. Adolescent Lifestyle Profile (ALP) [56,57]	Assess health-promoting behaviours	Adolescents	Child [a]	42	4-point Likert scale (never–routinely)	Total score; ranging from 42 to 168	NR
35. Childhood Family Mealtime Questionnaire (CFMQ) (reduced) [58]	Assess mealtime environment	Adolescents	Child [a]	22	5-point Likert scale (never–always)	NR	NR
36. Diet–Lifestyle Index [59]	Assess nutrition and lifestyle quality related to overweight and obesity	Adolescents	Child [a]	13	Likert-type scale (varying)	Total score; ranging from 11 to 57	NR
37. Shortened Health-Promoting Lifestyle Profile (HPLP) II [60]	Assess health-promoting behaviours	Adolescents	Child [a]	34	4-point Likert scale (never–routinely)	Subscale scores and total score	NR
38. Tool by Fernald et al. [61]	Assess health behaviour	Adolescents	Child [a]	16	NR	Total score; ranging from 0 to 3	NR
39. Tool by Hyun et al. [62]	Assess nutrition knowledge	Adolescents	Child [a]	20	2-point Likert scales (wrong–right)	Total score; ranging from 0 to 20	NR
40. Tool by Hyun et al. [62]	Assess dietary habits	Adolescents	Child [a]	9	5-point Likert scales (always–never)	Total score; ranging from 0 to 5	NR
41. VISA-TEEN [63]	Assess lifestyle	Adolescents	Child [a]	11	Combination of closed and open questions	Total score	NR

Note: Tools are sorted by target age. Abbreviations: NR, not reported; y, years. [a] Self-reported; [b] proxy-reported; [c] originally, 18 items were developed, but only 11 were validated.

Table 2. Addressed domains and items of lifestyle screening tools for children in the community setting.

Tool Name	Nutrition [a]	Physical Activity	Sedentary Behaviour/ Screen Time	Sleep	Hygiene	Environment	Other
1. NutricheQ [18–20]	✓						
2. Toddler Feeding Questionnaire (TFQ) [b] [21]	✓						
3. Toddler NutriSTEP [b] [22]	✓		Duration of watching TV or using the computer				Growth adequacy, child's weight status

Table 2. *Cont.*

Tool Name	Nutrition [a]	Physical Activity	Sedentary Behaviour/ Screen Time	Sleep	Hygiene	Environment	Other
4. Toddler Dietary Questionnaire (TDQ) [23]	✓						
5. Child Eating Behavior Questionnaire (CEBQ) [b] [24]	✓						
6. Nutrition Screening Tool for Every Preschooler (NutriSTEP) [25–27]	✓	Frequency of PA	Frequency and duration of watching TV, using computer and playing video games				Parental satisfaction of child's growth, child's weight status
7. Preschooler Dietary Questionnaire (PDQ) [b] [28]	✓						
8. Preschoolers Diet–Lifestyle Index (PDL-index) [29]	✓	Duration of moderate-to-vigorous PA	Duration of watching TV				
9. Healthy Kids [30]	✓	Preference for playing over watching TV	Duration of watching TV and playing video or computer games	Bedtime			
10. Tool by Das and Ghosh [b] [31]	✓						General knowledge on health and lifestyle
11. Start the Conversation 4–12 (STC-4-12) [32]	✓	Frequency and duration of sports, playing outside and being active, barriers and readiness to change regarding PA	Duration of screen time				
12. Healthy Families Survey [33]	✓	Duration of PA, child sees parent being physically active	Duration of watching TV and using other screens, availability of TV in child's bedroom				
13. Knowledge, Attitudes and Habits (KAH-) questionnaire [34]	✓	Frequency of playing active games, liking exercise, activities after school and during weekends, knowledge and attitudes towards PA	Activities after school and during weekends		Brushing teeth, washing hands, taking bath or shower		Knowledge, attitudes and habits regarding the human body and emotions
14. Parental Self-efficacy Questionnaire [35]	✓	Confidence regarding child being physically active and playing outside	Confidence regarding limiting amount of screen time				

Table 2. *Cont.*

Tool Name	Nutrition [a]	Physical Activity	Sedentary Behaviour/Screen Time	Sleep	Hygiene	Environment	Other
15. Tool by Chacko and Ganesan [36]	✓						
16. Food, Health and Choices questionnaire (FHC-Q) [37]	✓	Frequency of specific activities, medium PA and heavy PA	Frequency and duration of watching TV and playing video games				Self-determination, outcome expectations, self-efficacy, habit strength, goal intention, knowledge and social desirability regarding a healthy lifestyle
17. Healthy Eating and Physical Activity Self-Efficacy Questionnaire for Children (HEPASEQ-C) [38]	✓	Self-efficacy regarding PA					
18. Healthy Eating and Physical Activity Behavior Recall Questionnaire for Children (HEPABRQ-C) [38]	✓	Duration of PA					
19. Eating Behavior Questionnaire for School Children [b] [39]	✓				✓ Not further specified	✓ Not further specified	
20. Tool by Drouin and Winickoff [40]	✓				Recent dental care visit		Tobacco smoke exposure
21. Child Nutrition and Physical Activity (CNPA) Screening Tool [41]	✓	Frequency and duration of PA	Duration of media use, availability of media in child's bedroom				Perception, confidence and importance items on healthy choices
22. Electronic Kids Dietary Index (E-KINDEX) [42]	✓						
23. Family Health Behavior Scale (FHBS) [43]	✓	Duration of being physically active, PA with parents, playing outside, doing sports, preferring indoor activities over outdoor activities, parental PA with child					
24. Family Nutrition and Physical Activity (FNPA) screening tool [b] [44,45]	✓	Child's PA, family PA	Screen time behaviour and monitoring	Sleep duration		Healthy environment	

Table 2. *Cont.*

Tool Name	Nutrition [a]	Physical Activity	Sedentary Behaviour/ Screen Time	Sleep	Hygiene	Environment	Other
25. HABITS questionnaire [46]	✓	Frequency of playing outside	Duration of watching TV				
26. Healthy Living for Kids Survey (HLKS) [b] [47]	✓	Frequency and duration of 'hard', 'moderate' and 'mild' exercise, frequency of any activity to work up a sweat, self-efficacy for PA	Duration of screen time, number of TV shows/videos watched, self-efficacy for screen time				
27. HeartSmartKids (HSK) [b] [48] (HeartSmartKids, LLC, Boulder, US)	✓	Duration of active play or sports	Duration of watching TV and using other screens	✓ Not further specified			Anthropometric measures
28. Home Self-Administered Tool for Environmental Assessment of Activity and Diet (HomeSTEAD) [49]	✓						
29. Lifestyle Behavior Checklist (LBC) [50,51]	✓	Parental problems experiencing and confidence in dealing with child complaining about PA	Parental problem experience and confidence in dealing with child watching too much TV and playing too many computer games				Parental problems experiencing and confidence in dealing with child complaining about problems related to obesity
30. Pediatric Adapted Liking Survey (PALS) [52]	✓				Liking/disliking of brushing teeth, taking a bath, getting dressed		
31. Short-Form, Multicomponent Dietary Questionnaire (SF-FFQ4PolishChildren) [b] [53]	✓	Intensity of PA at school and leisure time	Duration of screen time				Family affluence, height, weight
32. Tool by Hendrie et al. [54]		Parental PA involvement, parental opportunity for PA role modelling, parental support of PA	Parental opportunity for screen time role modelling			See domain PA	

Table 2. *Cont.*

Tool Name	Nutrition [a]	Physical Activity	Sedentary Behaviour/ Screen Time	Sleep	Hygiene	Environment	Other
33. Tool by Huang et al. [55]		Child's self-efficacy regarding PA, home PA environment, sports facilities in neighbourhood, family and peer support for PA	Child's perceived enjoyment of screen-based behaviours with parents, parental role modelling regarding screen time, rules and guidance on screen-based behaviours, availability of electronic screens			Child's perceived neighbourhood safety, social environment in neighbourhood	
34. Adolescent Lifestyle Profile (ALP) [b] [56,57]	✓	At least: frequency and duration of vigorous PA, playing active games with friends					Health responsibility, interpersonal relations, stress management, personal growth
35. Childhood Family Mealtime Questionnaire (CFMQ) (reduced) [58]	✓						
36. Diet–Lifestyle Index [59]	✓	Duration of extracurricular sport activities	Duration of watching TV and playing electronic games				Obesity status of parents
37. Shortened Health-Promoting Lifestyle Profile (HPLP) II [b] [60]	✓	✓ Not further specified					Health responsibility, stress management
38. Tool by Fernald et al. [b] [61]	✓	At least: frequency and duration of PA	Duration of watching TV				Alcohol use, smoking
39. Tool by Hyun et al. [62]	✓						
40. Tool by Hyun et al. [62]	✓						
41. VISA-TEEN [63]	✓	Duration of moderate and intense PA	Duration of using internet or gaming	Sleep duration	Frequency of brushing teeth and washing hands		Amount of cigarettes smoked, frequency of consuming alcohol and using drugs

Note: Tools are sorted by target age. [a] Details on nutrition items are demonstrated in Table 3; [b] Specific items of screening tool not fully described.

Table 3. Addressed nutrition items of lifestyle screening tools for children in the community setting.

Tool Name	Consumption of Food Groups	Dietary Habits	Psychological Factors Associated with Nutrition	Other
1. NutricheQ [18–20]	Vegetables, fruits, milk, dairy products, sweetened beverages, fortified cereals, red meat instead of oily or dark fish, fast food, unhealthy snacks			Age moving to cow's milk, avoiding foods due to allergy or intolerance
2. Toddler Feeding Questionnaire (TFQ) [a] [21]		Parental indulgent and authoritative practices, not further specified		Food environment-related, not further specified
3. Toddler NutriSTEP [a] [22]	Vegetables and fruits, flavoured beverages, dairy and substitutes, grains, meat and alternatives, fast food	Eating while watching TV, eating episodes per day, child feeds him- or herself, drinking from bottle with a nipple		Food is expensive, problems with chewing or swallowing when eating, being hungry at mealtimes, child controls amount consumed
4. Toddler Dietary Questionnaire (TDQ) [23]	Vegetables, fruits, dairy, milk beverages, non-milk beverages, grains, white versus non-white bread, meat products, lean red meat, fish, hot potato products, snack products, sweet snacks, spreadable fats, vegemite-type spreads			
5. Child Eating Behavior Questionnaire (CEBQ) [a] [24]		Food fussiness, emotional overeating, emotional undereating, satiety responsiveness, slowness in eating, desire to drink, food responsiveness	Enjoyment of food	
6. Nutrition Screening Tool for Every Preschooler (NutriSTEP) [25–27]	Vegetables, fruits, dairy, grain products, meat or fish or poultry or alternatives, fast food, supplements	Eating while watching TV, eating episodes per day		Difficulty buying food because of costs, problems with chewing, swallowing, gagging or choking when eating, not hungry because of drinking all day, parental control of amount consumed
7. Preschooler Dietary Questionnaire (PDQ) [a] [28]	Vegetables, fruits, dairy, milk beverages, non-milk beverages, grains, white versus non-white bread, meat products, lean red meat, fish, hot potato products, snack products, sweet snacks, spreadable fats, vegemite-type spreads			
8. Preschoolers Diet–Lifestyle Index (PDL-index) [29]	Vegetables, fruits, sweets, dairy products, grains, red meat (products), white meat and legumes, fish and seafood, unsaturated fats			

Table 3. *Cont.*

Tool Name	Consumption of Food Groups	Dietary Habits	Psychological Factors Associated with Nutrition	Other
9. Healthy Kids [30]	Vegetables, fruits, sugar-sweetened beverages, dairy, unhealthy snacks	Parent and child eating together, removing fat from meat		
10. Tool by Das and Ghosh [a] [31]				Knowledge on healthy dietary habits, nutrients and child nutrition practice
11. Start the Conversation 4–12 (STC-4-12) [32]	Vegetables and fruits, sugar-sweetened beverages, milk type, unhealthy snacks, fast food		Barriers and readiness to change regarding healthy eating	
12. Healthy Families Survey [33]	Vegetables, fruits, sugar-sweetened beverages, healthy snacks, unhealthy snacks	Eating out, parent and child eating together, picky eating		Parental modelling and parent–child interactions regarding healthy eating, parental food resource management and shopping behaviours
13. Knowledge, Attitudes and Habits (KAH-) questionnaire [34]	Vegetables, fruits, pastries	Consuming breakfast, lunch and dinner, having mid-morning snack, trying new foods	Attitudes towards healthy and unhealthy eating	Knowledge on healthy and unhealthy eating
14. Parental Self-efficacy Questionnaire [35]			Confidence regarding intake of vegetables, fruits, fruit juice, sugary drinks, sweets, dairy, grains, meat and alternatives, sodium, fats and eating out, eating together, child making healthy choices	
15. Tool by Chacko and Ganesan [36]	Vegetables, green leafy vegetables, fruit, cereals, pulses and dahl and non-vegetarian food, milk and coffee and tea and flavoured milk and curd, junk food, food from street shops	Mid-morning and evening snack, meal skipping		
16. Food, Health and Choices questionnaire (FHC-Q) [37]	Vegetables, fruits, sugar-sweetened beverages, processed packaged snacks, fast food		Self-determination, outcome expectations, self-efficacy, habit strength, goal intention, knowledge and social desirability regarding a healthy diet	
17. Healthy Eating and Physical Activity Self-Efficacy Questionnaire for Children (HEPASEQ-C) [38]			Self-efficacy to adhere to recommendations and to choose the healthy option when in temptation	
18. Healthy Eating and Physical Activity Behavior Recall Questionnaire for Children (HEPABRQ-C) [38]	Vegetables, number of colours of vegetables, fruits, soda pop, dairy, healthy snacks	Choosing the healthy option when eating out		

Table 3. *Cont.*

Tool Name	Consumption of Food Groups	Dietary Habits	Psychological Factors Associated with Nutrition	Other
19. Eating Behavior Questionnaire for School Children [a] [39]		Food responsiveness, meal timings, eating problems, meal preparation		
20. Tool by Drouin and Winickoff [40]	Sugar-sweetened beverages			
21. Child Nutrition and Physical Activity (CNPA) Screening Tool [41]	Vegetables, fruits, sugar-sweetened beverages, milk, milk type, fast food	Consuming breakfast, dinner eaten with adult	Perception, confidence and importance items on a healthy diet	
22. Electronic Kids Dietary Index (E-KINDEX) [42]	Vegetables, fruits and fruit juices, sweets and junk food, soft drinks, milk, bread, cereals and grain foods, meat, salted and smoked meat food, fish and seafood, legumes, fried food, grilled food	Consuming breakfast, number of main meals and snacks, eating in fast food restaurants or other eating places, eating with family, eating alone, eating of healthy food, eating meals in afternoon school, eating foods because they are advertised, eating whatever food is prepared at home, parental insistence to eat all the food, eating when not hungry	Beliefs and attitudes regarding an (un)healthy diet, weight, dieting	
23. Family Health Behavior Scale (FHBS) [43]		Consuming breakfast, eating three meals a day, eating at table, staying seated at the table, eating at a routine time, asking for unhealthy snacks, eating when bored, emotional eating, eating frequently, sneaking of food	Being influenced to eat or offered unhealthy foods by others	Choices and teaching on healthy foods by parents
24. Family Nutrition and Physical Activity (FNPA) screening tool [a] [44,45]	Food choices, beverage choices	Family eating patterns, family eating habits	Restriction/rewarding	
25. HABITS questionnaire [46]	Vegetables, fruits, fruit juice, sugar-sweetened beverages, milk, water, fast food meals, unhealthy snacks	Eating while watching TV, eating three meals a day, eating extra meals or snacks		
26. Healthy Living for Kids Survey (HLKS) [a] [47]	Vegetables, fruits, low fat milk, whole wheat bread		Self-efficacy and nutritional intention for healthy eating	
27. HeartSmartKids (HSK) [a] [48] (HeartSmartKids, LLC, Boulder, US)	At least: vegetables and fruits, sugar-sweetened beverages (incl. juice), milk, unhealthy snacks	At least: consuming breakfast, eating at restaurants, eating while watching TV		
28. Home Self-Administered Tool for Environmental Assessment of Activity and Diet (HomeSTEAD) [49]		Parent and child eating together at table, eating while TV is on	Parental autonomy support, atmosphere during meals	Parental control and limit setting, eating area decoration
29. Lifestyle Behavior Checklist (LBC) [50,51]			Parental problems experiencing and confidence in dealing with child's eating habits (e.g., eats too quickly, yells about food, hides food)	

Table 3. *Cont.*

Tool Name	Consumption of Food Groups	Dietary Habits	Psychological Factors Associated with Nutrition	Other
30. Pediatric Adapted Liking Survey (PALS) [52]	[b] Vegetables, fruits, sugar-sweetened beverages, dairy, meat, fish, beans, peanut butter, unhealthy snacks (sweet, salty and fat)			
31. Short-Form, Multicomponent Dietary Questionnaire (SF-FFQ4PolishChildren) [a] [53]	Vegetables, fruits, sugar-sweetened beverages, energy drinks, juices, sweets, dairy, fish, fast food	Breakfast consumption, frequency of having two meals per day		Nutrition knowledge
34. Adolescent Lifestyle Profile (ALP) [a] [56,57]	At least: vegetables, fruits, sweets, low fat dairy, chicken or fish instead of beef	At least: consuming breakfast		
35. Childhood Family Mealtime Questionnaire (CFMQ) (reduced) [58]		Mealtime structure, mealtime communication	Family mealtime stress	Appearance weight control
36. Diet–Lifestyle Index [59]	Vegetables, fruits, sweets and added sugars, dairy type, wholegrain, breakfast cereals	Consuming breakfast, eating foods not prepared at home, eating episodes per day, removing visible fat from meat/poultry		
37. Shortened Health-Promoting Lifestyle Profile (HPLP) II [a] [60]	NR	NR	NR	NR
38. Tool by Fernald et al. [a] [61]	At least: vegetables, fruits			
39. Tool by Hyun et al. [62]				Nutrition knowledge, including general knowledge and knowledge regarding food composition, nutrients and diseases
40. Tool by Hyun et al. [62]	Vegetables, green and orange vegetables, seaweed, fruits, dairy, meat and fish and egg and beans	Consuming breakfast, eating adequate amounts, combining food groups at each meal		
41. VISA-TEEN [63]	Vegetables and fruit, soft drinks, dairy, grains and potatoes, red meats, chicken and fish and eggs, butter and sweets, liquid excluding soft drinks			

Notes: Tools are sorted by target age. We numbered the tools in Table 3 as in Table 1. As tool number 32 and 33 do not describe nutrition items, they have been omitted from Table 3. [a] Specific items of screening tool not fully described; [b] liking/disliking of food items is used as proxy for intake.

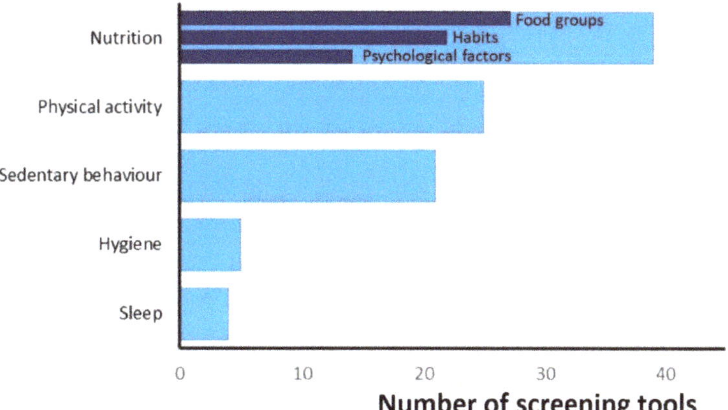

Figure 2. Prevalence of most frequently covered domains. N.B. The total number of covered domains exceeds the number of screening tools (*n* = 41) since most tools covered multiple domains.

3.2. Psychometric Properties

Table 4 demonstrates the validity and reliability outcomes of the included screening tools as illustrated by the different studies. For a total of 39 tools, psychometric properties were evaluated, whereas for two tools [36,61] they were not. The median sample size of the studies showing psychometric properties comprised 277 participants (IQR: 145, 486). Regarding reliability, Cronbach's α, as a measure of internal consistency, and the intraclass correlation coefficient (ICC), considering test–retest reliability, were assessed for 24 and 11 tools, respectively. Other measures of test–retest reliability, such as Cohen's kappa (κ, *n* = 4), Pearson's correlation coefficient (r, *n* = 4) and Spearman's rho (ρ, *n* = 2), were less evaluated. In general, internal consistency was moderate [64], but due to heterogeneity in the assessed concepts and tool aims, comparison between studies was not appropriate. Test–retest reliability was also highly variable, with eight tools clearly reaching cut-offs for 'sufficiency' based on ICC or κ [22,23,25,26,28,31,52,55,63,65]. Regarding validity, features of criterion validity were determined mostly. Criterion validity included sensitivity and specificity (*n* = 6, e.g., to detect nutritional risk or obesity) as well as concurrent validity (*n* = 31, e.g., association of tool score with body mass index (BMI)). Predictive validity was not assessed for any tool. Specifically, the 'NutricheQ' was tested for sensitivity, specificity, associations with food group intake and nutrient intake based on a 4-day weighed food diary, and associations with BMI z-scores [18–20]. The other screening tools were validated less extensively, usually comprising only one dimension of validity.

Table 4. Psychometric properties of lifestyle screening tools for children in the community setting.

Tool Name	Country	Sample Size	Age	Reliability	Criterion Validity
1. NutricheQ	Ireland [18]	$N = 371$	1–3 y	Internal consistency, $\alpha = 0.50$	Total score was associated with (4-day weighted food diary) intakes of fruits, vegetables, protein, dietary fibre, non-milk sugars, iron, vitamin D, zinc, calcium, riboflavin, niacin, folate, thiamine, phosphorous, potassium, carotene and retinol ($r = -0.390$–0.119, $p < 0.05$) A score > 4 (AUC = 76%) identified moderate risk with sensitivity = 83% and specificity = 48% A score > 8 (AUC = 85%) identified high risk with sensitivity = 70% and specificity = 80%
	Italy [19]	$N = 201$	1–3 y	Internal consistency, $\alpha = 0.83$ for Section 1 and $\alpha = 0.70$ for Section 2 ICC = 0.73 (95% CI [0.40, 0.89], $p = 0.0002$) for Section 1 and ICC = 0.55 (95% CI [0.13, 0.81], $p = 0.0074$) for Section 2	In Section 1, a score ≥ 4 identified toddlers with a poor iron intake (AUC = 0.678, $p = 0.001$) and a score of ≥ 2 identified toddlers exceeding the En% protein intake (AUC = 0.6024, $p = 0.009$). In Section 2, a score of ≥ 3 identified toddlers with poor fibre intake (AUC = 0.7028, $p < 0.0001$)
	Lebanon [20]	$N = 467$	1–3 y		Total score was associated with age and BMI ($r = 0.11$, $p = 0.021$ (for both)), and with fat ($\rho = 0.148$, $p = 0.039$) and fibre ($\rho = -0.137$, $p = 0.031$) intake AUC = 0.457 for correctly classifying toddlers into the high risk group based on their BMI z-score
2. Toddler Feeding Questionnaire (TFQ) [21]	United States	$N = 629$	3–5 y	Internal consistency, $\alpha = 0.66$ for indulgent subscale, $\alpha = 0.65$ for authoritative subscale, $\alpha = 0.48$ for environmental subscale	Indulgent subscale scores were correlated with the HEI-2010 ($\rho = -0.22$, $p < 0.001$), kcal/d ($\rho = 0.11$, $p = 0.011$), grams of fat/day ($\rho = 0.12$, $p = 0.008$), servings of vegetables ($\rho = -0.11$, $p = 0.01$), servings of desserts ($\rho = 0.13$, $p = 0.002$) and servings of sugary drinks ($\rho = 0.23$, $p < 0.001$) Authoritative subscale scores were correlated with the HEI-2010 ($\rho = 0.15$, $p < 0.001$), servings of vegetables ($\rho = 0.11$, $p = 0.011$), servings of desserts ($\rho = -0.15$, $p < 0.001$) and servings of sugary drinks ($\rho = -0.09$, $p < 0.039$) Environmental subscale scores were correlated with HEI-2010 ($\rho = -0.12$, $p = 0.004$), kcals/day ($\rho = 0.12$, $p = 0.007$), grams of fat/day ($\rho = 0.14$, $p = 0.001$), servings of desserts ($\rho = 0.13$, $p = 0.003$) and servings of sugary drinks ($\rho = 0.22$, $p < 0.001$)

Table 4. *Cont.*

Tool Name	Country	Sample Size	Age	Reliability	Criterion Validity
3. Toddler NutriSTEP [22]	Canada	$N = 200$	18–35 m	ICC = 0.951 ($p < 0.001$)	Total score was associated with dietician risk score ($\rho = 0.67$, $p < 0.000$) A score $\geq$ 21 identified moderate risk with sensitivity = 86% and specificity = 61% A score $\geq$ 26 identified high risk with sensitivity = 95% and specificity = 63%
4. Toddler Dietary Questionnaire (TDQ) [23]	Australia	$N = 111$	12–36 m	Total score ICC = 0.90 ($p < 0.001$) All children were classified into the same ($n = 83$, 75%) or adjacent ($n = 28$, 25%) dietary risk category during each administration Test–retest reliability for individual items, $\kappa_w = 0.40$–0.78	Total score and food frequency questionnaire risk score were associated (r = 0.71, $p < 0.001$) Classification analysis between the TDQ and food frequency questionnaire revealed that all the participants were classified into the same ($n = 88$, 79%) or adjacent ($n = 23$, 21%) dietary risk category
5. Child Eating Behavior Questionnaire (CEBQ) [24]	Sweden	$N = 1271$	3–8 y	Internal consistency: $\alpha = 0.73$	NR
6. Nutrition Screening Tool for Every Preschooler (NutriSTEP)	Canada [25]	$N = 269$	3–5 y	Total score ICC = 0.89 (95% CI [0.85, 0.92], $p < 0.001$)Test–retest reliability for individual items, $\kappa = 0.39$–1.0	Total score was associated with dietician risk rating (r = 0.48, $p = 0.01$) A score > 20 identified moderate risk with sensitivity = 53% and specificity = 79% A score > 25 identified high risk with sensitivity = 84% and specificity = 46%
	Canada [26]	$N = 63$ for internet use $N = 64$ for onscreen use	3–5 y	Internet use total score ICC = 0.91 (95% CI [0.90, 0.96]) Onscreen use total score ICC = 0.91 (95% CI [0.85, 0.95]) Test–retest reliability among risk categories, $\kappa = 0.58$ ($p = 0.000$) for internet use and $\kappa = 0.50$ ($p = 0.000$) for onscreen use	NR
	Iran [27]	$N = 192$	4–6 y	Test–retest reliability, r = 0.68 ($p < 0.001$)	Total score was associated with nutritionist risk score (r = 0.23, $p = 0.003$) and with healthy eating index (r = -0.16, $p = 0.03$) A score > 27 identified moderate risk with sensitivity = 41.7% and specificity = 85.7% A score > 31 identified high risk with sensitivity = 38.9% and specificity = 84.4%

Table 4. *Cont.*

Tool Name	Country	Sample Size	Age	Reliability	Criterion Validity
7. Preschooler Dietary Questionnaire (PDQ) [28]	Australia	$N = 74$	3–5 y	Total score ICC = 0.87 (95% CI [0.07, 2.95], $p = 0.040$)	Total score and food frequency questionnaire risk score were associated (r = 0.85, $p = 0.009$) PDQ scores were associated with the number of people per household ($\beta = -0.32$, 95% CI [-6.69, -0.59], $p = 0.020$), but not BMI z-score ($\beta = -0.09$, 95% CI [-0.02, -0.04], $p = 0.512$)
8. Preschoolers Diet–Lifestyle Index (PDL-index) [29]	Greece	$N = 2287$	2–5 y	NR	A 1/44 unit score increase was associated with an OR for obesity of 0.95 (95% CI [0.92, 0.98]) and an OR of 0.97 (95% CI [0.95, 0.99]) for overweight/obesity Correct classification rate for obesity = 85%, for overweight/obesity = 67% Sensitivity for obesity = 60%, for overweight/obesity = 55% Specificity for obesity and overweight/obesity = 52%
9. Healthy Kids [30]	United States	$N = 133$	2–5 y	Internal consistency, $\alpha = 0.76$ Test–retest reliability coefficient = 0.74 ($p \leq 0.01$)	The Healthy Kids scale score was inversely associated with BMI percentiles ($p = 0.02$)
10. Tool by Das and Ghosh [31]	India	$N = 134$	3–6 y	Internal consistency, $\alpha = 0.87$ Total score ICC = 0.77 ($p <0.01$)	NR
11. Start the Conversation 4–12 (STC-4-12) [32]	United States	$N = 115$	4–12 y	NR	Three of five queried dietary barriers were found to be significantly associated with at least one healthy eating behaviour Four of five queried barriers to PA were significantly associated with at least one PA-related behaviour
12. Healthy Families Survey [33]	United States	$N = 1376$	6–11 y	Internal consistency for subscales, $\alpha = 0.51$–0.77	NR
13. Knowledge, Attitudes and Habits (KAH-) questionnaire [34]	Spain	$N = 295$	6–7 y	Internal consistency, $\alpha = 0.79$	NR
14. Parental Self-efficacy Questionnaire [35]	United States	$N = 146$	6–11 y	Internal consistency, $\alpha = 0.94$ Test–retest reliability, r = 0.94 ($p < 0.001$)	NR
15. Tool by Chacko and Ganesan [36]	India	NR	6–17 y	NR	NR
16. Food, Health and Choices questionnaire (FHC-Q) [37]	United States	$N = 221$	9–11 y	Internal consistency: $\alpha = 0.77$–0.92 for behaviour scales and $\alpha = 0.44$–0.83 for psychosocial scales ICC = 0.59–0.81 for behaviours ($p < 0.001$) and ICC = 0.51–0.68 for continuous psychosocial determinants ($p <0.05$)	Correlation coefficients between the FHC-Q and reference questionnaires were all statistically significant ($p < 0.01$)

Table 4. *Cont.*

Tool Name	Country	Sample Size	Age	Reliability	Criterion Validity
17. Healthy Eating and Physical Activity Self-Efficacy Questionnaire for Children (HEPASEQ-C) [38]	United States	$N = 492$	9–13 y	Internal consistency, $\alpha = 0.75$	HEPASEQ-C was significantly correlated with HEPABRQ-C, r = 0.50 ($p = 0.000$)
18. Healthy Eating and Physical Activity Behavior Recall Questionnaire for Children (HEPABRQ-C) [38]	United States	$N = 492$	9–13 y	NR	HEPABRQ-C was significantly correlated with HEPASEQ-C, r = 0.50 ($p = 0.000$)
19. Eating Behavior Questionnaire for School Children [39]	India	$N = 462$	10–12 y	NR	No correlation between tool subscores and anthropometric measures (exact numerical data NR)
20. Tool by Drouin and Winickoff [40]	United States	$N = 626$	0–18 y	NR	Parents receiving the tool were not more likely to receive counselling or service delivery by clinicians than participants not screened No statistical difference in the proportion of parents reporting having taken steps towards correcting the behaviour in the parents that received the screening after one month follow-up
21. Child Nutrition and Physical Activity (CNPA) Screening Tool [41]	United States	$N = 2230$	2–18 y	Internal consistency: $\alpha = $ 'low', exact value NR	Both generated readiness to change and perception subscores were associated with weight status categories ($p < 0.001$)
22. Electronic Kids Dietary Index (E-KINDEX) [42]	Greece	$N = 622$	9–13 y	Internal consistency: $\alpha = 0.60$	Each 1 SD (i.e., 7.81 points) score increase was associated with a 2.31 ± 0.23 kg/m^2 decrease in BMI ($p < 0.001$), a 2.23 ± 0.35 decrease in calculated % body fat ($p < 0.001$) and a 2.16 ± 0.61 cm decrease in waist circumference ($p < 0.001$) Correct classification rate for excess body fat was 84% (95% CI [0.74, 0.94]) Sensitivity for overweight/obesity versus normal weight = 74%, for obesity versus normal weight/overweight = 61% Specificity for overweight/obesity versus normal weight = 46%, for obesity versus normal weight/overweight = 79%

Table 4. *Cont.*

Tool Name	Country	Sample Size	Age	Reliability	Criterion Validity
23. Family Health Behavior Scale (FHBS) [43]	United States	N = 233	5–12 y	Internal consistency: α = 0.86 Test–retest reliability coefficient = 0.85	FHBS was inversely associated with zBMI (r = −0.28, p< 0.01) Every unit increase was associated with an OR of 0.96 (95% CI [0.95, 0.99] for overweight/obesity (p < 0.01) Correct classification rate for weight classification = 63%
24. Family Nutrition and Physical Activity (FNPA) screening tool	United States [44]	N = 349	1st and 10th grade	NR	At both ages, the FNPA score was not significantly correlated with BMI% Only in first graders, scores in the lowest tertile were associated with higher odds for overweight/obesity compared to the highest tertile (OR = 2.49, 95% CI [1.17, 5.31])
	United States [44]	N = 19	2–5 y	NR	NR
25. HABITS questionnaire [46]	United States	N = 35	7–16 y	Internal consistency, α = 0.61 for dietary subscale and α = 0.59 for PA/sedentary behaviour subscale Test–retest reliability, κ = 0.27–0.78 for individual items of dietary subscale and κ = 0.29–0.48 for PA/sedentary behaviour subscale As a whole, the dietary subscale and PA/sedentary behaviour subscales had test–retest reliabilities of r = 0.94 and r = 0.87, respectively	NR
26. Healthy Living for Kids Survey (HLKS) [47]	United States	N = 88	9–12 y	Internal consistency for subscales, α = 0.63–0.80 Test–retest reliability for subscales: r = 0.37–0.78	NR
27. HeartSmartKids (HSK) [48] (HeartSmartKids, LLC, Boulder, US)	United States	N = 103	9–14 y	Test–retest reliability, ρ = 0.38–0.78	Each item of the HSK was significantly correlated with the HABITS, ρ = 0.21–0.65 (p <0.05)
28. Home Self-Administered Tool for Environmental Assessment of Activity and Diet (HomeSTEAD) [49]	United States	N = 129	3–12 y	Internal consistency for subscales, α = 0.62–0.93 Subscale ICC = 0.57–0.89	No statistically significant correlation between factor composite scores and child BMI z-scores

Table 4. *Cont.*

Tool Name	Country	Sample Size	Age	Reliability	Criterion Validity
29. Lifestyle Behavior Checklist (LBC)	Australia [50]	$N = 182$	4–11 y	Internal consistency, $\alpha = 0.97$ for Problem scale and $\alpha = 0.92$ for Confidence scale Test–retest reliability, $\rho = 0.87$ ($p < 0.001$) for Problem scale and $\rho = 0.66$ ($p < 0.001$) for Confidence scale	Correct classification rate for obesity was 91%
	The Netherlands [51]	$N = 273$	3–13 y	Internal consistency, $\alpha = 0.92$ for Problem scale and $\alpha = 0.98$ for Confidence scale Test–retest reliability, $\rho = 0.74$ ($p < 0.001$) for Problem scale and $\rho = 0.70$ ($p < 0.001$) for Confidence scale	Parents with healthy weight children scored lower on the Problem scale, F = 16.94 ($p < 0.001$), compared to those with overweight children The Problem scale was associated with nurturance ($\rho = -0.23$, $p < 0.01$), restrictiveness ($\rho = 0.14$, $p < 0.05$), psychological control ($\rho = 0.19$, $p < 0.01$) and BMI of child ($\rho = 0.21$, $p < 0.01$), mother ($\rho = 0.23$, $p < 0.01$) and father ($\rho = 0.14$, $p < 0.05$) The Confidence scale was associated with nurturance ($\rho = 0.14$, $p < 0.05$) and psychological control ($\rho = -0.22$, $p < 0.01$)
30. Pediatric Adapted Liking Survey (PALS) [52]	United States	$N = 144$	5–17 y	Internal consistency for subscales, $\alpha = 0.40$–0.72 ICC for individual items = 0.79–0.91	In girls, higher BMI was associated with greater preference for fat/sweet/salty foods ($\beta = 0.32$, 95% CI [0.14, 1.15], $p < 0.05$)
31. Short-Form, Multicomponent Dietary Questionnaire (SF-FFQ4PolishChildren) [53]	Poland	$N = 437$ children $N = 630$ adolescents	6–10 y 11–15 y	Test–retest reliability for consumption of food items and meals, $\kappa = 0.46$–0.81 in children, $\kappa = 0.30$–0.54 in adolescent's test–retest, and $\kappa = 0.27$–0.56 in adolescent's test and parent's retest Across study groups, test–retest reliability, $\kappa = 0.31$–0.72 for active/sedentary lifestyle items, $\kappa = 0.55$–0.93 for components of the Family Affluence Scale, $\kappa = 0.64$–0.67 for BMI categories, $\kappa = 0.36$ for the nutrition knowledge of adolescents and $\kappa = 0.62$ for the nutrition knowledge of children's parents	NR
32. Tool by Hendrie et al. [54]	Australia	$N = 106$	5–11 y	Internal consistency, $\alpha = 0.83$	The family activity environment was associated with children's fruit and vegetable intake assessed with a 24-h recall (r = 0.34, $p < 0.01$), PA assessed by the Children's Leisure Activity Study Survey (r = 0.27, $p < 0.01$) and screen time (r = -0.24, $p < 0.05$) assessed by a survey

Table 4. *Cont.*

Tool Name	Country	Sample Size	Age	Reliability	Criterion Validity
33. Tool by Huang et al. [55]	China	$N = 303$	9–14 y	Internal consistency for identified factors, $\alpha = 0.50$–0.86 Identified factor ICC = 0.82–0.89	Self-efficacy (r = 0.25, $p < 0.05$), home physical activity environment (r = 0.14, $p < 0.05$) and peer support (r = 0.25, $p < 0.05$) were associated with child-reported moderate-to-vigorous PA Family support for PA was associated with screen time (r = −0.22, $p < 0.05$)
34. Adolescent Lifestyle Profile (ALP)	United States [56]	$N = 207$	10–15 y	Internal consistency: $\alpha = 0.91$	ALP correlated with hope (r = 0.60, $p = 0.001$), self-efficacy (r = 0.47, $p = 0.001$) and self-esteem (r = 0.35, $p = 0.001$) scores
	Portugal [57]	$N = 236$	12–18 y	Internal consistency: $\alpha = 0.87$	NR
35. Childhood Family Mealtime Questionnaire (CFMQ) (reduced) [58]	United States	$N = 280$	11–15 y	Internal consistency for identified factors, $\alpha = 0.76$–0.82	Childhood mealtime communication was associated with physically active days ($\beta = 0.20$, 95% CI [0.07, 0.32], $p < 0.01$), fruits and vegetable intake ($\beta = 0.29$, 95% CI [0.15, 0.45], $p < 0.001$) and added sugar intake ($\beta = 0.23$, 95% CI [0.09, 0.37], $p < 0.001$) Childhood mealtime stress was associated with fruits and vegetable intake ($\beta = 0.26$, 95% CI [0.08, 0.45], $p < 0.01$) and added sugar intake ($\beta = 0.38$, 95% CI [0.21, 0.57], $p < 0.001$)
36. Diet–Lifestyle Index [59]	Greece	$N = 2008$	12–17 y	NR	The Diet–Lifestyle Index was inversely associated with BMI in boys ($\rho = -0.169$, $p < 0.001$) and girls ($\rho = -0.143$, $p < 0.001$) An 11/57 unit score increase was associated with an OR of 0.93 (95% CI [0.90, 0.96]) for overweight/obesity ($p < 0.001$) Correct classification rate for BMI category = 83% Sensitivity for overweight/obesity = 66%, specificity = 50%
37. Shortened Health-Promoting Lifestyle Profile (HPLP) II [60]	Iran	$N = 495$	14–18 y	Internal consistency, $\alpha = 0.86$	Total HPLP-II was associated with quality of life (r = 0.24, $p < 0.001$), self-efficacy (r = 0.48, $p < 0.001$) and demographic variables (data NR)
38. Tool by Fernald et al. [61]	United States	$N = 227$	Average 15 y	NR	NR
39. Tool by Hyun et al. [62]	Korea and China	$N = 406$	15–18 y	NR	Nutrition knowledge was associated with body shape satisfaction in Korean boys (r = 0.208, $p < 0.01$), not in Chinese boys

Table 4. *Cont.*

Tool Name	Country	Sample Size	Age	Reliability	Criterion Validity
40. Tool by Hyun et al. [62]	Korea and China	$N = 406$	15–18 y	NR	Healthy dietary habits were associated with body shape satisfaction in Chinese boys ($r = 0.210$, $p < 0.01$), not in Korean boys
41. VISA-TEEN [63]	Spain	$N = 419$	13–19 y	Internal consistency, $\alpha = 0.66$ Total score ICC = 0.86 (95% CI [0.82, 0.89])	Total VISA-TEEN score was associated with KIDSCREEN-10 ($r = 0.21$, $p < 0.001$) and self-rated health ($p < 0.001$)

Note: Tools are sorted by target age. Abbreviations: y, years; ICC, intraclass correlation coefficient; NR, not reported; HEI-2010, Healthy Eating Index 2010.

3.3. Implementation

A total of 35 tools calculated a subscore and/or total score. Six tools defined score cut-offs for the identification of risk [18–20,22,23,25–28,53]. Eighteen tools provided some form of a prospect of action following the answers given. Two of these tools [32,40] based their prospects of action on highlighted topics, whereas the other sixteen based prospects of action on tool scores. None of the tools for adolescents provided a prospect of action. The prospects of action could be intended for the health care professional, child or parent. It included counselling, education, a combination of these two, initiating the conversation about a healthy lifestyle or referring to a specialist for further examination, and/or treatment. Articles on the 'NutriSTEP', 'Start the Conversation 4–12', 'tool by Drouin and Winickoff', 'HeartSmartKids' (HeartSmartKids, LLC, Boulder, US) and 'Pediatric Adapted Liking Survey' described that their prospects of action are tailored to the answers given, but details on them were lacking [25–27,32,40,48,52]. The 'NutricheQ' was advised to be administered during regular growth check-ups [18–20]. Other tools did not describe recommendations for administering occasion or frequency. Despite being developed for out-of-hospital use, the intended target location of administering the tools was merely suggested. When administration methods were reported, it involved paper (n = 15) or online (n = 10) formats. The 'NutriSTEP' paper version was expanded by an internet and onscreen version in response to the interest of health care professionals [26] and the 'Food, Health and Choices questionnaire' used an audience response system to decrease administer burden [37]. Others did not describe their motivation for the choice of administration methods.

4. Discussion

The 41 lifestyle screening tools for children included in this review varied widely in their design, but items on nutrition, PA and sedentary behaviour/screen time were commonly addressed. Nutrition items predominantly covered the intake of specific food groups, dietary habits and psychological factors, such as (parental) beliefs and attitudes towards a healthy lifestyle. For most tools, one or more aspects of reliability and/or validity had been studied with varying results. Nearly half of the screening tools offered prospects of action, but none described the exact follow-up actions based on tool outcomes. Moreover, other features of implementation were sparse.

Most tools evaluated lifestyle determinants related to overweight and obesity. Considering overweight, domains related to energy balance, i.e., nutrition, PA and sedentary behaviour, were frequently evaluated. Compared to PA and sedentary behaviour/screen time, which mainly concerned frequency and duration, there was more variety in nutrition items, which reflects the versatility of this topic. The tools not only addressed the intake of foods directly related to energy intake, such as sugar-sweetened beverages and unhealthy snacks/fast food but also foods and dietary habits that might be more indirectly associated with weight status, such as fruits and vegetables, having breakfast and eating together at the table [66–68]. The concept of a balanced diet, characterised by adequate amounts and proportions of nutrients required for good health, is broader than energy balance alone. The 'NutricheQ' aimed to evaluate the risk of dietary imbalances in toddlers, with a particular focus on iron and vitamin D [18–20]. Next to iron and vitamin D, the total score of the 'NutricheQ' was associated with the intake of fruits, vegetables, protein, dietary fibre, non-milk sugars and other specific micronutrients [18], and its 18-item version score was also associated with BMI z-scores [20], indicating extensive dietary exploration. It could be proposed that screening tools addressing both dietary and energy balance may be most effective in screening for the risk of overall health problems, including overweight. This could for instance be conducted through the assessment of children's adherence to age-specific recommendations for commonly consumed food groups.

While there is emerging evidence on the importance of sleep on weight status and overall health [69,70], only four tools covered sleep. This finding accords with the results of Byrne et al., who conducted a systematic review on brief tools measuring obesity-related behaviours for children under five years of age [17]. Only two out of their twelve appraised tools covered sleep, indicating paucity [17]. Regarding the specific items on sleep, sleep duration was the most common in our results. A systematic review on sleep and childhood obesity supports the relevance of sleep duration on weight status but stated that associations with other dimensions, such as sleep quality and bedtime, need to be studied further [69]. The previous findings that shorter sleep duration in children is associated with unhealthy dietary habits and lower PA suggest a pathway from sleep deficiency to obesity and indicate that certain lifestyle behaviours might cluster in individuals [71,72].

The ten screening tools specifically developed for toddlers and preschoolers covered fewer domains than the tools for the other age groups; yet, all comprised nutrition. The early years of life form a critical window of opportunity for growth and development, in which proper nutrition is fundamental [1]. However, other lifestyle factors, such as PA, sedentary behaviour and sleep, have also been shown to affect health in toddlers and preschoolers [5–7]. An explanation for the lack of these domains in tools for toddlers and preschoolers might be that guidelines on these topics for this age group are not universally available. Howbeit, none of the reviewed articles clearly justified their choice of the exact items included. Depending on the aim of the lifestyle screening tool, it could be useful to base tool domains on clustering lifestyle behaviours in the target population to provide integrated follow-up advice. In addition, it might be valuable to study accurate indicators of an unhealthy lifestyle in advance. Furthermore, the accuracy of the questions should be optimized to obtain the desired information (e.g., the exact question to evaluate general vegetable intake).

In addition to lifestyle behaviours and habits, the included screening tools evaluated psychological factors related to lifestyle. Psychological factors, such as parental attitudes towards healthy eating and self-efficacy to adhere to recommendations, are important [73]. On the one hand, these perceptions can imply certain behaviours. On the other, they can map motivation and perceived barriers for behaviour change. As children's lifestyle behaviour is highly reliant on parental support behaviours [74], it is helpful to evaluate parental perceptions regarding lifestyle. When health care professionals gain an insight into parental indicators of behaviour change, they obtain cues for motivational interviewing to help parents and children shifting towards a healthier lifestyle.

Although 39 out of 41 screening tools had undergone some form of psychometric testing, the results were inconclusive and hardly comparable due to high heterogeneity in tool aim and study design. However, a number of tools, such as the 'NutricheQ', 'NutriSTEP' and Lifestyle Behavior Checklist [18–20,25–27,50,51], have been researched more thoroughly than others and may therefore have a more solid foundation for use in practice. Becker et al. [14] concluded in their review that no nutrition screening tool for children in the community setting provided enough evidence for moderate to high validity and reliability [14]. As the reliability and validity influence the effectiveness of screening tools, assessing these psychometric properties is crucial. Nevertheless, the interpretation of group-level validity and reliability for individual counselling should be performed with prudence [75]. Proper psychometric assessment should also take into account differences in socioeconomic status and language and fill the current gap in testing predictive validity. The lack of a gold standard for screening children's lifestyle impairs the validity testing of new lifestyle screening tools. Nonetheless, studying the association of validated dietary assessment methods and activity trackers with items of lifestyle screening tools could assess criterion validity. In addition, longitudinal studies addressing a common outcome of an unhealthy lifestyle, such as overweight, and applying identical intervention strategies could study the effectiveness of a new tool over another one or over a health care professional's clinical view.

Eighteen tools provided recommendations for actions to be taken based on the answers given. Overall, these recommendations for both children and parents were as general as 'receiving tips' or health care professionals 'offering counselling' or 'referring to a specialist', and are therefore open to interpretation. Neither of the tools that identified cut-offs for particular risk classifications defined clear follow-up actions according to the classification. This is in contrast with established nutrition screening tools for hospitalised children, which offer specific action points per identified risk group [76–79]. Defining risk score cut-offs corresponding with unambiguous follow-up steps, such as 'no action required', 'discuss lifestyle with parents and repeat screening in X weeks' and 'initiate further examination by a specialist', might strengthen the effectiveness of lifestyle screening tools. Considering the various domains of lifestyle, integrating subscores and cut-offs for different domains could pinpoint the areas that need attention and guide health care professionals to address these specifically.

With this review, we have created a hitherto lacking overview of the literature. Searching for screening tools encompassing lifestyle in the broadest sense of the term made our search strategy comprehensive and enabled the inclusion of tools that evaluate a broad variety of lifestyle determinants. Our additional focus explicitly on nutrition highlighted the importance of this topic within children's lifestyle.

Not preselecting specific lifestyle factors (except nutrition) in our search strategy could also be considered a limitation, as we may have missed articles on screening tools that only denote specific determinants (e.g., PA and screen time), without framing them in the context of lifestyle in general. Moreover, we might have missed certain screening tools due to publication bias. Another important concern was the definition of screening tools, which we predefined in our protocol as tools that assign a certain value to behaviour and/or characteristics and/or offer prospects of action to an individual. The ascertainment of screening tools was performed in duplicate and independently, but the lack of a universal definition may have hampered the robustness of our methods. As this review was conducted to provide an overview of all recent literature on lifestyle screening tools for children in the community setting, regardless of methodological quality and tool outcome, we did not include a quality or risk of bias assessment. However, we expect that the limitations of this review have not altered the main conclusions and that we gained clear insights into existing lifestyle screening tools for children.

Ideally, a balance exists between the set of items retrieving as much information as possible and convenience by the person completing the tool. Considering the association between questionnaire length and response burden [80], future studies should target the optimal number of items relative to the aim of the screening tool. Moreover, addressing aspects of implementation of a screening tool might contribute to fulfilling the potential of its usage. For example, studies that explore the most effective administration method (e.g., paper format, online or mobile application), setting (e.g., at home or at a clinic) and target group of health care professionals handling the results of the screening tool could detect vital features in making the screening tool advantageous. Finally, it is crucial to validate current and new lifestyle screening tools to identify children at risk as early as possible.

5. Conclusions

This systematic review shows that a fair variety exists in lifestyle screening tools for children in the community setting. The majority addressed dietary and/or lifestyle behaviours and habits related to overweight and obesity. Domains that were mostly covered included nutrition, PA and sedentary behaviour/screen time. Tool validation was, however, limited, and the availability of unambiguous prospects of actions following tool outcomes was lacking. Considering the importance of a healthy lifestyle during childhood, there is a need for an easy-to-administer lifestyle screening tool for children with distinct follow-up actions in order to improve a child's lifestyle at an early age.

Supplementary Materials: The following supporting information can be downloaded at: https://www.mdpi.com/xxx/s1, Supplementary file S1: Search strategy.

Author Contributions: K.J., C.v.R. and J.V.-K. conceptualized the study. A.K. and S.t.B. performed the search and article screening and selection. A.K., S.t.B., C.v.R. and L.E. interpreted the results. A.K. wrote the original draft of the manuscript. A.K., S.t.B., L.E., C.v.R., J.V.-K., E.S., H.R. and K.J. reviewed and edited all drafts. All authors have read and agreed to the published version of the manuscript.

Funding: This study was conducted as part of the larger project 'Nutrition and lifestyle screening tool for youth healthcare 2019–2022' and funded by the Dutch Ministry of Health, Welfare and Sport (VWS). The funding body had no role in the design and execution of the study nor in interpreting the data and writing the manuscript.

Institutional Review Board Statement: Not applicable.

Informed Consent Statement: Not applicable.

Data Availability Statement: Not applicable.

Conflicts of Interest: The authors declare no conflict of interest.

References

1. Prado, E.L.; Dewey, K.G. Nutrition and brain development in early life. *Nutr Rev.* **2014**, *72*, 267–284. [CrossRef] [PubMed]
2. Laitinen, T.T.; Pahkala, K.; Magnussen, C.G.; Viikari, J.S.; Oikonen, M.; Taittonen, L.; Mikkila, V.; Jokinen, E.; Hutri-Kahonen, N.; Laitinen, T.A.; et al. Ideal cardiovascular health in childhood and cardiometabolic outcomes in adulthood: The Cardiovascular Risk in Young Finns Study. *Circulation* **2012**, *125*, 1971–1978. [CrossRef] [PubMed]
3. World Health Organization. *Healthy Living: What Is a Healthy Lifestyle?* WHO Regional Office for Europe: Copenhagen, Denmark, 1999.
4. Kumar, S.; Kelly, A.S. Review of childhood obesity: From epidemiology, etiology, and comorbidities to clinical assessment and treatment. *Mayo Clin. Proc.* **2017**, *92*, 251–265. [CrossRef] [PubMed]
5. Carson, V.; Lee, E.Y.; Hewitt, L.; Jennings, C.; Hunter, S.; Kuzik, N.; Stearns, J.A.; Powley Unrau, S.; Poitras, V.J.; Gray, C.; et al. Systematic review of the relationships between physical activity and health indicators in the early years (0–4 years). *BMC Public Health* **2017**, *17* (Suppl. 5), 854.
6. Poitras, V.J.; Gray, C.E.; Janssen, X.; Aubert, S.; Carson, V.; Faulkner, G.; Goldfield, G.S.; Reilly, J.J.; Sampson, M.; Tremblay, M.S. Systematic review of the relationships between sedentary behaviour and health indicators in the early years (0–4 years). *BMC Public Health* **2017**, *17* (Suppl. 5), 868. [CrossRef]
7. Chaput, J.P.; Gray, C.E.; Poitras, V.J.; Carson, V.; Gruber, R.; Birken, C.S.; MacLean, J.E.; Aubert, S.; Sampson, M.; Tremblay, M.S. Systematic review of the relationships between sleep duration and health indicators in the early years (0–4 years). *BMC Public Health* **2017**, *17* (Suppl. 5), 855. [CrossRef]
8. Correia, M. Nutrition screening vs nutrition assessment: What's the difference? *Nutr. Clin. Pract.* **2018**, *33*, 62–72. [CrossRef]
9. Hartman, C.; Shamir, R.; Hecht, C.; Koletzko, B. Malnutrition screening tools for hospitalized children. *Curr. Opin. Clin. Nutr. Metab. Care* **2012**, *15*, 303–309. [CrossRef]
10. Huysentruyt, K.; Devreker, T.; Dejonckheere, J.; De Schepper, J.; Vandenplas, Y.; Cools, F. Accuracy of nutritional screening tools in assessing the risk of undernutrition in hospitalized children. *J. Pediatr. Gastroenterol. Nutr.* **2015**, *61*, 159–166. [CrossRef]
11. Joosten, K.F.; Hulst, J.M. Nutritional screening tools for hospitalized children: Methodological considerations. *Clin. Nutr.* **2014**, *33*, 1–5. [CrossRef]
12. Klanjsek, P.; Pajnkihar, M.; Marcun Varda, N.; Povalej Brzan, P. Screening and assessment tools for early detection of malnutrition in hospitalised children: A systematic review of validation studies. *BMJ Open* **2019**, *9*, e025444. [CrossRef] [PubMed]
13. Teixeira, A.F.; Viana, K.D. Nutritional screening in hospitalized pediatric patients: A systematic review. *J. Pediatr.* **2016**, *92*, 343–352. [CrossRef] [PubMed]
14. Becker, P.J.; Gunnell Bellini, S.; Wong Vega, M.; Corkins, M.R.; Spear, B.A.; Spoede, E.; Hoy, K.; Piemonte, T.A.; Rozga, M. Validity and reliability of pediatric nutrition screening tools for hospital, outpatient, and community settings: A 2018 evidence analysis center systematic review. *J. Acad. Nutr. Diet.* **2020**, *120*, 288–318.e2. [CrossRef] [PubMed]
15. Moher, D.; Liberati, A.; Tetzlaff, J.; Altman, D.G.; Group, P. Preferred reporting items for systematic reviews and meta-analyses: The PRISMA statement. *PLoS Med.* **2009**, *6*, e1000097.
16. Cochrane. Cochrane PICO Ontology. Available online: https://linkeddata.cochrane.org/pico-ontology (accessed on 15 December 2020).
17. Byrne, R.; Bell, L.; Taylor, R.W.; Mauch, C.; Mihrshahi, S.; Zarnowiecki, D.; Hesketh, K.D.; Ming Wen, L.; Trost, S.G.; Golley, R. Brief tools to measure obesity-related behaviours in children under 5 years of age: A systematic review. *Obes. Rev.* **2019**, *20*, 432–447.
18. Rice, N.; Gibbons, H.; McNulty, B.A.; Walton, J.; Flynn, A.; Gibney, M.J.; Nugent, A.P. Development and validation testing of a short nutrition questionnaire to identify dietary risk factors in preschoolers aged 12–36 months. *Food Nutr. Res.* **2015**, *59*, 27912. [CrossRef]

19. Morino, G.S.; Cinelli, G.; Di Pietro, I.; Papa, V.; Spreghini, N.; Manco, M. NutricheQ Questionnaire assesses the risk of dietary imbalances in toddlers from 1 through 3 years of age. *Food Nutr. Res.* **2015**, *59*, 29686.

20. Aramouny, E.; Sacy, R.; Chokr, I.; Joudy, B. Local validation study for NutricheQ tool in Lebanon. *J. Compr. Pediatrics* **2018**, *9*, e63573. [CrossRef]

21. Heerman, W.J.; Lounds-Taylor, J.; Mitchell, S.; Barkin, S.L. Validity of the toddler feeding questionnaire for measuring parent authoritative and indulgent feeding practices which are associated with stress and health literacy among Latino parents of preschool children. *Nutr. Res.* **2018**, *49*, 107–112. [CrossRef]

22. Randall Simpson, J.; Gumbley, J.; Whyte, K.; Lac, J.; Morra, C.; Rysdale, L.; Tufryer, M.; McGibbon, K.; Beyers, J.; Keller, H. Development, reliability, and validity testing of Toddler NutriSTEP: A nutrition risk screening questionnaire for children 18–35 months of age. *Appl. Physiol. Nutr. Metab.* **2015**, *40*, 877–886. [CrossRef]

23. Bell, L.K.; Golley, R.K.; Magarey, A.M. A short food-group-based dietary questionnaire is reliable and valid for assessing toddlers' dietary risk in relatively advantaged samples. *Br. J. Nutr.* **2014**, *112*, 627–637. [CrossRef] [PubMed]

24. Sandvik, P.; Ek, A.; Somaraki, M.; Hammar, U.; Eli, K.; Nowicka, P. Picky eating in Swedish preschoolers of different weight status: Application of two new screening cut-offs. *Int. J. Behav. Nutr. Phys. Act.* **2018**, *15*, 74. [PubMed]

25. Randall Simpson, J.A.; Keller, H.H.; Rysdale, L.A.; Beyers, J.E. Nutrition Screening Tool for Every Preschooler (NutriSTEP): Validation and test-retest reliability of a parent-administered questionnaire assessing nutrition risk of preschoolers. *Eur. J. Clin. Nutr.* **2008**, *62*, 770–780. [CrossRef] [PubMed]

26. Carducci, B.; Reesor, M.; Haresign, H.; Rysdale, L.; Keller, H.; Beyers, J.; Paquette-Duhaime, S.; O'Connor, A.; Simpson, J.R. NutriSTEP®is reliable for internet and onscreen use. *Can. J. Diet Pract. Res.* **2015**, *76*, 9–14. [CrossRef]

27. Mehdizadeh, A.; Vatanparast, H.; Khadem-Rezaiyan, M.; Norouzy, A.; Abasalti, Z.; Rajabzadeh, M.; Nematy, M. Validity and reliability of the persian version of nutrition screening tool for every preschooler (NutriSTEP®) in iranian preschool children. *J. Pediatr. Nurs.* **2020**, *52*, e90–e95. [CrossRef]

28. Bell, L.K.; Golley, R.K.; Mauch, C.E.; Mathew, S.M.; Magarey, A.M. Validation testing of a short food-group-based questionnaire to assess dietary risk in preschoolers aged 3–5 years. *Nutr. Diet.* **2019**, *76*, 642–645. [CrossRef]

29. Manios, Y.; Kourlaba, G.; Grammatikaki, E.; Androutsos, O.; Moschonis, G.; Roma-Giannikou, E. Development of a diet-lifestyle quality index for young children and its relation to obesity: The Preschoolers Diet-Lifestyle Index. *Public Health Nutr.* **2010**, *13*, 2000–2009. [CrossRef]

30. Townsend, M.S.; Ontai, L.; Shilts, M.K.; Styne, D.M.; Drake, C.; Lanoue, L. An obesity risk assessment tool for young children: Validity with BMI and nutrient values. *J. Nutr. Educ. Behav.* **2018**, *50*, 705–717. [CrossRef]

31. Das, N.; Ghosh, A. Psychometric validation of a nutrition knowledge questionnaire among parents of 3–6-Year-Old Asian Indian children in east Barddhaman district, West Bengal, India. *Indian J. Community Med.* **2020**, *45*, 130–134.

32. Jacobson Vann, J.C.; Finkle, J.; Ammerman, A.; Wegner, S.; Skinner, A.C.; Benjamin, J.T.; Perrin, E.M. Use of a tool to determine perceived barriers to children's healthy eating and physical activity and relationships to health behaviors. *J. Pediatr. Nurs.* **2011**, *26*, 404–415. [CrossRef]

33. Zemeir, L.A.; Butler, J.; Howard, D.E. Validity and reliability of the healthy families survey: A key component of the maryland Supplemental Nutrition Assistance Program–Education (SNAP-Ed) evaluation. *J. Nutr. Educ. Behav.* **2018**, *50*, 632–637. [CrossRef] [PubMed]

34. Santos-Beneit, G.; Sotos-Prieto, M.; Bodega, P.; Rodríguez, C.; Orrit, X.; Pérez-Escoda, N.; Bisquerra, R.; Fuster, V.; Penalvo, J.L. Development and validation of a questionnaire to evaluate lifestyle-related behaviors in elementary school children. *BMC Public Health* **2015**, *15*, 901.

35. Decker, J.W. Initial development and testing of a questionnaire of parental self-efficacy for enacting healthy lifestyles in their children. *J. Spec. Pediatr. Nurs.* **2012**, *17*, 147–158. [CrossRef]

36. Chacko, T.V.; Ganesan, S. A tool for quickly identifying gaps in diet of school children for nutritional educational interventions. *Indian J. Public Health* **2018**, *62*, 146–149. [PubMed]

37. Gray, H.L.; Koch, P.A.; Contento, I.R.; Bandelli, L.N.; Ang, I.; Di Noia, J. Validity and reliability of behavior and theory-based psychosocial determinants measures, using audience response system technology in urban upper-elementary schoolchildren. *J. Nutr. Educ. Behav.* **2016**, *48*, 437–452.e1. [CrossRef] [PubMed]

38. Lassetter, J.H.; Macintosh, C.I.; Williams, M.; Driessnack, M.; Ray, G.; Wisco, J.J. Psychometric testing of the healthy eating and physical activity self-efficacy questionnaire and the healthy eating and physical activity behavior recall questionnaire for children. *J. Spec. Pediatr. Nurs.* **2018**, *23*, e12207. [CrossRef] [PubMed]

39. Ayesha Fathima, A.; Hema, T.H.; Hemamalini, A.J. Development and validation of a questionnaire on eating behaviour for school children and its correlation with nutritional status. *Indian J. Public Health Res. Dev.* **2020**, *11*, 1–6.

40. Drouin, O.; Winickoff, J.P. Screening for behavioral risk factors is not enough to improve preventive services delivery. *Acad. Pediatr.* **2018**, *18*, 460–467. [CrossRef]

41. Nemec, K.B.; Keim, K.S.; Mullen, M.; Sowa, D.; Lui, K.J. Reliability and validity of the Child Nutrition and Physical Activity Screening Tool. *Top. Clin. Nutr.* **2018**, *33*, 106–117. [CrossRef]

42. Lazarou, C.; Panagiotakos, D.B.; Spanoudis, G.; Matalas, A.L. E-KINDEX: A dietary screening tool to assess children's obesogenic dietary habits. *J. Am. Coll. Nutr.* **2011**, *30*, 100–112. [CrossRef]

43. Moreno, J.P.; Kelley, M.L.; Landry, D.N.; Paasch, V.; Terlecki, M.A.; Johnston, C.A.; Foreyt, J.P. Development and validation of the Family Health Behavior Scale. *Int. J. Pediatr. Obes.* **2011**, *6*, e480–e486. [CrossRef] [PubMed]
44. Peyer, K.L.; Welk, G.J. Construct validity of an obesity risk screening tool in two age groups. *Int. J. Environ. Res. Public Health* **2017**, *14*, 419. [CrossRef] [PubMed]
45. Park, S.H.; Park, C.G.; McCreary, L.; Norr, K.F. Cognitive interviews for validating the family nutrition physical activity instrument for korean-american families with young children. *J. Pediatr. Nurs.* **2017**, *36*, 1–6. [CrossRef] [PubMed]
46. Wright, N.D.; Groisman-Perelstein, A.E.; Wylie-Rosett, J.; Vernon, N.; Diamantis, P.M.; Isasi, C.R. A lifestyle assessment and intervention tool for pediatric weight management: The HABITS questionnaire. *J. Hum. Nutr. Diet.* **2011**, *24*, 96–100. [CrossRef] [PubMed]
47. Quelly, S.B. Developing and testing adapted measures of children's self-efficacy, intentions, and behaviors associated with childhood obesity. *Child. Health Care* **2018**, *47*, 67–82. [CrossRef]
48. Gance-Cleveland, B.; Schmiege, S.; Aldrich, H.; Stevens, C.; Scheller, M. Reliability and validity of HeartSmartKids: A survey of cardiovascular risk factors in children. *J. Pediatr. Health Care* **2018**, *32*, 381–386. [CrossRef]
49. Vaughn, A.E.; Dearth-Wesley, T.; Tabak, R.G.; Bryant, M.; Ward, D.S. Development of a comprehensive assessment of food parenting practices: The Home Self-Administered Tool for Environmental Assessment of Activity and Diet Family Food Practices Survey. *J. Acad. Nutr. Diet.* **2017**, *117*, 214–227. [CrossRef]
50. West, F.; Sanders, M.R. The Lifestyle Behaviour Checklist: A measure of weight-related problem behaviour in obese children. *Int. J. Pediatr. Obes.* **2009**, *4*, 266–273. [CrossRef]
51. Gerards, S.M.; Hummel, K.; Dagnelie, P.C.; de Vries, N.K.; Kremers, S.P. Parental self-efficacy in childhood overweight: Validation of the Lifestyle Behavior Checklist in the Netherlands. *Int. J. Behav. Nutr. Phys Act.* **2013**, *10*, 7. [CrossRef]
52. Smith, S.R.; Johnson, S.T.; Oldman, S.M.; Duffy, V.B. Pediatric Adapted Liking Survey: A novel, feasible and reliable dietary screening in clinical practice. *Caries Res.* **2019**, *53*, 153–159. [CrossRef]
53. Kowalkowska, J.; Wadolowska, L.; Hamulka, J.; Wojtas, N.; Czlapka-Matyasik, M.; Kozirok, W.; Bronkowska, M.; Sadowska, J.; Naliwajko, S.; Dziaduch, I.; et al. Reproducibility of a short-form, multicomponent dietary questionnaire to assess food frequency consumption, nutrition knowledge, and lifestyle (SF-FFQ4PolishChildren) in polish children and adolescents. *Nutrients* **2019**, *11*, 2929. [CrossRef] [PubMed]
54. Hendrie, G.A.; Coveney, J.; Cox, D.N. Factor analysis shows association between family activity environment and children's health behaviour. *Aust. NZJ Public Health* **2011**, *35*, 524–529. [CrossRef] [PubMed]
55. Huang, Y.J.; Wong, S.H.; Salmon, J.; Hui, S.S. Reliability and validity of psychosocial and environmental correlates measures of physical activity and screen-based behaviors among Chinese children in Hong Kong. *Int. J. Behav. Nutr. Phys. Act.* **2011**, *8*, 16. [CrossRef] [PubMed]
56. Hendricks, C.; Murdaugh, C.; Pender, N. The Adolescent Lifestyle Profile: Development and psychometric characteristics. *J. Natl. Black Nurses Assoc.* **2006**, *17*, 1–5.
57. Sousa, P.; Gaspar, P.; Fonseca, H.; Hendricks, C.; Murdaugh, C. Health promoting behaviors in adolescence: Validation of the Portuguese version of the Adolescent Lifestyle Profile. *J. Pediatr.* **2015**, *91*, 358–365. [CrossRef] [PubMed]
58. Lebron, C.N.; Lee, T.K.; Prado, G.; St. George, S.M.; Pantin, H.; Messiah, S.E. Psychometric properties of an abbreviated Childhood Family Mealtime Questionnaire among overweight and obese Hispanic adolescents. *Appetite* **2019**, *140*, 169–179. [CrossRef]
59. Kosti, R.I.; Panagiotakos, D.B.; Mariolis, A.; Zampelas, A.; Athanasopoulos, P.; Tountas, Y. The Diet-Lifestyle Index evaluating the quality of eating and lifestyle behaviours in relation to the prevalence of overweight/obesity in adolescents. *Int. J. Food. Sci. Nutr.* **2009**, *60* (Suppl. 3), 34–47. [CrossRef]
60. Mohamadian, H.; Ghannaee, M.; Kortdzanganeh, J.; Meihan, L. Reliability and construct validity of the Iranian version of health-promoting lifestyle profile in a female adolescent population. *Int. J. Prev. Med.* **2013**, *4*, 42–49.
61. Fernald, D.H.; Froshaug, D.B.; Dickinson, L.M.; Balasubramanian, B.A.; Dodoo, M.S.; Holtrop, J.S.; Hung, D.Y.; Glasgow, R.E.; Niebauer, L.J.; Green, L.A. Common measures, better outcomes (COMBO): A field test of brief health behavior measures in primary care. *Am. J. Prev. Med.* **2008**, *35*, S414–S422. [CrossRef]
62. Hyun, H.; Lee, H.; Ro, Y.; Gray, H.L.; Song, K. Body image, weight management behavior, nutritional knowledge and dietary habits in high school boys in Korea and China. *Asia Pac. J. Clin. Nutr.* **2017**, *26*, 923–930.
63. Costa-Tutusaus, L.; Guerra-Balic, M. Development and psychometric validation of a scoring questionnaire to assess healthy lifestyles among adolescents in Catalonia. *BMC Public Health* **2016**, *16*, 89. [CrossRef] [PubMed]
64. Tavakol, M.; Dennick, R. Making sense of Cronbach's alpha. *Int. J. Med. Educ.* **2011**, *2*, 53–55. [CrossRef] [PubMed]
65. Prinsen, C.A.C.; Mokkink, L.B.; Bouter, L.M.; Alonso, J.; Patrick, D.L.; de Vet, H.C.W.; Terwee, C.B. COSMIN guideline for systematic reviews of patient-reported outcome measures. *Qual. Life Res.* **2018**, *27*, 1147–1157. [CrossRef] [PubMed]
66. Ledoux, T.A.; Hingle, M.D.; Baranowski, T. Relationship of fruit and vegetable intake with adiposity: A systematic review. *Obes. Rev.* **2011**, *12*, e143–e150. [CrossRef] [PubMed]
67. Monzani, A.; Ricotti, R.; Caputo, M.; Solito, A.; Archero, F.; Bellone, S.; Prodam, F. A systematic review of the association of skipping breakfast with weight and cardiometabolic risk factors in children and adolescents. What should we better investigate in the future? *Nutrients* **2019**, *11*, 387. [CrossRef]
68. Berge, J.M.; Wall, M.; Hsueh, T.F.; Fulkerson, J.A.; Larson, N.; Neumark-Sztainer, D. The protective role of family meals for youth obesity: 10-year longitudinal associations. *J. Pediatr.* **2015**, *166*, 296–301. [CrossRef]

69. Morrissey, B.; Taveras, E.; Allender, S.; Strugnell, C. Sleep and obesity among children: A systematic review of multiple sleep dimensions. *Pediatr. Obes.* **2020**, *15*, e12619. [CrossRef]
70. Matricciani, L.; Paquet, C.; Galland, B.; Short, M.; Olds, T. Children's sleep and health: A meta-review. *Sleep Med. Rev.* **2019**, *46*, 136–150. [CrossRef]
71. Córdova, F.V.; Barja, S.; Brockmann, P.E. Consequences of short sleep duration on the dietary intake in children: A systematic review and metanalysis. *Sleep Med. Rev.* **2018**, *42*, 68–84. [CrossRef]
72. Gazmararian, J.; Smith, J. Role of sleep duration and obesity-related health behaviors in young children. *Prev. Med. Rep.* **2020**, *20*, 101199. [CrossRef]
73. Xu, H.; Wen, L.M.; Rissel, C. Associations of parental influences with physical activity and screen time among young children: A systematic review. *J. Obes.* **2015**, *2015*, 546925. [CrossRef] [PubMed]
74. Golan, M.; Crow, S. Parents are key players in the prevention and treatment of weight-related problems. *Nutr. Rev.* **2004**, *62*, 39–50. [CrossRef]
75. De Moor, C.; Baranowski, T.; Cullen, K.W.; Nicklas, T. Misclassification associated with measurement error in the assessment of dietary intake. *Public Health Nutr.* **2003**, *6*, 393–399. [CrossRef] [PubMed]
76. Hulst, J.M.; Zwart, H.; Hop, W.C.; Joosten, K.F. Dutch national survey to test the STRONGkids nutritional risk screening tool in hospitalized children. *Clin. Nutr.* **2010**, *29*, 106–111. [CrossRef] [PubMed]
77. McCarthy, H.; Dixon, M.; Crabtree, I.; Eaton-Evans, M.J.; McNulty, H. The development and evaluation of the Screening Tool for the Assessment of Malnutrition in Paediatrics (STAMP©) for use by healthcare staff. *J. Hum. Nutr. Diet.* **2012**, *25*, 311–318. [CrossRef]
78. Gerasimidis, K.; Keane, O.; Macleod, I.; Flynn, D.M.; Wright, C.M. A four-stage evaluation of the Paediatric Yorkhill Malnutrition Score in a tertiary paediatric hospital and a district general hospital. *Br. J. Nutr.* **2010**, *104*, 751–756. [CrossRef]
79. White, M.; Lawson, K.; Ramsey, R.; Dennis, N.; Hutchinson, Z.; Soh, X.Y.; Matsuyama, M.; Doolan, A.; Todd, A.; Elliott, A.; et al. Simple Nutrition Screening Tool for Pediatric Inpatients. *JPEN J. Parenter. Enteral Nutr.* **2016**, *40*, 392–398. [CrossRef]
80. Rolstad, S.; Adler, J.; Rydén, A. Response burden and questionnaire length: Is shorter better? A review and meta-analysis. *Value Health* **2011**, *14*, 1101–1108. [CrossRef]

Review

The Impact on Dietary Outcomes of Celebrities and Influencers in Marketing Unhealthy Foods to Children: A Systematic Review and Meta-Analysis

Jessica Packer [1,*], Simon J. Russell [1], Gabriela Siovolgyi [1], Katie McLaren [1], Claire Stansfield [2], Russell M. Viner [1,†] and Helen Croker [1,†]

1. Population, Policy and Practice Research and Teaching Department, UCL Great Ormond Street Institute of Child Health, University College London, London WC1N 1EH, UK; s.russell@ucl.ac.uk (S.J.R.); gabrielansiovolgyi@gmail.com (G.S.); katie.mclaren@hotmail.co.uk (K.M.); r.viner@ucl.ac.uk (R.M.V.); h.croker@wcrf.org (H.C.)
2. EPPI-Centre, UCL Social Research Institute, University College London, London WC1H 0NR, UK; c.stansfield@ucl.ac.uk
* Correspondence: jessica.packer@ucl.ac.uk
† These authors contributed equally to this work.

Abstract: Celebrities, including influencers, are commonly used to market products that are high in fat, sugar, and salt (HFSS) to children but the impact on dietary outcomes has been unclear. The primary aim of this study was to systematically review the literature and quantify the impact of celebrities in HFSS marketing on children's dietary outcomes. We searched eight databases and included studies from all countries and languages published from 2009 until August 2021. Participants were defined as under 16 years, exposure was marketing for HFSS products with a celebrity, and the outcomes were dietary preference, purchasing behaviors, and consumption of HFSS products. We were able to conduct a meta-analysis for consumption outcomes. Seven articles met the inclusion criteria, of which three were included in the meta-analysis. Under experimental conditions, the use of celebrities in HFSS marketing compared to non-food marketing was found to significantly increase consumption of the marketed HFSS product by 56.4 kcals ($p = 0.021$). There was limited evidence on the impact on preference or purchase intentions and on the comparisons between use and non-use of celebrities and influencers.

Keywords: child and adolescent health; food marketing; obesity; policy research

Citation: Packer, J.; Russell, S.J.; Siovolgyi, G.; McLaren, K.; Stansfield, C.; Viner, R.M.; Croker, H. The Impact on Dietary Outcomes of Celebrities and Influencers in Marketing Unhealthy Foods to Children: A Systematic Review and Meta-Analysis. *Nutrients* **2022**, *14*, 434. https://doi.org/10.3390/nu14030434

Academic Editor: Zhiyong Zou

Received: 22 December 2021
Accepted: 17 January 2022
Published: 19 January 2022

Publisher's Note: MDPI stays neutral with regard to jurisdictional claims in published maps and institutional affiliations.

Copyright: © 2022 by the authors. Licensee MDPI, Basel, Switzerland. This article is an open access article distributed under the terms and conditions of the Creative Commons Attribution (CC BY) license (https://creativecommons.org/licenses/by/4.0/).

1. Introduction

Increased exposure to high in fat, sugar, and salt (HFSS) marketing is occurring simultaneously with the global childhood obesity epidemic [1]. The marketing of unhealthy foods is ubiquitous and is particularly impactful for children and young people (the term 'marketing' includes both advertising and packaging) [2]. Children are exposed to marketing throughout the food environment, including via television and other broadcast media, in shops and supermarkets, on the street, and, increasingly, online [3]. The majority of food marketing is for HFSS products, some of which is directly targeted at children [4,5]. Evidence suggests that children from ethnic minority and lower socioeconomic groups are disproportionately exposed to, and influenced by, food marketing [6]. Children with higher body weight have also been found to be disproportionately affected by screen advertising for HFSS products [7].

The need for enhanced regulations of commercial marketing is a priority on the global health and policy agenda, as highlighted in a recent WHO–UNICEF–Lancet child health Commission [8]. This follows on from the 2010 WHO recommendations, for policies to limit the effectiveness of HFSS food marketing to children by limiting its exposure and power (the creative content, design, and execution of the marketing message/impacted by techniques used) [9]. In the accompanying WHO implementation guidance, restricting the use of celebrities in HFSS product marketing was used as a specific example of how to reduce the power of marketing [10]. Celebrity endorsements, across a broad array of categories, including HFSS food products, have been shown to increase sales, bring about more positive attitudes to brand and product, and increase purchase intentions [11–14].

Celebrity endorsements are thought to work through the process of evaluative conditioning, where liking of a stimulus is the result of its pairing with other positive stimuli [15]. This process is mediated by the parasocial relationships children can form with celebrities (one-sided relationships between media users and celebrities), especially through social media interaction [16]. A cross-disciplinary review examining how celebrities influence patients' health-related behaviors found that they can help distinguish products and elicit herd behavior (economics); transfer positive characteristics to the endorsed products (marketing); activate brain regions associated with trust, creating positive associations and encoding memories (neuroscience); and cause positive reactions (psychology) [17]. This results in the possibility of celebrities having a substantial influence on people's health-related behaviors.

Current restrictions on the use of celebrities in marketing of HFSS products to children have been identified as an area of concern [18–20]. Whilst the UK, Ireland, Chile, Australia, Netherlands, Portugal, Spain, and Brazil restrict the use of celebrities in HFSS advertising to children, loopholes exist in the interpretation of defining celebrities and audience thresholds, and the implementation of regulations [19,21–24]. In the UK, for example, the use of celebrities 'popular with children' is restricted in broadcast and non-broadcast HFSS advertisements targeting pre- and primary school children (under 12 years) but no definition of what constitutes a celebrity 'popular with children' is provided [25,26]. Additionally, the regulation of restrictions varies between self-regulated or statutory legislation. Evidence has shown that self-regulation is broadly ineffective at limiting HFSS marketing to children [19,27–29]. An example of voluntary and self-regulated restrictions is the Spanish Publicidad, Actividad, Obesidad, Salud (PAOS) Code for food and drinks marketing to children, which prevents the participation, appearance, and exploitation of well-known and famous persons [30]. The scope of restrictions is mainly focused on broadcast marketing, with packaging, sponsorship, cinema, and in-store promotions commonly neglected [18,19]. Restrictions frequently apply only to pre-digital media and need to be updated to react to changes in marketing and media consumption, with digital marketing now accounting for the majority of UK advertising spend [18,31]. Children as young as 3–4 years old are increasingly switching their preference and usage from TV to online (e.g., YouTube), and YouTube is growing as the preferred platform [3]. This has led to a new type of celebrity, the 'influencer' (or YouTuber), defined as gaining fame by successfully branding themselves as experts on social media platforms [32], which has been identified as a new marketing source targeting children [33]. Experiments have shown that influencer marketing leads to greater purchase intentions due to participants identifying, relating to and trusting influencers more than other celebrities [32]. Evidence suggests that the integration of 'real-life' scenarios into social media marketing (use of advertised product in their daily lives) leads to greater positive brand effects (brand attitude, purchase intention, willingness to pay for a product, and feeling connected to the brand) compared to traditional commercial celebrity-endorsed advertising [34]. Sports celebrities are also of interest in our review, with research showing that their association with high-sugar products can foster beliefs in children that they are healthy and improve sports performance [35].

Content analyses show that food marketing featuring celebrities is particularly prominent for HFSS products [36–39]. This was consistent across television advertisements in the UK [36] and musician [37], athlete [38], and YouTube influencer endorsements in the US [39]. Celebrities are also used on HFSS product packaging [40]. Analysis of social media advertising exposure in children aged 7–16 years found that during 10 min of social media use, 72% were exposed to food advertising, primarily for HFSS products, of which 17% was embedded in celebrity generated content [41]. Embedded content is not explicitly advertising, adding to the difficulty children already face in recognizing online advertising and impacting their ability to understand the intent of advertising [42,43]. A longitudinal study, looking at the association between self-reported vlog (i.e., video weblogs) viewing and consumption of HFSS beverages or snacks at three time points, found a significant association between exposure and consumption of HFSS beverages 24 months later [44].

Despite the widespread use of celebrities in the marketing of HFSS products and given the evidence of their impact on dietary outcomes, there have been no systematic reviews to date examining their impact on preference, purchasing, and consumption outcomes in children. Reviews have reported that celebrities are a popular marketing tactic for promoting HFSS foods to children but impacts on outcomes have rarely been assessed [45,46]. One review examined the impact of food marketing tactics on children's attitudes, preferences, and consumption and included the use of endorsers but found only limited evidence relating to advertising with celebrities [47]. Another review found evidence that celebrity endorsements positively impact brand attitudes and purchase intentions but included only one study in adolescents [11]. Due to these gaps in the literature, we undertook a review to better understand how celebrities used in HFSS food marketing (advertising and packaging) impact on children's food preferences, purchasing behaviors, and consumption. Our secondary aims were to assess the differential impact of the type of celebrity (sports, YouTubers/influencers, or other), child characteristics (age and socio-economic status), format of advertisement (content within advertisements versus on packaging), and length of any effects (short- or long-term).

2. Materials and Methods

2.1. Protocol and Registration

The current systematic review and meta-analysis was conducted and reported in accordance with the PRISMA statement checklist and was registered with PROSPERO (CRD42019155037) [48].

2.2. Eligibility Criteria, Information Resources, and Search Strategy

To be eligible for inclusion, studies needed to be quantitative, experimental (randomized or non-randomized) with an advertising exposure featuring a celebrity and a comparison group (non-food advertisement, no exposure, healthy food advertisement), or "real-world" (longitudinal, interrupted time series, controlled before and after). Studies from 2009, in any country or language, were included, as well as both between-subject and within-subject designs. Studies before 2009 were excluded as they were likely to be of limited relevance due to rapid advancement of technology and celebrity/influencer culture. Participant criteria were children aged between 0 and 15 years, in line with UK advertising regulations. Any marketing modality (TV, online, internet/advergames, poster, packaging, digital advertising) with a celebrity/influencer was eligible. Outcomes were all related to the advertised HFSS food product and included consumption (measured dietary intake, ad libitum consumption), preferences (self-reported, like/dislike ratings), and purchasing/request (quantity of product purchased, pester intention).

Searches of the following electronic databases were conducted on 22 October 2019 and updated on 16 August 2021: Ovid MEDLINE, Cochrane Library, Scopus, PsycINFO, ProQuest (Central)—ASSIA, Web of Science—Social Science Citation Index and Emerging Sources Citation Index and Social Policy and Practice (see Supplementary Table S1 for full search details and Supplementary Table S2 for search strings). K.M., G.S., and J.P. conducted the searches, imported records into EndNoteX9 and EPPI-Reviewer 4 and removed duplicates. EPPI-Reviewer 4 was used for screening and for search management [49].

2.3. Study Selection

Exclusion criteria were date (published pre-2009), participant age group (over 16 years), study design (qualitative, content analyses, cross-sectional), publication type (reviews, dissertations), intervention (no HFSS marketing exposure with a celebrity), and outcome measure (no measure of food intake, consumption, choice, preference, purchase, purchase intention, or pestering).

Double screening of papers on title and abstract and full-text were independently completed by K.M., G.S., J.W., and J.P. Discrepancies between reviewers were mutually reconciled. Full-texts of relevant articles were acquired via library and web services, in addition to direct contact with authors. All papers eligible for screening were retrieved successfully.

2.4. Data Extraction and Items

Data were independently extracted and jointly reconciled by K.M., G.S., and J.P. Corresponding authors were contacted to request additional data, where required, for the meta-analysis. Four corresponding authors were contacted, of whom one responded with the required data.

Data extracted included study identification (authors, country, year of publication), target population (children and/or adolescents), sample group description (size of sample, age range, and mean age of participants), study description (study design, number of participants in each condition and assignment to conditions), intervention description (advertising medium, celebrity/influencer), comparison type (HFSS food advertisement vs. healthy food or non-food advertisement), test foods used, outcome type (consumption, preference, or purchasing), and outcome measures (kcals, kJ, grams, preference ratings, purchase request measures).

2.5. Assessment of Quality

The Cochrane risk of bias tool RoB 2 was used to assess bias in the included experimental studies [50]. Bias assessment was conducted independently by two reviewers with discrepancies reconciled.

2.6. Data Synthesis

Our primary analysis was meta-analysis, but where studies did not provide sufficient data, they were included in a narrative synthesis. For inclusion in the meta-analysis, experimental studies were required to have an appropriate comparison group, including no advertisement or non-food advertisement. We considered these comparison groups due to the ubiquity of using celebrities in marketing of both food and non-food products and to be consistent with previous research [20]. Due to a lack of studies measuring preference or purchase of a HFSS product, meta-analysis was only possible for consumption outcomes. Three articles were identified that measured HFSS consumption and reported/provided the mean values with standard deviations. The consumption outcomes were standardized to report the total energy content consumed (kcals), which required conversion from weight (grams) using published nutritional values of the consumed products. Further details about the standardization methods, and the rationale for the experimental conditions and outcomes, were included in the meta-analysis, is provided in a supplemental file (Supplementary Table S3). Due to heterogeneity in study design, including the type of celebrity (sport or influencer), the advertising exposure (online static image, TV, or YouTube clip), and the HFSS products advertised and consumed (crisps or chocolate biscuit), a DerSimonian-Laird random-effects model was used for meta-analysis. We presented the results graphically using forest plots. Analyses were conducted using Stata 16 (16.1, StataCorp LLC, College Station, TX, USA) [51].

3. Results

3.1. Study Selection

The search process is shown in the flowchart (Figure 1). The searches resulted in 414 articles, of which 294 were unique records following the removal of duplicates. After screening on title and abstract, 264 were excluded and 30 were screened on full-text. One article was found through screening of a related review. This led to the inclusion of seven studies, from seven reports. Three studies were included in the meta-analysis.

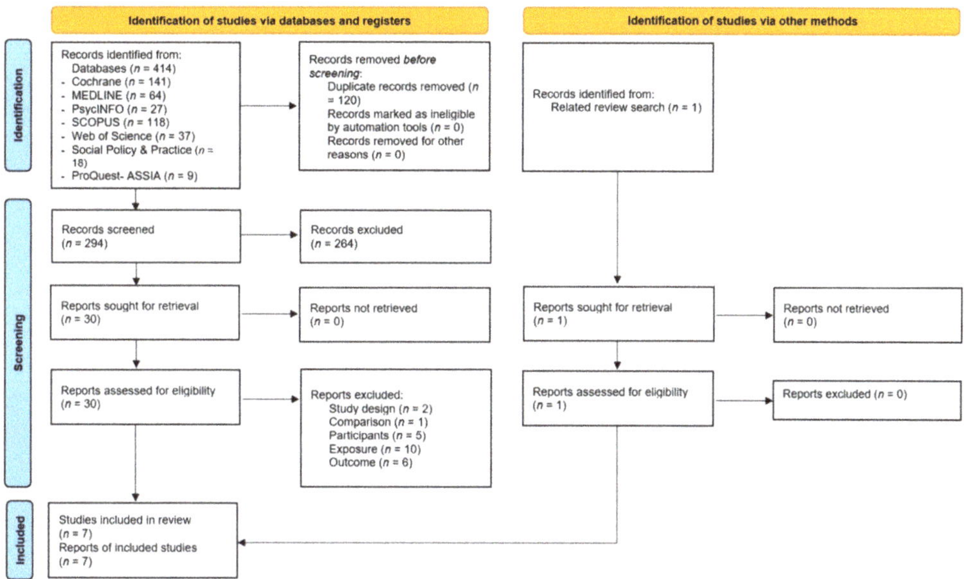

Figure 1. PRISMA screening flowchart.

3.2. Study Description and Results

A summary of study information is provided in Table 1.

Table 1. Descriptive table of experimental studies.

Author, Year, Country	Participants	Design	Advertising Intervention	Comparison	Outcome	Relevant Results	Risk of Bias
Boyland [20], 2013, UK	$N = 181$ Age range = 8–11 Mean age = 10.3	Experimental (school), between-subject, allocation not specified	20 min cartoon with 45 s TV advert for HFSS product (Walker's crisps) with sports celebrity endorser (Gary Lineker)	20 min cartoon with 45 s non-food advert; food advert with no endorser; or TV footage of endorser	Post-intervention, ad libitum consumption of potato crisps, labeled branded, and non-branded (grams)	Celebrity endorsed TV food adverts significantly increased intake of food, compared to food advert with no endorser and non-food advert.	Some concerns
Coates [52], 2019, UK EoI	$N = 151$ Age range = 9–11 Mean age = 10	Experimental (school), between-subject, random assignment	5 min YouTube video with 1 min influencer marketing (Zoella and PointlessBlog) segment of branded HFSS product (McVitie's chocolate biscuits), with and without disclosure	5 min YouTube video with 1 min influencer marketing segment of branded non-food product (Apple iPhone)	Post-intervention, ad libitum consumption of cookies (kcal), labeled, branded, and non-branded, 5 min	Influencer endorsed HFSS advert significantly increased intake of promoted food, compared to non-food advert	Low
Coates [53], 2019, UK SMI	$N = 176$ Age range = 9–11 Mean age = 10.5	Experimental (school), between-subject, random assignment	1 min viewing of mock Instagram profile of popular YouTube influencers (not stated due to copyright) with marketing of HFSS product (unbranded chocolate biscuits)	1 min viewing of mock Instagram with image of YouTube influencer marketing healthy product (banana) or non-food (sneakers)	Post-intervention, ad libitum consumption of unbranded HFSS products HFSS (candy, chocolate) and healthy (carrot, grapes) products (kcal), 10 min	Intake of HFSS products and overall snacks significantly increased following exposure to celebrity endorsement of HFSS products, compared to non-food condition.	Low

Table 1. *Cont.*

Study	Sample	Design	HFSS exposure	Control	Outcome measure	Results	Risk of bias
De Jans [54], 2021, Belgium	*N* = 190 Age range = 8–12 Mean age = 10.04	Experimental (classroom), between-subject, random	Instagram post of influencer (fictitious) promotion of HFSS snack (unbranded donuts) (either sedentary lifestyle versus athletic lifestyle)	Instagram post of influencer promotion of snack high in nutritional value (strawberries) (both (influencer lifestyle: sedentary versus athletic)	Snack choice between mini donut or a strawberry.	Children exposed to influencer promotion of the donut, chose the donut 52.2% (47/90) compared to 49.5% exposed to influencer promotion of non-HFSS product. Significance not tested.	Low
Dixon [55], 2014, Australia	*N* = 1302 Age range = 10–12 Mean age = 11	Experimental (online school), between-subject, random assignment	Packaging of HFSS products (cereal, cheese dips, chicken nuggets, ice cream, flavored milk, brands not stated) with sports celebrity endorsement (popular Australian male athletes, names not stated)	Packaging of same HFSS products with no celebrity endorsement (no promotion)	During-intervention, forced choice of randomly allocated HFSS exposure or comparable healthier food pack, on a computer	Celebrity endorsed HFSS products were significantly more likely to be chosen compared to control, in boys only. No significant difference in girls.	Low
Jain [56], 2011, India	*N* = 378 Age range = 13–17 Mean age = not stated	Experimental (school), between-subject, allocation not specified	5–10 min viewing of print advertisement of HFSS product (unbranded chocolate) with celebrity endorsement (Hindi actor, Aamir Khan)	5–10 min viewing of print adverts of HFSS product with no endorsement	Post-intervention, purchase intention product (scale NS)	Purchase intentions of HFSS product endorsed by a celebrity were significantly greater compared to control or character-endorsed HFSS product.	Some concerns
Ponce-Blandon [57], 2020, Spain	*N* = 421 Age range = 4–6 Mean age = 4.8	Experimental (education centers), between-subject, random assignment	8 min episode of cartoon (Caillou) with an advert for HFSS product (Príncipe Double Choc chocolate cookies) with sports celebrity (famous Spanish soccer player, name not stated)	No advert control and non-food advert control	Preference choice between advertised product (Príncipe Double Choc chocolate cookies) vs. similar non advertised product (Tosta Rica Chocoguay, Cuétara chocolate cream filled cookies)	Preference for the advertised product was not significantly different between the conditions.	Low

3.3. Participants

The age of participants across all studies ranged from 6–17 years. One study included participants aged 13–17, and results were not split by age [56]. The range for the three studies in the meta-analysis was 8 to 11 years, and the mean age was 10.3 years.

3.4. Settings

All studies were experimental using between-subjects designs. All but two [20,56] stated that subjects were randomly allocated. All of the studies were conducted in schools or education centers. Studies were conducted in the UK (*n* = 3, same research group), India (*n* = 1), Belgium (*n* = 1), Australia (*n* = 1), and Spain (*n* = 1).

3.5. Interventions

The HFSS marketing exposure varied: TV advertisements, embedded in cartoons (*n* = 2) [20,57], an online advertisement embedded in a YouTube clip (*n* = 1) [52]; static Instagram posts (*n* = 2) [53,54]; food product packaging (*n* = 1) [55]; and printed advertisements (*n* = 1) [56]. The celebrities featured influencers (*n* = 3), sports celebrities (*n* = 3), and a Hindi movie star (*n* = 1). The HFSS products included crisps, chocolate, chocolate biscuits, donuts, and sweetened cereals.

3.6. Outcomes

The outcome measures were ad libitum consumption of snacks immediately following the advertising exposure (from which total calories consumed could be calculated) and self-reported consumption intention (*n* = 3) [20,52,53], preference/snack choice between paired items (*n* = 3) [54,55,57], and self-reported purchase intention (*n* = 1) [56]. The HFSS products available for ad libitum consumption were crisps, chocolate biscuits, and jelly candy/chocolate buttons. Duration of snacking, when reported, was between 5–10 min.

3.7. Comparisons

Calorie consumption following viewing of advertisements of HFSS products with celebrities was compared to consumption following non-food advertisements with or without celebrity endorsements in the three studies included in the meta-analysis [20,52,53]. Of the studies not included in the meta-analysis, two studies compared marketing of a HFSS product with celebrity endorsement to the same HFSS product without celebrity endorsement packaging [55] or print advertisement [56]; one study compared TV advertisements with celebrity endorsement for HFSS product to non-exposure or non-food TV advertisements [57]; and one study compared Instagram posts with influencers for HFSS products to Instagram posts with influencers for non-HFSS food products [54].

3.8. Meta-Analysis

Three studies provided sufficient data on calorie consumption to be included in meta-analysis of consumption. We found that use of a celebrity in HFSS food marketing, compared to non-food marketing, resulted in significantly greater consumption of HFSS products, with a pooled effect size of 56.4 kcals (95% CI 8.50, 104.20; p = 0.021) (Figure 2). We found evidence of high heterogeneity (I^2 = 79.5%); Egger's regression analysis showed low risk of publication bias (p = 0.347); and trim and fill analysis suggested evidence of two missing studies (See Supplementary Figure S1).

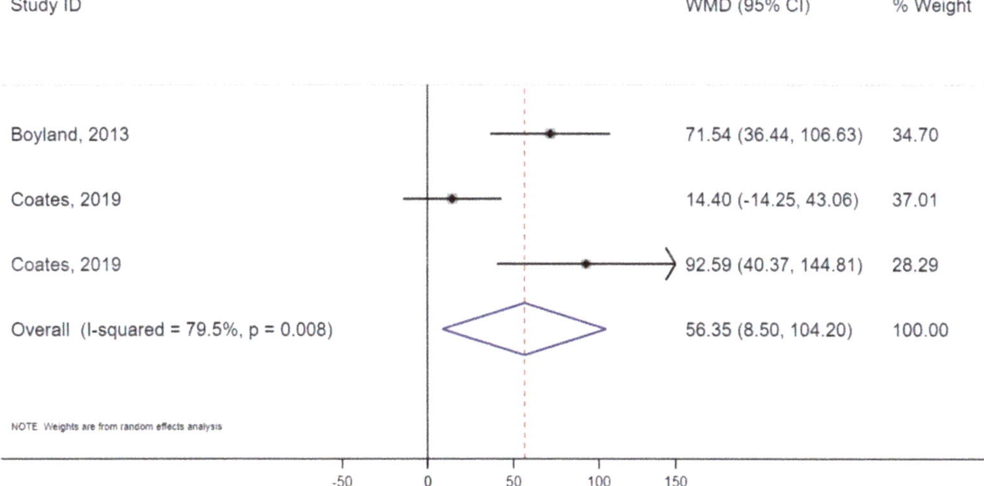

Figure 2. Forest plot showing mean difference (kcals) in total snack consumption of HFSS products between celebrity HFSS advertisement and non-food advertisement. Boyland, 2013; Coates, 2019; Coates, 2019.

3.9. Other Findings

One study measured purchase intentions and found purchase intentions for HFSS products were significantly greater in the celebrity HFSS advertisement group compared to non-food advertisement [56]. Preference was measured in three studies, with mixed results [55]. One found that when exposed to product packaging featuring an endorsement from a sports celebrity that product was chosen by participants significantly more when compared to the same product with no endorsement; however, this effect was only seen in boys [55]. Two studies found that preference for HFSS product was not significantly different with celebrity-endorsed HFSS product advertisement exposure compared to no advertisement exposure or non-food advertisement control [57] or non-HFSS food product [54]. The impact of SES was only reported in one study, which found no association between SES and food preference in response to HFSS advertising [55]. Impact of age was measured in three studies and was found to not be significantly associated with consumption (n = 2) [20,52] or consumption preference [57]. We were unable to assess the secondary aims of differential impact of celebrity type or format of advertisement, as the data did not support this. None of the included studies were longitudinal; therefore, long-term consequences of these effects could also not be assessed. Across all studies, there was little evidence relating to comparisons between use and non-use of celebrities and influencers (i.e., endorsed vs. non-endorsed) in HFSS food marketing.

3.10. Quality of Studies

The risk of bias across included studies was assessed as mostly low (see Supplementary Figure S2 for bias assessment).

4. Discussion

Our systematic review included the first meta-analysis examining the impact of celebrities used in the marketing of HFSS products to children and found evidence that marketing HFSS products with celebrities' influences children's calorie consumption. The meta-analysis showed that HFSS advertisements with a celebrity endorser, compared to a non-food advertisement, resulted in significantly greater calorie intake in children under experimental conditions. We found limited evidence that celebrities impact purchase intentions and mixed evidence that they impact preference outcomes. The findings from our review extend the findings of previous reviews, which have indicated that celebrities are persuasive HFSS marketing tools [11,45–47]. Our previous work showed that, following exposure to screen advertising for food, children consumed an additional 57 kcals [7] when compared to non-food advertisement exposure, which is consistent with the finding here of an additional 56.4 kcals. The impacts of advertising may be modest but over time can accrue to have substantial impacts on energy balance, body weight, and associated morbidities [58,59].

There was limited evidence that age did not impact on consumption outcomes but only one study examined the impact of SES and found no evidence. A review recently found that children from ethnic minority and low SES backgrounds are exposed to more HFSS advertising than children from higher SES and non-ethnic minority backgrounds [6]. This emphasizes the need for policy actions that addresses these inequalities.

Our data suggest there is potential for population-level interventions, including policy action such as enhanced regulations, to have an impact on HFSS consumption by children, even if effects are modest at an individual level. The use of celebrities in HFSS marketing is often not restricted or subject to weak regulations, and greater policy action has been recommended by WHO. The quantifiable impact of celebrities on children's dietary outcomes has not been previously evidenced in the literature [18,19]. Our findings suggest that tightening policies regulating HFSS marketing directed at children that contain celebrities may be effective in reducing children's calorie intake. Celebrity-endorsed HFSS brand advertising is frequently omitted from regulations, due to complexities in identifying advertised products and assessing if restrictions are applicable [60]. An additional concern with using celebrities in HFSS marketing is the knock-on effect of exposure to them in contexts outside of HFSS advertisements. This has been shown to increase consumption of the HFSS food (seeing Gary Lineker in Match of the Day led to children eating more Walkers crisps) [20]. Chile has implemented policies that comprehensively restrict use of celebrities, and these have been shown to be effective at reducing HFSS food marketing to children [19,60,61]. The UK government announced plans to introduce a pre-watershed ban on HFSS advertisements across television and on-demand program services, and a restriction on paid-for less healthy food and drink advertising online, which could be effective at overcoming some gaps and limiting exposure of celebrities in HFSS marketing to children [60,61]. These policies touch on areas where regulations could be strengthened including a standardization of approaches to achieve a consistent definition of celebrities, scope (programs and medias), audience thresholds (approved composition of the audience i.e., proportion of children), and enforcement. In the UK, only celebrities 'popular with children' are restricted (but this is not defined), allowing free use of celebrities with general appeal, such as Gary Lineker, and unrestricted endorsement opportunities for influencers [18,19]. Certain regulations (e.g., Ireland, UK, and the EU pledge) only apply if children make up between 20–50% of the viewing audience, but quantifying the audience demographics for broadcast and non-broadcast media is difficult to accurately assess [18]. This is especially true online, where user age restrictions on social media platforms are rarely followed by site users and children often use their parents' account or devices [3]. Restrictions specific to children's programs, mean that family programs popular with children and broadcast during peak children viewing times (6–9 p.m. in the UK) are out of scope in many jurisdictions but could be overcome by total bans on HFSS advertising, such as the proposed pre-watershed ban in the UK [3,29].

Limitations of this review include the small number of search results and limited number of studies eligible for inclusion in meta-analysis; therefore, care needs to be taken due to variability. The heterogeneity of the included studies was high, but the random effects model was used to account for differences. Findings from the meta-analysis should be interpreted with caution, as all studies were completed by the same research team at the University of Liverpool but conducted to a high standard. Further primary research would be beneficial in building the evidence base, especially digital marketing. We were unable to address the secondary aims, due to a lack of data. Despite our search including real-world studies, we were unable to identify any and therefore were unable to investigate the long-term impacts of celebrity endorsers on dietary outcomes. Beyond age, we were unable to assess the influence of SES on impact of advertising and in future would also examine the impact of weight status and ethnicity. We were unable to assess if there were any changes due to COVID-19, as no papers specifically mention this and most of the data was collected pre-2020. Due to the data available, we were unable to complete meta-analysis comparing use and non-use of celebrities and influencers in HFSS marketing and the impact on dietary outcomes (i.e., endorsed vs. non-endorsed). We suggest this as an area for future primary research; it is also a limitation about the difficulty of separating the effect of celebrity from marketing more generally. The impact of celebrity and influencers on promoting healthier food products was not the focus of this review and could be another area of future research. Evidence suggests food adverts do not appear to work in the same way for healthy food [62].

5. Conclusions

We found evidence that HFSS food marketing featuring celebrities or influencers increases children's food consumption, although this was from a limited number of studies. These findings suggest that limiting exposure of children to HFSS marketing including all celebrity types may have beneficial impacts upon dietary consumption. Further research on the impact of child characteristics such as SES, long-term impacts, and real-world studies would be beneficial to further inform the thinking of policy makers.

Supplementary Materials: The following supporting information can be downloaded at: https://www.mdpi.com/article/10.3390/nu14030434/s1, Table S1: Details of search, Table S2: Search history, Table S3: Rationale for meta-analysis inclusion and data processing, Figure S1: Trim and fill analysis, Figure S2: Bias assessment for experimental studies

Author Contributions: Conceptualization, J.P., H.C., R.M.V., S.J.R. and C.S.; methodology, J.P., H.C., C.S., G.S. and K.M.; software, J.P., H.C., C.S., G.S. and K.M.; validation, J.P., H.C., G.S. and K.M.; formal analysis, J.P. and H.C.; investigation, J.P., H.C., G.S. and K.M.; resources, J.P., H.C., G.S. and K.M.; data curation, J.P., H.C., G.S. and K.M.; writing—original draft preparation, J.P. and H.C.; writing—review and editing, J.P., H.C., C.S., S.J.R., R.M.V., G.S. and K.M.; visualization, J.P.; supervision, H.C. and R.M.V.; project administration, H.C. and R.M.V.; funding acquisition, R.M.V. All authors have read and agreed to the published version of the manuscript.

Funding: This report is independent research commissioned and funded by the National Institute for Health Research Policy Research Program. The views expressed in this publication are those of the authors and not necessarily those of the NHS, the National Institute for Health Research, the Department of Health and Social Care or its arm's length bodies, and other Government Departments.

Institutional Review Board Statement: Not applicable.

Informed Consent Statement: Not applicable.

Data Availability Statement: The data presented in this study are available in Boyland [20], 2013; Coates [52], 2019; Coates [53], 2019.

Acknowledgments: We thank Semina Michalopoulou and Jamie Wong for their assistance in bias assessment and screening, respectively.

Conflicts of Interest: The authors declare no conflict of interest.

References

1. Sahoo, K.; Sahoo, B.; Choudhury, A.K.; Sofi, N.Y.; Kumar, R.; Bhadoria, A.S. Childhood obesity: Causes and consequences. *J. Fam. Med. Prim. Care* **2015**, *4*, 187–192. [CrossRef]
2. Lapierre, M.A.; Fleming-Milici, F.; Rozendaal, E.; McAlister, A.R.; Castonguay, J. The Effect of Advertising on Children and Adolescents. *Pediatrics* **2017**, *140*, S152. [CrossRef]
3. Ofcom. Children and Parents: Media Use and Attitudes Report 2018. 2019. Available online: https://www.ofcom.org.uk/__data/assets/pdf_file/0024/134907/children-and-parents-media-use-and-attitudes-2018.pdf (accessed on 9 December 2021).
4. Griffith, R.; O'Connell, M.; Smith, K.; Stroud, R. *Children's Exposure to TV Advertising of Food and Drink*; The Institute for Fiscal Studies: London, UK, 2018; Available online: https://www.ifs.org.uk/uploads/publications/bns/BN238.pdf (accessed on 9 December 2021).
5. Boyland, E.J.; Whalen, R. Food advertising to children and its effects on diet: Review of recent prevalence and impact data. *Pediatric Diabetes* **2015**, *16*, 331–337. [CrossRef]
6. Backholer, K.; Gupta, A.; Zorbas, C.; Bennett, R.; Huse, O.; Chung, A.; Isaacs, A.; Golds, G.; Kelly, B.; Peeters, A. Differential exposure to, and potential impact of, unhealthy advertising to children by socio-economic and ethnic groups: A systematic review of the evidence. *Obes. Rev.* **2020**, *22*, e13144. [CrossRef] [PubMed]
7. Russell, S.J.; Croker, H.; Viner, R.M. The effect of screen advertising on children's dietary intake: A systematic review and meta-analysis. *Obes. Rev.* **2019**, *20*, 554–568. [CrossRef]
8. Clark, H.; Coll-Seck, A.M.; Banerjee, A.; Peterson, S.; Dalglish, S.L.; Ameratunga, S.; Balabanova, D.; Bhan, M.K.; Bhutta, Z.A.; Borrazzo, J.; et al. A future for the world's children? A WHO-UNICEF-Lancet Commission. *Lancet* **2020**, *395*, 605–658. [CrossRef]
9. WHO. *Set of Recommendations on the Marketing of Foods and Non-Alcoholic Beverages to Children*; World Health Organization: Geneva, Switzerland, 2010; Available online: https://apps.who.int/iris/bitstream/handle/10665/44416/9789241500210_eng.pdf (accessed on 9 December 2021).
10. WHO. *A Framework for Implementing the Set of Recommendations on the Marketing of Foods and Non-Alcoholic Beverages to Children*; World Health Organization: Geneva, Switzerland, 2012; Available online: https://www.who.int/dietphysicalactivity/MarketingFramework2012.pdf (accessed on 9 December 2021).
11. Bergkvist, L.; Zhou, K.Q. Celebrity endorsements: A literature review and research agenda. *Int. J. Advert.* **2016**, *35*, 642–663. [CrossRef]
12. Cuomo, M.T.; Foroudi, P.; Tortora, D.; Hussain, S.; Melewar, T.C. Celebrity Endorsement and the Attitude Towards Luxury Brands for Sustainable Consumption. *Sustainability* **2019**, *11*, 6791. [CrossRef]
13. Elberse, A.; Verleun, J. The Economic Value of Celebrity Endorsements. *J. Advert. Res.* **2012**, *52*, 149. [CrossRef]
14. Knoll, J.; Matthes, J. The effectiveness of celebrity endorsements: A meta-analysis. *J. Acad. Mark. Sci.* **2017**, *45*, 55–75. [CrossRef]
15. De Houwer, J.; Thomas, S.; Baeyens, F. Associative learning of likes and dislikes: A review of 25 years of research on human evaluative conditioning. *Psychol. Bull.* **2001**, *127*, 853–869. [CrossRef] [PubMed]
16. Aw, E.C.-X.; Labrecque, L.I. Celebrity endorsement in social media contexts: Understanding the role of parasocial interactions and the need to belong. *J. Consum. Mark.* **2020**, *37*, 895–908. [CrossRef]
17. Hoffman, S.J.; Tan, C. Biological, psychological and social processes that explain celebrities' influence on patients' health-related behaviors. *Arch. Public Health* **2015**, *73*, 3. [CrossRef]
18. WHO. *Evaluating Implementation of the WHO Set of Recommendations on the Marketing of Foods and Non-Alcoholic Beverages to Children*; World Health Organization: Geneva, Switzerland, 2018; Available online: https://www.euro.who.int/__data/assets/pdf_file/0003/384015/food-marketing-kids-eng.pdf (accessed on 9 December 2021).
19. Taillie, L.S.; Busey, E.; Stoltze, F.M.; Dillman Carpentier, F.R. Governmental policies to reduce unhealthy food marketing to children. *Nutr. Rev.* **2019**, *77*, 787–816. [CrossRef] [PubMed]
20. Boyland, E.J.; Harrold, J.A.; Dovey, T.M.; Allison, M.; Dobson, S.; Jacobs, M.C.; Halford, J.C. Food choice and overconsumption: Effect of a premium sports celebrity endorser. *J. Pediatr.* **2013**, *163*, 339–343. [CrossRef] [PubMed]

21. WCRF. *Restrict Food Advertising and Other Forms of Commercial Promotion*; WCRF: London, UK, 2019; Available online: https://policydatabase.wcrf.org/level_one?page=nourishing-level-one#step2=3 (accessed on 9 December 2021).

22. Hawkes, C. *Marketing Food to Children: Changes in the Global Regulatory Environment 2004–2006*; World Health Organization: Geneva, Switzerland, 2007; Available online: https://www.who.int/dietphysicalactivity/regulatory_environment_CHawkes07.pdf (accessed on 9 December 2021).

23. Australian Communications and Media Authority. Guide to the Children's Television Standards 2009. Available online: https://www.acma.gov.au/sites/default/files/2019-06/Previous-guide-to-the-Childrens-Television-Standards-2009.pdf (accessed on 9 December 2021).

24. WHO. *Marketing of Foods High in Fat, Salt and Sugar to Children: Update 2012–2013*; WHO Regional Office: Copenhagen, Denmark, 2013; Available online: http://www.euro.who.int/__data/assets/pdf_file/0019/191125/e96859.pdf (accessed on 9 December 2021).

25. Advertising Standards Authority. *The BCAP Code—The UK Code of Broadcast Advertising*; The Advertising Standards Authority: London, UK, 2018; Available online: https://www.asa.org.uk/static/846f25eb-f474-47c1-ab3ff571e3db5910/BCAP-Code-full.pdf (accessed on 9 December 2021).

26. Advertising Standards Authority. *The CAP Code—The UK Code of Non-Broadcast Advertising and Direct & Promotional Marketing*; The Advertising Standards Authority: London, UK, 2018; Available online: https://www.asa.org.uk/static/47eb51e7-028d-4509-ab3c0f4822c9a3c4/bbca4ed3-9b41-4c6b-8d13299664f119be/The-Cap-code.pdf (accessed on 9 December 2021).

27. Boyland, E.J.; Harris, J.L. Regulation of food marketing to children: Are statutory or industry self-governed systems effective? *Public Health Nutr.* **2017**, *20*, 761–764. [CrossRef] [PubMed]

28. Zhou, M.; Rincón-Gallardo Patiño, S.; Hedrick, V.E.; Kraak, V.I. An accountability evaluation for the responsible use of celebrity endorsement by the food and beverage industry to promote healthy food environments for young Americans: A narrative review to inform obesity prevention policy. *Obes. Rev.* **2020**, *21*, e13094. [CrossRef]

29. Department of Health and Social Care. Introducing Further Advertising Restrictions on TV and Online for Products High in Fat, Sugar and Salt (HFSS). Available online: https://assets.publishing.service.gov.uk/government/uploads/system/uploads/attachment_data/file/807378/hfss-advertising-consultation-10-april-2019.pdf (accessed on 9 December 2021).

30. León-Flández, K.; Rico-Gómez, A.; Moya-Geromin, M.; Romero-Fernández, M.; Bosqued-Estefania, M.J.; Damián, J.; López-Jurado, L.; Royo-Bordonada, M. Evaluation of compliance with the Spanish Code of self-regulation of food and drinks advertising directed at children under the age of 12 years in Spain, 2012. *Public Health* **2017**, *150*, 121–129. [CrossRef]

31. House of Lords. *UK Advertising in a Digital Age*; Authority of the House of Lords: London, UK, 2018; Available online: https://publications.parliament.uk/pa/ld201719/ldselect/ldcomuni/116/116.pdf (accessed on 9 December 2021).

32. Schouten, A.P.; Janssen, L.; Verspaget, M. Celebrity vs. Influencer endorsements in advertising: The role of identification, credibility, and Product-Endorser fit. *Int. J. Advert.* **2020**, *39*, 258–281. [CrossRef]

33. De Veirman, M.; Hudders, L.; Nelson, M.R. What Is Influencer Marketing and How Does It Target Children? A Review and Direction for Future Research. *Front. Psychol.* **2019**, *10*, 2685. [CrossRef]

34. Russell, C.A.; Rasolofoarison, D. Uncovering the power of natural endorsements: A comparison with celebrity-endorsed advertising and product placements. *Int. J. Advert.* **2017**, *36*, 761–778. [CrossRef]

35. Phillipson, L.J.; Jones, S.C. I eat Milo to make me run faster: How the use of sport in food marketing may influence the food beliefs of young Australians. *Proc. Aust. N. Z. Mark. Acad. Conf.* **2008**, 1–7. Available online: https://ro.uow.edu.au/cgi/viewcontent.cgi?article=3332&context=hbspapers (accessed on 9 December 2021).

36. Boyland, E.J.; Harrold, J.A.; Kirkham, T.C.; Halford, J.C. Persuasive techniques used in television advertisements to market foods to UK children. *Appetite* **2012**, *58*, 658–664. [CrossRef]

37. Bragg, M.A.; Miller, A.N.; Elizee, J.; Dighe, S.; Elbel, B.D. Popular Music Celebrity Endorsements in Food and Nonalcoholic Beverage Marketing. *Pediatrics* **2016**, *138*, e20153977. [CrossRef]

38. Bragg, M.A.; Yanamadala, S.; Roberto, C.A.; Harris, J.L.; Brownell, K.D. Athlete endorsements in food marketing. *Pediatrics* **2013**, *132*, 805–810. [CrossRef] [PubMed]

39. Alruwaily, A.; Mangold, C.; Greene, T.; Arshonsky, J.; Cassidy, O.; Pomeranz, J.L.; Bragg, M. Child Social Media Influencers and Unhealthy Food Product Placement. *Pediatrics* **2020**, *146*, e20194057. [CrossRef] [PubMed]

40. Hebden, L.; King, L.; Kelly, B.; Chapman, K.; Innes-Hughes, C. A menagerie of promotional characters: Promoting food to children through food packaging. *J. Nutr. Educ. Behav.* **2011**, *43*, 349–355. [CrossRef] [PubMed]

41. Potvin Kent, M.; Pauzé, E.; Roy, E.A.; de Billy, N.; Czoli, C. Children and adolescents' exposure to food and beverage marketing in social media apps. *Pediatr. Obes.* **2019**, *14*, e12508. [CrossRef] [PubMed]

42. Blades, M.; Oates, C.; Li, S. Children's recognition of advertisements on television and on Web pages. *Appetite* **2013**, *62*, 190–193. [CrossRef]

43. Rozendaal, E.; Buijzen, M.; Valkenburg, P. Comparing Children's and Adults' Cognitive Advertising Competences in the Netherlands. *J. Child. Media* **2010**, *4*, 77–89. [CrossRef]

44. Smit, C.R.; Buijs, L.; van Woudenberg, T.J.; Bevelander, K.E.; Buijzen, M. The Impact of Social Media Influencers on Children's Dietary Behaviors. *Front. Psychol.* **2020**, *10*, 2975. [CrossRef]

45. Jenkin, G.; Madhvani, N.; Signal, L.; Bowers, S. A systematic review of persuasive marketing techniques to promote food to children on television. *Obes. Rev.* **2014**, *15*, 281–293. [CrossRef] [PubMed]

46. Cairns, G.; Angus, K.; Hastings, G.; Caraher, M. Systematic reviews of the evidence on the nature, extent and effects of food marketing to children. A retrospective summary. *Appetite* **2013**, *62*, 209–215. [CrossRef] [PubMed]
47. Smith, R.; Kelly, B.; Yeatman, H.; Boyland, E. Food Marketing Influences Children's Attitudes, Preferences and Consumption: A Systematic Critical Review. *Nutrients* **2019**, *11*, 875. [CrossRef]
48. Moher, D.; Liberati, A.; Tetzlaff, J.; Altman, D.G. Preferred reporting items for systematic reviews and meta-analyses: The PRISMA statement. *BMJ* **2009**, *339*, b2535. [CrossRef] [PubMed]
49. Thomas, J.; Brunton, J.; Graziosi, S. *EPPI-Reviewer 4.0: Software for Research Synthesis*; Social Science Research Unit, Institute of Education, University of London: London, UK, 2010.
50. Sterne, J.A.C.; Savović, J.; Page, M.J.; Elbers, R.G.; Blencowe, N.S.; Boutron, I.; Cates, C.J.; Cheng, H.-Y.; Corbett, M.S.; Eldridge, S.M.; et al. RoB 2: A revised tool for assessing risk of bias in randomised trials. *BMJ* **2019**, *366*, l4898. [CrossRef]
51. StataCorp. *Stata Statistical Software: Release 16*; StataCorp LLC.: College Station, TX, USA, 2019.
52. Coates, A.E.; Hardman, C.A.; Halford, J.C.G.; Christiansen, P.; Boyland, E.J. The effect of influencer marketing of food and a "protective" advertising disclosure on children's food intake. *Pediatr. Obes.* **2019**, *14*, e12540. [CrossRef] [PubMed]
53. Coates, A.E.; Hardman, C.A.; Halford, J.C.G.; Christiansen, P.; Boyland, E.J. Social Media Influencer Marketing and Children's Food Intake: A Randomized Trial. *Pediatrics* **2019**, *143*, e20182554. [CrossRef] [PubMed]
54. De Jans, S.; Spielvogel, I.; Naderer, B.; Hudders, L. Digital food marketing to children: How an influencer's lifestyle can stimulate healthy food choices among children. *Appetite* **2021**, *162*, 105182. [CrossRef]
55. Dixon, H.; Scully, M.; Niven, P.; Kelly, B.; Chapman, K.; Donovan, R.; Martin, J.; Baur, L.A.; Crawford, D.; Wakefield, M. Effects of nutrient content claims, sports celebrity endorsements and premium offers on pre-adolescent children's food preferences: Experimental research. *Pediatr. Obes.* **2014**, *9*, e47–e57. [CrossRef]
56. Jain, V.; Roy, S.; Daswani, A.; Sudha, M. What really works for teenagers: Human or fictional celebrity? *Young Consum.* **2011**, *12*, 171–183. [CrossRef]
57. Ponce-Blandón, J.A.; Pabón-Carrasco, M.; Romero-Castillo, R.; Romero-Martín, M.; Jiménez-Picón, N.; Lomas-Campos, M.d.L.M. Effects of Advertising on Food Consumption Preferences in Children. *Nutrients* **2020**, *12*, 3337. [CrossRef]
58. Public Health England. *Calorie Reduction: The Scope and Ambition for Action*; Public Health England: London, UK, 2018. Available online: https://assets.publishing.service.gov.uk/government/uploads/system/uploads/attachment_data/file/800675/Calories_Evidence_Document.pdf (accessed on 9 December 2021).
59. Mytton, O.T.; Boyland, E.; Adams, J.; Collins, B.; O'Connell, M.; Russell, S.J.; Smith, K.; Stroud, R.; Viner, R.M.; Cobiac, L.J. The potential health impact of restricting less-healthy food and beverage advertising on UK television between 05.30 and 21.00 hours: A modelling study. *PLoS Med.* **2020**, *17*, e1003212. [CrossRef] [PubMed]
60. Department of Health and Social Care. New Advertising Rules to Help Tackle Childhood Obesity. 2021. Available online: https://www.gov.uk/government/news/new-advertising-rules-to-help-tackle-childhood-obesity (accessed on 9 December 2021).
61. Department of Health and Social Care. Introducing Further Advertising Restrictions on TV and Online for Products High in Fat, Salt and Sugar: Government Response-Consultation Outcome. Available online: https://www.gov.uk/government/consultations/further-advertising-restrictions-for-products-high-in-fat-salt-and-sugar/outcome/introducing-further-advertising-restrictions-on-tv-and-online-for-products-high-in-fat-salt-and-sugar-government-response (accessed on 9 December 2021).
62. Dovey, T.M.; Taylor, L.; Stow, R.; Boyland, E.J.; Halford, J.C. Responsiveness to healthy television (TV) food advertisements/commercials is only evident in children under the age of seven with low food neophobia. *Appetite* **2011**, *56*, 440–446. [CrossRef] [PubMed]

Review

Inositol Nutritional Supplementation for the Prevention of Gestational Diabetes Mellitus: A Systematic Review and Meta-Analysis of Randomized Controlled Trials

Jingshu Wei [†], Jie Yan [†] and Huixia Yang *

Department of Obstetrics and Gynecology, Peking University First Hospital, Beijing 100034, China;
1610301133@pku.edu.cn (J.W.); yanjie20120303@126.com (J.Y.)
* Correspondence: yanghuixia@bjmu.edu.cn
† These authors contributed equally to this work.

Abstract: This study was aimed at assessing the efficacy and safety of inositol nutritional supplementation during pregnancy for the prevention of GDM. PubMed, Embase, MEDLINE, and Cochrane library were systematically searched for randomized controlled trails (RCTs) in this field until May 2022. Primary outcomes were the incidence for GDM and plasma glucose levels by oral glucose tolerance test (OGTT). Pooled results were expressed as relative risk (RR) or mean difference (MD) with a 95% confidence interval (95% CI). Seven RCTs with 1321 participants were included in this study. Compared with the control group, 4 g myo-inositol (MI) supplementation per day significantly decreased the incidence of GDM (RR = 0.30, 95% CI (0.18, 0.49), $p < 0.00001$). It significantly decreased the plasma glucose levels of OGTT regarding fasting-glucose OGTT (MD = -4.20, 95% CI (-5.87, -2.54), $p < 0.00001$), 1-h OGTT (MD = -8.75, 95% CI (-12.42, -5.08), $p < 0.00001$), and 2-h OGTT (MD = -8.59, 95% CI (-11.81, -5.83), $p < 0.00001$). It also decreased the need of insulin treatment, and reduced the incidence of preterm delivery and neonatal hypoglycemia. However, no difference was observed between 1.1 g MI per day plus 27.6 mg D-chiro-inositol (DCI) per day and the control group regarding all evaluated results. In conclusion, 4 g MI nutritional supplementation per day during early pregnancy may reduce GDM incidence and severity, therefore may be a practical and safe approach for the prevention of GDM.

Keywords: inositol; myo-inositol; D-chiro-inositol; gestational diabetes mellitus; insulin resistance; randomized controlled trial; meta-analysis

Citation: Wei, J.; Yan, J.; Yang, H. Inositol Nutritional Supplementation for the Prevention of Gestational Diabetes Mellitus: A Systematic Review and Meta-Analysis of Randomized Controlled Trials. *Nutrients* **2022**, *14*, 2831. https://doi.org/10.3390/nu14142831

Academic Editor: Annunziata Lapolla

Received: 7 June 2022
Accepted: 6 July 2022
Published: 9 July 2022

Publisher's Note: MDPI stays neutral with regard to jurisdictional claims in published maps and institutional affiliations.

Copyright: © 2022 by the authors. Licensee MDPI, Basel, Switzerland. This article is an open access article distributed under the terms and conditions of the Creative Commons Attribution (CC BY) license (https://creativecommons.org/licenses/by/4.0/).

1. Introduction

Gestational diabetes mellitus (GDM) is one of the most common complications during pregnancy. It is defined as "diabetes diagnosed in the second or third trimester of pregnancy that was not clearly overt diabetes prior to gestation" according to the latest guideline of the American Diabetes Association (ADA) [1]. The International Diabetes Federation (IDF) has reported [2] that the global prevalence of GDM among pregnant women aged 20–49 years was 16.7% in 2021. According to the recommendations of the International Federation of Gynecology and Obstetrics and guidelines of various countries, a 75 g 2-h oral glucose tolerance test is suggested to be performed on all pregnant women between 24- and 28-weeks' gestation, in order to diagnose and effectively manage GDM [3]. Current evidence indicates that GDM is associated with not only higher risk of pregnancy complications, but also the long-term health of mothers and infants, such as increased risk of developing type 2 diabetes, cardiovascular disease, chronic kidney disease, and cancer for mothers and obesity, overweight, insulin resistance, and delayed neurocognitive development for their offspring [4,5]. Therefore, it is crucial to find safe, effective, and acceptable measures for the prevention of GDM, and nutritional supplementation has been considered as a possible intervention [6].

Inositol, or cyclohexane−1,2,3,4,5,6-hexol, is a polyol which naturally presents in animal and plant cells. Foods that are rich in inositol include cereals, legumes, oil, seeds, and nuts. Inositol exists under nine stereoisomeric forms, of which myo-inositol (MI) and D-chiro-inositol (DCI) are the most common forms in eukaryotic cells. Inositol serves as the structural basis for a variety of second messengers in eukaryotic cells, thus involved in the transduction of several endocrine signals, including follicle stimulating hormone (FSH), thyroid stimulating hormone (TSH), and insulin. By increasing glucose transporter type 4 (GLUT4) translocation to the cell membrane, MI and DCI act as insulin sensitizers, thereby decreasing insulin resistance and plasma glucose levels [7,8].

It has been reported that inositol may reduce hyperinsulinemia and restore ovarian function in women with polycystic ovarian syndrome (PCOS) [9]. As insulin resistance increasing during pregnancy is now recognized as the main pathophysiological mechanism of GDM, it is reasonable to hypothesize that inositol dietary supplementation may have a preventive effect on the development of GDM and its complications.

Over the past ten years, several randomized controlled trails (RCTs) have investigated the effectiveness of inositol supplementation for the prevention of GDM. A Cochran review in 2015 concluded the body of evidence of this topic was of low quality [10]. Since then, other new RCTs have provided more evidence on this topic. Therefore, this systematic review and meta-analysis is performed to assess and summarize the efficacy and safety of inositol nutritional supplementation during pregnancy for the prevention of GDM.

2. Materials and Methods

2.1. Study Design

This systematic review is organized in accordance with current recommendations of Preferred Reporting Items for Systematic Reviews and Meta-analyses (PRISMA) guidelines [11].

2.2. Search Strategy

A literature search was performed in the following databases: PubMed, Embase, MEDLINE, and Cochrane library from inception to May 2022. The search terms used were inositol, myo-inositol, D-chiro-inositol, gestational diabetes mellitus, gestational diabetes, OGTT, plasma glucose level, insulin resistance, randomized controlled trial, and clinical trial. Additional searches were performed among the reference lists of relevant review articles.

2.3. Study Selection

The study selection underwent a two-stage process. Firstly, the titles and abstracts were screened. The full tests of all relevant trials were then examined, based on the following inclusion and exclusion criteria. Duplicated studies were removed.

Inclusion criteria:

1. RCTs evaluating the effects of inositol (MI and/or DCI) for preventing GDM during pregnancy.

Exclusion criteria:

1. Pregnant women who were already diagnosed with type 1 diabetes (T1DM), type 2 diabetes mellitus (T2DM), or GDM.
2. Interventions included supplementation other than inositol and folic acid.
3. Did not assess incidence of GDM as outcome.
4. Pilots, protocols, observational studies, reviews, case reports, trials, comments, letters, news, notes, editorial, or conference abstracts.
5. Unable to access the full text of article.

2.4. Data Extraction

Data were extracted from the selected RCTs, including first author, year of publication, research site, number of participants, baseline characteristics of participants, detailed intervention protocols, outcome measures, and any reported adverse events. Outcomes that appeared at least in two included studies were extracted for further meta-analysis.

2.5. Outcome Measures

Primary outcomes were the incidence for GDM and plasma glucose levels (fasting glucose, 1-h glucose, and 2-h glucose) of the oral glucose tolerance test (OGTT). Secondary outcomes included maternal health outcomes, delivery outcomes, and neonatal health outcomes. Addtionally, side effects associated with the intervention were also evaluated.

Maternal health outcomes measures:

--Need of insulin treatment (%).

--Hypertensive disorders (%).

Delivery outcomes measures:

--Preterm delivery (%): the birth of a baby at earlier than 37 weeks gestational age.

--Cesarean section (%).

--Shoulder dystocia (%).

Neonatal health outcomes measures:

--Birth weight (g).

--Birth weight (%).

--Macrosomia (%): birth weight $\geq$ 4000 g.

--Neonatal hypoglycemia (%): plasma glucose level of less than 45 mg/dl.

--Neonatal Intensive Care Unit admission (NICU admission) (%).

2.6. Risk of Bias and Quality Assessment

The risk of bias was assessed within studies in seven domains, following the criteria outlined in the Cochrane Handbook for Systematic Reviews of Interventions [12]. Discrepancies were resolved by group discussion.

The quality of evidence was evaluated by using the Grading of Recommendations Assessment, Development, and Evaluation (GRADE) approach as outlined in the GRADE handbook [13].

2.7. Data Analysis

The meta-analysis estimated the risk ratio (RR) with the 95% confidence interval (CI) for dichotomous variables, and mean difference (MD) with the 95% CI for continuous variables. Heterogeneity was assessed using I^2 statistics, with an I^2 between 50 and 90% indicating substantial heterogeneity [11]. All statistical analyses were performed using Review Manager version 5.4.1 (The Cochrane Collaboration, Oxford, UK).

3. Results

3.1. Literature Search

The initial search in electronic databases and reference lists of review articles identified 23 potentially relevant RCTs. The full text of nine RCTs was retrieved after screaming titles and abstracts, of which seven RCTs were finally included in the meta-analysis, with a total of 1321 patients. The detailed process of the literature search and selection is shown via a flowchart in Figure 1.

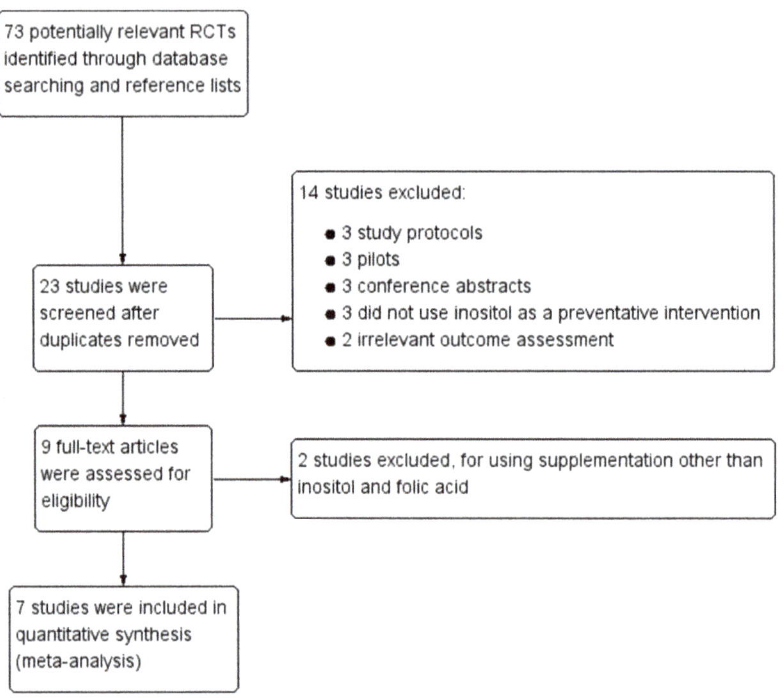

Figure 1. Flowchart of study search and selection.

3.2. Characteristics of Included Studies

The main characteristics of the included studies are summarized in Table 1.

Table 1. The main characteristics of the included studies. BMI: body mass index. GA: gestational age. FPG: fast plasma glucose. RG: random glucose.

Study	Country	Inclusion Criteria	Interventions	Inositol Group	Control Group
Matarrelli 2013 [14]	Italy	(1) Prepregnancy BMI < 35 kg/m² (2) FPG ≥ 5.1 mmol/L and <7.0 mmol/L in the first or early second trimester (3) Single pregnancy	Group A: 2 g MI plus 200 μg folic acid twice a day Group B: 200 μg folic acid twice a day	Group A (*n* = 35) Age: 33.0 ± 4.9 BMI: 23.5 ± 3.4 FPG: 5.4 ± 0.3	Group B (*n* = 38) Age: 33.8 ± 4.7 BMI: 24.7 ± 4.2 FPG: 5.4 ± 0.3
D'Anna 2013 [15]	Italy	(1) GA: 12~13 w (2) First-degree relatives affected by T2DM (3) Prepregnancy BMI < 30 kg/m² (4) FPG < 126 mg/dl, RG < 200 mg/dl (5) Single pregnancy (6) Caucasian race	Group A: 2 g MI plus 200 μg folic acid twice a day Group B: 200 μg folic acid twice a day	Group A (*n* = 99) Age: 31.0 ± 5.3 BMI: 22.8 ± 3.1	Group B (*n* = 98) Age: 31.6 ± 5.6 BMI: 23.6 ± 3.1
D'Anna 2015 [16]	Italy	(1) GA: 12~13 w (2) Prepregnancy BMI ≥ 30 kg/m² (3) FPG < 126 mg/dl, RG < 200 mg/dl (4) Single pregnancy (5) Caucasian race	Group A: 2 g MI plus 200 μg folic acid twice a day Group B: 200 μg folic acid twice a day	Group A (*n* = 107) Age: 30.9 (18~44) BMI: 33.8 (30.0~46.9) FPG: 83.1 ± 8.5	Group B (*n* = 107) Age: 31.7 (19~43) BMI: 33.8 (30.0~46.0) FPG: 82.3 ± 10.6
Santamaria 2016 [17]	Italy	(1) GA: 12~13 w (2) Prepregnancy BMI ≥ 25 kg/m², <30 kg/m² (3) FPG < 126 mg/dl, RG < 200 mg/dl (4) Single pregnancy (5) Caucasian race	Group A: 2 g MI plus 200 μg folic acid twice a day Group B: 200 μg folic acid twice a day	Group A (*n* = 95) Age: 32.1 ± 4.8 BMI: 26.9 ± 1.3 FPG: 81.09 ± 8..03	Group B (*n* = 102) Age: 32.7 ± 5.3 BMI: 27.1 ± 1.3 FPG: 78.63 ± 6.15
Farren 2017 [18]	Ireland	(1) GA: 10~16 w (2) First-degree relatives affected by T1DM/T2DM (3) FPG < 126 mg/dl, RG < 200 mg/dl (4) Single pregnancy	Group A: 1.1 g MI, 27.6 mg DCI plus 400 μg folic acid per day Group B: 400 μg folic acid per day	Group A (*n* = 120) Age: 31.1 ± 5.1 BMI: 26.0 ± 5.3	Group B (*n* = 120) Age: 31.5 ± 5.0 BMI: 26.2 ± 5.5

Table 1. *Cont.*

Study	Country	Inclusion Criteria	Interventions	Inositol Group	Control Group
Celentano 2020 [19]	Italy	(1) Prepregnancy BMI < 35 kg/m² (2) FPG ≥ 5.1 mmol/L and <7.0 mmol/L in the first trimester (3) Single pregnancy	Group A: 2 g MI plus 200 µg folic acid twice a day Group B: 500 mg DCI plus 400 µg folic acid per day Group C: 0.55 g MI, 13.8 mg DCI plus 200 µg folic acid twice a day Group D: 400 µg folic acid per day	Group A ($n = 39$) Age: 33.1 ± 4.9 BMI: 23.5 ± 3.4 FPG: 5.4 ± 0.3 Group B ($n = 32$) Age: 34.4 ± 3.7 BMI: 24.4 ± 4.9 FPG: 5.3 ± 0.2 Group C ($n = 34$) Age: 34.1 ± 4.2 BMI: 23.5 ± 4.6 FPG: 5.4 ± 0.3	Group D ($n = 52$) Age: 33.9 ± 4.9 BMI: 24.4 ± 4.1 FPG: 5.4 ± 0.3
Vitale 2021 [20]	Italy	(1) GA: 12~13 w (2) Prepregnancy BMI ≥ 25 kg/m², <30 kg/m² (3) FPG < 126 mg/dl, RG < 200 mg/dl (4) Single pregnancy (5) Caucasian race	Group A: 2 g MI plus 200 µg folic acid twice a day Group B: 200 µg folic acid twice a day	Group A ($n = 110$) Age: 27.18 ± 6.03 BMI: 27.00 ± 1.49 FPG: 82.20 ± 12.12	Group B ($n = 113$) Age: 27.95 ± 4.90 BMI: 26.68 ± 1.56 FPG: 83.10 ± 14.10

The seven included studies were published between 2013 and 2021, six of which were conducted in Italy, the other one in Ireland. The sample sizes ranged from 73 to 240. Two studies were double-center studies [16,17] and the remaining five were carried out in a single center [14,15,18–20].

Two studies [17,19] focused on overweight non-obese women, whose pre-pregnancy body mass index (BMI) was between 25 and 30 kg/m², while one study [16] focused on obese women, whose pre-pregnancy BMI was over 30 kg/m². Two studies [15,18] only included patients with a family history of T2DM. Two studies [14,19] focused on pregnant women with elevated plasma glucose levels in the first trimester.

Six studies [14–17,19,20] compared the effect of 4 g MI per day with the placebo; two studies [18,19] compared the effect of 1.1 g MI plus 27.6 mg DCI with the placebo; and one study [19] investigated the effect of 500 mg DCI versus the placebo. The placebo was 200 µg folic acid twice a day (or 400 µg per day) in all studies. The duration of intervention in all studies was from trial entry to delivery.

In all studies, the diagnosis of GDM was based on the Hyperglycemia and Adverse Pregnancy Outcome (HAPO) study [21] and the International Association of Diabetes and Pregnancy Study Groups (IADPSG) recommendations [22].

3.3. Risk of Bias of Included Studies

The risk of bias of the included studies is shown in Figures 2 and 3.

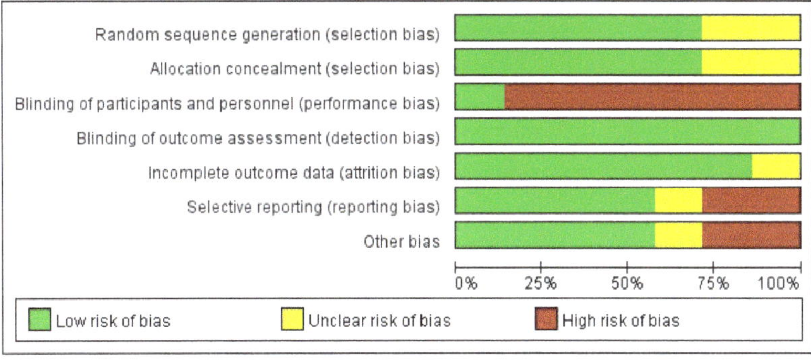

Figure 2. Review of authors' judgement about each risk of bias item presented as percentages across all included studies.

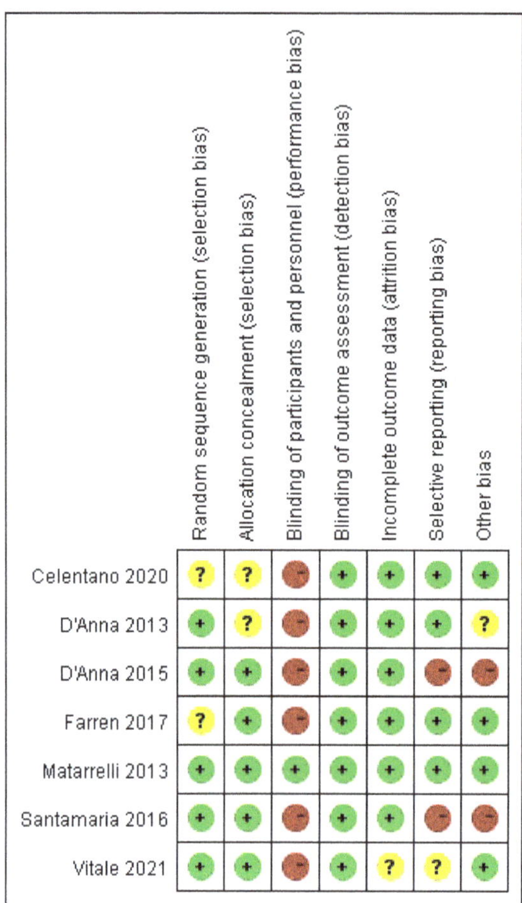

Figure 3. Review of authors' judgement about each risk of bias item for each included study [14–20].

3.3.1. Selection Bias

All but two studies used a computer-generated random sequence. Farren 2017 and Celentano 2020 stated the participants were randomized but did not provide further information on the method (unclear risk of bias).

D'Anna 2013 and Celentano 2020 did not report the method of allocation concealment (unclear risk of bias), while the remaining studies used an adequate allocation strategy (four with sealed envelopes and one with central allocation).

3.3.2. Performance Bias

Matarrelli 2013 is the only study which was double-blinded for both participants and healthcare providers. The others were open-label studies and therefore were assessed as having a high risk of bias.

3.3.3. Detection Bias

The outcomes evaluated in these studies were either objective measurements of laboratory values or established facts, which were unlikely to be influenced by the blinding strategy. Therefore, all studies were assessed as having a low risk of bias.

3.3.4. Attrition Bias

None of the seven studies had a drop-out rate over 20%. However, Vitale 2021 used a protocol treatment analysis, therefore was assessed as having an unclear risk of bias.

3.3.5. Reporting Bias

D'Anna 2015 and Santamaria 2016 were assessed as having a high risk of bias because the outcomes were changed after protocol registration. Vitale 2021 was assessed as having an unclear risk of bias because the detailed results of several secondary outcomes were not reported. The remaining studies were assessed as having a low risk of bias.

3.3.6. Other Bias

D'Anna 2015 and Santamaria 2016 were assessed as having a high risk of bias because a baseline difference in the rate of family history of T2DM between groups was reported, which might affect the effect evaluation. D'Anna 2013 stated that an intention-to-treat analysis was performed and did not show results different from those of per-protocol analysis, but only published the results of per-protocol analysis. Therefore, it was assessed as having an unclear risk of bias. The remaining studies were assessed as having a low risk of bias.

3.4. Effects of Intervention

3.4.1. Primary Outcomes

All seven studies evaluated the incidence of GDM and OGTT plasma glucose levels.

Compared with the control group, 4 g MI supplementation per day significantly decreased the incidence of GDM (RR = 0.30, 95% CI (0.18, 0.49), $p < 0.00001$; six trials, $n = 995$), with substantial heterogeneity among studies ($I^2 = 52\%$, $p = 0.06$). No difference was observed between 1.1 g MI plus 27.6 mg DCI supplementation per day and the control group ($p = 0.74$; two trials, $n = 326$) (Figure 4).

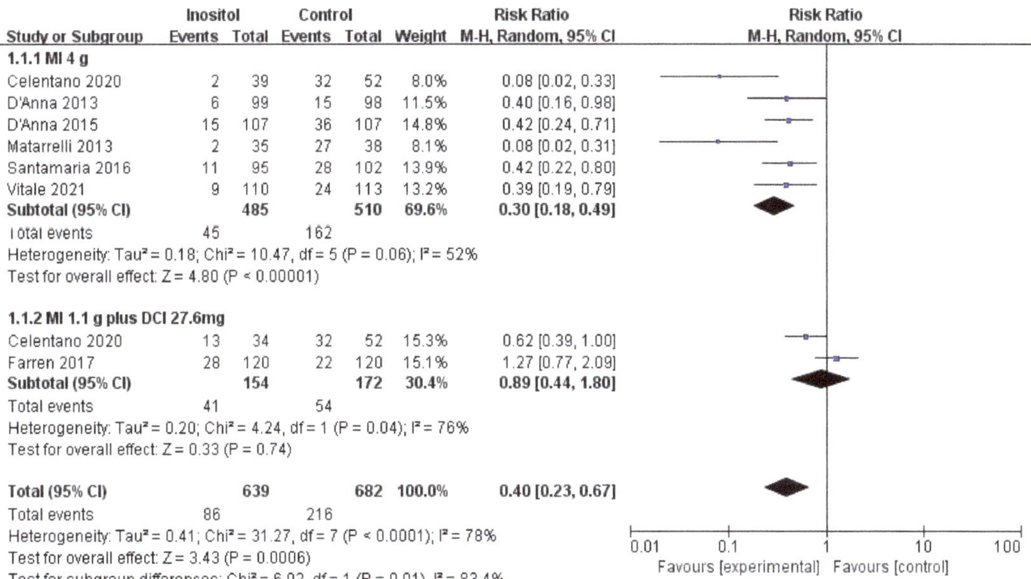

Figure 4. Inositol vs. control: incidence of GDM [14–20].

Compared with the control group, 4 g MI supplementation per day significantly decreased the OGTT plasma glucose levels regarding fasting-glucose OGTT (MD = −4.20,

95% CI (-5.87, -2.54), $p < 0.00001$, $I^2 = 50\%$; six trials, $n = 995$) (Figure 5), 1-h OGTT (MD $= -8.75$, 95% CI [-12.42, -5.08], $p < 0.00001$, $I^2 = 27\%$; six trials, $n = 995$) (Figure 6), and 2-h OGTT (MD $= -8.59$, 95% CI (-11.81, -5.83), $p < 0.00001$, $I^2 = 44\%$; six trials, $n = 995$) (Figure 7). No difference was observed between 1.1 g MI plus 27.6 mg DCI supplementation per day and the control group ($p = 0.45/0.29/0.48$; two trials, $n = 326$).

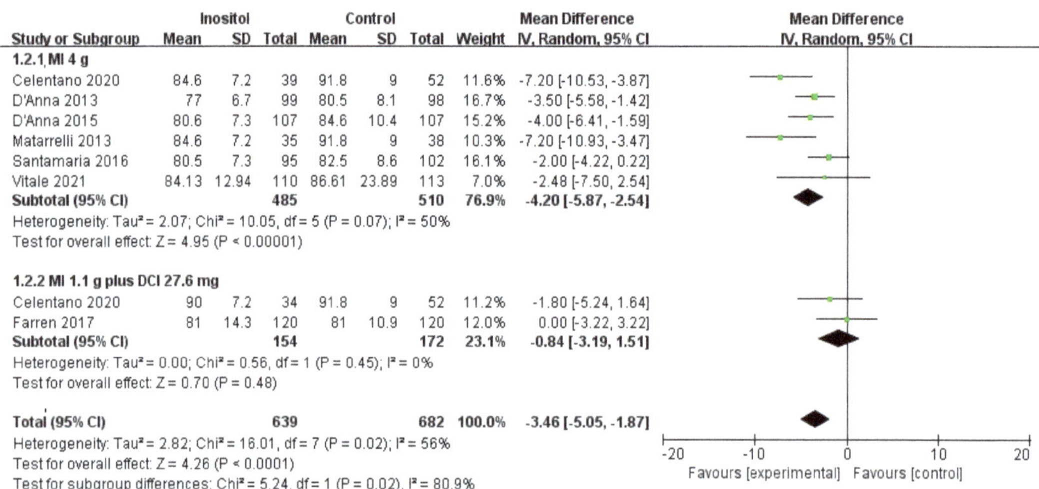

Figure 5. Inositol vs. control: fasting-glucose OGTT [14–20].

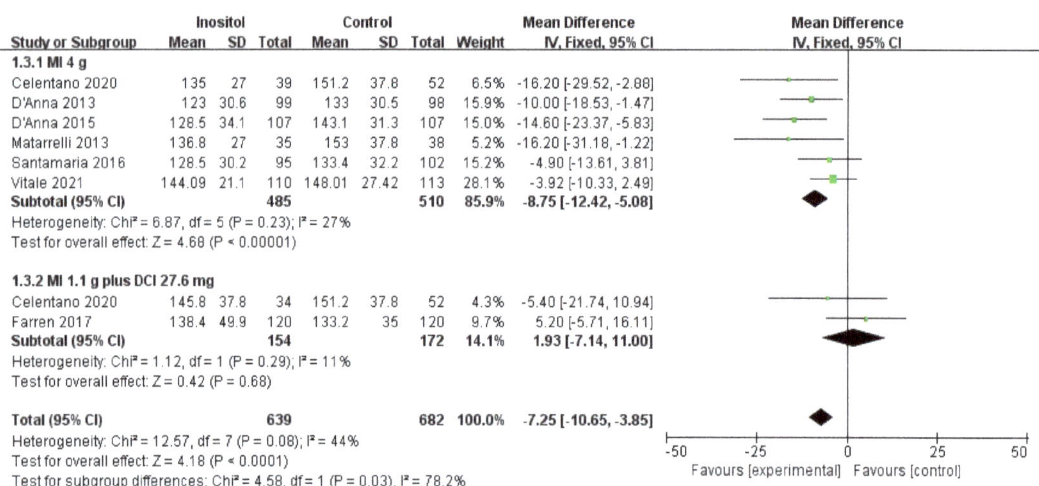

Figure 6. Inositol vs. control: 1-h OGTT [14–20].

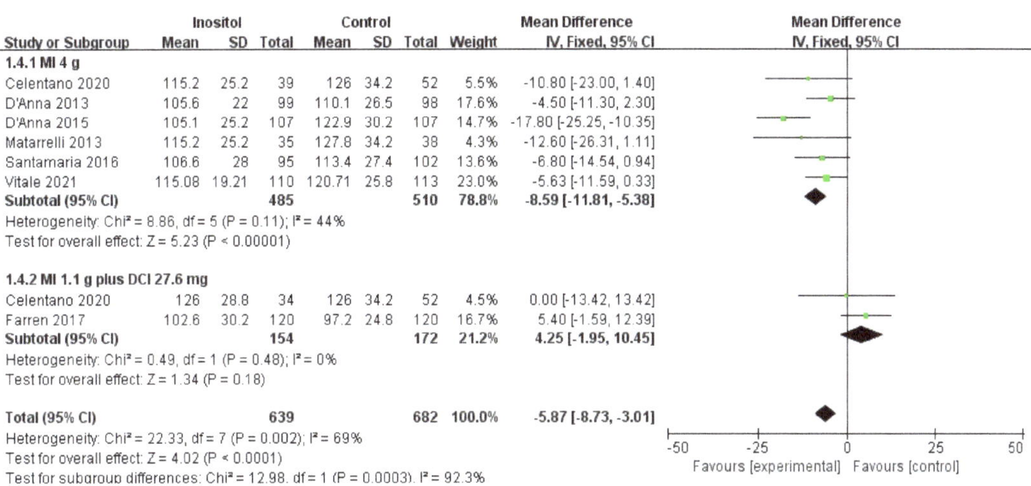

Figure 7. Inositol vs. control: 2-h OGTT [14–20].

3.4.2. Secondary Outcomes

Four studies evaluated patients' need of insulin treatment. Compared with the control group, 4 g MI supplementation per day significantly decreased the need of insulin treatment (RR = 0.27, 95% CI (0.11, 0.66), p = 0.004, I^2 = 0%; four trials, n = 562).

Five studies evaluated the incidence of hypertensive disorders. No difference in the incidence of hypertensive disorders was observed between 4 g MI supplementation per day and the control group or 1.1 g MI plus 27.6 mg DCI supplementation per day and the control group (respectively, p = 0.06, four trials, n = 686; p = 0.23, two trials, n = 320).

Five studies evaluated the incidence of preterm delivery. Compared with the control group, 4 g MI supplementation per day significantly decreased the incidence of preterm delivery (RR = 0.39, 95% CI (0.18, 0.82), p = 0.01, I^2 = 0%; four trials, n = 686). No difference was observed between 1.1 g MI plus 27.6 mg DCI supplementation per day and the control group (p = 0.13; two trials, n = 320).

Five studies evaluated the cesarean section rate. No difference in the cesarean section rate was observed between 4 g MI supplementation per day and the control group or 1.1 g MI plus 27.6 mg DCI supplementation per day and the control group (respectively, p = 0.14, four trials, n = 686; p = 0.57, two trials, n = 320).

Three studies evaluated the incidence of shoulder dystocia. No difference in the incidence of shoulder dystocia was observed between 4 g MI supplementation per day and the control group (p = 0.48, three trials, n = 595).

Six studies evaluated neonatal birth weight. No difference in birth weight was observed between 4 g MI supplementation and the control group or 1.1 g MI plus 27.6 mg DCI supplementation per day and the control group (respectively, p = 0.12, five trials, n = 759; p = 0.88, two trials, n = 320). However, compared with the control group, 4 g MI supplementation per day significantly decreased birth weight in percentiles (MD = −14.86, 95% CI (−21.38, −8.35), p < 0.00001, I^2 = 0%; two trials, n = 164).

Three studies evaluated the incidence of macrosomia. No difference in the incidence of macrosomia was observed between 4 g MI supplementation per day and the control group (p = 0.22, three trials, n = 595).

Six studies evaluated the incidence of neonatal hypoglycemia. Compared with the control group, 4 g MI supplementation per day significantly decreased the incidence of neonatal hypoglycemia (RR = 0.15, 95% CI (0.03, 0.66), p = 0.01, I^2 = 0%; five trials, n = 759). No difference was observed between 1.1 g MI plus 27.6 mg DCI supplementation per day and the control group (p = 0.53; two trials, n = 320).

Four studies evaluated the incidence of NICU admission. No difference in the NICU admission rate was observed between 4 g MI supplementation per day and the control group or 1.1 g MI plus 27.6 mg DCI supplementation per day and the control group (respectively, $p = 0.08$, three trials, $n = 489$; $p = 0.75$, two trials, $n = 320$).

3.5. Side Effects

A total of 1321 patients from seven studies were included in the meta-analysis, among which no side effect was reported.

3.6. Overall Quality of Evidence

The overall quality of evidence was rated as very low or very low for all the outcomes evaluated (Table 2).

Table 2. Summary of finding: inositol for the prevention of GDM.

Inositol for the Prevention of GDM						
Patient or Population: Women in Early Pregnancy Who Were at Risk of GDM (Those with Pre-Existing T1DM/T2DM Excluded) **Intervention: MI 4 g/MI 1.1 g Plus DCI 27.6 mg**						
Outcomes	Illustrative comparative risks * (95% CI)		Relative effect (95% CI)	No of Participants (studies)	Quality of the evidence (GRADE)	Comments
	Assumed risk	Corresponding risk				
	Control	Inositol				
GDM rate	Study population		RR 0.4 (0.23 to 0.67)	1321 (7 studies)	⊕⊖⊖⊖ very low [1,2,3,4]	
	317 per 1000	127 per 1000 (73 to 212)				
	Moderate					
	306 per 1000	122 per 1000 (70 to 205)				
Insulin treatment	Study population		RR 0.27 (0.11 to 0.66)	562 (4 studies)	⊕⊕⊖⊖ low [1,2,4,5]	
	84 per 1000	23 per 1000 (9 to 56)				
	Moderate					
	106 per 1000	29 per 1000 (12 to 70)				
Hypertensive disorders	Study population		RR 0.43 (0.2 to 0.91)	1006 (5 studies)	⊕⊖⊖⊖ very low [1,2,4]	
	42 per 1000	18 per 1000 (8 to 38)				
	Moderate					
	30 per 1000	13 per 1000 (6 to 27)				
Preterm delivery	Study population		RR 0.4 (0.22 to 0.74)	1006 (5 studies)	⊕⊖⊖⊖ very low [1,2,4]	
	69 per 1000	27 per 1000 (15 to 51)				
	Moderate					
	63 per 1000	25 per 1000 (14 to 47)				
Cesarean section	Study population		RR 0.89 (0.77 to 1.03)	1006 (5 studies)	⊕⊖⊖⊖ very low [1,2,4,6]	
	448 per 1000	398 per 1000 (345 to 461)				
	Moderate					
	471 per 1000	419 per 1000 (363 to 485)				
Shoulder dystocia	Study population		RR 0.58 (0.12 to 2.68)	595 (3 studies)	⊕⊖⊖⊖ very low [1,2,4,6]	
	13 per 1000	8 per 1000 (2 to 35)				
	Moderate					
	10 per 1000	6 per 1000 (1 to 27)				

Table 2. *Cont.*

Inosamic for the Prevention of GDM					
Patient or Population: Women in Early Pregnancy Who Were at Risk of GDM (Those with Pre-Existing T1DM/T2DM Excluded)					
Intervention: MI 4 g/MI 1.1 g Plus DCI 27.6 mg					
Macrosomia	Study population		RR 0.35 (0.06 to 1.92)	595 (3 studies)	⊕⊖⊖⊖ very low [1,2,3,4,6]
	56 per 1000	20 per 1000 (3 to 107)			
	Moderate				
	49 per 1000	17 per 1000 (3 to 94)			
Neonatal hypoglycemia	Study population		RR 0.62 (0.32 to 1.18)	1079 (6 studies)	⊕⊖⊖⊖ very low [1,4,6]
	46 per 1000	29 per 1000 (15 to 54)			
	Moderate				
	10 per 1000	6 per 1000 (3 to 12)			
NICU admission	Study population		RR 0.53 (0.23 to 1.21)	809 (4 studies)	⊕⊖⊖⊖ very low [1,2,4,6]
	37 per 1000	20 per 1000 (9 to 45)			
	Moderate				
	39 per 1000	21 per 1000 (9 to 47)			

⊕ Upgrade. ⊖ Downgrade. [1] Downgraded due to the high risk of performance bias and unclear risk of reporting bias. [2] Downgraded because the research location and race of participants were limited. [3] Downgraded due to high heterogeneity. [4] Publication bias due to "positive results" is strongly suspected. [5] Upgraded because the RR/*p* value is low. [6] Downgraded due to wide CI crossing the line of no effect. * The basis for the assumed risk (e.g., the median control group risk across studies) is provided in footnotes. The corresponding risk (and its 95% confidence interval) is based on the assumed risk in the comparison group and the relative effect of the intervention (and its 95% CI).

4. Discussion

The incidence of GDM has been increasing worldwide over the past years. As current evidence indicates that GDM is associated with higher risk of pregnancy complications and long-term maternal and fetal adverse outcomes, researchers have been focused on finding more practical measures to prevent GDM. Clinical trials and observational studies have shown that exercise can be a non-invasive therapeutic option [23,24], while others have focused on diet control and nutritional supplementation.

Inositol is a polyol which widely exists in eukaryotic cells. It is the structural basis of various secondary messengers, especially of inositol triphosphates, phosphatidylinositol phosphate lipids, and inositol glycans. Therefore, inositol is involved in insulin signal transduction through regulating the phosphatidylinositol 3-kinase (PI3K)/protein kinase B (AKT) signal pathway and increasing GLUT4 translocation to the cell membrane, which has been supported by the latest animal experiments in type 2 diabetic db/db mice [25]. It is natural to consider inositol as an effective insulin sensitizer to prevent GDM, the main pathophysiological mechanism of which is believed to be insulin resistance.

The present meta-analysis included seven RCTs with a total number of 1321 participants. In all, 485 patients received the intervention of 4 g MI supplementation per day, and 154 patients received the intervention of 1.1 g MI plus 27.6 mg DCI supplementation per day. The results suggest that compared with the control group (placebo), 4 g MI supplementation per day during pregnancy significantly reduced incidence of GDM, plasma glucose levels of OGTT, and the need of insulin treatment. It was also associated with significantly reduced incidence of preterm delivery, lower birth weight (in percentiles), and reduced incidence of neonatal hypoglycemia. The incidence of hypertensive disorders and NICU admission rate also reduced, although not statistically significant. It showed no remarkable impact on the incidence of cesarean section, shoulder dystocia, and macrosomia. It is noted that no difference was observed between 1.1 g MI plus 27.6 mg DCI supplementation per day and the control group in all outcomes. The dosing regimen for MI+DCI was extrapolated from previous studies investigating the effects of inositol in women with PCOS [9]. However, this dosage seemed inadequate in terms of preventing GDM.

The present meta-analysis included RCTs that used inositol as the only intervention. There were also studies investigating the effect of inositol plus other nutritional supplementation. Dell'Edera 2017 [26] suggested that a combination of 1.75 g MI, 250 mg DCI, plus zinc and methylsulfonylmethane might prevent the onset of GDM in pregnant women with glucose intolerance. Godfrey 2021 [27], however, found the combination of 4 g MI plus vitamin D/B6/B12, riboflavin, zinc, and probiotics was not effective in an international, multi-center, double-blind trial. Previous studies have suggested that vitamin D supplementation could possibly reduce the risk of GDM [28], while the effect of probiotics remained unclear [29]. Whether the other components of the intervention played an additional role is yet to be investigated. Melvasi 2014 [30] and Melvasi 2017 [31] focused on the improvement of maternal metabolic profile, suggesting that the intervention of 2 g MI plus 400 mg DCI, and several other components could improve glycemic and lipidic parameters in pregnant women. In addition, 400 µg folic acid supplementation per day was given to all participants as a placebo control in the RCTs included in this meta-analysis. Several cohort studies have reported that folic acid supplementation in early pregnancy [32] and prolonged duration [33] of it may increase the risk of GDM. This is unlikely to affect the results of the present meta-analysis but should be taken into consideration when conducting further clinical trials in the future.

Inositol is included in the list of compounds that are 'generally recognized as safe', according to United States Food and Drug Administration, which means it has been proven to be free of side effects and is safe for use in pregnancy [34]. Among the studies included in the present meta-analysis, no adverse event associated with inositol supplementation was reported, and previous reviews declared the same conclusion [35]. Therefore, as a safe molecule at usual administered dosage, inositol could be a potential choice of nutritional supplementation for preventing GDM.

Previous systematic reviews and meta-analyses mostly focused on the effects of MI and suggested that MI supplementation during pregnancy shows the potential in preventing GDM, with low quality previous evidence [10,36–38]. The main conclusion is similar between the present study and previous studies. However, the presented meta-analysis has included all existing RCTs with the intervention of inositol only (both MI and DCI) on the topic up to May 2022. Three aspects of a secondary outcome were evaluated, including ten subitems. A more comprehensive evaluation of inositol's effect on GDM prevention was performed by the present study.

Still, several limitations should be taken into consideration. Six studies were conducted in Italy with Caucasian women. The lack of a blinding method in most studies lead to an overall high risk of performance bias, and the risk of reporting bias was unclear. Moreover, there was substantial heterogeneity when performing a sensitivity analysis of some outcomes. Therefore, the overall quality of evidence was rated as low or very low for all the outcomes evaluated. Further studies are required in different countries and ethnic groups. The dosage of inositol and intervention time need to be further evaluated. Besides, long-term outcomes of both mothers and their offspring should be included in the evaluation.

5. Conclusions

Inositol nutritional supplementation of 4 g MI per day during early pregnancy may reduce GDM incidence and severity. It may also decrease the need of insulin treatment, and reduce the incidence of preterm delivery and neonatal hypoglycemia. Besides, it is reported to be safe for use in pregnancy. Therefore, inositol nutritional supplementation may be a practical and safe approach for the prevention of GDM. Further studies are required to apply inositol into more clinical practice.

Funding: This research was funded by the National Key Research and Development Program of China (No. 2021YFC2700700) and Peking University First Hospital Medicine Fund of Fostering Young Scholars under grant (No. 2021CR01).

Institutional Review Board Statement: Not applicable.

Informed Consent Statement: Not applicable.

Conflicts of Interest: The authors declare no conflict of interest.

References

1. American Diabetes Association Professional Practice Committee. 2. Classification and Diagnosis of Diabetes: Standards of Medical Care in Diabetes-2022. *Diabetes Care* **2022**, *45*, S17–S38. [CrossRef] [PubMed]
2. International Diabetes Federation. *IDF Diabetes Atlas*, 10th ed.; International Diabetes Federation: Belgium, Brussels, 2021.
3. Tsakiridis, I.; Giouleka, S.; Mamopoulos, A.; Kourtis, A.; Athanasiadis, A.; Filopoulou, D.; Dagklis, T. Diagnosis and Management of Gestational Diabetes Mellitus: An Overview of National and International Guidelines. *Obstet. Gynecol. Surv.* **2021**, *76*, 367–381. [CrossRef] [PubMed]
4. Yan, J.; Yang, H. Gestational diabetes in China: Challenges and coping strategies. *Lancet Diabetes Endocrinol.* **2014**, *2*, 930–931. [CrossRef]
5. Shou, C.; Wei, Y.M.; Wang, C.; Yang, H.X. Updates in Long-term Maternal and Fetal Adverse Effects of Gestational Diabetes Mellitus. *Matern. Fetal Med.* **2019**, *1*, 91–94.
6. Griffith, R.J.; Alsweiler, J.; Moore, A.E.; Brown, S.; Middleton, P.; Shepherd, E.; Crowther, C.A. Interventions to prevent women from developing gestational diabetes mellitus: An overview of Cochrane Reviews. *Cochrane Database Syst. Rev.* **2020**, *6*, CD012394. [CrossRef] [PubMed]
7. Croze, M.L.; Soulage, C.O. Potential role and therapeutic interests of myo-inositol in metabolic diseases. *Biochimie* **2013**, *95*, 1811–1827. [CrossRef]
8. Dinicola, S.; Unfer, V.; Facchinetti, F.; Soulage, C.O.; Greene, N.D.; Bizzarri, M.; Laganà, A.S.; Chan, S.Y.; Bevilacqua, A.; Pkhaladze, L.; et al. Inositols: From Established Knowledge to Novel Approaches. *Int. J. Mol. Sci.* **2021**, *22*, 10575. [CrossRef]
9. Unfer, V.; Nestler, J.E.; Kamenov, Z.A.; Prapas, N.; Facchinetti, F. Effects of Inositol(s) in Women with PCOS: A Systematic Review of Randomized Controlled Trials. *Int. J. Endocrinol.* **2016**, *2016*, 1849162. [CrossRef]
10. Crawford, T.J.; Crowther, C.A.; Alsweiler, J.; Brown, J. Antenatal dietary supplementation with myo-inositol in women during pregnancy for preventing gestational diabetes. *Cochrane Database Syst. Rev.* **2015**, *2015*, CD011507. [CrossRef]
11. Liberati, A.; Altman, D.G.; Tetzlaff, J.; Mulrow, C.; Gøtzsche, P.C.; Ioannidis, J.P.; Clarke, M.; Devereaux, P.J.; Kleijnen, J.; Moher, D. The PRISMA statement for reporting systematic reviews and meta-analyses of studies that evaluate healthcare interventions: Explanation and elaboration. *BMJ* **2009**, *339*, b2700. [CrossRef]
12. Higgins, J.P.T.; Thomas, J.; Chandler, J.; Cumpston, M.; Li, T.; Page, M.J.; Welch, V.A. (Eds.) Cochrane Handbook for Systematic Reviews of Interventions Version 6.2 (Updated February 2021). *Cochrane.* 2021. Available online: www.training.cochrane.org/handbook (accessed on 4 May 2022).
13. Balshem, H.; Helfand, M.; Schünemann, H.J.; Oxman, A.D.; Kunz, R.; Brozek, J.; Vist, G.E.; Falck-Ytter, Y.; Meerpohl, J.; Norris, S.; et al. GRADE guidelines: 3. Rating the quality of evidence. *J. Clin. Epidemiol.* **2011**, *64*, 401–406. [CrossRef] [PubMed]
14. Matarrelli, B.; Vitacolonna, E.; D'Angelo, M.; Pavone, G.; Mattei, P.A.; Liberati, M.; Celentano, C. Effect of dietary myo-inositol supplementation in pregnancy on the incidence of maternal gestational diabetes mellitus and fetal outcomes: A randomized controlled trial. *J. Matern. Fetal Neonatal Med.* **2013**, *26*, 967–972. [CrossRef] [PubMed]
15. D'Anna, R.; Scilipoti, A.; Giordano, D.; Caruso, C.; Cannata, M.L.; Interdonato, M.L.; Corrado, F.; Di Benedetto, A. myo-Inositol supplementation and onset of gestational diabetes mellitus in pregnant women with a family history of type 2 diabetes: A prospective, randomized, placebo-controlled study. *Diabetes Care* **2013**, *36*, 854–857. [CrossRef]
16. D'Anna, R.; Di Benedetto, A.; Scilipoti, A.; Santamaria, A.; Interdonato, M.L.; Petrella, E.; Neri, I.; Pintaudi, B.; Corrado, F.; Facchinetti, F. Myo-inositol Supplementation for Prevention of Gestational Diabetes in Obese Pregnant Women: A Randomized Controlled Trial. *Obstet. Gynecol.* **2015**, *126*, 310–315. [CrossRef]
17. Santamaria, A.; Di Benedetto, A.; Petrella, E.; Pintaudi, B.; Corrado, F.; D'Anna, R.; Neri, I.; Facchinetti, F. Myo-inositol may prevent gestational diabetes onset in overweight women: A randomized, controlled trial. *J. Matern. Fetal Neonatal Med.* **2016**, *29*, 3234–3237. [CrossRef] [PubMed]
18. Farren, M.; Daly, N.; McKeating, A.; Kinsley, B.; Turner, M.J.; Daly, S. The Prevention of Gestational Diabetes Mellitus with Antenatal Oral Inositol Supplementation: A Randomized Controlled Trial. *Diabetes Care* **2017**, *40*, 759–763. [CrossRef]
19. Celentano, C.; Matarrelli, B.; Pavone, G.; Vitacolonna, E.; Mattei, P.A.; Berghella, V.; Liberati, M. The influence of different inositol stereoisomers supplementation in pregnancy on maternal gestational diabetes mellitus and fetal outcomes in high-risk patients: A randomized controlled trial. *J. Matern. Fetal Neonatal Med.* **2020**, *33*, 743–751. [CrossRef]
20. Vitale, S.G.; Corrado, F.; Caruso, S.; Di Benedetto, A.; Giunta, L.; Cianci, A.; D'Anna, R. Myo-inositol supplementation to prevent gestational diabetes in overweight non-obese women: Bioelectrical impedance analysis, metabolic aspects, obstetric and neonatal outcomes—A randomized and open-label, placebo-controlled clinical trial. *Int. J. Food Sci. Nutr.* **2021**, *72*, 670–679. [CrossRef]

21. HAPO Study Cooperative Research Group. Hyperglycemia and Adverse Pregnancy Outcome (HAPO) Study: Associations with neonatal anthropometrics. *Diabetes* **2009**, *58*, 453–459. [CrossRef]
22. International Association of Diabetes and Pregnancy Study Groups Consensus Panel; Metzger, B.E.; Gabbe, S.G.; Persson, B.; Buchanan, T.A.; Catalano, P.A.; Damm, P.; Dyer, A.R.; Leiva, A.D.; Hod, M.; et al. International association of diabetes and pregnancy study groups recommendations on the diagnosis and classification of hyperglycemia in pregnancy. *Diabetes Care* **2010**, *33*, 676–682. [CrossRef]
23. Wang, C.; Wei, Y.; Zhang, X.; Zhang, Y.; Xu, Q.; Sun, Y.; Su, S.; Zhang, L.; Liu, C.; Feng, Y.; et al. A randomized clinical trial of exercise during pregnancy to prevent gestational diabetes mellitus and improve pregnancy outcome in overweight and obese pregnant women. *Am. J. Obstet. Gynecol.* **2017**, *216*, 340–351. [CrossRef] [PubMed]
24. Wang, C.; Guelfi, K.J.; Yang, H.X. Exercise and its role in gestational diabetes mellitus. *Chronic Dis. Transl. Med.* **2016**, *2*, 208–214. [CrossRef] [PubMed]
25. Fan, C.; Liang, W.; Wei, M.; Gou, X.; Han, S.; Bai, J. Effects of D-Chiro-Inositol on Glucose Metabolism in db/db Mice and the Associated Underlying Mechanisms. *Front. Pharmacol.* **2020**, *11*, 354. [CrossRef] [PubMed]
26. Dell'Edera, D.; Sarlo, F.; Allegretti, A.; Simone, F.; Lupo, M.G.; Epifania, A.A. The influence of D-*chiro*-inositol and D-*myo*-inositol in pregnant women with glucose intolerance. *Biomed. Rep.* **2017**, *7*, 169–172. [CrossRef]
27. Godfrey, K.M.; Barton, S.J.; El-Heis, S.; Kenealy, T.; Nield, H.; Baker, P.N.; Chong, Y.S.; Cutfield, W.; Chan, S.Y. Myo-Inositol, Probiotics, and Micronutrient Supplementation from Preconception for Glycemia in Pregnancy: NiPPeR International Multicenter Double-Blind Randomized Controlled Trial. *Diabetes Care* **2021**, *44*, 1091–1099. [CrossRef]
28. Chan, K.Y.; Wong, M.; Pang, S. Lo, K.K.H. Dietary supplementation for gestational diabetes prevention and management: A meta-analysis of randomized controlled trials. *Arch. Gynecol. Obstet.* **2021**, *303*, 1381–1391. [CrossRef]
29. Davidson, S.J.; Barrett, H.L.; Price, S.A.; Callaway, L.K.; Dekker Nitert, M. Probiotics for preventing gestational diabetes. *Cochrane Database Syst. Rev.* **2021**, *4*, CD009951. [CrossRef]
30. Malvasi, A.; Casciaro, F.; Minervini, M.M.; Kosmas, I.; Mynbaev, O.A.; Pacella, E.; Monti Condesnitt, V.; Creanza, A.; Di Renzo, G.C.; Tinelli, A. Myo-inositol, D-chiro-inositol, folic acid and manganese in second trimester of pregnancy: A preliminary investigation. *Eur. Rev. Med. Pharmacol. Sci.* **2014**, *18*, 270–274.
31. Malvasi, A.; Kosmas, I.; Mynbaev, O.A.; Sparic, R.; Gustapane, S.; Guido, M.; Tinelli, A. Can trans resveratrol plus d-chiro-inositol and myo-inositol improve maternal metabolic profile in overweight pregnant patients? *Clin. Ter.* **2017**, *168*, e240–e247. [CrossRef]
32. Zhu, B.; Ge, X.; Huang, K.; Mao, L.; Yan, S.; Xu, Y.; Huang, S.; Hao, J.; Zhu, P.; Niu, Y.; et al. Folic Acid Supplement Intake in Early Pregnancy Increases Risk of Gestational Diabetes Mellitus: Evidence from a Prospective Cohort Study. *Diabetes Care* **2016**, *39*, e36–e37. [CrossRef]
33. Huang, L.; Yu, X.; Li, L.; Chen, Y.; Yang, Y.; Yang, Y.; Hu, Y.; Zhao, Y.; Tang, H.; Xu, D.; et al. Duration of periconceptional folic acid supplementation and risk of gestational diabetes mellitus. *Asia Pac. J. Clin. Nutr.* **2019**, *28*, 321–329. [CrossRef] [PubMed]
34. Facchinetti, F.; Appetecchia, M.; Aragona, C.; Bevilacqua, A.; Bezerra Espinola, M.S.; Bizzarri, M.; D'Anna, R.; Dewailly, D.; Diamanti-Kandarakis, E.; Hernández Marín, I.; et al. Experts' opinion on inositols in treating polycystic ovary syndrome and non-insulin dependent diabetes mellitus: A further help for human reproduction and beyond. *Expert. Opin. Drug Metab. Toxicol.* **2020**, *16*, 255–274. [CrossRef] [PubMed]
35. Facchinetti, F.; Cavalli, P.; Copp, A.J.; D'Anna, R.; Kandaraki, E.; Greene, N.D.E.; Unfer, V. Experts Group on Inositol in Basic and Clinical Research. An update on the use of inositols in preventing gestational diabetes mellitus (GDM) and neural tube defects (NTDs). *Expert. Opin. Drug Metab. Toxicol.* **2020**, *16*, 1187–1198. [CrossRef] [PubMed]
36. Guo, X.; Guo, S.; Miao, Z.; Li, Z.; Zhang, H. Myo-inositol lowers the risk of developing gestational diabetic mellitus in pregnancies: A systematic review and meta-analysis of randomized controlled trials with trial sequential analysis. *J. Diabetes Complicat.* **2018**, *32*, 342–348. [CrossRef]
37. Vitagliano, A.; Saccone, G.; Cosmi, E.; Visentin, S.; Dessole, F.; Ambrosini, G.; Berghella, V. Inositol for the prevention of gestational diabetes: A systematic review and meta-analysis of randomized controlled trials. *Arch. Gynecol. Obstet.* **2019**, *299*, 55–68. [CrossRef]
38. Liu, Q.; Liu, Z. The efficacy of myo-inositol supplementation to reduce the incidence of gestational diabetes: A meta-analysis. *Gynecol. Endocrinol.* **2022**, *38*, 450–454. [CrossRef]

 **nutrients**

Article

Anthropometric Parameters and Mediterranean Diet Adherence in Preschool Children in Split-Dalmatia County, Croatia—Are They Related?

Dora Bučan Nenadić [1,2,*], Ela Kolak [1,2,*], Marija Selak [1,2], Matea Smoljo [2], Josipa Radić [3,4], Marijana Vučković [3], Bruna Dropuljić [2], Tanja Pijerov [2] and Dora Babić Cikoš [2]

1 Department of Nutrition and Dietetics, University Hospital Centre Split, 21000 Split, Croatia; marija.selak3@gmail.com
2 Croatian Association of Nutritionists, 10000 Zagreb, Croatia; matea.cigic@gmail.com (M.S.); brunatripicic@gmail.com (B.D.); pijerovtanja@gmail.com (T.P.); dorababic1@gmail.com (D.B.C.)
3 Department of Nephrology and Dialysis, University Hospital Centre Split, 21000 Split, Croatia; josiparadic1973@gmail.com (J.R.); mavuckovic@kbsplit.hr (M.V.)
4 Department of Internal Medicine, University of Split School of Medicine, 21000 Split, Croatia
* Correspondence: dorabucan@gmail.com (D.B.N.); elakolak93@gmail.com (E.K.)

Abstract: Obesity is a rapidly growing problem in European countries, Croatia being among them. According to the latest CroCOSI data, every third child in Croatia aged 8.0–8.9 years is overweight or obese. The Mediterranean diet (MeDi) and its impact on nutritional status and health has been the focus of recent research. Therefore, the aim of this cross-sectional, observational study was to determine the nutritional status and adherence to the MeDi of preschool children in Split, Croatia. We included 598 preschool children aged 3 to 7 years and, for each child, parents completed a lifestyle questionnaire and the Mediterranean Diet Quality Index (KIDMED) in order to assess adherence to the MeDi. The anthropometric assessment included the measurement of weight, height, mid-upper arm circumference (MUAC), waist circumference (WC) and the z-score was calculated. According to the z-score, 420 (70.2%) children had a healthy body weight with 54 (9%) underweight and 124 (20.8%) overweight or obese children. Almost half (49%) of the study participants had a low KIDMED index score, indicating a low MeDi adherence, 37% had an average score, while only 14% had high MeDi compliance. Statistically significant negative correlations between MUAC and WC and the consumption of a second daily serving of fruit ($p = 0.04$) as well as a daily serving of vegetables ($p = 0.03$) were found. In conclusion, low compliance to the MeDi principles in preschool children is concerning. Considering the beneficial effects of the MeDi on overall health, further education, and the adoption of healthy eating habits in preschool children in this Mediterranean region are required.

Keywords: preschool children; nutritional status; eating habits; Mediterranean diet; Split-Dalmatia County

Citation: Bučan Nenadić, D.; Kolak, E.; Selak, M.; Smoljo, M.; Radić, J.; Vučković, M.; Dropuljić, B.; Pijerov, T.; Babić Cikoš, D. Anthropometric Parameters and Mediterranean Diet Adherence in Preschool Children in Split-Dalmatia County, Croatia—Are They Related? *Nutrients* **2021**, *13*, 4252. https://doi.org/10.3390/nu13124252

Academic Editor: Zhiyong Zou

Received: 14 November 2021
Accepted: 25 November 2021
Published: 26 November 2021

Publisher's Note: MDPI stays neutral with regard to jurisdictional claims in published maps and institutional affiliations.

Copyright: © 2021 by the authors. Licensee MDPI, Basel, Switzerland. This article is an open access article distributed under the terms and conditions of the Creative Commons Attribution (CC BY) license (https://creativecommons.org/licenses/by/4.0/).

1. Introduction

For preschool children, a healthy nourishment characterized by an adequate and balanced nutritional intake alone can ensure proper physical growth as well as cognitive and emotional development, a feeling of satiety and sufficient physical ability [1]. In this preschool period, children shape their dietary patterns by developing independent feeding skills and forming their own food preferences [2]. In addition to the direct effect on growth and development, inadequate nutrition can affect the occurrence of certain chronic non-communicable diseases in childhood such as cardiovascular diseases, obstructive sleep apnea and psychosocial problems [3]. Given that eating habits acquired in childhood might continue into adulthood [4] and increase the risk of the development of type 2 diabetes mellitus, arterial hypertension and arteriosclerosis especially in children with persistent obesity [5,6], it is important to provide children with quality nutrition and allow them to adopt proper eating habits [1]. On the other hand, overnutrition, which leads to children becoming overweight and obese because of their excessive intake of energy dense food and a sedentary lifestyle, is a rapidly growing problem in European countries [3] and as such in Croatia. According to the European Childhood Obesity Surveillance Initiative in Croatia (CroCOSI, 2018–2019), one in three Croatian children (35.0%) aged 8.0–8.9 years were overweight and obese [7]. In the past few years, nutrition research has focused on the MeDi and its impact on nutritional status and health. Considered as one of the healthiest dietary patterns, MeDi is based on high intake of fresh fruits, vegetables, legumes, whole grains, dairy products, fish and olive oil; moderate intake of poultry, eggs and wine, and finally, a low intake of red and processed meat and sweets [8]. Greater adherence to the MeDi is associated with a significant reduction of risk in developing chronic morbidities [9] and an inverse relationship with obesity [10]. In order to assess the compliance to the principles of the MeDi in the pediatric population, the use of the KIDMED index is suggested [11]. As the growth and nutritional assessment by health professionals become less frequent after the first year of a child's life, dietary habits are most often guided by parents, caregivers, or kindergarten professionals. Therefore, the nutritional status and dietary habits of preschool children should be given special attention. A departure from the traditional MeDi has been observed in the general population [12], and according to recent research the same drift has occurred in preschool children [13]. Therefore, the main aim of this study was to determine nutritional status and adherence to the MeDi in preschool children from Split-Dalmatia County, Croatia.

2. Materials and Methods

2.1. Study Design and Population

Five hundred and ninety-eight preschool children aged 3 to 7 years were included in this cross-sectional study. The study was carried out in randomly chosen kindergartens in Split, Split-Dalmatia County, Croatia between March and October 2019. The research was conducted as part of a series of lectures for educators and parents as well as educational workshops for children about healthy eating habits for this population at the initiative of the Croatian Association of Nutritionists. Oral and written informed consent was obtained from parents of each study participant as well as from kindergarten management. Only one child per household was included in the study. Children who met one of the following criteria were excluded from the study: (1) refused to participate; (2) had psychological disorders that prevent them from participating in regular activities; (3) lack of parental approval; (4) in cases where parents had not fully completed or returned questionnaires.

2.2. Lifestyle Questionnaire

Parents who had consented to their child participating in the study were asked to complete a lifestyle questionnaire to provide insight into the characteristics of the population. The questionnaire included demographic information such as date of birth and sex as well as questions about the duration of stay at the kindergarten, possible present food allergies, number of meals and the use of dietary supplements.

2.3. The Mediterranean Diet Quality Index (KIDMED)

For each study participant, parents were also asked to complete the KIDMED index. The KIDMED index was created to assess adherence to the MeDi for children and young people from two (2) to twenty four (24) years of age based on the principles that reflect Mediterranean eating habits and those that undermine them [14]. The index itself consists of sixteen questions that can be answered with the response of yes or no. Questions with negative connotations were assigned a value of -1 and those with a positive connotation were attributed a value of $+1$ [15]. The sum of all values range from zero (0) to twelve (12) and are therefore classified into the following three levels: >8 which indicates the optimal MeDi; 4–7, which means adjustment is needed to improve food intake according to the MeDi principles; $\leq$3 i.e., very low quality of nutrition according to the MeDi [11].

2.4. Anthropometric Measurements

Each child underwent an anthropometric assessment at the kindergarten while wearing light clothes. Height was measured using a stadiometer and weight was measured to the nearest tenth decimal with a calibrated Omron BF511 diagnostic scale (Omron, Kyoto, Japan). Non-stretchable, flexible body-measuring tape was used to measure the mid-upper arm circumference (MUAC) and waist circumference (WC). For each study participant, a BMI-to-age z-score was calculated with WHO AnthroPlus software [16]. Because of its ability to describe nutritional status, including at the extreme ends of the distribution and the ability to derive summary statistics, this classification system is recommended by WHO (for, i.e., means and SDs of z-scores (WHO, 1995)). Z-scores are derived using the exact age in days for the WHO standards and months for the WHO reference of 2007 [16]. Anthropometric measurements were performed in a playful manner to lessen the anxiety levels of children. No child was forced to participate in anthropometric measuring if unwilling, regardless of parental consent.

2.5. Statistical Analysis

A database was created with the information obtained from the questionnaires and anthropometric assessment were produced for a statistical analysis using IBM SPSS Statistics for Windows, Version 21.0 (IBM Corp, Armonk, NY, USA) software. The categorical data are represented by absolute and relative frequencies (*n* and %). Numerical data were described by arithmetic mean and standard deviation (SD) in cases of normal distribution following, and, in other cases, median and interquartile range boundaries were used. The variance of the category variables was tested by the Chi-square test. A point-biserial correlation was used to measure the strength and direction of the association that exists between one continuous variable and one dichotomous variable. The significance level was set at $p < 0.05$, with a 95% confidence level.

3. Results

Out of the 598 children involved in this study, 310 (51.8%) were boys with a median weight of 21.2 kg. According to the z-score, 420 (70.2%) children had healthy body weight with 54 (9%) children found to be underweight and 124 (20.8%) found to be overweight or obese. The basic characteristics and anthropometric measurements of the study population are shown in Table 1. Nearly 90% of participants were enrolled in a full day kindergarten programme which provides two main meals (breakfast and lunch) as well as morning and afternoon snacks. An example of a menu from kindergartens in which the research was conducted is shown in Table S1.

Almost half (49%) of the study participants had a low KIDMED score, suggesting low MeDi adherence; 37% had an average score while only 14% had high MeDi compliance. The highest compliance to the KIDMED components was determined related to points distributed for eating breakfast, the consumption of dairy products for breakfast, the daily intake of fruit and juice and eating fast food less than once a week. The total adherence to the KIDMED score and its components is shown in Figure 1.

There were no statistically significant differences in total KIDMED score regarding sex (Figure 2) nor its components.

The most apparent differences with respect to age in responses to the KIDMED questionnaire were found for the consumption of vegetables once a day ($p < 0.02$) and dairy product consumption for breakfast ($p = 0.04$). There was no statistically significant difference between the total KIDMED score and age groups, as demonstrated in Figure 3.

Differences in adherence to the MeDi, according to the nutritional status are shown in Figure 4. Statistically significant differences were not found except for a borderline significance in fish consumption two or three times a week ($p < 0.04$) for undernourished children, and dairy product consumption with a frequency of two or more times a day ($p < 0.03$) in obese children.

Table 1. Basic characteristic and anthropometric measurements of the study population.

Sex [*n* (%)]	
Boys	310 (51.8)
Girls	288 (48.2)
Age (years) [Mean (SD)]	5 (1)
Age groups [*n* (%)]	
<4 years	15 (2.5)
4–5 years	290 (48.5)
6–7 years	293 (49)
Body height (m) [Mean (SD)]	1.17 (0.09)
Body weight (kg) [Median (IQR)] 12–40.5	21.2 (18.6–24.3)
Waist circumference (cm) [Median (IQR)] 43–76	54 (51–56)
Mid-upper arm circumference (cm) [Median (IQR)] 13–24	17.7 (16.5–19)
z-score [Median (IQR)]	0.25 (−0.40–0.92)
Nutritional status regarding z-score [*n* (%)]	
≤ -1.04	54 (9)
$> -1.04, < 1.04$	420 (70.2)
≥ 1.04	124 (20.7)
Supplements [*n* (%)]	124 (20.7)
Food allergies [*n* (%)]	>35 (5.9)
Number of meals [*n* (%)]	
3–4/day	117 (19.6)
4–5/day	396 (66.2)
>5/day	85 (14.2)
Length of kindergarten stay [*n* (%)]	
Half a day programme (5 h)	8 (1.3)
Half a day programme (6 h)	53 (8.9)
Half a day programme (6 h; lunch included)	19 (3.2)
Full day (10 h)	518 (86.6)

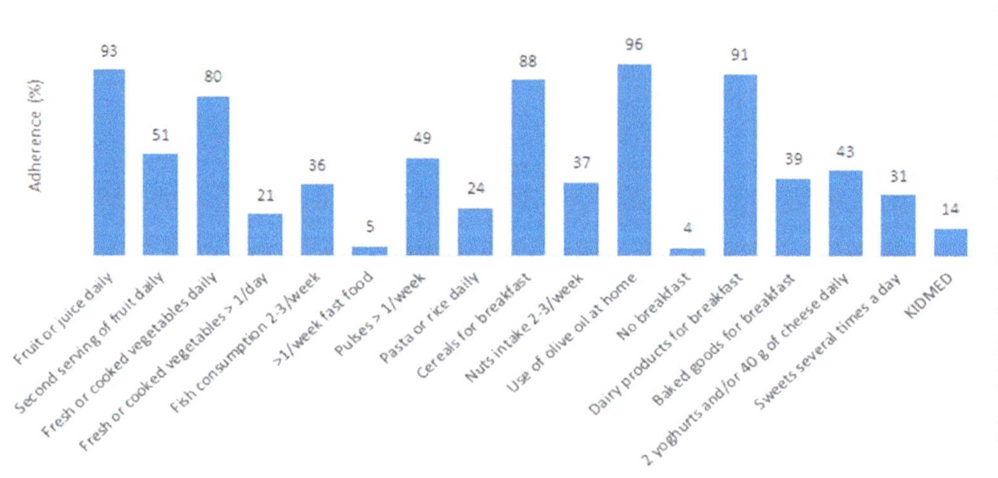

Figure 1. Adherence to the MeDi and its components for total study population.

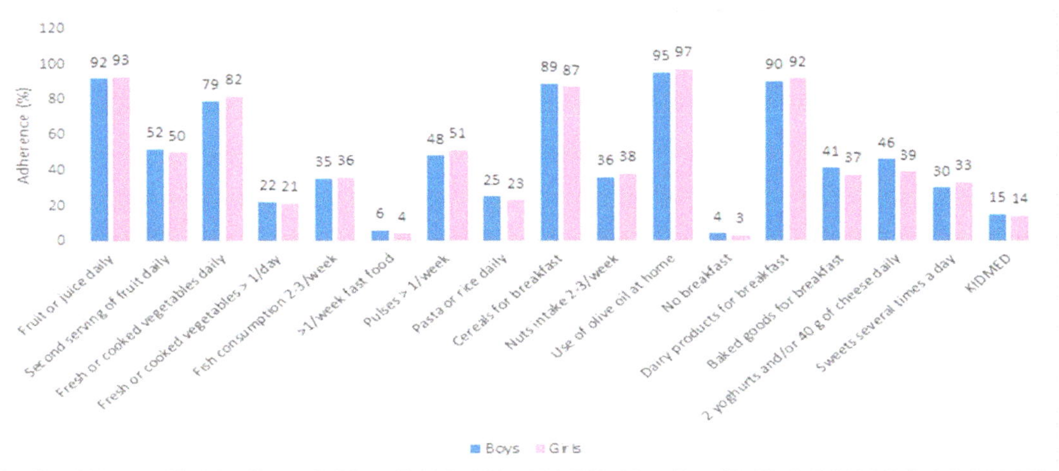

Figure 2. Adherence to the MeDi and its components regarding sex.

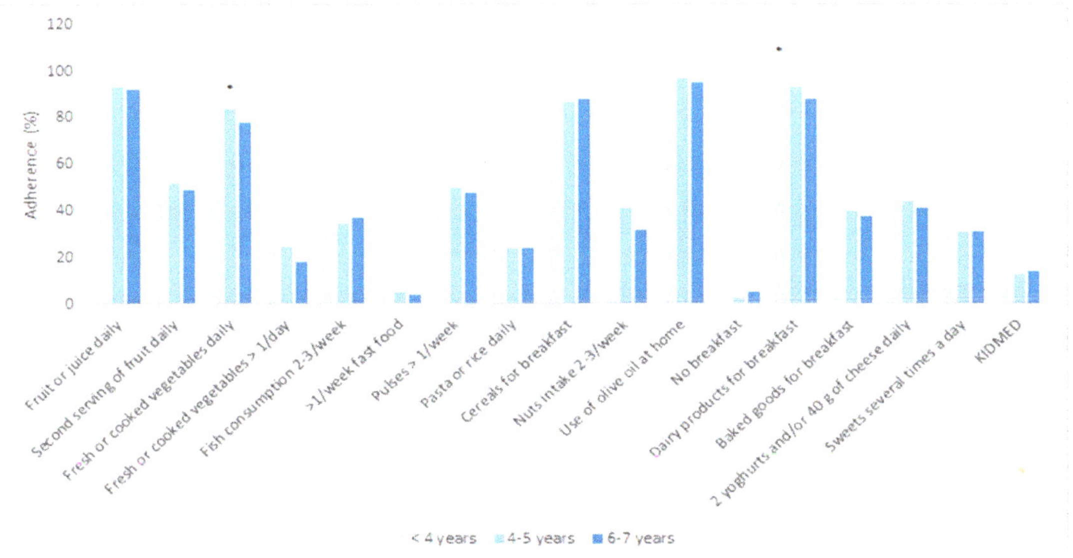

Figure 3. Adherence to the MeDi and its components regarding age. * $p < 0.05$; p-values were obtained with Chi-square test.

A correlation between MUAC and WC and components of the KIDMED index was observed. As shown in Table 2, children who consumed a second daily serving of fruit ($p = 0.04$) as well as a daily serving of vegetables ($p = 0.03$) had a significantly smaller WC, whereas children who had a vegetable intake of more than once a day ($p = 0.01$) had a significantly smaller MUAC.

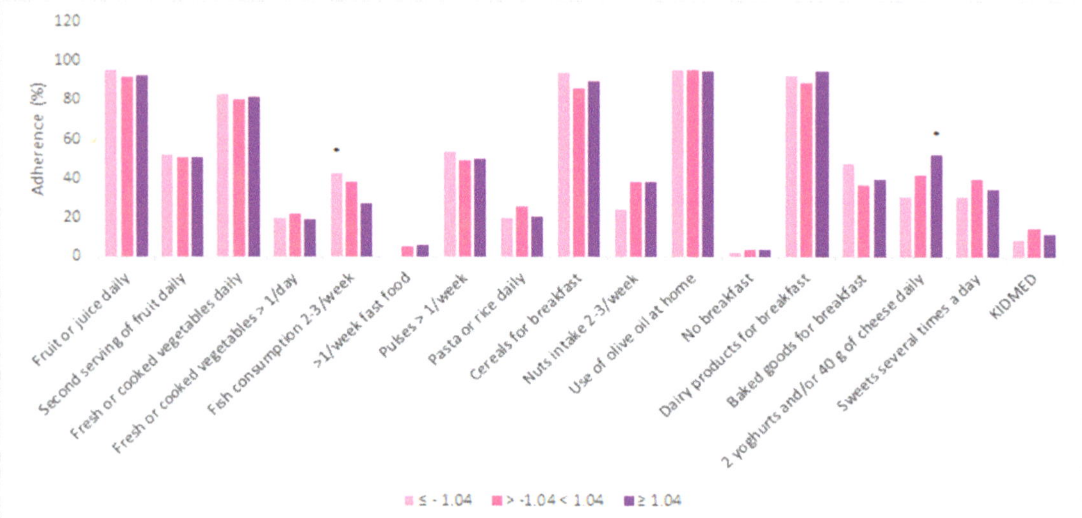

Figure 4. Adherence to the MeDi and its components regarding z-score. * $p < 0.05$; p-values were obtained with Chi-square test.

Table 2. Correlation between KIDMED index and waist and mid-upper arm circumference (only statistically significant values shown).

	WC	MUAC
Second serving of fruit daily	−0.085 (0.04 *)	−0.050 (0.23 *)
Fresh or cooked vegetables > 1/day	−0.091 (0.03 *)	−0.101 (0.01 *)

Abbreviations: WC—waist circumference, MUAC—mid-upper arm circumference * p-values were obtained with Point biserial correlation.

4. Discussion

Out of total study participants, 20.8% were overweight or obese, which was expected considering the increase in the prevalence of obesity in the pediatric population globally and in Croatia. Data from the World Health Organization indicate that approximately 32.8 million children under the age of 5 were overweight or obese in 2019 [17]. According to the latest results of the CROCosi survey from 2019, the percentage of children who are overweight or obese is 36.9%—the highest in the Adriatic region, and 36.9%, of which 23.1% of children were overweight and 13.8% were obese [7]. Almost half of the preschool children included in this study presented a low adherence to the MeDi, underlining low adherence in the general population. These results are in accordance with the study from Kolčić et al. conducted with adults, in which only 23% of the participants from Southern Dalmatia adhered to the principles of the MeDi [12]. As shown in Figure 1, a low intake of fish, nuts, and dairy products most likely contributes to the poor adherence to the MeDi in children, as observed in this study. In previous studies, the adherence to the MeDi, calculated using the KIDMED index, has shown poor results among children [18] and adolescents [19] from the European Union, suggesting the necessity of nutritional intervention in this particular population. On the contrary, the results of an earlier study involving preschool children from Split-Dalmatia County indicate an exceptionally high adherence to the MeDi with only 6% out of 260 children having a low KIDMED score [20]. The difference in the results can be explained by the smaller number of respondents in the above-mentioned study, as well as parental overestimation or underestimation of the quality of a child's nutrition.

Our results did not show a significant difference in adherence to the MeDi between girls and boys, which is consistent with the results of previous studies from Spain and Turkey [21,22].

A difference in the total KIDMED score with regard to age was not found in our study, but a statistically higher intake of dairy products for breakfast was noticed in the youngest participants while statistically higher daily intake of vegetables was noticed in participants aged 4 to 5 years. The Portuguese Eat Mediterranean (EM) programme demonstrated that participants aged from 6 to 9 years old (68.9%) were the most adherent to the MeDi, whereas lower adherence to the MeDi was found in children aged 2 to 5 years (59.0%) and the lowest adherence was found for adolescents older than 15 years (46.0%) [23]. If we consider that most of our study participants were enrolled in a 10 h kindergarten program that provides breakfast and lunch as well as morning and afternoon snacks, it could be said that the diet of these children was uniform and, therefore, age-related differences should be minimal. Additionally, food fussiness (picky eating) and food neophobia are characteristic for children aged 2 to 5 years [24]. The most common food group that causes aversion is vegetables, because of their bitter taste and unique smell [25].

No statistically significant differences were found in the adherence to the total KIDMED score regarding the z-score values, except a borderline significance for fish consumption two or three times per week in underweight children and dairy consumption twice or more per day in overweight and obese children. In contrast to our results, previous studies found that a low adherence to the MeDi is associated with a high prevalence of overweight and obesity in children from the Mediterranean countries of the European Union [26–28]. Fish consumption was inversely associated with the z-score, which can be explained by the lower energy value of fish, as well as higher vegetable intake that is traditionally offered as a side dish to fish. Nevertheless, the evidence presented in the overview by Naveed et al. suggests that a diet rich in fish, vegetables and fruit, especially berries, is linked to better cognitive function and academic performance in children and adolescents [29]. Regarding dairy products intake and obesity rates, Dougkas et al. demonstrated that milk and dairy products do not have an impact on the development of childhood obesity, despite being a significant contribution to children's nutrient intake. Although there may be some mechanisms that could explain the connection between dairy intake and adiposity based on appetite, lipid absorption and even intestinal microflora, there is still no evidence that suggests benefits of limiting dairy products for the purpose of preventing obesity in this particular population [30].

No significant correlations were determined between the total KIDMED score and WC as well as MUAC, unlike EnKid [14] and LLAL [31] studies, where the KIDMED score was inversely correlated to the WC. Regarding individual components of the KIDMED questionnaire, significant negative correlations were noticed for both WC and MUAC and vegetable consumption, as well as WC and the second daily fruit serving.

There are some limitations to this study. The KIDMED questionnaire was self-administered by the parents and data about household income and parental education are lacking. In addition, the study was conducted in one urban area but included a high number of age specific participants which is why the obtained results do not reflect the adherence to the MeDi for the entire Split-Dalmatia County. Further research should include participants from a larger geographical region, as well as a wider age range, to assess compliance to the MeDi principles.

5. Conclusions

The results of this study conducted with preschool children showed low adherence to the principles of the Mediterranean diet, which is characterized by a high presence of plant-based food as well as fish and dairy products. As such, the MeDi has been shown to have beneficial effects on body composition and obesity prevention in all age groups, which further emphasizes the importance of the timely implementation of this dietary pattern in the pediatric population, especially now, when the childhood obesity epidemic has reached its peak. Healthy eating habits play a major role in the growth and development of children, but also in preventing the development of chronic diseases in later life. Considering the favourable effects of the MeDi on overall health, further intervention studies could be useful for educating children and their caregivers about the benefits of healthy eating habits and of adopting them to reduce the risk of developing chronic non-communicable diseases in this population.

Supplementary Materials: The following are available online at https://www.mdpi.com/article/10.3390/nu13124252/s1, Table S1: An example of a menu from kindergartens.

Author Contributions: D.B.N. and E.K. contributed equally to this paper. Conceptualization, D.B.N., E.K.; Data curation, E.K., M.S. (Matea Smoljo), B.D. and T.P.; Funding acquisition, D.B.N. and D.B.C.; Investigation, E.K., M.S. (Matea Smoljo), B.D. and T.P.; Methodology, D.B.N., E.K., M.S. (Marija Selak) and M.S. (Matea Smoljo); Project administration, D.B.N., M.S. (Matea Smoljo) and D.B.C.; Resources, D.B.N., E.K., M.S. (Marija Selak) and D.B.C.; Supervision, J.R. and D.B.C.; Validation, M.S. (Marija Selak), M.S. (Matea Smoljo), J.R., M.V., B.D., T.P. and D.B.C.; Visualization, E.K.; Writing—original draft, D.B.N. and E.K.; Writing—review and editing, M.S. (Marija Selak), M.S. (Matea Smoljo), J.R., M.V., B.D., T.P. and D.B.C. All authors have read and agreed to the published version of the manuscript.

Funding: This research is a part of project "We Grow Healthy" by Croatian Association of Nutritionists funded by City of Split, Croatia (Field of social welfare and health care for the realization of the right to financial support within the available funds from the budget of the City of Split for 2018 with projections for 2019–2020 (Class: 022-05/18-03/993; Ur.no 2181/01-07-02/04-18-1)).

Institutional Review Board Statement: The study was approved by the Board of the Preschool Institution-Cvit Mediterana, Split, Croatia, (Class: 601-01/21-01/0002; Ur. no 2181-01-21-0007).

Informed Consent Statement: Informed consent was obtained from all parents of children involved in the study.

Data Availability Statement: Raw data can be found at corresponding authors via e-mail: elakolak93@gmail.com (E.K.); dorabucan@gmail.com (D.B.N.).

Conflicts of Interest: The authors declare no conflict of interest. The funders had no role in the design of the study; in the collection, analyses, or interpretation of data; in the writing of the manuscript, or in the decision to publish the results.

References

1. Kim, S.Y.; Cha, S.M. Evaluation of dietary behaviors of preschool children in seoul and gyeonggi-do associated with the level of parents' health consciousness: Using nutrition quotient for preschoolers (nq-p). *Nutr. Res. Pract.* **2021**, *15*, 248–265. [CrossRef]
2. Scaglioni, S.; De Cosmi, V.; Ciappolino, V.; Parazzini, F.; Brambilla, P.; Agostoni, C. Factors influencing children's eating behaviours. *Nutrients* **2018**, *10*, 706. [CrossRef] [PubMed]
3. Steenbergen, E.; Krijger, A.; Verkaik-kloosterman, J.; Elstgeest, L.E.M.; Ter Borg, S.; Joosten, K.F.M.; van Rossum, C.T.M. Evaluation of nutrient intake and food consumption among dutch toddlers. *Nutrients* **2021**, *13*, 1531. [CrossRef] [PubMed]
4. Llewellyn, A.; Simmonds, M.; Owen, C.G.; Woolacott, N. Childhood obesity as a predictor of morbidity in adulthood: A systematic review and meta-analysis. *Obes. Rev.* **2016**, *17*, 56–67. [CrossRef] [PubMed]
5. Ryder, J.R.; Jacobs, D.R.; Sinaiko, A.R.; Kornblum, A.P.; Steinberger, J. Longitudinal Changes in Weight Status from Childhood and Adolescence to Adulthood. *J. Pediatr.* **2019**, *214*, 187–192.e2. [CrossRef]
6. Juonala, M.; Magnussen, C.G.; Berenson, G.S.; Venn, A.; Burns, T.L.; Sabin, M.A.; Raitakari, O.T. Childhood Adiposity, Adult Adiposity, and Cardiovascular Risk Factors. *N. Engl. J. Med.* **2011**, *365*, 1876–1885. [CrossRef]
7. Musić Milanović, S.; Lang Morović, M.; Križan, H. *European Childhood Obesity Initiative, Croatia 2018/2019 (CroCOSI)*; Croatian Institute of Public Health: Zagreb, Croatia, 2021.
8. Bach-Faig, A.; Berry, E.M.; Lairon, D.; Reguant, J.; Trichopoulou, A.; Dernini, S.; Medina, F.X.; Battino, M.; Belahsen, R.; Miranda, G.; et al. Mediterranean diet pyramid today. Science and cultural updates. *Public Health Nutr.* **2011**, *14*, 2274–2284. [CrossRef]
9. Sofi, F.; Abbate, R.; Gensini, G.F.; Casini, A. Accruing evidence on benefits of adherence to the Mediterranean diet on health: An updated systematic review and meta-analysis. *Am. J. Clin. Nutr.* **2010**, *92*, 1189–1196. [CrossRef]
10. Jennings, A.; Welch, A.; Van Sluijs, E.M.F.; Griffin, S.J.; Cassidy, A. Diet quality is independently associated with weight status in children aged 9–10 years. *J. Nutr.* **2011**, *141*, 453–459. [CrossRef] [PubMed]

11. Serra-Majem, L.; Ribas, L.; Ngo, J.; Ortega, R.M.; García, A.; Pérez-Rodrigo, C.; Aranceta, J. Food, youth and the Mediterranean diet in Spain. Development of KIDMED, Mediterranean Diet Quality Index in children and adolescents. *Public Health Nutr.* **2004**, *7*, 931–935. [CrossRef]

12. Kolčić, I.; Relja, A.; Gelemanović, A.; Miljković, A.; Boban, K.; Hayward, C.; Rudan, I.; Polašek, O. Mediterranean diet in the southern Croatia—Does it still exist? *Croat. Med. J.* **2016**, *57*, 415–424. [CrossRef]

13. Pereira-Da-Silva, L.; Rêgo, C.; Pietrobelli, A. The diet of preschool children in the Mediterranean countries of the European Union: A systematic review. *Int. J. Environ. Res. Public Health* **2016**, *13*, 572. [CrossRef] [PubMed]

14. Schröder, H.; Mendez, M.A.; Gomez, S.F.; Fíto, M.; Ribas, L.; Aranceta, J.; Serra-Majem, L. Energy density, diet quality, and central body fat in a nationwide survey of young Spaniards. *Nutrition* **2013**, *29*, 1350–1355. [CrossRef]

15. Costarelli, V.; Koretsi, E.; Georgitsogianni, E. Health-related quality of life of Greek adolescents: The role of the Mediterranean diet. *Qual. Life Res.* **2013**, *22*, 951–956. [CrossRef] [PubMed]

16. WHO. AnthroPlus Software. Available online: https://www.who.int/tools/growth-reference-data-for-5to19-years/application-tools (accessed on 12 May 2021).

17. WHO. Obesity and Overweight. Available online: https://www.who.int/news-room/fact-sheets/detail/obesity-and-overweight (accessed on 12 May 2021).

18. Labayen Goñi, I.; Arenaza, L.; Medrano, M.; García, N.; Cadenas-Sanchez, C.; Ortega, F.B. Associations between the adherence to the Mediterranean diet and cardiorespiratory fitness with total and central obesity in preschool children: The PREFIT project. *Eur. J. Nutr.* **2018**, *57*, 2975–2983. [CrossRef] [PubMed]

19. Farajian, P.; Risvas, G.; Karasouli, K.; Pounis, G.D.; Kastorini, C.M.; Panagiotakos, D.B.; Zampelas, A. Very high childhood obesity prevalence and low adherence rates to the Mediterranean diet in Greek children: The GRECO study. *Atherosclerosis* **2011**, *217*, 525–530. [CrossRef] [PubMed]

20. Salcin, L.O.; Karin, Z.; Damjanovic, V.M.; Ostojic, M.; Vrdoljak, A.; Gilic, B.; Sekulic, D.; Lang-Morovic, M.; Markic, J.; Sajber, D. Physical activity, body mass, and adherence to the mediterranean diet in preschool children: A cross-sectional analysis in the split-dalmatia county (Croatia). *Int. J. Environ. Res. Public Health* **2019**, *16*, 3237. [CrossRef]

21. Cabrera, S.G.; Fernández, N.H.; Hernández, C.R.; Nissensohn, M.; Román-Viña, B.; Serra-Majem, L. Test KIDMED; prevalencia de la Baja Adhesión a la Dieta Mediterránea en Niños y Adolescentes; Revisión Sistemática. *Nutr. Hosp.* **2015**, *32*, 2390–2399. [CrossRef]

22. Sahingoz, S.A.; Sanlier, N. Compliance with Mediterranean Diet Quality Index (KIDMED) and nutrition knowledge levels in adolescents. A case study from Turkey. *Appetite* **2011**, *57*, 272–277. [CrossRef]

23. Rito, A.I.; Dinis, A.; Rascôa, C.; Maia, A.; Mendes, S.; Stein-Novais, C.; Lima, J. Mediterranean Diet Index (KIDMED) Adherence, Socioeconomic Determinants, and Nutritional Status of Portuguese Children: The Eat Mediterranean Program. *Port. J. Public Health* **2018**, *36*, 141–149. [CrossRef]

24. Ventura, A.K.; Worobey, J. Early influences on the development of food preferences. *Curr. Biol.* **2013**, *23*, R401–R408. [CrossRef] [PubMed]

25. Gibson, E.L.; Wardle, J.; Watts, C.J. Fruit and vegetable consumption, nutritional knowledge and beliefs in mothers and children. *Appetite* **1998**, *31*, 205–228. [CrossRef] [PubMed]

26. Agnoli, C.; Sieri, S.; Ricceri, F.; Giraudo, M.T.; Masala, G.; Assedi, M.; Panico, S.; Mattiello, A.; Tumino, R.; Giurdanella, M.C.; et al. Adherence to a Mediterranean diet and long-term changes in weight and waist circumference in the EPIC-Italy cohort. *Nutr. Diabetes* **2018**, *8*, 22. [CrossRef] [PubMed]

27. Tognon, G.; Moreno, L.A.; Mouratidou, T.; Veidebaum, T.; Molnár, D.; Russo, P.; Siani, A.; Akhandaf, Y.; Krogh, V.; Tornaritis, M.; et al. Adherence to a Mediterranean-like dietary pattern in children from eight European countries. The IDEFICS study. *Int. J. Obes.* **2014**, *38*, S108–S114. [CrossRef]

28. Kanellopoulou, A.; Giannakopoulou, S.P.; Notara, V.; Antonogeorgos, G.; Rojas-Gil, A.P.; Kornilaki, E.N.; Konstantinou, E.; Lagiou, A.; Panagiotakos, D.B. The association between adherence to the Mediterranean diet and childhood obesity; the role of family structure: Results from an epidemiological study in 1728 Greek students. *Nutr. Health* **2021**, *27*, 39–47. [CrossRef]

29. Naveed, S.; Lakka, T.; Haapala, E.A. An overview on the associations between health behaviors and brain health in children and adolescents with special reference to diet quality. *Int. J. Environ. Res. Public Health* **2020**, *17*, 953. [CrossRef]

30. Dougkas, A.; Barr, S.; Reddy, S.; Summerbell, C.D. A critical review of the role of milk and other dairy products in the development of obesity in children and adolescents. *Nutr. Res. Rev.* **2019**, *32*, 106–127. [CrossRef]

31. Mazaraki, A.; Tsioufis, C.; Dimitriadis, K.; Tsiachris, D.; Stefanadi, E.; Zampelas, A.; Richter, D.; Mariolis, A.; Panagiotakos, D.; Tousoulis, D.; et al. Adherence to the Mediterranean diet and albuminuria levels in Greek adolescents: Data from the Leontio Lyceum ALbuminuria (3L study). *Eur. J. Clin. Nutr.* **2011**, *65*, 219–225. [CrossRef]

Article

Association between Fruit Consumption and Lipid Profile among Children and Adolescents: A National Cross-Sectional Study in China

Jieyu Liu [1,2], Yanhui Li [1,2], Xinxin Wang [3], Di Gao [1,2], Li Chen [1,2], Manman Chen [1,2], Tao Ma [1,2], Qi Ma [1,2], Ying Ma [1,2], Yi Zhang [1,2], Jun Jiang [4], Zhiyong Zou [1,2], Xijie Wang [5,6,*], Yanhui Dong [1,2,*] and Jun Ma [1,2]

[1] Institute of Child and Adolescent Health, School of Public Health, Peking University, Beijing 100191, China; jieyulynne@163.com (J.L.); yanhui_lyh@163.com (Y.L.); gaodi1993@163.com (D.G.); clcl@bjmu.edu.cn (L.C.); 1911210173@pku.edu.cn (M.C.); 1610306216@pku.edu.cn (T.M.); 18702110295@163.com (Q.M.); mypku232@163.com (Y.M.); 1710306140@pku.edu.cn (Y.Z.); harveyzou2002@bjmu.edu.cn (Z.Z.); majunt@bjmu.edu.cn (J.M.)

[2] National Health Commission Key Laboratory of Reproductive Health, Peking University, Beijing 100191, China

[3] School of Public Health and Management, Ningxia Medical University, Yinchuan 750004, China; wangxinxin291314@163.com

[4] Department of Plant Science and Landscape Architecture, University of Maryland, College Park, MD 20742, USA; sherryj@umd.edu

[5] Vanke School of Public Health, Tsinghua University, Beijing 100084, China

[6] Institute for Healthy China, Tsinghua University, Beijing 100084, China

* Correspondence: xijie_wang@126.com (X.W.); dongyanhui@bjmu.edu.cn (Y.D.); Tel.: +86-10-62782199 (X.W.); +86-10-82801624 (Y.D.); Fax: +86-10-82801178 (Y.D.)

Citation: Liu, J.; Li, Y.; Wang, X.; Gao, D.; Chen, L.; Chen, M.; Ma, T.; Ma, Q.; Ma, Y.; Zhang, Y.; et al. Association between Fruit Consumption and Lipid Profile among Children and Adolescents: A National Cross-Sectional Study in China. *Nutrients* **2022**, *14*, 63. https://doi.org/10.3390/nu14010063

Academic Editor: Stephen Ives

Received: 5 November 2021
Accepted: 22 December 2021
Published: 24 December 2021

Publisher's Note: MDPI stays neutral with regard to jurisdictional claims in published maps and institutional affiliations.

Copyright: © 2021 by the authors. Licensee MDPI, Basel, Switzerland. This article is an open access article distributed under the terms and conditions of the Creative Commons Attribution (CC BY) license (https://creativecommons.org/licenses/by/4.0/).

Abstract: To investigate associations between fruit consumption and lipid profiles, and to further explore a satisfactory level of frequency and daily fruit intake for children and adolescents. A national sample of 14,755 children and adolescents aged 5–19 years from seven provinces in China were recruited. Fasting blood samples were collected to test the lipid profile. Information regarding fruit consumption and other characteristics was collected by questionnaires. Logistic regression models adjusting for confounding covariates were applied to calculate the odds ratio (OR) and 95% confidence interval (95% CI). Participants who consumed fruits for 6–7 days per week had lower risks of high triglycerides (OR: 0.66, 95% CI: 0.58–0.75), dyslipidemia (OR: 0.77, 95% CI: 0.68–0.86), and hyperlipidemia (OR: 0.72, 95% CI: 0.63–0.81), compared to fruit consumption of 0–2 days per week. Risks of high triglycerides, dyslipidemia and hyperlipidemia of those who consumed fruits for 0.75–1.5 servings each day also decreased, compared to the insufficient fruit intake. The combined effects of high frequency and moderate daily intake of fruit on lipid disorders did not change essentially. The associations were more evident in girls, younger children and those whose families had higher educational levels. Moderate fruit consumption was associated with lower odds of lipid disorders, predominantly in girls, younger participants, and those came from higher-educated families. These findings supported the health effect of moderate fruit intake frequently to improve the childhood lipid profiles.

Keywords: fruit consumption; lipid; dyslipidemia; children and adolescents; China

1. Introduction

Dyslipidemia, characterized by abnormal blood lipids and lipoprotein levels, is a recognized risk factor for cardiovascular disease [1]. Previous studies have observed a high prevalence of lipid disorders during both childhood and adolescence in China [2,3]. Observed from the 10-year trend of serum lipids, the pooled prevalence of dyslipidemia in Chinese children and adolescents was 28.9% in 2014 [2], while in 2015, it increased to 31.6% in a nationally representative pediatric population with 129,426 participants [4]. It showed a younger-age trend during recent decades, and could be progressive into adulthood. Growing evidence indicated that, in children and adolescents, higher concentrations of low density lipoprotein-cholesterol (LDL-C), as well as lower concentrations of high density lipoprotein-cholesterol (HDL-C) were associated with higher risk of atherosclerosis in later life [5]. As such, screening children and adolescents for lipid profiles, which usually included the measurements of serum concentrations of total cholesterol (TC), triglycerides (TG), HDL-C and LDL-C, might have the potential to identify the affected subjects, reduce the burdens of long-term cholesterol through intervention, and postpone or prevent cardiovascular events during adulthood. The challenge was to maintain appropriate lipid levels at the right time, most commonly by early behavioral and lifestyle interventions for high-risk children and adolescents.

Previous work revealed that the nutritional and diet factors were important determinants for the development of childhood dyslipidemia [1,6], and therefore could be important targets for prevention strategies. Among the multiple diet factors, vegetables and fruits are important sources of a healthy diet. Notably, low fruit consumption is considered to be the fifth leading contributor to the global disease burden [7]. Growing observational evidence suggested that the fruit consumption might parallel the decrease in the risks of obesity, diabetes mellitus, and cardiovascular events in both childhood and adulthood [8,9]. Pan and his colleagues found a significant inverse association between healthy eating index score of fruits and the risk of metabolic syndrome (MS) among US adolescents, suggesting that a fruit-rich diet could exert beneficial effects in prevention of MS [10]. However, inadequate fruit consumption of children and adolescents worldwide despite the generally higher preference for consumption of fruits than vegetables [11], interventions to encourage fruit consumption during childhood and adolescence might therefore be an effective strategy in reducing disease burden.

So far, evidence for the effects of fruit consumption on lipid disorders had been limited to date. In 2013, a randomized controlled trial demonstrated no significant effect between increased fruits and vegetables consumption and healthier blood levels of TC [12]. In contrast, a cross-sectional study in Macao, China indicated that students who consumed less fruits suffered from a higher rate of low HDL-C level and elevated TG [13]. Besides, a fruits- and vegetables-rich diet was associated with a healthier metabolic profile, reflected by low concentrations of TC and LDL-C [14]. However, these studies only determined the consumption frequency of fruits, such as consuming fruits every day, not focused on the specific frequencies or average daily intake amount. Currently, the dietary guidelines for fruit intake were mostly based on adults, with the recommended daily fruit intake of 400 g [15], while there was no consensus regarding daily fruit consumption for children and adolescents [16], especially aimed at reducing adverse lipid profiles. Different from children and adolescents, adults could accumulate more complex environmental effects, while children might be more sugar-sensitive than adults. Guidelines based on findings in adults may therefore lead to ambiguity about fruit recommendations for pediatric cardiovascular and lipid health. In addition, fruit also contains a large amount of fructose, the accumulation of which is detrimental. Whether a causal link exists between natural sources of fructose present in fruits and the development of lipid disorders continues to be contested [17]. For these reasons, exploring the exact frequency and amount of fruit for children and adolescents to prevent lipid disorders was essential.

Given the role that fruit consumption played in pediatric health, we hereby investigated the relationship between fruit consumption and lipid profiles among children and adolescents aged 5–19 years, based on data of a large cross-sectional survey, which was conducted in seven provinces in China. The aim of the present study was to provide evidence on a satisfactory frequency and daily amount of fruit consumption to prevent deleterious lipid events for children and adolescents.

2. Materials and Methods

2.1. Study Population

Data in this study came from the baseline of a multi-centered, cluster-controlled trial, aiming to prevent obesity in children and adolescents from seven provinces or cities of China (Hunan, Ningxia, Hunan, Chongqing, Liaoning, Shanghai, and Guangzhou; registration number: NCT02343588). The full trial protocol, including sampling procedure and measurements has been published previously in detail [18]. In brief, a multi-stage cluster random sampling method was used to determine the original study population. Firstly, several regions were randomly chosen from each province. Approximately 12–16 schools covering primary schools, junior high schools, and middle high schools were then randomly chosen from each region. In each school, two classes were randomly selected from each grade. All the students and their parents from selected classes were invited for participation. Among the original surveyed population of 16,637 participants aged 5–19 years whose physical examination and blood samples were available, 1862 participants were excluded from the present analysis because of missing information on fruit consumption, making the final sample size of 14,755. The study has been approved by the ethical committee of the Peking University (number: IRB0000105213034). Written informed consent was obtained from both students and their parents or legal guardian in both waves.

2.2. Questionnaire

The children's questionnaire was performed in order to collect basic information and lifestyle behaviors. Besides, the parental self-administrated questionnaire included information about residence area, monthly household income, parental body mass index (BMI), and parental educational attainment. Both parental and children's questionnaires of children grade 1–3 were reported by parents. Children above the fourth grade would fill in children's questionnaire by themselves instructed by the class teacher.

Data regarding consumption of fruits and other eating behaviors were collected. As previously published [18–21], the frequency (days) and amount (serving per day) of dietary behaviors, including the total consumption of fruits, vegetables, meat and sugar-sweetened beverages (SSBs) over the past 7 days, were investigated. Participants were asked "How many days, over the past 7 days, have you eaten fruit/vegetables/meat or drunk SSB? How many servings in one day?" [18–21]. Previous studies determined the consumption frequency of fruits as consuming fruits every day [13], in order to avoid extreme few samples in each group, we categorized the frequency of those dietary behaviors into three groups of "0–2 days/week", "3–5 days/week", and "6–7 days/week". To better understand the intake of fruit/vegetable, one serving was defined as the size of an ordinary adult's closed fist (Supplementary Figure S1) and roughly equaled a medium-sized apple or orange (≈200 g) [22], which has been described in detail elsewhere [21]. As set in the questionnaire, SSB included Coca-Cola, Sprite, orange juice, Nutrition Express, and Red Bull [20]. One serving of SSB was determined as a canned beverage (approximately 250 mL), while one portion of meat equaled the size of an adult's palm (approximately 100 g) [23]. The daily dietary intake was calculated as: average daily intake = (days of consumption × servings in those days)/7.

According to the Dietary Guidelines for Chinese School-age Children 2016 [24], inadequate daily fruit intake was defined as eating fruits less than 150 g each day, and the recommended daily intake was approximately 250–350 g for the pediatric population. The amount of daily fruit intake was therefore categorized into groups of "<0.75 servings/day (approximately <150 g)", "0.75–1.5 servings/day (≥0.75 and <1.5 servings/day) (approximately 150–300 g)", and "≥1.5 servings/day (approximately ≥300 g)".

Information about the child's physical activity was collected with the International Physical Activity Questionnaire-Short Form (IPAQ-SF) [25], which is more suitable for population surveillance and large-scale studies [26]. Instructed by trained project members, all recruited participants were asked to report their frequency (days) and duration (hours and minutes) of moderate to vigorous-intensity physical activities

(MVPA) over the past 7 days, and the average time for MVPA per day was calculated as: average daily time = (days of physical activity × duration in those days)/7.

Parents were asked to report their height (cm) and weight (kg), while body mass index (BMI) was calculated as the weight (kg) divided by the square of the height (m^2). According to the criteria established by the Working Group on Obesity in China (WGOC) for Chinese adults [27], BMI cut-offs of 24 and 28 kg/m^2 were applied to categorize each parent into the categories of normal, overweight and obesity. Parental educational attainment was surveyed and grouped into "primary school or below", "secondary or equivalent" and "junior college or above". In addition, the residence area was divided into "rural" and "urban", and monthly household income was defined with the sum of monthly income (in CNY) of all household members and then classified into <5000, or ≥5000 CNY.

2.3. Anthropometric Measurements

Anthropometric measurements were conducted by trained project members according to standardized procedure, and the measuring instruments were similar at all study sites. Children were required to stand straight in light clothing and without shoes for the measurements. Height was measured using the portable stadiometer with 0.1 cm precision, weight was measured to the nearest 0.1 kg by a Lever type weight scale. Every indicator was measured twice, and the average level of the two measurements was calculated for final analyses. BMI was calculated as body weight (kg) divided by height (m) squared.

2.4. Blood Sample Collection and Detection

Venous blood sample was obtained in the morning after an overnight (at least 8 h) fasting. Children were asked to rest for at least 10 min before blood sample collection. Blood specimens were then transported in a chilled insulated container immediately, and then centrifuged at 3000 rpm for 10 min and then frozen at −80 °C. All plasma samples were transported by air in dry ice to the laboratory in Beijing, where the samples were stored at −80 °C before laboratory detections. All the biochemical analyses were conducted at a biomedical analyses company accredited by Peking University [18]. Lipid profiles were measured with an autoanalyzer (TBA-120FR, Toshiba, Tokyo, Japan), with TC and TG assayed by enzymatic method, while LDL-C and HDL-C measured by clearance method.

2.5. Definition of Abnormal Lipid Profile and Overweight/Obesity

Following the 2011 Expert Panel on Integrated Guidelines for Cardiovascular Health and Risk Reduction in Children and Adolescents [28], high TG was defined as TG ≥ 1.13 mmol/L for children aged 9 years or younger, and TG ≥ 1.47 mmol/L for adolescents aged 10 years or older. High TC was considered as TC ≥ 5.18 mmol/L, high LDL-C was regarded as LDL-C ≥ 3.37 mmol/L, and low HDL-C referred to HDL-C ≤ 1.04 mmol/L. Because non-HDL-C levels showed better prediction of persistent dyslipidemia, the prevalence of high non-HDL-C levels was also presented, as was calculated as the difference between TC level and the HDL-C level. High non-HDL-C were determined as non-HDL-C ≥ 3.76 mmol/L [28]. A participant with one or more abnormal lipid levels (high LDC-C, low HDL-C, high TG, or high TC) was defined as having dyslipidemia. In addition, hyperlipidemia was defined with the presence of high TG or high TC levels [28].

In accordance with the guideline of the Working Group on Obesity in China, participants with age- and sex-specific BMI < 85th percentile were considered as non-overweight/ obese, and those with ≥85th percentile BMI were considered as overweight/obese [29].

2.6. Statistical Analysis

Data were expressed as mean ± SD for continuous variables and number (%) for categorical variables. We used Bonferroni multiple comparison methods to examine the differences in prevalence of adverse lipid profiles between groups. The significance of Bonferroni formula was set at α = 0.017. To understand whether some demographics, dietary or lifestyle factors could modify the associations, two separate logistic regression models were applied to estimate the odds ratio (OR) and 95% confidence interval (95% CI) between fruit consumption and lipid profile. In model 1, age and residence area were adjusted. In model 2, several additional confounding factors, including sex (boy, girl), BMI values, ethnicity (Han, Hui, Tibetan, Mongolian and Other), monthly household income (<5000, or $\geq$5000 CNY), parental weight status (normal, overweight or obesity), parental educational attainment (primary school or below, secondary or equivalent, junior college or above), vegetable consumption, sugar-sweetened beverages consumption, meat consumption and physical activity were included. Additionally, previous studies had shown that lipid levels were dependent on age and sexual maturation [30], and could be influenced by the difference of socioeconomic support mainly determined by the degree of their parents' education [31]. Besides, BMI values were closely related with lipid health. Therefore, stratified analyses were conducted according to sex, age, parental educational attainment and BMI values.

All analyses were performed using SAS software (version 9.4; SAS Institute, Inc., Cary, NC, USA). A two-sided p value < 0.05 was considered statistically significant.

3. Results

3.1. General Characteristics

A total of 7420 boys and 7335 girls were included in the final analysis. Table 1 showed the characteristics of the study population. The mean age was 11.15 $\pm$ 3.29 years old and their average BMI value was 18.74 $\pm$ 3.85 kg/m^2. The majority of the study population was Han ethnicity (92.56%), and most of their parents belonged to the normal weight status (paternal: 56.88%; maternal: 79.49%). In addition, 35.33% of the children's fathers received junior college or above education, 58.33% for secondary or equivalent and 6.34% for primary school or below. A similar trend was observed for maternal educational attainment. It was worth noting that most of them had a monthly household income of less than 5000 yuan. Besides, the average concentrations of TC, TG, LDL-C, HDL-C and non-HDL-C were 3.89 $\pm$ 0.88, 1.09 $\pm$ 0.77, 2.01 $\pm$ 0.69, 1.90 $\pm$ 1.35 and 1.99 $\pm$ 1.56 mmol/L, respectively.

Table 1. Baseline characteristic of included population.

Characteristics	Total Population (n = 14,755)	Boys (n = 7420)	Girls (n = 7335)	p-Value
Age, year	11.15 $\pm$ 3.29	11.13 $\pm$ 3.25	11.16 $\pm$ 3.33	0.533
Weight, kg	42.27 $\pm$ 15.91	44.09 $\pm$ 17.46	40.45 $\pm$ 13.94	<0.01
Height, m	1.48 $\pm$ 0.17	1.49 $\pm$ 0.18	1.46 $\pm$ 0.15	<0.01
BMI, kg/m^2	18.74 $\pm$ 3.85	19.04 $\pm$ 4.04	18.45 $\pm$ 3.63	<0.01
Residence area (n, %)				0.701
Urban	7860 (53.27%)	3941 (53.11%)	3919 (53.43%)	
Rural	6895 (46.73%)	3479 (46.89%)	3416 (46.57%)	
Ethnicity (n, %)				0.066
Han	13,657 (92.56%)	6876 (92.67%)	6781 (92.45%)	
Hui	528 (3.58%)	248 (3.34%)	280 (3.82%)	
Tibetan	33 (0.22%)	14 (0.19%)	19 (0.26%)	
Mongolian	187 (1.27%)	110 (1.48%)	77 (1.05%)	
Other	350 (2.37%)	172 (2.32%)	178 (2.43%)	
Paternal weight status, n (%)				0.451
Normal	8393 (56.88%)	4252 (57.30%)	4141 (56.46%)	
Overweight	4834 (32.76%)	2395 (32.28%)	2439 (33.25%)	
Obesity	1528 (10.36%)	773 (10.42%)	755 (10.29%)	
Maternal weight status, n (%)				0.002
Normal	11,729 (79.49%)	5983 (80.63%)	5746 (78.34%)	
Overweight	2462 (16.69%)	1176 (15.85%)	1286 (17.53%)	
Obesity	564 (3.82%)	261 (3.52%)	303 (4.13%)	
Paternal educational attainment, n (%)				0.229
Primary school or below	936 (6.34%)	451 (6.08%)	485 (6.61%)	
Secondary or equivalent	8606 (58.33%)	4371 (58.91%)	4235 (57.74%)	

Table 1. *Cont.*

Characteristics	Total Population (*n* = 14,755)	Boys (*n* = 7420)	Girls (*n* = 7335)	*p*-Value
Junior college or above	5213 (35.33%)	2598 (35.01%)	2615 (35.65%)	
Maternal educational attainment, *n* (%)				0.385
Primary school or below	1277 (8.65%)	619 (8.34%)	658 (8.97%)	
Secondary or equivalent	8429 (57.13%)	4246 (57.22%)	4183 (57.03%)	
Junior college or above	5049 (34.22%)	2555 (34.43%)	2494 (34.00%)	
Monthly household income, *n* (%)				0.220
<5000 yuan	12,402 (84.05%)	6264 (84.42%)	6138 (83.68%)	
≥5000 yuan	2353 (15.95%)	1156 (15.58%)	1197 (16.32%)	
Frequency of consumption (days per week)				
Fruit	4.93 ± 2.11	4.75 ± 2.19	5.11 ± 2.02	<0.01
Vegetables	6.02 ± 1.82	5.99 ± 1.85	6.06 ± 1.79	0.011
Sugar-sweetened beverages	1.63 ± 1.78	1.84 ± 1.91	1.41 ± 1.62	<0.01
Meat	5.01 ± 2.24	5.21 ± 2.18	4.80 ± 2.28	<0.01
Average daily consumption (servings per day)				
Fruit	1.33 ± 1.14	1.32 ± 1.19	1.33 ± 1.08	0.567
Vegetables	1.87 ± 1.50	1.89 ± 1.53	1.85 ± 1.46	0.148
Sugar-sweetened beverages	0.46 ± 0.84	0.56 ± 0.96	0.36 ± 0.68	<0.01
Meat	1.24 ± 1.30	1.39 ± 1.42	1.09 ± 1.15	<0.01
Frequency of physical activity, days/week	3.35 ± 2.53	3.47 ± 2.55	3.23 ± 2.50	<0.01
Average daily physical activity, hours and minutes	0.37 ± 0.82	0.42 ± 0.89	0.32 ± 0.74	<0.01
TC, mmol/L	3.89 ± 0.88	3.83 ± 0.87	3.94 ± 0.89	<0.01
TG, mmol/L	1.09 ± 0.77	1.04 ± 0.74	1.14 ± 0.79	<0.01
LDL-C, mmol/L	2.01 ± 0.69	1.98 ± 0.67	2.03 ± 0.70	<0.01
HDL-C, mmol/L	1.90 ± 1.35	1.90 ± 1.38	1.90 ± 1.33	0.993
Non-HDL-C, mmol/L	1.99 ± 1.56	1.94 ± 1.58	2.04 ± 1.54	<0.01

Abbreviation: BMI, body mass index; TC, total cholesterol; TG, triglycerides; LDL-C, low density lipoprotein-cholesterol; HDL-C, high density lipoprotein-cholesterol.

Furthermore, girls tended to consume fruits and vegetables more frequently per week ($p < 0.05$), while boys were more likely to consume sugar-sweetened beverages and meat ($p < 0.01$). As for the average daily consumption, intakes of fruits ($p = 0.567$) and vegetables ($p = 0.148$) were not significantly different among sex-specific groups, but boys consumed more sugar-sweetened beverages and meat daily (both $p < 0.01$). Of note, boys were prone to exercise more frequently and for longer duration (both $p < 0.01$).

3.2. The Prevalence of Abnormal Lipid Profile

The prevalence of abnormal lipid profile was present in Supplementary Table S1. Compared with participants consuming fruits for only 0–2 days/week, the prevalence of high TG, low HDL-C, dyslipidemia, and hyperlipidemia tended to be low among those who consumed fruits more frequently. Notably, similar lower prevalence for these lipid disorders was observed in the group of average daily fruit intake of 0.75–1.5 servings, compared to those with daily fruit consumption of ≥1.5 servings. Similar trends of prevalence of adverse lipid profiles were found for boys and girls, respectively.

3.3. Association between Fruit Consumption and Lipid Profile

The associations between fruit consumption and the odds of abnormal lipid profile were present in Table 2, high frequency of fruit intake per week was associated with lower possibility of unfavorable lipid profiles. In the basically adjusted model 1, the ORs (95% CI) of frequent fruit consumption (6–7 days/week) were 0.74 (0.66–0.83), 0.83 (0.75–0.92) and 0.79 (0.70–0.88) for the risk of high TG, dyslipidemia and hyperlipidemia, respectively. After controlling for potential covariates included in model 2, the results did not change essentially.

Table 2. Multivariate odds ratios (OR) and 95% confidence intervals (CI) of fruit consumption and lipid profile (n = 14,755).

Lipid Profile	Frequency of Fruit Consumption			Average Daily Fruit Consumption		
	0–2 Days/Week	3–5 Days/Week	6–7 Days/Week	<0.75 Servings/Day	0.75–1.5 Servings/Day	≥1.5 Servings/Day
			Model 1 [1]			
High TC	1 (Reference)	1.09 (0.86–1.38)	1.17 (0.93–1.46)	1 (Reference)	1.01 (0.85–1.20)	0.93 (0.78–1.12)
High TG	1 (Reference)	0.88 (0.79–0.99) *	0.74 (0.66–0.83) **	1 (Reference)	0.92 (0.83–1.01)	1.09 (0.99–1.21)
High LDL-C	1 (Reference)	1.14 (0.84–1.55)	1.13 (0.84–1.53)	1 (Reference)	0.93 (0.75–1.17)	0.86 (0.67–1.09)
Low HDL-C	1 (Reference)	0.96 (0.84–1.11)	0.88 (0.76–1.01)	1 (Reference)	0.94 (0.83–1.05)	1.06 (0.94–1.19)
High non-HDL-C	1 (Reference)	0.99 (0.78–1.26)	1.05 (0.83–1.32)	1 (Reference)	1.07 (0.89–1.28)	0.90 (0.74–1.10)
Dyslipidemia	1 (Reference)	0.97 (0.87–1.07)	0.83 (0.75–0.92) **	1 (Reference)	0.93 (0.86–1.01)	1.05 (0.96–1.15)
Hyperlipidemia	1 (Reference)	0.92 (0.82–1.03)	0.79 (0.70–0.88) **	1 (Reference)	0.92 (0.84–1.01)	1.07 (0.97–1.17)
			Model 2 [2]			
High TC	1 (Reference)	1.13 (0.88–1.45)	1.18 (0.92–1.50)	1 (Reference)	1.01 (0.85–1.22)	0.94 (0.77–1.14)
High TG	1 (Reference)	0.84 (0.74–0.96) *	0.66 (0.58–0.75) **	1 (Reference)	0.87 (0.78–0.97) **	1.00 (0.90–1.13)
High LDL-C	1 (Reference)	1.23 (0.89–1.71)	1.09 (0.79–1.51)	1 (Reference)	0.92 (0.73–1.17)	0.80 (0.62–1.03)
Low HDL-C	1 (Reference)	1.02 (0.88–1.19)	0.88 (0.76–1.03)	1 (Reference)	0.91 (0.80–1.03)	1.00 (0.87–1.14)
High non-HDL-C	1 (Reference)	1.07 (0.83–1.38)	1.05 (0.82–1.35)	1 (Reference)	1.08 (0.89–1.31)	0.87 (0.70–1.07)
Dyslipidemia	1 (Reference)	0.95 (0.85–1.06)	0.77 (0.68–0.86) **	1 (Reference)	0.88 (0.81–0.97) **	0.98 (0.89–1.08)
Hyperlipidemia	1 (Reference)	0.88 (0.78–1.00)	0.72 (0.63–0.81) **	1 (Reference)	0.88 (0.80–0.97) **	1.00 (0.90–1.11)

[1] Model 1: adjusted for age and residence area. [2] Model 2: additionally adjusted for sex, BMI values, ethnicity, incomes, parental educational attainment, parental weight, vegetable consumption, sugar-sweetened beverages consumption, meat consumption and physical activity. TC, total cholesterol; TG, triglycerides; LDL-C, low density lipoprotein-cholesterol; HDL-C, high density lipoprotein-cholesterol. * $p < 0.05$, ** $p < 0.01$.

In addition, adjusting for factors in model 2, participants who consumed fruits moderately for 0.75–1.5 servings each day were estimated to have 0.87 times (OR: 0.87, 95% CI: 0.78–0.97), 0.88 times (OR: 0.88, 95% CI: 0.81–0.97) and 0.88 times (OR: 0.88, 95% CI: 0.80–0.97) lower odds of high TG, dyslipidemia and hyperlipidemia compared with their counterparts with insufficient fruit consumption each time. There were no statistically significant associations between fruit consumption over 1.5 servings/day and risks of high TC, high LDL-C, low HDL-C and high non-HDL-C.

Apart from this, consuming fruits for 6–7 days/week combined with moderate intake of 0.75–1.5 servings each day could improve adverse lipid profiles (Table 3). Children and adolescents who consumed fruits moderately of 0.75–1.5 servings/day for 6–7 days/week had lower risks of high TG, low HDL-C, dyslipidemia and hyperlipidemia (all $p < 0.05$).

Table 3. Combined effects of frequencies and daily intake of fruit consumption on adverse lipid profile (n = 14,755).

Lipid Profile	Unselected Population	6–7 Days/Week (n = 6829)		
		<0.75 Servings/Day (n = 61)	0.75–1.5 Servings/Day (n = 3036)	≥1.5 Servings/Day (n = 3732)
High TC	1 (Reference)	0.48 (0.11–2.04)	1.10 (0.92–1.32)	1.01 (0.85–1.21)
High TG	1 (Reference)	0.78 (0.36–1.68)	0.68 (0.61–0.77) **	0.94 (0.85–1.05)
High LDL-C	1 (Reference)	0.71 (0.15–3.23)	1.03 (0.81–1.31)	0.90 (0.71–1.13)
Low HDL-C	1 (Reference)	0.86 (0.35–2.09)	0.83 (0.73–0.96) *	0.97 (0.86–1.09)
High non-HDL-C	1 (Reference)	0.75 (0.21–2.73)	1.12 (0.92–1.36)	0.91 (0.75–1.09)
Dyslipidemia	1 (Reference)	0.69 (0.36–1.33)	0.76 (0.69–0.83) **	0.95 (0.87–1.04)
Hyperlipidemia	1 (Reference)	0.68 (0.33–1.41)	0.72 (0.65–0.81) **	0.96 (0.88–1.06)

Adjusted in Model 2: including age, residence area, sex, BMI values, ethnicity, incomes, parental educational attainment, parental weight, vegetable consumption, sugar-sweetened beverages consumption, meat consumption and physical activity. TC, total cholesterol; TG, triglycerides; LDL-C, low density lipoprotein-cholesterol; HDL-C, high density lipoprotein-cholesterol. * $p < 0.05$, ** $p < 0.01$.

3.4. Stratified Analyses

The results of stratified analyses were similar with the main results (Figures 1,2 and S1). The more frequent of fruit intake was associated with lower odds of high TG, dyslipidemia and hyperlipidemia both in boys and girls, while moderately consuming fruits each day could only lower the likelihood of lipid disorders among girls (Figure 1). Notably, compared to the reference group, significantly lower likelihoods of high TG and hyperlipidemia were observed among all participants aged 5–19 years old who consumed fruits more frequently, but only 5–14-year-old children and adolescents had a lower risk of dyslipidemia ($p < 0.01$). However, when considering consuming fruits moderately of 0.75–1.5 servings per day, only 5-to-9-year-old children were estimated to have 0.80 times (OR: 0.80, 95% CI: 0.66–0.96), 0.79 times (OR: 0.79, 95% CI: 0.68–0.92) and 0.77 times (OR: 0.77, 95% CI: 0.66–0.91) lower odds of high TG, dyslipidemia and hyperlipidemia, respectively (Figure 2).

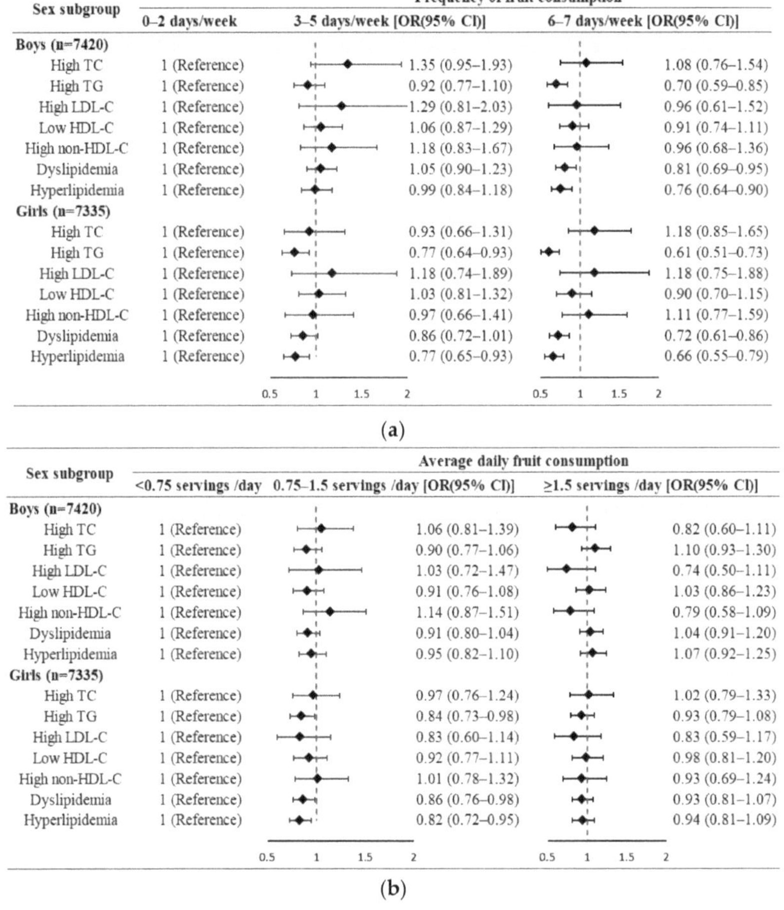

Sex subgroup	Frequency of fruit consumption		
	0–2 days/week	3–5 days/week [OR(95% CI)]	6–7 days/week [OR(95% CI)]
Boys (n=7420)			
High TC	1 (Reference)	1.35 (0.95–1.93)	1.08 (0.76–1.54)
High TG	1 (Reference)	0.92 (0.77–1.10)	0.70 (0.59–0.85)
High LDL-C	1 (Reference)	1.29 (0.81–2.03)	0.96 (0.61–1.52)
Low HDL-C	1 (Reference)	1.06 (0.87–1.29)	0.91 (0.74–1.11)
High non-HDL-C	1 (Reference)	1.18 (0.83–1.67)	0.96 (0.68–1.36)
Dyslipidemia	1 (Reference)	1.05 (0.90–1.23)	0.81 (0.69–0.95)
Hyperlipidemia	1 (Reference)	0.99 (0.84–1.18)	0.76 (0.64–0.90)
Girls (n=7335)			
High TC	1 (Reference)	0.93 (0.66–1.31)	1.18 (0.85–1.65)
High TG	1 (Reference)	0.77 (0.64–0.93)	0.61 (0.51–0.73)
High LDL-C	1 (Reference)	1.18 (0.74–1.89)	1.18 (0.75–1.88)
Low HDL-C	1 (Reference)	1.03 (0.81–1.32)	0.90 (0.70–1.15)
High non-HDL-C	1 (Reference)	0.97 (0.66–1.41)	1.11 (0.77–1.59)
Dyslipidemia	1 (Reference)	0.86 (0.72–1.01)	0.72 (0.61–0.86)
Hyperlipidemia	1 (Reference)	0.77 (0.65–0.93)	0.66 (0.55–0.79)

(**a**)

Sex subgroup	Average daily fruit consumption		
	<0.75 servings /day	0.75–1.5 servings /day [OR(95% CI)]	≥1.5 servings /day [OR(95% CI)]
Boys (n=7420)			
High TC	1 (Reference)	1.06 (0.81–1.39)	0.82 (0.60–1.11)
High TG	1 (Reference)	0.90 (0.77–1.06)	1.10 (0.93–1.30)
High LDL-C	1 (Reference)	1.03 (0.72–1.47)	0.74 (0.50–1.11)
Low HDL-C	1 (Reference)	0.91 (0.76–1.08)	1.03 (0.86–1.23)
High non-HDL-C	1 (Reference)	1.14 (0.87–1.51)	0.79 (0.58–1.09)
Dyslipidemia	1 (Reference)	0.91 (0.80–1.04)	1.04 (0.91–1.20)
Hyperlipidemia	1 (Reference)	0.95 (0.82–1.10)	1.07 (0.92–1.25)
Girls (n=7335)			
High TC	1 (Reference)	0.97 (0.76–1.24)	1.02 (0.79–1.33)
High TG	1 (Reference)	0.84 (0.73–0.98)	0.93 (0.79–1.08)
High LDL-C	1 (Reference)	0.83 (0.60–1.14)	0.83 (0.59–1.17)
Low HDL-C	1 (Reference)	0.92 (0.77–1.11)	0.98 (0.81–1.20)
High non-HDL-C	1 (Reference)	1.01 (0.78–1.32)	0.93 (0.69–1.24)
Dyslipidemia	1 (Reference)	0.86 (0.76–0.98)	0.93 (0.81–1.07)
Hyperlipidemia	1 (Reference)	0.82 (0.72–0.95)	0.94 (0.81–1.09)

(**b**)

Figure 1. Sex-specific analysis of the associations between fruit consumption and lipid profile (adjusted for age, BMI values, residence area, ethnicity, incomes, parental educational attainment, parental weight, vegetable consumption, sugar-sweetened beverages consumption, meat consumption and physical activity) (**a**) Frequency of fruit consumption; (**b**) Average daily fruit consumption. TC, total cholesterol; TG, triglycerides; LDL-C, low density lipoprotein-cholesterol; HDL-C, high density lipoprotein-cholesterol; diamond symbol, point estimates; red dashed line, invalid line.

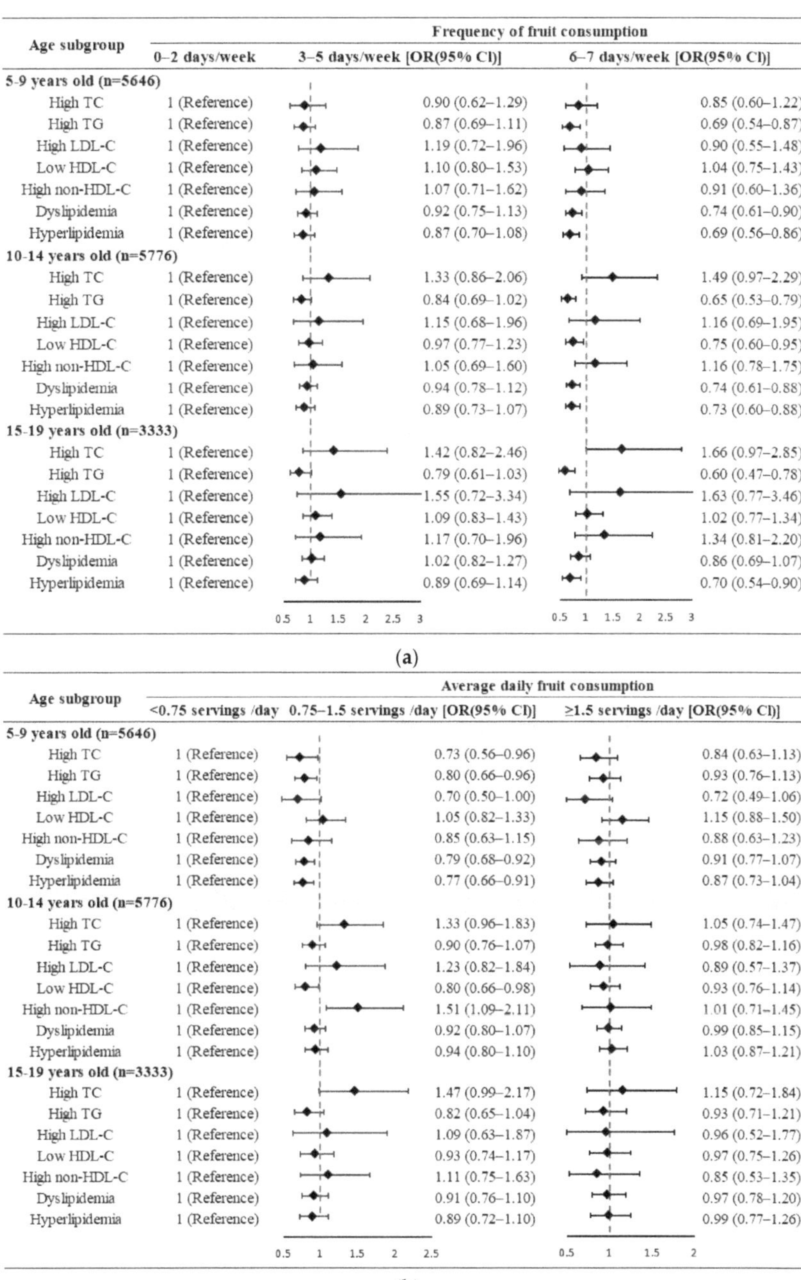

Figure 2. Age-specific analysis of the associations between fruit consumption and lipid profile (adjusted for sex, BMI values, residence area, ethnicity, incomes, parental educational attainment, parental weight, vegetable consumption, sugar-sweetened beverages consumption, meat consumption and physical activity) (**a**) Frequency of fruit consumption; (**b**) Average daily fruit consumption. TC, total cholesterol; TG, triglycerides; LDL-C, low density lipoprotein-cholesterol; HDL-C, high density lipoprotein-cholesterol; diamond symbol, point estimates; red dashed lines, invalid line.

When the study sample was restricted according to parental educational attainment, the main associations did not change essentially, but were more significantly pronounced among participants from families with higher parental education levels, compared with those from lower educational families (Supplementary Figure S2). Notably, the results of the two subgroups presented similar according to BMI values (Supplementary Figure S3).

4. Discussion

Based on the national representative sample, our findings suggested significant inverse associations between high frequency of fruit consumption per week and moderate daily fruit intake with the odds of adverse lipid profiles. The associations were more pronounced among girls, younger participants and those whose families had a higher educational background, when presented similar by BMI subgroups. These findings supported the health effect of moderate fruit intake frequently to improve the childhood lipid profile.

Our findings were in accordance with previous studies in direction. A case-control study in Jordanians revealed that consumption of banana could reduce the odds of cardiovascular events to about 44% and 62% during adulthood when consuming 1–2 and 3–6 servings per week, respectively [32]. Growing reviews and meta-analyses found consistent associations between fruit consumption and cardiovascular events, diabetes, cancer, and other chronic diseases [33,34]. Among a pediatric population, a diet rich in dietary fiber from fruits was beneficial to a healthy cardiovascular profile, regardless of European children's weight [35]. Since there was evidence supporting that eating behaviors of children and adolescents were likely to persist into adulthood [36], encouraging healthy eating habits among school-aged children might therefore represent an effective primary prevention strategy for reducing the risk of chronic diseases. As recommended by ISPAD Clinical Practice Consensus Guidelines 2018 [37], intake of a variety of fruit should be encouraged among children and adolescents with diabetes, which could be particularly useful in helping to reduce lipid levels.

Although WHO recommends 5 servings (400 g) of fruits and vegetables per day for the overall population [15], among the pediatric population, studies rarely applied the criteria of 5 servings per day since only few children and adolescents met this recommendation [16,38], thus it was unlikely that any significant association between the recommended levels and lipid health would be detected. Therefore, the absence of a specific frequency and daily amount of fruit consumption for preventing pediatrics' lipid disorders was an important limitation. In the present study, daily fruit consumption of 0.75–1.5 servings (approximately 150–300 g) for 6–7 days/week was crucial for Chinese pediatrics' lipid health. We speculated that the continued increase in fruit intake would not always promote lipid health, which might be explained by the following reasons: Fructose is a kind of sugar found in fruits, the excess accumulation of which might be positively associated with the level of plasma TG and atherogenic indices [39]. Besides, it could disrupt the utilization of dietary copper, while copper deficiency could produce high TC and high TG [40]. Since there was no consensus regarding the method of assessing fruit consumption in children and adolescents [41], additional studies with high methodological quality are needed to develop informed guidelines for fruit consumption among pediatric population and to better promote lipid health.

However, the exact mechanisms by which increased fruit consumption reduced abnormal lipid profile were not fully understood. As for these inverse associations, antioxidant compounds and polyphenols found in fruit (e.g., vitamin C, carotenoids, and flavonoids) could prevent the oxidation of cholesterol in the arteries [42]. In addition, those compounds could reduce systemic inflammation through cellular signaling processes, therefore preventing atherosclerosis and cardiovascular disease [43]. Apart from this, several components of fruit have cholesterol lowering properties, particularly dietary fiber, as well as high amounts of water and low amounts of saturated fat, which are associated with reduced energy density, hunger control, and satiety [44]. Taken together, the evidence accumulated so far does support that fruit consumption may reduce the odds of lipid disorders.

Subgroup analyses suggested that the associations between moderate fruit consumption each day and the odds of lipid disorders were more pronounced in girls. A previous study conducted in Korea also emphasized the sex differences, and concluded that low fruit intakes were significantly associated with overweight and obesity among adolescent girls but not in boys [45]. To be noted, we found more frequent of fruit/vegetable consumption and less intake of SSBs among girls; in this context, different behavioral factors may influence the associations in girls and boys. Another potential behavioral explanation was that higher fruit consumption might reflect their health consciousness, consequently, a lower intake of sugar- and fat-rich products, especially in girls.

An interesting age effect was also present; this possibility indicated the potential implications of different lifestyle behaviors in school and preschool children. Our present study indicated these associations were more evident in those aged 5–14 years old. One of the reasons might be that younger children lived healthier lifestyles, while older children and adolescents were under heavier academic pressure, which often led them to overconsume SSBs and meat, as well as to reduce their outdoor activities (Supplementary Table S2). Notably, few studies investigated the association between fruit intakes and lipid profile among different age groups. Therefore we could not discuss an age-specific association between low fruit intakes and adverse lipid profile based on the previous findings.

Targeting the family environment for the promotion of lipid health among children and adolescents is important. Our findings were in agreement with previous surveys which concluded that family education could potentially modify the relationships between dietary behaviors and risks of chronic diseases [30]. It was necessary to influence the fruit consumption preferences of their parents, in order to increase the fruit intake of children [46]. Higher parental educational background may be related to higher household income and thus greater availability of healthy foods, increased nutrition knowledge and higher motivation to follow a healthy lifestyle. In addition, we proposed that health and education programs should consider these findings and be implemented widely to make the public aware of the importance of moderate fruit consumption for the pediatric population. In addition, the similarity of subgroup analysis by BMI showed that the relationship between fruit consumption and lipid profiles was independent of BMI values, to a certain extent. Children and adolescents should consume fruits moderately each day to maintain lipid health, regardless of their weight status.

Our study had strengths of the national representative sample from seven provinces in China, and we focused on 5–19-year-old children and adolescents. However, there were also limitations that should be taken into account when interpreting the findings. Firstly, since 92.56% of the study population was of Han ethnicity, our results may not be applicable to other ethnic groups. Secondly, we only relied on self-reported dietary intake, which could lead to a certain degree of recall bias. However, in this process, we carried out strict quality control to ensure the reliability. In addition, the dietary data recall of 7 days might not represent long-term dietary behaviors; thus, in future studies, a pilot study using food tracking, such as photo tracking, could be used to accurately assess the frequency and amount of fruit and other dietary intake in individuals. Thirdly, this was a cross-sectional survey and thus could not generate casual relationships of fruit consumption and lipid profile. A randomized controlled trial was desired to confirm our results. Fourth, measuring fruit consumption was complex, with interactions of different fruit compounds, cultural and socioeconomic conditions [47]. Since it was difficult to collect accurate and detailed information on intake of various types of food in a large population survey, we could not identify the specific type of fruits consumed by the participants, and different fruits might have various influences on lipid profiles. However, as for food and diet information, both parent and child questionnaire of children grade 1–3 were reported by parents. In addition, trained project members interpreted all the questionnaires in detail. Appropriate guidance would be given by these project members as effectively as possible. The questionnaires would be rechecked by 3% within one week for the same participants [18]. Therefore, the quality of diet information was guaranteed largely. Besides, other important dietary factors, such as macronutrient and energy intakes, were not available in the present study. In order to minimize possible influence of these factors, we included vegetable, meat and SSBs consumption in the adjusted model. Fifth, the IPAQ-SF questionnaire used in this study was not completely suitable for the assessment of physical activity of children and adolescents. However, due to the open-ended questions of IPAQ-SF, and appropriate guidance would be given by project members, to a certain extent, the IPAQ-SF could reflect the levels of physical activity of children and adolescents. Further studies with more information regarding confounding factors and eating behaviors were desired in the future.

5. Conclusions

Our findings confirmed that a diet rich in fruit could be effective for improving lipid health. Given the reason that the basis of dietary intake recommendations is related with its health implications, moderate fruit consumption of approximately 150–300 g each day for 6–7 days/week could be beneficial to lipid health for children and adolescents. Monitoring dietary habits early in childhood and adolescence might have a positive impact on lipid health and overall quality of life. For the pediatric population, indicators of family circumstances should also be applied to identify target groups for interventions aimed at promoting eating habits and lipid health.

Supplementary Materials: The following are available online at https://www.mdpi.com/article/10.3390/nu14010063/s1, Figure S1. The size of one serving of fruit or vegetable (as the size of a fist of an ordinary adult). Figure S2: Parental educational attainment-specific analysis of the associations between fruit consumption and lipid profile (adjusted for age, residence area, sex, BMI values, ethnicity, incomes, parental weight, vegetable consumption, sugar-sweetened beverages consumption, meat consumption and physical activity). Figure S3: BMI-specific analysis of the associations between fruit consumption and lipid profile (adjusted for age, residence area, sex, ethnicity, incomes, parental weight, parental educational attainment, vegetable consumption, sugar-sweetened beverages consumption, meat consumption and physical activity). Table S1: Prevalence of the abnormal lipid profile (n, %). Table S2: Difference of demographics and lifestyle behaviors among age groups.

Author Contributions: Conceptualization, J.L.; Data curation, Y.L., D.G., L.C., M.C. and Y.M.; Formal analysis, J.L. and X.W. (Xinxin Wang); Funding acquisition, Y.D. and J.M.; Methodology, J.L., X.W. (Xinxin Wang), D.G., L.C. and M.C.; Project administration, Y.D. and J.M.; Resources, Z.Z. and J.M.; Software, X.W. (Xinxin Wang) and L.C.; Supervision, Z.Z., X.W. (Xijie Wang), Y.D. and J.M.; Validation, Z.Z., X.W. (Xijie Wang), Y.D. and J.M.; Visualization, Y.L., D.G., L.C., Y.M., Y.Z., J.J., Z.Z., X.W. (Xijie Wang) and Y.D.; Writing—original draft, J.L.; Writing—review and editing, Y.L., X.W. (Xinxin Wang), D.G., M.C., T.M., Q.M., Y.Z., J.J. and X.W. (Xijie Wang). All authors have read and agreed to the published version of the manuscript.

Funding: This research was funded by the Research Special Fund for Public Welfare Industry of Health of the Ministry of Health of China (1147 Project; Grant No. 20122010 to J.M.), China Postdoctoral Science Foundation (BX20200019 and 2020M680266 to Y.D.) and National Natural Science Foundation of China (82103865 to Y.D.).

Institutional Review Board Statement: This study was conducted according to the guidelines laid down in the Declaration of Helsinki and all procedures involving students were approved by the ethics committee of the Peking University (IRB0000105213034).

Informed Consent Statement: All participants and their parents signed informed consents voluntarily.

Data Availability Statement: The data supporting the conclusions of this article will be made available from the corresponding author upon request.

Acknowledgments: The authors would like to acknowledge the support from all the team members and the participating students, teachers, parents and local education and health staff in the programs.

Conflicts of Interest: The authors declare no conflict of interest.

References

1. Kit, B.K.; Kuklina, E.; Carroll, M.D.; Ostchega, Y.; Freedman, D.S.; Ogden, C.L. Prevalence of and Trends in Dyslipidemia and Blood Pressure Among US Children and Adolescents, 1999–2012. *JAMA Pediatr.* **2015**, *169*, 272–279. [CrossRef]
2. Ding, W.; Cheng, H.; Yan, Y.; Zhao, X.; Chen, F.; Huang, G.; Hou, D.; Mi, J. 10-Year Trends in Serum Lipid Levels and Dyslipidemia Among Children and Adolescents From Several Schools in Beijing, China. *J. Epidemiol.* **2016**, *26*, 637–645. [CrossRef] [PubMed]
3. He, H.; Pan, L.; Du, J.; Liu, F.; Jin, Y.; Ma, J.; Wang, L.; Jia, P.; Hu, Z.; Shan, G. Prevalence of, and biochemical and anthropometric risk factors for, dyslipidemia in children and adolescents aged 7 to 18 years in China: A cross-sectional study. *Am. J. Hum. Biol.* **2019**, *31*, e23286. [CrossRef] [PubMed]
4. Ding, W.; Dong, H.; Mi, J. Prevalence of dyslipidemia in Chinese children and adolescents:a Meta-analysis. *Zhonghua Liu Xing Bing Xue Za Zhi = Zhonghua Liuxingbingxue Zazhi* **2015**, *36*, 71–77. [PubMed]
5. Lozano, P.; Henrikson, N.; Morrison, C.; Dunn, J.; Nguyen, M.; Blasi, P.; Whitlock, E. Lipid Screening in Childhood and Adolescence for Detection of Multifactorial Dyslipidemia: Evidence Report and Systematic Review for the US Preventive Services Task Force. *JAMA Pediatr.* **2016**, *316*, 634–644. [CrossRef]
6. Rauber, F.; Campagnolo, P.; Hoffman, D.; Vitolo, M. Consumption of ultra-processed food products and its effects on children's lipid profiles: A longitudinal study. *Nutr. Metab. Cardiovasc. Dis.* **2015**, *25*, 116–122. [CrossRef]
7. Lim, S.S.; Vos, T.; Flaxman, A.D.; Danaei, G.; Shibuya, K.; Adair-Rohani, H.; AlMazroa, M.A.; Amann, M.; Anderson, H.R.; Andrews, K.G.; et al. A comparative risk assessment of burden of disease and injury attributable to 67 risk factors and risk factor clusters in 21 regions, 1990–2010: A systematic analysis for the Global Burden of Disease Study 2010. *Lancet* **2012**, *380*, 2224–2260. [CrossRef]
8. Van Duyn, M.; Pivonka, E. Overview of the health benefits of fruit and vegetable consumption for the dietetics professional: Selected literature. *J. Am. Diet. Assoc.* **2000**, *100*, 1511–1521. [CrossRef]
9. LeDoux, T.A.; Hingle, M.D.; Baranowski, T. Relationship of fruit and vegetable intake with adiposity: A systematic review. *Obes. Rev.* **2010**, *12*, e143–e150. [CrossRef]
10. Pan, Y.; Pratt, C. Metabolic syndrome and its association with diet and physical activity in US adolescents. *J. Am. Diet. Assoc.* **2008**, *108*, 276–286. [CrossRef]
11. Rosi, A.; Paolella, G.; Biasini, B.; Scazzina, F. SINU Working Group on Nutritional Surveillance in Adolescents. Dietary habits of adolescents living in North America, Europe or Oceania: A review on fruit, vegetable and legume consumption, sodium intake, and adherence to the Mediterranean Diet. *Nutr. Metab. Cardiovasc. Dis.* **2019**, *29*, 544–560. [CrossRef]
12. John, J.; Ziebland, S.; Yudkin, P.; Roe, L.; Neil, H. Effects of fruit and vegetable consumption on plasma antioxidant concentrations and blood pressure: A randomised controlled trial. *Lancet* **2002**, *359*, 1969–1974. [CrossRef]
13. Keung, V.; Lo, K.; Cheung, C.; Tam, W.; Lee, A. Changes in dietary habits and prevalence of cardiovascular risk factors among school students in Macao, China. *Obes. Res. Clin. Pract.* **2019**, *13*, 541–547. [CrossRef] [PubMed]
14. Cardoso Chaves, O.; Franceschini, S.C.; Machado Rocha Ribeiro, S.; Ferreira Rocha Sant Ana, L.; Garçon De Faria, C.; Priore, S.E. Anthropometric and biochemical parameters in adolescents and their relationship with eating habits and household food availability. *Nutr. Hosp.* **2013**, *28*, 1352–1356. [PubMed]
15. World Health Organization. *Diet, Nutrition, and the Prevention of Chronic Diseases. Report of the Joint WHO/FAO Expert Consultation;* (WHO Technical Report Series, no. 916); World Health Organization: Geneva, Switzerland, 2002.
16. Collese, T.S.; Nascimento-Ferreira, M.V.; De Moraes, A.C.F.; Rendo-Urteaga, T.; Bel-Serrat, S.; Moreno, L.A.; Carvalho, H.B. Role of fruits and vegetables in adolescent cardiovascular health: A systematic review. *Nutr. Rev.* **2017**, *75*, 339–349. [CrossRef] [PubMed]
17. Sharma, S.P.; Chung, H.J.; Kim, H.J.; Hong, S.T. Paradoxical Effects of Fruit on Obesity. *Nutrients* **2016**, *8*, 633. [CrossRef]
18. Chen, Y.; Ma, L.; Ma, Y.; Wang, H.; Luo, J.; Zhang, X.; Luo, C.; Wang, H.; Zhao, H.; Pan, D.; et al. A national school-based health lifestyles interventions among Chinese children and adolescents against obesity: Rationale, design and methodology of a randomized controlled trial in China. *BMC Public Health* **2015**, *15*, 210. [CrossRef]
19. Dong, Y.H.; Zou, Z.Y.; Yang, Z.P.; Wang, Z.H.; Jing, J.; Luo, J.Y.; Zhang, X.; Luo, C.Y.; Wang, H.; Zhao, H.P.; et al. Association between high birth weight and hypertension in children and adolescents: A cross-sectional study in China. *J. Hum. Hypertens.* **2017**, *31*, 737–743. [CrossRef]
20. Gui, Z.-H.; Zhu, Y.-N.; Cai, L.; Sun, F.-H.; Ma, Y.-H.; Jing, J.; Chen, Y.-J. Sugar-Sweetened Beverage Consumption and Risks of Obesity and Hypertension in Chinese Children and Adolescents: A National Cross-Sectional Analysis. *Nutrients* **2017**, *9*, 1302. [CrossRef] [PubMed]
21. Yang, Y.; Dong, B.; Zou, Z.; Wang, S.; Dong, Y.; Wang, Z.; Ma, J. Association between Vegetable Consumption and Blood Pressure, Stratified by BMI, among Chinese Adolescents Aged 13–17 Years: A National Cross-Sectional Study. *Nutrients* **2018**, *10*, 451. [CrossRef]
22. Ho, S.-Y.; Wong, B.Y.-M.; Lo, W.-S.; Mak, K.-K.; Thomas, G.N.; Lam, T.H. Neighbourhood food environment and dietary intakes in adolescents: Sex and perceived family affluence as moderators. *Int. J. Pediatr. Obes.* **2010**, *5*, 420–427. [CrossRef]
23. Hicks, T.M.; Knowles, S.O.; Farouk, M. Global Provisioning of Red Meat for Flexitarian Diets. *Front. Nutr.* **2018**, *5*, 50. [CrossRef] [PubMed]
24. The Chinese Nutrition Society. *Dietary Guidelines for Chinese School-Age Children (2016);* People's Medical Publishing House: Beijing, China, 2016.

25. Lee, P.H.; Macfarlane, D.J.; Lam, T.H.; Stewart, S.M. Validity of the international physical activity questionnaire short form (IPAQ-SF): A systematic review. *Int. J. Behav. Nutr. Phys. Act.* **2011**, *8*, 115. [CrossRef] [PubMed]
26. Craig, C.L.; Marshall, A.L.; Sjöström, M.; Bauman, A.E.; Booth, M.L.; Ainsworth, B.E.; Pratt, M.; Ekelund, U.; Yngve, A.; Sallis, J.F.; et al. International Physical Activity Questionnaire: 12-Country Reliability and Validity. *Med. Sci. Sports Exerc.* **2003**, *35*, 1381–1395. [CrossRef]
27. Zhou, B.-F. Effect of body mass index on all-cause mortality and incidence of cardiovascular diseases–report for meta-analysis of prospective studies open optimal cut-off points of body mass index in Chinese adults. *Biomed. Environ. Sci.* **2002**, *15*, 245–252. [PubMed]
28. Expert Panel on Integrated Guidelines for Cardiovascular Health and Risk Reduction in Children and Adolescents; National Heart, Lung, and Blood Institute. Expert panel on integrated guidelines for cardiovascular health and risk reduction in children and adolescents: Summary report. *Pediatrics* **2011**, *128* (Suppl. 5), S213–S256. [CrossRef]
29. Group of China Obesity Task Force. Body mass index reference norm for screening overweight and obesity in Chinese children and adolescents. *Zhonghua Liu Xing Bing Xue Za Zhi* **2004**, *25*, 97–102.
30. Spinneker, A.; Egert, S.; González-Gross, M.; Breidenassel, C.; Albers, U.; Stoffel-Wagner, B.; Huybrechts, I.; Manios, Y.; Venneria, E.; Molnar, D.; et al. Lipid, lipoprotein and apolipoprotein profiles in European adolescents and its associations with gender, biological maturity and body fat—The HELENA Study. *Eur. J. Clin. Nutr.* **2012**, *66*, 727–735. [CrossRef]
31. Kocaoglu, B.; Manios, Y.; Dimitriou, M.; Kolotourou, M.; Keskin, Y.; Sur, H.; Hayran, O.; Manios, Y. Parental educational level and cardiovascular disease risk factors in schoolchildren in large urban areas of Turkey: Directions for public health policy. *BMC Public Health* **2005**, *5*, 13. [CrossRef]
32. Tayyem, R.F.; Al-Bakheit, A.; Hammad, S.S.; Al-Shudifat, A.-E.; Azab, M.; Bawadi, H. Fruit and vegetable consumption and cardiovascular diseases among Jordanians: A case-control study. *Central Eur. J. Public Health* **2020**, *28*, 208–218. [CrossRef]
33. Wu, Y.; Zhang, D.; Jiang, X.; Jiang, W. Fruit and vegetable consumption and risk of type 2 diabetes mellitus: A dose-response meta-analysis of prospective cohort studies. *Nutr. Metab. Cardiovasc. Dis.* **2015**, *25*, 140–147. [CrossRef] [PubMed]
34. Wang, X.; Ouyang, Y.; Liu, J.; Zhu, M.; Zhao, G.; Bao, W.; Hu, F.B. Fruit and vegetable consumption and mortality from all causes, cardiovascular disease, and cancer: Systematic review and dose-response meta-analysis of prospective cohort studies. *BMJ* **2014**, *349*, g4490. [CrossRef] [PubMed]
35. Larrosa, S.; Luque, V.; Grote, V.; Closa-Monasterolo, R.; Ferré, N.; Koletzko, B.; Verduci, E.; Gruszfeld, D.; Xhonneux, A.; Escribano, J. Fibre Intake Is Associated with Cardiovascular Health in European Children. *Nutrients* **2020**, *13*, 12. [CrossRef] [PubMed]
36. Craigie, A.M.; Lake, A.; Kelly, S.; Adamson, A.; Mathers, J.C. Tracking of obesity-related behaviours from childhood to adulthood: A systematic review. *Maturitas* **2011**, *70*, 266–284. [CrossRef]
37. Smart, C.E.; Annan, F.; Higgins, L.A.; Jelleryd, E.; Lopez, M.; Acerini, C.L. ISPAD Clinical Practice Consensus Guidelines 2018: Nutritional management in children and adolescents with diabetes. *Pediatr. Diabetes* **2018**, *19* (Suppl. S27), 136–154. [CrossRef]
38. Panagiotopoulos, C.; Nguyen, D.; Smith, J. Cardiovascular risk factors and health behaviours in elementary school-age Inuvialuit and Gwich'in children. *Paediatr. Child Health* **2014**, *19*, 256–260. [CrossRef]
39. Czerwonogrodzka-Senczyna, A.; Rumińska, M.; Majcher, A.; Credo, D.; Jeznach-Steinhagen, A.; Pyrżak, B. Fructose Consumption and Lipid Metabolism in Obese Children and Adolescents. *Adv. Exp. Med. Biol.* **2019**, *1153*, 91–100. [CrossRef]
40. Klevay, L.M. Metabolic interactions among dietary cholesterol, copper, and fructose. *Am. J. Physiol. Endocrinol. Metab.* **2010**, *298*, E138–E139. [CrossRef]
41. Collins, C.E.; Watson, J.; Burrows, T. Measuring dietary intake in children and adolescents in the context of overweight and obesity. *Int. J. Obes.* **2010**, *34*, 1103–1115. [CrossRef]
42. Asplund, K. Antioxidant vitamins in the prevention of cardiovascular disease: A systematic review. *J. Intern. Med.* **2002**, *251*, 372–392. [CrossRef]
43. Joseph, S.V.; Edirisinghe, I.; Burton-Freeman, B.M. Fruit Polyphenols: A Review of Anti-inflammatory Effects in Humans. *Crit. Rev. Food Sci. Nutr.* **2016**, *56*, 419–444. [CrossRef] [PubMed]
44. Rolls, B.J.; Ello-Martin, J.A.; Tohill, B.C. What can intervention studies tell us about the relationship between fruit and vegetable consumption and weight management? *Nutr. Rev.* **2004**, *62*, 1–17. [CrossRef] [PubMed]
45. You, J.; Choo, J. Adolescent Overweight and Obesity: Links to Socioeconomic Status and Fruit and Vegetable Intakes. *Int. J. Environ. Res. Public Health* **2016**, *13*, 307. [CrossRef] [PubMed]
46. Groele, B.; Głąbska, D.; Gutkowska, K.; Guzek, D. Mother's Fruit Preferences and Consumption Support Similar Attitudes and Behaviors in Their Children. *Int. J. Environ. Res. Public Health* **2018**, *15*, 2833. [CrossRef]
47. Corella, D.; Ordovás, J. Biomarkers: Background, classification and guidelines for applications in nutritional epidemiology. *Nutr. Hosp.* **2015**, *31* (Suppl. S3), 177–188. [PubMed]

Article

Caregiver Perceptions of Child Diet Quality: What Influenced Their Judgment

Lijing Shao [1,†], Yan Ren [1,†], Yanming Li [1], Mei Yang [2], Bing Xiang [2], Liping Hao [3], Xuefeng Yang [3] and Jing Zeng [1,2,*]

[1] School of Public Health, Wuhan University of Science and Technology, Wuhan 430065, China; 202009703046@wust.edu.cn (L.S.); 202009703044@wust.edu.cn (Y.R.); liyanming@wust.edu.cn (Y.L.)

[2] School Research Center for Woman and Child Health, Wuhan University of Science and Technology, Wuhan 430065, China; yangmei88@wust.edu.cn (M.Y.); Xiangbing@wust.edu.cn (B.X.)

[3] Department of Nutrition and Food Hygiene, Tongji Medical College, Huazhong University of Science and Technology, Wuhan 430065, China; haolp@mails.tjmu.edu.cn (L.H.); xxyxf@mails.tjmu.edu.cn (X.Y.)

* Correspondence: zengjing@wust.edu.cn; Tel.: +86-137-9702-0322

† These authors contributed equally to this work.

Citation: Shao, L.; Ren, Y.; Li, Y.; Yang, M.; Xiang, B.; Hao, L.; Yang, X.; Zeng, J. Caregiver Perceptions of Child Diet Quality: What Influenced Their Judgment. *Nutrients* 2022, 14, 125. https://doi.org/10.3390/nu14010125

Academic Editor: Zhiyong Zou

Received: 12 December 2021
Accepted: 24 December 2021
Published: 28 December 2021

Publisher's Note: MDPI stays neutral with regard to jurisdictional claims in published maps and institutional affiliations.

Copyright: © 2021 by the authors. Licensee MDPI, Basel, Switzerland. This article is an open access article distributed under the terms and conditions of the Creative Commons Attribution (CC BY) license (https://creativecommons.org/licenses/by/4.0/).

Abstract: This study aimed at assessing the correctness of a caregiver's perception of their child's diet status and to determine the factors which may influence their judgment. 815 child-caregiver pairs were recruited from two primary schools. 3-day 24-h recall was used to evaluate children's dietary intake, Chinese Children Dietary Index (CCDI) was used to evaluate the dietary quality. Multivariate logistic regression models were used to explore the factors that could influence the correctness of caregiver's perception. In the current study, 371 (62.1%) children with "high diet quality" and 35 (16.1%) children with "poor diet quality" were correctly perceived by their caregivers. Children who were correctly perceived as having "poor diet quality" consumed less fruits and more snacks and beverages than those who were not correctly perceived ($p < 0.05$). Obese children were more likely to be correctly identified as having "poor diet quality" (OR = 3.532, $p = 0.040$), and less likely to be perceived as having "high diet quality", even when they had a balanced diet (OR = 0.318, $p = 0.020$). Caregivers with a high level of education were more likely to correctly perceive children's diet quality (OR = 3.532, $p = 0.042$). Caregivers in this study were shown to lack the ability to correctly identify their children's diet quality, especially amongst children with a "poor diet quality". Obesity, significantly low consumption of fruits or high consumption of snacks can raise caregivers' awareness of "poor diet quality".

Keywords: child; diet quality; caregiver perception; CCDI

1. Introduction

Childhood obesity is a major health problem both in the developed and developing countries [1]. In 2015, amongst the 20 most populous countries, China ranked first in terms of the number of obese children [2]. The national prevalence estimates of obesity and overweight among Chinese children were increased to 10.4% for children younger than 6 years and 19% for children aged 6 to 17 years in 2019 [3], and this trend will only continue to increase with the continuous development of the economy.

Diet quality plays an important role in weight control [4,5]. Diets containing adequate fruits and vegetables and less energy-dense nutrient-poor (EDNP) foods can not only help people control their weight [6–8], but also reduce the risk of negative health outcomes and all-cause mortality [9–13]. Previous studies have revealed that more than 60% of Chinese school-age children failed to meet the recommendations issued by Chinese Dietary Guidelines regarding fruit and vegetable intake [14–16], which undoubtedly will have a detrimental impact on their development. Given that dietary habits developed in childhood can be carried into adulthood and impact long-term health outcomes [17,18], early detection of poor diet quality in children and taking timely action is of great importance.

In order to cultivate healthy eating habits at an early stage, researchers and governments have started to develop nutrition education activities targeting school-age children. However, children (especially young children) have little autonomy over their food choices, with almost all foods being provided by their caregivers. Thus, attitudes of caregivers are crucial to the effectiveness of these nutrition interventions [19,20]. If they are unable to recognize unhealthy habits and help their child to change these, the effect of health education will be minimal. Previous research found parents with correct perception of their child's overweight status were more likely to make changes to their children's lifestyles and participate in healthy lifestyle behaviors with their children [21,22]. However, it also found that those who failed to recognize their child's weight status were less motivated to address the problem [21,22]. Thus, we have reason to believe that caregivers with correct perception of their children's poor diet quality could also be more willing to participate in nutrition promotion activities with their children, which could greatly improve the effectiveness of health interventions.

However, limited information is available concerning the caregiver's perception of their child's diet quality, with previous studies focusing mainly on caregiver's perception of their child's weight status. It appears common for caregivers to underestimate their child's weight; a meta-analysis indicated that nearly half of parents underestimated their children's overweight/obese status and a significant minority underestimated their children's normal weight status [21]. Furthermore, Reyes et al. found that 50% of caregivers of children aged 2–18 years underestimated their children's weight status [23]. Boys' weight status was more likely to be underestimated compared to girls [24], potentially due to different perceptions of an ideal body shape for boys and girls. Girls were more likely to be correctly perceived as overweight, whilst overweight boys however were more likely to be regarded as strong rather than overweight.

Caregiver's correct perception of children's health issues is crucial to the maintenance of children's health. It has already been reported that caregiver's perception of children's weight status was sub optimal, potentially indicating poor health literacy among caregivers, therefore it could also be assumed that their perception of children's diet quality could require improvement. However, there is still a lack of research regarding the perception of caregivers of children's diet quality. Therefore, in the current study, we aimed to examine the correctness of caregiver's perception of their child's diet quality and to investigate the factors which may influence their judgment.

2. Materials and Methods

2.1. Study Population and Ethical Statement

Participants were recruited from two primary schools using cluster sampling in Hongshan district, Wuhan, China in April 2016. Considering that the children's diet needed to be self-reported and the cognitive ability of students in lower grades is limited, in the current study we selected students in grades 3rd to 6th and their caregivers as subjects. Initially, 1132 eligible child-caregiver pairs were recruited. Of these, 317 were excluded due to incomplete information or not providing consent. Therefore, this analysis was based on a final sample of 815 child–caregiver pairs. Figure 1 provides an overview of the recruitment process. This study was conducted according to the guidelines of the Declaration of Helsinki and all procedures involving human subjects were approved by the Ethics Committee of Wuhan University of Science and Technology (No. 201519). All parents gave written informed consent and children gave assent.

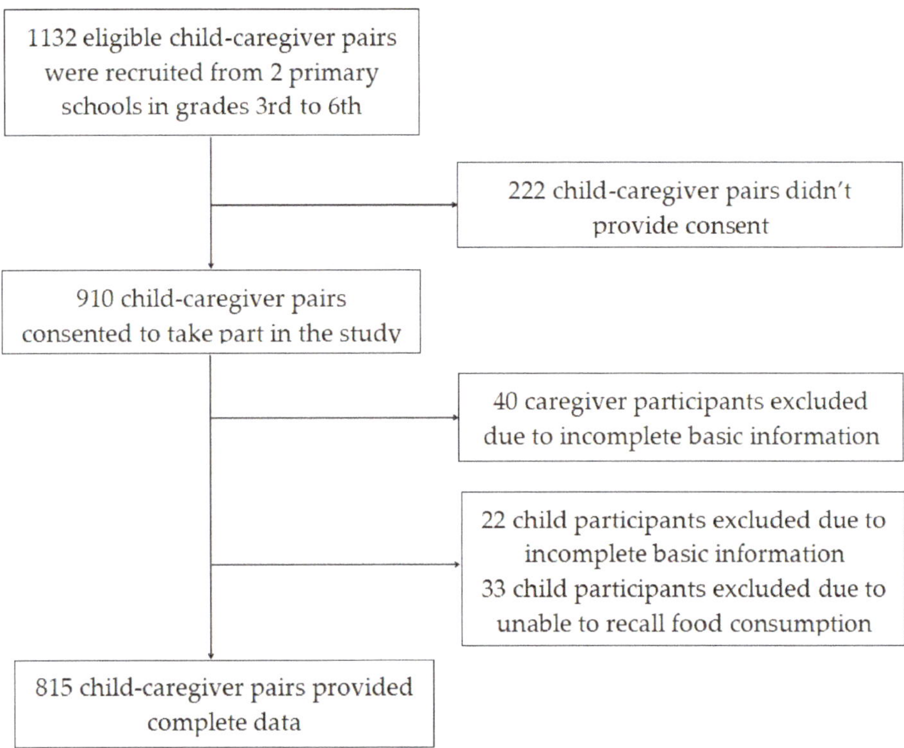

Figure 1. Flow diagram of subject recruitment.

2.2. Demographic Characteristics

Data relating to socio-demographic status and caregiver's perception were collected using customised, self-rated questionnaires. Information on age, gender, frequency and location of meals were collected for children, whilst information on age, gender, relationship to the children, annual household income and level of education was collected for caregivers. Members of the research team including faculty and postgraduates of the school of public health helped students fill out the questionnaire in the school setting. Caregivers' questionnaires were brought home by the students; the primary caregivers of the children were then asked to fill in the questionnaire, and after filling it in, the questionnaires were brought back to school by the students the following days.

2.3. Diet Survey and Evaluation

3-day 24-h recall was used to evaluate children's dietary intake. The children were asked to recall all food and beverages consumed in the past 24 h for three consecutive days, the investigators went into the classroom after lunch, and using food size reference models recorded all the food consumed by each student in the past 24 h (including drinks, snacks, inter-class meals, etc.). Data relating to eating behaviors were collected through interviews. Children's dietary reference values issued by Chinese Dietary Guidelines vary by age and sex [25], so food intakes were converted to food density (g/1000 kcal) for the purpose of comparison (except for snacks and beverages). The daily dietary intake of calories and nutrients were calculated using the China Food Composition 2004 and were presented as an average intake over the 3-day period.

The Chinese Children Dietary Index (CCDI) was used to evaluate the diet quality of the children involved in the study based on the Chinese dietary intake recommendations

(Chinese Dietary Guidelines and Chinese Dietary Reference Intakes) and health-promoting behaviors. It was developed by Guo Cheng et al., though it is not widely used at present, the validity of CCDI in evaluating the dietary status of Chinese school-age children has been verified in several studies [14,26,27].

The CCDI contains 16 items in four sections. The highest score for each item is 10 points, therefore the CCDI has a total score of 160 points, with higher scores representing better diet quality. The scoring scheme is based on the amounts and types of nutrients or foods that the children consumed, and whether they exhibited health-promoting behaviors. The first part evaluates the intake of 8 food groups (grains, vegetables, fruits, dairy and dairy products, soybean and its products, fish and shrimp, meat and eggs), water and sugar-sweetened beverages. The second part evaluates intakes of vitamin A, fatty acids and dietary fiber. The third part evaluates food diversity and the final part evaluates dietary behaviors, including breakfast, dinner and energy intake. Criteria relating to maximum and minimum score of food and nutrients were derived from recent age- and sex-specific dietary reference values [25,28], and the criteria for food diversity and diet behaviors were derived from the Chinese Dietary Guidelines. Details relating to these have been published previously [14].

Since sedentary behaviors or water intake were not assessed in this study, the CCDI was slightly modified (Table 1). The item "drinking water" was removed, and the score of "energy balance" was determined only by total energy intake, rather than total energy intake and sedentary behaviors. "Fatty acid" was changed to total fat intake, and the cut-offs were derived from the Chinese Dietary Guidelines. Following modification, scores ranged from 0 to 150. Scores above 60% were considered acceptable, therefore a score of over 90 points was considered "high diet quality" and a score below 90 points was considered "poor diet quality".

2.4. Caregiver Perception

Caregiver's perception of their child's diet was assessed using the question 'How would you describe your child's diet quality?' and they were given three choices: high diet quality, poor diet quality, and unknown (reason is needed). The accuracy of this perception was assessed by comparing the caregiver's perception with the child's actual diet quality. Caregivers who had a different perception of their child's diet quality compared to what their child's diet quality actually were deemed to have incorrectly perceived their child's diet quality.

2.5. Covariates

In this study, the primary caregiver was defined as the person who takes care of the child or prepares food for the child most often. The caregiver's level of education was defined as the highest degree that the primary caregiver had completed at the time of the survey. Family income refers to the average annual household income, including but not limited to wages, self-employed income and agricultural income.

Data relating to the children's weight and height were obtained from a physical examination made by Wuhan ChangeDong Hospital; height was measured with a precision of 0.1 cm and weight was measured with a precision of 0.1 kg. This information was then used to calculate Body Mass Index (BMI) and obesity was defined according to the BMI cutoffs points issued by the National Health Commission of the People's Republic of China [29].

Table 1. Adjusted scoring system for the components of the CCDI, a measure of overall diet quality for Chinese school-aged children.

CCDI Component	Range of Score (Points)	Criteria for Maximum Score	Criteria for Minimum Score
Food Groups [a]			
Grains [b]	0–10	140–160 g/1000 kcal	0 or >320 g/1000 kcal
Vegetables [c]	0–10	≥175 g/1000 kcal	0 g/1000 kcal
Fruits	0–10	≥110 g/1000 kcal	0 g/1000 kcal
Dairy and dairy products	0–10	≥110 g/1000 kcal	0 g/1000 kcal
Soybeans and its products	0–10	≥17 g/1000 kcal	0 g/1000 kcal
Meat	0–10	25–35 g/1000 kcal	0 or >70 g/1000 kcal
Fish and shrimp	0–10	≥30 g/1000 kcal	0 g/1000 kcal
Eggs	0–10	12.5–22.5 g/1000 kcal	0 or 45 g/1000 kcal
SSBs [d]	0–10	0 mL/day	≥1 serving/day
Nutrients			
Vitamin A [e]	0–10	≥100% RNI/day	0% RNI/day
Fat [f]	0–10	20–30% E/day	0% or >60% E/day
Dietary fiber	0–10	≥14 g/1000 kcal	0 g/1000 kcal
Diet variety	0–10	>1 serving of food from each of these groups (grains, vegetables, fruits, dairy/beans, and meat/fish/eggs)	<1 serving of food from each of these groups (grains, vegetables, fruits, dairy/beans, and meat/fish/eggs)
Behaviors			
Breakfast and dinner	0–10	Eating breakfast and having dinner with parents regularly	Skipping breakfast and not having dinner with parents regularly
Energy balance	0–10	0.9 EER ≤ EI ≤ 1.1 EER	EI = 0 or EI ≥ 2.2 EER
CCDI total score	0–150		

Abbreviations: CCDI, Chinese Children Dietary Index; SSBs, sugar-sweetened beverages; RNI, recommended nutrient intakes; E, energy; EER, estimated energy requirement; EI, energy intake. [a] To characterize diet quality, consumption of food groups were expressed on a per −1000-calorie basis in the CCDI. [b] Given that grains, meat, and eggs should be consumed moderately, consumption between the lowest and highest recommended amount per 1000 kcal according to the Chinese Dietary Guidelines (2007) was chosen as the standard for the maximum score. [c] Vegetables, fruits, dairy and dairy products, soybeans and its products, and fish and shrimp should be consumed sufficiently. The lowest recommended amount per 1000 kcal according to the Chinese Dietary Guidelines (2007) was chosen as the standard for the maximum score for these food groups. [d] SSBs were defined as beverages with added sugar, such as lemonades, fruit drinks, ice teas, soft drinks (soda pop), sport drinks, tea and coffee drinks, and sweetened milks. One serving is 250 mL. [e] RNI of vitamin A: 500 μgRAE/day (children aged 7 to 10 years), 630 μgRAE/day (girls aged 11 to 12 years), 670 μgRAE/day (boys aged 11 to 12 years). [f] Consumption of fat within the AMDR (Acceptable Macronutrient Distribution Range) was chosen as the standard for the maximum score.

2.6. Statistical Analysis

Descriptive statistics included frequency and percentages for categorical variables and median (P25, P75) for continuous variables that were not normally distributed. Rank sum test for continuous variables with a non-normal distribution and chi-square tests for categorical variables were used. Variables found to be statistically significant in univariate analysis and variables that are thought to be associated with diet quality (such as gender, family income, et al.) were included in multivariate logistic regression, to explore the factors that could influence the correctness of caregiver's perception. Statistical analyses for this study were performed using Stata (version 13.0; StataCorp, College Station, TX, USA). Differences were considered significant if $p < 0.05$.

3. Results

3.1. Participant Characteristics

Demographic characteristics of children based on their diet quality can be seen in Table 2. More than half of the children were migrants (57.9%). Most caregivers were not well educated and had only completed middle school (51.8%). Half of the families had an average annual income lower than 50,000 CNY (50.7%). 26.7% of the children were classed as having a poor-quality diet, with boys and overweight/obese children being more likely to have a poor-quality diet ($p < 0.05$) (Table 2).

Table 2. Characteristics of study participants [a].

Characteristic		Child Diet Status		χ^2	*p*-Value
		High Diet Quality (*n* = 597)	**Poor Diet Quality** (*n* = 218)		
Gender	Boys	243 (40.7%)	114 (52.3%)	8.714	0.003
	Girls	354 (59.3%)	104 (47.7%)		
Primary caregiver	Mother	400 (67.0%)	151 (69.3%)	1.664	0.645
	Father	102 (17.1%)	37 (17.0%)		
	Grandparents	63 (10.5%)	23 (10.5%)		
	Others	32 (5.4%)	7 (3.2%)		
Family income	<50,000¥	298 (49.9%)	115 (52.7%)	0.514	0.473
	≥50,000¥	299 (50.1%)	103 (47.3%)		
Caregiver's educational level [b]	Primary school	44 (7.4%)	21 (9.6%)	3.646	0.302
	Middle school	258 (43.2%)	99 (45.4%)		
	High school	231 (38.7%)	83 (38.1%)		
	College	64 (10.7%)	15 (6.9%)		
Weight status	Normal weight	493 (82.6%)	164 (75.2%)	6.496	0.039
	Overweight	73 (12.2%)	34 (15.6%)		
	Obesity	31 (5.2%)	20 (9.2%)		
Caregiver's perception	High diet quality	371 (62.1%)	131 (60.1%)	3.012	0.222
	Poor diet quality	69 (11.6%)	35 (16.1%)		
	Unknown	157 (26.3%)	52 (23.8%)		

[a] Data are presented as counts (percentages). [b] Caregiver's educational level: represented as the highest degree of the primary caregiver.

3.2. Caregiver Perception

Among the 597 children with "high diet quality", 371 (62.1%) were correctly perceived by their caregivers, whilst among the 218 children with "poor diet quality", just 35 (16.1%) were correctly perceived (Table 2). Whether in the "high diet quality" group or the "low diet quality" group, most parents believe that their children had a high diet quality ($p > 0.05$).

Of the 815 caregivers, 209 (25.6%) caregivers could not make a clear judgment on their child's diet. These caregivers did not exhibit any significant differences in relation to educational level, family income and their child's gender, weight status and diet scores when compared to caregivers with clear judgment ($p > 0.05$). Therefore, in further analysis only those child-caregiver pairs in which the caregiver had given a clear judgment were included ($n = 606$) (Table 3).

Table 3. The correctness of caregiver's perception of diet status of children in with differing diet quality.

Characteristic		High Diet Quality [a] (*n* = 440)			Poor Diet Quality [a] (*n* = 166)		
		Total	Correct [b]	*p*	Total	Correct [b]	*p*
Gender	Boys	170	142 (83.5%)	0.718	95	20 (21.0%)	0.991
	Girls	270	229 (84.8%)		71	15 (21.1%)	
Primary caregiver	Mother	303	250 (82.5%)	0.178	114	24 (21.0%)	0.889
	Father	67	57 (85.1%)		27	5 (18.5%)	
	Grandparents/Others	70	64 (91.4%)		25	6 (24.0%)	
Family income	<50,000¥	214	173 (80.8%)	0.051	83	16 (19.3%)	0.568
	≥50,000¥	226	198 (87.6%)		83	19 (22.9%)	
Caregiver's educational level [c]	Primary school	34	25 (73.5%)	0.096	18	4 (22.2%)	0.923
	Middle/High school	352	297 (84.4%)		136	29 (21.3%)	
	College	54	49 (90.7%)		12	2 (16.7%)	
Weight status	Normal weight	365	312 (85.5%)	0.069	121	22 (18.2%)	0.012
	Overweight	54	45 (83.3%)		29	5 (17.2%)	
	Obesity	21	14 (66.7%)		16	8 (50.0%)	

[a] High diet quality (or poor diet quality) represents the child's diet status which assessed by CCDI. [b] Correct means the caregiver's recognition is consistent with the child's diet status which assessed by CCDI. [c] Caregiver's educational level represented as the highest degree of the primary caregiver.

3.3. Children's Daily Food Intake

Table 4 shows that children with "poor diet quality" had lower intakes of vegetables, fruits, fish, eggs, beans and milk ($p < 0.01$), and higher intakes of grains, snacks and beverages ($p < 0.01$). Children who were correctly perceived by their caregivers as having "poor diet quality" consumed less fruits than those who were not correctly perceived, and they were more likely to consume snacks and beverages than the other groups ($p < 0.05$).

Table 4. Food consumed by each diet quality group according to diet status and caregiver's perception (M (P25, P75)).

Food Group	High Diet Quality [a]				Poor Diet Quality [a]			
	Correct [b] (*n* = 371)	Incorrect [b] (*n* = 69)	Z	*p*	Correct [b] (*n* = 35)	Incorrect [b] (*n* = 131)	Z	*p*
Grain (g/1000 kcal)	158.0 (145.0, 173.3)	157.5 (140.6, 172.4)	0.783	0.434	170.5 (160.4, 185.5)	167.5 (154.8, 178.3)	1.045	0.296
Vegetables (g/1000 kcal)	102.8 (85.0, 122.7)	103.1 (85.0, 125.4)	0.123	0.902	95.6 (77.4, 106.6)	93.6 (81.0, 107.3)	−0.107	0.915
Fruit (g/1000 kcal)	106.8 (68.3, 150.7)	117.1 (87.2, 145.6)	−0.902	0.367	20.2 (0.0, 61.4)	48.3 (0.0, 93.2)	−2.380	0.017
Meat (g/1000 kcal)	25.8 (18.5, 35.4)	26.8 (19.1, 33.0)	0.317	0.752	24.8 (15.2, 42.6)	24.3 (14.9, 36.1)	0.424	0.672
Fish (g/1000 kcal)	4.9 (0.0, 13.6)	6.3 (0.0, 16.7)	−1.312	0.190	0.0 (0.0, 8.7)	2.8 (0.0, 8.9)	−0.431	0.667
Egg (g/1000 kcal)	11.6 (2.0, 20.4)	13.6 (5.0, 20.2)	−0.260	0.795	0.0 (0.0, 9.9)	0.0 (0.0, 9.6)	0.992	0.321
Beans (g/1000 kcal)	4.4 (0.0, 10.3)	3.1 (0.0, 9.9)	0.630	0.529	0.0 (0.0, 6.2)	0.0 (0.0, 5.0)	0.353	0.724
Milk (g/1000 kcal)	41.2 (0.0, 97.1)	41.7 (0.0, 109.9)	−0.496	0.620	0.0 (0.0, 31.7)	0.0 (0.0, 35.0)	−0.220	0.826
Snacks (g/day)	18.5 (12.8, 24.0)	20.4 (16.7, 26.9)	−1.765	0.078	30.5 (19.0, 38.5)	19.7 (13.5, 28.2)	3.841	<0.001
Beverages (g/day)	34.4 (0.0, 59.6)	30.8 (0.0, 69.4)	0.312	0.755	71.8 (0.0, 160.7)	46.0 (0.0, 76.4)	2.052	0.040

[a] High diet quality/poor diet quality represents child diet status. [b] Correct/incorrect represents whether the caregiver's recognition is consistent with the child's diet status.

3.4. Influence Factors of Caregiver's Perception of Diet

Migration, child's gender, caregiver's association with the child, family income and breakfast habits did not affect the correctness of the caregiver's perception of the child having poor diet quality. Caregivers were more likely to identify "poor diet quality" among children who consumed less fruits (OR = 0.989, $p = 0.031$) or those eat more snacks (OR = 1.074, $p = 0.004$). Obese children were more likely to be correctly identified as having "poor diet quality" (OR = 3.532, $p = 0.040$), and these children were also more likely to be perceived as having "poor diet quality" even when they had a balanced diet (OR = 0.318, $p = 0.020$). Moreover, caregivers with a high level of education were more likely to correctly perceive children's "high diet quality" (OR = 3.532, $p = 0.042$) (Table 5).

Table 5. Regression model of the relationship between caregiver's perception and other socio-demographic predictors, and intake of some kinds of food [a].

High Diet Quality Group			
Characteristic		OR (95% CI)	*p*-Value
Family income	<50,000¥	Reference	
	≥50,000¥	1.555 (0.914, 2.645)	0.103
Caregiver's educational level [b]	Primary school	Reference	
	Middle/High school	1.932 (0.844, 4.427)	0.119
	College	3.532 (1.046, 11.925)	0.042
Weight status	Normal weight	Reference	
	Overweight	0.858 (0.392, 1.877)	0.701
	Obesity	0.318 (0.121, 0.836)	0.020

Poor Diet Quality Group			
Characteristic		OR (95% CI)	*p*-Value
Weight status	Normal weight	Reference	
	Overweight	1.109 (0.357, 3.443)	0.858
	Obesity	3.532 (1.056, 11.805)	0.040
Fruits intake		0.989 (0.980, 0.999)	0.031
Snacks intake		1.074 (1.023, 1.129)	0.004
Beverages intake		1.001 (0.995, 1.008)	0.640

[a] According to the dietary status of children, two logistic regression models were established respectively, the dependent variables were the correctness of the caregivers' judgment. [b] Caregiver's educational level: represented as the highest degree of the primary caregiver.

4. Discussion

The aim of the current study was to examine the correctness of caregiver's perception of their child's diet quality and to investigate the potential factors that may influence this. To our knowledge, this is one of the first studies to pair caregivers with their child to examine caregiver's perception of child diet quality, given that previous studies have focused on caregiver's perception of their child's weight status [21–24]. In the present study, our results showed that it is common for caregivers to be unable to identify the diet quality of their child. This was particularly noticeable in children with "poor diet quality", with less than a fifth of children with a poor-quality diet being correctly perceived by their caregivers. This is worrying to note; if caregivers are not aware of their child's poor diet quality, they might be less likely to change their dietary behaviors, and this could lead to an increased risk of obesity, malnutrition or chronic diseases in the future.

Children in the "poor diet quality" group had lower intakes of several kinds of foods apart from grains and meat, and their snacks and beverages intakes were significantly higher. However, it was only extremely low intakes of fruits or high intakes of snacks that caregivers seemed to be aware of, they did not take other food types (including vegetables, fish, beans, eggs, and milk) into consideration. According to our survey, children with "poor diet quality" had significantly lower intakes of fish, eggs, beans and milk, with nearly half of them having never consumed these foods over a 3-day period, which may suggest that these kinds of foods were not frequently served as part of school lunches. A previous study found that the supply of milk, beans, fish and eggs at school lunches in Shanghai, China, did not reach the recommended level [30]. Although Shanghai is a first-tier city in China with high standard of living, the schools investigated in this study were average public primary schools in Wuhan, which are less likely to provide such foods. These foods are therefore more likely to be prepared by caregivers at breakfast or dinner time. The lower intake of these foods may be due to the caregivers not understanding the benefits of them, or not regularly preparing these foods for their children. Therefore, they did not take these foods into consideration when judging their children's diet quality. Most of the children in this study were eating a certain quantity of vegetables, however in those with "poor diet quality", their consumption was significantly lower. A potential reason why caregivers could not make accurate judgments on this insufficient intake might because they were not aware of the number of vegetables that children should be eating. Therefore, they would not be able to know whether their children's consumption matched the recommended number while making the judgment.

Generally, highly educated caregivers have more access to correct health information [22,31] and are more likely to make accurate judgments about their children's diet. This was highlighted in this study; caregivers with a college degree or above were able to make more accurate judgments relating to children with "high diet quality", but not "poor diet quality". Caregivers play an integral role in shaping children's eating behavior, and their attitude towards diet can influence children's eating habits [32]. Compared to the caregivers with children in the "high diet quality" group, those with children in the "poor diet quality" group might be less concerned about diet quality and therefore were not as aware of their children's diet habits, which may have led to their children having a poor-quality diet. Therefore, even if they did have adequate nutrition knowledge, their lack of concern and understanding of children's diet could prevent them from making accurate judgments.

Obese children with both "high diet quality" or "low diet quality" were more likely to be perceived as having a "poor diet quality", even when their diet quality was in fact adequate. Caregivers were aware that improper diet can lead to obesity, however they may have ignored the influence of genetics, sports, mental health and other factors [33], which indicates that caregivers may not have comprehensive health knowledge.

Regardless of level of education and family income, the inability to correctly identify poor diet quality appears to be a common phenomenon. It has been suggested previously that the action of parents' judging their child's actual weight is not a cognitive task, but an emotional evaluation [34]. This may also be the case for diet judgment; caregivers tend to believe that their children have a "high quality diet" when they are unsure whether their children's diet is up to the standard or not. This tendency may also lead caregivers to be less motivated when participating in diet promotion courses [21,22]. Thus, it is necessary for them to be taught how to judge diet accurately.

Therefore, in a future nutrition intervention program, caregivers should firstly be educated about the importance of a healthy diet for children, which may improve their awareness of their child's diet. There is the potential to develop tailored nutrition education for caregivers, so that they can have adequate knowledge relating to nutrition, be aware of the standards of a healthy diet for children and learn cooking skills, so that they can spot any deficiencies in children's diet and correct them timely. Finally, caregivers should be informed that causes of obesity are multifactorial, such as dietary, exercise and genetics et al. They should understand that diet is not the only factor that can cause obesity, and similarly, obesity does not necessarily mean the child has an unhealthy diet.

Strengths and Limitations

There are several limitations in this study. Firstly, as the participants were recruited from two local schools, the sample is not nationally representative and therefore the results of this study cannot be generalized to other samples. Secondly, in order to improve participant compliance, the 3-day dietary recalls were performed during school hours, therefore the data may not represent children's diet on the weekends. At the same time, best practice for 24-h recalls is three unannounced days, not three consecutive days, so this may lead to bias caused by changes of food behaviors. Thirdly, since there are no widely accepted indices to measure diet quality in Chinese children, CCDI was used to assess children's diet quality. Although not widely used, the effectiveness of CCDI has been proved in another study in children of the same age [14]. Finally, although caregivers' perceptions of children's diet quality were obtained by asking only one question, which may not be comprehensive, similar one question-based methods have been used previously in research relating to caregivers' perception of children's weight status [23,24]. Despite these limitations, this study is one of the first to examine caregiver's perception of child diet quality and the factors that can influence this, and these results have highlighted the inadequate health awareness among parents or other caregivers and can provide direction for future health interventions.

5. Conclusions

Caregivers in this study were shown to lack the ability to correctly identify their children's diet quality, especially amongst children with a "poor diet quality". Obesity, significantly low consumption of fruits or high consumption of snacks can raise caregivers' awareness of "poor diet quality". However, if the child was not classed as obese or they did not show any special rejection/preference for fruits and snacks, their poor diet quality was less likely to be noticed. Caregivers' judgment was also influenced by children's consumption of certain types of food and their body type, rather than the child's overall diet. Such information may be valuable for the prevention of obesity and malnutrition among children through improving caregivers' awareness of child diet quality.

Author Contributions: Conceptualization, J.Z.; investigation, L.S., Y.R. and Y.L.; Formal analysis, L.S. and M.Y.; Project administration, B.X.; Writing—Original draft, L.S. and Y.R.; Writing—Review & Editing, J.Z., L.H. and X.Y. All authors have read and agreed to the published version of the manuscript.

Funding: This work was supported by National Social Science Foundation of China (NSSFC) (No. 20BSH066).

Institutional Review Board Statement: This study was conducted according to the guidelines of the Declaration of Helsinki, and approved by the Ethics Committee of Wuhan University of Science and Technology (No. 201519).

Informed Consent Statement: Informed consent was obtained from all subjects involved in the study.

Data Availability Statement: The database used in the study was available from the corresponding author upon reasonable request.

Acknowledgments: Authors are grateful to the teachers, caregivers and children participated in the study, and the investigators who took great passion to collect data.

Conflicts of Interest: The authors declare no conflict of interest. The funders had no role in the design of the study; in the collection, analyses, or interpretation of data; in the writing of the manuscript, or in the decision to publish the results.

References

1. Afşhin, A.; Forouzanfar, M.H.; Reitsma, M.B.; Sur, P.; Estep, K.; Lee, A.; Marczak, L.; Mokdad, A.H.; Moradi-Lakeh, M.; Naghavi, M.; et al. Health effects of overweight and obesity in 195 countries over 25 years. *N. Engl. J. Med.* **2017**, *377*, 13–27. [PubMed]
2. Zhang, J.; Wang, H.; Wang, Z.; Du, W.; Su, C.; Zhang, J.; Jiang, H.; Jia, X.; Huang, F.; Ouyang, Y.; et al. Prevalence and stabilizing trends in overweight and obesity among children and adolescents in China, 2011–2015. *BMC Public Health* **2018**, *18*, 571. [CrossRef]
3. The State Council Information Office of the People's Republic of China. Press Briefing for the Report on Chinese Residents' Chronic Diseases and Nutrition 2020 (In Chinese). Available online: http://www.gov.cn/xinwen/2020-12/24/content_5572983.htm (accessed on 2 June 2021).
4. Magriplis, E.; Michas, G.; Petridi, E.; Chrousos, G.P.; Roma, E.; Benetou, V.; Cholopoulos, N.; Micha, R.; Panagiotakos, D.; Zampelas, A. Dietary Sugar Intake and Its Association with Obesity in Children and Adolescents. *Children* **2021**, *8*, 676. [CrossRef]
5. Sui, Z.; Wong, W.K.; Louie, J.C.Y.; Rangan, A. Discretionary food and beverage consumption and its association with demographic characteristics, weight status, and fruit and vegetable intakes in Australian adults. *Public Health Nutr.* **2017**, *20*, 274–281. [CrossRef] [PubMed]
6. Bucher Della Torre, S.; Keller, A.; Laure Depeyre, J.; Kruseman, M. Sugar-sweetened beverages and obesity risk in children and adolescents: A systematic analysis on how methodological quality may influence conclusions. *J. Acad. Nutr. Diet.* **2016**, *116*, 638–659. [CrossRef] [PubMed]
7. Smethers, A.D.; Rolls, B.J. Dietary Management of Obesity: Cornerstones of Healthy Eating Patterns. *Med. Clin. N. Am.* **2018**, *102*, 107–124. [CrossRef] [PubMed]
8. Hwang, S.B.; Park, S.; Jin, G.R.; Jung, J.; Park, H.; Lee, S.; Shin, S.; Lee, B.-H. Trends in Beverage Consumption and Related Demographic Factors and Obesity among Korean Children and Adolescents. *Nutrients* **2020**, *12*, 2651. [CrossRef]
9. Shin, J.Y.; Kim, J.Y.; Kang, H.T.; Han, K.H.; Shim, J.Y. Effect of fruits and vegetables on metabolic syndrome: A systematic review and meta-analysis of randomized controlled trials. *Int. J. Food Sci. Nutr.* **2015**, *66*, 416–425. [CrossRef]
10. Aune, D.; Giovannucci, E.; Boffetta, P.; Fadnes, L.T.; Keum, N.; Norat, T.; Greenwood, D.C.; Riboli, E.; Vatten, L.J.; Tonstad, S. Fruit and vegetable intake and the risk of cardiovascular disease, total cancer and all-cause mortality-a systematic review and dose-response meta-analysis of prospective studies. *Int. J. Epidemiol.* **2017**, *46*, 1029–1056. [CrossRef]
11. Miller, V.; Mente, A.; Dehghan, M.; Rangarajan, S.; Zhang, X.; Swaminathan, S.; Dagenais, G.; Gupta, R.; Mohan, V.; Lear, S.; et al. Prospective Urban Rural Epidemiology (PURE) study investigators. Fruit, vegetable, and legume intake, and cardiovascular disease and deaths in 18 countries (PURE): A prospective cohort study. *Lancet* **2017**, *390*, 2037–2049. [CrossRef]
12. Park, H.A. Fruit Intake to Prevent and Control Hypertension and Diabetes. *Korean J. Fam. Med.* **2021**, *42*, 9–16. [CrossRef]
13. Zhan, J.; Liu, Y.-J.; Cai, L.-B.; Xu, F.-R.; Xie, T.; He, Q.-Q. Fruit and vegetable consumption and risk of cardiovascular disease: A meta-analysis of prospective cohort studies. *Crit. Rev. Food Sci. Nutr.* **2017**, *57*, 1650–1663. [CrossRef]
14. Cheng, G.; Duan, R.; Kranz, S.; Libuda, L.; Zhang, L. Development of a dietary index to assess overall diet quality for Chinese school-aged children: The Chinese children dietary index. *J. Acad. Nutr. Diet.* **2016**, *116*, 608–617. [CrossRef]
15. Zhang, J.; Wang, D.; Eldridge, A.L.; Huang, F.; Ouyang, Y.; Wang, H.; Zhang, B. Urban-Rural Disparities in Energy Intake and Contribution of Fat and Animal Source Foods in Chinese Children Aged 4–17 Years. *Nutrients* **2017**, *9*, 526. [CrossRef]
16. Liu, H.; Cai, L.; Li, Y.; Rong, S.; Li, W.; Zeng, J. Nutrition quality of school lunch for children aged 9-12 years in Wuhan City and its relationship with physical development. *Wei Sheng Yan Jiu* **2019**, *48*, 403–412.
17. Operamolla, P.; Rogai, S.; Scipione, F.; D'Andrea, M.; Spadea, A.; Poscia, A. Unhealthy dietary habits among toddlers in Rome (Italy): A cross sectional study. *Ig. Sanita Pubbl.* **2018**, *74*, 349–357. [PubMed]
18. Mikkila, V.; Rasanen, L.; Raitakari, O.T.; Pietinen, P.; Viikari, J. Consistent dietary patterns identified from childhood to adulthood: The cardiovascular risk in Young Finns Study. *Br. J. Nutr.* **2005**, *93*, 923–931. [PubMed]
19. DeJesus, J.M.; Gelman, S.A.; Viechnicki, G.B.; Appugliese, D.P.; Miller, A.L.; Rosenblum, K.L.; Lumeng, J.C. An investigation of maternal food intake and maternal food talk as predictors of child food intake. *Appetite* **2018**, *127*, 356–363. [CrossRef] [PubMed]
20. Sharp, G.; Pettigrew, S.; Wright, S.; Pratt, I.S.; Blane, S.; Biagioni, N. Potential in-class strategies to increase children's vegetable consumption. *Public Health Nutr.* **2017**, *20*, 1491–1499. [CrossRef] [PubMed]
21. Lundahl, A.; Kidwell, K.M.; Nelson, T.D. Parental underestimates of child weight: A meta-analysis. *Pediatrics* **2014**, *133*, e689–e703. [CrossRef]
22. Tompkins, C.L.; Seablom, M.; Brock, D.W. Parental perception of child's body weight: A systematic review. *J. Child Fam. Stud.* **2014**, *24*, 1384–1391. [CrossRef]

23. Reyes, I.; Higgins, M. Parental perception of child's body mass index and health within primary care. *J. Am. Assoc. Nurse Pract.* **2017**, *29*, 375–383. [CrossRef] [PubMed]
24. Alshahrani, A.; Shuweihdi, F.; Swift, J.; Avery, A. Underestimation of overweight weight status in children and adolescents aged 0–19 years: A systematic review and meta-analysis. *Obes. Sci. Pract.* **2021**, *7*, 760–796. [CrossRef] [PubMed]
25. Ge, K. The transition of Chinese dietary guidelines and food guide pagoda. *Asia Pac. J. Clin. Nutr.* **2011**, *20*, 439–446.
26. Duan, R.; Qiao, T.; Chen, Y.; Chen, M.; Xue, H.; Zhou, X.; Yang, M.; Liu, Y.; Zhao, L.; Libuda, L.; et al. The overall diet quality in childhood is prospectively associated with the timing of puberty. *Eur. J. Nutr.* **2021**, *60*, 2423–2434. [CrossRef]
27. Qiao, T.; Duan, R.; Cheng, G. Revision of the Chinese children dietary index. *Acta Nutr. Sin.* **2019**, *41*, 105–109.
28. Chinese Nutrition Society. *Dietary Reference Intakes for Chinese, Version 2013*; Science Press: Beijing, China, 2014.
29. National Health Commission of the People's Republic of China. *Screening for Overweight and Obesity among School-Age Children and Adolescents*; Standards Press: Beijing, China, 2018.
30. Huang, Z.; Gao, R.; Zhang, Y. Investigation on school lunch and nutritional status in primary schools and middle. *J. Shanghai Jiaotong Univ. (Med. Sci.)* **2017**, *37*, 106–109.
31. Wuerlajiafu, B.; Tuerxunjiang, M.; Shen, J. The Status Quo of Nutrition Knowledge, Attitude and Behavior of Residents in One District of Hami City in Xinjiang. *Food Nutr. China* **2019**, *25*, 75–78.
32. Faith, M.S.; Van Horn, L.; Appel, L.J.; Burke, L.E.; Carson, J.A.S.; Franch, H.A.; Jakicic, J.M.; Kral, T.V.E.; Odoms-Young, A.; Wansink, B.; et al. Evaluating parents and adult caregivers as "agents of change" for treating obese children: Evidence for parent behavior change strategies and research gaps: A scientific statement from the American Heart Association. *Circulation* **2012**, *125*, 1186–1207. [CrossRef]
33. Milica, M.; Goran, B.; Evans, G.W.; Lausevic, D.; Asanin, B.; Samardzic, M.; Terzic, N.; Pantovic, S.; Jaksic, M.; Boljevic, J. Prevalence of and contributing factors for overweight and obesity among Montenegrin school children. *Eur. J. Public Health* **2015**, *5*, 833.
34. Noor Shafina, M.N.; Abdul Rasyid, A.; Anis Siham, Z.A.; Nor Izwa, M.K.; Jamaluddi, M. Parental perception of children's weight status and sociodemographic factors associated with childhood obesity. *Med. J. Malays.* **2020**, *75*, 221–225.

Article

Dietary Fat and Polyunsaturated Fatty Acid Intakes during Childhood Are Prospectively Associated with Puberty Timing Independent of Dietary Protein

Yujie Xu [1], Jingyuan Xiong [2], Wanke Gao [1], Xiaoyu Wang [3], Shufang Shan [3], Li Zhao [4] and Guo Cheng [3,*]

[1] Department of Nutrition and Food Safety, West China School of Public Health and West China Fourth Hospital, Sichuan University, Chengdu 610041, China; jayee1217@126.com (Y.X.); gaowanke_gwk@126.com (W.G.)

[2] Healthy Food Evaluation Research Center, West China School of Public Health and West China Fourth Hospital, Sichuan University, Chengdu 610041, China; jzx0004@tigermail.auburn.edu

[3] Laboratory of Molecular Translational Medicine, Center for Translational Medicine, Key Laboratory of Birth Defects and Related Diseases of Women and Children, Ministry of Education, Department of Pediatrics, West China Second University Hospital, Sichuan University, Chengdu 610041, China; wangxiaoyu12168@163.com (X.W.); shan103163@163.com (S.S.)

[4] Department of Health Policy and Management, West China School of Public Health and West China Fourth Hospital, Sichuan University, Chengdu 610041, China; zhaoli@scu.edu.cn

* Correspondence: gcheng@scu.edu.cn; Tel.: +86-28-85502220

Citation: Xu, Y.; Xiong, J.; Gao, W.; Wang, X.; Shan, S.; Zhao, L.; Cheng, G. Dietary Fat and Polyunsaturated Fatty Acid Intakes during Childhood Are Prospectively Associated with Puberty Timing Independent of Dietary Protein. *Nutrients* **2022**, 14, 275. https://doi.org/10.3390/nu14020275

Academic Editor: Zhiyong Zou

Received: 25 October 2021
Accepted: 7 January 2022
Published: 10 January 2022

Publisher's Note: MDPI stays neutral with regard to jurisdictional claims in published maps and institutional affiliations.

Copyright: © 2022 by the authors. Licensee MDPI, Basel, Switzerland. This article is an open access article distributed under the terms and conditions of the Creative Commons Attribution (CC BY) license (https://creativecommons.org/licenses/by/4.0/).

Abstract: Dietary fat and fat quality have been inconsistently associated with puberty timing. The aim of this study was to investigate the prospective associations of dietary fat, saturated fatty acid (SFA), polyunsaturated fatty acid (PUFA), and monounsaturated fatty acid (MUFA) with puberty timing. Using longitudinal data from China Health and Nutrition Survey (CHNS) and Southwest China Childhood Nutrition and Growth (SCCNG) Study, we analyzed dietary data, anthropometric measurements, and potential confounders. Dietary intakes were assessed by 3-day 24-h recalls. Age at Tanner stage 2 for breast/genital development (B2/G2) and age at menarche/voice break (M/VB) were used as puberty development markers. Cox proportional hazard regression models were used to estimate the relevance of dietary intake of total fat, SFA, PUFA, and MUFA on puberty timing. Among 3425 girls and 2495 boys, children with higher intakes of total fat and PUFA were more likely to reach their B2/G2 or M/VB at an earlier age. Associations were not attenuated on additional adjustment for childhood dietary protein intake. However, higher intakes of SFA or MUFA were not independently associated with puberty development. A higher intake of dietary fat and PUFA in prepuberty was associated with earlier puberty timing, which was independent of dietary protein intake.

Keywords: dietary fat; polyunsaturated fatty acid; monounsaturated fatty acid; puberty timing

1. Introduction

Puberty represents the transition stage from childhood to adulthood [1]. There has been a global secular trend towards earlier puberty timing in the past decades [2–4]. Such a trend is concerning because individuals with an earlier puberty onset are at a higher risk of diabetes, cardiovascular diseases, hormone-related cancers, and all cause-mortality in later life [5,6]. Identifying modifiable factors of puberty timing is thus critical to public health ramifications, and nutrition is the most relevant to determine the timing of puberty among these modifiable influences [7].

Observational studies have demonstrated that dietary intakes of food groups or nutrients during childhood are associated with puberty timing [8–10]. Although dietary fat has been reported to alter sex hormones in humans [11,12], prospective associations regarding dietary fat and the onset of puberty are limited and conflicting. A prospective study has found an inverse association between childhood total fat intake and age at puberty onset [13], whereas other studies have reported no association [14,15], or positive association [16]. From a public health standpoint, it is more meaningful to investigate the possible effect of fat quality, i.e., saturated fatty acid (SFA), polyunsaturated fatty acid (PUFA), and monounsaturated fatty acid (MUFA), other than total fat [7], but the relationship between different types of fatty acids and puberty timing varies: some studies found that higher consumption of PUFA was associated with early puberty onset [8,17], while others reported a lack of an association [16,18]; no association between SFA intake with puberty timing was observed [8,16,18]; inverse [15], positive [17] or null association [16] were reported between MUFA intake and age at puberty onset. Notably, most of these studies were limited mainly to girls and examined solely a single self-report puberty marker [8,13–18], usually age at menarche, which highlights the need to consider other puberty timing traits reflecting different sex hormone pathways [19]. The measurement of different pubertal markers covering the range from earlier (Tanner stage 2 for breast development in girls or genital development in boys (B2/G2)) to later stages (menarche in girls or voice break in boys (M/VB)) of pubertal development might provide more comprehensive pubertal data. Furthermore, those previous studies were conducted in Caucasian from Western countries [8,13–18], including the British [8], Germany [13], America [14,17], Canada [16], and the Netherlands [18]; little is known about the impact of dietary fat and specific types of fatty acids on puberty timing in Chinese children. Given the secular trend of earlier puberty has been observed in both Chinese boys and girls [20,21], investigation on the determinants of puberty timing in this population is warranted.

Currently, the obesity epidemic has received large attention, and body composition in childhood has turned out to be associated with puberty timing [22]. In recent years, China has witnessed a secular increase in childhood obesity [23]. Owing to the two trends that coincide in China, the impact of dietary fat on puberty timing beyond body composition needs to be investigated. In addition, fat-rich foods, such as red meat, beans, and peanuts, are also the predominant nutritional source of dietary protein, and whose effect on the timing of sexual maturation has been largely confirmed [10]. However, few studies handled appropriately the independent association of dietary fat on the timing of puberty. Hence, it is important to examine whether a potential effect of dietary fat or fatty acids on puberty onset is mediated by childhood dietary protein intake.

Using longitudinal data from the China Health and Nutrition Survey (CHNS) and the Southwest China Childhood Nutrition and Growth (SCCNG) Study, we thus aimed to investigate the prospective associations between dietary fat and fatty acids intake and earlier puberty markers (B2/G2) or later puberty markers (M/VB) in Chinese boys and girls. Moreover, we also examined whether these associations were independent of pre-pubertal dietary protein intake.

2. Materials and Methods

2.1. Study Sample

We used nationally representative data from the CHNS (1997–2015) and the SCCNG studies.

The CHNS is an ongoing, large-scale, and longitudinal study conducted in China. Fifteen provinces were selected, and a multi-stage random cluster sampling method was performed in the ten waves between 1989 to 2015. A detailed protocol of the cohort has been published elsewhere [24]. Since data on B2 in girls or G2 and VB in boys were not available in the CHNS, we only included girls with menarche information in the present study. Based on girls with available data of the occurrence of menarche, our analysis considered 2405 girls aged 6–13 years that were recruited between 1997 to 2015. Of these participants, 1433 girls with complete dietary assessment were included for baseline. Then, 173 girls with incomplete information on household income, anthropometry, or other potential confounders, and 19 girls with implausible energy intakes (<600 kcal/day and >4000 kcal/day) [25] were excluded. Finally, 1240 CHNS girls were eligible in the present analysis.

The SCCNG study was started in March 2013 and covered three provinces (Sichuan province, Guizhou province, and Chongqing province) in southwest China, aiming to investigate the development and nutritional status of Chinese children. The study procedure was described in detail elsewhere [26]. The yearly recruitment from 2013 onward enrolled children aged 6–8 years who were cooperative and voluntary. At the first examination, we collected information on sociodemographic issues, dietary intake and eating behaviors, physical activity and sedentary behaviors, anthropometry, and pubertal development. From then on, the participants were followed up for assessment on nutritional status and growth at regular intervals until the age of 15. Detailed information on the assessments of anthropometry and puberty status were obtained every year, while data of dietary intake and physical activity were collected biennially. The study was approved by the Ethics Committee of the Sichuan University, and all the participants gave their written confirmed consent before enrollment. Between January 2013 and December 2019, 5439 children aged 6–8 years had baseline dietary information and completed at least 2 follow-up assessments. Since we were interested in the prospective relevance of dietary fat intake on puberty timing, 389 children who had already reached B2/G2 at baseline were excluded. Then, 268 participants with implausible energy intakes (<600 kcal/day and >4000 kcal/day) [25] and 102 participants with incomplete information on potential confounders were further excluded. A final sample of 4680 children (2185 girls and 2495 boys) was included in this analysis.

Hence, we included 5920 children (1240 girls from CHNS, 4680 children from SCCNG) in the present study; 1752 girls reached B2, 1897 girls (732 girls from CHNS, 1165 girls from SCCNG) experienced menarche, and 1239 boys reached G2, 831 boys experienced voice break. The flow chart of the sample selection was shown in Figure 1.

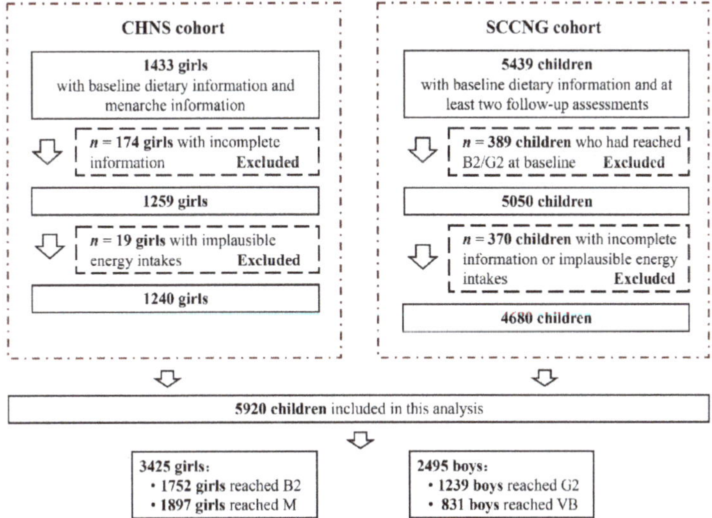

Figure 1. Flow chart for the study sample in our study. CHNS, China Health and Nutrition Survey; SCCNG, Southwest China Childhood Nutrition and Growth; B2, Tanner stage 2 for breast development; G2, Tanner stage for genital development; M, menarche; VB, voice break.

2.2. Nutrition Assessment

Nutritional data of children in CHNS and SCCNG were collected by trained investigators via 3-day 24-h dietary recalls. In CHNS, when children were 12 years or older, they were asked to recall their consumption of all foods and beverages. For children younger than 12 years, dietary intake data from school were provided by themselves while the information on food consumption at home was provided by their parents. In SCCNG, children aged 9 years or older were asked to recall their consumption of all foods and beverages. For children younger than 9 years, parents provided the information on food consumption at home, while children provided the dietary intake information from school themselves.

Details on recipes and brands of all food items reported were inquired. Food models, standard tableware including bowls, plates, and glasses, and picture aids were provided to enhance the accuracy of portion size estimates [27]. In the SCCNG study, a designed photo book that contains photos of snacks and beverages and pictures of the commonly used commercial packaging to improve the accuracy of diet recall was also given to children.

Nutrition data of children that were collected via 3-day 24-h dietary recalls recorded their consumption of foods and beverages, including categories and quantities. Then we transformed this original information to energy and specific nutrient intake by calculating the nutrient content in every food item based on the Chinese food composition (FCT). In CHNS, FCT 1991 was used in the 1997 and 2000 dietary survey; FCT 2002, FCT 2004, and 2009 (combined) were used in the 2004, 2006, 2009, and 2011 dietary survey, respectively. While in SCCNG, dietary intake data were converted into energy and nutrient intake data using the continuously updated in-house nutrient database based on NCCW software (version 11.0, 2014), which reflects the FCT.

2.3. Assessment of Puberty Onset

In the CHNS study, girls aged 8 years or older were asked whether their menarche had already occurred and the detail on the month and year of the first menstrual period were recorded in each survey. If there were different reported menarche ages for one girl, we only included the first reported menarche age in the panel data for analysis to reduce potential recall bias. In the SCCNG study, pubertal maturation for breast (girls) and pubic hair (girls and boys) stages were assessed at each examination by investigators according to the standardized criteria of Tanner stages. Testicular volume was assessed by comparative palpation with the Prader orchidometer. If the testicular volumes of the two testes were not equal, the volume of the larger one was recorded. Testicular volume less than 1 mL was recorded as 1 mL. In addition, during the annual physical examination, children were asked whether menarche (girls) or voice break (boys) had already occurred, and the respective month and year were recorded.

For this analysis, age at B2 in girls and the initiation of G2 in boys as well as age at menarche in girls and age at voice break in boys were considered.

2.4. Anthropometry

In CHNS and SCCNG study, anthropometric measurements of the participants were performed at each visit by trained research assistants according to standard procedures, with the subjects dressed lightly and barefoot. Height and weight were measured to the nearest 0.1 cm and 0.1 kg, respectively. Body mass index (BMI) was calculated as weight divided by the square of height (kg/m^2). All anthropometric measurements were performed twice, and the averages were calculated for each child.

For this analysis, sex- and age-independent BMI SDS and age-specific BMI Z-score were calculated for each children using the equation by Cole et al. [28] based on a Chinese reference population [29]. Overweight was defined according to the International Obesity Task Force (IOTF) BMI cut-offs for children, which correspond to an adult BMI of 25 kg/m^2 [28].

2.5. Covariates

In both CHNS and SCNNG study, detailed information about pregnancy and infancy (i.e., children's birth weight, exclusive breastfeeding duration, the timing of complementary feeding), and family characteristics (i.e., place of residence, household income, family size, smoking in the household, parental age, parental occupation, and parental education levels) were collected using structured questionnaires.

2.6. Statistical Analysis

SAS® procedures (version 9.4, SAS Inc., Cary, NC, USA) were used for all data analyses. All analyses were performed with a significance level at $p < 0.05$.

Dietary fat intakes were expressed as sex- and age-specific residuals from the regression of dietary fat intakes on energy intake. To examine the potential associations of dietary fat intakes with puberty timing, their distribution was grouped into tertiles (T1–T3), with the lowest tertile serving as the reference group. T1 indicates the lowest tertile who have a lowest intake of dietary fat, while T3 indicates the highest tertile with those having the highest dietary fat consumption.

Kolmogorov-Smirnov and Shapiro-Wilk tests were conducted to test the normality of the data. The continuous variables were normally distributed and presented as means (SD). Differences in anthropometric data, socio-demographic data, and nutritional intake between genders were tested using ANOVA test for normally distributed continuous variables, Kruskal-Wallis test for not normally distributed continuous variables, Chi-square test was used for categorical variables, followed by Student-Newman-Keuls tests or Dunn's post hoc tests. Since different pubertal markers were used in boys and girls, statistical models, as well as descriptive tables, were stratified by sex.

To investigate the prospective relevance of dietary fat intakes at baseline with age at B2/G2 or M/VB, Cox proportional hazard regression models were used. Censoring occurred at the age of reaching B2/G2 and M/VB or age at last follow-up if the puberty events had not been reported. In the basic models, the tertiles of dietary fat intakes at baseline were the principle independent fixed effects. Cox regression models were adjusted for birth year, age at baseline, body composition (Z-scores of BMI, overweight (Y/N)), family income, parental educational level, mother's age at menarche (only from SCCNG population), and total energy intake at baseline and dietary protein intake (residuals) at baseline. Each potential confounder was initially considered separately and included if it was associated with both the dietary index and indicators of puberty timing and substantially altered the estimate by more than 10%. Thus, birth year, family income level, energy intake at baseline, and mother's age at menarche (for B2, VB, and G2 model) were adjusted for in model 2. In the final model, Z-scores of BMI at baseline and dietary protein intake (residuals) at baseline were considered. Hazard ratios (HRs) and 95% confidence intervals (CIs) were estimated by comparing the second and third tertiles to the first tertile in these models.

3. Results

This study analyzed data on 5920 children (3425 girls and 2495 boys). There were no significant differences in age, or the BMI SDS at baseline among these 5920 children.

Table 1 summarized anthropometric, parental, and nutritional characteristics, and puberty timing in girls and boys. The mean age at baseline was 7.0 (0.8) years for girls from SCCNG and CHNS, and 7.1 (0.8) years for boys from SCCNG. Most puberty traits occurred earlier in girls than boys. Among girls from SCCNG and CHNS, 1748 (51.0%) had reached B2 at a mean (SD) age of 9.2 (1.4) years, and 1897 (55.4%) had experienced menarche at age of 12.6 (0.7) years. Among boys from SCCNG, 1233 (49.4%) had reached G2 aged 11.2 (1.1) years, and 829 (33.2%) had experienced voice break at a mean age of 13.7 (1.0) years. Compared to boys from SCCNG, girls from SCCNG and CHNS had a lower prevalence of overweight (10.1% of the girls and 11.9% of the boys) and were more likely to have a highly educated father and mother and grow up in a family with high income. As for childhood dietary energy and macronutrients intakes, boys from SCCNG consumed more total energy, dietary fat, and more dietary SFA, MUFA, and PUFA compared with girls from SCCNG and CHNS.

The associations of childhood dietary fat intakes at baseline with early and late markers of puberty development were presented in Table 2. According to the Cox proportional hazard regression model adjusted for birth year, family income level, energy intake, BMI Z-score, and dietary protein intake at baseline, girls with a higher intake of dietary fat were more likely to reach their B2 (adjusted HR = 1.13, 95% CI, 1.07 to 1.21, p for trend = 0.03) or experience their menarche (adjusted HR = 1.17, 95% CI, 1.11 to 1.23, *p* for trend = 0.01) earlier than those with a lower intake of dietary fat. Similarly, boys who consumed more dietary fat experienced their G2 (adjusted HR = 1.09, 95% CI, 1.03 to 1.15, *p* for trend = 0.03) and voice break (adjusted HR = 1.12, 95% CI, 1.07 to 1.16, *p* for trend = 0.03) at an earlier age compared with boys with a low dietary fat intake. In both genders, these associations were independent of dietary protein intake in childhood.

Table 1. Characteristics [1] of participants in our study.

	Girls from SCCNG (*n* = 2185)	Girls from SCCNG and CHNS (*n* = 3425)	Boys from SCCNG (*n* = 2495)
Age at baseline (years)	7.1 (0.7)	7.0 (0.8)	7.1 (0.8)
Age at Tanner stage B2 [2] (years, *n* = 1748)	9.1 (1.3)	9.2 (1.4)	−
Age at Tanner stage G2 [2] (years, *n* = 1233)	−	−	11.2 (1.1)
Age at menarche (years, *n* = 1894)	12.9 (0.8)	12.6 (0.7)	−
Age at voice break (years, *n* = 829)	−	−	13.7 (1.0)
BMI SDS at baseline (kg/m^2)	0.1 (0.8)	−0.1 (0.7)	0.2 (0.8) [#]
Overweight [3] (*n* (%))	251 (11.5)	345 (10.1)	299 (11.9) [#]
High physical activity (*n* (%))	572 (26.2)	791 (23.1)	671 (26.9)
Parental data at baseline			
High family income [4] (*n* (%))	478 (21.9)	688 (20.1)	578 (23.2) [#]
High paternal educational level [5] (*n* (%))	463 (21.2)	671 (19.6)	646 (25.9) [#*]
High maternal educational level [5] (*n* (%))	412 (18.9)	602 (17.6)	489 (19.6) [#]
Mother's age at menarche (years)	12.3 (1.1)	−	12.3 (1.0)
Nutritional data [6]			
Total energy intake (kcal/day)	1762 (237)	1729 (262)	1928 (267) [#*]
Fat (g/day)	51.2 (16.9)	50.3 (14.6)	57.6 (15.8) [#]
Fat (% of energy)	26.9 (7.5)	26.2 (7.6)	26.9 (7.6)
Saturated fatty acids (g/day)	14.6 (3.0)	13.2 (5.6)	16.9 (5.6) [#]
Saturated fatty acids (% of energy)	7.5 (1.8)	6.9 (1.6)	7.9 (1.6) [#]
Monounsaturated fatty acids (g/day)	18.7 (4.2)	17.4 (4.6)	19.4 (6.2) [#]
Monounsaturated fatty acids (% of energy)	9.6 (2.7)	9.1 (2.8)	9.1 (4.2)
Polyunsaturated fatty acids (g/day)	14.1 (4.1)	12.8 (2.7)	17.7 (5.2) [#*]
Polyunsaturated fatty acids (% of energy)	7.2 (2.1)	6.7 (1.7)	8.3 (2.7) [#*]
Carbohydrate (% of energy)	59.3 (7.2)	60.5 (8.4)	58.9 (8.0)
Protein (% of energy)	13.8 (2.4)	13.3 (2.1)	14.2 (2.3)

[1] Values are means (SD) or frequency; test for difference between the groups was performed, using ANOVA test for normal distributed continuous variables, Kruskal-Wallis test for not normally distributed continuous variables, followed by Student-Newman-Keuls tests or Dunn's post hoc tests, and Chi-square test for categorical variables. *p* < 0.05 between girls from SCCNG and girls from SCCNG and CHNS, [#] *p* < 0.05 between girls from SCCNG and CHNS and boys from SCCNG, [*] *p* < 0.05 between boys from SCCNG and girls from SCCNG; [2] Tanner stage 2 for breast development; [3] definition according to the International Obesity Task Force (IOTF) [28]; [4] average annual income of family income in each survey year was inflated to values in 2015 by adjusting for consumer price index and then divided into five groups (yuan): low (≤5000), middle (5000–10,000), and high (>10,000) [30]; [5] school education at least 12 years; [6] mean values of dietary data at baseline using 3-day 24 h recall.

Furthermore, we examined the effect of the three dietary fatty acids, i.e., SFA, MUFA, and PUFA, on the puberty timing in girls and boys. Table 3 revealed significant associations of childhood PUFA intake with early and late puberty markers in both genders: girls with a higher PUFA intake had approximately 11% higher HR to reach their B2 (adjusted HR = 1.11, 95% CI, 1.05 to 1.17, *p* for trend = 0.02) or 13% higher HR to experience their menarche (adjusted HR = 1.13, 95% CI, 1.08 to 1.20, *p* for trend = 0.03) than girls with a lower PUFA intake. Similarly, compared to boys with a lower PUFA intake, boys who consumed more PUFA had approximately 8% higher HR to reach G2 (adjusted HR = 1.08, 95% CI, 1.03 to 1.13, *p* for trend = 0.03) or 10% higher HR to experience voice break (adjusted HR = 1.10, 95% CI, 1.06 to 1.15, *p* for trend = 0.02). However, no association of SFA intake with age at B2/G2 or M/VB was found in any model for both genders (Table S1). Moreover, the associations between MUFA intakes in childhood and pubertal development in our participants were presented in Table S2: higher MUFA intake in girls was associated with earlier B2 (adjusted HR = 1.10, 95% CI, 1.04 to 1.18, *p* for trend = 0.04) and earlier menarche (adjusted HR = 1.11, 95% CI, 1.06 to 1.17, *p* for trend = 0.05) in a model adjusted for parental and childhood characteristics. Nevertheless, further adjustment for dietary protein intakes eliminated these associations, indicating that the effect of MUFA, unlike PUFA, on puberty timing was not independent of dietary protein intake. In contrast, no association of MUFA with puberty onset was seen in boys in any model.

Table 2. Association [1] of dietary fat intake in childhood with puberty timing.

	Dietary Fat Intake at Baseline			
Girls	**T1** [2]	**T2** [2]	**T3** [2]	p_{trend} [3]
Age at Tanner stage B2 (*n* = 2185)				
Unadjusted model:	1	1.12 (1.06, 1.19)	1.18 (1.13, 1.24)	0.04
Model 2 [4]:	1	1.13 (1.05, 1.18)	1.17 (1.11, 1.22)	0.03
Final model [5]:	1	1.11 (1.03, 1.19)	1.13 (1.07, 1.21)	0.03
Age at menarche (*n* = 3425)				
Unadjusted model:	1	1.15 (1.09, 1.20)	1.20 (1.15, 1.26)	0.03
Model 2 [6]:	1	1.17 (1.12, 1.22)	1.22 (1.17, 1.27)	0.02
Final model [5]:	1	1.13 (1.06, 1.19)	1.17 (1.11, 1.23)	0.01
Boys	**T1** [7]	**T2** [7]	**T3** [7]	p_{trend} [3]
Age at Tanner stage G2 (*n* = 2495)				
Unadjusted model:	1	1.09 (1.04, 1.13)	1.12 (1.08, 1.17)	0.04
Model 2 [6]:	1	1.11 (1.05, 1.17)	1.15 (1.10, 1.21)	0.04
Final model [5]:	1	1.07 (1.03, 1.12)	1.09 (1.03, 1.15)	0.03
Age at voice break (*n* = 2495)				
Unadjusted model:	1	1.11 (1.06, 1.18)	1.14 (1.09, 1.19)	0.04
Model 2 [6]:	1	1.12 (1.08, 1.17)	1.16 (1.11, 1.21)	0.03
Final model [5]:	1	1.08 (1.02, 1.13)	1.12 (1.07, 1.16)	0.03

[1] Values are models adjusted hazard ratios (95% CI), HR = hazard ratio; [2] values are min-max in tertiles in girls for age at Tanner stage B2: T1 (12.7–39.2), T2 (39.6–64.7), and T3 (64.9–81.7); values are min-max in tertiles in girls for age at menarche: T1 (11.6–37.7), T2 (38.1–62.3), and T3 (62.5–76.3); [3] *p* for trend across tertiles were performed by including dietary fat intake at baseline as continuous variables; [4] adjusted for birth year, family income level and energy intake at baseline; [5] additionally adjusted for BMI Z-scores at baseline and dietary protein intake (residual) at baseline; [6] adjusted for birth year, family income level, energy intake at baseline and mother's age at menarche; [7] values are min-max in tertiles in boys: T1 (15.8–47.2), T2 (47.3–69.5), and T3 (69.7–93.6).

Table 3. Association [1] of dietary polyunsaturated fatty acid (PUFA) in childhood with puberty timing.

	Dietary PUFA at Baseline			
Girls	**T1** [2]	**T2** [2]	**T3** [2]	p_{trend} [3]
Age at Tanner stage B2 (*n* = 2185)				
Unadjusted model:	1	1.08 (1.03, 1.12)	1.13 (1.08, 1.19)	0.04
Model 2 [4]:	1	1.09 (1.04, 1.13)	1.14 (1.09, 1.20)	0.03
Final model [5]:	1	1.07 (1.02, 1.11)	1.11 (1.05, 1.17)	0.02
Age at menarche (*n* = 3425)				
Unadjusted model:	1	1.11 (1.07, 1.16)	1.18 (1.12, 1.22)	0.04
Model 2 [6]:	1	1.13 (1.08, 1.19)	1.17 (1.12, 1.21)	0.03
Final model [5]:	1	1.09 (1.06, 1.13)	1.13 (1.08, 1.20)	0.03
Boys	**T1** [7]	**T2** [7]	**T3** [7]	p_{trend} [3]
Age at Tanner stage G2 (*n* = 2495)				
Unadjusted model:	1	1.08 (1.03, 1.13)	1.11 (1.05, 1.18)	0.04
Model 2 [6]:	1	1.09 (1.03, 1.14)	1.12 (1.05, 1.20)	0.03
Final model [5]:	1	1.06 (1.02, 1.11)	1.08 (1.03, 1.13)	0.03
Age at voice break (*n* = 2495)				
Unadjusted model:	1	1.09 (1.04, 1.13)	1.12 (1.06, 1.18)	0.03
Model 2 [6]:	1	1.11 (1.06, 1.17)	1.13 (1.08, 1.19)	0.03
Final model [5]:	1	1.07 (1.03, 1.12)	1.10 (1.06, 1.15)	0.02

[1] Values are models adjusted hazard ratios (95% CI), HR = hazard ratio; [2] values are min-max in tertiles in girls for age at Tanner stage B2: T1 (4.3–8.2), T2 (8.3–15.6), and T3 (15.7–20.1); values are min-max in tertiles in girls for age at menarche: T1 (4.1–7.5), T2 (7.6–13.9), and T3 (14.0–18.6); [3] *p* for trend across tertiles were performed by including dietary fat intake at baseline as continuous variables; [4] adjusted for birth year, family income level and energy intake at baseline; [5] additionally adjusted for Z-scores of BMI at baseline and dietary protein intake (residual) at baseline; [6] adjusted for birth year, family income level, energy intake at baseline and mother's age at menarche; [7] values are min-max in tertiles in boys: T1 (3.9–13.1), T2 (13.2–18.3), and T3 (18.4–21.6).

4. Discussion

We observed that higher childhood habitual total fat and PUFA intakes were associated with subsequent earlier timing of puberty, which was independent of pre-puberty dietary protein intake. However, MUFA or SFA was not associated with any puberty markers.

Our findings with total fat and earlier puberty are comparable with previous cohort studies conducted in Germany [13] and America [31]. Data from 261 German girls aged 8–15 years showed that girls with higher consumption of total fat were associated with accelerated menarche [13]. Based on data from 64 Caucasian American boys, childhood fat and animal protein intakes were found to be predictive of earlier age at peak height velocity [31]. This may lie in the fact that fat intake has the potential to accelerate the maturation of the hypothalamic–pituitary–gonadal axis and result in differences in hormone concentrations [32]. However, other prospective cohort studies that conducted in girls from the USA [17,33–35], Canada [15,16,36], Greece [36], and New Zealand [18] or boys from Australia [37] and the UK [38] reported inconsistent conclusions with our analysis. The reasons behind the discrepancies are likely to be multifactorial. The differences in genetic background, sample size, and dietary assessment may partly lead to inconsistent results. Our study is the first report of the Asian population with 3425 girls and 2495 boys, while the participants of others ranged from 67 to 2299. Larger sample size was achieved in our study and thus might have greater statistical power to reveal diet–puberty associations than most previous studies. Moreover, main food sources of dietary fat, such as red meat and milk, are also the predominant resources of dietary protein and have been demonstrated to be inversely associated with puberty timing [9,39]. However, the influence of dietary protein on the association of dietary fat with puberty timing has not been excluded appropriately in these previous analyses [9,15–18,33–35,37–40]. Since dietary protein intake is associated with puberty timing [10], we adjusted for childhood dietary protein intake in the Cox model. In this way, we could figure out the independent impact of dietary fat on puberty development, and we believe that our findings are helpful to build public health initiatives of dietary fat intake.

We also extended previous evidence of associations between dietary fat quality and puberty timing [8,15–18,40]. Our findings with PUFA and SFA are broadly consistent with conclusions of an earlier meta-analysis in 2020 which reported an inverse association of PUFA and null association of SFA with menarche in girls [10] based on cohorts in the USA [17] and Canada [15,16,36]. Another cohort of 3872 UK girls [41] and a cohort including 63 Caucasian girls [18] also demonstrated the inverse association of PUFA intake and null association of SFA intake with age at B2, respectively. Nevertheless, most of the previous reports [15–18,40] on associations between PUFA or SFA and puberty timing did not adjust for child body composition, which was associated with puberty development [22]. In our analysis, we have overcome this common limitation and presented more convincing conclusions after adjusting for parental characteristics, child body composition, and other potential confounders. Randomized trials in boys and girls also confirmed the null association between SFA and puberty timing [42,43]. In addition, we described associations of higher PUFA intake and higher odds of earlier puberty onset in boys, which was inconsistent with the only existing data from Avon Longitudinal Study of Parents and Children (ALSPC) cohort boys from the UK [41]. The reasons for the difference may be as follows. On one hand, the average timing of G2 (mean (SD), 8.7 (1.6) years) reported in the ALSPC cohort was unusually earlier compared to that reported in non-Hispanic white (10.1 (2.1) years) and Hispanic (10.0 (1.8) years) boys [44], which indicated that the subsample in ALSPC cohort was not a good representation of the total Caucasian boys, and the null association of PUFA based on these data should be considered with caution; while boys in our study entered G2 at a mean age of 11.2 (1.1) and close to that (10.55 (95% confidential intervals, 10.27 to 10.79)) reported in 18,807 urban Chinese boys [20], thus our investigation on diet-puberty association had a good representative sample. On the other hand, the mean PUFA intake (11.0 (1.3) g/day vs. 17.7 (5.2) g/day) and age at G2 (8.7 (1.6) years vs. 11.2 (1.1) years) of participants, nutritional assessment instrument (3-day food diaries vs. 3-day 24-h recalls), and statistical method (multivariable linear regression model vs. Cox proportional hazard regression model) were quite different between ALSPC cohort and our study, which could partly interpret this discrepancy as well. For MUFA, we found that there is no association with puberty onset in girls and boys, which has been reported in existing cohorts [16,41].

Furthermore, except Cheng TS et al. [41], most of the previous studies [8,15–18] have not ruled out the effect of dietary protein intake on the association between SFA, PUFA, or MUFA and puberty timing. We considered dietary protein intake at baseline in the multivariable model using the residual method and figured out the inverse association of PUFA, which was independent of dietary protein intake. Similar to our results, Cheng TS et al. reported the independent effect of PUFA using an isocaloric substitution model [41]. However, the exact mechanisms elucidating the association between PUFA intakes and puberty timing remain to be determined. A higher intake of PUFA may affect steroidogenic machinery and mammary gland development to accelerate growth and reproductive progress [45]. In vitro, PUFA modulates adrenal steroidogenesis and acts on the production of adrenal androgen, which eventually stimulates gonadotropin-releasing hormone neurons that are required for puberty onset [46]. In vivo, an increased ratio of n-6 to n-3 PUFA (5:1) modulated the reproductive function in female zebrafish [47]. Our findings failed to draw a conclusion on specific PUFA, and the optimal ratio of PUFA merited further investigation. We also found that there were no independent association of SFA with the puberty timing, but interestingly, the inverse association between MUFA intake and puberty in girls was changed in the model additionally adjusting for dietary protein, i.e., this association was mediated by dietary protein intake. Specific associations between SFA, MUFA, or PUFA and puberty timing suggest that there may be differences in the origin, metabolism, and biological activity of fatty acids [48]. Children usually get most of their dietary protein from animal foods, like red meat, milk, and eggs, which are also rich in SFA and MUFA, but relatively less in PUFA [49]. Similar data reported that children consumed most of their PUFA from fish, beans, and peanuts, but these foods have a relatively low concentration of SFA and MUFA [50]. This might partly explain why we observed an independent association for PUFA, but not for SFA or MUFA, and the previous association observed of MUFA and puberty timing may be driven by the dietary protein [15,17]. Meanwhile, MUFA can be sourced from the diet or endogenously synthesized from SFA [49], and Stearoyl-CoA desaturase (SCD) is the rate-limiting enzyme required for this synthesis. The expression of SCD1 gene has been reported to relate to the ratio of SFA and MUFA in vivo [51,52], and the polymorphisms of SCD1 gene are associated with SCD1 enzyme activity [53]. Thus, genetic variation of SCD1 gene among our participants would influence their endogenous MUFA levels. In aggregate, the interaction in SFA and MUFA metabolism may thus partly complicate the null association of SFA or MUFA in our study, although MUFA could theoretically stimulate mammary gland development. Hence, more studies that focus on the association of sex-specific individual fatty acids with pubertal development and the underlying pathway are needed.

Our study has several strengths. In contrast to most of the previous studies, we examined the puberty data from both girls and boys and provided a representable sample of Asian children. We also assessed multiple pubertal timing traits in both genders, showing a global view of puberty development. The prospective design, repeated follow-up for measurement on anthropometric, pubertal, and dietary data, as well as necessary adjustment for potential cofounders both in children and their parents were other strengths. Among these confounders, adjustment for dietary protein intake was the major strength in our analysis.

Some limitations should be acknowledged. In the CHNS, data on parental pubertal (mother's age at menarche) characteristics were not available, so we could not adjust for genetic influences; meanwhile, only the year of menarche occurrence in girls was considered in this study, therefore potential bias could not be avoided. However, our models have included a larger number of major potential confounders both in children and in their parents than most previous studies [15–18,33–36,40], which could partly overcome this potential limitation.

In conclusion, our analysis suggested that children with higher intakes of dietary fat and PUFA in prepuberty would enter their puberty at an earlier age. This association was independent of childhood dietary protein intake.

Supplementary Materials: The following supporting information can be downloaded at: https://www.mdpi.com/article/10.3390/nu14020275/s1, Table S1: Association of dietary saturated fatty acid (SFA) in childhood with puberty timing, Table S2: Association of dietary monounsaturated fatty acid (MUFA) in childhood with puberty timing.

Author Contributions: G.C. conceived the project. Y.X. and J.X. performed the analyses and wrote the manuscript. X.W. and W.G. performed the initial data analyses. L.Z., X.W., S.S. and J.X. coordinated the study centers. G.C. supervised the study. All authors have read and agreed to the published version of the manuscript.

Funding: This research was funded by National Key R&D Program of China, grant number 2020YFC2006300.

Institutional Review Board Statement: The study was conducted according to the guidelines of the Declaration of Helsinki, and approved by Ethics Committee of Sichuan University.

Informed Consent Statement: Informed consent was obtained from all subjects.

Data Availability Statement: Data from CHNS described in the manuscript will be made publicly and freely available without restriction at China Health and Nutrition Survey. Available online: https://www.cpc.unc.edu/projects/china/data/datasets/index.html (accessed on 1 January 2021). And data from SCCNG will be made available upon request pending approval by the corresponding author.

Conflicts of Interest: The authors have no conflicts of interest to disclose.

References

1. Patton, G.C.; Viner, R. Pubertal transitions in health. *Lancet* **2007**, *369*, 1130–1139. [CrossRef]
2. Euling, S.Y.; Herman-Giddens, M.E.; Lee, P.A.; Selevan, S.G.; Juul, A.; Sorensen, T.I.; Dunkel, L.; Himes, J.H.; Teilmann, G.; Swan, S.H. Examination of US puberty-timing data from 1940 to 1994 for secular trends: Panel findings. *Pediatrics* **2008**, *121* (Suppl. 3), S172–S191. [CrossRef] [PubMed]
3. Ersoy, B.; Balkan, C.; Gunay, T.; Onag, A.; Egemen, A. Effects of different socioeconomic conditions on menarche in Turkish female students. *Early Hum. Dev.* **2004**, *76*, 115–125. [CrossRef] [PubMed]
4. Prentice, S.; Fulford, A.J.; Jarjou, L.M.; Goldberg, G.R.; Prentice, A. Evidence for a downward secular trend in age of menarche in a rural Gambian population. *Ann. Hum. Biol.* **2010**, *37*, 717–721. [CrossRef]
5. Day, F.R.; Elks, C.E.; Murray, A.; Ong, K.K.; Perry, J.R. Puberty timing associated with diabetes, cardiovascular disease and also diverse health outcomes in men and women: The UK Biobank study. *Sci. Rep.* **2015**, *5*, 11208. [CrossRef]
6. Chen, X.; Liu, Y.; Sun, X.; Yin, Z.; Li, H.; Liu, X.; Zhang, D.; Cheng, C.; Liu, L.; Liu, F.; et al. Age at menarche and risk of all-cause and cardiovascular mortality: A systematic review and dose-response meta-analysis. *Menopause* **2018**, *26*, 670–676. [CrossRef] [PubMed]
7. Villamor, E.; Jansen, E.C. Nutritional Determinants of the Timing of Puberty. *Annu. Rev. Public Health* **2016**, *37*, 33–46. [CrossRef]
8. Rogers, I.S.; Northstone, K.; Dunger, D.B.; Cooper, A.R.; Ness, A.R.; Emmett, P.M. Diet throughout childhood and age at menarche in a contemporary cohort of British girls. *Public Health Nutr.* **2010**, *13*, 2052–2063. [CrossRef]
9. Jansen, E.C.; Marín, C.; Mora-Plazas, M.; Villamor, E. Higher Childhood Red Meat Intake Frequency Is Associated with Earlier Age at Menarche. *J. Nutr.* **2015**, *146*, 792–798. [CrossRef]
10. Nguyen, N.T.K.; Fan, H.-Y.; Tsai, M.-C.; Tung, T.-H.; Huynh, Q.T.V.; Huang, S.-Y.; Chen, Y.C. Nutrient Intake through Childhood and Early Menarche Onset in Girls: Systematic Review and Meta-Analysis. *Nutrients* **2020**, *12*, 2544. [CrossRef]
11. Hämäläinen, E.; Adlercreutz, H.; Puska, P.; Pietinen, P. Diet and serum sex hormones in healthy men. *J. Steroid Biochem.* **1984**, *20*, 459–464. [CrossRef]
12. Wu, A.H.; Pike, M.C.; Stram, D.O. Meta-analysis: Dietary fat intake, serum estrogen levels, and the risk of breast cancer. *J. Natl. Cancer Inst.* **1999**, *91*, 529–534. [CrossRef] [PubMed]
13. Merzenich, H.; Boeing, H.; Wahrendorf, J. Dietary fat and sports activity as determinants for age at menarche. *Am. J. Epidemiol.* **1993**, *138*, 217–224. [CrossRef]
14. Meyer, F.; Moisan, J.; Marcoux, D.; Bouchard, C. Dietary and physical determinants of menarche. *Epidemiology* **1990**, *1*, 377–381. [CrossRef] [PubMed]
15. Moisan, J.; Meyer, F.; Gingras, S. A nested case-control study of the correlates of early menarche. *Am. J. Epidemiol.* **1990**, *132*, 953–961. [CrossRef] [PubMed]

16. Koo, M.M.; Rohan, T.E.; Jain, M.; McLaughlin, J.R.; Corey, P.N. A cohort study of dietary fibre intake and menarche. *Public Health Nutr.* **2002**, *5*, 353–360. [CrossRef]
17. Maclure, M.; Travis, L.B.; Willett, W.; MacMahon, B. A prospective cohort study of nutrient intake and age at menarche. *Am. J. Clin. Nutr.* **1991**, *54*, 649–656. [CrossRef]
18. de Ridder, C.M.; Thijssen, J.H.; Van't Veer, P.; van Duuren, R.; Bruning, P.F.; Zonderland, M.L.; Erich, W.B. Dietary habits, sexual maturation, and plasma hormones in pubertal girls: A longitudinal study. *Am. J. Clin. Nutr.* **1991**, *54*, 805–813. [CrossRef]
19. Berenbaum, S.A.; Beltz, A.M.; Corley, R. The importance of puberty for adolescent development: Conceptualization and measurement. *Adv. Child Dev. Behav.* **2015**, *48*, 53–92. [CrossRef]
20. Ma, H.-M.; Chen, S.-K.; Chen, R.-M.; Zhu, C.; Xiong, F.; Li, T.; Wang, W.; Liu, G.-L.; Luo, X.-P.; Liu, L.; et al. Pubertal development timing in urban Chinese boys. *Int. J. Androl.* **2011**, *34*, e435–e445. [CrossRef]
21. Ma, H.-M.; Du, M.-L.; Luo, X.-P.; Chen, S.-K.; Liu, L.; Chen, R.-M.; Zhu, C.; Xiong, F.; Li, T.; Wang, W.; et al. Onset of Breast and Pubic Hair Development and Menses in Urban Chinese Girls. *Pediatrics* **2009**, *124*, e269–e277. [CrossRef]
22. Wagner, I.V.; Sabin, M.; Pfäffle, R.W.; Hiemisch, A.; Sergeyev, E.; Körner, A.; Kiess, W. Effects of obesity on human sexual development. *Nat. Rev. Endocrinol.* **2012**, *8*, 246–254. [CrossRef]
23. Jia, P.; Xue, H.; Zhang, J.; Wang, Y. Time Trend and Demographic and Geographic Disparities in Childhood Obesity Prevalence in China-Evidence from Twenty Years of Longitudinal Data. *Int. J. Env. Res. Public Health* **2017**, *14*, 369. [CrossRef]
24. Popkin, B.M.; Du, S.; Zhai, F.; Zhang, B. Cohort Profile: The China Health and Nutrition Survey—monitoring and understanding socio-economic and health change in China, 1989-2011. *Int. J. Epidemiol.* **2010**, *39*, 1435–1440. [CrossRef]
25. Mervish, N.A.; Teitelbaum, S.L.; Pajak, A.; Windham, G.C.; Pinney, S.M.; Kushi, L.H.; Biro, F.M.; Wolff, M.S. Peripubertal dietary flavonol and lignan intake and age at menarche in a longitudinal cohort of girls. *Pediatr Res.* **2017**, *82*, 201–208. [CrossRef]
26. Yang, M.Z.; Xue, H.M.; Pan, J.; Libuda, L.; Muckelbauer, R.; Yang, M.; Quan, L.; Cheng, G. High protein intake along with paternal part-time employment is associated with higher body fat mass among girls from South China. *Eur. J. Nutr.* **2018**, *57*, 1845–1854. [CrossRef]
27. Zhai, F.Y.; Du, S.F.; Wang, Z.H.; Zhang, J.G.; Du, W.W.; Popkin, B.M. Dynamics of the Chinese diet and the role of urbanicity, 1991-2011. *Obes. Rev.* **2014**, *15* (Suppl. 1), 16–26. [CrossRef] [PubMed]
28. Cole, T.J.; Bellizzi, M.C.; Flegal, K.M.; Dietz, W.H. Establishing a standard definition for child overweight and obesity worldwide: International survey. *BMJ* **2000**, *320*, 1240–1243. [CrossRef]
29. Li, H.; Ji, C.Y.; Zong, X.N.; Zhang, Y.Q. Body mass index growth curves for Chinese children and adolescents aged 0 to 18 years. *Zhonghua Er Ke Za Zhi* **2009**, *47*, 493–498.
30. National Bureau of Statistics of China. *China Statistical Yearbook*; China Statistics Press: Beijing, China, 2020.
31. Alimujiang, A.; Colditz, G.A.; Gardner, J.D.; Park, Y.; Berkey, C.S.; Sutcliffe, S. Childhood diet and growth in boys in relation to timing of puberty and adult height: The Longitudinal Studies of Child Health and Development. *Cancer Causes Control.* **2018**, *29*, 915–926. [CrossRef]
32. Lemarchand-Béraud, T.; Zufferey, M.-M.; Reymond, M.; Rey, I. Maturation of the Hypothalamo-Pituitary-Ovarian Axis in Adolescent Girls. *J. Clin. Endocrinol. Metab.* **1982**, *54*, 241–246. [CrossRef] [PubMed]
33. Berkey, C.S.; Gardner, J.D.; Frazier, A.L.; Colditz, G.A. Relation of childhood diet and body size to menarche and adolescent growth in girls. *Am. J. Epidemiol.* **2000**, *152*, 446–452. [CrossRef]
34. Koprowski, C.; Ross, R.K.; Mack, W.J.; Henderson, B.E.; Bernstein, L. Diet, body size and menarche in a multiethnic cohort. *Br. J. Cancer* **1999**, *79*, 1907–1911. [CrossRef]
35. Kissinger, D.G.; Sanchez, A. The association of dietary factors with the age of menarche. *Nutr. Res.* **1987**, *7*, 471–479. [CrossRef]
36. Petridou, E.; Syrigou, E.; Toupadaki, N.; Zavitsanos, X.; Willett, W.; Trichopoulos, D. Determinants of age at menarche as early life predictors of breast cancer risk. *Int. J. Cancer* **1996**, *68*, 193–198. [CrossRef]
37. Cheng, H.L.; Raubenheimer, D.; Steinbeck, K.; Baur, L.; Garnett, S. New insights into the association of mid-childhood macronutrient intake to pubertal development in adolescence using nutritional geometry. *Br. J. Nutr.* **2019**, *122*, 274–283. [CrossRef]
38. Cheng, T.S.; Sharp, S.J.; Brage, S.; Emmett, P.M.; Forouhi, N.G.; Ong, K.K. Longitudinal associations between prepubertal childhood total energy and macronutrient intakes and subsequent puberty timing in UK boys and girls. *Eur. J. Nutr.* **2021**. [CrossRef]
39. Ramezani Tehrani, F.; Moslehi, N.; Asghari, G.; Gholami, R.; Mirmiran, P.; Azizi, F. Intake of dairy products, calcium, magnesium, and phosphorus in childhood and age at menarche in the Tehran Lipid and Glucose Study. *PLoS ONE* **2013**, *8*, e57696. [CrossRef]
40. Moisan, J.; Meyer, F.; Gingras, S. Diet And Age At Menarche. *Cancer Causes Control.* **1990**, *1*, 149–154. [CrossRef]
41. Cheng, T.S.; Day, F.R.; Perry, J.R.B.; Luan, J.; Langenberg, C.; Forouhi, N.G.; Wareham, N.J.; Ong, K.K. Prepubertal Dietary and Plasma Phospholipid Fatty Acids Related to Puberty Timing: Longitudinal Cohort and Mendelian Randomization Analyses. *Nutrients* **2021**, *13*, 1868. [CrossRef] [PubMed]
42. Sadov, S.; Virtanen, H.E.; Main, K.M.; Andersson, A.M.; Juul, A.; Jula, A.; Raitakari, O.T.; Pahkala, K.; Niinikoski, H.; Toppari, J. Low-saturated-fat and low-cholesterol diet does not alter pubertal development and hormonal status in adolescents. *Acta Paediatr.* **2019**, *108*, 321–327. [CrossRef] [PubMed]
43. Niinikoski, H.; Lagstrom, H.; Jokinen, E.; Siltala, M.; Ronnemaa, T.; Viikari, J.; Raitakari, O.T.; Jula, A.; Marniemi, J.; Nanto-Salonen, K.; et al. Impact of repeated dietary counseling between infancy and 14 years of age on dietary intakes and serum lipids and lipoproteins: The STRIP study. *Circulation* **2007**, *116*, 1032–1040. [CrossRef] [PubMed]

44. Herman-Giddens, M.E.; Steffes, J.; Harris, D.; Slora, E.; Hussey, M.; Dowshen, S.A.; Wasserman, R.; Serwint, J.R.; Smitherman, L.; Reiter, E.O. Secondary sexual characteristics in boys: Data from the Pediatric Research in Office Settings Network. *Pediatrics* **2012**, *130*, e1058–e1068. [CrossRef]
45. Anderson, B.M.; MacLennan, M.B.; Hillyer, L.M.; Ma, D.W.L. Lifelong exposure to n-3 PUFA affects pubertal mammary gland development. *Appl. Physiol. Nutr. Metab.* **2014**, *39*, 699–706. [CrossRef]
46. Sampath, H.; Ntambi, J.M. Polyunsaturated fatty acid regulation of genes of lipid metabolism. *Annu. Rev. Nutr.* **2005**, *25*, 317–340. [CrossRef] [PubMed]
47. Fowler, L.A.; Dennis-Cornelius, L.N.; Dawson, J.A.; Barry, R.J.; Davis, J.L.; Powell, M.L.; Yuan, Y.; Williams, M.B.; Makowsky, R.; D'Abramo, L.R.; et al. Both Dietary Ratio of n-6 to n-3 Fatty Acids and Total Dietary Lipid Are Positively Associated with Adiposity and Reproductive Health in Zebrafish. *Curr. Dev. Nutr.* **2020**, *4*, nzaa034. [CrossRef]
48. Tvrzicka, E.; Kremmyda, L.S.; Stankova, B.; Zak, A. Fatty acids as biocompounds: Their role in human metabolism, health and disease—A review. Part 1: Classification, dietary sources and biological functions. *Biomed. Pap. Med. Fac. Univ. Palacky Olomouc Czech Repub.* **2011**, *155*, 117–130. [CrossRef]
49. Michas, G.; Micha, R.; Zampelas, A. Dietary fats and cardiovascular disease: Putting together the pieces of a complicated puzzle. *Atherosclerosis* **2014**, *234*, 320–328. [CrossRef] [PubMed]
50. Yaméogo, C.W.; Cichon, B.; Fabiansen, C.; Rytter, M.J.H.; Faurholt-Jepsen, D.; Stark, K.D.; Briend, A.; Shepherd, S.; Traoré, A.S.; Christensen, V.B.; et al. Correlates of whole-blood polyunsaturated fatty acids among young children with moderate acute malnutrition. *Nutr. J.* **2017**, *16*, 44. [CrossRef]
51. Karahashi, M.; Ishii, F.; Yamazaki, T.; Imai, K.; Mitsumoto, A.; Kawashima, Y.; Kudo, N. Up-regulation of stearoyl-CoA desaturase 1 increases liver MUFA content in obese Zucker but not Goto-Kakizaki rats. *Lipids* **2013**, *48*, 457–467. [CrossRef] [PubMed]
52. Alim, M.A.; Fan, Y.P.; Wu, X.P.; Xie, Y.; Zhang, Y.; Zhang, S.L.; Sun, D.X.; Zhang, Y.; Zhang, Q.; Liu, L.; et al. Genetic effects of stearoyl-coenzyme A desaturase (SCD) polymorphism on milk production traits in the Chinese dairy population. *Mol. Biol. Rep.* **2012**, *39*, 8733–8740. [CrossRef] [PubMed]
53. Martín-Núñez, G.M.; Cabrera-Mulero, R.; Rojo-Martínez, G.; Gómez-Zumaquero, J.M.; Chaves, F.J.; de Marco, G.; Soriguer, F.; Castaño, L.; Morcillo, S. Polymorphisms in the SCD1 gene are associated with indices of stearoyl CoA desaturase activity and obesity: A prospective study. *Mol. Nutr. Food Res.* **2013**, *57*, 2177–2184. [CrossRef] [PubMed]

Article

Soy Food Intake Associated with Obesity and Hypertension in Children and Adolescents in Guangzhou, Southern China

Xiaotong Wang [1,†], Tongtong He [2,†], Suhua Xu [1], Hailin Li [1], Miao Wu [1], Zongyu Lin [1], Fenglian Huang [1] and Yanna Zhu [1,*]

[1] Guangdong Provincial Key Laboratory of Food, Nutrition and Health, Department of Maternal and Child Health, School of Public Health, Sun Yat-sen University, Guangzhou 510080, China; wangxt27@mail2.sysu.edu.cn (X.W.); xush59@mail2.sysu.edu.cn (S.X.); lihlin29@mail2.sysu.edu.cn (H.L.); wumiao6@mail2.sysu.edu.cn (M.W.); linzy35@mail2.sysu.edu.cn (Z.L.); huangflian@mail2.sysu.edu.cn (F.H.)

[2] Department of Nutrition, School of Public Health, Sun Yat-sen University, Guangzhou 510080, China; hett5@mail2.sysu.edu.cn

* Correspondence: zhuyn3@mail.sysu.edu.cn; Tel.: +86-020-8733-0663

† These authors contributed equally to this work.

Abstract: The associations between soy food intake and cardio-metabolic risk factors in children remain unclear due to limited evidence. We aim to explore soy food intake and its association with the risks of obesity and hypertension in Chinese children and adolescents. A total of 10,536 children and adolescents aged 7–18 years (5125 boys and 5411 girls) were enrolled in a cross-sectional study in Guangzhou City, southern China. Data on demographic characteristics and dietary consumption were collected using self-reported questionnaires, and anthropometric characteristics were measured. Obesity, abdominal obesity, and hypertension were defined using Chinese criteria for children and adolescents. A multiple logistic regression model was applied to estimate the association between soy food intake and obesity and hypertension. Roughly 39.5% of the participants consumed soy food more than three times per week. The mean amounts of liquid and solid soy food intake were 0.35 ± 0.54 cups/day and 0.46 ± 0.63 servings/day, respectively. The adjusted odds ratios (OR) of hypertension among those with high liquid soy food intake and a high frequency of all soy food intake (more than three times/week) were 0.79 (95% confidence interval (CI), 0.67–0.94), and 0.83 (95% CI, 0.70–0.97) compared to those with no intake. Additionally, the adjusted OR of obesity among those with high solid soy food intake and a high frequency of all soy food intake were 1.34 (95% CI, 1.09–1.63) and 1.30 (95% CI, 1.07–1.58), respectively. In conclusion, 39.5% of southern Chinese children and adolescents had high soy food intake (more than three times/week), which was significantly associated with a lower prevalence of hypertension and a greater prevalence of obesity.

Keywords: soy food; obesity; hypertension; children and adolescents

Citation: Wang, X.; He, T.; Xu, S.; Li, H.; Wu, M.; Lin, Z.; Huang, F.; Zhu, Y. Soy Food Intake Associated with Obesity and Hypertension in Children and Adolescents in Guangzhou, Southern China. *Nutrients* **2022**, *14*, 425. https://doi.org/10.3390/nu14030425

Academic Editor: María-José Castro

Received: 11 December 2021
Accepted: 16 January 2022
Published: 18 January 2022

Publisher's Note: MDPI stays neutral with regard to jurisdictional claims in published maps and institutional affiliations.

Copyright: © 2022 by the authors. Licensee MDPI, Basel, Switzerland. This article is an open access article distributed under the terms and conditions of the Creative Commons Attribution (CC BY) license (https:// creativecommons.org/licenses/by/ 4.0/).

1. Introduction

The prevalence of childhood cardio-metabolic risk factors, specifically obesity and hypertension, are sharply increasing and constitute a major public health threat in China [1–3]; these risk factors may persist into adulthood and are the leading causes of cardiovascular disease (CVD) [4,5]. Many studies have reported that a habitual diet is one of the most effective factors in regulating weight and blood pressure (BP) [6–9]. The consumption of soy food, one of the common habitual diet components, as a valuable source of isoflavones, phytosterols, lecithin, poly-unsaturated fatty acids, dietary fibers, and high-quality protein has attracted significant attention [10] and has demonstrated many health benefits, including a reduction of CVD risk, making soy food worthy of further examination [11–13].

Several studies have reported that habitual soy food intake lowers the risk of high BP [14–22]. However, some studies observed no significant changes in BP after the consumption of soy food [7,23]. Current epidemiological findings suggest certain effects of soy supplementation on weight control and the prevention of obesity [6,24]. For instance, Ruscica et al. documented significant changes in median percentage of body weight (-1.5%, $p = 0.005$) and body mass index (BMI) (-1.5%, $p = 0.05$) in the soy group [24]. Moreover, another study showed that soy protein is at least as good as other protein sources for weight loss during low-calorie dietary interventions in older adults [11]. Nevertheless, some studies have also shown that the protective effect of legumes on weight remains insufficient [25–27]. In addition, a systematic review and meta-analysis suggested that soy had no statistically significant effect on weight in the general population, yet an obesogenic effect of soy was observed in obese subjects (BMI ≥ 30 kg/m^2) and participants whose daily soy food intake included at least 40 g of soy protein [28].

However, most of these studies have focused on Western populations or older adults; whether the same relationship exists in children remains obscure. Furthermore, there is little information on the association between soy food intake and cardio-metabolic risk factors in children and adolescents; in particular, a better understanding of soy food intake status and its impact on obesity and hypertension, as well as the underlying influencing factors, among children and adolescents is needed. This study, therefore, sought to describe the status of soy food intake and investigate the association of soy food intake with the risk of obesity and hypertension in Chinese children and adolescents aged 7–18 years.

2. Materials and Methods

2.1. Study Sample

The data were obtained from cross-sectional health and nutrition surveys deployed among children and adolescents in Guangzhou, southern China, in 2018. Fourteen elementary schools and 15 middle schools were randomly selected through multiple cluster sampling, and two classes of each grade from the selected schools were randomly chosen. In total, 13,979 children and adolescents aged 7–18 years old were involved in the present study and asked to undergo anthropometric measurements and questionnaire assessments. Participants were excluded if they did not undergo anthropometric measurements or if their completed questionnaire lacked information on dietary intake or demographic characteristics ($n = 3443$). Finally, the study sample consisted of 10,536 participants (48.6% boys). Informed consent forms were voluntarily signed by all participants and their legal guardians, and the study was approved by the Ethical Committee of School of Public Health, Sun Yat-sen University (Number (2018)005).

2.2. Anthropometric Measurements

All participants underwent anthropometric measurements, including height, weight, waist circumference, and blood pressure, which were collected by trained clinicians and nurses according to a uniform standard. Height was measured to the nearest 0.1 cm with the use of a fixed stadiometer (Yilian TZG; Yangzhou, Jiangsu, China) and weight was measured to the nearest 0.1 kg with the use of a lever scale (Hengxing RGT-140; Yixing, Jiangsu, China). Body mass index (BMI) was calculated as the weight divided by the height squared (kg/cm^2). Waist circumference (WC) was measured to an accuracy of 0.1 cm at 1 cm above umbilicus using a flexible tape in the standing position. Systolic blood pressure (SBP) and diastolic blood pressure (DBP) were measured on the mid-upper right arm using a mercury sphygmomanometer (Yutu XJ1ID; Shanghai, China) after participants sat quietly for at least 5 min. All indices were measured twice and the average numbers were recorded.

2.3. Questionnaire Assessment

The self-reported questionnaire in this study was developed based on previously tested and validated questions, including those on demographic information (examination date, sex, birth date, age, and family income) and dietary intake (fruits, vegetables, meat, sugar-sweetened beverages (SSBs), milk, fried food, and soy food intake). The children and adolescents were asked to answer the questionnaire on their own if they were over 9 years old, and children younger than 9 years old were required to bring the questionnaire to their legal guardians. Trained staff and teachers interpreted all of the questions so that the participants understood the questions correctly. The researchers performed quality control when the questionnaires were collected from each class.

To investigate soy food intake and frequency, every participant was required to answer the following questions: "How many times have you consumed soy food in the last week (including soybeans, tofu, soy milk, fermented bean curd, and other products made from soybean)?" and "How many cups of liquid soy food or servings of solid soy food did you consume each time? (Milliliters is equal to one cup, the size of an adult's palm is equal to one serving)". The consumption of fruit and vegetables was evaluated using the following two questions: "How many days have you eaten fruits/vegetables in the last week?" and "How many servings of fruits/vegetables did you eat each day? The size of an adult's fist is equal to one serving". For meat or meat product consumption, the following two questions were asked: "How many days have you eaten meat or meat products in the last week?" and "How many servings of meat or meat products did you eat last week? The size of an adult's palm is equal to one serving". For SSB consumption, the following two questions were asked: "How many days did you drink SSBs in the last week?" and "How many servings of SSB did you drink each day? Two hundred fifty milliliters is equal to one serving". For this study, SSBs included Coca-Cola (Coca-Cola Company, Atlanta, GA, USA), orange juice, Red Bull (Red Bull GmbH, Fuschl, Austria), and other beverages containing sugar. For milk consumption, the following questions were asked: "How many times did you drink milk in the last week?" and "How many milliliters of milk did you drink each time? Milk includes fresh milk, yogurt, milk powder, and other dairy products". Finally, data on fried food consumption were collected by asking the following question: "How many times have you eaten fried food in the last week?". For this study, fried food included fried chicken, fried dough sticks, fried chips, and so on. For demographic information, family income was divided into the following five RMB/month groups: 2000 or below, 2000 to 5000, 5000 to 8000, 8000 or above, and don't know or no answer.

2.4. Definitions

Overweight and obesity were defined by BMIs in at least the 85th percentile and at least the 95th percentile according to the age- and sex-specific cutoff points recommended by the Group of China Obesity Task Force [29]. Abdominal obesity was defined by a WC of at least the 90th age- and sex-specific percentile cutoff point for Chinese children and adolescents [30]. Individuals with SBP and/or DBP values in at least the 90th percentile but less than the 95th percentile for their age and sex were considered to be pre-hypertensive, while those with values in at least the 95th percentile were considered to have hypertension [31].

2.5. Statistical Analysis

The data were entered into the EpiData 3.0 software (The EpiData Association, Odense, Denmark). All statistical tests were performed using the Statistical Package for the Social Sciences version 21.0 software (IBM Corporation, Armonk, NY, USA). The participants were classified into three groups according to tertile cutoff points of soy food intake, as follows: (1) liquid soy food intake (none = 0 cups/day, medium = 0–0.4 cups/day, high > 0.4 cups/day); (2) solid soy food intake (none = 0–0.1429 servings/day, medium = 0.1429–0.4286 servings/day, high > 0.4286 servings/day); (3) total soy food intake frequency (none = 0–1 times/week, medium = 1–3 times/week, high > 3 times/week). The average soy food intake by age is shown using mean ± and standard deviation values, and the characteristics of the participants stratified by soy food intake are presented as mean ± standard error values for quantitative variables or numbers and percentages for the categorical variables. The *t* test was used to compare differences in soy food intake between boys and girls. When assessing the differences of the characteristics among the three groups of soy food intake, a one-way analysis of variance (ANOVA) was applied for quantitative variables, and a non-parametric test was applied if the normality test and homogeneity test of variances were unsatisfied; meanwhile, a χchi-squared test was applied for the categorical variables. Bonferroni's test was used to analyze pairwise comparisons. The association of soy food intake with obesity and hypertension was evaluated by using multivariate logistic regression in four models (model 1, unadjusted; model 2, adjusted for age and sex; model 3, model 2 plus adjustments for family income and overweight, obesity, and abdominal obesity, and family income and BMI for blood pressure; model 4, model 3 plus adjustments for dietary intakes (intake frequency of fried food and SSBs)). The results were reported with an odds ratio (OR) and corresponding 95% confidence interval (95% CI) values, and p values of less than 0.05 were considered to be statistically significant.

3. Results

3.1. Soy Food Intake in Chinese Children and Adolescents According to Age

In this cross-sectional study, a total of 10,536 Chinese children and adolescents (5125 boys and 5411 girls) aged 7–18 years old were included. Table 1 shows the status of soy food intake according to age and sex. The mean liquid and solid soy food intakes of boys and girls were 0.35 ± 0.54 cups/day and 0.46 ± 0.63 servings/day, respectively. The overall frequency of all soy products was 2.41 ± 2.00 times/week among participants. The servings and frequency of soy food intake were higher in boys compared to those in girls (all $p < 0.05$). The trend showed that boys and girls tended to have more soy food intake as they grew up (both $p < 0.001$), especially in the middle of adolescence.

Table 1. Soy food intake in Chinese children and adolescents according to age ($n = 10,536$).

Age (y)	Liquid Soy Food (Cups/Day) Mean ± SD			Solid Soy Food (Servings/Day) Mean ± SD			All Soy Food (Times/Week) Mean ± SD		
	Total $N = 10,536$	Boys $n = 5125$	Girls $n = 5411$	Total $N = 10,536$	Boys $n = 5125$	Girls $n = 5411$	Total $N = 10,536$	Boys $n = 5125$	Girls $n = 5411$
7	0.26 ± 0.44	0.26 ± 0.44	0.27 ± 0.46	0.33 ± 0.47	0.33 ± 0.44	0.34 ± 0.51	2.01 ± 1.69	1.95 ± 1.60	2.08 ± 1.79
8	0.27 ± 0.38	0.25 ± 0.31	0.29 ± 0.44	0.35 ± 0.51	0.35 ± 0.46	0.36 ± 0.55	2.05 ± 1.65	2.06 ± 1.69	2.04 ± 1.60
9	0.31 ± 0.47	0.34 ± 0.57	0.28 ± 0.34 *	0.47 ± 0.70	0.48 ± 0.74	0.47 ± 0.66	2.23 ± 1.87	2.26 ± 1.97	2.19 ± 1.75
10	0.35 ± 0.51	0.38 ± 0.55	0.32 ± 0.46 *	0.52 ± 0.66	0.55 ± 0.69	0.49 ± 0.63	2.39 ± 2.02	2.49 ± 2.23	2.27 ± 1.76
11	0.38 ± 0.61	0.38 ± 0.64	0.37 ± 0.59	0.48 ± 0.69	0.52 ± 0.77	0.44 ± 0.58	2.34 ± 2.09	2.38 ± 2.26	2.29 ± 1.86
12	0.43 ± 0.56	0.48 ± 0.63	0.39 ± 0.49 *	0.49 ± 0.61	0.51 ± 0.69	0.46 ± 0.52	2.50 ± 1.96	2.61 ± 1.14	2.40 ± 1.77
13	0.44 ± 0.71	0.46 ± 0.82	0.42 ± 0.61	0.49 ± 0.65	0.53 ± 0.77	0.47 ± 0.54	2.59 ± 2.07	2.62 ± 2.29	2.57 ± 1.90
14	0.33 ± 0.43	0.36 ± 0.47	0.30 ± 0.40	0.46 ± 0.60	0.52 ± 0.66	0.42 ± 0.56 *	2.51 ± 2.06	2.63 ± 2.11	2.41 ± 2.02
15	0.38 ± 0.57	0.43 ± 0.62	0.35 ± 0.52 *	0.50 ± 0.65	0.53 ± 0.72	0.47 ± 0.58	2.71 ± 2.14	2.92 ± 2.28	2.53 ± 2.01 *
16	0.40 ± 0.61	0.43 ± 0.67	0.37 ± 0.56	0.50 ± 0.64	0.55 ± 0.68	0.47 ± 0.61	2.72 ± 2.18	2.98 ± 2.45	2.51 ± 1.91 *
17	0.37 ± 0.58	0.42 ± 0.64	0.34 ± 0.52	0.49 ± 0.66	0.54 ± 0.74	0.46 ± 0.59	2.65 ± 2.04	2.82 ± 2.08	2.52 ± 2.00 *
18	0.33 ± 0.51	0.38 ± 0.56	0.28 ± 0.44	0.44 ± 0.66	0.43 ± 0.52	0.46 ± 0.76	2.52 ± 2.20	2.76 ± 2.45	2.31 ± 1.94
Total	0.35 ± 0.54	0.37 ± 0.58	0.33 ± 0.50 *	0.46 ± 0.63	0.48 ± 0.67	0.44 ± 0.59 *	2.41 ± 2.00	2.48 ± 2.13	2.32 ± 1.86 *

Abbreviation: SD, standard deviation. Liquid soy food (cups/day) = liquid soy food (cups/time) * intake of liquid soy food (times/week)/7. Solid soy food (servings/day) = solid soy food (servings/time) * intake of solid soy food (times/week)/7. * $p < 0.05$, t test compared the differences between boys and girls in soy food intake.

As shown in Figure 1A1–C3, about 39.5% of participants consumed soy food more than three times per week. Boys tended to increase their liquid soy food intake as they grew up ($p < 0.01$). In general, the proportion of those with higher soy food intake frequency tended to increase with age (all $p < 0.001$), while there was no significant change in solid soy food intake with age.

3.2. Characteristics of Participants by Soy Food Intake Levels

The characteristics of the participants according to the tertiles of the liquid and solid servings and the frequency of soy food intake are described in Table 2, including sex, anthropometry, family income, and dietary intake. A high soy food intake and high soy food frequency were associated with higher weight, BMI, QC, and SBP (all $p < 0.001$). Meanwhile, the high servings and high frequency groups showed greater dietary intake, including fruits, vegetables, meat, milk, SSBs, and fried food. No significant differences were found in terms of family income and DBP level among the groups stratified by soy food intake or frequency.

Figure 1. *Cont.*

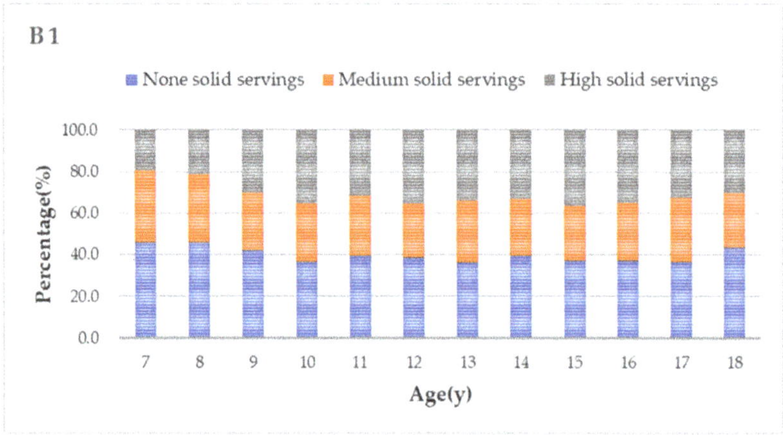

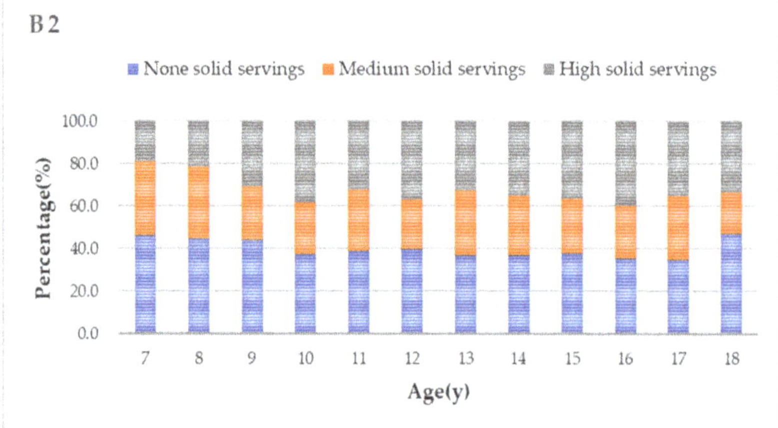

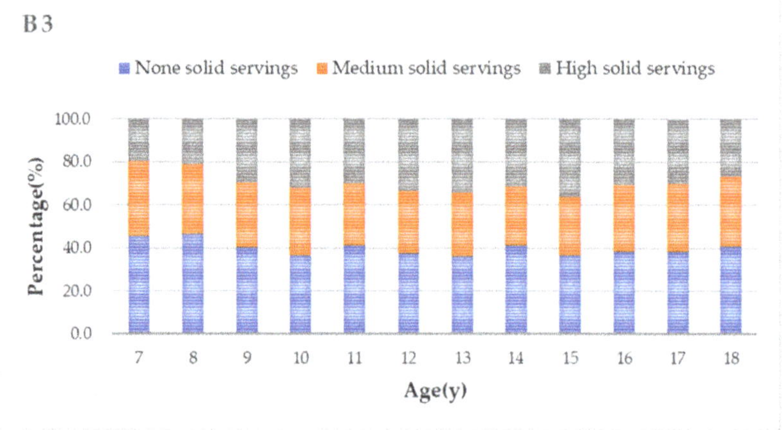

Figure 1. *Cont.*

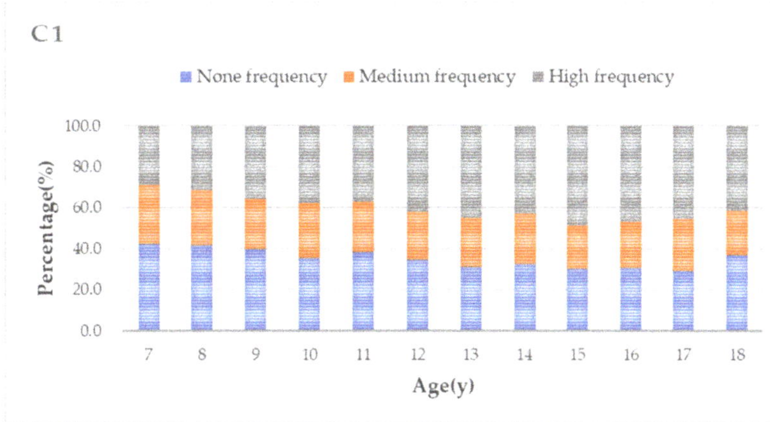

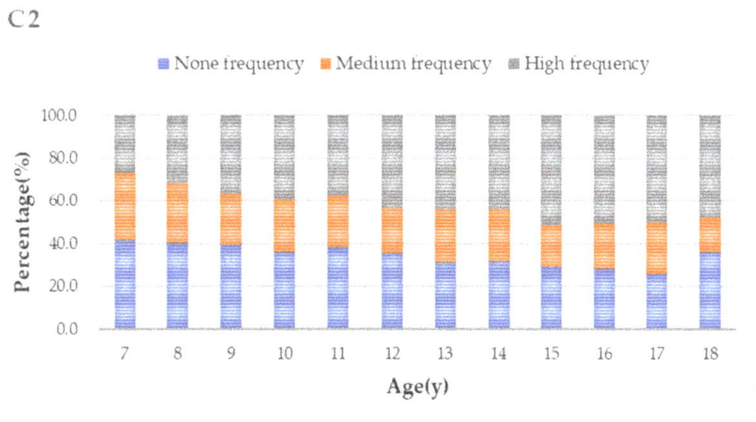

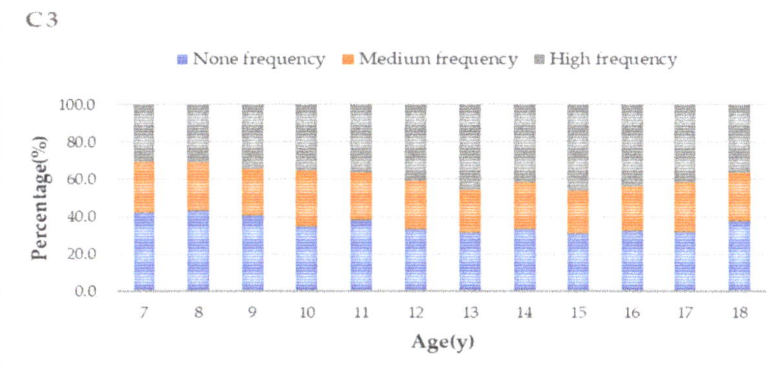

Figure 1. Soy food intake in Chinese children and adolescents according to age 1 for total, 2 for males and 3 for females; (**A1–A3**) present data on liquid soy food; (**B1–B3**) present data on solid soy food; (**C1–C3**) present data on the intake frequency of all soy food. (**A1,B1,C1**) show data for the total population; (**A2,B2,C2**) show data for the boys; (**A3,B3,C3**) show data for the girls.

Table 2. Characteristics of participants stratified by soy food intake level.

Variables	Liquid Soy Food (Cups/Day)				Solid Soy Food (Servings/Day)				All Soy Food (Times/Week)			
	None [1,†]	Medium [1,†]	High [1,†]	*p* Value	None [2,†]	Medium [2,†]	High [2,†]	*p* Value	None [3,†]	Medium [3,†]	High [3,†]	*p* Value
Sex, %				0.010 [b]				0.011 [b]				0.127
Male	37.3	28.5 *	34.2		40.0	28.3	31.7		35.1	24.4	40.4	
Female	35.6	31.2 *	32.2		40.1	30.5	29.4		36.0	25.5	38.5	
Family income, RMB/month, %				0.320				0.009 [b]				0.038 [b]
2000 or below	35.5	29.8	34.6	0.880	44.9	25.1 *	30.1	0.052	39.7	21.1 *	39.2	0.076
2000–5000	34.4	30.8	34.8 *	0.061	38.7	29.6	31.7	0.219	34.3	26.1	39.6	0.233
5000–8000	36.5	30.9	32.6	0.416	38.9	31.2	29.9	0.115	36.6	26.2	37.2 *	0.046 [b]
≥8000	37.2	29.6	33.2	0.525	39.8	30.3	29.9	0.376	34.9	25.0	40.0	0.640
Don't know or no answer	37.7	28.4	33.6	0.214	42.4	26.8 *	30.9	0.009 [b]	35.9	23.2	40.9	0.098
Anthropometry, mean ± SE												
Weight (kg)	42.62 ± 0.23	39.53 ± 0.26 *	43.77 ± 0.25 *	<0.001 [a]	41.21 ± 0.23	41.14 ± 0.26	44.08 ± 0.26 *	<0.001 [a]	40.51 ± 0.24	40.83 ± 0.28	44.30 ± 0.23 *	<0.001 [a]
BMI (kg/m²)	18.30 ± 0.06	17.78 ± 0.06 *	18.52 ± 0.06 *	<0.001 [a]	18.03 ± 0.05	18.07 ± 0.06	18.61 ± 0.07 *	<0.001 [a]	17.94 ± 0.06	17.99 ± 0.07	18.63 ± 0.06 *	<0.001 [a]
WC (cm)	64.11 ± 0.16	62.52 ± 0.18 *	64.57 ± 0.16	<0.001 [a]	63.30 ± 0.15	63.30 ± 0.17	64.91 ± 0.18 *	<0.001 [a]	63.04 ± 0.16	63.22 ± 0.19	64.84 ± 0.15 *	<0.001 [a]
SBP (mmHg)	105.97 ± 0.19	104.91 ± 0.21 *	106.24 ± 0.19	<0.001 [a]	105.35 ± 0.18	105.55 ± 0.21	106.44 ± 0.21 *	<0.001 [a]	105.05 ± 0.19	105.63 ± 0.23	106.43 ± 0.18 *	<0.001 [a]
DBP (mmHg)	65.81 ± 0.13	65.65 ± 0.15	65.89 ± 0.13	0.471	65.71 ± 0.12	65.62 ± 0.15	66.05 ± 0.14	0.082	65.59 ± 0.13	65.93 ± 0.16	65.87 ± 0.13	0.171
Dietary conditions, mean ± SE												
Fruit, servings/day	1.06 ± 0.02	1.09 ± 0.02 *	1.25 ± 0.02 *	<0.001 [c]	1.05 ± 0.02	1.08 ± 0.02 *	1.28 ± 0.02 *	<0.001 [c]	1.07 ± 0.02	1.09 ± 0.02 *	1.22 ± 0.02 *	<0.001 [c]
Vegetables, servings/day	1.84 ± 0.02	1.82 ± 0.03	2.01 ± 0.03 *	<0.001 [c]	1.76 ± 0.02	1.79 ± 0.02 *	2.17 ± 0.03 *	<0.001 [c]	1.79 ± 0.03	1.90 ± 0.03 *	1.98 ± 0.02 *	<0.001 [c]
Meat, servings/day	1.71 ± 0.02	1.59 ± 0.02 *	1.79 ± 0.03 *	<0.001 [c]	1.64 ± 0.02	1.53 ± 0.02	1.94 ± 0.03 *	<0.001 [c]	1.67 ± 0.02	1.65 ± 0.03	1.76 ± 0.02 *	<0.001 [c]
Milk, servings/day	0.61 ± 0.01	0.59 ± 0.01	0.72 ± 0.01 *	<0.001 [c]	0.60 ± 0.01	0.63 ± 0.01 *	0.71 ± 0.10 *	<0.001 [c]	0.58 ± 0.01	0.61 ± 0.01 *	0.71 ± 0.01 *	<0.001 [c]
SSBs, servings/day	0.21 ± 0.01	0.25 ± 0.01 *	0.32 ± 0.01 *	<0.001 [c]	0.24 ± 0.01	0.24 ± 0.01 *	0.31 ± 0.01 *	<0.001 [c]	0.24 ± 0.01	0.24 ± 0.01 *	0.29 ± 0.01 *	<0.001 [c]
Fried food times/week	0.62 ± 0.02	0.82 ± 0.02 *	1.00 ± 0.02 *	<0.001 [c]	0.67 ± 0.02	0.81 ± 0.02 *	0.99 ± 0.03 *	<0.001 [c]	0.68 ± 0.02	0.77 ± 0.02 *	0.95 ± 0.02 *	<0.001 [c]

Abbreviations: BMI, body mass index; DBP, diastolic blood pressure; SBP, systolic blood pressure; SE, standard error; SSB, sugar-sweetened beverage; WC, waist circumference; ANOVA, analysis of variance. [†] None [1], 0 cups/day; medium [1], 0–0.4 cups/day; high [1], >0.4 cups/day. None [2], 0–0.1429 servings/day; medium [2], 0.1429–0.4286 servings/day; high [2], >0.4286 servings/day. None [3], 0–1 times/week; medium [3], 1–3 times/week; high [3], >3 times/week. * $p < 0.05$, compared with no intake (χ2 test or one-way ANOVA test to discriminate pairwise comparisons). [a] $p < 0.05$, boys vs. girls, assessed by one-way ANOVA. [b] $p < 0.05$, boys vs. girls, assessed by the χ2 test for categorical variables. [c] $p < 0.05$, boys vs. girls, assessed by the Kruskal–Wallis H test.

3.3. The Distribution of Obesity and Blood Pressure Stratified by Soy Food Intake

The distributions of obesity and blood pressure according to soy food intake are shown in Table 3. Compared with no servings, a high liquid soy food intake was associated with a lower prevalence of normal BMI and a higher prevalence of hypertension (both $p < 0.05$), and children who consumed a high number of servings of solid soy food were more likely to be obese ($p = 0.032$). In addition, the prevalence of pre-hypertension was higher in the high-frequency group ($p < 0.05$). However, soy food intake showed no association with abdominal obesity.

3.4. ORs (95% CIs) for Obesity and Hypertension across Soy Food Intake

The association between the risks of obesity, abdominal obesity, and hypertension in children and adolescents across the soy food intake categories is shown in Table 4. The crude OR of hypertension to high servings of liquid soy food intake compared to no liquid servings was 0.840 (95% CI, 0.717–0.984). After adjustment for age, sex, BMI, family income, and dietary intake, compared with no consumption, the OR for hypertension in the high servings of liquid soy food intake and high frequency of the all soy food intake group were 0.790 (95% CI, 0.666–0.937) and 0.828 (95% CI, 0.704–0.974), respectively. The crude OR of obesity to high solid servings compared to no solid servings was 1.287 (95% CI, 1.062–1.559), and the adjusted OR was 1.335 (95% CI, 1.091–1.633). The adjusted OR for obesity in the high frequency of the all soy food intake group was 1.302 (95% CI, 1.070–1.583). However, no significant correlation was found between soy food intake and abdominal obesity.

Table 3. The distribution of obesity and blood pressure stratified by soy food intake (%).

	Total	Liquid Soy Food (Cups/Day)				Solid Soy Food (Servings/Day)				All Soy Food (Times/Week)			
		None [1,†]	Medium [1,†]	High [1,†]	p Value	None [2,†]	Medium [2,†]	High [2,†]	p Value	None [3,†]	Medium [3,†]	High [3,†]	p Value
BMI					0.004 [a]				0.059				0.105
Normal, %	70.6	72.0	68.8 *	70.7	0.013 [a]	71.2	70.5	69.8	0.381	70.5	70.8	70.4	0.935
Overweight, %	9.5	9.5	9.0	9.9	0.448	9.4	8.8	10.3	0.139	9.6	8.6	10.0	0.138
Obesity, %	6.2	5.8	6.5	6.3	0.414	5.5	6.5	6.9 *	0.032 [a]	5.7	6.2	6.7	0.175
Abdominal obesity					0.639				0.143				0.074
No, %	86.1	86.5	86.0	85.7		86.4	86.6	85.1		86.2	87.2	85.3	
Yes, %	13.9	13.5	14.0	14.3		13.6	13.4	14.9		13.8	12.8	14.7	
Hypertension HTN					0.038 [a]				0.263				0.116
Normal, %	75.6	75.3	75.3	76.3	0.569	75.9	76.1	74.9	0.528	76.0	75.2	75.6	0.750
Pre-hypertension, %	14.5	14.5	13.9	15.1	0.395	14.0	14.0	15.6	0.095	13.7	14.4	15.3 *	0.142
Hypertension, %	9.9	10.2	10.8	8.7 *	0.010 [a]	10.1	9.9	9.4	0.600	10.3	10.5	9.1	0.114

Abbreviations: BMI, body mass index; DBP, diastolic blood pressure; SBP, systolic blood pressure. [†] None [1], 0 cups/day; medium [1], 0–0.4 cups/day; high [1], >0.4 cups/day. None [2], 0–0.1429 servings/day; medium [2], 0.1429–0.4286 servings/day; high [2], >0.4286 servings/day. None [3], 0–1 times/week; medium [3], 1–3 times/week; high [3], >3 times/week. * $p < 0.05$, compared with no intake (χ2 test to discriminate pairwise comparisons). [a] $p < 0.05$, boys vs. girls, assessed by the χ2 test for categorical variables.

Table 4. Odds ratios (95% confidence intervals) for obesity and hypertension across soy food intake groups.

	Overweight OR (95% CI)	Obesity OR (95% CI)	Abdominal Obesity OR (95% CI)	Pre-Hypertension OR (95% CI)	Hypertension OR (95% CI)
Liquid soy food (cups/day) [a]					
Model 1					
No servings	1	1	1	1	1
Medium servings	0.987 (0.837–1.164)	1.179 (0.968–1.436)	1.044 (0.911–1.197)	0.960 (0.837–1.101)	1.057 (0.906–1.235)
High servings	1.057 (0.904–1.235)	1.112 (0.917–1.348)	1.064 (0.933–1.214)	1.028 (0.903–1.171)	0.840 (0.717–0.984) *
Model 2					
No servings	1	1	1	1	1
Medium servings	0.994 (0.841–1.175)	1.141 (0.934–1.394)	1.026 (0.894–1.177)	1.098 (0.954–1.263)	1.076 (0.920–1.257)
High servings	1.071 (0.915–1.253)	1.154 (0.949–1.402)	1.071 (0.939–1.223)	1.000 (0.876–1.142)	0.839 (0.716–0.983) *
Model 3					
No servings	1	1	1	1	1
Medium servings	1.009 (0.852–1.239)	1.163 (0.950–1.425)	1.027 (0.893–1.180)	1.123 (0.973–1.297)	1.116 (0.949–1.312)
High servings	1.072 (0.913–1.259)	1.175 (0.964–1.431)	1.079 (0.944–1.234)	0.985 (0.860–1.129)	0.804 (0.681–0.948) *
Model 4					
No servings	1	1	1	1	1
Medium servings	1.067 (0.896–1.272)	1.131 (0.918–1.394)	1.032 (0.893–1.192)	1.091 (0.941–1.265)	1.112 (0.941–1.313)
High servings	1.123 (0.951–1.327)	1.194 (0.975–1.461)	1.124 (0.979–1.291)	0.958 (0.832–1.102)	0.790 (0.666–0.937) *
Solid soy food (servings/day) [b]					
Model 1					
No servings	1	1	1	1	1
Medium servings	0.944 (0.802–1.112)	1.195 (0.981–1.455)	0.984 (0.859–1.127)	0.997 (0.871–1.141)	0.975 (0.835–1.140)
High servings	1.113 (0.952–1.301)	1.287 (1.062–1.559) *	1.118 (0.980–1.274)	1.129 (0.991–1.286)	0.942 (0.806–1.102)
Model 2					
No servings	1	1	1	1	1
Medium servings	0.961 (0.815–1.134)	1.220 (1.000–1.488)	0.986 (0.861–1.130)	1.005 (0.876–1.154)	0.981 (0.839–1.146)
High servings	1.123 (0.959–1.315)	1.365 (1.123–1.659) *	1.137 (0.997–1.297)	1.027 (0.898–1.173)	0.934 (0.798–1.093)
Model 3					
No servings	1	1	1	1	1
Medium servings	0.974 (0.823–1.152)	1.212 (0.992–1.481)	0.974 (0.848–1.118)	1.000 (0.868–1.151)	0.980 (0.834–1.152)
High servings	1.132 (0.964–1.330)	1.330 (1.092–1.620) *	1.123 (0.983–1.283)	0.997 (0.87–1.143)	0.866 (0.735–1.020)
Model 4					
No servings	1	1	1	1	1
Medium servings	1.008 (0.848–1.199)	1.189 (0.967–1.461)	0.968 (0.840–1.116)	0.974 (0.842–1.127)	0.985 (0.834–1.163)
High servings	1.158 (0.981–1.367)	1.335 (1.091–1.633) *	1.147 (1.000–1.315)	1.006 (0.874–1.158)	0.867 (0.732–1.027)
All soy food (times/week) [c]					
Model 1					
No frequency	1	1	1	1	1
Medium frequency	0.887 (0.743–1.059)	1.086 (0.878–1.343)	0.912 (0.787–1.056)	1.060 (0.917–1.225)	1.027 (0.871–1.211)
High frequency	1.042 (0897–1.212)	1.180 (0.980–1.422)	1.076 (0.948–1.221)	1.120 (0.986–1.271)	0.891 (0.767–1.036)
Model 2					
No frequency	1	1	1	1	1
Medium frequency	0.894 (0.748–1.069)	1.104 (0.891–1.368)	0.916 (0.790–1.061)	1.043 (0.899–1.209)	1.027 (0.871–1.212)
High frequency	1.061 (0.910–1.236)	1.281 (1.060–1.548) *	1.103 (0.971–1.253)	0.972 (0.853–1.108)	0.881 (0.757–1.025)
Model 3					
No frequency	1	1	1	1	1
Medium frequency	0.884 (0.737–1.061)	1.119 (0.902–1.389)	0.898 (0.773–1.043)	1.059 (0.911–1.232)	1.034 (0.871–1.227)
High frequency	1.062 (0.909–1.241)	1.256 (1.036–1.521) *	1.089 (0.957–1.239)	0.938 (0.821–1.073)	0.827 (0.706–0.968) *
Model 4					
No frequency	1	1	1	1	1
Medium frequency	0.896 (0.743–1.081)	1.089 (0.871–1.361)	0.887 (0.760–1.036)	1.072 (0.918–1.252)	1.051 (0.881–1.254)
High frequency	1.087 (0.926–1.277)	1.302 (1.070–1.583) *	1.119 (0.981–1.278)	0.937 (0.816–1.075)	0.828 (0.704–0.974) *

* $p < 0.05$. [a] None [1], 0 cups/day; medium [1], 0–0.4 cups/day; high [1], > 0.4 cups/day; [b] None [2], 0–0.1429 servings/day; medium [2], 0.1429–0.4286 servings/day; high [2], > 0.4286 servings/day. [c] None [3], 0–1 times/week; medium [3], 1–3 times/week; high [3], >3 times/week. Model 1 is without adjustment; model 2 is adjusted for age and sex; model 3 is additionally adjusted for family income for overweight, obesity, and abdominal obesity and family income and BMI for blood pressure; model 4 is additionally adjusted for dietary intakes (intake frequency of fried food and SSB).

4. Discussion

With the sharply increasing prevalence of childhood obesity and hypertension in recent years [32], several studies have explored the effect of habitual diet in regulating weight and

BP. However, the association between soy food intake and cardio-metabolic risk factors in children and adolescents remains unclear due to limited evidence. Using a cross-sectional survey of 10,536 children and adolescents aged 7–18 years old, we analyzed the daily soy food intake and determined that high soy food intake (>3 times/week) was correlated with a higher prevalence of obesity and lower prevalence of hypertension independent of family income and consumption of meat and fried foods. However, no association between soy food and abdominal obesity was observed.

In this study, the mean intakes of liquid and solid soy food were 0.35 ± 0.54 cups/day and 0.46 ± 0.63 servings/day, respectively, with 39.5% of participants consuming soy food more than 3 times per week. The overall frequency of all soy products was 2.41 times per week among participants. Hsiao reported a mean daily soy food intake of 150.1 g among children aged 8–9 years old in Taiwan [33], which is similar to the intake estimated by our study. Daily soy food intake was less than 61 g for 7–14 years old participants in Japan [34], but higher than 40 g of soymilk in 4–14-year-olds in Hong Kong [35]. Nevertheless, Leung et al. reported that, in Hong Kong, young boys had approximately 10 soy-containing meals per week [35], which was much higher than the consumption in our study. In addition, a survey of almost 200 12-year-olds in Beijing found that 41% normally included soymilk or tofu at breakfast [36]. These findings revealed that soy food intake by children was quite uneven across Asian countries [37], even in China. This can be partly explained by regional differences in dietary patterns and a lack of information about soy consumption by children relative to adults.

The present study has shown that habitual soy food intake lowers the risk of high BP. A meta-analysis of 12 trials indicates that the ingestion of at least 25 g of soy protein per day has BP-lowering effects [22], which is consistent with our findings. Additionally, a prospective cohort study in Japan demonstrated that the intake of fermented soy products showed a protecting effect of BP in healthy populations [21]. Soy and some of its constituents mitigate hypertension through the effects of vasodilation [10], which may be caused by isoflavones [38]. However, clinical evidence supporting the underlying mechanisms remains controversial. There are also some studies showing that no significant changes in BP could be observed after the consumption of soy food [7,23].

A growing body of literature suggests soy supplementation affects weight control and obesity prevention [6,24,39–41]. For instance, in a 12-week randomized controlled trial, Beavers et al. proposed that soy protein is at least as good as other protein sources for weight loss during low-calorie dietary interventions in older adults [11]. Another study recorded significant changes in the median percentage of body weight (-1.5%, $p = 0.005$) and BMI (-1.5%, $p = 0.05$) in the soy group [24]. There are also studies reporting that the protective effect of legumes, including soy food, beans, and soy dietary supplements, on weight is still insufficient [25–27]. Moreover, some studies present conflicting results. For example, a systematic review and meta-analysis suggested that soy showed no statistically significant effect on weight in the general population, but an obesogenic effect of soy was observed in obese subjects (BMI ≥ 30 kg/m^2) and participants whose daily soy food intake included at least 40 g of soy protein [28], which may be explained by the fact that obese people have less control over their diet and tend to consume more amounts of various foods in daily life compared to people in the normal weight range. Our findings, in part, were consistent with this previous study in that greater soy food intake (>3 times/week) was associated with an increased risk of obesity. In summary, these controversial findings might be due to the weight-loss property of isoflavones [42] and the weight-gain effect of a high amount of soy consumption. Briefly, soy and isoflavones have different impacts on weight status [28].

However, no significant associations were observed between soy food intake and abdominal obesity in this study or in previous studies [28] conducted in children.

Our study has several strengths. First, the current study was based on a large sample of children and adolescents across a wide age range living in Guangzhou City, which is the third-largest city in China and the largest city in southern China. It has a large immigrant population; this could assure the reliability of the data and widen the application of our results. Second, we analyzed the association between the consumption of soy food and cardio-metabolic risk factors in a pediatric population stratified by age and sex. Still, several limitations of the study should not be overlooked. First, as our study applied a cross-sectional analysis, the causality of the relationships observed could hardly be determined with certainty. Second, self-reported questionnaires were used to collect the data on dietary information of the previous week; hence, the data might be affected by recall bias. Finally, although we controlled for several dietary and other potential confounding factors, residual confounding by unknown or unmeasured factors might have also been present.

5. Conclusions

In conclusion, 39.5% of 10,536 children and adolescents aged 7–18 years living in Guangzhou City, southern China, had high soy food intake (>3 times/week), and this was significantly associated with a lower prevalence of hypertension and higher prevalence of obesity. Our results may provide a new diet intervention method to reduce the risk of hypertension in childhood and associated cardiovascular diseases in adulthood. However, large prospective studies need to be carried out to confirm the present findings, which can be useful in clarifying the mechanisms and mitigating the current epidemics of childhood obesity and hypertension.

Author Contributions: Conceptualization, T.H. and Y.Z.; data curation, X.W.; formal analysis, T.H.; funding acquisition, Y.Z.; investigation, H.L., M.W., Z.L. and F.H.; methodology, X.W.; project administration, Y.Z.; resources, Y.Z.; software, X.W. and T.H.; supervision, Y.Z.; validation, S.X.; writing—original draft, T.H.; writing—review and editing, X.W. and Y.Z. All authors have read and agreed to the published version of the manuscript.

Funding: The study was funded by a grant from the Research Fund for the Guangdong Provincial Natural Science Foundation (Grant No. 2021A1515010439), and the Sanming Project of Medicine in Shenzhen (Grant No. SZSM201803061). The funders had no role in the study design, data collection and analysis, decision to publish, or manuscript preparation.

Institutional Review Board Statement: The study was conducted according to the guidelines of the Declaration of Helsinki, and approved by the Ethical Committee of School of Public Health, Sun Yat-sen University (Number (2018)005).

Informed Consent Statement: Informed consent forms were voluntarily signed by all participants and their legal guardians.

Data Availability Statement: All data generated or analyzed during this study are included in this article. Further inquiries can be directed to the corresponding author.

Acknowledgments: We thank the participants, doctors, and nurses involved in this study.

Conflicts of Interest: The authors declare no conflict of interest. The funders had no role in the design of the study; data collection, analyses, or interpretation of data; in the writing of the manuscript; or in the decision to publish the results.

References

1. Sun, J.; Buys, N.; Shen, S. Dietary patterns and cardiovascular disease-related risks in chinese older adults. *Front. Public Health* **2013**, *1*, 48. [CrossRef] [PubMed]
2. Takahira, M.; Noda, K.; Fukushima, M.; Zhang, B.; Mitsutake, R.; Uehara, Y.; Ogawa, M.; Kakuma, T.; Saku, K. Randomized, double-blind, controlled, comparative trial of formula food containing soy protein vs. milk protein in visceral fat obesity FLAVO study. *Circ. J.* **2011**, *75*, 2235–2243. [CrossRef]
3. Xi, B.; Bovet, P.; Hong, Y.M.; Zong, X.; Chiolero, A.; Kim, H.S.; Zhang, T.; Zhao, M. Recent blood pressure trends in adolescents from China, Korea, Seychelles and the United States of America, 1997–2012. *J. Hypertens.* **2016**, *34*, 1948–1958. [CrossRef] [PubMed]
4. Jahan-Mihan, A.; Smith, C.E.; Hamedani, A.; Anderson, G.H. Soy protein–based compared with casein-based diets fed during pregnancy and lactation increase food intake and characteristics of metabolic syndrome less in female than male rat offspring. *Nutr. Res.* **2011**, *31*, 644–651. [CrossRef]

5. Dietz, W.H. Health consequences of obesity in youth: Childhood predictors of adult disease. *Pediatrics* **1998**, *101*, 518–525. [CrossRef] [PubMed]
6. Anderson, J.W.; Fuller, J.; Patterson, K.; Blair, R.; Tabor, A. Soy compared to casein meal replacement shakes with energy-restricted diets for obese women: Randomized controlled trial. *Metabolism* **2007**, *56*, 280–288. [CrossRef]
7. Mohammadifard, N.; Salehi-Abargouei, A.; Salas-Salvadó, J.; Guasch-Ferré, M.; Humphries, K.; Sarrafzadegan, N. The effect of tree nut, peanut, and soy nut consumption on blood pressure: A systematic review and meta-analysis of randomized controlled clinical trials. *Am. J. Clin. Nutr.* **2015**, *101*, 966–982. [CrossRef]
8. McCarty, M.F. Vegan Proteins May Reduce Risk of Cancer, Obesity, and Cardiovascular Disease by Promoting Increased Glucagon Activity. *Med. Hypotheses* **1999**, *53*, 459–485. [CrossRef]
9. Hu, F.B. Do functional foods have a role in the prevention of cardiovascular disease? *Circulation* **2011**, *124*, 538–540. [CrossRef]
10. Ramdath, D.; Padhi, E.; Sarfaraz, S.; Renwick, S.; Duncan, A. Beyond the Cholesterol-Lowering Effect of Soy Protein: A Review of the Effects of Dietary Soy and Its Constituents on Risk Factors for Cardiovascular Disease. *Nutrients* **2017**, *9*, 324. [CrossRef] [PubMed]
11. Beavers, K.M.; Gordon, M.M.; Easter, L.; Beavers, D.P.; Hairston, K.G.; Nicklas, B.J.; Vitolins, M.Z. Effect of protein source during weight loss on body composition, cardiometabolic risk and physical performance in abdominally obese, older adults: A pilot feeding study. *J. Nutr. Health Aging* **2015**, *19*, 87–95. [CrossRef] [PubMed]
12. Sharifi-Zahabi, E.; Entezari, M.H.; Maracy, M.R. Effects of Soy Flour Fortified Bread Consumption on Cardiovascular Risk Factors According toAPOE Genotypes in Overweight and Obese Adult Women: A Cross-over Randomized Controlled Clinical Trial. *Clin. Nutr. Res.* **2015**, *4*, 225. [CrossRef]
13. Lee, M.; Kim, I. Soy protein and obesity. *Nutrients* **2000**, *16*, 459–460. [CrossRef]
14. Pan, A.; Franco, O.H.; Ye, J.; Demark-Wahnefried, W.; Ye, X.; Yu, Z.; Li, H.; Lin, X. Soy protein intake has sex-specific effects on the risk of metabolic syndrome in middle-aged and elderly Chinese. *J. Nutr.* **2008**, *138*, 2413–2421. [CrossRef] [PubMed]
15. Simao, A.N.; Lozovoy, M.A.; Simao, T.N.; Dichi, J.B.; Matsuo, T.; Dichi, I. Nitric oxide enhancement and blood pressure decrease in patients with metabolic syndrome using soy protein or fish oil. *Arq. Bras. Endocrinol. Metabol.* **2010**, *54*, 540–545. [CrossRef]
16. Palanisamy, N.; Viswanathan, P.; Ravichandran, M.K.; Anuradha, C.V. Renoprotective and blood pressure-lowering effect of dietary soy protein via protein kinase C beta II inhibition in a rat model of metabolic syndrome. *Can. J. Physiol. Pharmacol.* **2010**, *88*, 28–37. [CrossRef]
17. Hermsdorff, H.H.; Zulet, M.A.; Abete, I.; Martinez, J.A. A legume-based hypocaloric diet reduces proinflammatory status and improves metabolic features in overweight/obese subjects. *Eur. J. Nutr.* **2011**, *50*, 61–69. [CrossRef] [PubMed]
18. Hecker, K.D. Effects of dietary animal and soy protein on cardiovascular disease risk factors. *Curr. Atheroscler. Rep.* **2001**, *3*, 471–478. [CrossRef] [PubMed]
19. Bakhtiary, A.; Yassin, Z.; Hanachi, P.; Rahmat, A.; Ahmad, Z.; Jalali, F. Effects of soy on metabolic biomarkers of cardiovascular disease in elderly women with metabolic syndrome. *Arch. Iran. Med.* **2012**, *15*, 462–468.
20. He, J.; Gu, D.; Wu, X.; Chen, J.; Duan, X.; Chen, J.; Whelton, P.K. Effect of soybean protein on blood pressure: A randomized, controlled trial. *Ann. Intern. Med.* **2005**, *143*, 1–9. [CrossRef] [PubMed]
21. Nozue, M.; Shimazu, T.; Sasazuki, S.; Charvat, H.; Mori, N.; Mutoh, M.; Sawada, N.; Iwasaki, M.; Yamaji, T.; Inoue, M.; et al. Fermented Soy Product Intake Is Inversely Associated with the Development of High Blood Pressure: The Japan Public Health Center-Based Prospective Study. *J. Nutr.* **2017**, *147*, 1749–1756. [CrossRef]
22. Kou, T.; Wang, Q.; Cai, J.; Song, J.; Du, B.; Zhao, K.; Ma, Y.; Geng, B.; Zhang, Y.; Han, X.; et al. Effect of soybean protein on blood pressure in postmenopausal women: A meta-analysis of randomized controlled trials. *Food Funct.* **2017**, *8*, 2663–2671. [CrossRef] [PubMed]
23. Teede, H.J.; Giannopoulos, D.; Dalais, F.S.; Hodgson, J.; McGrath, B.P. Randomised, controlled, cross-over trial of soy protein with isoflavones on blood pressure and arterial function in hypertensive subjects. *J. Am. Coll. Nutr.* **2006**, *25*, 533–540. [CrossRef]
24. Ruscica, M.; Pavanello, C.; Gandini, S.; Gomaraschi, M.; Vitali, C.; Macchi, C.; Morlotti, B.; Aiello, G.; Bosisio, R.; Calabresi, L.; et al. Effect of soy on metabolic syndrome and cardiovascular risk factors: A randomized controlled trial. *Eur. J. Nutr.* **2018**, *57*, 499–511. [CrossRef] [PubMed]
25. Williams, P.G.; Grafenauer, S.J.; O'Shea, J.E. Cereal grains, legumes, and weight management: A comprehensive review of the scientific evidence. *Nutr. Rev.* **2008**, *66*, 171–182. [CrossRef] [PubMed]
26. Newby, P.K. Plant foods and plant-based diets: Protective against childhood obesity? *Am. J. Clin. Nutr.* **2009**, *89*, 1572S–1587S. [CrossRef] [PubMed]
27. Carmignani, L.O.; Pedro, A.O.; Da, C.L.; Pinto-Neto, A.M. The effect of soy dietary supplement and low dose of hormone therapy on main cardiovascular health biomarkers: A randomized controlled trial. *Rev. Bras. Ginecol. Obstet.* **2014**, *36*, 251–258. [CrossRef]
28. Akhlaghi, M.; Zare, M.; Nouripour, F. Effect of Soy and Soy Isoflavones on Obesity-Related Anthropometric Measures: A Systematic Review and Meta-analysis of Randomized Controlled Clinical Trials. *Adv. Nutr.* **2017**, *8*, 705–717. [CrossRef] [PubMed]
29. Group of China Obesity Task Force. Body mass index reference norm for screening overweight and obesity in Chinese children and adolescents. *Zhonghua Liu Xing Bing Xue Za Zhi* **2004**, *25*, 97–102.
30. Ma, G.S.; Ji, C.Y.; Ma, J.; Mi, J.; Sung, R.Y.; Xiong, F.; Yan, W.L.; Hu, X.Q.; Li, Y.P.; Du, S.M.; et al. Waist circumference reference values for screening cardiovascular risk factors in Chinese children and adolescents. *Biomed. Environ. Sci.* **2010**, *23*, 21–31. [CrossRef]

31. Mi, J.; Wang, T.; Meng, L.; Zhu, G.; Han, S.; Zhong, Y.; Liu, G.; Wan, Y.; Xiong, F.; Shi, J.; et al. Development of blood pressure reference standards for Chinese children and adolescents. *Chin. J. Evid. Based Pediatr.* **2010**, *5*, 4–14.

32. Messina, M. Soy and Health Update: Evaluation of the Clinical and Epidemiologic Literature. *Nutrients* **2016**, *8*, 754. [CrossRef] [PubMed]

33. Hsiao, L.P. Soy consumption in Taiwanese children in Taipei. *J. Nutr.* **2000**, *130*, 705S.

34. Messina, M.; Nagata, C.; Wu, A.H. Estimated Asian Adult Soy Protein and Isoflavone Intakes. *Nutr. Cancer* **2006**, *55*, 12. [CrossRef] [PubMed]

35. Leung, S.S.F.; Lee, R.H.Y.; Luo, H.Y.; Kam, C.W.; Yuen, M.P.; Hjelm, M.; Lee, S.H. Growth and nutrition of Chinese vegetarian children in HongKong. *J. Paediatr. Child Health* **2001**, *37*, 247–253. [CrossRef]

36. Wang, D.; Shi, Y.; Chang, C.; Stewart, D.; Ji, Y.; Wang, Y.; Harris, N. Knowledge, attitudes and behaviour regarding nutrition and dietary intake of seventh-grade students in rural areas of Mi Yun County, Beijing, China. *Environ. Health Prev. Med.* **2014**, *19*, 179–186. [CrossRef] [PubMed]

37. Messina, M.; Rogero, M.M.; Fisberg, M.; Waitzberg, D. Health impact of childhood and adolescent soy consumption. *Nutr. Rev.* **2017**, *75*, 500–515. [CrossRef]

38. Liu, X.X.; Li, S.H.; Chen, J.Z.; Sun, K.; Wang, X.J.; Wang, X.G.; Hui, R.T. Effect of soy isoflavones on blood pressure: A meta-analysis of randomized controlled trials. *Nutr. Metab. Cardiovasc. Dis.* **2012**, *22*, 463–470. [CrossRef] [PubMed]

39. Allison, D.B.; Gadbury, G.; Schwartz, L.G.; Murugesan, R.; Kraker, J.L.; Heshka, S.; Fontaine, K.R.; Heymsfield, S.B. A Novel Soy-Based Meal Replacement Formula for Weight Loss among Obese Individuals: A Randomized Controlled Clinical Trial. *Eur J. Clin. Nutr.* **2003**, *57*, 514–522. [CrossRef] [PubMed]

40. Hu, X.; Gao, J.; Zhang, Q.; Fu, Y.; Li, K.; Zhu, S.; Li, D. Soy fiber improves weight loss and lipid profile in overweight and obese adults: A randomized controlled trial. *Mol. Nutr. Food Res.* **2013**, *57*, 2147–2154. [CrossRef]

41. Liao, F.; Shieh, M.; Yang, S.; Lin, S.; Chien, Y. Effectiveness of a soy-based compared with a traditional low-calorie diet on weight loss and lipid levels in overweight adults. *Nutrition* **2007**, *23*, 551–556. [CrossRef] [PubMed]

42. Tan, J.; Huang, C.; Luo, Q.; Liu, W.; Cheng, D.; Li, Y.; Xia, Y.; Li, C.; Tang, L.; Fang, J.; et al. Soy Isoflavones Ameliorate Fatty Acid Metabolism of Visceral Adipose Tissue by Increasing the AMPK Activity in Male Rats with Diet-Induced Obesity (DIO). *Molecules* **2019**, *24*, 2809. [CrossRef] [PubMed]

 *nutrients*

Article

An Assessment of the Nutritional Value of the Preschool Food Rations for Children from the Wroclaw District, Poland—The Case of a Big City

Agnieszka Orkusz

Department of Biotechnology and Food Analysis, Wroclaw University of Economics and Business, 53-345 Wroclaw, Poland; agnieszka.orkusz@ue.wroc.pl; Tel.: +48-713680480

Abstract: The evaluation of nutrition is an essential element of preventing chronic diseases and can be used to determine nutritional recommendations. A child spends about 7–8 h a day in a kindergarten; therefore, meals served there should be balanced appropriately to ensure the full psychophysical development of the young organism. At preschool age, children develop eating habits that can have life-long effects. Based on 10-day menus, the study aimed to estimate the energy and nutritional value of children's diets at four randomly selected kindergartens in the Wroclaw district, Poland. In total, 80 menus were analyzed (40 for summer and 40 for autumn). The data from kindergartens were analyzed based on the Diet 6D computer program. Regardless of the kindergarten, the analyzed food rations showed irregularities related to excessive supplies (in reference to the dietary recommendations) of sucrose, fiber, salt, magnesium, and vitamin A. The preschool food rations did not cover demands with respect to PUFA n-3, PUFA-n-6, calcium, and vitamin D. The observed irregularities confirm the need to monitor the content of energy and nutrients in preschool menus to be able to correct any dietary errors.

Keywords: child; diet; kindergarten; menus; nutritional errors

Citation: Orkusz, A. An Assessment of the Nutritional Value of the Preschool Food Rations for Children from the Wroclaw District, Poland—The Case of a Big City. *Nutrients* **2022**, *14*, 460. https://doi.org/10.3390/nu14030460

Academic Editor: Iain A. Brownlee

Received: 10 December 2021
Accepted: 19 January 2022
Published: 20 January 2022

Publisher's Note: MDPI stays neutral with regard to jurisdictional claims in published maps and institutional affiliations.

Copyright: © 2022 by the author. Licensee MDPI, Basel, Switzerland. This article is an open access article distributed under the terms and conditions of the Creative Commons Attribution (CC BY) license (https://creativecommons.org/licenses/by/4.0/).

1. Introduction

The proper nutrition of pre-school children guarantees their optimal health [1,2]. Because a child spends about 7–8 h a day in kindergarten, the food served there should be balanced appropriately to provide energy and nutrients in the right amounts and proportions [3]. The collective nutrition of children and adolescents within the education system must meet the appropriate requirements for given age groups, adhering to the current nutrition standards for the Polish population [4]. The following standards were adopted in the latest 2020 Polish nutrition standards: Estimated Energy Requirement (EER); Estimated Average Requirement (EAR); Recommended Dietary Allowance (RDA); Adequate Intake (AI), and Reference Intake ranges for macronutrients (RIs) [4]. In planning mass nutrition in kindergartens, the Recommended Dietary Allowance (RDA) level should be applied. This guarantees that the planned nutrition will cover the average demand of the group of preschoolers at the EAR level [5]. In addition, the Tolerable Upper Intake Level (UL) has been developed for vitamins and mineral ingredients. The UL is the maximum level of nutrient intake from all sources (food, drinking water, dietary supplements) that does not cause adverse health effects in almost all people in each group. Consumption of nutrients above this level may cause adverse health effects. The UL value is a guideline level that should not be pursued as a standard but should be observed to ensure that it is not exceeded [4]. A child receives at least 3 meals daily in kindergarten. Therefore, it is assumed that the daily preschool ration should cover 75% of energy and nutrient needs [2,5–10].

Proper nutrition in kindergarten is also crucial for shaping the correct eating habits of children and can prevent the later development of many diet-related diseases (obesity, type

2 diabetes, cardiovascular diseases) [11]. The statistics related to obesity among children and adolescents are worrying across the world, including Poland. Data from the World Health Organization indicate an upward trend in obesity among children aged 5–9 and 10–19 [12]. Over 10 years, boys showed an increase in obesity of 6.3% and 4.8%, respectively, while girls showed an increase of 3.0% and 1.7%. It is estimated that, by 2025, if the rate of increase in obesity continues to grow at a similar pace, about 268 million children aged 5–17 years will be overweight, of whom 91 million will be obese [13]. In 2025, 38 million children will have hepatic steatosis or a buildup of fat in the liver, 27 million will have hypertension, 12 million will suffer from impaired glucose tolerance, and 4 million will have type 2 diabetes [14]. A lot of studies show that obesity in childhood is strongly associated with obesity in adulthood [15–18]. Most children who gained excess weight before the age of 6 remain overweight during puberty [18]. It has been shown that the earlier the increase in body fat occurs, the greater the final weight of the child is [15,19].

Many authors point to improper nutrition of children in kindergartens in Poland [7,20–31]. Meals are improperly balanced, both in terms of energy and nutritional value. It also happens that they are prepared from low-quality ingredients. This situation may be due to the insufficient knowledge and skills of the people responsible for preschool nutrition, which suggests the need to implement nutritional education for staff [2]. Perhaps a regular assessment of menus would help to avoid basic mistakes that occur when composing them.

Based on 10-day menus, the study aimed at estimating the energy and nutritional value of children's diets in four randomly selected kindergartens in the Wroclaw district, Poland.

2. Materials and Methods

2.1. Data

The research was conducted in 2019. The menus from the 4 randomly selected preschool institutions (designated as: A, B, C, D) located in the Wroclaw district were assessed. There were 5–7 pre-school groups in the facilities, to which children aged 4 to 6 were assigned. Each group consisted of about 25 children. The kindergartens had their own kitchens where meals were prepared (the services of catering companies were not used). The 10-day menus from the summer and autumn seasons were assessed in each of 4 kindergartens. In total, 80 menus were analyzed (40 menus for summer and 40 menus for autumn).

To assess the energy value and the content of nutrients in the food rations, such as proteins, fats, carbohydrates (including glucose, fructose, sucrose, lactose, starch), fiber, minerals (sodium (Na), potassium (K), calcium (Ca), phosphorus (P), magnesium (Mg), iron (Fe), zinc (Zn), copper (Cu), and manganese (Mn)), vitamins (thiamine (B_1), riboflavin (B_2), niacin (B_3), pyridoxine (B_6), and vitamins A, C, D, and E), cholesterol, and fatty acids (saturated, monounsaturated, polyunsaturated), the Diet 6D computer program [32] was implemented. The salt content was also calculated. The Diet 6D program has proved to be the best in Poland so far. It offers the possibility of calculating energy intake and as many as 89 nutrients. The Diet program contains a food composition database based on the Polish food composition tables. The portion size was estimated based on the number of products, meals, and drinks expressed in grams specified in the menus. The calculations considered losses resulting from technological processes. Therefore, for the energy value, total protein, fats, carbohydrates, calcium, and iron, losses equal to 10% were used, and for vitamins A and C, losses equivalent to 20% and 55%, respectively, were used [33]. Then all values for 10-day menus were totaled. In this way, for each kindergarten, the average values of energy and nutrients were estimated. Next, the obtained results were compared with the applicable standards and recommendations for children aged 4–6 years [4]. Energy standards were set at the level of average demand (Estimated Energy Requirement). The Reference Intake ranges, expressed as a percentage of the energy requirement, were defined for fat and carbohydrates. The standards for protein, selected minerals (calcium, phosphorus, magnesium, iron, zinc, copper), and vitamins (thiamine, riboflavin, niacin, pyridoxine, vitamins A and C) were set as per the Recommended Dietary

Allowance. Standards for other nutrients (sodium, potassium, manganese, vitamins D and E, fiber, n-3 PUFA, and n-6 PUFA) were set as per the level of sufficient consumption (Adequate Intake) [4].

To assess the diet of children aged 4–6 years, with a bodyweight of 19 kg and moderate physical activity, a daily energy requirement of 1400 kcal was adopted in accordance with the standards [4]. For children up to the age of 9, the standards are the same for boys and girls. Their differentiation by gender is applied in the older age groups. The energy standards for children up to 6 years of age include the same levels of physical activity. The norms for energy, protein, and fat were established based on body weight, which is understood as a reference body weight value, not the actual weight of a child. The data from studies representative of the Polish population were used to determine the reference body weight values for preschool children (3–6 years) [34]. The reference body weight value for children aged 4–6 years, based on the median, is 19 kg. The energy and nutrient content of the preschool diet was compared to 75% of the daily norm for children aged 4–6 [2,5–10]. According to this, the recommended value for energy was 1050 kcal [4]. The consumption values differing by ±10% from the standard values were assumed to be correct. The content of minerals (Mg, Zn, Cu) and vitamins (A, B_6) was also compared to 75% of the Tolerable Upper Intake Level [35]. Various groups of experts have established UL levels for individual vitamins and minerals. UL values proposed by the European Food Safety Authority were used in the study [35]. It was assumed that the children consumed all served food.

2.2. Statistical Analysis

Two-way variance analysis (ANOVA) was used to evaluate changes in energy value and nutrients in kindergarten menus. The effect of such factors as the kindergarten (A, B, C, D), the season (summer and autumn), and a couple of interactions between them were analyzed. When a significant main effect or interaction was identified, the mean values were further analyzed using the Tukey's multiple range test. The results were presented as a mean ± standard deviation. The experimental data were analyzed using Statistica version 13.3 (StatSoft Inc., Kraków, Poland) [36]. Differences with a probability level <0.05 were considered significant.

2.3. Ethical Statement

Ethical review and approval were waived for this study as the research was based on an evaluation of the preschool menus based on the computer program Diet 6D.

3. Results

Based on the analysis of variance results, it was shown that for energy and nutrients (except for copper, manganese, cholesterol, and PUFA n-3), the main effect was significant for the type of kindergarten for the adopted significance level of $p < 0.05$. For most nutrients, the season factor was not found to be significant. A two-way interaction effect occurred only with carbohydrates and starch (Table 1).

Based on the results of this research, it was shown that the energy value of meals differed significantly depending on the kindergarten, both in summer and in autumn (Table 1). The energy value of meals served in summer was correct in kindergartens A, B and C, while in autumn, this was the case only in kindergartens A and B. The energy value of menus in the kindergarten D was higher than the Estimated Energy Requirement, regardless of the season (Table 2).

Table 1. Mean values ($\pm$ standard deviation) of energy value and nutrient content in preschool menus.

Energy and Nutrients	Season (S)	Kindergarten (K)				Significant Effects		
		A	B	C	D	K	S	K × S
Energy (kcal)	Summer	945.4 [a] ± 205.7	1024.4 [a] ± 148.9	1059.8 [a] ± 138.3	1397.3 [b] ± 13.5	***	**	ns
	Autumn	1121.8 [a] ± 116.7	1029.4 [a] ± 108.5	1217.7 [a] ± 238.7	1397.1 [b] ± 6.6			
Protein (g)	Summer	30.7 [a] ± 6.7	35.8 [a] ± 3.6	36.2 [a] ± 8.0	52.3 [b] ± 9.6	***	*	ns
	Autumn	40.5 [a] ± 9.6	37.5 [a] ± 4.8	39.1 [a] ± 10.0	55.4 [b] ± 14.4			
Fat (g)	Summer	29.4 [ab] ± 13.3	24.6 [a] ± 4.4	23.6 [a] ± 4.5	40.7 [b] ± 14.5	***	ns	ns
	Autumn	30.1 [ab] ± 7.5	29.6 [ab] ± 7.5	22.8 [a] ± 7.8	42.8 [b] ± 12.9			
Carbohydrates (g)	Summer	146.7 [a] ± 27.3	171.5 [ab] ± 33.5	183.7 [ab] ± 29.8	212.6 [b] ± 33.1	***	ns	*
	Autumn	181.0 [ab] ± 26.7	159.4 [a] ± 24.4	222.2 [b] ± 54.8	205.8 [ab] ± 28.4			
Sodium (mg)	Summer	1763.3 [a] ± 332.5	1747.5 [a] ± 382.2	2100.5 [a] ± 422.6	2261.3 [a] ± 755.9	**	ns	ns
	Autumn	2172.6 [ab] ± 386.7	1756.5 [a] ± 282.0	1974.3 [ab] ± 378.0	2448.2 [b] ± 574.5			
Potassium (mg)	Summer	1928.1 ± 374.5	1652.0 ± 350.7	2085.4 ± 455.0	2070.4 ± 589.4	**	ns	ns
	Autumn	2337.7 [a] ± 550.3	1577.2 [b] ± 321.7	1988.1 [ab] ± 400.0	2313.3 [ab] ± 535.5			
Calcium (mg)	Summer	250.4 [a] ± 102.6	401.9 [ab] ± 130.3	373.2 [ab] ± 107.4	473.2 [b] ± 201.4	***	ns	ns
	Autumn	364.0 [a] ± 148.9	380.6 [a] ± 110.2	365.9 [a] ± 111.3	622.3 [b] ± 222.8			
Phosphorus (mg)	Summer	476.6 [a] ± 121.3	652.8 [ab] ± 82.2	653.0 [ab] ± 163.5	821.2 [b] ± 171.1	***	**	ns
	Autumn	643.8 [a] ± 191.7	679.4 [a] ± 141.3	675.3 [a] ± 123.8	1022.6 [b] ± 257.2			
Calcium/Phosphorus	Summer	0.5 ± 0.2	0.6 ± 0.2	0.6 ± 0.1	0.6 ± 0.1	ns	ns	ns
	Autumn	0.6 ± 0.1	0.6 ± 0.1	0.5 ± 0.1	0.6 ± 0.2			
Magnesium (mg)	Summer	142.0 ± 35.3	170.4 ± 30.0	182.2 ± 46.5	196.8 ± 51.0	**	*	ns
	Autumn	194.9 ± 49.7	188.9 ± 57.3	179.5 ± 37.5	259.5 ± 113.8			
Iron (mg)	Summer	5.3 ± 0.7	5.9 ± 0.8	6.2 ± 1.5	7.3 ± 1.3	*	*	ns
	Autumn	7.0 [ab] ± 1.8	6.2 [a] ± 1.4	6.5 [a] ± 1.1	9.8 [b] ± 1.7			
Zinc (mg)	Summer	3.9 [a] ± 0.6	5.1 [ab] ± 0.5	4.7 [ab] ± 1.1	6.1 [b] ± 1.7	***	*	ns
	Autumn	5.4 ± 1.9	5.4 ± 1.3	5.0 ± 1.0	6.8 ± 1.7			
Copper (mg)	Summer	0.7 ± 0.1	0.9 ± 0.2	0.8 ± 0.2	0.9 ± 0.3	ns	ns	ns
	Autumn	0.9 ± 0.2	0.9 ± 0.2	0.8 ± 0.1	1.0 ± 0.3			
Manganese (mg)	Summer	2.3 ± 0.3	2.7 ± 0.6	2.5 ± 0.8	2.8 ± 1.2	ns	ns	ns
	Autumn	2.8 ± 1.4	3.0 ± 0.4	2.5 ± 0.5	3.5 ± 1.7			
Vitamin B$_1$ (mg)	Summer	0.6 ± 0.2	0.6 ± 0.2	0.5 ± 0.2	0.8 ± 0.3	*	ns	ns
	Autumn	0.8 ± 0.3	0.6 ± 0.1	0.7 ± 0.2	0.7 ± 0.3			
Vitamin B$_2$ (mg)	Summer	0.6 [a] ± 0.1	0.9 [ab] ± 0.3	0.8 [ab] ± 0.2	1.0 [b] ± 0.2	***	*	ns
	Autumn	0.9 [a] ± 0.3	0.9 [ab] ± 0.1	0.8 [a] ± 0.3	1.2 [b] ± 0.3			
Niacin (mg)	Summer	7.8 ± 3.3	5.4 ± 1.6	8.2 ± 3.8	10.4 ± 3.6	***	ns	ns
	Autumn	9.7 [ab] ± 2.5	5.4 [a] ± 1.6	8.4 [ab] ± 2.9	12.0 [b] ± 8.5			
Vitamin C (mg)	Summer	77.5 ± 27.6	51.9 ± 12.6	67.0 ± 29.8	48.2 ± 20.6	***	ns	ns
	Autumn	98.1 [a] ± 70.0	37.4 [b] ± 16.4	62.0 [ab] ± 22.9	55.0 [ab] ± 22.6			
Vitamin A (µg)	Summer	857.4 ± 309.8	689.6 ± 281.1	758.9 ± 426.2	827.9 ± 277.4	*	ns	ns
	Autumn	975.3 [ab] ± 370.7	631.2 [a] ± 346.1	801.6 [ab] ± 289.0	1210.3 [b] ± 702.4			
Vitamin E (mg)	Summer	4.0 ± 1.0	3.7 ± 0.7	3.4 ± 0.7	4.7 ± 1.0	***	ns	ns
	Autumn	4.0 [a] ± 0.7	4.3 [ab] ± 1.6	3.7 [a] ± 1.1	5.5 [b] ± 1.5			
Vitamin B$_6$ (mg)	Summer	1.0 ± 0.2	0.8 ± 0.2	1.1 ± 0.3	1.2 ± 0.4	***	ns	ns
	Autumn	1.2 [a] ± 0.2	0.8 [b] ± 0.3	1.0 [ab] ± 0.3	1.2 [a] ± 0.5			
Vitamin D (µg)	Summer	0.8 ± 0.4	1.0 ± 0.7	0.8 ± 0.6	1.8 ± 0.7	*	ns	ns
	Autumn	1.1 ± 0.8	1.2 ± 1.0	0.8 ± 0.4	1.3 ± 0.9			
Glucose (g)	Summer	10.6 [a] ± 2.8	5.3 [b] ± 1.4	7.5 [ab] ± 2.3	6.0 [b] ± 3.5	***	ns	ns
	Autumn	10.4 [a] ± 3.2	5.0 [b] ± 2.2	7.1 [ab] ± 2.2	5.8 [b] ± 1.8			
Sucrose (g)	Summer	31.6 ± 8.3	40.1 ± 17.4	51.6 ± 13.7	34.1 ± 17.9	**	*	ns
	Autumn	40.1 ± 7.2	43.0 ± 12.4	61.4 ± 33.2	43.5 ± 12.2			
Fructose (g)	Summer	13.3 [a] ± 3.4	6.9 [b] ± 2.7	9.2 [ab] ± 2.6	6.5 [b] ± 3.2	***	ns	ns
	Autumn	12.2 [a] ± 4.7	6.8 [b] ± 2.0	8.7 [ab] ± 4.1	7.7 [ab] ± 2.6			
Lactose (g)	Summer	2.7 [a] ± 2.6	8.1 [b] ± 3.9	6.2 [ab] ± 4.8	7.7 [b] ± 2.7	***	ns	ns
	Autumn	4.9 [a] ± 3.8	8.6 [ab] ± 2.5	6.9 [ab] ± 4.1	10.3 [b] ± 3.7			
Starch (g)	Summer	62.0 [a] ± 12.6	89.1 [ab] ± 45.5	82.7 [a] ± 21.4	124.6 [b] ± 24.1	***	ns	*
	Autumn	76.3 ± 27.3	76.0 ± 18.3	106.2 ± 34.8	100.6 ± 14.9			
Fibre (g)	Summer	14.7 ± 2.3	13.2 ± 2.5	16.1 ± 4.7	14.5 ± 3.1	*	ns	ns
	Autumn	17.6 ± 2.7	12.8 ± 2.1	16.3 ± 2.6	16.7 ± 6.4			
Cholesterol (mg)	Summer	142.6 ± 73.0	165.4 ± 125.6	120.4 ± 88.2	184.2 ± 92.6	ns	ns	ns
	Autumn	187.2 ± 169.9	163.9 ± 101.4	108.3 ± 78.1	183.6 ± 67.5			
SFA (g)	Summer	15.2 [ab] ± 8.8	11.6 [ab] ± 2.4	10.6 [a] ± 2.4	20.4 [b] ± 10.3	***	ns	ns
	Autumn	13.9 [ab] ± 5.0	13.3 [ab] ± 3.3	10.7 [a] ± 5.4	21.0 [b] ± 7.5			
MUFA (g)	Summer	9.6 [ab] ± 3.8	7.7 [a] ± 1.7	7.2 [a] ± 2.2	12.7 [b] ± 4.6	***	ns	ns
	Autumn	10.7 [ab] ± 3.4	9.7 [ab] ± 3.2	6.8 [a] ± 2.7	13.2 [b] ± 4.8			

Table 1. *Cont.*

Energy and Nutrients	Season (S)	Kindergarten (K)				Significant Effects		
		A	B	C	D	K	S	K × S
PUFA (g)	Summer	2.2 [a] ± 0.4	3.2 [ab] ± 0.5	3.6 [ab] ± 3.0	4.4 [b] ± 0.9	***	ns	ns
	Autumn	2.8 ± 0.6	4.1 ± 1.5	3.5 ± 1.6	4.6 ± 1.1			
PUFA/SFA	Summer	0.2 ± 0.1	0.3 ± 0.1	0.4 ± 0.6	0.3 ± 0.2	*	ns	ns
	Autumn	0.2 ± 0.1	0.3 ± 0.2	0.4 ± 0.4	0.2 ± 0.1			
PUFA n-3 (g)	Summer	0.4 ± 0.1	0.4 ± 0.1	0.6 ± 0.5	0.6 ± 0.2	ns	ns	ns
	Autumn	0.5 ± 0.1	0.5 ± 0.4	0.5 ± 0.3	0.7 ± 0.3			
PUFA n-6 (g)	Summer	1.8 [a] ± 0.4	2.8 [ab] ± 0.4	3.0 [ab] ± 2.6	3.8 [b] ± 0.8	***	ns	ns
	Autumn	2.3 ± 0.6	3.6 ± 1.4	2.9 ± 1.5	3.8 ± 1.0			
PUFA n-6/n-3	Summer	4.4 ± 1.0	7.4 ± 1.1	4.6 ± 1.4	7.1 ± 2.7	***	ns	ns
	Autumn	4.7 [a] ± 1.8	8.6 [b] ± 4.3	6.1 [ab] ± 2.8	5.7 [ab] ± 2.2			
Salt (g)	Summer	4.4 ± 0.8	4.4 ± 1.0	5.3 ± 1.1	5.7 ± 1.9	**	ns	ns
	Autumn	5.4 [ab] ± 1.0	4.4 [a] ± 0.7	4.9 [ab] ± 0.9	6.1 [b] ± 1.4			

Significant effects: *** $p < 0.001$; ** $0.001 \leq p < 0.01$; * $0.01 \leq p < 0.05$; ns—not significant; K—kindergarten; S—season. a–b: Means. Different letters in the same row indicate differences at $p < 0.05$ in view of the kindergarten. SFA—saturated fatty acids, MUFA—monounsaturated fatty acids, PUFA—polyunsaturated fatty acids.

Table 2. Energy and nutrients in preschool food rations constituting 75% of the Polish norms and recommendations and the UL established by the EFSA.

Energy and Nutrients	Norm	75% of the Daily Requirement	Season	Norm Realization (%)			
				Kindergarten A	B	C	D
Energy (kcal)	1400 [1]	1050 [1]	Summer	90.0 [a]	97.6 [a]	100.9 [a]	133.1 [b]
			Autumn	106.8 [a]	98.0 [a]	116.0 [a]	133.1 [b]
Protein (g)	21 [2]	15.8 [2]	Summer	195.0 [a]	227.4 [a]	229.7 [a]	332.3 [b]
			Autumn	257.2 [a]	238.2 [a]	248.1 [a]	351.9 [b]
Fat (g)	47 [3]	35.3 [3]	Summer	83.4 [ab]	69.7 [a]	66.9 [a]	115.4 [b]
			Autumn	85.3 [ab]	84.1 [ab]	64.6 [a]	121.5 [b]
Carbohydrates (g)	227.5 [3]	170.6 [3]	Summer	86.0 [a]	100.5 [ab]	107.7 [ab]	124.6 [b]
			Autumn	106.1 [ab]	93.4 [a]	130.2 [b]	120.6 [ab]
Sodium (mg)	1000 [4]	750 [4]	Summer	235.1	233.0	280.1	301.5
			Autumn	289.7 [ab]	234.2 [a]	263.2 [ab]	326.4 [b]
Potassium (mg)	1100 [4]	825 [4]	Summer	233.7	200.2	252.8	251.0
			Autumn	283.4 [a]	191.2 [b]	241.0 [ab]	280.[ab]
Calcium (mg)	1000 [2]	750 [2]	Summer	33.4 [a]	53.6 [ab]	49.8 [ab]	63.1 [b]
			Autumn	48.5 [a]	50.7 [a]	48.8 [a]	83.0 [b]
Phosphorus (mg)	500 [2]	375 [2]	Summer	127.1 [a]	174.1 [ab]	174.1 [ab]	219.0 [b]
			Autumn	171.7 [a]	181.2 [a]	180.1 [a]	272.7 [b]
Magnesium (mg)	130 [2]/250 [5]	97.5 [2]/187.5 [5]	Summer	145.6/75.7	174.8/90.9	186.9/97.2	201.8/105.0
			Autumn	199.9/104.0	193.7/100.8	184.1/95.7	266.1/138.4
Iron (mg)	10.0 [2]	7.5 [2]	Summer	70.0	78.9	82.2	97.4
			Autumn	93.5 [ab]	82.7 [a]	86.8 [a]	130.2 [b]
Zinc (mg)	5.0 [2]/10.0 [5]	3.8 [2]/7.5 [5]	Summer	104.8/52.4	124.8/68.3	124.8/62.4	161.5/80.8
			Autumn	144.0/72.0	136.6/71.3	133.0/66.5	180.5/90.3
Copper (mg)	0.4 [2]/2.0 [5]	0.3 [2]/1.5 [5]	Summer	228.9/46.0	281.8/56.6	274.8/54.7	291.5/58.0
			Autumn	285.7/57.3	283.5/56.7	268.0/53.3	331.2/66.0
Manganese (mg)	1.5 [4]	1.1 [4]	Summer	203.3	242.5	223.4	245.6
			Autumn	244.2	265.4	223.6	305.3
Vitamin B$_1$ (mg)	0.6 [2]	0.5 [2]	Summer	130.7	122.1	119.0	173.6
			Autumn	169.1	122.1	145.0	158.7
Vitamin B$_2$ (mg)	0.6 [2]	0.5 [2]	Summer	143.0 [a]	203.5 [ab]	179.3 [ab]	225.0 [b]
			Autumn	195.8 [a]	200.3 [ab]	184.8 [a]	267.3 [b]
Niacin (mg)	8.0 [2]	6.0 [2]	Summer	130.7	89.9	136.6	173.4
			Autumn	162.3 [ab]	90.5 [a]	140.1 [ab]	199.1 [b]
Vitamin C (mg)	50.0 [2]	37.5 [2]	Summer	206.7	138.5	178.7	128.4
			Autumn	261.5 [a]	99.7 [b]	165.2 [ab]	146.7 [ab]
Vitamin A (µg)	450 [2]/1100 [5]	337.5 [2]/825 [5]	Summer	254.0/103.9	204.3/83.6	224.9/92.0	245.3/100.4
			Autumn	289.0 [ab]/118.2	187.0 [a]/76.5	237.5 [ab]/97.2	358.6 [b]/146.7
Vitamin E (mg)	6.0 [4]	4.5 [4]	Summer	88.0	83.1	75.0	104.9
			Autumn	87.7 [a]	94.4 [ab]	82.1 [a]	123.0 [b]

Table 2. *Cont.*

Energy and Nutrients	Norm	75% of the Daily Requirement	Season	Norm Realization (%)			
				Kindergarten A	B	C	D
Vitamin B$_6$ (mg)	0.6 [2]/7.0 [5]	0.5 [2]/5.3 [5]	Summer	223.3/19.1	180.9/15.4	238.2/20.4	263.3/22.5
			Autumn	273.6 [a]/23.4	169.7 [b]/14.5	223.1 [ab]/19.1	275.0 [a]/23.6
Vitamin D (µg)	15.0 [4]	11.3 [4]	Summer	6.9	8.5	7.2	15.6
			Autumn	9.9	10.5	7.2	11.7
Glucose (g)			Summer	40.2 [a]	20.3 [b]	28.7 [ab]	23.0 [b]
			Autumn	39.8 [a]	19.2 [b]	26.9 [ab]	21.9 [b]
Sucrose (g)	* 10% energy value	26.3	Summer	120.2	152.8	196.7	130.0
			Autumn	152.8	163.7	234.0	165.8
Fructose (g)			Summer	50.6 [a]	26.3 [b]	35.2 [ab]	24.7 [b]
			Autumn	46.3 [a]	25.9 [b]	33.1 [ab]	29.1 [ab]
Fiber (g)	14 [4]	10.5 [4]	Summer	140.4	125.3	153.2	137.6
			Autumn	167.6	122.1	154.8	158.6
Cholesterol (mg)	300	225	Summer	63.4	73.5	53.5	81.9
			Autumn	83.2	72.9	48.1	81.6
SFA (g)	15.6	11.7	Summer	130.1 [ab]	98.9 [ab]	90.5 [a]	174.7 [b]
			Autumn	119.2 [ab]	113.8 [ab]	91.3 [a]	179.8 [b]
PUFA n-3 (g)	1.0 [4]	0.8 [4]	Summer	54.8	49.4	82.4	76.3
			Autumn	63.5	68.4	64.6	95.5
PUFA n-6 (g)	6.2 [4]	4.7 [4]	Summer	39.0 [a]	60.1 [ab]	64.2 [ab]	81.2 [b]
			Autumn	49.0	77.1	61.0	81.8
Salt (g)	2.5	1.9	Summer	234.7	232.5	279.3	301.1
			Autumn	289.1 [ab]	233.5 [a]	262.8 [ab]	325.0 [b]

* 10% energy value of food ratio; [1] Estimated Energy Requirement (EER); [2] Recommended Dietary Allowance (RDA); [3] Reference Intake ranges (RIs); [4] Adequate Intake (AI); [5] Tolerable Upper Intake Level (UL). SFA—saturated fatty acids, MUFA—monounsaturated fatty acids, PUFA—polyunsaturated fatty acids. a–b: Means. Different letters in the same row indicate differences at $p < 0.05$ in view of the kindergarten.

The protein content in the analyzed preschool rations ranged from 30.7 g to 52.3 g in summer and from 37.5 g to 55.4 g in autumn, respectively (Table 1). This amount of protein exceeded the recommended level of consumption (intake above 75% of RDA) by two or even three times (Table 2). The menus from kindergarten D had a significantly higher content of this nutrient compared to the other kindergartens (Table 2).

The food ration of the children who spend part of the day in kindergarten should provide 35.3 g of fat. In this study, this amount was exceeded by 15.4% in summer and 21.5% in autumn in kindergarten D (Table 2). In the remaining kindergartens, the fat content was lower than recommended. From the nutritional point of view, it is also important to identify the type of fatty acids in the diet. In each kindergarten, the children's diet was characterized by too low n-3 and n-6 polyunsaturated fatty acid content. An excess of unsatisfactory saturated fatty acids was found in kindergartens A and D in both seasons (Table 2).

The carbohydrate content differed significantly between preschools. The menus were balanced in terms of carbohydrate content in kindergartens B and C in summer and A and B in autumn (Table 2). An excessive supply of carbohydrates in summer and autumn was recorded in kindergarten D (Table 2). In the assessment of the supply of added sugars (sucrose, fructose, glucose), it was assumed that they should constitute no more than 10% of the diet's energy. Regardless of the kindergarten, the sucrose content exceeded the recommended intake (Table 2). The proportion of carbohydrates from each group in the children's meals was inappropriate (Table 3). The children's diet was characterized by a too low proportion of starch-containing products. The dietary fiber content was above the recommended level in each facility (Table 2).

The analyzed menus, compared to the norms, covered the children's needs for minerals (such as sodium, potassium, calcium, phosphorus, magnesium, calcium, iron, zinc, copper, manganese) to varying degrees. Regardless of the kindergarten, a significant calcium inadequacy was noted in the studied menus (Table 2). Its content in summer ranged from 250.4 mg to 473.2 mg, and in autumn, from 364.0 mg to 622.3 mg (with the recommended standard of 750 mg) (Table 1).

Table 3. Percentage of carbohydrate groups in total carbohydrate content in kindergarten menus.

Type of Carbohydrates	Season	Kindergatren			
		A	B	C	D
Glucose	Summer	7.2	3.1	4.1	2.8
	Autumn	5.8	2.9	3.2	2.8
Sucrose	Summer	21.5	23.4	28.2	16.1
	Autumn	22.2	27.0	27.6	21.1
Fructose	Summer	9.1	4.0	5.0	3.1
	Autumn	6.7	4.3	3.9	3.7
Lactose	Summer	1.8	4.8	3.4	3.6
	Autumn	2.7	5.4	3.1	5.0
Starch	Summer	42.3	52.0	45.0	58.6
	Autumn	42.2	47.7	47.8	48.9

The iron content in kindergartens B and C summer and autumn menus was also below 75% of the RDA. The daily supply of this mineral exceeded 75% of RDA in kindergarten D in autumn (Table 2). At the same time, it should be emphasized that UL standards for iron have not been determined [35]. The content of the remaining analyzed minerals (sodium, potassium, phosphorus, magnesium, zinc, copper, and manganese) in all the analyzed menus was too high according to Polish standards (RDA, AI) (Table 2). It should also be noted that the intake of magnesium was higher than 75% of the UL in the fall in the kindergartens A and B and in institution D in both seasons, and for vitamin A in kindergartens A and D. The UL levels for zinc, copper, and vitamin B_6 were not exceeded.

In the analyzed preschool menus there were irregularities in the vitamin content. A vitamin D deficiency (intake lower than 75% of AI) and an excess of vitamins A, B_1, B_2, and B_6 (intakes above 75% of RDA) were demonstrated in each preschool (Table 2). The supply of niacin and vitamins C and E varied depending on the kindergarten (Table 2).

In all menus, there were irregularities in the salt content. The content in preschool rations ranged from 232.5 g to 325.0 g, respectively (Table 1). This amount of salt was twice or even three times higher than the recommended intake level (Table 2).

The all-day food ration in the kindergarten consisted of 3 meals: breakfast, lunch, and afternoon tea. The intervals between meals were adequate at 4 and 3 h, respectively.

Regardless of the kindergarten, breakfast was usually served with wheat bread, butter, and a high-protein product of animal origin (cold cuts of meat, eggs, cheese) or an item with high carbohydrate content such as jam (most often strawberry). Children were also served milk with corn flakes or other cereals. To drink, tea, grain coffee, or cocoa were provided. The drinks were sweetened with sugar. Each breakfast was served with a piece of fruit (most often apples, bananas) or a vegetable (most often cucumbers or tomatoes).

Lunch always consisted of two dishes: soup and a second course. The most frequently served soups included tomato, potato, barley, or chicken, and cabbage. They were prepared based on meat and vegetable stock. They were not enriched with fresh dill or parsley. The second course usually consisted of a protein product of animal origin—usually meat, a bulk product—mainly potatoes and less often pasta or rice, as well as cooked or raw vegetables. Compote with sugar or fruit and vegetable drinks were usually served with lunch.

Afternoon tea usually consisted of products rich in carbohydrates (buns, jelly, bread with chocolate spread, honey) and a warm drink (tea, milk, or milk drinks with added sugar). Sometimes sandwiches were served (with pate, pork sirloin, sausage, or jam).

The meals were prepared in a variety of techniques. The products were fried, boiled, stewed, or baked. The meals were composed with contrasts in taste, smell, and color to encourage children to eat. Combining two dishes of a similar nature (sweet, sour, or similar in color and texture) was avoided during one meal. No seasonality of products was found in the menus. The disadvantage of the analyzed menus was the lack of whole grain products, the limited quantities of fish and legumes, and the servings of meat products with high fat content (mortadella, sausages, and cheese). Spices were added in limited amounts.

4. Discussion

The statistical analysis of the obtained results showed that the energy value of the studied menus differed regarding the season and the kindergarten. In autumn, a favorable tendency of an energy increase in menus was observed when the body's energy expenditure related to the need to maintain the body temperature at about 37 °C increased, with a simultaneous decrease in the surrounding temperature. The energy values of the kindergarten meals, except for facility D and C (in autumn), were correct. Preschool meals should cover the energy and nutrient needs of children to a greater extent than home meals. However, they should not exceed the recommendations because they lead to, among other issues, excess weight and obesity. The results of this study do not differ from the data in the literature on the nutrition of children in kindergartens [20,23,27,37–39], which report too high energy supply as compared to the recommendations.

Kindergarten D also reported excess fat in the children's diet, which probably came from chocolate and confectionery products, which provided 10 to 15 g of fat per 100 g of product [32] and were also a source of unfavorable saturated fatty acids. The consumption of products rich in fat, including saturated fatty acids, is associated with the risk of ischemic heart disease and the development of cancer, mainly colorectal and breast cancer [40–42].

The average protein content of the evaluated food rations from the analyzed kindergartens was more than double the RDA norm. Similar results in preschool diets were also obtained by the other authors [22,27,30,43]. Such a protein supply was determined by the high consumption of meat and processed meat products. So far, an UL standard has not been defined for protein but attempts are underway to determine the upper limit of consumption. However, it is known that high protein consumption may be associated with hypercalciuria, resulting in osteoporosis, acidosis, and an increased risk of kidney stones [4,44]. Excess dietary protein content in childhood contributes to an increased risk of obesity [45].

An analysis of the data in the literature shows that there is a lack of studies on the intake of specific groups of carbohydrates, including fructose, which is extremely important because of its negative impact on human health [46–48]. The increased consumption of products (sweetened soft drinks, candies) to which fructose is added (especially as a glucose–fructose syrup [4]) during the production process was observed. The highest fructose content was found in the meals of kindergarten A. Excessive fructose intake is believed to be one of the factors responsible for the steady increase in excess weight and obesity problems in modern societies [49]. This is especially true in children and adolescents [50].

One of the tools to support the improvement of the nutrition of children and adolescents in educational institutions was the Regulation of the Minister of Health of 26 July 2016 on the groups of food intended for sale to children and adolescents within the educational system and the requirements for food served as part of the collective nutrition of children and adolescents in educational settings (Journal of Laws of 2016, item 1154) [51]. The provisions of the regulation covered, among others, requirements for collective nutrition in kindergartens, school establishments, boarding schools, and boarding houses. The regulation requires the inclusion of more vegetables and fruit so that they are present every day in each meal, with the inclusion of more fish in the weekly diet while limiting the addition of sugar in drinks prepared onsite to a maximum of 10 g per 250 mL of drink. This was designed to enable easier compliance with the standards' recommendation to limit sugars (sucrose, fructose, glucose) added to food during its manufacture and production, which should provide no more than 10% of energy.

Despite the above-mentioned regulation, in each kindergarten the amount of sugar added to food during its production process exceeded recommendations. Sucrose content ranged from 31.6 g to 61.4 g, corresponding to 120% and 234.0% of the standard value, which is a disturbing phenomenon. It is well known that an excessive intake of sucrose and simple sugars contributes to the occurrence of many chronic non-communicable diseases, including excess weight and obesity problems, insulin resistance, diabetes, metabolic syndrome, tooth decay, cancer, and non-alcoholic fatty liver disease [52–55]. Other authors have also noted high sucrose content in the diet of preschool children [24,27].

It should be noted that from 1 January 2021, regulations entered into force in Poland introducing taxes on sweetened and energy drinks (ACT of 14 February 2020 on the amendment of certain acts in connection with the promotion of pro-health consumer choices) [56]. Perhaps this regulation will reduce the sale and consumption of sweetened beverages and improve the health indicators of the Polish society.

The mineral content of the study rations varied widely. In all the evaluated dietary rations from the studied kindergartens, the calcium content was too low compared to the standard. The low supply of calcium in the analyzed dietary rations was caused by the low supply of milk and milk products (e.g., cottage cheese, kefir, buttermilk), which represent the main sources of calcium in the diet. Adequate calcium content is particularly desirable in the diet of small children because calcium is an important building material for bones and teeth [57,58]; it is also involved in muscle contraction, blood coagulation, conduction of nerve impulses, activation of enzymes, and the reduction in the permeability of cell membranes, as well as supporting the proper functioning of the cardiovascular system. Calcium inadequacy in children's diets is a common problem [28–30,59–61]. The utilization of calcium from the diet is hindered by a too-high phosphorus content [62]. An adequate calcium/phosphorus ratio is a very significant factor in calcium phosphate metabolism. A too low calcium/phosphorus ratio (indicating a high dietary phosphorus intake) stimulates parathormone secretion and enhances calcium resorption from the bones. The calcium-to-phosphorus ratio was abnormal in the children's diets, ranging between 0.5 to 0.6 regardless of the preschool institution and a season.

It should also be mentioned that the factors increasing calcium absorption include, among others, vitamin D, the amount of which in the analyzed diets was significantly below the norm. Insufficient intake of calcium and vitamin D raises serious concerns about the proper condition of children's skeletons. The results of this research indicate that it is necessary to educate the employees of preschool institutions responsible for children's nutrition. Social campaigns promoting the importance of calcium intake and indicating food products and their quantities needed to meet the child's demand for this mineral may be a way to draw public attention to the consequences of calcium deficiency in children.

Based on the research results, it was found that, regardless of the kindergarten, the sodium content in the children's diet was very high (intake higher than 75% of AI). Such high sodium intake indicates the need to limit the addition of salt to dishes and the elimination of processed products with a high salt content (e.g., cold meat products). The results of sodium consumption converted into table salt exceeded the recommendations by more than 2–3 times. There is no doubt that salt consumption in Poland should be reduced [2,3,6]. It is advisable to develop proper eating habits from an early age to reduce the risk of developing cardiovascular diseases, arterial hypertension, and cancer of the stomach and throat at a later age [63,64]. Many studies report excessive sodium and table salt intake in children's diets, suggesting that it is a common phenomenon [30,59,65–67].

Regardless of the kindergarten, the analyzed food rations also contained too high intakes of potassium, copper, manganese, and magnesium compared to the Polish norms. It should be noted that EFSA experts concluded that there is currently insufficient scientific research on the negative effects of the excess of sodium, potassium, and manganese on the human body [35]. Consequently, it is currently impossible to establish ULs for these components. The UL for copper intake was not exceeded. In this situation, the intake of potassium, copper, and manganese is satisfactory. The intake of magnesium was higher than 75% of UL. Doses of magnesium above 250 mg/day have laxative properties [4].

The analyzed food rations covered the children's iron needs to varying degrees. In kindergartens B and C, iron deficiency was demonstrated, while in kindergarten D an excess was found. At the same time, it should be emphasized that UL standards for iron have not been determined [35]. It is well known that iron deficiency promotes the development of anemia [68] and may reduce a child's immunity [69]. At the same time, its excess increases the risk of cancer, gastric disorders, atherosclerosis, strokes, and Alzheimer's and Parkinson's disease [70–72]. Literature data also show large variations in iron consumption among children, usually insufficient [28,30,39,67,73] or exceeding the norm [22,43].

Regardless of the kindergarten, the children's food rations showed an excess of vitamins A, B_1, B_2, B_6, and C (except for kindergarten B) and vitamin D deficiency. The study results do not differ from the results of the studies by the other authors who found excessive intake of vitamin A in preschool food rations compared to the norms [28,30,43]. An excessive supply of vitamin A manifests itself in irritability, headaches, loss of appetite, dry skin, and digestive tract disorders, which are accompanied by enlargement of the liver and spleen [1]. Since the UL values have not been established for vitamins B_1, B_2 and C and the UL value standard for vitamin B_6 was not exceeded, their intake level can be assessed positively.

Children's food rations were characterized by a deficient supply of vitamin D, whose deficiency may result in the negative calcium balance and impaired bone mineralization, resulting in rickets in children [74]. Deficiency of this vitamin was due to a very low intake of fish, which was served sporadically, regardless of the preschool. Vitamin D is found in smaller amounts in butter, yellow cheese, egg yolk, and meat, which were served to the children in meals. However, with the deficient intake of fish, the other listed sources of the vitamin did not cover the children's vitamin D requirements.

Numerous studies indicate irregularities in the nutrition of children in kindergartens. The most frequent recurring problem is an excessive supply of fat, vitamin A, salt, and calcium, and vitamin D deficiency, which is also confirmed by the results of this study.

It should be mentioned that the data on the energy and nutritional value of children diet were obtained from the tables of composition and nutritional value [32]. The tables' values are averaged data, so the energy and nutrient content of a product consumed by children may differ from the values given in the tables. The content can depend on many factors, both genetic and environmental. The limitation of this study is the lack of data regarding the real food consumption of children, food waste on the plate, and the real amount of salt added to dishes. Consequently, the estimated content of energy and nutrients in the daily food rations of children may differ from their actual consumption. Thus, the observed deficiencies of the selected nutrients in the children's diets are hazardous, as their actual consumption may be even lower. It is essential to be aware that preschool nutrition represents only a part of the diet. It is not known what foods products the children received at home and whether their parents introduced supplements into their diet. It is worth mentioning that the menu data were acquired through 2019 only. However, they are still valid, as during 2020 and 2021 there were lockdowns, and kindergartens were closed. Therefore, there were more important issues to combat than kindergarten menu adjustment to the proper nutrition schemes.

5. Conclusions

The adequate nutrition of children is necessary to ensure their proper development. The evaluation of preschool food rations is important to prevent nutrition errors in this group. The study results indicate that the analyzed preschools did not fully implement the requirements set out in the Regulation of the Ministry of Health of 26 July 2016 [51]. Not every kindergarten meal included vegetable or fruit. More than two portions of fried food were served per week. The children's diet did not contain nuts. Fish was served sporadically. One of the disadvantages of the analyzed menus was the presence of sugary foods. Chocolate, honey, and cornflakes with added sugar were often served for breakfast. The afternoon snack was served with confectionery products such as buttered buns and puff pastries. As is commonly known, sweets in a child's diet should be limited to a minimum, and such an institution as a kindergarten should provide children with proper models concerning nutrition. No seasonality of products was noticed in the menus. The same fruit and vegetables were served in summer and autumn. The assessment of menus, regardless of the kindergarten, showed that they were arranged incorrectly, which indicates the need for modification of their composition. Therefore, it is necessary to carry out nutritional education among kindergarten staff. To ensure the health of Polish children and thus the health of future society, it is necessary to correct dietary mistakes as soon as possible, since diet-related diseases may develop from childhood.

Funding: This research received no external funding.

Institutional Review Board Statement: Not applicable.

Informed Consent Statement: Not applicable.

Data Availability Statement: Not applicable.

Acknowledgments: The author thanks Joanna Harasym for manuscript editing consultation and Dorota Matysiak, an employee of the Main Library of the Wroclaw University of Economics and Business, for help in searching for literature to prepare this work.

Conflicts of Interest: The author declares no conflict of interest.

References

1. Sadowska, J.; Radziszewska, M.; Krzymuska, A. Evaluation of nutrition manner and nutritional status of pre-school children. *Acta Sci. Pol. Technol. Aliment.* **2010**, *9*, 105–115.
2. Myszkowska-Ryciak, J.; Harton, A. Nutrition-related practices in kindergartens in the context of changes to legal regulations on foodstuffs used in canteen menus for children. *Rocz. Panstw. Zakl. Hig.* **2018**, *69*, 31–36.
3. Merkiel, S.; Chalcarz, W. Preschool diets in children from Piła, Poland, require urgent intervention as implied by high risk of nutrient inadequacies. *J. Health Popul. Nutr.* **2016**, *35*, 11. [CrossRef]
4. Jarosz, M.; Rychlik, E.; Stoś, K.; Charzewska, J. *Nutrition Standards for the Population of Poland and Their Application*; National Institute of Public Health—National Institute of Hygiene: Warsaw, Poland, 2020.
5. Charzewska, J. *Recommendations for Nutrition Implementers in the Field of Proper Nutrition of Children in Kindergartens*; Food and Nutrition Institute: Warsaw, Poland, 2011.
6. Merkiel, S.; Chalczarz, W. Review of the studies on nutrition in polish preschool children. Part 1. Preschool menus. *Rocz. Panstw. Zakl. Hig.* **2016**, *67*, 223–235. [PubMed]
7. Stroba, M.; Malinowska-Borowska, J. Assessment of calcium and vitamin d content in meals served for preschool children attending to kindergartens in some Silesian cities. *Rocz. Panstw. Zakl. Hig.* **2021**, *72*, 83–88.
8. Myszkowska-Ryciak, J.; Harton, A. Evaluation of implementation of dietary recommendations and guidelines in selected kindergartens from Poznan area. *Probl. Hig. Epidemiol.* **2018**, *99*, 7–11.
9. Klemarczyk, W.; Strucińska, M.; Weker, H.; Więch, M. Dietary assessment of children in vegetarian kindergarten. Pediatria współczesna. *Gastroenterol. Hepatol. I Zyw. Dziecka* **2005**, *7*, 243–246.
10. Dymkowska Malesa, M.; Szparaga, A. Evaluation of intake of selected vitamins and minerals in the diets of preschool children from Koszalin. *Nowa Pediatr.* **2013**, *3*, 106–110.
11. Stempel, P.; Galczak-Kondraciuk, A.; Czeczelewski, J.; Kołdej, M. Assessment of nutritional behaviour of children aged 3–7 from selected kindergartens in Biala Podlaska county. *Rocz. Panstw. Zakl. Hig.* **2018**, *69*, 289–297. [PubMed]
12. World Health Organization. Prevalence of Obesity among Children and Adolescents, BMI > +2 Standard Deviations above the Median (Crude Estimate) (%). Available online: https://www.who.int/data/gho/data/indicators/indicator-details/GHO/prevalence-of-obesity-among-children-and-adolescents-bmi-2-standard-deviations-above-the-median-(crude-estimate)-(-) (accessed on 10 December 2021).
13. Lobstein, T.; Jackson-Leach, R. Planning for the worst: Estimates of obesity and comorbidities in school-age children in 2025. *Pediatr. Obes.* **2016**, *11*, 321–325. [CrossRef]
14. WHO. Reducing Childhood Obesity in Poland Final WEB.pdf (who.int). Available online: https://www.euro.who.int/__data/assets/pdf_file/0011/350030/Reducing-childhood-obesity-in-Poland_final_WEB.pdf (accessed on 10 December 2021).
15. Noczyńska, A.; Zubkiewicz-Kucharska, A. The influence of birth weight and parents' weight on overweight and obesity prevalence among their children. *Pediatr. Endocrinol.* **2014**, *13*, 23–30. [CrossRef]
16. Stąpor, N.; Kapczuk, I.; Krzewska, A.; Sieniawska, J.; Rakuś-Kwiatosz, A.; Piątek, D.; Bąk, K.; Szydełko, J.; Beń-Skowronek, I. What is the difference in lifestyle in slim and overweight children? *Pediatr. Endocrinol.* **2016**, *15*, 29–35. [CrossRef]
17. Llewellyn, A.; Simmonds, M.; Owen, C.G.; Woolacott, N. Childhood obesity as a predictor of morbidity in adulthood: A systematic review and meta-analysis. *Obes. Rev.* **2016**, *17*, 56–67. [CrossRef] [PubMed]
18. Batko, B.; Kowal, M.; Szwajca, M.; Pilecki, M. Relationship between biopsychosocial factors, body mass and body composition in preschool children. *Psychiatr. Psychol. Klin.* **2020**, *20*, 164–173. [CrossRef]
19. Rolland-Cachera, M.F.; Deheeger, M.; Bellisle, F.; Sempé, M.; Guilloud-ataille, M.; Patois, E. Adiposity rebound in children: A simple indicator for predicting obesity. *Am. J. Clin. Nutr.* **1984**, *39*, 129–135. [CrossRef]

20. Grajeta, H.; Ilow, R.; Prescha, A.; Regulska-Ilow, B.; Biernat, J. Evaluation of energy and nutritive value of nursery schools' meals. *Rocz. Panstw. Zakl. Hig.* **2003**, *54*, 417–425.
21. Leszczyńska, T.; Sikora, E.; Kręcina, K.; Pysz, K. Meals served in nursery schools and their share in meeting the recommended daily demand for energy and nutrients exemplified by one selected canteen. *Żywność Nauka Technol. Jakość* **2007**, *6*, 327–334.
22. Michota-Katulska, E.; Zegan, M. Comperative analysis of children of nutrition in kindergartens in the traditional system and catering. *Med. Rodz.* **2014**, *4*, 166–169.
23. Czech, A.; Kęska, A. Nutrient content in preschool rations in spring and autumn. *Żyw. Człow. Metab.* **2007**, *34*, 561–571.
24. Rogalska-Niedźwiedź, M.; Charzewska, J.; Chabros, E.; Chwojnowska, Z.; Wajszczyk, Z.; Zachrewicz, E. Nutrition of 4-year-old children from rural and urban environments. *Probl. Hig. Epidemiol.* **2008**, *89*, 80–84.
25. Kowieska, A.; Biel, W.; Chalaba, A. Characteristics of the nutrition of preschool children. *Żyw. Człow. Metab.* **2009**, *36*, 179–184.
26. Dymkowska-Malesa, M.; Skibniewska, K.A. Meals served at nursery schools and their share in meeting the recommended daily demand for nutrients and energy Bromat. *Chem. Toksykol.* **2011**, *44*, 374–379.
27. Frąckiewicz, J.; Ring-Andrzejczuk, K.; Gronowska-Senger, A. Energy and selected nutrients content in pre-school children diet of Warsowia district. *Roczn. Panstw. Zakl. Hig.* **2011**, *62*, 181–185.
28. Orkusz, A.; Olech, A. Evaluation of the nutritional value of nursery school meals. *Eng. Sci. Technol.* **2014**, *2*, 77–87. [CrossRef]
29. Orkusz, A.; Włodarczyk, A. Assessment of preschool children's decades menus. *Eng. Sci. Technol.* **2014**, *1*, 72–81. [CrossRef]
30. Kwiecień, M.; Winiarska-Mleczan, A.; Danek-Majewska, A.; Kiczorowska, B.; Olcha, M. Assesment of nutritional value of feed rations for preschool children with particular emphasis on mineral content. *Probl. Hig. Epidemiol.* **2015**, *96*, 742–745.
31. Wrzochal, A.; Gładyś-Jakubczyk, A.; Edyta Suliga, E. Evaluation of diet in preschool-age children with Down syndrome—Preliminary examination. *Med. Stud.* **2019**, *35*, 128–138. [CrossRef]
32. *Computer Program Diet 6D, Independent Laboratory of Epidemiology and Nutrition Standards*; Institute of Food and Nutrition: Warsaw, Poland, 2018.
33. Kunachowicz, H.; Nadolna, I.; Przygoda, B.; Iwanow, K. *Tables of Composition and Nutritional Value of Food*; PZWL: Warsaw, Poland, 2005.
34. Kułaga, Z.; Grajda, A.; Gurzkowska, B.; Góźdź, M.; Wojtyło, M.; Świąder, A.; Różdżyńska-Świątkowska, A.; Litwin, M. Polish 2012 growth references for preschool children. *Eur. J. Pediatr.* **2013**, *172*, 753–761. [CrossRef] [PubMed]
35. European Food Safety Authority (EFSA). Summary of Tolerable Upper Intake Levels. September 2018; pp. 1–4. Available online: https://www.efsa.europa.eu/sites/default/files/assets/UL_Summary_tables.pdf (accessed on 10 December 2021).
36. *Statistica, Version 13.3*; StatSoft Inc.: Kraków, Poland, 2013.
37. Kłos, A.; Bertrand, J. Nutrition of children in selected military kindergartens in Warsaw. *Lek. Wojsk.* **1999**, *5*, 275–279.
38. Hamułka, J.; Wawrzyniak, A. Nutritional assessment of decadal menus of children aged 1–6 years. *Bromat. Chem. Toksykol.* **2003**, *36*, 7–11.
39. Orkusz, A.; Janczar-Smuga, M.; Frysiak, D. Assessment of nutrition of children aged 4–6 years basing on decade menus. *Zesz. Probl. Post. Nauk. Roln.* **2018**, *594*, 37–47. [CrossRef]
40. Enos, R.T.; Velázquez, K.T.; McClellan, J.L.; Cranford, T.L.; Nagarkatti, M.; Nagarkatti, P.S.; Davis, J.M.; Murphy, E.A. High-fat diets rich in saturated fat protect against azoxymethane/dextran sulfate sodium-induced colon cancer. *Am. J. Physiol. Gastrointest. Liver Physiol.* **2016**, *310*, G906–G919. [CrossRef] [PubMed]
41. Nettleton, J.A.; Brouwer, I.A.; Geleijnse, J.M.; Hornstra, G. Saturated Fat Consumption and Risk of Coronary Heart Disease and Ischemic Stroke: A Science Update. *Ann. Nutr. Metab.* **2017**, *70*, 26–33. [CrossRef] [PubMed]
42. Sieri, S.; Chiodini, P.; Agnoli, C.; Pala, V.; Berrino, F.; Trichopoulou, A.; Benetou, V.; Vasilopoulou, E.; Sánchez, M.-J.; Chirlaque, M.-D.; et al. Dietary Fat Intake and Development of Specific Breast Cancer Subtypes. *JNCI* **2014**, *106*, dju068. [CrossRef]
43. Orkusz, A.; Hapanowicz, K. Assessment of the energy and nutritional values of meals in selected kindergarten in Wroclaw. *Eng. Sci. Technol.* **2016**, *4*, 85–94. [CrossRef]
44. Delimaris, J. Adverse effects associated with protein intake above the recommended dietary allowance for adults. *ISRN Nutr.* **2013**, *2013*, 1–6. [CrossRef] [PubMed]
45. Campbell, K.J.; Abott, G.M.P.H.; Zheng, M.; Mc Naughton, S.A. Early Life Protein Intake: Food Sources, Correlates, and Tracking across the First 5 Years of Life. *J. Acad. Nutr. Diet.* **2017**, *117*, 1188–1197. [CrossRef]
46. Gaby, A.R. Adverse effects of dietary fructose. *Altern. Med. Rev.* **2005**, *10*, 294–306. [PubMed]
47. Federico, A.; Rosato, V.; Masarone, M.; Torre, P.; Dallio, M.; Romeo, M.; Persico, M. The Role of Fructose in Non-Alcoholic Steatohepatitis: Old Relationship and New Insights. *Nutrients* **2021**, *13*, 1314. [CrossRef]
48. Stricker, S.; Rudloff, S.; Geier, A.; Steveling, A.; Roeb, E.; Zimmer, K.P. Fructose consumption-free sugars and their health effects. *Dtsch. Arztebl. Int.* **2021**, *5*, 71–78. [CrossRef]
49. Okręglicka, K.; Pardecki, M.; Jagielska, A.; Tyszko, P.Z. Metabolic effects of excessive dietary fructose intake. *Med. Og. Nauk. Zdr.* **2017**, *23*, 165–170. [CrossRef]
50. Kretowicz, M.; Goszka, G.; Brymora, A.; Flisiński, M.; Odrowąż-Sypniewska, G.; Manitius, J. Is there a relationship among daily fructose intake, blood pressure and uric acid level in chronic kidney disease patients without diabetes? *Arter. Hypertens.* **2011**, *15*, 341–346.
51. Regulation of the Polish Minister of Health of 26 July 2016 on the Groups of Foodstuffs Intended for Sale to Children and Adolescents in Units of the Educational System and the Requirements to Be Met by Foodstuffs Used as a Part of Collective Nutrition of Children and Adolescents in These Units (Dz.U. from 2016 poz. 1154; 01.08.2016). Available online: https://isap.sejm.gov.pl/isap.nsf/DocDetails.xsp?id=WDU20160001154 (accessed on 10 December 2021).

52. Sartorius, B.; Sartorius, K.; Aldous, C.; Madiba, T.E.; Stefan, C.; Noakes, T. Carbohydrate intake, obesity, metabolic syndrome and cancer risk? A two-part systematic review and meta-analysis protocol to estimate attributability. *BMJ Open* **2016**, *6*, e009301. [CrossRef] [PubMed]

53. Stanhope, K.L. Sugar consumption, metabolic disease and obesity: The state of the controversy. *Crit. Rev. Clin. Lab. Sci.* **2016**, *53*, 52–67. [CrossRef] [PubMed]

54. Sheiham, A.; James, W.P. A new understanding of the relationship between sugars, dental caries, and fluoride use: Implications for limits on sugars consumption. *Public Health Nutr.* **2014**, *17*, 2176–2184. [CrossRef]

55. Delli Bovi, A.P.; Di Michele, L.; Laino, G.; Vajro, P. Obesity and Obesity Related Diseases, Sugar Consumption and Bad Oral Health: A Fatal Epidemic Mixtures: The Pediatric and Odontologist Point of View. *Transl. Med. UniSa.* **2017**, *16*, 11–16.

56. Polish Act of February 14, 2020 on the Amendment of Certain Acts in Connection with the Promotion of Pro-Health Consumer Choices. (Dz.U. from 2020 poz. 1492; 31.08.2020). Available online: https://isap.sejm.gov.pl/isap.nsf/DocDetails.xsp?id=WDU20200001492 (accessed on 10 December 2021).

57. Bueno, A.L.; Czepielewski, M.A. The importance for growth of dietary intake of calcium and vitamin D. *J. Pediatr.* **2008**, *84*, 386–394. [CrossRef]

58. Dermience, M.; Lognay, G.; Mathieu, F.; Goyens, P. Effects of thirty elements on bone metabolism. *J. Trace Elem. Med. Biol.* **2015**, *32*, 86–106. [CrossRef]

59. Merkiel, S.; Chalcarz, W. Dietary intake in 6-year-old children from southern Poland. Part 2—Vitamin and mineral intakes. *BMC Pediatrics* **2014**, *14*, 310. [CrossRef]

60. Pettifor, J.M. Calcium and vitamin D metabolism in children in developing countries. *Ann. Nutr. Metab.* **2014**, *64*, 15–22. [CrossRef]

61. Merkiel, S.; Chalcarz, W. Analysis of mineral intake in preschool children from Turek. *Med. Rodz.* **2016**, *1*, 7–13.

62. Zhu, K.; Prince, R.L. Calcium and bone. *Clin. Biochem.* **2012**, *45*, 936–942. [CrossRef] [PubMed]

63. Strnad, M. Sol. Salt and cancer. *Acta Med. Croatica.* **2010**, *64*, 159–161.

64. Hu, J.; La Vecchia, C.; Morrison, H.; Negri, E.; Mery, L. Canadian Cancer Registries Epidemiology Research Group. Salt processed meat and the risk of cancer. *Eur. J. Cancer Prev.* **2011**, *20*, 132–139. [CrossRef]

65. Huybrechts, I.; De Henauw, S. Energy and nutrient intakes by pre-school children in Flanders-Belgium. *Br. J. Nutr.* **2007**, *98*, 600–610. [CrossRef] [PubMed]

66. Huybrechts, I.; Maes, L.; Vereecken, C.; De Keyzer, W.; De Bacquer, D. High dietary supplements intakes among Flemish preschools. *Appetite* **2010**, *54*, 340–345. [CrossRef]

67. Wielgos, B.; Leszczyńska, T.; Kopeć, A.; Cieślik, E.; Piątkowska, E.; Pysz, M. Assesment of intake of minerals with daily diets by children aged 10–12 years from Malopolska region. *Rocz. Panstw. Zakl. Hig.* **2012**, *63*, 329–337.

68. Domellöf, M.; Thorsdottir, I.; Thorstensen, K. Health effects of different dietary iron intakes: A systematic literature review for the 5th Nordic Nutrition Recommendations. *Food Nutr. Res.* **2013**, *57*, 21667. [CrossRef]

69. Singh, M. Role of micronutrients for physical growth and mental development. *Indian J. Pediatr.* **2004**, *71*, 59–62. [CrossRef]

70. Tsai, C.J.; Leitzmann, M.F.; Willett, W.C.; Giovannucci, E.L. Heme and non-heme iron consumption and risk of gallstone disease in men. *Am. J. Clin. Nutr.* **2007**, *85*, 518–522. [CrossRef]

71. Ndayisaba, A.; Kaindlstorfer, C.; Wenning, G.K. Iron in Neurodegeneration—Cause or Consequence? *Front. Neurosci.* **2019**, *13*, 180. [CrossRef] [PubMed]

72. Zhang, W.; Iso, H.; Ohira, T.; Date, C.; Tanabe, N.; Kikuchi, S.; Tamakoshi, A. JACC Study Group. Associations of dietary iron intake with mortality from cardiovascular disease. *J. Epidemiol.* **2012**, *22*, 484–493. [CrossRef] [PubMed]

73. Sadowska, J.; Krzymuska, A. The estimation of complementation of the nursery school food rations by parents of pre-school children. *Bromat. Chem. Toksykol.* **2010**, *52*, 203–211.

74. Malhotra, K.; Baggott, P.J.; Livingstone, J. Vitamin D in the foot and ankle—A review of the literature. *J. Am. Podiatr. Med. Assoc.* **2020**, *110*, 1–11. [CrossRef] [PubMed]

Article

Exploring the Associations between Single-Child Status and Childhood High Blood Pressure and the Mediation Effect of Lifestyle Behaviors

Rui Deng [1], Ke Lou [1], Siliang Zhou [1], Xingxiu Li [1], Zhiyong Zou [1], Jun Ma [1], Bin Dong [1,*] and Jie Hu [2]

[1] Institute of Child and Adolescent Health, School of Public Health, Peking University Health Science Center, Beijing 100191, China; dengrui@stu.pku.edu.cn (R.D.); louke@pku.edu.cn (K.L.); slzhou@pku.edu.cn (S.Z.); 1610306239@pku.edu.cn (X.L.); harveyzou2002@bjmu.edu.cn (Z.Z.); majunt@bjmu.edu.cn (J.M.)

[2] Menzies Health Institute, Griffith University, Nathan, QLS 4111, Australia; jie.hu@griffith.edu.au

* Correspondence: bindong@bjmu.edu.cn

Abstract: Background: This study aimed to assess the association between single-child status and childhood high blood pressure (HBP) and to explore the role of lifestyle behaviors in this relationship. Methods: This study used data from a cross-sectional survey of 50,691 children aged 7~18 years in China. Linear and logistic regression models were used to assess the relationship between single-child status and HBP, and interactions between single-child status and lifestyle behaviors were also evaluated. Mediation analysis was conducted to detect the mediation effect of lifestyle behaviors. Results: Of the participants enrolled, 67.2% were single children and 49.4% were girls. Non-single children were associated with a greater risk of HBP, especially in girls (OR = 1.11, 95%CI: 1.03~1.19). Meat consumption and sedentary behavior mediated 58.9% of the association between single-child status and HBP ($p < 0.01$). When stratified by sleeping duration, non-single girls of insufficient sleep and hypersomnia showed a higher risk of HBP ($p < 0.05$) than single-child peers, but not in those with adequate sleep. Conclusion: Findings suggest that non-single children had an increased risk of HBP, and keeping healthy lifestyle behaviors could help to mitigate the adverse impact in non-single children.

Keywords: children; blood pressure; single-child status; lifestyle behaviors

Citation: Deng, R.; Lou, K.; Zhou, S.; Li, X.; Zou, Z.; Ma, J.; Dong, B.; Hu, J. Exploring the Associations between Single-Child Status and Childhood High Blood Pressure and the Mediation Effect of Lifestyle Behaviors. *Nutrients* **2022**, *14*, 500. https://doi.org/10.3390/nu14030500

Academic Editor: Manuel Sánchez Santos

Received: 5 January 2022
Accepted: 18 January 2022
Published: 24 January 2022

Publisher's Note: MDPI stays neutral with regard to jurisdictional claims in published maps and institutional affiliations.

Copyright: © 2022 by the authors. Licensee MDPI, Basel, Switzerland. This article is an open access article distributed under the terms and conditions of the Creative Commons Attribution (CC BY) license (https://creativecommons.org/licenses/by/4.0/).

1. Introduction

Hypertension is the leading risk for global disability-adjusted life-years (DALYs), accounting for 9.2% of DALYs for men and 7.8% of DALYs for women in 2015 [1]. A Chinese national survey [2] conducted from 2012 to 2015 showed that an estimated 244.5 million adults had hypertension. Evidence has suggested that childhood high blood pressure (HBP) could track to adulthood [3] and increase the risk of cardiovascular disease events in adulthood [4]. A meta-analysis [5] showed that the pooled prevalence of elevated BP among Chinese children and adolescents was 9.8% (95%CI: 7.9~11.9). Thus, preventing hypertension during childhood and identifying children and adolescents with an increased risk of HBP is of far-reaching public health significance.

Currently, countries [6] are struggling with population aging, and various policies have been proposed. In October 2015, China abolished the one-child policy, which lasted for 35 years, and subsequently implemented the universal two-child policy [7]. Soon after, in May 2021, the Chinese government further announced the policy that one couple is allowed to have three children. These new fertility policies may contribute to a large increase in non-single-child families in the future. Thus far, many research studies have investigated the influence of single-child status on health issues, including obesity [8–10], poor vision [11], anxiety and depression [12]. However, cardio-metabolic disorders and sing-child status are seldom studied.

Considering the progressing demographic transition in China, we wondered whether the rising proportion of non-single children would change the risk of HBP in the Chinese population, from both an immediate and long-term perspective. Furthermore, previous studies have suggested that lifestyle behaviors might contribute to the differences in body mass index (BMI) between single and non-single children [13]. It is unclear whether lifestyle behaviors could also mediate or modify the relationship between single-child status and HBP.

Using data collected from nationwide Chinese children and adolescents, this study aimed to examine the association of single-child status with childhood HBP and to further investigate the role of lifestyle behaviors in the relationship between single-child status and risk of HBP.

2. Materials and Methods

2.1. Study Design and Participants

This study used data from the baseline cross-sectional survey of a national school program from seven provinces or cities of China (Guangdong, Shanghai, Hunan, Ningxia, Chongqing, Liaoning, and Tianjin) in 2013. A detailed description of the study design is reported elsewhere [14]. Briefly, using the multistage cluster random sampling method, a total of 93 primary and secondary schools were randomly selected, and 53,532 children and adolescents were involved in the present study. Exclusion criteria included: (a) participants aged out of 7~18 years old (n = 250); (b) participants who were not a singleton birth (n = 1577); (c) participants with a missing value for blood pressure (n = 1014). Thus, a total of 50,691 participants (94.7%) were involved in analyses. The study has been approved by the Ethics Committee of Peking University (No. IRB0000105213034). All participating students and their parents have signed the informed consent.

2.2. Anthropometric Measurements

All participants underwent anthropometric and clinical measurements by qualified medical physicians from medical establishments in schools. Height was measured two times using a portable stadiometer (model TZG, China) to the nearest 0.1 cm without shoes. Weight was measured two times using a lever-type weight scale (model RGT-140, China) to the nearest 0.1 kg in light clothing and without shoes. The average of two repeated measures was calculated for height and weight. BMI was calculated as weight in kilograms divided by the square of height in meters (kg/m^2) and then converted into age- and sex-specific BMI z scores according to the World Health Organization (WHO) growth references [15]. BP was obtained by Mercury sphygmomanometer (model XJ11D, Shanghai Medical Instruments Co, Ltd, Shanghai, China) and stethophone (model TZ-1, Shanghai Medical Instruments Co, Ltd, Shanghai, China) from the right arm using the appropriate cuffs. Prior to the first reading, children were asked to sit quietly for at least 5 min. Systolic blood pressure (SBP) was determined by the onset of the first Korotkoff sound, and diastolic blood pressure (DBP) was determined by the fifth Korotkoff sound. BP was measured twice with a 5 min gap between replicates, and the averages of two readings for SBP and DBP were calculated for each participant.

2.3. Questionnaires Data Collection

Questionnaires were designed for students and their parents, separately. Students of grade four or older ($\geq$10 years) finished the student questionnaire independently in a classroom, while children of grade 1~3 (7~9 years) answered the questionnaire with the help of parents at home. Child self-reported lifestyle behaviors were obtained by the student questionnaire, while obstetric information (e.g., gestational age, delivery mode and birth weight) were obtained from the parental questionnaire. Moreover, information on family history of hypertension was also obtained from the parental questionnaire, and children were considered as having a family history of hypertension when either father or mother self-reported that he/she had been diagnosed with hypertension by medical physicians. Both father's and mother's education degrees were collected, and the higher one was defined as the parental highest education degree.

2.4. Definitions

Single-child status was defined using the question "Is your child the only child?" in the parental questionnaire, and children were assigned to "single children" or "non-single children" groups accordingly. As suggested by the American Academy of Pediatrics [16], high blood pressure (HBP) was defined if child blood pressure was $\geq$90th percentile or 120/80 mmHg. Lifestyle behaviors, including diet behaviors, physical activity, sedentary behavior and sleeping duration, were investigated in the current study. Diet behaviors were identified as frequencies (days per week) of consuming meat, fruit, vegetables and beverages during the past 7 days. Based on child self-reported frequency (days) and duration (minutes per day) of moderate and vigorous-intensity activity over the past 7 days, physical activity was defined with reference to the International Physical Activity Questionnaire [17]. Sedentary behavior (h) referred to the total time spent sitting or lying down at school or home, including watching television, playing computer games and doing homework. Sleeping duration per day was recorded, and students were divided into three groups (enough sleep ($\geq$9 h), insufficient sleep (7~8 h) and hypersomnia (<7 h)) according to the recommendation of National Sleep Foundations [18].

2.5. Statistical Analysis

Continuous variables were reported as mean $\pm$ standard deviation, and categorical variables were reported as frequency (percentage). Differences between single children and non-single children were tested by t test for continuous variables and χ^2 test for categorical variables. Linear regression models were used to evaluate the association between single-child status and BP z score, while logistic regression models were used to test the relationship between single-child status and risk of HBP by sex. To explore the potential factors mediating the association of single-child status with HBP, a mediation analysis proposed by Karlson and colleagues, was conducted [19]. This method decomposed the total effect of a variable into direct effect and indirect effect and calculated the percentage of the main association explained by the mediators. Moreover, the interaction terms of single-child status and each lifestyle behavioral factor, such as interaction between single-child status and frequency of meat consumption, were also tested. All analyses were performed with R studio 1.2.5033 and Stata software version 14 (StataCorp, College Station, TX, USA). A two-sided $p < 0.05$ was considered statistically significant.

3. Results

Characteristics of participants are shown in Table 1 by sex and single-child status. Among 50,691 participants, 67.2% were single children and 49.4% were girls. Non-single children were associated with higher z scores of BP, while single children were related to a greater BMI z score in both boys and girls ($p < 0.05$). A significant difference in HBP prevalence was observed in single and non-single girls. In both sexes, single children more frequently ate meat, fruit and vegetables, and they spent more time sitting, lying, and sleeping, compared to those children with siblings.

The associations between single-child status and BP levels were assessed by sex, and non-single children generally had higher BP levels compared with single children (Table 2). When stratified by sex, similar relationships were detected in girls after adjusting for confounders (for SBP: $\beta = 0.073$, 95%CI: 0.043~0.102, $p < 0.001$; for DBP: $\beta = 0.062$, 95% CI: 0.032~0.091, $p < 0.001$; for HBP: OR = 1.11, 95% CI: 1.03~1.19, $p < 0.05$), but not in boys.

Table 1. Characteristics of participants involved according to single-child status and by sex (n = 50,691).

	Boys			Girls		
	Single Children	Non-Single Children	p Value	Single Children	Non-Single Children	p Value
Number of children	18,397 (71.7)	7264 (28.3)		15,691 (62.7)	9339 (37.3)	
Age, years	10.59 ± 3.2	10.62 ± 3.2	<0.001 ***	10.65 ± 3.27	11.22 ± 3.32	<0.001 ***
Height, cm	146.7 ± 18.3	145.3 ± 18.1	<0.001 ***	144.1 ± 15.5	144.9 ± 15.1	<0.001 ***
Weight, kg	42.7 ± 17.3	40.4 ± 15.7	<0.001 ***	39.2 ± 14.1	39.7 ± 13.5	<0.001 ***
BMI z score	0.48 ± 1.4	0.25 ± 1.3	<0.001 ***	0.07 ± 1.2	−0.04 ± 1.1	<0.001 ***
SBP z score	0.07 ± 1.0	0.11 ± 1.0	0.002 **	−0.10 ± 1.0	0.05 ± 1.0	<0.001 ***
DBP z score	0.04 ± 1.0	0.09 ± 1.0	<0.001 ***	−0.11 ± 1.0	0.01 ± 1.0	<0.001 ***
HBP	5647 (30.7)	2237 (30.8)	0.887	3247 (20.7)	2223 (23.8)	<0.001 ***
Breastfeeding	15,138 (83.5)	6273 (87.7)	<0.001 ***	12,878 (82.9)	8187 (89.1)	<0.001 ***
Birth weight, g	3359.9 ± 497.8	3406.7 ± 533.1	<0.001 ***	3273.7 ± 472.8	3306.3 ± 499.8	<0.001 ***
Caesarean section	8814 (48.6)	2104 (29.5)	<0.001 ***	7410 (47.8)	2416 (26.4)	<0.001 ***
Gestational age, week	39.7 ± 1.2	39.9 ± 1.1	<0.001 ***	39.8 ± 1.21	39.9 ± 1.06	<0.001 ***
Meat consumption per week			<0.001 ***			<0.001 ***
0~1 day(s)	1127 (6.4)	630 (9.0)		1105 (7.3)	1019 (11.3)	
2~3 days	3164 (17.9)	1608 (23.1)		3151 (20.9)	2648 (29.4)	
4~5 days	3013 (17.1)	1340 (19.2)		2692 (17.8)	1749 (19.4)	
6~7 days	10,358 (58.6)	3387 (48.6)		8157 (54.0)	3598 (39.9)	
Fruits consumption per week			<0.001 ***			<0.001 ***
0~1 days	1362 (7.7)	633 (9.1)		703 (4.7)	532 (5.9)	
2~3 days	3674 (20.8)	1753 (25.3)		2612 (17.3)	1927 (21.4)	
4~5 days	4232 (24.0)	1830 (26.4)		3658 (24.2)	2417 (26.9)	
6~7 days	8373 (47.5)	2725 (39.3)		8119 (53.8)	4124 (45.8)	
Vegetable consumption per week			<0.001 ***			<0.001 ***
0~1 day(s)	775 (4.4)	360 (5.2)		563 (3.7)	391 (4.3)	
2~3 days	1408 (8.0)	648 (9.3)		1029 (6.8)	857 (9.5)	
4~5 days	1989 (11.3)	938 (13.5)		1522 (10.1)	1132 (12.6)	
6~7 days	13,496 (76.4)	5021 (72.1)		11,990 (79.4)	6632 (73.6)	
Beverage consumption per week			0.502			0.071
0 day	5237 (29.8)	2035 (29.3)		5477 (36.4)	3148 (35.0)	
1~2 days	7796 (44.4)	3137 (45.2)		6724 (22.8)	4111 (45.7)	
3 days and more	4530 (25.8)	1766 (25.5)		2835 (18.9)	1745 (19.4)	
Physical activity			0.051			<0.001 ***
Low intensity	2533 (15.9)	932 (15.2)		2457 (18.0)	1477 (18.6)	
Median intensity	4970 (31.1)	1847 (30.0)		5354 (39.1)	2874 (36.2)	
High intensity	8454 (53.0)	3369 (54.8)		5870 (42.9)	3580 (45.1)	

Table 1. *Cont.*

	Boys			Girls		
	Single Children	**Non-Single Children**	***p* Value**	**Single Children**	**Non-Single Children**	***p* Value**
Sedentary behavior, h	5.54 ± 3.6	5.25 ± 3.7	<0.001 ***	5.90 ± 3.7	5.64 ± 3.7	<0.001 ***
Sleeping duration			0.020 *			<0.001 ***
Adequate sleep	5908 (35.6)	2391 (36.7)		4931 (34.2)	3006 (35.2)	
Insufficient sleep	10,275 (61.9)	3938 (60.4)		9264 (64.2)	5324 (62.4)	
Hypersomnia	403 (2.4)	191 (2.9)		242 (1.7)	203 (2.4)	
Having a family history of hypertension	1157 (6.6)	495 (7.2)	0.091	1063 (7.0)	662 (7.4)	0.278
Parental highest education degree			<0.001 ***			<0.001 ***
Illiteracy or elementary school	342 (1.9)	496 (7.1)		246 (1.6)	573 (6.2)	
Junior high school	4639 (26.0)	3542 (51.0)		3108 (20.4)	4662 (51.9)	
Senior high school	5248 (29.4)	1928 (27.8)		4454 (29.2)	2466 (27.4)	
Technical secondary school/ Junior college	3596 (20.1)	586 (8.4)		3492 (22.9)	771 (8.6)	
Undergraduate or above	4033 (22.6)	394 (5.7)		3930 (25.8)	516 (5.7)	

Note: Continuous variables were expressed by mean values ± standard deviations, and categorical variables were expressed by frequency value (percentages, %). BMI, body mass index; HBP, high blood pressure; SBP, systolic blood pressure; DBP, diastolic blood pressure. * $p < 0.05$, ** $p < 0.01$, *** $p < 0.001$.

Table 2. Associations between single-child status and childhood BP.

	Model 1		Model 2	
	β/OR (95% CI)	***p* Value**	**β/OR (95% CI)**	***p* Value**
SBP z score				
Total				
Single children	0 (ref)		0 (ref)	
Non-single children	0.123 (0.084~0.121)	<0.001 ***	0.037 (0.016–0.058)	<0.001 ***
Boys				
Single children	0 (ref)		0 (ref)	
Non-single children	0.043 (0.016~0.070)	0.002 **	0.005 (−0.025~0.035)	0.735
Girls				
Single children	0 (ref)		0 (ref)	
Non-single children	0.161 (0.135~0.186)	<0.001 ***	0.073 (0.043~0.102)	<0.001 ***
DBP z score				
Total				
Single children	0 (ref)		0 (ref)	
Non-single children	0.095 (0.076~0.113)	<0.001 ***	0.035 (0.014~0.056)	0.001 **

Table 2. *Cont.*

	Model 1		Model 2	
	β/OR (95% CI)	*p* Value	β/OR (95% CI)	*p* Value
Boys				
Single children	0 (ref)		0 (ref)	
Non-single children	0.055 (0.029~0.082)	<0.001 ***	0.013 (−0.017~0.043)	0.404
Girls				
Single children	0 (ref)		0 (ref)	
Non-single children	0.137 (0.111~0.163)	<0.001 ***	0.062 (0.032~0.091)	<0.001 ***
HBP				
Total				
Single children	1 (ref)		1 (ref)	
Non-single children	1.08 (1.03~1.12)	0.001 **	1.00 (0.95~1.05)	0.884
Boys				
Single children	1 (ref)		1 (ref)	
Non-single children	1.00 (0.94~1.06)	0.956	1.00 (0.93~1.07)	0.991
Girls				
Single children	1 (ref)		1 (ref)	
Non-single children	1.20 (1.13~1.27)	<0.001 ***	1.11 (1.03~1.19)	0.008 **

Model 1 adjusted for age and sex if the variable was not the grouping variable. Model 2 adjusted for age, birth weight, gestational age, breastfeeding, delivery mode, family history of hypertension, parental highest education degree, BMI z score if the variable was not the grouping variable. BP, blood pressure; SBP, systolic blood pressure; DBP, diastolic blood pressure; HBP, high blood pressure. ** $p < 0.01$, *** $p < 0.001$.

Mediation analysis was conducted to identify the mediation effects of lifestyle behaviors on the relationship between single-child status and BP level. In girls (Table 3), frequency of meat consumption mediated about one-third of the association between single-child status and SBP z score (mediation effect = 0.017, 95% CI: 0.013~0.021, $p < 0.001$, mediation ratio = 30.31%) and DBP z score (mediation effect = 0.016, 95% CI: 0.012~0.020, $p < 0.001$, mediation ratio = 31.40%). The corresponding mediation effect for sedentary behavior was 0.002 ($p < 0.01$) with the mediation proportion of 4.02% for SBP and 4.18% for DBP. When HBP was the outcome variable, the mediation proportions of consuming meat and sedentary behavior were 46.98% and 11.87%, with a mediation effect of 0.022 (95% CI: 0.015~0.028, $p < 0.001$) and 0.004 (95%CI: 0.001~0.007, $p < 0.01$), respectively. Similar patterns were also found in boys, especially the mediation effect of consuming meat (Supplementary Material Table S3).

When interaction terms of single-child status and different lifestyle behaviors were analyzed, a significant interaction effect of sleeping duration and single-child status was observed (p values < 0.05) (for girls: Supplementary Materials Tables S1 and S2; for boys: Supplementary Materials Tables S4 and S5). Among various sleeping duration groups (Table 4), non-single girls of insufficient sleep had higher BP z scores (for SBP: β = 0.110, 95% CI: 0.072~0.151, $p < 0.001$; for DBP: β = 0.098, 95% CI: 0.058~0.137, $p < 0.001$) than their single peers, with the risk of HBP raised by 15% (OR = 1.15, 95%CI: 1.04~1.27, $p < 0.05$). Similar patterns were also observed in children with hypersomnia, in which SBP z scores and risk of HBP in non-single girls increased by 0.220 (95%CI: 0.019~0.420, $p < 0.05$) and 88% (OR = 1.88, 95% CI: 1.08~3.32, $p < 0.05$), respectively. However, in girls with adequate sleep, the BP levels were similar regardless of single-child status. The above analyses were repeated in boys, and similar patterns were observed, although the associations were mostly not statistically significant.

Table 3. Mediation of lifestyle behaviors on the association between single-child status and BP in girls ($n = 25{,}030$).

Mediator	n	Outcome	Total Association		Direct Association		Indirect Association		Mediation Proportion
			β (95% CI)	p	β (95% CI)	p	β (95% CI)	p	
Meat consumption per week	20,053	SBP z score	0.059 (0.025~0.087)	<0.001 ***	0.039 (0.008~0.070)	0.014 **	0.017 (0.013~0.021)	<0.001 ***	30.31%
	20,054	DBP z score	0.050 (0.003~0.065)	0.002 **	0.034 (0.003~0.065)	0.032 *	0.016 (0.012~0.020)	<0.001 ***	31.40%
	29,957	HBP	0.046 (−0.030~0.121)	0.234	0.024 (−0.051~0.100)	0.529	0.022 (0.015~0.028)	<0.001 ***	46.98%
Fruits consumption per week	20,049	SBP z score	0.056 (0.024~0.087)	<0.001 ***	0.058 (0.027~0.089)	<0.001 ***	—	—	—
	20,050	DBP z score	0.050 (0.019~0.081)	0.002 **	0.053 (0.021~0.083)	0.001 **	—	—	—
	20,053	HBP	0.045 (−0.031~0.120)	0.247	0.050 (−0.026~0.125)	0.195	—	—	—
Vegetable consumption per week	20,053	SBP z score	0.057 (0.026~0.088)	<0.001 ***	0.057 (0.026~0.088)	<0.001 ***	—	—	—
	20,054	DBP z score	0.051 (0.020~0.082)	0.001 **	0.052 (0.021~0.083)	0.001 **	—	—	—
	20,057	HBP	0.049 (−0.030~0.125)	0.200	0.052 (−0.023~0.127)	0.176	—	—	—
Beverage consumption per week	19,980	SBP z score	0.056 (0.024~0.087)	<0.001 ***	0.055 (0.024~0.087)	0.001 **	0.000 (−0.000~0.001)	0.304	—
	19,981	DBP z score	0.049 (0.018~0.080)	0.002 **	0.049 (0.017~0.080)	0.002 **	0.000 (−0.001~0.001)	0.364	—
	19,984	HBP	0.046 (−0.029~0.122)	0.231	0.045 (−0.030~0.121)	0.239	0.001 (−0.001~0.002)	0.411	—
Physical activity	18,181	SBP z score	0.065 (0.032~0.098)	<0.001 ***	0.065 (0.032~0.098)	<0.001 ***	—	—	—
	18,182	DBP z score	0.059 (0.026~0.092)	<0.001 ***	0.059 (0.026~0.092)	<0.001 ***	0.000 (−0.000~0.000)	0.895	—

Table 3. *Cont.*

Mediator	n	Outcome	Total Association		Direct Association		Indirect Association		Mediation Proportion
			β (95% CI)	p	β (95% CI)	p	β (95% CI)	p	
	18,185	HBP	0.055 (−0.024~0.134)	0.173	0.055 (−0.024~0.134)	0.174	0.000 (−0.002~0.002)	0.962	—
Sedentary behavior	18,256	SBP z score	0.052 (0.020~0.085)	0.002 **	0.050 (0.017~0.083)	0.003 **	0.002 (0.001~0.004)	0.004 **	4.02%
	18,257	DBP z score	0.044 (0.011~0.076)	0.009 **	0.042 (0.009~0.075)	0.012 *	0.002 (0.001~0.003)	0.007 **	4.18%
	18,260	HBP	0.036 (−0.043~0.115)	0.373	0.032 (−0.047~0.110)	0.432	0.004 (0.001~0.007)	0.008 **	11.87%
Sleeping duration	19,211	SBP z score	0.054 (0.220~0.086)	0.001 **	0.054 (0.022~0.086)	0.001 **	—	—	—
	19,212	DBP z score	0.048 (0.016~0.080)	0.003 **	0.048 (0.016~0.080)	0.003 **	0.000 (−0.000~0.000)	0.902	—
	19,215	HBP	0.046 (−0.031~0.123)	0.243	0.046 (−0.031~0.123)	0.243	0.000 (−0.000~0.000)	0.978	—

Note: The independent variable in the model was single-child status and the reference group was "single children". Adjusted for age, birth weight, gestational age, breastfeeding, delivery mode, family history of hypertension, parental highest education degree, BMI z score. "—" represents the negative value of indirect association. BP, blood pressure; SBP, systolic blood pressure; DBP, diastolic blood pressure; HBP, high blood pressure. * $p < 0.05$, ** $p < 0.01$, *** $p < 0.001$.

Table 4. Associations between single-child status with childhood BP stratified by sex and sleeping duration.

Sleeping Duration	SBP z Score		DBP z Score		HBP	
	β (95% CI)	*p* Value	β (95% CI)	*p* Value	OR (95% CI)	*p* Value
Girls						
Adequate sleep						
Single children	0 (ref)		0 (ref)		1 (ref)	
Non-single children	−0.007 (−0.060~0.046)	0.794	−0.003 (−0.057~0.052)	0.927	0.99 (0.86~1.13)	0.860
Insufficient sleep						
Single children	0 (ref)		0 (ref)		1 (ref)	
Non-single children	0.110 (0.072~0.151)	<0.001 ***	0.098 (0.058~0.137)	<0.001 ***	1.15 (1.04~1.27)	0.006 **
Hypersomnia						
Single children	0 (ref)		0 (ref)		1 (ref)	
Non-single children	0.220 (0.019~0.420)	0.032 *	0.149 (−0.071~0.369)	0.184	1.88 (1.08~3.32)	0.026 *
Boys						
Adequate sleep						
Single children	0 (ref)		0 (ref)		1 (ref)	
Non-single children	−0.025 (−0.078~0.028)	0.351	−0.049 (−0.103~0.005)	0.074	0.88 (0.78~1.00)	0.058
Insufficient sleep						
Single children	0 (ref)		0 (ref)		1 (ref)	
Non-single children	0.020 (−0.021~0.061)	0.336	0.056 (0.015~0.098)	0.008 **	1.06 (0.96~1.17)	0.236
Hypersomnia						
Single children	0 (ref)		0 (ref)		1 (ref)	
Non-single children	0.031 (−0.163~0.225)	0.755	−0.070 (−0.278~0.137)	0.505	1.12 (0.70~1.76)	0.639

Adjusted for age, birth weight, gestational age, breastfeeding, delivery mode, family history of hypertension, parental highest education degree, BMI z score. BP, blood pressure; SBP, systolic blood pressure; DBP, diastolic blood pressure; HBP, high blood pressure. * $p < 0.05$, ** $p < 0.01$, *** $p < 0.001$.

4. Discussion

In the present study, non-single girls were positively associated with HBP in those with insufficient sleeping duration or hypersomnia, but not in those with adequate sleep. These associations were independent of BMI. Furthermore, lifestyle behaviors, such as meat consumption and sedentary behavior, mediated 34.3%~58.9% of the association between single-child status and BP levels. These findings suggested that adherence to a healthy lifestyle could relieve the adverse impact on BP in non-single children.

Our findings are contrary to some previous research. Kelishadi and colleagues [20] found that single Iranian children had an increased risk of SBP (OR = 1.58; 95% CI: 1.17~2.14). Considering that single children were all firstborns, studies [21,22] found that firstborns had higher blood pressure compared with later-born children. However, these studies did not take child BMI into account. It has been established that single children were associated with higher risk of obesity [23–25], which was also supported by our study. Besides, BMI was also strongly associated with HBP [26,27]. In the present study, BMI was identified as a confounder, rather than a mediator, in the associations between single-child status and BP, because BMI did not show its mediation effect in the mediation analysis. After controlling for BMI, we concluded that non-single children had a greater risk of HBP than single peers. In previous studies, results of elevated BP levels in single children also disappeared after adjusting for BMI [20] and weight status [22].

In this study, lifestyle behaviors were investigated, with meat consumption and sedentary behavior identified as mediators. Multiple studies have shed light on dietary behaviors (such as meat consumption and eating outside) and sedentary time as the mechanism underlying the association between single-child status and their health outcomes [9,20,28,29]. A systematic literature review [30] showed that peers or siblings often had a negative influence on children's healthy eating behavior; for example, children may increase consumption of energy-dense foods such as meat because their siblings may tease them about eating fruits and vegetables [31]. Meanwhile, studies have showed that higher meat consumption is related to greater risk of HBP [32,33]. For example, a cohort study [34] found that participants who consumed ≥ 5 servings of red meat had a 17% higher rate of hypertension compared with those who consumed <1 serving/wk. Although these results could support our findings, the mechanism remains highly speculative, and more studies are needed to provide further evidence.

Our study additionally found that the adverse effect related to single-child status could be modified by improving lifestyle behaviors, such as sufficient sleep. Quist and colleagues reviewed [35] the effect of sleeping on childhood health outcomes and concluded that inadequate sleep was associated with HBP. Another study conducted [36] on American children showed that both short and long sleeping duration was associated with an elevated risk of hypertension, with a stronger association for the short sleeping duration. In our study, we demonstrated that non-single children with adequate sleep were not related to a greater HBP risk, while those of both short or long sleeping duration showed a raised HBP risk. These findings suggested that interventions of a modifiable lifestyle, such as sleeping, could be helpful in relieving the risk of HBP in non-single children.

Furthermore, we found that the association between single-child status and the risk of HBP was only statistically significant among girls. Similarly, a study [37] conducted within the Pelotas 2004 cohort also observed lower SBP and DBP in first-born females, and another study [38] involving over 1 million Swedish men found no birth order differences in BP. In our study, the possible explanation for the gender differences among the Chinese might be partly attributed to the Chinese traditional preference for sons [39]. Chinese boys may have less discrepancy in lifestyle behaviors with their brothers because they equally enjoy the family's attention and resources, which were more than that for girls [40,41]. As a result, there are limited differences in BP between single and non-single Chinese boys.

Several limitations should be taken into consideration when interpreting the findings. First, causality is difficult to be inferred from this cross-sectional study. Second, some factors, such as maternal and paternal BMI and parent–child interaction, were not investigated. In addition, lifestyle behaviors were collected from self-reported questionnaires, and instrumental measurements are needed to obtain more accurate and objective data, such as ActiGraph GT3X accelerometers. Furthermore, BP measurements were taken in one visit for each individual, which may overestimate the prevalence of HBP. Finally, as this study was conducted in the Chinese population, studies in other countries are warranted to clarify the generalizability of our findings.

5. Conclusions

To summarize, this study found that non-single children were associated with a greater risk of HBP than their single-child peers, especially in girls and those with short or prolonged sleeping duration. This association was potentially mediated by lifestyle behaviors, such as meat consumption and sedentary behavior. These findings suggested that keeping a healthy lifestyle could mitigate the disadvantageous impact on BP profile among non-single children and adolescents. More studies are desired to clarify the mechanisms underlying the observations that this study presented.

Supplementary Materials: The following supporting information can be downloaded at: https://www.mdpi.com/article/10.3390/nu14030500/s1, Table S1: Moderation of lifestyle behaviors on the association between single-child status and BP z score in girls (*n* = 25,030), Table S2: Moderation of lifestyle behaviors on the association between single-child status and HBP in girls (*n* = 25,030), Table S3: Mediation of lifestyle behaviors on the association between single-child status and BP in boys (*n* = 25,661), Table S4: Moderation of lifestyle behaviors on the association between single-child status and BP z score in boys (*n* = 25,661), Table S5: Moderation of lifestyle behaviors on the association between single-child status and HBP in boys (*n* = 25,661).

Author Contributions: Conceptualization, R.D.; Data curation, S.Z., K.L. and X.L.; Formal analysis, R.D.; Funding acquisition, B.D. and J.M.; Investigation, B.D., Z.Z. and J.M.; Methodology, R.D. and B.D.; Project administration, B.D., Z.Z. and J.M.; Resources, B.D., Z.Z. and J.M.; Software, R.D. and B.D.; Supervision, B.D., Z.Z., J.H. and J.M.; Validation, B.D., Z.Z., J.H. and J.M.; Visualization, R.D. and B.D.; Writing—original draft, R.D.; Writing—review and editing, B.D. and J.H. All authors have read and agreed to the published version of the manuscript.

Funding: The present research was supported by funding of the National Natural Science Foundation of China (Grant No. 81903344 to B.D.) and the Young Researcher Personal Project of Beijing to Bin Dong.

Institutional Review Board Statement: This study was conducted according to the guidelines laid down in the Declaration of Helsinki, and all procedures involving students were approved by the Ethics Committee of Peking University (no. IRB0000105213034).

Informed Consent Statement: All participants and their parents signed informed consent voluntarily.

Data Availability Statement: The data supporting the conclusions of this article will be made available from the corresponding author upon request.

Acknowledgments: We would like to thank all the schoolteachers and doctors for their assistance in data collection and all students and their guardians for participating in the study.

Conflicts of Interest: The authors declare no conflict of interest related to this manuscript.

References

1. GBD 2015 Risk Factors Collaborators. Global, regional, and national comparative risk assessment of 79 behavioural, environmental and occupational, and metabolic risks or clusters of risks, 1990–2015: A systematic analysis for the Global Burden of Disease Study 2015. *Lancet* **2016**, *388*, 1659–1724. [CrossRef]
2. Wang, Z.; Chen, Z.; Zhang, L.; Wang, X.; Hao, G.; Zhang, Z.; Shao, L.; Tian, Y.; Dong, Y.; Zheng, C.; et al. Status of Hypertension in China: Results from the China Hypertension Survey, 2012–2015. *Circulation* **2018**, *137*, 2344–2356. [CrossRef] [PubMed]
3. Chen, X.; Wang, Y. Tracking of blood pressure from childhood to adulthood: A systematic review and meta-regression analysis. *Circulation* **2008**, *117*, 3171–3180. [CrossRef] [PubMed]
4. Yang, L.; Magnussen, C.G.; Yang, L.; Bovet, P.; Xi, B. Elevated Blood Pressure in Childhood or Adolescence and Cardiovascular Outcomes in Adulthood. *Hypertension* **2020**, *75*, 948–955. [CrossRef]
5. Wang, L.; Song, L.; Liu, B.; Zhang, L.; Wu, M.; Cao, Z.; Wang, Y. Trends and Status of the Prevalence of Elevated Blood Pressure in Children and Adolescents in China: A Systematic Review and Meta-analysis. *Curr. Hypertens. Rep.* **2019**, *21*, 88. [CrossRef]
6. Parsons, A.J.Q.; Gilmour, S. An evaluation of fertility- and migration-based policy responses to Japan's ageing population. *PLoS ONE* **2018**, *13*, e0209285. [CrossRef]
7. Zeng, Y.; Hesketh, T. The effects of China's universal two-child policy. *Lancet* **2016**, *388*, 1930–1938. [CrossRef]
8. Haugaard, L.K.; Ajslev, T.A.; Zimmermann, E.; Ängquist, L.; Sørensen, T.I.A. Being an Only or Last-Born Child Increases Later Risk of Obesity. *PLoS ONE* **2013**, *8*, e56357. [CrossRef]
9. Min, J.; Xue, H.; Wang, V.H.C.; Li, M.; Wang, Y. Are single children more likely to be overweight or obese than those with siblings? The influence of China's one-child policy on childhood obesity. *Prev. Med.* **2017**, *103*, 8–13. [CrossRef]
10. Mosli, R.H.; Miller, A.L.; Peterson, K.E.; Kaciroti, N.; Rosenblum, K.; Baylin, A.; Lumeng, J.C. Birth order and sibship composition as predictors of overweight or obesity among low-income 4- to 8-year-old children. *Pediatr. Obes.* **2016**, *11*, 40–46. [CrossRef]
11. Zhao, L.; Zhou, M. Do only children have poor vision? Evidence from China's One-Child Policy. *Health Econ.* **2018**, *27*, 1131–1146. [CrossRef]
12. Cao, Y.; Huang, L.; Si, T.; Wang, N.Q.; Qu, M.; Zhang, X.Y. The role of only-child status in the psychological impact of COVID-19 on mental health of Chinese adolescents. *J. Affect. Disord.* **2021**, *282*, 316–321. [CrossRef]
13. Tian, X.; von Cramon-Taubadel, S. Are only children in China more likely to be obese/overweight than their counterparts with siblings? *Econ. Hum. Biol.* **2020**, *37*, 100847. [CrossRef]

14. Chen, Y.; Ma, L.; Ma, Y.; Wang, H.; Luo, J.; Zhang, X.; Luo, C.; Wang, H.; Zhao, H.; Pan, D.; et al. A national school-based health lifestyles interventions among Chinese children and adolescents against obesity: Rationale, design and methodology of a randomized controlled trial in China. *BMC Public Health* **2015**, *15*, 210. [CrossRef]

15. De Onis, M.; Onyango, A.W.; Borghi, E.; Siyam, A.; Nishida, C.; Siekmann, J. Development of a WHO growth reference for school-aged children and adolescents. *Bull. World Health Organ.* **2007**, *85*, 660–667. [CrossRef]

16. Flynn, J.T.; Kaelber, D.C.; Baker-Smith, C.M.; Blowey, D.; Carroll, A.E.; Daniels, S.R.; de Ferranti, S.D.; Dionne, J.M.; Falkner, B.; Flinn, S.K.; et al. Clinical Practice Guideline for Screening and Management of High Blood Pressure in Children and Adolescents. *Pediatrics* **2017**, *140*, e20171904. [CrossRef]

17. Fan, M.; Lyu, J.; He, P. Chinese guidelines for data processing and analysis concerning the International Physical Activity Questionnaire. *Zhonghua Liu Xing Bing Xue Za Zhi* **2014**, *35*, 961–964. (In Chinese)

18. Hirshkowitz, M.; Whiton, K.; Albert, S.M.; Alessi, C.; Bruni, O.; DonCarlos, L.; Hazen, N.; Herman, J.; Adams Hillard, P.J.; Katz, E.S.; et al. National Sleep Foundation's updated sleep duration recommendations: Final report. *Sleep Health* **2015**, *1*, 233–243. [CrossRef]

19. Kohler, U.; Karlson, K.B.; Holm, A. Comparing Coefficients of Nested Nonlinear Probability Models. *Stata J.* **2011**, *11*, 420–438. [CrossRef]

20. Kelishadi, R.; Qorbani, M.; Rezaei, F.; Motlagh, M.E.; Djalalinia, S.; Ziaodini, H.; Taheri, M.; Ochi, F.; Shafiee, G.; Aminaei, T.; et al. Is single-child family associated with cardio-metabolic risk factors: The CASPIAN-V study. *BMC Cardiovasc. Disord.* **2018**, *18*, 109. [CrossRef]

21. Ayyavoo, A.; Savage, T.; Derraik, J.G.; Hofman, P.L.; Cutfield, W.S. First-born children have reduced insulin sensitivity and higher daytime blood pressure compared to later-born children. *J. Clin. Endocrinol. Metab.* **2013**, *98*, 1248–1253. [CrossRef] [PubMed]

22. Wells, J.C.K.; Hallal, P.C.; Reichert, F.F.; Dumith, S.C.; Menezes, A.M.; Victora, C.G. Associations of Birth Order with Early Growth and Adolescent Height, Body Composition, and Blood Pressure: Prospective Birth Cohort from Brazil. *Am. J. Epidemiol.* **2011**, *174*, 1028–1035. [CrossRef] [PubMed]

23. Datar, A. The more the heavier? Family size and childhood obesity in the U.S. *Soc. Sci. Med.* **2017**, *180*, 143–151. [CrossRef] [PubMed]

24. Li, M.; Xue, H.; Wang, W.; Wen, M.; Wang, Y. Increased obesity risks for being an only child in China: Findings from a nationally representative study of 19,487 children. *Public Health* **2017**, *153*, 44–51. [CrossRef]

25. Hesketh, T. Health effects of family size: Cross sectional survey in Chinese adolescents. *Arch. Dis. Child.* **2003**, *88*, 467–471. [CrossRef]

26. Chen, Z.; Smith, M.; Du, H.; Guo, Y.; Clarke, R.; Bian, Z.; Collins, R.; Chen, J.; Qian, Y.; Wang, X.; et al. Blood pressure in relation to general and central adiposity among 500,000 adult Chinese men and women. *Int. J. Epidemiol.* **2015**, *44*, 1305–1319. [CrossRef]

27. Dong, Y.; Ma, J.; Song, Y.; Ma, Y.; Dong, B.; Zou, Z.; Prochaska, J.J. Secular Trends in Blood Pressure and Overweight and Obesity in Chinese Boys and Girls Aged 7 to 17 Years from 1995 to 2014. *Hypertension* **2018**, *72*, 298–305. [CrossRef]

28. Gao, D.; Li, Y.; Yang, Z.; Ma, Y.; Chen, M.; Dong, Y.; Zou, Z.; Ma, J. The Association Between Single-Child Status and Risk of Abdominal Obesity: Result from a Cross-Sectional Study of China. *Front. Pediatr.* **2021**, *9*, 697047. [CrossRef]

29. Hunsberger, M.; Formisano, A.; Reisch, L.A.; Bammann, K.; Moreno, L.; De Henauw, S.; Molnar, D.; Tornaritis, M.; Veidebaum, T.; Siani, A.; et al. Overweight in singletons compared to children with siblings: The IDEFICS study. *Nutr. Diabetes* **2012**, *2*, e35. [CrossRef]

30. Ragelienė, T.; Grønhøj, A. The influence of peers' and siblings' on children's and adolescents' healthy eating behavior. A systematic literature review. *Appetite* **2020**, *148*, 104592. [CrossRef]

31. Povey, R.; Cowap, L.; Gratton, L. "They said I'm a square for eating them" Children's beliefs about fruit and vegetables in England. *Brit Food J.* **2016**, *118*, 2949–2962. [CrossRef]

32. Farajian, P.; Panagiotakos, D.B.; Risvas, G.; Micha, R.; Tsioufis, C.; Zampelas, A. Dietary and lifestyle patterns in relation to high blood pressure in children: The GRECO study. *J. Hypertens.* **2015**, *33*, 1174–1181. [CrossRef]

33. Oude Griep, L.M.; Seferidi, P.; Stamler, J.; Van Horn, L.; Chan, Q.; Tzoulaki, I.; Steffen, L.M.; Miura, K.; Ueshima, H.; Okuda, N.; et al. Relation of unprocessed, processed red meat and poultry consumption to blood pressure in East Asian and Western adults. *J. Hypertens.* **2016**, *34*, 1721–1729. [CrossRef]

34. Lajous, M.; Bijon, A.; Fagherazzi, G.; Rossignol, E.; Boutron-Ruault, M.C.; Clavel-Chapelon, F. Processed and unprocessed red meat consumption and hypertension in women. *Am. J. Clin. Nutr.* **2014**, *100*, 948–952. [CrossRef]

35. Quist, J.S.; Sjödin, A.; Chaput, J.-P.; Hjorth, M.F. Sleep and cardiometabolic risk in children and adolescents. *Sleep Med. Rev.* **2016**, *29*, 76–100. [CrossRef]

36. Makarem, N.; Shechter, A.; Carnethon, M.R.; Mullington, J.M.; Hall, M.H.; Abdalla, M. Sleep Duration and Blood Pressure: Recent Advances and Future Directions. *Curr. Hypertens. Rep.* **2019**, *21*, 33. [CrossRef]

37. Howe, L.D.; Hallal, P.C.; Matijasevich, A.; Wells, J.C.; Santos, I.S.; Barros, A.J.D.; Lawlor, D.A.; Victora, C.G.; Smith, G.D. The association of birth order with later body mass index and blood pressure: A comparison between prospective cohort studies from the United Kingdom and Brazil. *Int. J. Obes.* **2014**, *38*, 973–979. [CrossRef]

38. Jelenkovic, A.; Silventoinen, K.; Tynelius, P.; Myrskylä, M.; Rasmussen, F. Association of birth order with cardiovascular disease risk factors in young adulthood: A study of one million Swedish men. *PLoS ONE* **2013**, *8*, e63361. [CrossRef]

39. Zhou, C.; Wang, X.L.; Zhou, X.D.; Hesketh, T. Son preference and sex-selective abortion in China: Informing policy options. *Int. J. Public Health* **2012**, *57*, 459–465. [CrossRef]
40. Zhai, L.; Dong, Y.; Bai, Y.; Wei, W.; Jia, L. Trends in obesity, overweight, and malnutrition among children and adolescents in Shenyang, China in 2010 and 2014: A multiple cross-sectional study. *BMC Public Health* **2017**, *17*, 151. [CrossRef]
41. Zhang, J.; Wang, H.; Wang, Z.; Du, W.; Su, C.; Zhang, J.; Jiang, H.; Jia, X.; Huang, F.; Ouyang, Y.; et al. Prevalence and stabilizing trends in overweight and obesity among children and adolescents in China, 2011–2015. *BMC Public Health* **2018**, *18*, 571. [CrossRef]

Article

Association of Maternal Dietary Patterns during Gestation and Offspring Neurodevelopment

Siyuan Lv [1,2,†], Rui Qin [1,3,†], Yangqian Jiang [1,3], Hong Lv [1,3,4], Qun Lu [1,5], Shiyao Tao [1,3], Lei Huang [1,5], Cong Liu [1,3], Xin Xu [1,5], Qingru Wang [1], Mei Li [1,6], Zhi Li [1,6], Ye Ding [5], Ci Song [3], Tao Jiang [7], Hongxia Ma [1,3,4], Guangfu Jin [1,3], Yankai Xia [1,6], Zhixu Wang [5], Shanshan Geng [2,*], Jiangbo Du [1,3,4,*], Yuan Lin [1,4,5,*] and Zhibin Hu [1,3,4]

[1] State Key Laboratory of Reproductive Medicine, Nanjing Medical University, Nanjing 211166, China; lvsiyuan_njmu@163.com (S.L.); qinrui@njmu.edu.cn (R.Q.); jiangyangqian23@163.com (Y.J.); lvhong1203@njmu.edu.cn (H.L.); luqun@njmu.edu.cn (Q.L.); taoshiyao_njmu@163.com (S.T.); leihuang@njmu.edu.cn (L.H.); lc1454732687@126.com (C.L.); xx3411251996@163.com (X.X.); wangqingru_0228@163.com (Q.W.); limei_njmu@163.com (M.L.); lizhi@njmu.edu.cn (Z.L.); hongxiama@njmu.edu.cn (H.M.); guangfujin@njmu.edu.cn (G.J.); yankaixia@njmu.edu.cn (Y.X.); zhibin_hu@njmu.edu.cn (Z.H.)

[2] Department of Toxicology and Nutritional Science, School of Public Health, Nanjing Medical University, Nanjing 211166, China

[3] Department of Epidemiology, Center for Global Health, School of Public Health, Nanjing Medical University, Nanjing 211166, China; songci@njmu.edu.cn

[4] Suzhou Municipal Hospital, Gusu School, Nanjing Medical University, Suzhou 215002, China

[5] Department of Maternal, Child and Adolescent Health, Center for Global Health, School of Public Health, Nanjing Medical University, Nanjing 211166, China; dingye@njmu.edu.cn (Y.D.); zxwang@njmu.edu.cn (Z.W.)

[6] Key Laboratory of Modern Toxicology of Ministry of Education, School of Public Health, Nanjing Medical University, Nanjing 211166, China

[7] Department of Biostatistics, School of Public Health, Nanjing Medical University, Nanjing 211166, China; tao.chiang0923@njmu.edu.cn

[*] Correspondence: gss9814@njmu.edu.cn (S.G.); dujiangbo@njmu.edu.cn (J.D.); yuanlin@njmu.edu.cn (Y.L.); Tel.: +86-025-86868453 (S.G.); +86-025-86868317 (J.D.); +86-025-86868471 (Y.L.)

[†] These authors contributed equally to this work.

Citation: Lv, S.; Qin, R.; Jiang, Y.; Lv, H.; Lu, Q.; Tao, S.; Huang, L.; Liu, C.; Xu, X.; Wang, Q.; et al. Association of Maternal Dietary Patterns during Gestation and Offspring Neurodevelopment. *Nutrients* **2022**, *14*, 730. https://doi.org/10.3390/nu14040730

Academic Editor: Zhiyong Zou

Received: 3 January 2022
Accepted: 31 January 2022
Published: 9 February 2022

Publisher's Note: MDPI stays neutral with regard to jurisdictional claims in published maps and institutional affiliations.

Copyright: © 2022 by the authors. Licensee MDPI, Basel, Switzerland. This article is an open access article distributed under the terms and conditions of the Creative Commons Attribution (CC BY) license (https://creativecommons.org/licenses/by/4.0/).

Abstract: The health effects of diet are long term and persistent. Few cohort studies have investigated the influence of maternal dietary patterns during different gestational periods on offspring's health outcomes. This study investigated the associations between maternal dietary patterns in the mid- and late-gestation and infant's neurodevelopment at 1 year of age in the Jiangsu Birth Cohort (JBC) Study. A total of 1178 mother–child pairs were available for analysis. A semiquantitative food frequency questionnaire (FFQ) was used to investigate dietary intake at 22–26 and 30–34 gestational weeks (GWs). Neurodevelopment of children aged 1 year old was assessed using Bayley-III Screening Test. Principal component analysis (PCA) and Poisson regression were used to extract dietary patterns and to investigate the association between dietary patterns and infant neurodevelopment. After adjusting for potential confounders, the maternal 'Aquatic products, Fresh vegetables and Homonemeae' pattern in the second trimester was associated with a lower risk of being non-competent in cognitive and gross motor development, respectively (cognition: aRR = 0.84; 95% CI 0.74–0.94; gross motor: aRR = 0.80; 95% CI 0.71–0.91), and the similar pattern, 'Aquatic products and Homonemeae', in the third trimester also showed significant association with decreased risk of failing age-appreciate cognitive and receptive communication development (cognition: aRR = 0.89; 95% CI 0.80–0.98; receptive communication: aRR = 0.91; 95% CI 0.84–0.99). Notably, adherence to the dietary pattern with relatively high aquatic and homonemeae products in both trimesters demonstrated remarkable protective effects on child neurodevelopment with the risk of being non-competent in cognitive and gross motor development decreasing by 59% (95% CI 0.21–0.79) and 63% (95% CI 0.18–0.77), respectively. Our findings suggested that adherence to the 'Aquatic products and Homonemeae' dietary pattern during pregnancy may have optimal effects on offspring's neurodevelopment.

Keywords: birth cohort; prospective study; maternal dietary pattern; neurodevelopment

1. Introduction

Early neurodevelopmental delays have negative impacts on children's daily life and school performance, which has become a pressing public health concern [1,2]. Fetal neurodevelopment begins approximately 22 days after conception and develops rapidly in the second and third trimesters [3]. Existing evidence has demonstrated the prominent benefits on fetal neurodevelopment from maternal dietary patterns during pregnancy in a reasonable amount of food and nutrients [4–10]. In contrast, the adverse influences of an unhealthy diet during pregnancy on the neurodevelopment of offspring remains after birth and even for long term. For instance, a randomized-controlled trial demonstrated that the maternal diet of two servings of cod per week during pregnancy might have adverse effects on infants' neurodevelopment [11]. A prospective birth cohort study found that maternal intake of excess folic acid supplements during pregnancy was associated with adverse psychomotor development in children after the first year of life [12]. The potential mechanisms include the alteration of epigenetics of fetal genome and the regulation of fetal brain-derived neurotrophic factor (BDNF) levels by maternal nutrition [13–15].

Indeed, considering the complexity of dietary status during pregnancy, individual foods and nutrients cannot represent the overall nutritional status. Dietary patterns, proven to be simple, useful, comprehensive and complementary, have been well-accepted and increasingly used by nutritionists to assess the overall dietary status [16,17]. Epidemiological data on dietary patterns during pregnancy and neurodevelopment in offspring after birth are currently sparse. One study conducted within the Avon Longitudinal Study of Parents and Children (ALSPAC) cohort suggested that children born to mothers who were prone to the 'Meat and Potatoes' pattern and the 'White bread and Coffee' pattern in the third trimester had lower verbal performance and full-scale intelligence quotient (IQ) at school age as compared to those born to mothers whose dietary pattern was prone to 'Fruit and Vegetables' [18]. Additionally, the Generation R Study suggested that both low adherence to a Mediterranean diet and high adherence to a Traditionally Dutch diet during the first trimester were associated with an increased risk of child externalizing problems [19]. Given the considerable heterogeneity in dietary patterns and habits between European and Chinese populations, it is of public health significance to assess the effects of Chinese maternal dietary patterns over gestation on offspring neurodevelopment.

Therefore, in the present study, we aimed to determine the main maternal dietary patterns across gestation in a prospective birth cohort in Jiangsu, China, and further to investigate the maternal dietary patterns in relation to infant neurodevelopment at 1 year of age.

2. Materials and Methods

2.1. Study Population

The present study is a prospective cohort study within the ongoing Jiangsu Birth Cohort (JBC) Study, which recruits and follows up families receiving assisted reproduction or conceiving naturally in the Women's Hospital of Nanjing Medical University and the Suzhou Affiliated Hospital of Nanjing Medical University to investigate the heterogeneity of assisted versus natural pregnancies in perinatal outcomes and children health, and to systematically assess potential influencing factors for child health and wellbeing. Details of the JBC Study have been published elsewhere [20]. Starting in April 2017, women were asked to complete the semiquantitative food frequency questionnaire (FFQ) in the first (10–14 weeks), second (22–26 weeks), and third (30–34 weeks) trimester of gestation. They were contacted again for children's follow-up at 42 days, 6 months, and 1 year. The initial study and subsequent follow-ups of the women and their children were approved by the Human Research Ethics Committee of Nanjing Medical University. Written informed consent was obtained from all participants. The ethical approval code for the project is NJMUIRB (2017) 002.

By September 2020, a total of 2493 singleton children in the cohort reached 1 year of age. We excluded children from the present study if (1) their mothers did not complete two FFQs (one in the second trimester, and one in the third trimester) (n = 440); (2) they did not complete the Bayley-III Screening Test (n = 462); (3) they completed the Bayley-III Screening Test aged above 12.5 months (n = 411); (4) their mothers reported daily energy intake <500 kcal/d or $\geq$5000 kcal/d (n = 2). Finally, a total of 1178 mother–infant pairs were included in the analysis. A detailed inclusion and exclusion flow chart was shown in Figure 1.

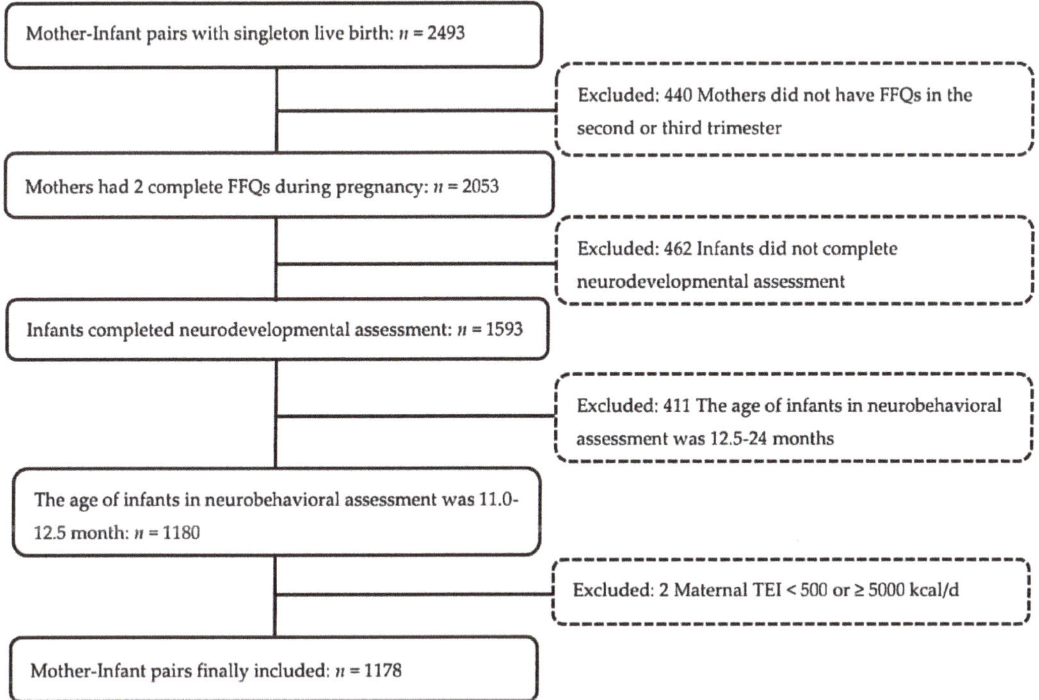

Figure 1. Participants flow chart. (Abbreviations: FFQ = food frequency questionnaire; TEI = total energy intake).

2.2. Dietary Assessment

Dietary intake information was assessed during the second and the third trimester, using the semiquantitative FFQ. The FFQ was administered by well-trained interviewers with food models and food maps [21] when women received face-to-face follow-ups. In the FFQ, the frequency and amounts of 25 food items were recorded. The validity of the FFQ have been reported in Table S1. The daily nutrient intake was calculated by multiplying the daily frequency of consumption, the amount per serving, and the nutrient per gram, which were obtained through the Chinese Food Composition Table [22]. Additionally, the dietary data used in this paper were converted logarithmically first, and then the residual method [23] was used for energy correction.

2.3. Neurodevelopment Evaluation

At the 1-year follow-up, children's neurodevelopment was assessed by licensed psychologists using the Bayley-III Screening Test, accompanied by their primary caregiver. The Bayley-III Screening Test is one of the most commonly used screening tools for assessing neurodevelopmental function, for which the effectiveness and reliability has been proved [24]. The Bayley-III Screening Test consists of five domains: cognition, expressive communication, receptive communication, fine motor, and gross motor. The children who received the test were categorized as 'at risk', 'emerging' or 'competent' according to the cutoff points for specific age [25]. In this study, we classified 'at risk' and 'emerging' as 'non-competent'. Additionally, in order to ensure the validity and reliability of the neurodevelopment evaluation, the JBC Study have taken a series of measures. Specifically, all the psychologists have undergone rigorous training by one appointed developmental neuropsychologist. After the approval of the guardians, the whole process of assessments was filmed. All psychologists were not aware of the dietary status of mothers.

2.4. Potential Confounders

All procedures were conducted according to standardized protocols. Maternal and child socio-demographic and lifestyle characteristics from early pregnancy to 1 year after birth were obtained via face-to-face or telephone questionnaires and medical records. Mode of conception (spontaneous pregnancy (SP)/assisted reproductive technology (ARTP)), maternal age, maternal pre-pregnancy BMI, parity (primipara/multipara), and folic acid supplements intake (mg/day) were collected at recruitment. We obtained information on maternal hypertension (chronic or pregnancy-induced hypertension, yes/no), maternal diabetes (pre-pregnancy or gestational diabetes, yes/no), gestational week at delivery, and infant sex (male/female) from electronic medical records (EMR). At the 1-year follow-up, we collected details on the duration of breastfeeding (<6/6–12 months) and child age when receiving the Bayley assessment.

2.5. Statistical Analysis

Dietary patterns were extracted by principal component analysis (PCA), which turned a large number of related variables into a small set of unrelated variables and maximized the explained variance. The foods listed in the food frequency questionnaire were first grouped into 16 food groups based on the similarity of the nutritional content of food. The intake of these food groups (g/day) were log transformed and then were adjusted using the residual method for total energy intake [23]. According to the scree plot and interpretability, we selected the eigenvalue of >1.3 criterion and determined the number of components [26]. We named the selected components based on factor loadings above 0.5 (Table S2), and the factor scores for each pattern and for each individual were derived from the component's factor loadings for the food groups and the calculated food intake. Finally, four dietary patterns were extracted during the second and third trimesters, respectively.

After adjusting for potential confounders, Poisson regression analysis was conducted to investigate the association between different dietary patterns scores and categorical outcomes of infant's neurodevelopment. The intraclass correlation coefficient (ICC) and 95% CI were calculated by dietary pattern scores to assess temporal variability of dietary patterns during pregnancy. Additionally, continuous variables were expressed in terms of mean and standard deviation, and classified variables were expressed in terms of frequency (n) and percentage (%). All hypothesis testing was conducted assuming a 0.05 significance level and a two-sided alternative hypothesis. All statistical analyses were performed using R software (Version 3.6.1, R Foundation for Statistic Computing, Vienna, Austria. URL https://www.R-project.org/ (accessed on 30 January 2022)).

3. Results

3.1. Basic Characteristics and Neurodevelopmental Assessment

Descriptive characteristics of 1178 mother–infant pairs were presented in Table 1. In our study population, the mean age of mothers at delivery was 31 years old. One in five women were overweight or obese before pregnancy, and approximately 30% of them were complicated with chronic diabetes or gestational diabetes during gestation. Given that the cohort was originally designed to investigate the heterogeneity of assisted versus natural pregnancies in perinatal outcomes and child health, we had forty percent mothers conceived after assisted reproduction (n = 471) included in the study. Less than 1% of women smoked or drank during pregnancy. With regard to the infants included, 3.8% (n = 45) were born preterm and 2.3% (n = 27) had a low birth weight under 2500 g. The prevalence of non-competent neurodevelopment among 1178 infants ranged from 4.1% (fine motor) to 17.5% (receptive language). Additionally, we compared the characteristics of mothers who completed the FFQ in both the second and the third trimesters with those mothers who did not complete the FFQs as requested. As shown in Table S3, women who did not complete the FFQs were more likely to be those who conceived after ART, and delivered preterm and LBW babies. No differences were observed in maternal age, pre-pregnancy BMI, parity and prevalence of diabetes or hypertension during pregnancy.

Table 1. Descriptive characteristics of baseline demographic and lifestyle factors of 1178 mother–infant pairs *.

	Overall
Maternal age at delivery (years)	31.01 (3.92)
Maternal pre-pregnancy BMI (kg/m^2)	21.68 (2.98)
<18.5	127 (10.8)
18.5–23.9	807 (68.5)
24–27.9	189 (16.0)
≥28	46 (3.9)
Diabetes [a]	339 (28.8)
Hypertension [b]	60 (5.1)
Mode of conception	
SP	707 (60.0)
ARTP	471 (40.0)
Primipara	915 (77.7)
Tobacco use during pregnancy	1 (0.1)
Alcohol intake during pregnancy	5 (0.4)
Preterm birth	45 (3.8)
Infant sex	
Male	612 (52.0)
Female	566 (48.0)
LBW (<2500 g) [c]	27 (2.3)
Duration of breastfeeding, months	
<6	406 (34.5)
6–12	754 (64.0)
Bayley-III screening test scale	
Non-competent in cognition	142 (12.1)
Non-competent in receptive communication	206 (17.5)
Non-competent in expressive communication	62 (5.3)
Non-competent in fine motor skills	48 (4.1)
Non-competent in gross motor skills	131 (11.1)

* Data are presented as mean (SD) or n (%). Abbreviations: BMI = body mass index; SP = spontaneous pregnancy; ARTP = assisted reproductive technology pregnancy. [a] Diabetes includes chronic and gestational diabetes. [b] Hypertension includes chronic, gestational and pre-eclampsia. [c] Low birth weight.

3.2. Maternal Dietary Patterns across Gestation

In the second and the third trimester, four dietary patterns were extracted, respectively. The 'Aquatic products and Homonemeae' and 'Nut' patterns were observed in both trimesters. The 'Aquatic products and Homonemeae' pattern, which was characterized by high consumption of high dietary fiber and high-quality protein, explained 9.7% and 10.0% of the variation in the dietary data in the second and third trimesters, respectively. The 'Nut' pattern, mainly associated with consumption of nut, explained 8.6% and 8.7% of the variation in the second and third trimesters, respectively. In addition, the 'Haslet, Beans, Shells and Molluscs' pattern and 'Sweets' pattern were determined in the second trimester, and 'Pome, Berry and Melon fruits' pattern and 'Citrus' pattern were observed in the third trimester, which reflected the variation in the dietary pattern in the mid and late gestation. Detailed factor loadings for the food groups with different dietary patterns were shown in Table S2. In the evaluation of temporal variability of dietary across pregnancy, the 'Aquatic products and Homonemeae' dietary pattern showed high consistency (ICC > 0.40) from mid- to late gestation in our study population, while the 'Nut' pattern fluctuated significantly (Table S4).

3.3. Maternal Dietary Patterns and Neurodevelopment in Infants

Among the 1178 infants, the multivariable associations of maternal dietary patterns in the second and the third trimester with the risk of nonoptimal neurodevelopment in infants were presented in Figure 2 and Table S5. A higher adherence score of 'Aquatic products, Fresh vegetables and Homonemeae' food consumption in the second trimester was significantly associated with decreased risk in infants being non-competent in cognitive development (aRR = 0.84; 95% CI 0.75–0.93) and in gross motor development (aRR = 0.82; 95% CI 0.73–0.91) after adjusting for conventional covariates including mode of conception, maternal age, maternal pre-pregnancy BMI, parity, gestational week at delivery, hypertension or diabetes during pregnancy, infant sex, and duration of breastfeeding (Model 1). After the further adjustments for folic acid supplementation and the scores for the rest dietary patterns (Model 2), the associations remained steady. Notably, similar maternal dietary pattern, 'Aquatic products and Homonemeae' pattern in the third trimester, also demonstrated optimal effects on infant's neurodevelopment with higher adherence score being associated with lower risk of nonoptimal cognitive development (Model 2: aRR = 0.89; 95% CI 0.80–0.98) and receptive communication development (Model 2: aRR = 0.91; 95% CI 0.84–0.99). In addition, the adherence score of the 'Nut' pattern in the second trimester was observed associated with decreased risk of infants being non-competent in expressive communication (Model 2: aRR = 0.79; 95% CI 0.66–0.94), while the association diminished in the third trimester. We did not observe the rest dietary patterns in mid- and later gestation associated with infant's neurodevelopment in any of the five domains.

3.4. Adherence to 'Aquatic Products and Homonemeae' Pattern and Infant Neurodevelopment

To assess the adherence to the 'Aquatic products and Homonemeae' pattern from mid- to late pregnancy in relation to neurodevelopment in infants, we categorized participants into tertiles according to their scores of dietary patterns in each trimester. As compared to the infants whose mothers at the lowest tertile of 'Aquatic products and Homonemeae' pattern in both trimesters, those whose mothers at the highest tertile in both trimesters demonstrated 59% (Model 2: aRR = 0.41; 95% CI 0.21–0.79; *p* = 0.008) and 63% (Model 2: aRR = 0.37; 95% CI 0.18–0.77; *p* = 0.008) decreased risk of being non-competent in cognitive and gross motor development after adjusting for potential confounders, respectively (Table 2).

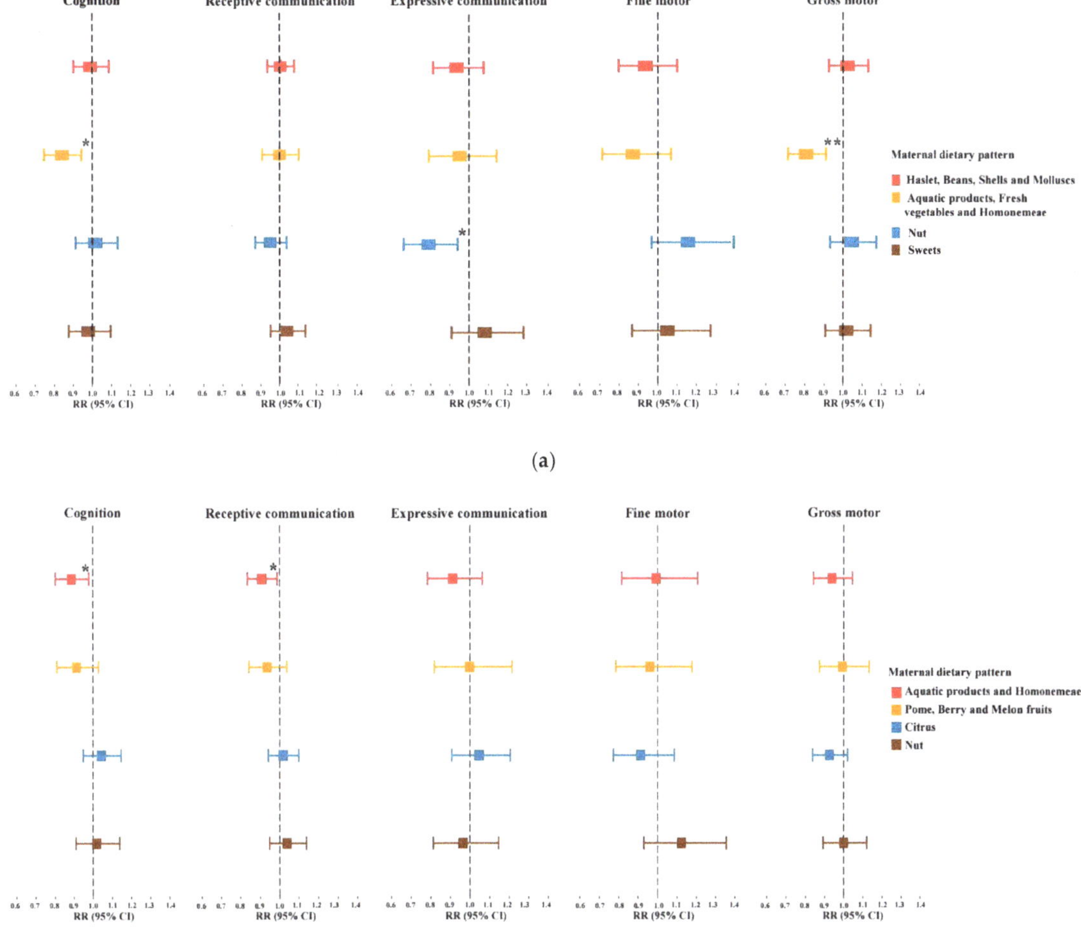

Figure 2. Adjusted associations of dietary patterns in the second and third trimesters with categorical outcomes of neurodevelopment in infants, respectively. Analyses were adjusted for mode of conception, maternal age, maternal pre-pregnancy body mass index (BMI), parity, hypertension, diabetes, gestational week at delivery, infant sex, duration of breastfeeding, principal component score of maternal dietary intake, folic acid supplements. * $p < 0.05$, ** $p < 0.001$. (**a**), Second trimester; (**b**), Third trimester. Abbreviations: RR = Risk ratio; 95% CI = 95% confidence intervals.

Table 2. Multivariable associations of adherence to the 'Aquatic products and Homonemeae' pattern in the second and third trimesters and categorical outcomes of neurodevelopment in infants.

		Cognition		Receptive Communication		Expressive Communication		Fine Motor		Gross Motor	
		RR (95% CI)	*p*	RR (95% CI)	*p*	RR (95% CI)	*p*	RR (95% CI)	*p*	RR (95% CI)	*p*
Model 1 Lowest tertile in both trimesters	*n* = 196	Ref		Ref		Ref		Ref		Ref	

Table 2. *Cont.*

		Cognition		Receptive Communication		Expressive Communication		Fine Motor		Gross Motor	
		RR (95% CI)	*p*	RR (95% CI)	*p*	RR (95% CI)	*p*	RR (95% CI)	*p*	RR (95% CI)	*p*
Highest tertile in both trimesters Model 2	*n* = 180	0.40 (0.21, 0.78)	0.007	0.75 (0.47, 1.20)	0.232	0.84 (0.37, 1.93)	0.682	0.45 (0.16, 1.31)	0.142	0.37 (0.18, 0.76)	0.007
Lowest tertile in both trimesters	*n* = 196	Ref		Ref		Ref		Ref		Ref	
Highest tertile in both trimesters	*n* = 180	0.41 (0.21, 0.79)	0.008	0.77 (0.48, 1.23)	0.266	0.86 (0.38, 1.98)	0.724	0.46 (0.16, 1.33)	0.150	0.37 (0.18, 0.77)	0.008

Model 1: adjusted for mode of conception, maternal age, maternal pre-pregnancy BMI, parity, hypertension, diabetes, gestational week at delivery, infant sex, duration of breastfeeding. Model 2: adjusted for mode of conception, maternal age, maternal pre-pregnancy BMI, parity, hypertension, diabetes, gestational week at delivery, infant sex, duration of breastfeeding, folic acid supplements. Statistically significant differences are highlighted in bold. Abbreviations: Ref = Reference group, which is the group of mothers with the lowest tertile of dietary pattern scores in both trimesters.

4. Discussion

It is well known that children's neurodevelopment is affected by various factors such as mother's lifestyle [27], environmental pollution [28], food contamination with chemical compounds [29], and pregnancy leisure physical activity [30]. Notably, maternal dietary intake has been extensively studied due to its greater controllability. This prospective birth cohort study conducted in Jiangsu province, China, investigated maternal dietary patterns across mid- to late gestation in relation to infant neurodevelopment. Our results demonstrated that the maternal dietary pattern of 'Aquatic products and Homonemeae' in the second and the third trimester was associated with a lower risk of being non-competent in cognitive development in infants at 1 year of age. Particularly, adherence to this pattern in both trimesters reduced the risk by 59% and 63% for nonoptimal development in cognition and gross motor skills, respectively.

The use of dietary patterns to consider the interactions of various foods and nutrients simultaneously is becoming increasingly common. Western dietary patterns, such as the 'DASH dietary pattern' and the 'Mediterranean-type diet', have been linked to lower risk of maternal and infant diseases [31,32], while the 'Western dietary pattern' can increase the risk of adverse health outcomes [33]. The Chinese dietary patterns extracted by a data-driven method reflected the local food consumption habits well. The Guangzhou Cohort Study demonstrated that a maternal diet with lots of milk and few vegetables during pregnancy may be associated with an increased risk of preterm delivery [34]. While no previous study has examined the relation between maternal dietary patterns in China during pregnancy and subsequent children neuropsychological function.

The neuroprotective effects of the 'Aquatic, Fresh vegetables and Homonemeae' pattern may result from the biological properties of the nutrients contained. Aquatic products are good sources of high-quality protein, omega-3 (ω-3) fatty acids and minerals such as iodine. Previous studies have demonstrated that omega-3 polyunsaturated fatty acid (ω-3 PUFA) may play key roles in neural development. The ALSPAC cohort study reported that the encouragement of maternal seafood consumption during the third trimester may benefit the neurodevelopment of children aged 6 months to 8 years [35]. In addition, several meta-analyses showed that children with attention deficit hyperactivity disorder (ADHD) have lower blood levels of ω-3 PUFA [36,37], and two observational studies suggested that higher dietary intake of fish or ω-3 PUFA can reduce the risk of psychosis [37,38]. Existing animal experiments showed that a diet rich in omega-3 reduced inflammation in the hypothalamus in rats [39]. In addition, maternal iodine has been associated with children neurodevelopment. For instance, the Southampton Women cohort reported that maternal urinary iodine concentration was positively correlated with cognitive function in children aged 6–7 years [40]. It has been demonstrated that iodine deficiency in mothers leads to insufficient thyroid hormone production, which further affects neuronal proliferation and migration in the cortex and hippocampus of their offspring [41,42]. In an animal experiment, maternal mild iodine deficiency promoted developmental hypothyroxinaemia, which further led to highly retarding the ability of offspring's hippocampal axonal growth in rats [43]. Homonemeae, with well-balanced essential amino acids and rich iodine [44], has been known for its neuroprotective effects. Furthermore, evidence from cell experiments suggested that dimethylsulfoniopropionate (DMSP) produced by homonemeae could strengthen antioxidant defense in mammalian nerve cells [45]. In addition, fresh vegetables are rich in folate, vitamin C, beta carotene (provitamin A), and other nutrients. More fresh vegetables in a mother's diet have been suggested in relation to better neural development in offspring [18,46]. Given the difficulty of collecting dietary data repeatedly, the combined effects of maternal dietary over gestation are less studied in prior studies. A notable finding of the present study is the remarkable protective effects we observed of adherence to the 'Aquatic products, Fresh vegetables and Homonemeae' dietary pattern on child neurodevelopment, which indicates the cumulative effect of dietary nutrients on neural development.

The advantages of this study should be elaborated. First, the use of dietary patterns captured the overall quality of the diet and were good proxy for dietary status over time [47]. In addition, abundant data on maternal and child characteristics were collected from a prospective and longitudinal cohort study to control potential confounding factors. Finally, a range of approaches were adopted to guarantee the validity and reliability of outcome measures. The study also has some limitations. First, recall bias may be caused by FFQ because it requires participants to recall their food intake over a period of time. However, we compared the estimated daily intake using FFQ with the mean of 3-day 24 h dietary recall (24HR) and proved that FFQ had comparable validity in a pre-survey. Second, residual confounding may occur due to other unmeasured confounding factors. In our analysis, factors related to fetal neurodevelopment such as maternal folic acid supplements and other dietary patterns, were considered as much as possible. Third, though with available maternal dietary data, we did not determine and assess the dietary pattern in the first trimester. A large number of women experience nausea and vomiting in their early pregnancy and thus the dietary pattern might not be well correlated to actual food and nutrient consumption.

5. Conclusions

In the prospective and longitudinal birth cohort study, our results suggested the cumulative effects of maternal diet on infant neurodevelopment. These findings highlighted the optimal effects of adherence to the 'Aquatic products and Homonemeae' dietary pattern during pregnancy and might provide add-on evidence to nutrition counseling for pregnant women.

Supplementary Materials: The following supporting information can be downloaded at: https://www.mdpi.com/article/10.3390/nu14040730/s1, Table S1: Pearson correlation coefficients between FFQ and 3-day 24 h completed by 141 pregnant women in the second trimester; Table S2: dietary patterns identified by PCA and factor loadings of food groups/items included in each dietary pattern for the pregnant woman in her second and third trimesters (n = 1178); Table S3: comparison between characteristics of eligible mothers who completed the FFQ at both trimesters and those who did not complete as requested; Table S4: distribution of similar dietary pattern scores across gestation; Table S5: multivariable associations between maternal dietary patterns in the second and third trimesters and categorical outcomes of neurodevelopment in infants.

Author Contributions: Y.L., J.D. and S.G. initiated, conceived, and designed the study. Z.H. supervised the whole study. S.L. and R.Q. contributed to the data management and statistical analysis. Data were collected by S.L., R.Q., Y.J., H.L., Q.L., S.T., L.H., C.L., X.X., Q.W., M.L., Z.L., Y.D., Y.L., S.G., S.L. and R.Q. prepared the manuscript. J.D., T.J., C.S., H.M., G.J., Y.X. and Z.W. proofread the manuscript. All authors have read and agreed to the published version of the manuscript.

Funding: This manuscript and research were supported by the National Key Research & Development (R&D) Program of China (2021YFC2700705), the National Natural Science Foundation of China (No. 82003527 and No. 82003415) and the Natural Science Program (Major Program) of Higher Education of Jiangsu, China (No. 20KJA310003).

Institutional Review Board Statement: The study was approved by the Human Research Ethics Committee of Nanjing Medical University. The ethical approval code for the project is NJMUIRB (2017) 002.

Informed Consent Statement: Written informed consent was obtained from all participants.

Data Availability Statement: All data analyzed or generated during this study are available from the corresponding authors on reasonable request.

Acknowledgments: Thanks to all the families who participated in this study and to the entire Jiangsu Birth Cohort team.

Conflicts of Interest: The authors declare no conflict of interest.

References

1. Perou, R.; Bitsko, R.H.; Blumberg, S.J.; Pastor, P.; Ghandour, R.M.; Gfroerer, J.C.; Hedden, S.L.; Crosby, A.E.; Visser, S.N.; Schieve, L.A.; et al. Mental health surveillance among children—United States, 2005–2011. *MMWR Suppl.* **2013**, *62*, 1–35. [PubMed]
2. Soni, A. Top Five Most Costly Conditions among Children, Ages 0–17, 2012: Estimates for the U.S. Civilian Noninstitutionalized Population. In *Statistical Brief (Medical Expenditure Panel Survey (US))*; Agency for Healthcare Research and Quality (US): Rockville, MD, USA, 2001.
3. Thompson, R.A.; Nelson, C.A. Developmental science and the media. Early brain development. *Am. Psychol.* **2001**, *56*, 5–15. [CrossRef] [PubMed]
4. Prado, E.L.; Dewey, K.G. Nutrition and brain development in early life. *Nutr. Rev.* **2014**, *72*, 267–284. [CrossRef]
5. Carlson, S.E. Docosahexaenoic acid supplementation in pregnancy and lactation. *Am. J. Clin. Nutr.* **2009**, *89*, 678s–684s. [CrossRef] [PubMed]
6. Jensen, C.L.; Voigt, R.G.; Prager, T.C.; Zou, Y.L.; Fraley, J.K.; Rozelle, J.C.; Turcich, M.R.; Llorente, A.M.; Anderson, R.E.; Heird, W.C. Effects of maternal docosahexaenoic acid intake on visual function and neurodevelopment in breastfed term infants. *Am. J. Clin. Nutr.* **2005**, *82*, 125–132. [CrossRef] [PubMed]
7. Zeisel, S.H. Is maternal diet supplementation beneficial? Optimal development of infant depends on mother's diet. *Am. J. Clin. Nutr.* **2009**, *89*, 685s–687s. [CrossRef] [PubMed]
8. Tofail, F.; Persson, L.A.; El Arifeen, S.; Hamadani, J.D.; Mehrin, F.; Ridout, D.; Ekström, E.C.; Huda, S.N.; Grantham-McGregor, S.M. Effects of prenatal food and micronutrient supplementation on infant development: A randomized trial from the Maternal and Infant Nutrition Interventions, Matlab (MINIMat) study. *Am. J. Clin. Nutr.* **2008**, *87*, 704–711. [CrossRef] [PubMed]
9. Christian, P.; Murray-Kolb, L.E.; Khatry, S.K.; Katz, J.; Schaefer, B.A.; Cole, P.M.; Leclerq, S.C.; Tielsch, J.M. Prenatal micronutrient supplementation and intellectual and motor function in early school-aged children in Nepal. *JAMA* **2010**, *304*, 2716–2723. [CrossRef]
10. Li, Q.; Yan, H.; Zeng, L.; Cheng, Y.; Liang, W.; Dang, S.; Wang, Q.; Tsuji, I. Effects of maternal multimicronutrient supplementation on the mental development of infants in rural western China: Follow-up evaluation of a double-blind, randomized, controlled trial. *Pediatrics* **2009**, *123*, e685–e692. [CrossRef]
11. Markhus, M.W.; Hysing, M.; Midtbø, L.K.; Nerhus, I.; Næss, S.; Aakre, I.; Kvestad, I.; Dahl, L.; Kjellevold, M. Effects of Two Weekly Servings of Cod for 16 Weeks in Pregnancy on Maternal Iodine Status and Infant Neurodevelopment: Mommy's Food, a Randomized-Controlled Trial. *Thyroid Off. J. Am. Thyroid Assoc.* **2021**, *31*, 288–298. [CrossRef]
12. Valera-Gran, D.; García de la Hera, M.; Navarrete-Muñoz, E.M.; Fernandez-Somoano, A.; Tardón, A.; Julvez, J.; Forns, J.; Lertxundi, N.; Ibarluzea, J.M.; Murcia, M.; et al. Folic acid supplements during pregnancy and child psychomotor development after the first year of life. *JAMA Pediatrics* **2014**, *168*, e142611. [CrossRef] [PubMed]
13. Waterland, R.A.; Jirtle, R.L. Early nutrition, epigenetic changes at transposons and imprinted genes, and enhanced susceptibility to adult chronic diseases. *Nutr. Burbank Los Angeles Cty. Calif.* **2004**, *20*, 63–68. [CrossRef] [PubMed]
14. MacLennan, N.K.; James, S.J.; Melnyk, S.; Piroozi, A.; Jernigan, S.; Hsu, J.L.; Janke, S.M.; Pham, T.D.; Lane, R.H. Uteroplacental insufficiency alters DNA methylation, one-carbon metabolism, and histone acetylation in IUGR rats. *Physiol. Genom.* **2004**, *18*, 43–50. [CrossRef]
15. Wu, A.; Ying, Z.; Gomez-Pinilla, F. The interplay between oxidative stress and brain-derived neurotrophic factor modulates the outcome of a saturated fat diet on synaptic plasticity and cognition. *Eur. J. Neurosci.* **2004**, *19*, 1699–1707. [CrossRef] [PubMed]
16. Cespedes, E.M.; Hu, F.B. Dietary patterns: From nutritional epidemiologic analysis to national guidelines. *Am. J. Clin. Nutr.* **2015**, *101*, 899–900. [CrossRef] [PubMed]
17. Hu, F.B. Dietary pattern analysis: A new direction in nutritional epidemiology. *Curr. Opin. Lipidol.* **2002**, *13*, 3–9. [CrossRef]

18. Freitas-Vilela, A.A.; Pearson, R.M.; Emmett, P.; Heron, J.; Smith, A.; Emond, A.; Hibbeln, J.R.; Castro, M.B.T.; Kac, G. Maternal dietary patterns during pregnancy and intelligence quotients in the offspring at 8 years of age: Findings from the ALSPAC cohort. *Matern. Child Nutr.* **2018**, *14*, e12431. [CrossRef] [PubMed]

19. Steenweg-de Graaff, J.; Tiemeier, H.; Steegers-Theunissen, R.P.; Hofman, A.; Jaddoe, V.W.; Verhulst, F.C.; Roza, S.J. Maternal dietary patterns during pregnancy and child internalising and externalising problems. The Generation R Study. *Clin. Nutr. Edinb. Scotl.* **2014**, *33*, 115–121. [CrossRef]

20. Lv, H.; Diao, F.; Du, J.; Chen, T.; Meng, Q.; Ling, X.; Li, H.; Song, C.; Xi, Q.; Jiang, Y.; et al. Assisted reproductive technology and birth defects in a Chinese birth cohort study. *Lancet Reg. Health. West. Pac.* **2021**, *7*, 100090. [CrossRef]

21. Ding, Y.; Yang, Y.; Li, F.; Shao, Y.; Sun, Z.; Zhong, C.; Fan, P.; Li, Z.; Zhang, M.; Li, X.; et al. Development and validation of a photographic atlas of food portions for accurate quantification of dietary intakes in China. *J. Hum. Nutr. Diet.* **2021**, *34*, 604–615. [CrossRef]

22. Yang, Y. China food composition tables. 6th edition. *Beijing China Peking Univ. Med. Press* **2018**, *11*, 1452.

23. Sugiura, H.; Uchiyama, M.; Omoto, M.; Sasaki, K.; Uehara, M. Prevalence of infantile and early childhood eczema in a Japanese population: Comparison with the disease frequency examined 20 years ago. *Acta Derm.-Venereol.* **1997**, *77*, 52–53. [PubMed]

24. Yang, Q.; Shu-Zhen, L.I.; Liu, L.L.; Lin, S.R.; Hua, J.J. Predictive Validity of Bayley Scale of Infant Development-3rd Edition-Screening Test in Early Term Children. *Chin. J. Child Health Care* **2018**, *26*, 729–732.

25. Bayley, N. *Bayley Scales of Infant and Toddler Development*, 3rd ed.; Harcourt Assessment: San Antonio, TX, USA, 2006.

26. Cattell, R.B. The Scree Test For The Number Of Factors. *Multivar. Behav. Res.* **1966**, *1*, 245–276. [CrossRef]

27. Polańska, K.; Muszyński, P.; Sobala, W.; Dziewirska, E.; Merecz-Kot, D.; Hanke, W. Maternal lifestyle during pregnancy and child psychomotor development—Polish Mother and Child Cohort study. *Early Hum. Dev.* **2015**, *91*, 317–325. [CrossRef]

28. Dórea, J.G. Environmental exposure to low-level lead (Pb) co-occurring with other neurotoxicants in early life and neurodevelopment of children. *Environ. Res.* **2019**, *177*, 108641. [CrossRef] [PubMed]

29. Gil, A.; Gil, F. Fish, a Mediterranean source of n-3 PUFA: Benefits do not justify limiting consumption. *Br. J. Nutr.* **2015**, *113* (Suppl. 2), S58–S67. [CrossRef] [PubMed]

30. Álvarez-Bueno, C.; Cavero-Redondo, I.; Sánchez-López, M.; Garrido-Miguel, M.; Martínez-Hortelano, J.A.; Martínez-Vizcaíno, V. Pregnancy leisure physical activity and children's neurodevelopment: A narrative review. *BJOG Int. J. Obstet. Gynaecol.* **2018**, *125*, 1235–1242. [CrossRef]

31. Bouwland-Both, M.I.; Steegers-Theunissen, R.P.; Vujkovic, M.; Lesaffre, E.M.; Mook-Kanamori, D.O.; Hofman, A.; Lindemans, J.; Russcher, H.; Jaddoe, V.W.; Steegers, E.A. A periconceptional energy-rich dietary pattern is associated with early fetal growth: The Generation R study. *BJOG Int. J. Obstet. Gynaecol.* **2013**, *120*, 435–445. [CrossRef]

32. Polanska, K.; Kaluzny, P.; Aubert, A.M.; Bernard, J.Y.; Duijts, L.; El Marroun, H.; Hanke, W.; Hébert, J.R.; Heude, B.; Jankowska, A.; et al. Dietary Quality and Dietary Inflammatory Potential During Pregnancy and Offspring Emotional and Behavioral Symptoms in Childhood: An Individual Participant Data Meta-analysis of Four European Cohorts. *Biol. Psychiatry* **2021**, *89*, 550–559. [CrossRef]

33. Englund-Ögge, L.; Brantsæter, A.L.; Sengpiel, V.; Haugen, M.; Birgisdottir, B.E.; Myhre, R.; Meltzer, H.M.; Jacobsson, B. Maternal dietary patterns and preterm delivery: Results from large prospective cohort study. *BMJ* **2014**, *348*, g1446. [CrossRef]

34. Lu, M.S.; He, J.R.; Chen, Q.; Lu, J.; Wei, X.; Zhou, Q.; Chan, F.; Zhang, L.; Chen, N.; Qiu, L.; et al. Maternal dietary patterns during pregnancy and preterm delivery: A large prospective cohort study in China. *Nutr. J.* **2018**, *17*, 71. [CrossRef] [PubMed]

35. Hibbeln, J.R.; Davis, J.M.; Steer, C.; Emmett, P.; Rogers, I.; Williams, C.; Golding, J. Maternal seafood consumption in pregnancy and neurodevelopmental outcomes in childhood (ALSPAC study): An observational cohort study. *Lancet* **2007**, *369*, 578–585. [CrossRef]

36. Hawkey, E.; Nigg, J.T. Omega-3 fatty acid and ADHD: Blood level analysis and meta-analytic extension of supplementation trials. *Clin. Psychol. Rev.* **2014**, *34*, 496–505. [CrossRef] [PubMed]

37. Tesei, A.; Crippa, A.; Ceccarelli, S.B.; Mauri, M.; Molteni, M.; Agostoni, C.; Nobile, M. The potential relevance of docosahexaenoic acid and eicosapentaenoic acid to the etiopathogenesis of childhood neuropsychiatric disorders. *Eur. Child Adolesc. Psychiatry* **2017**, *26*, 1011–1030. [CrossRef] [PubMed]

38. Schlögelhofer, M.; Amminger, G.P.; Schaefer, M.R.; Fusar-Poli, P.; Smesny, S.; McGorry, P.; Berger, G.; Mossaheb, N. Polyunsaturated fatty acids in emerging psychosis: A safer alternative? *Early Interv. Psychiatry* **2014**, *8*, 199–208. [CrossRef] [PubMed]

39. de Mello, A.H.; Schraiber, R.B.; Goldim, M.P.S.; Garcez, M.L.; Gomes, M.L.; de Bem Silveira, G.; Zaccaron, R.P.; Schuck, P.F.; Budni, J.; Silveira, P.C.L.; et al. Omega-3 Fatty Acids Attenuate Brain Alterations in High-Fat Diet-Induced Obesity Model. *Mol. Neurobiol.* **2019**, *56*, 513–524. [CrossRef] [PubMed]

40. Robinson, S.M.; Crozier, S.R.; Miles, E.A.; Gale, C.R.; Calder, P.C.; Cooper, C.; Inskip, H.M.; Godfrey, K.M. Preconception Maternal Iodine Status Is Positively Associated with IQ but Not with Measures of Executive Function in Childhood. *J. Nutr.* **2018**, *148*, 959–966. [CrossRef]

41. Zimmermann, M.B. The effects of iodine deficiency in pregnancy and infancy. *Paediatr. Perinat. Epidemiol.* **2012**, *26* (Suppl. 1), 108–117. [CrossRef]

42. Williams, G.R. Neurodevelopmental and neurophysiological actions of thyroid hormone. *J. Neuroendocrinol.* **2008**, *20*, 784–794. [CrossRef]

43. Wei, W.; Wang, Y.; Wang, Y.; Dong, J.; Min, H.; Song, B.; Teng, W.; Xi, Q.; Chen, J. Developmental hypothyroxinaemia induced by maternal mild iodine deficiency delays hippocampal axonal growth in the rat offspring. *J. Neuroendocrinol.* **2013**, *25*, 852–862. [CrossRef] [PubMed]
44. Zimmermann, M.B. The Importance of Adequate Iodine during Pregnancy and Infancy. *World Rev. Nutr. Diet.* **2016**, *115*, 118–124. [PubMed]
45. Wichmann, H.; Brinkhoff, T.; Simon, M.; Richter-Landsberg, C. Dimethylsulfoniopropionate Promotes Process Outgrowth in Neural Cells and Exerts Protective Effects against Tropodithietic Acid. *Mar. Drugs* **2016**, *14*, 89. [CrossRef] [PubMed]
46. Freitas-Vilela, A.A.; Smith, A.D.; Kac, G.; Pearson, R.M.; Heron, J.; Emond, A.; Hibbeln, J.R.; Castro, M.B.; Emmett, P.M. Dietary patterns by cluster analysis in pregnant women: Relationship with nutrient intakes and dietary patterns in 7-year-old offspring. *Matern. Child Nutr.* **2017**, *13*, e12353. [CrossRef] [PubMed]
47. Hoffmann, K.; Schulze, M.B.; Schienkiewitz, A.; Nöthlings, U.; Boeing, H. Application of a new statistical method to derive dietary patterns in nutritional epidemiology. *Am. J. Epidemiol.* **2004**, *159*, 935–944. [CrossRef] [PubMed]

Article

Multilevel Analysis of the Nutritional and Health Status among Children and Adolescents in Eastern China

Ting Tian [1], Yuanyuan Wang [1,2], Wei Xie [1], Jingxian Zhang [1], Yunlong Ni [1], Xianzhen Peng [3], Guiju Sun [2], Yue Dai [1] and Yonglin Zhou [1,*]

[1] Institute of Food Safety and Assessment, Jiangsu Provincial Center for Disease Control and Prevention, Nanjing 210009, China; jstt@jscdc.cn (T.T.); wyypro@foxmail.com (Y.W.); jscdcxiewei@sina.com (W.X.); z18252063009@163.com (J.Z.); nyl2008@163.com (Y.N.); 18915999341@163.com (Y.D.)

[2] Department of Nutrition and Food Hygiene, School of Public Health, Southeast University, Nanjing 210009, China; gjsun@seu.edu.cn

[3] Department of Public Health and Preventive Medicine, Kangda College of Nanjing Medical University, Lianyungang 222000, China; xianzhenpeng@njmu.edu.cn

* Correspondence: jsepipublic@sina.com

Citation: Tian, T.; Wang, Y.; Xie, W.; Zhang, J.; Ni, Y.; Peng, X.; Sun, G.; Dai, Y.; Zhou, Y. Multilevel Analysis of the Nutritional and Health Status among Children and Adolescents in Eastern China. *Nutrients* **2022**, *14*, 758. https://doi.org/10.3390/nu14040758

Academic Editor: Zhiyong Zou

Received: 12 January 2022
Accepted: 6 February 2022
Published: 11 February 2022

Publisher's Note: MDPI stays neutral with regard to jurisdictional claims in published maps and institutional affiliations.

Copyright: © 2022 by the authors. Licensee MDPI, Basel, Switzerland. This article is an open access article distributed under the terms and conditions of the Creative Commons Attribution (CC BY) license (https://creativecommons.org/licenses/by/4.0/).

Abstract: We aimed to identify multiple nutritional health problems and the relevant factors among children and adolescents aged 7–17 years. This study was part of the China Nutrition and Health Surveillance of Children and Lactating Mothers in 2016–2017, conducted in Jiangsu Province in eastern China. After sampling, 3025 school-age children and adolescents were enrolled into this study. Demographic information collections and anthropometric measurements were conducted by trained local Center for Disease Control and Prevention (CDC) staff. Venous blood in the amount of 6 mL was drawn from each participant in the morning and used for testing biochemical and nutritional indicators. Multivariate logistic regression analysis and Poisson regression analysis were used for overnutrition- and undernutrition-related disorders to test relevant personal, parental, and household factors. The prevalence of wasting, overweight, and obesity was 5.5%, 14.8%, and 12.7%, respectively. Metabolic syndrome (MetS) was prevalent among 5.1% of participants. Among the study participants, 29.5% had hyperuricemia. The overall prevalence of high low-density lipoprotein (LDL) and high total cholesterol (TC) of all participants was 4.8% and 7.4%, respectively. 0.9% of the participants had vitamin A deficiency (VAD) and 14.6% had marginal vitamin A deficiency; 25.1% had vitamin D deficiency (VDD) and 54.5% had inadequate vitamin D levels. Anemia was present in 4.0% of all participants. The prevalence of zinc deficiency was 4.8%. Demographic characteristics, behavioral characteristics, parents' characteristics, and family characteristics were associated with these multiple malnutrition disorders. The double burdens of malnutrition, which includes overnutrition- and undernutrition-related diseases, were prevalent among the school-age children and adolescents in Jiangsu Province in eastern China. There were various factors related to different nutritional problems. Thus, health education focusing on behavior intervention and nutrition education are necessary in containing nutritional problems among children.

Keywords: nutritional status; children and adolescents; malnutrition; overnutrition

1. Introduction

Nutrition is essential for children's neurodevelopment and reproductive development and is associated with linear growth [1]. Improving children's nutrition is urgently needed for global development and for reducing the burden of noncommunicable diseases (NCDs) [2]. Nutritional conditions in the early stages of life may have both short-term and longer lasting effects, including an increased risk of NCDs and other chronic diseases [3]. Cardiovascular diseases, which are the most common causes of mortality among adults worldwide, are rooted in childhood, underlying the urgency of identifying and intervening with at-risk children [4,5].

Further, approximately one-half of all childhood deaths globally are caused directly or indirectly by malnutrition [6]. Various malnutrition diseases, including undernutrition- and overnutrition-related disorders, are highly prevalent among children and adolescents in many countries [7,8]. For example, vitamin A deficiency (VAD) remains a public health problem in China, where the prevalence of VAD in children was 5.16% in 2015 [9]. A high prevalence of vitamin D deficiency (VDD) has also been reported in infants, children, and adolescents from diverse countries around the world [10]. The overall prevalence of anemia was 8.8% among children in Hunan Province in China [11] and 17.49% among school-age children in Latin America and the Caribbean [12]. Zinc is an essential trace element for growth and development in children, but zinc deficiency is a serious nutritional problem in China. In 2012, the China Nutrition and Health Survey showed a 6.8% prevalence of zinc deficiency in 3407 school-age children [13].

In addition to undernutrition, overnutrition is a serious problem. The results of five consecutive national surveys from 1995 to 2004, as part of the Chinese National Survey on Students' Constitution and Health, showed that the main growth problems had changed from undernutrition to overnutrition in the past 20 years [14]. The nutrition issues of children and adolescents have become more important in view of the global epidemic of obesity, metabolic syndrome (MetS), and other metabolic risk factors. Worldwide, more than 100 million children are obese [15].

One predictor of future health risk is Metabolic syndrome (MetS), a cluster of cardio-vascular risk factors that include abdominal obesity, which is mainly measured by high waist circumference (WC), hypertension, high fasting triglycerides (TG), low high-density lipoprotein (HDL) cholesterol, and high fasting blood glucose (FBG) [16]. The recent prevalence of MetS in American children and adolescents was 9.83%, according to the results of National Health and Nutrition Examination Survey (NHANES) [17]. Data from the China Health and Nutrition Surveys (CHNS) showed that the prevalence of MetS in Chinese children and adolescents aged 7–18 years was 3.37% [18]. Meanwhile, recent evidence suggests that hyperuricemia is at a high level in children and adolescents in both China and America, with the prevalence ranging from 10.1% to 26.02% [19–21].

With sustained economic development, the living standards of Chinese residents, and especially residents of Jiangsu Province, which is a relatively developed area in the country, have continuously improved. However, China still faces many nutritional problems, such as the coexistence of undernutrition- and overnutrition-related disorders among children and adolescents and the lack of clarity regarding the sociodemographic factors that are related to these nutritional problems. Sociodemographic differences play important roles in nutrition status, especially among children and adolescents [22–24]. To date, there have been few studies conducted in Jiangsu Province to describe comprehensively the nutritional status of school-age children and adolescents. Thus, this study aimed to identify the multiple nutritional health problems among children and adolescents aged 7–17 years in Jiangsu Province in China, and the relevant factors contributing to those problems, based on provincial representative data.

2. Materials and Methods

2.1. Study Participants

This study was part of the China Nutrition and Health Surveillance of Children and Lactating Mothers in 2016–2017. We selected school-age children and adolescents from 12 survey sites, including urban and rural areas in Jiangsu Province in eastern China. The representative provincial samples for accessing the nutritional and health status of school-aged participants were randomly selected by multistage stratified cluster random sampling methods. The study design and sampling methods were described by Dongmei Yu et al. in an article introducing the China Nutrition and Health Survey from 1982–2017 [25].

In the first stage of the study, 12 survey sites were randomly picked from the whole province, including two survey sites in large cities, eight sites in medium cities, and two sites in generally rural counties. In the second stage, two townships/subdistricts were randomly picked from each survey site. In the third stage, two communities or villages of township/subdistricts were randomly selected. In the last stage, 70 school-age children and adolescents were randomly picked from the selected communities or villages. There were at least 280 children aged 7–17 years, with equal numbers of males and females, selected in each of the survey sites. Replacements were allowed only from the same village/neighborhood and similar households as those of the original participants. The replacement rate had to be less than 35%. After sampling, 3360 school-aged children and adolescents were enrolled into the study. In the final analysis, 3025 participants were included, after excluding participants who were unable to provide key information, such as sociodemographic characteristics, lab-testing information, or anthropometric results (n = 335).

2.2. Ethics Approval and Consent to Participate

This study was performed according to the Declaration of Helsinki and approved by the ethical committee of the China Center for Disease Control and Prevention (the China CDC); the corresponding ethical approval number was 201614. All of the participants agreed to take part in this study and signed the required informed consent form.

2.3. Data Collection and Measurements

Interviews were based on household and individually structured questionaries provided by the China CDC project group, to collect demographic information and additional information concerning physical activities, screen time, parents, household sizes, and other details. Interviews were conducted by a strictly trained investigator in a face-to-face manner.

Screen time refers to the time using an electronic screen every day. A low level of screen time was defined as <2 h and a high level was defined as ≥2 h. The assessment of physical activities was based on reaching a level of coarsened breathing or a heart rate rise for more than 30 min. An exercise level of 0–3 days per week was defined as a low physical activity level; an exercise level for more than 4 days per week was defined as a high physical activity level.

Anthropometric measurements including height, weight, WC, and blood pressure were conducted by trained local CDC staff. All equipment used in the measurements were selected according to the guidance of the national project group. Height and weight were measured without shoes or coats. A stadiometer (TZG, Nantong, China) was used for height measurement, with an accuracy to 0.1 cm. The equipment used for weight measurement was an electric scale (TANITA HD-390, Tokyo, Japan) and the results were accurate to 0.1 kg. A soft tape was used to measure WC and the participants were measured in fasting states, at the midpoint between the bottom of the rib cage and the uppermost border of the iliac crest, following exhalation. An electronic sphygmomanometer (Omron HBP 1300, Kyoto, Japan) was used for the measurement of blood pressure. Each participant was measured three times for blood pressure, in 5 min intervals, and the average value was recorded.

Laboratory sample collection and tests included the collection of participants' fingertip blood for testing hemoglobin by a Hemocue 201$^+$ Hemoglobin Analyzer (HemoCue, Angelholm, Sweden). In addition, 6 mL of fasting venous blood was drawn from participants in the morning and used for testing biochemical and nutritional indicators. Blood glucose, lipids, uric acid, serum vitamin A, vitamin D, and zinc, etc., were detected according to the standard methods provided by the China CDC project group.

2.4. Diagnostic Criteria and Definitions

The Chinese "screening standard for malnutrition of school-age children and adolescents" was applied to screen wasting among children by age-specific values [26].

The Chinese "screening for overweight and obesity among school-age children and adolescents" was used to identify overweight and obesity by age-specific values [27]

MetS was diagnosed according to the modified criteria of the National Cholesterol Education Program-Adult Treatment Panel III (NCEP-ATP III) for children and adolescents aged 7–17 years [28]. When participants had more than three of the following five components, they were diagnosed as MetS:

(1)   Abdominal obesity, where WC $\geq$ age- and gender-specific 90th percentile [29];
(2)   Elevated triglyceride (TG), where TG $\geq$ 1.24 mmol/L;
(3)   Low high-density lipoprotein (HDL), where HDL $\leq$ 1.03 mmol/L;
(4)   Elevated blood pressure, where systolic blood pressure (SBP) or diastolic blood pressure (DBP) $\geq$ 90th percentile for gender, age, and height [30];
(5)   Elevated fasting blood glucose (FBG), where glucose $\geq$ 6.1 mmol/L.

The following diagnostic criteria and definitions were applied in this study:

Elevated total cholesterol (TC): TC $\geq$ 5.18 mmol/L [31]; elevated low high-density lipoprotein (LDL): LDL $\geq$ 3.36 mmol/L [31];

Hyperuricemia: serum uric acid (SUA) $\geq$ 357 μmol/L [21];

Vitamin A deficiency (VAD), marginal deficiency in vitamin A (VAMD), and vitamin A sufficiency were referred to serum retinol levels less than 0.2 μg/mL, 0.2–0.3 μg/mL, and more than 0.3 μg/mL, respectively [32];

Vitamin D deficiency (VDD), inadequacy of vitamin D (VDI), and vitamin D sufficiency were defined by serum 25(OH)D levels less than 12 ng/mL, 12–20 ng/mL, and more than 20 ng/mL, respectively [33];

Anemia: for participants aged 5–11 years, hemoglobin < 115 g/L; for participants aged 12–14 years, hemoglobin < 120 g/L; for male participants aged 15–17 years, hemoglobin < 130 g/L; for female participants aged 15–17 years, hemoglobin < 120 g/L [34];

Zinc deficiency in children and adolescents was defined as serum zinc concentrations < 76.5 μg/dL [35].

The number of overnutrition-related disorders included a count of five MetS components (abdominal obesity, high TG, low HDL, elevated blood pressure, and elevated FBG) and hyperuricemia, high LDL, high TC. The number of undernutrition-related disorders included a count of prevalent vitamin A insufficiency, vitamin D deficiency, anemia, and zinc deficiency.

2.5. Statistical Analysis

All statistical analysis was conducted on R software (Version 4.1.0, developed by R Development Core Team, originated from the Statistics Department of the University of Auckland, Auckland, New Zealand). According to the normality of the continuous variables, they were displayed as either median (interquartile range, IQR) or average (standard deviation, SD). Pearson's chi-square test was used to compare the distribution of different characteristics and metabolic risk factors between gender groups. Multivariate logistic regression analysis was applied to find the related factors for multiple nutritional problems in the research populations.

In order to understand the relevant factors for children's and adolescents' nutritional status, we conducted the multivariate logistic regression analysis by including the following potential factors in the analysis: participants' demographic characteristics, behavioral characteristics, parents' characteristics, and family characteristics. Multivariate Poisson regression analysis was carried out for the overnutrition- and undernutrition-related disorders to test the participants' personal, parental, and household factors. A two-sided $p < 0.05$ was utilized to define statistical significance.

3. Results

3.1. Characteristics of Research Populations

Of the 3025 school-age children and adolescents (7–17 years old) in the study, there were 50.2% males and 49.8% females; 34.6% of the participants were from urban areas and 65.4% were from rural areas. As displayed in Table 1, the distribution of most characteristics, such as age, age group, living area, physical activity level, screen time, age of mother, age of father, education of mother, and education of father, were not significantly different between male and female participants (all $p > 0.05$). There was a remarkable difference in household size between genders ($p = 0.009$). When it came to anthropometric indexes, male participants had higher height, weight, and SBP than female participants (all $p < 0.05$), except for DBP, which was not different between males and females (Table 2). There were noticeable differences in many biochemical indices. Male participants had higher levels of FBG, serum uric acid, vitamin D, and serum zinc than did females (all $p < 0.05$). The TG, TC, LDL, and HDL of female participants were at higher levels than they were males (all $p < 0.05$). However, there was no significant difference in vitamin A level between genders ($p = 0.963$), as shown in Table 2.

Table 1. Basic characteristics of the research population.

Variables	Males (%) $n = 1520$	Females (%) $n = 1505$	All $n = 3025$	t or χ^2 Values	p Value
Age (years)	11.4 ± 3.0	11.4 ± 3.0	11.4 ± 3.0	0.074	0.941
Age group				0.089	0.765
7–12 years	990 (65.1)	988 (65.6)	1978 (65.4)		
13–17 years	530 (34.9)	517 (34.4)	1047 (34.6)		
Living area				0.001	0.982
Urban	1272 (83.7)	1259 (83.7)	2531 (83.7)		
Rural	248 (16.3)	246 (16.3)	494 (16.3)		
Physical activity				0.613	0.434
Low	907 (59.7)	919 (61.1)	1826 (60.4)		
High	613 (40.3)	586 (38.9)	1199 (39.6)		
Screen time				3.432	0.064
Low	1320 (86.8)	1340 (89.0)	2660 (87.9)		
High	200 (13.2)	165 (11.0)	365 (12.1)		
Age of mother				0.170	0.680
≤37 (median)	783 (51.5)	764 (50.8)	1547 (51.1)		
≥38	737 (48.5)	741 (49.2)	1478 (48.9)		
Age of father				0.389	0.533
≤39 (median)	806 (53.0)	781 (51.9)	1587 (52.5)		
≥40	714 (47.0)	724 (48.1)	1438 (47.5)		
Education of mother				1.900	0.387
Primary school and below	226 (14.9)	206 (13.7)	432 (14.3)		
Middle school	1007 (66.3)	989 (65.7)	1996 (66.0)		
University and above	287 (18.9)	310 (20.6)	597 (19.7)		
Education of father				0.908	0.635
Primary school and below	134 (8.8)	119 (7.9)	253 (8.4)		
Middle school	1027 (67.6)	1034 (68.7)	2061 (68.1)		
University and above	359 (23.6)	352 (23.4)	711 (23.5)		
Household size				6.907	0.009
≤4 members	924 (60.8)	844 (56.1)	1768 (58.4)		
≥5 members	596 (39.2)	661 (43.9)	1257 (41.6)		

As noted above, screen time refers to the time of using an electronic screen every day. A low level of screen time is defined as <2 h and a high level of screen time is defined as ≥2 h. The assessment of physical activities was based on reaching a level of coarsened breathing or a heart rate rise for more than 30 min. An exercise level of 0–3 days per week was defined as a low physical activity level; an exercise level of more than 4 days per week was defined as a high physical activity level.

Table 2. Nutritional and biochemistry indicators for the research population.

Variables	Males Median (IQR)	Females Median (IQR)	All Median (IQR)	Z Values	*p* Value
Anthropometric measurements					
Height (cm)	152.1 (138.4, 168.0)	152.0 (138.0, 160.3)	152.1 (138.1, 162.6)	−5.826	<0.001
Weight (kg)	45.0 (32.6, 59.4)	42.2 (30.9, 52.1)	43.5 (31.7, 55.3)	−6.791	<0.001
WC (cm)	66.4 (59.1, 75.1)	62.5 (56.1, 68.5)	64.0 (57.7, 71.5)	−11.076	<0.001
SBP (mmHg)	116.8 (108.3, 125.5)	113.0 (105.3, 120.5)	115.0 (106.7, 122.7)	−9.233	<0.001
DBP (mmHg)	67.3 (62.0, 73.3)	67.7 (62.7, 73.3)	67.7 (62.3, 73.3)	−1.155	0.248
Biochemical indexes					
FBG (mmol/L)	5.29 (5.00, 5.59)	5.16 (4.90, 5.45)	5.22 (4.94, 5.52)	−7.172	<0.001
TG (mmol/L)	0.77 (0.60, 1.06)	0.83 (0.65, 1.09)	0.80 (0.62, 1.06)	−5.747	<0.001
TC (mmol/L)	3.95 (3.50, 4.49)	4.03 (3.63, 4.54)	4.00 (3.59, 4.52)	−3.282	0.001
LDL (mmol/L)	2.16 (1.80, 2.55)	2.21 (1.88, 2.59)	2.19 (1.84, 2.57)	−2.448	0.014
HDL (mmol/L)	1.56 (1.30, 1.84)	1.60 (1.37, 1.87)	1.59 (1.34, 1.86)	−2.801	0.005
Serum uric acid (μmol/L)	336.0 (276.0, 405.0)	297.0 (254.0, 339.0)	310.0 (260.0, 369.0)	−13.106	<0.001
Vitamin A (μg/mL)	0.38 (0.32, 0.43)	0.38 (0.32, 0.43)	0.38 (0.32, 0.43)	−0.046	0.963
Vitamin D (ng/mL)	16.1 (12.8, 20.1)	14.5 (11.4, 18.1)	15.3 (12.0, 19.1)	−8.368	<0.001
Zinc (μg/dL)	94.0 (87.0, 102.0)	93.0 (85.0, 99.0)	93.0 (86.0, 101.0)	−4.386	<0.001

IQR = interquartile range; WC = waist circumference; SBP = systolic blood pressure; DBP =: diastolic blood pressure; FBG = fasting blood glucose; TG = triglyceride; TC = total cholesterol; LDL = low-density lipoprotein; HDL = high-density lipoprotein.

3.2. Nutritional and Health Statuses of School-Age Children and Adolescents

The nutritional and health statuses of school-age children and adolescents are displayed in Figure 1 and Table 3. Figure 1 is presented according to the sequences of the prevalence of the relevant nutritional and health problems. The prevalence of wasting, overweight, and obesity were 5.5%, 14.8%, and 12.7%, respectively. MetS was prevalent among 5.1% of the participants. The prevalence of the five MetS components, including abdominal obesity, high TG, low HDL, elevated blood pressure, and elevated FBG, were 18.7%, 15.0%, 4.1%, 41.3%, and 3.5%, respectively. Among the study participants, 29.5% had hyperuricemia. The prevalence of high LDL among all participants was 4.8%. The prevalence of high TC among all participants was 7.4%; 0.9% of the participants were VAD, 14.6% had marginal vitamin D deficiency, and 84.5% were sufficient in vitamin A levels; 25.1% of the participants had VDD, 54.5% had inadequate VDD, and 20.4% were sufficient in vitamin D levels. The prevalence of anemia among all participants was 4.0%. The prevalence of zinc deficiency was 4.8%. There were significant differences between genders in the prevalence of weight groups, MetS, hyperuricemia, high TC, and vitamin D conditions, as shown in Table 3.

After age group stratification, we compared the nutritional status within each age group and between the age groups, as shown in the Supplementary Materials, Table S1. The proportions for weight groups, vitamin A, and vitamin D were significantly different between participants in the 7–12 years age group and the 13–17 years age group ($p < 0.001$). The 13–17 years age group had a significant higher possibility of hyperuricemia and anemia than did the 7–12 years age group (both $p < 0.05$).

3.3. Related Factors Associated with Various Nutritional Diseases

In Table 4, the multivariate logistic regression of multiple nutritional diseases (MetS, hyperuricemia, vitamin A insufficiency, vitamin D deficiency, anemia, and zinc deficiency) are displayed. Table 4 includes the most common factors of demographic characteristics, behavioral characteristics, parents' characteristics, and family characteristics. These common factors were considered in the multivariate logistic analysis. The adjusted odds ratios (AORs), along with their 95% confidence intervals (95% CI) and p values, are portrayed in Table 4.

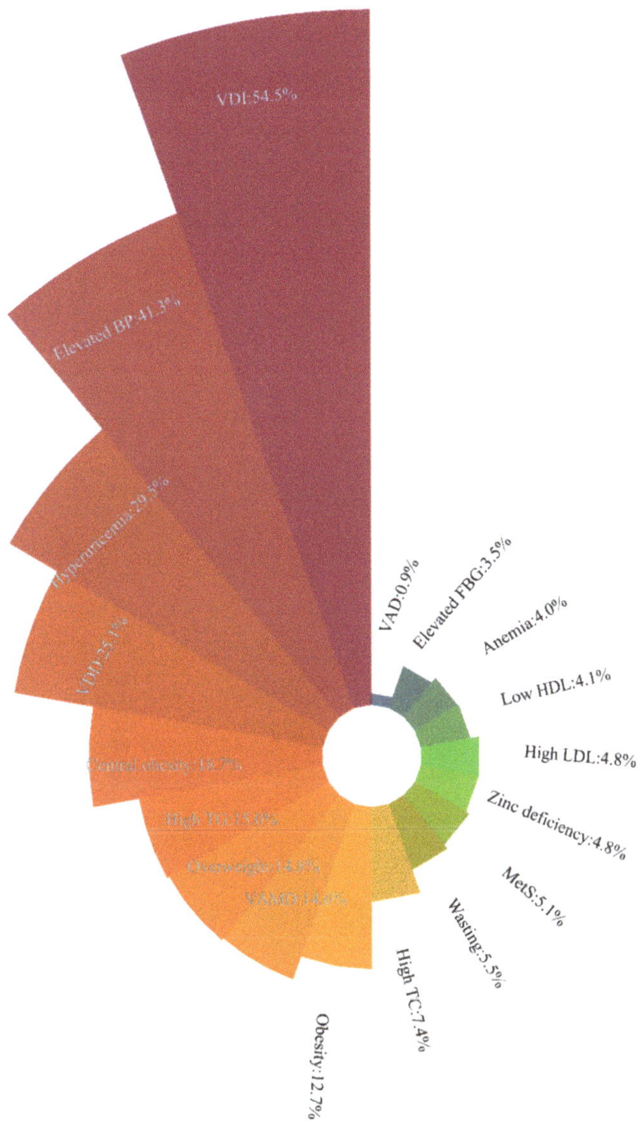

Figure 1. Sequence of prevalence of the relevant nutritional and health problems among school-age children and adolescents. VDI = vitamin D inadequacy; elevated BP = elevated blood pressure; VDD = vitamin D deficiency; high TG = high triglyceride; VAMD = vitamin A marginal deficiency; high TC = high total cholesterol; MetS = metabolic syndrome; high LDL = high low-density lipoprotein; low HDL = low high-density lipoprotein; elevated FBG = elevated fasting blood glucose; VAD = vitamin A deficiency. MetS and its five components (central obesity, high TG, low HDL, elevated BP, and elevated FBG) were diagnosed according to the modified criteria of the National Cholesterol Education Program-Adult Treatment Panel III (NCEP-ATP III) for children and adolescents aged 7–17 years.

Table 3. Prevalence of various categories of nutritional status among school-age children and adolescents in Jiangsu.

Nutritional Status	Males n (%)	Female n (%)	Total n (%)	χ^2 Values	p Value
Weight groups				102.298	<0.001
Wasting	96 (6.3)	70 (4.7)	166 (5.5)		
Normal weight	893 (58.8)	1136 (45.5)	2029 (67.1)		
Overweight	271 (17.8)	176 (11.7)	447 (14.8)		
Obesity	260 (17.1)	123 (8.2)	383 (12.7)		
MetS				6.215	0.013
No	1427 (93.9)	1443 (95.9)	2870 (94.9)		
Yes	93 (6.1)	62 (4.1)	155 (5.1)		
Abdominal obesity				18.075	<0.001
No	1190 (78.3)	1269 (84.3)	2459 (81.3)		
Yes	330 (21.7)	236 (15.7)	566 (18.7)		
High TG				1.403	0.236
No	1304 (85.8)	1268 (84.3)	2572 (85.0)		
Yes	216 (14.2)	237 (15.7)	453 (15.0)		
Low HDL				3.160	0.075
No	95.3 (1448)	1453 (96.5)	2901 (95.9)		
Yes	72 (4.7)	52 (3.5)	124 (4.1)		
Elevated BP				0.159	0.690
No	887 (58.4)	889 (59.1)	1179 (58.7)		
Yes	633 (41.6)	616 (40.9)	1249 (41.3)		
Elevated FBG				15.868	<0.001
No	1446 (95.1)	1472 (97.8)	2918 (96.5)		
Yes	74 (4.9)	33 (2.2)	107 (3.5)		
Hyperuricemia				194.297	<0.001
No	897 (59.0)	1236 (82.1)	2133 (70.5)		
Yes	623 (41.0)	269 (17.9)	892 (29.5)		
High LDL				0.432	0.511
No	1451 (95.5)	1429 (95.0)	2880 (95.2)		
Yes	69 (4.5)	76 (5.0)	145 (4.8)		
High TC				7.727	0.005
No	1427 (93.9)	1373 (91.2)	2800 (92.6)		
Yes	93 (6.1)	132 (8.8)	225 (7.4)		
Vitamin A				0.049	0.976
Sufficiency	1285 (84.5)	1272 (84.5)	2557 (84.5)		
Marginal deficiency	222 (14.6)	219 (14.6)	441 (14.6)		
Deficiency	13 (0.9)	14 (0.9)	27 (0.9)		
Vitamin D				68.349	<0.001
Sufficiency	383 (25.2)	233 (15.5)	616 (20.4)		
Inadequacy	835 (54.9)	815 (54.2)	1650 (54.5)		
Deficiency	302 (19.9)	457 (30.4)	759 (25.1)		
Anemia				31.863	<0.001
No	1490 (98.0)	1415 (94.0)	2905 (96.0)		
Yes	30 (2.0)	90 (6.0)	120 (4.0)		
Zinc deficiency				3.417	0.065
No	1458 (95.9)	1422 (94.5)	2880 (95.2)		
Yes	62 (4.1)	83 (5.5)	145 (4.8)		

MetS = metabolic syndrome; high TG = high triglyceride; low HDL = low high-density lipoprotein; elevated BP = elevated blood pressure; elevated FBG = elevated fasting blood glucose; high LDL = high low-density lipoprotein; high TC = high total cholesterol.

We found that age groups and weight groups were significantly associated with MetS. Compared to the 7–12 years age group, the 13–17 years age group was more prone to have MetS (AOR = 1.79, 95% CI: 1.17–2.73, p = 0.007). Participants in the overweight groups were more likely to have MetS (OR = 7.47, 95% CI: 4.26–13.11, p < 0.001), compared to those in the reference group (in terms of wasting and normal weight). Furthermore, participants in the obesity group were more likely to have MetS than were those in the reference group (OR = 41.14, 95% CI: 25.05–67.56, p < 0.001).

Table 4. Multivariate logistic regression analysis of various nutritional disorders among school-age children and adolescents in Jiangsu Province.

Characteristics	Metabolic Syndrome		Hyperuricemia		Vitamin A Insufficiency		Vitamin D Deficiency		Anemia		Zinc Deficiency	
	AOR (95%CI)	p Value	AOR (95%CI)	p Value	AOR (95%CI)	p Value	AOR (95%CI)	p Value	AOR (95%CI)	p Value	AOR (95%CI)	p Value
Sex												
Males (reference)	1.00	–	1.00	–	1.00	–	1.00	–	1.00	–	1.00	–
Females	1.20 (0.83–1.74)	0.331	0.30 (0.25–0.36)	<0.001	0.91 (0.74–1.12)	0.378	1.80 (1.51–2.15)	<0.001	2.95 (1.93–4.51)	<0.001	1.33 (0.94–1.88)	0.102
Age group												
7–12 years (reference)	1.00	–	1.00	–	1.00	–	1.00	–	1.00	–	1.00	–
13–17 years	1.79 (1.17–2.73)	0.007	5.09 (4.13–6.27)	<0.001	0.34 (0.26–0.45)	<0.001	2.53 (2.09–3.08)	<0.001	1.82 (1.20–2.77)	0.005	0.94 (0.63–1.41)	0.774
Living area												
Urban	1.00	–	1.00	–	1.00	–	1.00	–	1.00	–	1.00	–
Rural	0.76 (0.43–1.34)	0.345	0.97 (0.75–1.25)	0.796	1.24 (0.95–1.62)	0.116	1.14 (0.89–1.44)	0.298	1.43 (0.89–2.31)	0.142	1.80 (1.21–2.68)	0.004
Weight group												
Others (reference)	1.00	–	1.00	–	1.00	–	1.00	–	1.00	–	1.00	–
Overweight	7.47 (4.26–13.11)	<0.001	1.94 (1.53–2.46)	<0.001	0.63 (0.46–0.87)	<0.001	1.02 (0.80–1.31)	0.877	0.71 (0.39–1.28)	0.255	0.78 (0.46–1.32)	0.357
Obesity	41.14 (25.05–67.56)	<0.001	4.38 (3.38–5.67)	<0.001	0.37 (0.25–0.55)	0.005	0.85 (0.64–1.13)	0.269	0.37 (0.15–0.93)	0.034	0.81 (0.45–1.44)	0.466
Physical activity												
Low	1.00	–	1.00	–	1.00	–	1.00	–	1.00	–	1.00	–
High	1.31 (0.91–1.88)	0.146	1.19 (0.99–1.43)	0.061	0.93 (0.75–1.16)	0.528	0.69 (0.58–0.83)	<0.001	1.07 (0.73–1.56)	0.746	1.25 (0.89–1.77)	0.204
Screen time												
Low	1.00	–	1.00	–	1.00	–	1.00	–	1.00	–	1.00	–
High	0.60 (0.33–1.10)	0.097	0.91 (0.69–1.19)	0.487	1.45 (1.09–1.94)	0.011	1.40 (1.09–1.80)	0.008	1.25 (0.73–2.13)	0.420	0.78 (0.46–1.32)	0.360
Age of mother												
<36	1.00	–	1.00	–	1.00	–	1.00	–	1.00	–	1.00	–
≥37	0.93 (0.55–1.55)	0.775	1.46 (1.13–1.89)	0.004	0.78 (0.58–1.06)	0.113	1.13 (0.88–1.45)	0.349	1.34 (0.78–2.29)	0.290	0.88 (0.53–1.45)	0.621
Age of father												
<36	1.00	–	1.00	–	1.00	–	1.00	–	1.00	–	1.00	–
≥37	0.97 (0.58–1.63)	0.901	0.94 (0.73–1.23)	0.668	1.07 (0.79–1.45)	0.662	1.03 (0.80–1.33)	0.788	0.75 (0.44–1.27)	0.285	0.82 (0.49–1.36)	0.443
Education of mother												
Primary school and below	1.00	–	1.00	–	1.00	–	1.00	–	1.00	–	1.00	–
Middle school	1.08 (0.58–2.03)	0.804	0.94 (0.71–1.25)	0.693	0.64 (0.47–0.88)	0.005	0.68 (0.52–0.89)	0.004	1.27 (0.70–2.29)	0.438	0.95 (0.58–1.56)	0.847
University and above	1.09 (0.47–2.50)	0.847	1.23 (0.82–1.83)	0.315	0.55 (0.35–0.89)	0.014	0.58 (0.40–0.86)	0.006	1.03 (0.43–2.45)	0.947	0.48 (0.20–1.16)	0.104
Education of father												
Primary school and below	1.00	–	1.00	–	1.00	–	1.00	–	1.00	–	1.00	–
Middle school	0.92 (0.45–1.87)	0.817	0.86 (0.61–1.22)	0.397	1.10 (0.74–1.64)	0.626	1.17 (0.84–1.62)	0.364	1.06 (0.51–2.20)	0.886	0.72 (0.41–1.28)	0.268
University and above	0.58 (0.24–1.40)	0.228	0.63 (0.41–0.97)	0.035	0.73 (0.43–1.22)	0.223	0.97 (0.63–1.48)	0.883	1.03 (0.51–2.20)	0.942	0.47 (0.20–1.09)	0.078
Household size												
≤4 members	1.00	–	1.00	–	1.00	–	1.00	–	1.00	–	1.00	–
≥5 members	1.20 (0.83–1.73)	0.327	0.93 (0.77–1.12)	0.465	1.20 (0.97–1.47)	0.088	1.25 (1.05–1.49)	0.014	1.10 (0.75–1.62)	0.609	1.30 (0.92–1.83)	0.136

With respect to hyperuricemia, a participant's sex, age group, weight group, age of the mother, and education of the father were relevant factors. Females had significantly less likelihood of hyperuricemia than did males (OR = 0.30, 95% CI: 0.25–0.36, $p < 0.001$). Participants in the 13–17 years age group had a higher possibility of hyperuricemia than did those in the 7–12 years age group (OR = 5.09, 95% CI: 4.13–6.27, $p < 0.001$). The overweight and obesity groups were both higher in the prevalence of hyperuricemia than were other groups (OR = 1.94, 95% CI: 1.53–2.46, and OR = 4.38, 95% CI: 3.38–5.67, respectively). Compared to the participants whose mothers were younger than 37 years, the participants whose mothers were older than 37 years were more likely to have hyperuricemia (OR = 1.46, 95% CI: 1.13–1.89, $p = 0.004$). Participants whose father's education was university and above were less likely to have hyperuricemia than were those whose father's education was primary school and below (OR = 0.63, 95% CI: 0.41–0.97, $p = 0.035$).

When focusing on vitamin A insufficiency, a participant's age group, weight group, screen time, and education of the mother were the relevant factors. Participants in the 13–17 years age group were less likely to have a prevalence of vitamin A insufficiency, when compared with the those in the 7–12 years age group (OR = 0.34, 95% CI: 0.26–0.45, $p < 0.001$). Participants who were overweight had less risk of vitamin A insufficiency than did those in the reference group (in terms of wasting and normal weight, OR = 0.63, 95% CI: 0.46–0.87). Obesity subjects also had less risk of vitamin A insufficiency than did those in the reference group (OR = 0.37, 95% CI:0.25–0.55, $p = 0.005$). In addition, participants with higher screen time were prone to have vitamin A insufficiency (OR = 1.45, 95% CI: 1.09–1.94, $p = 0.011$). Children and adolescents whose mothers had higher education levels were less likely to have vitamin A insufficiency, compared to those whose mothers had primary school education or below; the OR were 0.64 and 0.55, respectively.

With reference to VDD, a participant's sex, age group, physical activity, screen time, education of the mother, and household size were the relevant factors. Females were more likely to have VDD than males, with AOR = 1.80, 95% CI: 1.51–2.15, $p < 0.001$. The 13–17 years participants had a higher risk of being VDD than did the 7–12 years participants (OR = 2.53, 95% CI:2.09–3.08, $p < 0.001$). Remarkably, participants with a high level of physical activity were less likely to have VDD (AOR = 0.69, 95% CI: 0.58–0.83, $p < 0.001$). Subjects with high screen time were associated with an increased risk of VDD (AOR = 1.40, 95% CI: 1.09–1.80, $p = 0.008$). In addition, VDD was related to participants' mothers' education levels (middle school vs. primary school and below: AOR = 0.68, 95% CI: 0.52–0.89; university and above vs. primary school and below: AOR = 0.58, 95% CI: 0.40–0.86). Those participants whose household size was ≥5 members were more prone to have VDD than were those whose household size was ≤4 members (AOR = 1.25, 95% CI = 1.05–1.49, $p = 0.014$).

When considering anemia, the sex, age group, and obesity of participants were the relevant factors. Sex was associated with anemia (females were more likely to have anemia, with AOR = 2.95, 95% CI = 1.95–4.51, $p < 0.001$), as well the age group (13–17 years: AOR = 1.82, 95% CI: 1.20–2.77, $p = 0.005$). Inversely, obesity groups were less likely to show prevalence in anemia (AOR = 0.37, 95% CI: 0.15–0.93, $p = 0.034$). For zinc deficiency, participants living in the rural area had a higher risk than did those who lived in urban areas (AOR = 1.80, 95% CI: 1.21–2.68, $p = 0.004$).

3.4. The Relevant Factors of Overnutrition- and Undernutrtion-Related Disorders

The Poisson regression analysis took the amount of overnutrition-related disorders (i.e., the prevalent five MetS components and hyperuricemia, high LDL, and high TC) and undernutrition-related disorders (i.e., vitamin A insufficiency, VDD, anemia, and zinc deficiency) as dependent variables to test the personal, parental, and household factors. We found that the factors of sex, age group, and weight group were related to overnutrition-related disorders, seen in Table 5. Females were less likely inclined to overnutrition-related disorders (AOR = 0.81, 95% CI: 0.51–0.85, $p < 0.05$). The older age group (13–17 years) and the overweight and obesity participants were more prone to have a variety of overnutrition-related disorders (AOR = 1.44, 1.66 and 2.77, respectively, all $p < 0.05$). With respect to undernutrition-related disorders, we found that a participant's sex, age, living area, weight group, physical activity, screen time, education of mother, and household size were the relevant factors. Females, older age group (13–17 years), rural areas, higher screen time, and large household size ($\geq$5 members) were significantly and positively associated with the occurrence of undernutrition-related disorders (all AOR > 1). Conversely, participants who were overweight or obese, or who had a high level of physical activity or a mother with higher education, were less prone to clustering in undernutrition-related disorders. (All AOR < 1 and all $p < 0.05$).

Table 5. Multivariate Poisson regression analysis of overnutrition- and undernutrition-related disorders among school-age children and adolescents in Jiangsu Province.

Characteristics	Overnutrition-Related Disorders		Undernutrition-Related Disorders	
	IRR (95%CI)	*p* Value	IRR (95%CI)	*p* Value
Sex				
Males (reference)	1.00	–	1.00	–
Females	0.81 (0.51–0.85)	<0.001	1.33 (1.20–1.48)	<0.001
Age group				
7–12 years (reference)	1.00	–	1.00	–
13–17 years	1.44 (1.35–1.54)	<0.001	1.17 (1.04–1.32)	0.008
Living area				
Urban	1.00	–	1.00	–
Rural	1.00 (0.91–1.08)	0.939	1.21 (1.06–1.38)	0.005
Weight group				
Others (reference)	1.00	–	1.00	–
Overweight	1.66 (1.53–1.79)	<0.001	0.86 (0.73–1.00)	0.05
Obesity	2.77 (2.58–2.97)	<0.001	0.67 (0.55–0.81)	<0.001
Physical activity				
Low	1.00	–	1.00	–
High	1.05 (0.99–1.12)	0.095	0.89 (0.80–0.99)	0.027
Screen time				
Low	1.00	–	1.00	–
High	0.94 (0.86–1.03)	0.202	1.23 (1.06–1.42)	0.005
Age of mother				
$\leq$37	1.00	–	1.00	–
$\geq$38	1.07 (0.98–1.16)	0.122	0.99 (0.86–1.15)	0.946
Age of father				
$\leq$36	1.00	–	1.00	–
$\geq$37	0.98 (0.90–1.07)	0.658	0.99 (0.85–1.15)	0.852
Education of mother				
Primary school and below	1.00	–	1.00	–
Middle school	0.99 (0.90–1.08)	0.759	0.81 (0.70–0.94)	0.006
University and above	1.03 (0.91–1.18)	0.631	0.68 (0.54–0.89)	0.001
Education of father				
Primary school and below	1.00	–	1.00	–
Middle school	1.01 (0.90–1.13)	0.913	1.04 (0.86–1.26)	0.689
University and above	0.90 (0.77–1.04)	0.135	0.83 (0.65–1.07)	0.147
Household size				
$\leq$4 members	1.00	–	1.00	–
$\geq$5 members	1.02 (0.96–1.09)	0.435	1.17 (1.05–1.29)	0.004

Overnutrition-related disorders: the prevalent five MetS components and hyperuricemia, high LDL, and high TC. Undernutrition-related disorders: prevalent vitamin A insufficiency, vitamin D deficiency, anemia, and zinc deficiency. IRR: incidence rate rations. 95%CI: 95% confidence interval.

4. Discussion

The present study illustrated the multiple nutritional statuses of school-age children and adolescents in Jiangsu Province, China. The presence of the double burden of malnutrition was common among the children and adolescents. To be more specific, overnutrition-related diseases such as overweight, obesity, MetS, hyperuricemia, high LDL and high TC, and undernutrition-related diseases (vitamin A insufficiency, VDD, anemia, and zinc deficiency) were prevalent in the 7–17 years age group of participants. Demographic characteristics, behavior characteristics, parents' characteristics, and family characteristics were associated with different malnutrition disorders.

In this study, 5.5% of school-age children and adolescents in Jiangsu Province showed prevalence in wasting, according to the Chinese standard, and males were more prone to wasting than were females (6.3% vs. 4.7%). Based on data from 2014 Chinese National Surveys on Students Constitution and Health, wasting was present in 5.5% of Chinese children and adolescents, which was consistent with the findings in our study [36]. In present study, the prevalence of overweight and obesity was 14.8% and 12.7%, respectively. The prevalence of both overweight and obesity was higher in males than in females. Results from the 2015 China Health and Nutrition Survey (CHNS) showed that, generally, overweight prevalence was 15.43% and obesity prevalence was 11.06% among children and adolescents aged 6 to 17 years [37]. Data from NHANES indicated that obesity prevalence among America youth aged 2–19 years, from 2015–2016, was 18.5% [38].

Overweight and obesity have become a serious nutritional problem in children all around the world [39,40]. The prevalence of MetS in this study's participants was 5.1%, while the results of the CHNS yielded an overall 3.37% prevalence of MetS [18]. The Jiangsu Province's level of prevalence of MetS in school-age children and adolescents was higher than the national average. This indicated that urgent measures must be taken to contain MetS. Among school-age children and adolescents, 29.5% had hyperuricemia ($\geq$357 µmol/L). The overall prevalence of hyperuricemia among American adults had increased from 18.0% in 1988 to 21.4% in 2008 [41], which suggested a high prevalence level of hyperuricemia in a large number of children and adolescents.

In our study, we found VAD and marginal deficiency in vitamin A were prevalent, respectively, in 0.9% and 14.6% of the participants. Similarly, the national results of the China Nutrition and Health Surveillance of Children and Lactating Mothers, in 2016–2017, showed that the prevalence of VAD and the marginal deficiency of vitamin A was 0.96% and 14.71%, respectively [42]. The prevalence of VAD and marginal deficiency in vitamin A among school-age children and adolescents in Jiangsu Province were at the national average level. Based on these findings, vitamin A marginal deficiency was the main form of vitamin A insufficiency in Chinese children. The prevalence of VDD and inadequacy of vitamin D were 25.1% and 54.5%, respectively. In spite of different diagnostic criteria, VDD was very common among school-age children [43,44]. Sufficient sunlight, dietary vitamin D and vitamin D supplements are recommended for children.

Our study indicated that the presence of anemia in children and adolescents aged 7–17 years was 4.0%. Anemia was more common in girls than in boys, at 6.0% vs. 2.0%, respectively. These results were consistent with data from the Chinese National Nutrition and Health Survey (CNNHS) [45], where 4.8% of children had zinc deficiency and the prevalence of zinc deficiency was not different between genders. Zinc is an essential microelement for growth and development in children, but zinc deficiency is still a serious nutritional problem, especially in school-age children [13]. These findings indicated that multiple overnutrition-related diseases and undernutrition-related diseases coexisted in the surveyed areas.

In addition, we analyzed the roles of different factors, such as demographic characteristics, behavioral characteristics, parents' characteristics, and family characteristics in the development of different malnutrition diseases. Age and weight are related to MetS. Participating adolescents over 13 years, obese participants, and overweight participants are more likely to suffer from MetS. Other research has also shown that age and obesity can increase the risk of developing MetS [46]. These results indicate that more attention should be paid to obese and overweight children, and to adolescents aged 13–17 years.

Based on our study, age, weight and the age of mothers were positively associated with hyperuricemia. Females and children whose mothers had high education levels were less likely to have hyperuricemia. The level of uric acid increased with age in children aged 5–14 years in southeast China [47]. One study showed that boys had higher mean serum uric acid concentrations than girls [19], and this difference was consistent with our results. Age and education of the mother were new influencing factors for hyperuricemia in children. These results support the view that family factors, especially factors pertaining to mothers, may have some influence on the development of hyperuricemia in children.

When focusing on vitamin A insufficiency, age, weight, screen time, and the education of mothers are relevant factors. Older, overweight, and obese children, and those whose mothers have higher education levels, are less likely to be insufficient in vitamin A. Similarly, a meta-analysis reported that a prevalence of VAD and marginal VAD both decreased with increasing age in Chinese children [9]. Furthermore, a Chinese nationally representative study reported that the average of serum retinol concentration was notably higher in over-nutrified than non-over nutrified children and adolescents [48].

The screen time of children was positively associated with vitamin A insufficiency. This can be understood by the fact that vitamin A is the key nutrient implicated in maintaining the function of normal vision; prolonged eye use may consume vitamin A [49,50]. Vitamin D, with its influence on calcium and phosphorus homeostasis, as well as bone health and immunity health, is the essential micronutrient in the growth and development of children [51]. VDD is associated with numerous adverse health issues, such as MetS [52], dyslipidemia [53], and cardiovascular diseases [54]. In our study, sex, age, physical activity, screen time, the education of mothers, and household sizes were related to VDD. The factors of female gender, age more than 13 years, high screen time, and large household size were positively associated with VDD in this population.

One Chinese study investigated VDD among 460,537 children and concluded that the factor of female gender contributed to VDD and vitamin D insufficiency [55]. Girls and boys have different fat distributions and girls have a higher physiological level of estrogens that play certain roles in vitamin D metabolism [56]. Older children have a higher risk of VDD, which can be demonstrated by the facts that muscle mass increases after puberty and muscle is a storage site for vitamin D metabolites [57,58]. Higher screen time means less sun exposure, which can affect vitamin D synthesis [59]. Meanwhile, family factors, such as education of mothers and household sizes, may have some roles in the development of VDD.

In our research, the anemia-related factors were sex, age, and weight. Our results accorded with the CNNHS in 2010–2012, which analyzed the factors associated with anemia in Chinese children and adolescents [15]. That study found that girls and older children were less likely to have anemia and that obese children had a significantly lower prevalence of anemia than did those who had normal or low body weight, which could be due to a better dietary nutritional intake in overweight or obese people [45].

The present study found that children living in rural areas were more likely to be zinc deficient than were those who lived in urban areas. A systematic review of the literature came to the same conclusion [60]. Zinc deficiency may result from low zinc dietary intake and/or the reduction in bioavailability of zinc. Consumption of foods with high phytate content (e.g., rice and wheat) in the rural areas may lead to low zinc absorption and reduced bioavailability [61].

Moreover, the Poisson regression analysis results showed that girls seemed to have a higher risk of undernutrition-related disorders and less risk of overnutrition-related disorders than boys, which was consistent with previous evidence [62]. There is evidence that girls are more vulnerable than boys to disadvantage in nutrient intake, due to China's "son preference" norm [63]. Novel findings in our study revealed that the ages of children, and adolescents aged 13–17 years, were related to multiple overnutrition and undernutrition disorders. In this study, adolescents aged 13–17 years had a higher prevalence of overweight, MetS, and hyperuricemia. On the other side, adolescents aged 13–17 years had a higher prevalence of VDD and anemia. Therefore, we should pay special attention to the various nutritional and health problems of adolescents in puberty and late puberty. Children's caregivers in poor areas of China have a lack of feeding information and cannot give their children complementary foods in a proper way, including proper starting times, frequency, and quality [64]. Lower education among rural caregivers was another risk factor in children's stunting. This result was consistent with previous studies [65].

Furthermore, childhood obesity is linked to serious complications in adulthood, which include an increased risk of ill health and premature death [66]. The high prevalence and dramatically rising trend in childhood obesity suggest that aggressive approaches in prevention and treatment should be pursued to reduce the attendant health and social consequences of obesity. One previous study indicated that children and adolescents aged 6–17 who participate in moderate to vigorous physical activities are likely to experience multiple positive health outcomes [67]. Thus, providing sufficient physical activity time is essential for children's and adolescents' health. Screen time, including the daily use of mobile phones, TV watching, and playing computer and video games, was related to the nutritional status of children [68], and especially to vitamin A insufficiency and VDD. For children's nutrition and vision health, it is necessary to strictly limit everyday screen time. In addition, domestic factors have influences on children's nutritional health. For example, the significant positive effects of a family economy on the weight of children and adolescents have been gradually and persistently increasing [69].

It is noted that this study does have some limitations. First, this study did not take into consideration dietary factors and some other influences, such as environmental factors or genetic susceptibility; however, these factors may raise issues affecting different nutritional and health problems among children and adolescents. In future studies, we will further explore the roles of these factors in the nutritional status of children. Second, this study, as a cross-sectional design, only provides clues on the relationships between nutritional problems and relevant factors; thus, further studies are needed to verify these relationships. However, through rigorous surveys and scientific sampling methods, a large regional representative sample of participants among eastern Chinese children and adolescents can present findings on multiple nutritional issues.

5. Conclusions

The presence of the double burdens of malnutrition are common among school-age children and adolescents in Jiangsu Province in eastern China. Demographic characteristics, behavior characteristics, parents' characteristics, and family characteristics were associated with multiple malnutrition disorders. Health education, focusing on behavioral intervention and nutritional education of children, and on their living environments, may be a practical strategy in preventing multiple malnutrition disorders in school-age children and adolescents.

Supplementary Materials: The following supporting information can be downloaded at https://www.mdpi.com/article/10.3390/nu14040758/s1: Table S1: Prevalence of various nutritional status in Jiangsu Province children and adolescents aged 7–12 years and 13–17 years.

Author Contributions: Y.Z. and Y.D. participated in the conceptualization and methodology of the study; T.T. and Y.W. carried out the surveys and experiments; J.Z. and W.X. contributed to the acquisition and interpretation of data; X.P. performed the statistical analysis; T.T. and Y.N. wrote the manuscript and prepared the tables; W.X. prepared the figures; G.S. supervised the work. All authors have read and agreed to the published version of the manuscript.

Funding: This research was funded by the National Natural Science Foundation of China (No.81872618), the sub-project of the National Key R&D Program of China (No.2016YFC1305201), and Philosophy and Social Science Research Projects in Universities of Jiangsu Province (No.2020SJA2439).

Institutional Review Board Statement: This study was carried out in accordance with the Declaration of Helsinki and approved by the ethical committee of the China Center for Disease Control and Prevention. The ethical approval number was 201614.

Informed Consent Statement: Informed consent was obtained from all participants involved in the study.

Data Availability Statement: The datasets used and/or analyzed during this study are available from the corresponding author on reasonable request.

Acknowledgments: The authors are grateful to the Chinese Center for Disease Control and Prevention for its support in this study's design and field work. We also appreciate the cooperation and compliance of all of the study's participants.

Conflicts of Interest: The authors declare that they have no conflicts of interest.

References

1. Millward, D.J. Nutrition, infection and stunting: The roles of deficiencies of individual nutrients and foods, and of inflammation, as determinants of reduced linear growth of children. *Nutr. Res. Rev.* **2017**, *30*, 50–72. [CrossRef] [PubMed]
2. Williams, A.M.; Suchdev, P.S. Assessing and Improving Childhood Nutrition and Growth Globally. *Pediatric Clin. N. Am.* **2017**, *64*, 755–768. [CrossRef] [PubMed]
3. Stephenson, J.; Heslehurst, N.; Hall, J.; Schoenaker, D.; Hutchinson, J.; Cade, J.E.; Poston, L.; Barrett, G.; Crozier, S.R.; Barker, M.; et al. Before the beginning: Nutrition and lifestyle in the preconception period and its importance for future health. *Lancet* **2018**, *391*, 1830–1841. [CrossRef]
4. Genovesi, S.; Parati, G. Cardiovascular Risk in Children: Focus on Pathophysiological Aspects. *Int. J. Mol. Sci.* **2020**, *21*, 6612. [CrossRef] [PubMed]
5. Sommer, A.; Twig, G. The Impact of Childhood and Adolescent Obesity on Cardiovascular Risk in Adulthood: A Systematic Review. *Curr. Diabetes Rep.* **2018**, *18*, 91. [CrossRef] [PubMed]
6. Bhutta, Z.A.; Berkley, J.A.; Bandsma, R.; Kerac, M.; Trehan, I.; Briend, A. Severe childhood malnutrition. *Nat. Rev. Dis. Primers* **2017**, *3*, 17067. [CrossRef]
7. Jahan, I.; Muhit, M.; Hardianto, D.; Laryea, F.; Amponsah, S.K.; Chhetri, A.B.; Smithers-Sheedy, H.; McIntyre, S.; Badawi, N.; Khandaker, G. Epidemiology of Malnutrition among Children with Cerebral Palsy in Low- and Middle-Income Countries: Findings from the Global LMIC CP Register. *Nutrients* **2021**, *13*, 3637. [CrossRef] [PubMed]
8. Perez-Escamilla, R.; Bermudez, O.; Buccini, G.S.; Kumanyika, S.; Lutter, C.K.; Monsivais, P.; Victora, C. Nutrition disparities and the global burden of malnutrition. *BMJ* **2018**, *361*, k2252. [CrossRef]
9. Song, P.; Wang, J.; Wei, W.; Chang, X.; Wang, M.; An, L. The Prevalence of Vitamin A Deficiency in Chinese Children: A Systematic Review and Bayesian Meta-Analysis. *Nutrients* **2017**, *9*, 1285. [CrossRef]
10. Huh, S.Y.; Gordon, C.M. Vitamin D deficiency in children and adolescents: Epidemiology, impact and treatment. *Rev. Endocr. Metab. Disord.* **2008**, *9*, 161–170. [CrossRef]
11. Li, H.; Xiao, J.; Liao, M.; Huang, G.; Zheng, J.; Wang, H.; Huang, Q.; Wang, A. Anemia prevalence, severity and associated factors among children aged 6-71 months in rural Hunan Province, China: A community-based cross-sectional study. *BMC Public Health* **2020**, *20*, 989. [CrossRef] [PubMed]
12. Iglesias Vázquez, L.; Valera, E.; Villalobos, M.; Tous, M.; Arija, V. Prevalence of Anemia in Children from Latin America and the Caribbean and Effectiveness of Nutritional Interventions: Systematic Review and Meta-Analysis. *Nutrients* **2019**, *11*, 183. [CrossRef] [PubMed]
13. Liu, X.; Piao, J.; Zhang, Y.; He, Y.; Li, W.; Yang, L.; Yang, X. Assessment of Zinc Status in School-Age Children from Rural Areas in China Nutrition and Health Survey 2002 and 2012. *Biol. Trace Elem. Res.* **2017**, *178*, 194–200. [CrossRef]
14. Dong, Y.; Jan, C.; Ma, Y.; Dong, B.; Zou, Z.; Yang, Y.; Xu, R.; Song, Y.; Ma, J.; Sawyer, S.M.; et al. Economic development and the nutritional status of Chinese school-aged children and adolescents from 1995 to 2014: An analysis of five successive national surveys. The lancet. *Diabetes Endocrinol.* **2019**, *7*, 288–299. [CrossRef]
15. Afshin, A.; Forouzanfar, M.H.; Reitsma, M.B.; Sur, P.; Estep, K.; Lee, A.; Marczak, L.; Mokdad, A.H.; Moradi-Lakeh, M.; GBD 2015 Obesity Collaborators; et al. Health Effects of Overweight and Obesity in 195 Countries over 25 Years. *N. Engl. J. Med.* **2017**, *377*, 13–27. [CrossRef]
16. DeBoer, M.D. Assessing and Managing the Metabolic Syndrome in Children and Adolescents. *Nutrients* **2019**, *11*, 1788. [CrossRef]
17. Lee, A.M.; Gurka, M.J.; DeBoer, M.D. Trends in Metabolic Syndrome Severity and Lifestyle Factors among Adolescents. *Pediatrics* **2016**, *137*, e20153177. [CrossRef]
18. Song, P.; Yu, J.; Chang, X.; Wang, M.; An, L. Prevalence and Correlates of Metabolic Syndrome in Chinese Children: The China Health and Nutrition Survey. *Nutrients* **2017**, *9*, 79. [CrossRef]
19. Li, N.; Zhang, S.; Li, W.; Wang, L.; Liu, H.; Li, W.; Zhang, T.; Liu, G.; Du, Y.; Leng, J. Prevalence of hyperuricemia and its related risk factors among preschool children from China. *Sci. Rep.* **2017**, *7*, 9448. [CrossRef]
20. Kumar, J.; Gupta, A.; Dev, K.; Kumar, S.; Kataria, D.; Gul, A.; Abbas, M.; Jamil, A.; Shahid, S.; Memon, S. Prevalence and Causes of Hyperuricemia in Children. *Cureus* **2021**, *13*, e15307. [CrossRef]
21. Ford, E.S.; Li, C.; Cook, S.; Choi, H.K. Serum concentrations of uric acid and the metabolic syndrome among US children and adolescents. *Circulation* **2007**, *115*, 2526–2532. [CrossRef]

22. Mia, M.N.; Rahman, M.S.; Roy, P.K. Sociodemographic and geographical inequalities in under- and overnutrition among children and mothers in Bangladesh: A spatial modelling approach to a nationally representative survey. *Public Health Nutr.* **2018**, *21*, 2471–2481. [CrossRef]

23. Álvarez, C.; Guzmán-Guzmán, I.P.; Latorre-Román, P.Á.; Párraga-Montilla, J.; Palomino-Devia, C.; Reyes-Oyola, F.A.; Paredes-Arévalo, L.; Leal-Oyarzún, M.; Obando-Calderón, I.; Cresp-Barria, M.; et al. Association between the Sociodemographic Characteristics of Parents with Health-Related and Lifestyle Markers of Children in Three Different Spanish-Speaking Countries: An Inter-Continental Study at OECD Country Level. *Nutrients* **2021**, *13*, 2672. [CrossRef] [PubMed]

24. Zach, A.; Meyer, N.; Hendrowarsito, L.; Kolb, S.; Bolte, G.; Nennstiel-Ratzel, U.; Stilianakis, N.I.; Herr, C.; GME Study Group. Association of sociodemographic and environmental factors with the mental health status among preschool children-Results from a cross-sectional study in Bavaria, Germany. *Int. J. Hyg. Environ. Health* **2016**, *219*, 458–467. [CrossRef] [PubMed]

25. Yu, D.; Zhao, L.; Zhang, J.; Yang, Z.; Yang, L.; Huang, J.; Fang, H.; Guo, Q.; Xu, X.; Ju, L.; et al. China Nutrition and Health Surveys (1982-2017). *China CDC Wkly.* **2021**, *3*, 193–195. [CrossRef]

26. National Health and Family Planning Commission of People's Republic of China. *Screening Standard for Malnutrition of School-Age Children and Adolescents*; WS/T 456-2014; National Health and Family Planning Commission of People's Republic of China: Beijing, China, 2014.

27. National Health and Family Planning Commission of People's Republic of China. *Screening for Overweight and Obesity among School-Age Children and Adolescents*; WST 586-2018; National Health and Family Planning Commission of People's Republic of China: Beijing, China, 2018.

28. Cook, S.; Weitzman, M.; Auinger, P.; Nguyen, M.; Dietz, W.H. Prevalence of a metabolic syndrome phenotype in adolescents: Findings from the third National Health and Nutrition Examination Survey, 1988–1994. *Arch. Pediatrics Adolesc. Med.* **2003**, *157*, 821–827. [CrossRef] [PubMed]

29. Song, P.; Li, X.; Gasevic, D.; Flores, A.B.; Yu, Z. BMI, Waist Circumference Reference Values for Chinese School-Aged Children and Adolescents. *Int. J. Environ. Res. Public Health* **2016**, *13*, 589. [CrossRef] [PubMed]

30. Dong, Y.; Ma, J.; Song, Y.; Dong, B.; Wang, Z.; Yang, Z.; Wang, X.; Prochaska, J.J. National Blood Pressure Reference for Chinese Han Children and Adolescents Aged 7 to 17 Years. *Hypertension* **2017**, *70*, 897–906. [CrossRef]

31. Lee, J.M.; Okumura, M.J.; Davis, M.M.; Herman, W.H.; Gurney, J.G. Prevalence and determinants of insulin resistance among U.S. adolescents: A population-based study. *Diabetes Care* **2006**, *29*, 2427–2432. [CrossRef]

32. World Health Organization. *Global Prevalence of Vitamin A Deficiency in Populations at Risk 1995–2005: WHO Global Database on Vitamin A Deficiency*; WHO: Geneva, Switzerland, 2009.

33. Del Valle, H.B.; Yaktine, A.L.; Taylor, C.L. *Dietary Reference Intakes for Calcium and Vitamin D*; National Academies Press (US): Washington, DC, USA, 2011.

34. World Health Organization. *Iron deficiency anaemia: Assessment, Prevention, and Control*; A guide for Programme Managers; WHO: Geneva, Switzerland, 2001.

35. Liu, J.; Hanlon, A.; Ma, C.; Zhao, S.R.; Cao, S.; Compher, C. Low blood zinc, iron, and other sociodemographic factors associated with behavior problems in preschoolers. *Nutrients* **2014**, *6*, 530–545. [CrossRef] [PubMed]

36. Dong, Y.H.; Wang, Z.H.; Yang, Z.G.; Wang, X.J.; Chen, Y.J.; Zou, Z.Y.; Ma, J. Epidemic status and secular trends of malnutrition among children and adolescents aged 7-18 years from 2005 to 2014 in China. *Beijing da xue xue bao. Yi xue ban J. Peking Univ. Health Sci.* **2017**, *49*, 424–432.

37. Hu, X.; Jiang, H.; Wang, H.; Zhang, B.; Zhang, J.; Jia, X.; Wang, L.; Wang, Z.; Ding, G. Intraindividual Double Burden of Malnutrition in Chinese Children and Adolescents Aged 6-17 Years: Evidence from the China Health and Nutrition Survey 2015. *Nutrients* **2021**, *13*, 3097. [CrossRef] [PubMed]

38. Hales, C.M.; Fryar, C.D.; Carroll, M.D.; Freedman, D.S.; Ogden, C.L. Trends in Obesity and Severe Obesity Prevalence in US Youth and Adults by Sex and Age, 2007-2008 to 2015-2016. *JAMA* **2018**, *319*, 1723–1725. [CrossRef] [PubMed]

39. Yang, L.; Bovet, P.; Ma, C.; Zhao, M.; Liang, Y.; Xi, B. Prevalence of underweight and overweight among young adolescents aged 12-15 years in 58 low-income and middle-income countries. *Pediatric Obes.* **2019**, *14*, e12468. [CrossRef]

40. Wang, Y.; Beydoun, M.A.; Min, J.; Xue, H.; Kaminsky, L.A.; Cheskin, L.J. Has the prevalence of overweight, obesity and central obesity levelled off in the United States? Trends, patterns, disparities, and future projections for the obesity epidemic. *Int. J. Epidemiol.* **2020**, *49*, 810–823. [CrossRef]

41. Zhu, Y.; Pandya, B.J.; Choi, H.K. Prevalence of gout and hyperuricemia in the US general population: The National Health and Nutrition Examination Survey 2007–2008. *Arthritis Rheum.* **2011**, *63*, 3136–3141. [CrossRef] [PubMed]

42. Wang, R.; Zhang, H.; Hu, Y.C.; Chen, J.; Yang, Z.; Zhao, L.; Yang, L. Serum Vitamin A Nutritional Status of Children and Adolescents Aged 6-17 Years—China, 2016-2017. *China CDC Wkly.* **2021**, *3*, 189–192. [CrossRef] [PubMed]

43. Wang, S.; Shen, G.; Jiang, S.; Xu, H.; Li, M.; Wang, Z.; Zhang, S.; Yu, Y. Nutrient Status of Vitamin D among Chinese Children. *Nutrients* **2017**, *9*, 319. [CrossRef]

44. Wang, L.L.; Wang, H.Y.; Wen, H.K.; Tao, H.Q.; Zhao, X.W. Vitamin D status among infants, children, and adolescents in southeastern China. *J. Zhejiang Univ. Sci. B* **2016**, *17*, 545–552. [CrossRef] [PubMed]

45. Wu, J.; Hu, Y.; Li, M.; Chen, J.; Mao, D.; Li, W.; Wang, R.; Yang, Y.; Piao, J.; Yang, L.; et al. Prevalence of Anemia in Chinese Children and Adolescents and Its Associated Factors. *Int. J. Environ. Res. Public Health* **2019**, *16*, 1416. [CrossRef] [PubMed]

46. Xu, H.; Li, Y.; Liu, A.; Zhang, Q.; Hu, X.; Fang, H.; Li, T.; Guo, H.; Li, Y.; Xu, G.; et al. Prevalence of the metabolic syndrome among children from six cities of China. *BMC Public Health* **2012**, *12*, 13. [CrossRef] [PubMed]
47. Dai, C.; Wang, C.; Xia, F.; Liu, Z.; Mo, Y.; Shan, X.; Zhou, Y. Age and Gender-Specific Reference Intervals for Uric Acid Level in Children Aged 5-14 Years in Southeast Zhejiang Province of China: Hyperuricemia in Children May Need Redefinition. *Front. Pediatrics* **2021**, *9*, 560720. [CrossRef] [PubMed]
48. Yang, C.; Chen, J.; Liu, Z.; Yun, C.; Li, Y.; Piao, J.; Yang, X. Association of Vitamin A Status with Overnutrition in Children and Adolescents. *Int. J. Environ. Res. Public Health* **2015**, *12*, 15531–15539. [CrossRef]
49. Tanumihardjo, S.A. Vitamin A: Biomarkers of nutrition for development. *Am. J. Clin. Nutr.* **2011**, *94*, 658S–665S. [CrossRef]
50. Zhong, M.; Kawaguchi, R.; Kassai, M.; Sun, H. Retina, retinol, retinal and the natural history of vitamin A as a light sensor. *Nutrients* **2012**, *4*, 2069–2096. [CrossRef] [PubMed]
51. Brett, N.R.; Lavery, P.; Agellon, S.; Vanstone, C.A.; Goruk, S.; Field, C.J.; Weiler, H.A. Vitamin D Status and Immune Health Outcomes in a Cross-Sectional Study and a Randomized Trial of Healthy Young Children. *Nutrients* **2018**, *10*, 680. [CrossRef] [PubMed]
52. Wimalawansa, S.J. Associations of vitamin D with insulin resistance, obesity, type 2 diabetes, and metabolic syndrome. *J. Steroid Biochem. Mol. Biol.* **2018**, *175*, 177–189. [CrossRef] [PubMed]
53. Kelishadi, R.; Farajzadegan, Z.; Bahreynian, M. Association between vitamin D status and lipid profile in children and adolescents: A systematic review and meta-analysis. *Int. J. Food Sci. Nutr.* **2014**, *65*, 404–410. [CrossRef]
54. Zittermann, A.; Gummert, J.F.; Börgermann, J. The role of vitamin D in dyslipidemia and cardiovascular disease. *Curr. Pharm. Des.* **2011**, *17*, 933–942. [CrossRef] [PubMed]
55. Yang, C.; Mao, M.; Ping, L.; Yu, D. Prevalence of vitamin D deficiency and insufficiency among 460,537 children in 825 hospitals from 18 provinces in mainland China. *Medicine* **2020**, *99*, e22463. [CrossRef]
56. Li, H.; Huang, T.; Xiao, P.; Zhao, X.; Liu, J.; Cheng, H.; Dong, H.; Morris, H.A.; Mi, J.; China Child and Adolescent Cardiovascular Health (CCACH) Collaboration Group. Widespread vitamin D deficiency and its sex-specific association with adiposity in Chinese children and adolescents. *Nutrition* **2020**, *71*, 110646. [CrossRef]
57. Borrud, L.G.; Flegal, K.M.; Looker, A.C.; Everhart, J.E.; Harris, T.B.; Shepherd, J.A. *Body Composition Data for Individuals 8 Years of Age and Older: U.S. Population, 1999–2004*; Vital and Health Statistics Series 11; Data from the National Health Survey; U.S. Department of Health and Human Services: Washington, DC, USA, 2010; pp. 1–87.
58. Hazell, T.J.; DeGuire, J.R.; Weiler, H.A. Vitamin D: An overview of its role in skeletal muscle physiology in children and adolescents. *Nutr. Rev.* **2012**, *70*, 520–533. [CrossRef]
59. Alfredsson, L.; Armstrong, B.K.; Butterfield, D.A.; Chowdhury, R.; de Gruijl, F.R.; Feelisch, M.; Garland, C.F.; Hart, P.H.; Hoel, D.G.; Jacobsen, R.; et al. Insufficient Sun Exposure Has Become a Real Public Health Problem. *Int. J. Environ. Res. Public Health* **2020**, *17*, 5014. [CrossRef]
60. Wong, A.Y.; Chan, E.W.; Chui, C.S.; Sutcliffe, A.G.; Wong, I.C. The phenomenon of micronutrient deficiency among children in China: A systematic review of the literature. *Public Health Nutr.* **2014**, *17*, 2605–2618. [CrossRef]
61. Ma, G.; Li, Y.; Jin, Y.; Zhai, F.; Kok, F.J.; Yang, X. Phytate intake and molar ratios of phytate to zinc, iron and calcium in the diets of people in China. *Eur. J. Clin. Nutr.* **2007**, *61*, 368–374. [CrossRef]
62. Zhang, N.; Bécares, L.; Chandola, T. Patterns and Determinants of Double-Burden of Malnutrition among Rural Children: Evidence from China. *PLoS ONE* **2016**, *11*, e0158119. [CrossRef]
63. Ning, M.; Chang, H.H. Migration decisions of parents and the nutrition intakes of children left at home in rural China. *Agric. Econ.* **2013**, *59*, 467–477. [CrossRef]
64. Yang, W.; Li, X.; Li, Y.; Zhang, S.; Liu, L.; Wang, X.; Li, W. Anemia, malnutrition and their correlations with socio-demographic characteristics and feeding practices among infants aged 0-18 months in rural areas of Shaanxi province in northwestern China: A cross-sectional study. *BMC Public Health* **2012**, *12*, 1127. [CrossRef] [PubMed]
65. Jiang, Y.; Su, X.; Wang, C.; Zhang, L.; Zhang, X.; Wang, L.; Cui, Y. Prevalence and risk factors for stunting and severe stunting among children under three years old in mid-western rural areas of China. *Child Care Health Dev.* **2015**, *41*, 45–51. [CrossRef]
66. Smith, J.D.; Fu, E.; Kobayashi, M.A. Prevention and Management of Childhood Obesity and Its Psychological and Health Comorbidities. *Annu. Rev. Clin. Psychol.* **2020**, *16*, 351–378. [CrossRef] [PubMed]
67. Chen, P.; Wang, D.; Shen, H.; Yu, L.; Gao, Q.; Mao, L.; Jiang, F.; Luo, Y.; Xie, M.; Zhang, Y.; et al. Physical activity and health in Chinese children and adolescents: Expert consensus statement (2020). *Br. J. Sports Med.* **2020**, *54*, 1321–1331. [CrossRef] [PubMed]
68. Chen, J.L.; Esquivel, J.H.; Guo, J.; Chesla, C.A.; Tang, S. Risk factors for obesity in preschool-aged children in China. *Int. Nurs. Rev.* **2018**, *65*, 217–224. [CrossRef] [PubMed]
69. Gan, X.; Xu, W.; Yu, K. Economic Growth and Weight of Children and Adolescents in Urban Areas: A Panel Data Analysis on Twenty-Seven Provinces in China, 1985-2014. *Child. Obes.* **2020**, *16*, 86–93. [CrossRef] [PubMed]

Article

Insufficient Fruit and Vegetable Intake and Low Potassium Intake Aggravate Early Renal Damage in Children: A Longitudinal Study

Menglong Li [1], Nubiya Amaerjiang [1], Ziang Li [1], Huidi Xiao [1], Jiawulan Zunong [1], Lifang Gao [1], Sten H. Vermund [2] and Yifei Hu [1,*]

[1]   Department of Child, Adolescent Health and Maternal Care, School of Public Health, Capital Medical University, Beijing 100069, China; limenglong_ph@163.com (M.L.); 13693617970@163.com (N.A.); 15622764530@163.com (Z.L.); xhd19988023@163.com (H.X.); jiawulan@foxmail.com (J.Z.); lifanggao@ccmu.edu.cn (L.G.)
[2]   Yale School of Public Health, Yale University, New Haven, CT 06510-3201, USA; sten.vermund@yale.edu
*   Correspondence: huyifei@yahoo.com or huyifei@ccmu.edu.cn; Tel.: +86-10-83911747

Abstract: Insufficient fruit and vegetable intake (FVI) and low potassium intake are associated with many non-communicable diseases, but the association with early renal damage in children is uncertain. We aimed to identify the associations of early renal damage with insufficient FVI and daily potassium intake in a general pediatric population. We conducted four waves of urine assays based on our child cohort (PROC) study from October 2018 to November 2019 in Beijing, China. We investigated FVI and other lifestyle status via questionnaire surveys and measured urinary potassium, β_2-microglobulin (β_2-MG), and microalbumin (MA) excretion to assess daily potassium intake and renal damage among 1914 primary school children. The prevalence of insufficient FVI (<4/d) was 48.6% (95% CI: 46.4%, 50.9%) and the estimated potassium intake at baseline was 1.63 ± 0.48 g/d. Short sleep duration, long screen time, lower estimated potassium intake, higher β_2-MG and MA excretion were significantly more frequent in the insufficient FVI group. We generated linear mixed effects models and observed the bivariate associations of urinary β_2-MG and MA excretion with insufficient FVI ($\beta = 0.012$, 95% CI: 0.005, 0.020; $\beta = 0.717$, 95% CI: 0.075, 1.359), and estimated potassium intake ($\beta = -0.042$, 95% CI: -0.052, -0.033; $\beta = -1.778$, 95% CI: -2.600, -0.956), respectively; after adjusting for age, sex, BMI, SBP, sleep duration, screen time and physical activity. In multivariate models, we observed that urinary β_2-MG excretion increased with insufficient FVI ($\beta = 0.011$, 95% CI: 0.004, 0.018) and insufficient potassium intake (<1.5 g/d) ($\beta = 0.031$, 95% CI: 0.023, 0.038); and urinary MA excretion increased with insufficient FVI ($\beta = 0.658$, 95% CI: 0.017, 1.299) and insufficient potassium intake ($\beta = 1.185$, 95% CI: 0.492, 1.878). We visualized different quartiles of potassium intake showing different renal damage with insufficient FVI for interpretation and validation of the findings. Insufficient FVI and low potassium intake aggravate early renal damage in children and underscores that healthy lifestyles, especially adequate FVI, should be advocated.

Keywords: fruit and vegetable intake; potassium intake; renal damage; children; China

Citation: Li, M.; Amaerjiang, N.; Li, Z.; Xiao, H.; Zunong, J.; Gao, L.; Vermund, S.H.; Hu, Y. Insufficient Fruit and Vegetable Intake and Low Potassium Intake Aggravate Early Renal Damage in Children: A Longitudinal Study. *Nutrients* **2022**, *14*, 1228. https://doi.org/10.3390/nu14061228

Academic Editor: Lutz Schomburg

Received: 18 February 2022
Accepted: 12 March 2022
Published: 14 March 2022

Publisher's Note: MDPI stays neutral with regard to jurisdictional claims in published maps and institutional affiliations.

Copyright: © 2022 by the authors. Licensee MDPI, Basel, Switzerland. This article is an open access article distributed under the terms and conditions of the Creative Commons Attribution (CC BY) license (https://creativecommons.org/licenses/by/4.0/).

1. Introduction

Insufficient fruit and vegetable intake (FVI) is associated with almost all major non-communicable diseases [1], including emerging chronic kidney disease (CKD) in the general population [2]. Adequate FVI plays an important role in ensuring good development in children and adolescents, reducing risk of obesity and micronutrient deficiencies and improving health status in adulthood [3,4]. The unmet recommended standard of FVI in terms of composition and frequency among children ages 6–17 years in China has gained attention in efforts to optimize childhood diets [5,6].

Dietary intake affects the acid-base balance in the human body. Plant-based foods including fruits and vegetables are rich in mineral cations (such as potassium [7]) and bicarbonate precursors with alkalizing effects [8]. Insufficient FVI is associated with increased acidity and kidney function declines in CKD patients [8]. High consumption of fruits and vegetables may have a buffering effect on metabolic acid load [9]. This is postulated as having a potentially protective role in avoiding renal damage. One study showed that increased FVI intake for 8 weeks was associated with a decrease in net endogenous acid production (NEAP) [10].

Potassium is a major factor in NEAP, and estimated NEAP is negatively correlated with potassium intake [8]. Studies among adults have shown that higher dietary acid load is associated with CKD, and higher intake of potassium is negatively associated with CKD [11]. Potassium is freely filtered at the glomerulus, with approximately 60–80% of the filtered potassium reabsorbed in the proximal tubule and 10% excreted in the distal tubule [12]. Increased β_2-microglobulin (β_2-MG) and microalbumin (MA) in the urine are markers of early renal damage [13,14], and are used to assess a worse status of renal proximal tubular and glomerular function, respectively [15]. A prospective, population-based cohort study conducted in adults with normal renal function showed that low urinary potassium excretion was associated with an increased risk of CKD [16] and another young adult study suggested that higher potassium intake may help prevent renal damage [17].

The effects of FVI and potassium intake on renal damage have not been studied among school-aged healthy children, and we sought to determine their associations using longitudinal data. We hypothesized that insufficient FVI and lower potassium intake may aggravate early renal damage in children.

2. Materials and Methods

2.1. Study Design and Participants

This study was nested within the PROC study conducted in Beijing enrolling 1914 children aged 6-8 years in six non-boarding primary schools (detailed elsewhere [18]). We conducted 4 waves of repeated urine assays for the same cohort. The single baseline urine was obtained in October–November 2018, and three subsequent urine assays were obtained from each child within a one-week span in November 2019 at the one-year follow-up visit. All 1914 participants provided at least one urine sample and were included in this study. Figure 1 illustrates the procedures and the number of children in each frequency category of urine collection.

2.2. Anthropometric Measurements

Anthropometric measurements were conducted by trained staff from October to November 2018 for the baseline survey and the anthropometric indicators in November 2019 were estimated using the measurements of the baseline and the following visit in September 2020. In short, the standing height was the average of two measurements and rounded to 0.1 cm. The weight of participants was measured in light clothes and rounded to 0.1 kg (detailed elsewhere [18]). Body mass index (BMI) was calculated as weight in kilograms divided by height in meter squared (kg/m^2).

2.3. Blood Pressure Measurements

We performed three blood pressure measurements on the same day in June 2019 to estimate mean blood pressure by using an electronic sphygmomanometer (OMRON HBP-1300, Dalian, China). Systolic blood pressure (SBP) and diastolic blood pressure (DBP) were calculated as the average of the last two measurements.

2.4. Urine Collection and Measurements

Urine collection was conducted in four waves. The first wave was a baseline fasting urine assay in the morning in 2018. Wave 2 of 24 h urine samples were collected from Sunday morning to the following Monday morning, and waves 3–4 including two more fasting urine samples were collected on Wednesday and Friday morning within 1 week in November 2019. Urinalysis (including glucose [GLU], protein [PRO], bilirubin [BIL], urobilinogen [URO], pH, specific gravity [SG], blood [BLD], ketones [KET], nitrites [NIT], leukocytes [LEU], color tone, and turbidity) of wave 1 were performed using an automatic clinical chemistry analyzer Arkray AUTION MAX AX-4030 (Osaka, Japan) with AUTION Sticks. Urinalysis in wave 2 was performed by three instruments: URIT-500B (Guangxi, China), Dirui H-800 (Guangdong, China) and Combi Scan 500 (Lichtenfels, Germany), with corresponding urinalysis sticks to cope with many samples needing to be studied in a short time frame. Urine chemical tests (sodium, potassium, β_2-MG and MA) of wave 1 and (sodium, potassium, creatinine, uric acid, β_2-MG and MA) of waves 2–4 were performed using the automatic clinical chemistry analyzer Beckman Coulter AU5800 and AU680 (Osaka, Japan), respectively.

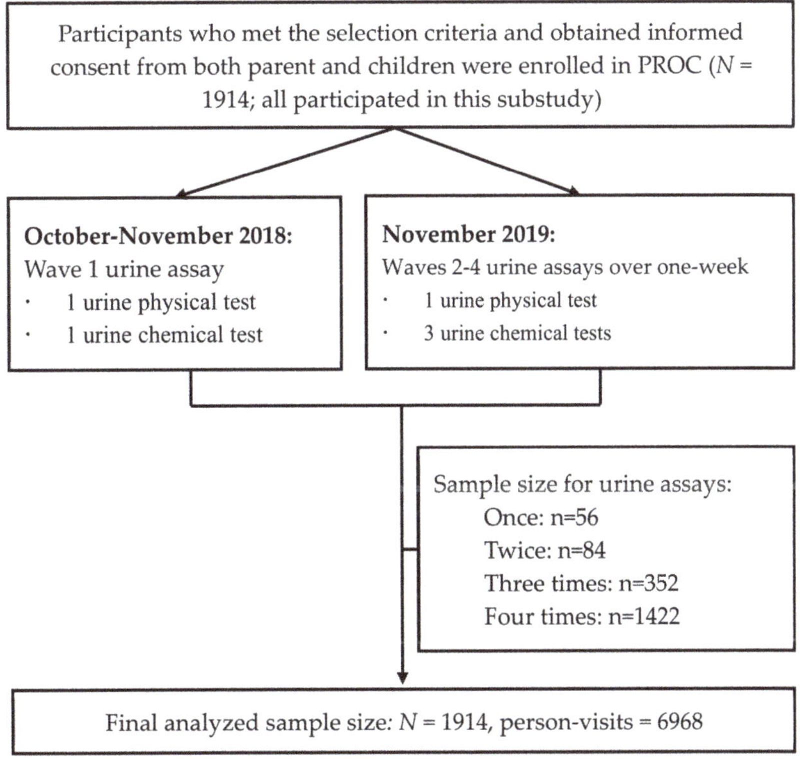

Figure 1. Flowchart of the procedure and frequency of urine collection for the study.

Based on spot urinary potassium excretion, we calculated the estimated 24 h urinary potassium excretion [19] and the estimated potassium intake [20]:

$$24\text{hUK} = 39.1 \times 5.2 \times \left[\frac{0.1 \times \text{SpotUK}}{\text{SpotUCr}} \left(-4.72 \times \text{Age} + 7.58 \times \text{Weight} + 6.09 \times \text{Height} - 64.50 \right) \right]^{0.5}$$

$$24\text{hKIntake} = 1.3 \times 24\text{hUK}/1000 \ (\text{g/d})$$

We defined the abnormal cutoffs value as <1.5 g/d [21] for insufficient potassium intake, >0.2 mg/L for elevated β_2-MG and tubular damage [22] and ≥20 mg/L for elevated MA and glomerular damage [23].

2.5. Lifestyle Covariates and Definition

FVI, sleep duration, screen time, and physical activity were reported by parents. FVI was assessed via the 16-item Mediterranean Diet Quality Index in children and adolescents (KIDMED) [24] and grouped as sufficient ($\geq$4/d) and insufficient (<4/d). Short sleep was defined as sleep duration <10 h/d; long screen time was defined as computer/cell phone screen time $\geq$2 h/d; insufficient physical activity was defined as <1 h/d.

2.6. Statistical Analysis

The main outcome indicators were the longitudinal urinary β_2-MG and MA concentrations. Descriptive statistics are presented according to FVI status. Categorical variables are presented as counts and percentages. Continuous variables are described as the mean $\pm$ standard deviation (SD) or as median and interquartile range (IQR). Multiple imputations were performed for variables with missing values; thus, 50 complete datasets were obtained. Independent *t*-tests, the Mann–Whitney U test and the χ^2 test were performed to compare the difference between sufficient and insufficient FVI groups. Linear mixed effects models were generated to determine the associations and coefficients with 95% confidence interval (95% CI) of renal damage indicators with FVI and urinary indicators, while the weekday and intra-wave of the urine assays were included as random effects. Sankey diagrams were used to visualize the associations of elevated β_2-MG and MA with FVI status and potassium intake quartiles. The results in Table 1 are based on the first imputed dataset. We generated Table 2; Table 3 via valid statistical inferences of the parameters based on 50 datasets using PROC MIANALYZE. A two-tailed *p* value of 0.05 was used to determine statistical significance. All data were analyzed using Statistical Analysis System V.9.4 (SAS Institute Inc., Cary, NC, USA).

Table 1. Descriptive characteristics of 6–9 year old children categorized by sufficiency of fruit and vegetable intake or not in Beijing, China (*N* = 1914).

Factors	Total	Sufficient FVI ($\geq$4/d)	Insufficient FVI (<4/d)	*p*
Sex [1]				0.018
Boy [n (%)]	956 (50.0)	465 (47.3)	491 (52.7)	
Girl [n (%)]	958 (50.0)	518 (52.7)	440 (47.3)	
Age (year) [2]	6.6 $\pm$ 0.3	6.6 $\pm$ 0.3	6.6 $\pm$ 0.3	0.09
Height (cm) [2]	122.5 $\pm$ 5.3	122.3 $\pm$ 5.4	122.6 $\pm$ 5.3	0.25
Weight (kg) [2]	24.8 $\pm$ 5.9	24.7 $\pm$ 5.9	24.8 $\pm$ 5.9	0.72
BMI (kg/m^2) [2]	16.4 $\pm$ 2.9	16.4 $\pm$ 2.9	16.4 $\pm$ 2.9	0.94
SBP (mmHg) [2]	101 $\pm$ 8	101 $\pm$ 8	101 $\pm$ 9	0.27
DBP (mmHg) [2]	56 $\pm$ 6	56 $\pm$ 6	56 $\pm$ 6	0.93
Short sleep (<10 h/d) [1]	1441 (75.3)	706 (71.8)	735 (78.9)	<0.001
Long screen time ($\geq$2 h/d) [1]	95 (5.0)	37 (3.8)	58 (6.2)	0.013
Insufficient physical activity (<1 h/d) [1]	1451 (75.8)	739 (75.2)	712 (76.5)	0.51
Spot urinary potassium excretion (mmol/L) [2]				
Wave 1	27.36 $\pm$ 12.53	27.41 $\pm$ 12.54	27.30 $\pm$ 12.53	0.85
Wave 2	30.42 $\pm$ 12.30	30.34 $\pm$ 12.47	30.50 $\pm$ 12.12	0.77
Wave 3	26.59 $\pm$ 12.86	26.82 $\pm$ 13.00	26.36 $\pm$ 12.74	0.45
Wave 4	28.00 $\pm$ 13.56	28.26 $\pm$ 13.45	27.73 $\pm$ 13.68	0.46
Estimated 24 h urinary potassium excretion (mg/d) [2]				
Wave 1	1253.5 $\pm$ 370.3	1272.4 $\pm$ 371.1	1233.6 $\pm$ 368.7	0.022
Wave 2	1363.1 $\pm$ 260.2	1383.4 $\pm$ 267.0	1342.4 $\pm$ 251.5	<0.001
Wave 3	1116.7 $\pm$ 276.5	1134.8 $\pm$ 270.8	1097.9 $\pm$ 281.3	0.005
Wave 4	1114.4 $\pm$ 266.6	1131.0 $\pm$ 263.3	1097.3 $\pm$ 269.1	0.017
Estimated 24 h potassium intake (g/d) [2]				
Wave 1	1.63 $\pm$ 0.48	1.65 $\pm$ 0.48	1.60 $\pm$ 0.48	0.022

Table 1. *Cont.*

Factors	Total	Sufficient FVI ($\geq$4/d)	Insufficient FVI (<4/d)	p
Wave 2	1.77 $\pm$ 0.34	1.80 $\pm$ 0.35	1.75 $\pm$ 0.33	<0.001
Wave 3	1.45 $\pm$ 0.36	1.48 $\pm$ 0.35	1.43 $\pm$ 0.37	0.005
Wave 4	1.45 $\pm$ 0.35	1.47 $\pm$ 0.34	1.43 $\pm$ 0.35	0.016
Spot urinary β_2-MG excretion (mg/L) [3]				
Wave 1	0.08 (0.04–0.13)	0.07 (0.04–0.12)	0.08 (0.05–0.13)	0.011
Wave 2	0.15 (0.12–0.18)	0.14 (0.12–0.18)	0.15 (0.12–0.19)	0.066
Wave 3	0.16 (0.13–0.21)	0.16 (0.13–0.20)	0.16 (0.13–0.21)	0.19
Wave 4	0.16 (0.13–0.21)	0.16 (0.13–0.21)	0.16 (0.13–0.21)	0.52
Spot urinary MA excretion (mg/L) [3]				
Wave 1	9.13 (6.45–12.56)	8.85 (6.36–12.18)	9.28 (6.53–12.96)	0.019
Wave 2	6.70 (6.10–8.50)	6.60 (6.10–8.30)	6.70 (6.10–8.90)	0.027
Wave 3	7.00 (6.20–9.10)	6.90 (6.20–9.00)	7.00 (6.20–9.40)	0.35
Wave 4	6.80 (6.00–8.90)	6.70 (6.00–8.80)	6.85 (6.00–9.10)	0.16

(FVI: fruit and vegetable intake, BMI: body mass index, SBP: systolic blood pressure, DBP: diastolic blood pressure, β_2-MG: β_2-microglobulin, MA: microalbumin.). [1] Comparison by FVI status using χ^2 test. [2] Mean and standard deviation (SD) compared by FVI status using independent t-test. [3] Median and interquartile ranges (IQR) compared by FVI status using the Mann–Whitney U test.

Table 2. Bivariate associations of renal damage indicators, FVI and potassium indicators using linear mixed effects models among 6–9 year old children, Beijing, China.

Dependent Variable	Independent Variable	Model 1 Estimate (95%CI)	p	Model 2 Estimate (95%CI)	p	Model 3 Estimate (95%CI)	p
Estimated pota-ssium intake	Insufficient FVI	−0.049 (−0.071, −0.027)	<0.001	−0.049 (−0.071, −0.027)	<0.001	−0.050 (−0.072, −0.027)	<0.001
	Spot urinary potassium	0.013 (0.012, 0.014)	<0.001	0.013 (0.012, 0.014)	<0.001	0.013 (0.012, 0.014)	<0.001
β_2-MG	Insufficient FVI	0.011 (0.004, 0.018)	0.003	0.011 (0.004, 0.018)	0.004	0.012 (0.005, 0.020)	<0.001
	Estimated potassium intake	−0.043 (−0.052, −0.034)	<0.001	−0.042 (−0.052, −0.033)	<0.001	−0.042 (−0.052, −0.033)	<0.001
MA	Insufficient FVI	0.616 (−0.020, 1.253)	0.058	0.669 (0.034, 1.304)	0.039	0.717 (0.075, 1.359)	0.029
	Estimated potassium intake	−1.837 (−2.656, −1.019)	<0.001	−1.753 (−2.574, −0.933)	<0.001	−1.778 (−2.600, −0.956)	<0.001

Model 1: unadjusted; Model 2: adjusting for age, sex, BMI; Model 3: adjusting for age, sex, BMI, SBP, sleep duration, screen time and physical activity. (FVI: fruit and vegetable intake, BMI: body mass index, SBP: systolic blood pressure, β_2-MG: β_2-microglobulin, MA: microalbumin. All models included two random effects: the weekday and wave of the urine assay.).

Table 3. Multivariable associations of renal damage indicators with insufficient FVI and insufficient potassium intake using linear mixed effects models among 6–9 year old children, Beijing, China.

Dependent Variable	Independent Variable	Model 1 Estimate (95%CI)	p	Model 2 Estimate (95%CI)	p	Model 3 Estimate (95%CI)	p
β_2-MG	Intercept	0.142 (0.104, 0.180)	<0.001	0.090 (−0.004, 0.184)	0.061	0.043 (−0.058, 0.145)	0.40
	Insufficient FVI	0.009 (0.002, 0.017)	0.010	0.009 (0.002, 0.017)	0.011	0.011 (0.004, 0.018)	0.003
	Insufficient potassium intake	0.031 (0.023, 0.039)	<0.001	0.031 (0.023, 0.038)	<0.001	0.031 (0.023, 0.038)	<0.001
MA	Intercept	8.940 (7.911, 9.969)	<0.001	6.614 (−0.662, 13.890)	0.075	2.228 (−5.736, 10.193)	0.58
	Insufficient FVI	0.558 (−0.077, 1.193)	0.085	0.612 (−0.022, 1.246)	0.059	0.658 (0.017, 1.299)	0.044
	Insufficient potassium intake	1.255 (0.563, 1.947)	<0.001	1.171 (0.479, 1.863)	<0.001	1.185 (0.492, 1.878)	<0.001

Model 1: unadjusted; Model 2: adjusting for age, sex, BMI; Model 3: adjusting for age, sex, BMI, SBP, sleep duration, screen time and physical activity. FVI: fruit and vegetable intake, BMI: body mass index, SBP: systolic blood pressure, β_2-MG: β_2-microglobulin, MA: microalbumin. All models included two random effects: the weekday and wave of the urine assay.

3. Results

3.1. Sociodemographic Characteristics

A total of 1914 children aged 6.6 ± 0.3 years old at the baseline were enrolled in this study. The prevalence of insufficient FVI is 48.6% (95% CI: 46.4%, 50.9%), and more common in boys (51.4%) than girls (45.9%). In the insufficient FVI group, short sleep and long screen time were significantly more prevalent, and urinary β_2-MG and MA excretion of wave 1 were significantly higher than that in the sufficient FVI group. In wave 1–4, estimated 24 h urinary potassium excretion and estimated potassium intake were significantly lower in the insufficient FVI group. There were no significant differences in height, weight, BMI, SBP, DBP, prevalence of insufficient physical activity, and spot urinary potassium excretion between the two FVI groups (Table 1).

3.2. Binary Regression of Renal Damage Indicators, FVI, and Potassium Indicators

The unadjusted model 1 and adjusting for age, sex, and BMI model 2 suggested that estimated potassium intake was associated with insufficient FVI and urinary potassium excretion, urinary β_2-MG and MA excretion were associated with insufficient FVI and estimated potassium intake, respectively. The coefficients of these associations remained stable and significant (all $p \leq 0.05$) after adjusting for age, sex, BMI, SBP, sleep duration, screen time and physical activity (model 3). The results in model 3 showed that higher estimated potassium intake was negatively associated with insufficient FVI ($\beta = -0.050$, 95% CI: -0.072, -0.027), and positively associated with higher urinary potassium excretion ($\beta = 0.013$, 95% CI: 0.012, 0.014); higher urinary β_2-MG excretion was positively associated with insufficient FVI ($\beta = 0.012$, 95% CI: 0.005, 0.020), and negatively associated with higher estimated potassium intake ($\beta = -0.042$, 95%CI: -0.052, -0.033); higher urinary MA excretion was positively associated with insufficient FVI ($\beta = 0.717$, 95% CI: 0.075, 1.359), and negatively associated with higher estimated potassium intake ($\beta = -1.778$, 95% CI: -2.600, -0.956) (Table 2).

3.3. Multivariable Regression of Renal Damage Indicators with Insufficient FVI and Insufficient Potassium Intake

The multivariable analysis (unadjusted model 1 and adjusting for age, sex, and BMI model 2) showed that urinary β_2-MG and MA excretion increased with insufficient FVI and insufficient potassium intake. After adjusting for age, sex, BMI, SBP, sleep duration, screen time and physical activity (model 3), the urinary β_2-MG excretion increased with insufficient FVI ($\beta = 0.011$, 95% CI: 0.004, 0.018) and insufficient potassium intake ($\beta = 0.031$, 95% CI: 0.023, 0.038); urinary MA excretion increased with insufficient FVI ($\beta = 0.658$, 95% CI: 0.017, 1.299) and insufficient potassium intake ($\beta = 1.185$, 95% CI: 0.492, 1.878) (Table 3).

3.4. Visualization of Early Renal Damage with Insufficient FVI and Quantiles of Potassium Intake

Tracking the proportion of children with elevated β_2-MG (Figure 2a) or elevated MA (Figure 2b), they were more likely to appear in the lower quantile of estimated potassium intake. Lower potassium intake contributed a larger proportion of both indicators of renal damage.

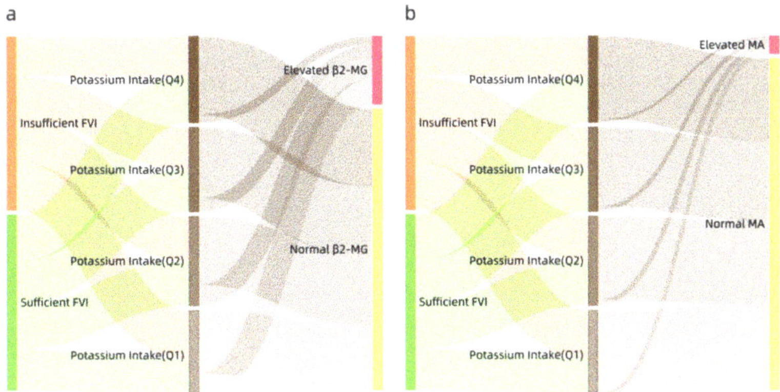

Figure 2. Sankey diagrams of renal damage with estimated potassium intake; elevated β_2-MG (**a**) or elevated MA (**b**) urinary excretion were more likely to originate from lower quantile of potassium intake.

4. Discussion

This longitudinal study with large sample size assessed the associations between FVI, estimated potassium intake and early renal damage among a general population of school-aged children in urban China. We found that children with insufficient FVI and low potassium intake were more likely to have increased urinary β_2-MG and MA excretion, after controlling for potential sociodemographic and lifestyle confounders, including age, sex, BMI, SBP, sleep duration, screen time and physical activity. These findings highlight the potential importance of adequate FVI intake and potassium intake to prevent early renal damage in children.

The prevalence of insufficient FVI (<4/d) was 48.6% (95% CI: 46.4–50.9%), which was lower than previous surveys, i.e., 58.0% of urban Chinese children in 1982 and 72.7% in 2012 (both reported prevalence corresponds to the WHO recommendation of 80%, i.e., <4 servings/d instead of 5 servings/d) [6]. The difference in prevalence may be due to many factors, including economic status [25], time period differences, and methodologies. For instance, we used KIDMED surveys that focus on frequency of fruit and vegetable consumption each week rather than quantities of consumption. Even so, we could observe significant differences of estimated 24 h urinary potassium excretion and potassium intake in the insufficient FVI group. Furthermore, the longitudinal negative association we observed between estimated potassium intake and insufficient FVI validated such findings. We also found children with insufficient FVI are more likely to have short sleep [26] and long screen time [27]. The clustering of healthy lifestyle features suggests that a set of needed interventions should include increased FVI alongside shorter screen time and enough sleep to prevent early renal damage [27–30].

In both the unadjusted analysis and those that adjusted for age, sex, BMI, SBP, sleep duration, screen time, and physical activity, we observed consistent longitudinal positive associations between insufficient FVI (<4/d) and urinary β_2-MG excretion, but not urinary MA excretion. This finding suggests that insufficient FVI leads to earlier renal tubular stress than glomerular stress, and evidence suggests tubular damage precede glomerular damage [31]. Clinical management practices in CKD patients with glomerular or tubular abnormalities include dietary intervention with base-producing foods such as fruits and vegetables to increase potassium absorption [9] and reduce metabolic acidosis [32]. We observed a longitudinal negative association between urinary β_2-MG or MA excretion and estimated potassium intake. Many studies have evaluated the effect of potassium intake on renal damage, and most results reached a consensus that insufficient potassium intake was associated with an increased risk of CKD in different populations [11,16,17], while our result provided the same evidence in the general pediatric population.

We found urinary β_2-MG and MA excretion increased with insufficient FVI (<4/d) and insufficient potassium intake (<1.5 g/d) in multivariable models, adjusting for age, sex, BMI, SBP, sleep duration, screen time, and physical activity. This finding suggests a consistent synergistic effect of FVI and potassium intake on early renal damage, i.e., at least 4/d FVI and 1.5 g/d potassium intake may mitigate or prevent glomerular or tubular damage. Our data support a conclusion that school-aged children should have no less than 4 kinds of FVI daily, even independent of the quantity of fruit and vegetable consumption. Moreover, the Sankey diagrams show that children with renal tubular or glomerular damage more likely originate from the lower quantile of the potassium intake group. The comprehensive analysis of the association between FVI, potassium intake and early renal damage leads to the robust conclusion that insufficient FVI and low potassium intake are associated with early renal damage and have a synergistic effect.

The major strength of this study was its use of multiple measures and longitudinal data from a population-based healthy children cohort in China. The imputation method for estimating missing data can maximize the use of information and avoid reporting bias. We selected appropriate linear mixed effects models and covariates to adjust for associations between renal outcomes and independent variables, especially SBP and lifestyle factors such as sleep duration, screen time, and physical activity. Therefore, our conclusions that insufficient FVI and low potassium intake aggravate early renal damage can be placed into a more relevant public health context. This study was limited by not considering other renal function indicators. Urine β_2-MG, MA, and potassium were tested via different machines due to the limited capacity of each testing site, but these effects were minimized by longitudinal data.

5. Conclusions

In conclusion, we present the longitudinal effects of FVI and potassium intake on early renal damage in school-aged children in China. We found that insufficient FVI and low potassium intake aggravate early renal damage in children. Our findings underscore the necessity of advocating more FVI for children to protect renal function from early damage in the context of other healthy lifestyle choices such as enough sleep, less screen time, and more physical activity.

Author Contributions: Y.H. conceptualized and designed the study; M.L., Z.L., H.X. and J.Z. carried out the survey; M.L. performed statistical analysis of the data; N.A. checked the data analysis process; M.L. and N.A. drafted the manuscript; S.H.V., L.G. and Y.H. edited, helped interpret and revise the manuscript. All authors were involved in writing the paper. All authors have read and agreed to the published version of the manuscript.

Funding: This research was funded by the National Natural Science Foundation of China (Y.H., grant No.82073574) and the Beijing Natural Science Foundation (Y.H., grant No.7202009).

Institutional Review Board Statement: The protocol of the child cohort designed to study sensitization, puberty, obesity and cardiovascular risk (PROC) was reviewed and approved by the Ethics Committee of Capital Medical University (No. 2018SY82), applying the guidance of the Declaration of Helsinki, later amendments, and comparable international ethical standards.

Informed Consent Statement: Informed consent was obtained from all participants' parents.

Data Availability Statement: The data that support the findings of this study are not publicly available but are available from the corresponding author on reasonable request.

Acknowledgments: We gratefully acknowledge the staff of the Shunyi District Center for Disease Control and Prevention, the Shunyi District Education Commission, and teachers of six primary school of Shunyi District, Beijing for their support and assistance to the field work. We give special thanks to all the study participants and parents for their contribution.

Conflicts of Interest: The authors declare no conflict of interest related to this work. The funders had no role in the study design; in the collection, analyses, or interpretation of data; in the writing of the manuscript, or in the decision to publish the results.

References

1. Msambichaka, B.; Eze, I.C.; Abdul, R.; Abdulla, S.; Klatser, P.; Tanner, M.; Kaushik, R.; Geubbels, E.; Probst-Hensch, N. Insufficient Fruit and Vegetable Intake in a Low- and Middle-Income Setting: A Population-Based Survey in Semi-Urban Tanzania. *Nutrients* **2018**, *10*, 222. [CrossRef] [PubMed]
2. Banerjee, T.; Carrero, J.J.; McCulloch, C.; Burrows, N.R.; Siegel, K.R.; Morgenstern, H.; Saran, R.; Powe, N.R. Dietary Factors and Prevention: Risk of End-Stage Kidney Disease by Fruit and Vegetable Consumption. *Am. J. Nephrol.* **2021**, *52*, 356–367. [CrossRef] [PubMed]
3. Hodder, R.K.; O'Brien, K.M.; Tzelepis, F.; Wyse, R.J.; Wolfenden, L. Interventions for increasing fruit and vegetable consumption in children aged five years and under. *Cochrane Database Syst. Rev.* **2020**, *5*, Cd008552. [CrossRef] [PubMed]
4. Lin, Y.C.; Fly, A.D. USDA Fresh Fruit and Vegetable Program Is More Effective in Town and Rural Schools Than Those in More Populated Communities. *J. Sch. Health* **2016**, *86*, 769–777. [CrossRef]
5. Xu, P.P.; Hu, X.Q.; Pan, H.; Yang, T.T.; Li, L.; Cao, W.; Gan, Q.; Xu, J.; Zhang, Q. The status of vegetables and fruits consumption of children aged 6 to 17-year-old from 2010 to 2012, China. *Chin. J. Prev. Med.* **2018**, *52*, 552–555. [CrossRef]
6. Xu, J.; Wang, L.-L.; Yang, T.-T.; Xu, P.-P.; Li, L.; Cao, W.; Gan, Q.; Pan, H.; Wang, H.-L.; Hu, X.-Q.; et al. Changes of the dietary intake of vegetable and fruit among Chinese children aged 6–17 in 1982 and 2012. *Chin. J. Dis. Control Prev.* **2021**, *25*, 509–514. [CrossRef]
7. Woodruff, R.C.; Zhao, L.; Ahuja, J.K.C.; Gillespie, C.; Goldman, J.; Harris, D.M.; Jackson, S.L.; Moshfegh, A.; Rhodes, D.; Sebastian, R.S.; et al. Top Food Category Contributors to Sodium and Potassium Intake—United States, 2015–2016. *MMWR Morb. Mortal. Wkly. Rep.* **2020**, *69*, 1064–1069. [CrossRef] [PubMed]
8. Toba, K.; Hosojima, M.; Kabasawa, H.; Kuwahara, S.; Murayama, T.; Yamamoto-Kabasawa, K.; Kaseda, R.; Wada, E.; Watanabe, R.; Tanabe, N.; et al. Higher estimated net endogenous acid production with lower intake of fruits and vegetables based on a dietary survey is associated with the progression of chronic kidney disease. *BMC Nephrol.* **2019**, *20*, 421. [CrossRef] [PubMed]
9. Carrero, J.J.; González-Ortiz, A.; Avesani, C.M.; Bakker, S.J.L.; Bellizzi, V.; Chauveau, P.; Clase, C.M.; Cupisti, A.; Espinosa-Cuevas, A.; Molina, P.; et al. Plant-based diets to manage the risks and complications of chronic kidney disease. *Nat. Rev. Nephrol.* **2020**, *16*, 525–542. [CrossRef]
10. Gunn, C.A.; Weber, J.L.; Coad, J.; Kruger, M.C. Increasing fruits and vegetables in midlife women: A feasibility study. *Nutr. Res.* **2013**, *33*, 543–551. [CrossRef]
11. Ko, B.-J.; Chang, Y.; Ryu, S.; Kim, E.M.; Lee, M.Y.; Hyun, Y.Y.; Lee, K.-B. Dietary acid load and chronic kidney disease in elderly adults: Protein and potassium intake. *PLoS ONE* **2017**, *12*, e0185069. [CrossRef] [PubMed]
12. Malnic, G.; Giebisch, G.; Muto, S.; Wang, W.; Bailey, M.A.; Satlin, L.M. Chapter 49—Regulation of K+ Excretion. In *Seldin and Giebisch's the Kidney*, 5th ed.; Alpern, R.J., Moe, O.W., Caplan, M., Eds.; Academic Press: Cambridge, MA, USA, 2013; pp. 1659–1715. [CrossRef]
13. Wu, H.Y.; Huang, J.W.; Peng, Y.S.; Hung, K.Y.; Wu, K.D.; Lai, M.S.; Chien, K.L. Microalbuminuria screening for detecting chronic kidney disease in the general population: A systematic review. *Ren. Fail.* **2013**, *35*, 607–614. [CrossRef] [PubMed]
14. Kadam, P.; Yachha, M.; Srivastava, G.; Pillai, A.; Pandita, A. Urinary beta-2 microglobulin as an early predictive biomarker of acute kidney injury in neonates with perinatal asphyxia. *Eur. J. Pediatr.* **2022**, *181*, 281–286. [CrossRef] [PubMed]
15. Ji, T.T.; Tan, N.; Lu, H.Y.; Xu, X.Y.; Yu, Y.Y. Early renal injury indicators can help evaluate renal injury in patients with chronic hepatitis B with long-term nucleos(t)ide therapy. *World J. Clin. Cases* **2020**, *8*, 6306–6314. [CrossRef] [PubMed]
16. Kieneker, L.M.; Bakker, S.J.; de Boer, R.A.; Navis, G.J.; Gansevoort, R.T.; Joosten, M.M. Low potassium excretion but not high sodium excretion is associated with increased risk of developing chronic kidney disease. *Kidney Int.* **2016**, *90*, 888–896. [CrossRef] [PubMed]
17. Elfassy, T.; Zhang, L.; Raij, L.; Bibbins-Domingo, K.; Lewis, C.E.; Allen, N.B.; Liu, K.J.; Peralta, C.A.; Odden, M.C.; Zeki Al Hazzouri, A. Results of the CARDIA study suggest that higher dietary potassium may be kidney protective. *Kidney Int.* **2020**, *98*, 187–194. [CrossRef]
18. Li, M.; Shu, W.; Zunong, J.; Amaerjiang, N.; Xiao, H.; Li, D.; Vermund, S.H.; Hu, Y. Predictors of non-alcoholic fatty liver disease in children. *Pediatr. Res.* **2021**. [CrossRef]
19. Meng, L.; Ma, L.; Sun, C. The Estimating Modole Building of 24-h Urine Excretion from the Casual Urinary Electrolytic Concentrations Based on Clinical Big Data. *Eur. J. Pediatr.* **2016**, *175*, 1866–1867. [CrossRef]
20. World Health Organization. *Guideline: Potassium Intake for Adults and Children*; World Health Organization: Geneva, Switzerland, 2012.
21. Chinese Society of Nutrition. *Dietary Guidelines for Chinese Residents*, 1st ed.; People's Medical Publishing House: Beijing, China, 2016; p. 336.
22. Barton, K.T.; Kakajiwala, A.; Dietzen, D.J.; Goss, C.W.; Gu, H.; Dharnidharka, V.R. Using the newer Kidney Disease: Improving Global Outcomes criteria, beta-2-microglobulin levels associate with severity of acute kidney injury. *Clin. Kidney J.* **2018**, *11*, 797–802. [CrossRef]

23. Shatat, I.F.; Qanungo, S.; Hudson, S.; Laken, M.A.; Hailpern, S.M. Changes in Urine Microalbumin-to-Creatinine Ratio in Children with Sickle Cell Disease over Time. *Front. Pediatr.* **2016**, *4*, 106. [CrossRef]
24. Serra-Majem, L.; Ribas, L.; Ngo, J.; Ortega, R.M.; García, A.; Pérez-Rodrigo, C.; Aranceta, J. Food, youth and the Mediterranean diet in Spain. Development of KIDMED, Mediterranean Diet Quality Index in children and adolescents. *Public Health Nutr.* **2004**, *7*, 931–935. [CrossRef] [PubMed]
25. Angeles-Agdeppa, I.; Lenighan, Y.M.; Jacquier, E.F.; Toledo, M.B.; Capanzana, M.V. The Impact of Wealth Status on Food Intake Patterns in Filipino School-Aged Children and Adolescents. *Nutrients* **2019**, *11*, 2910. [CrossRef] [PubMed]
26. Pengpid, S.; Peltzer, K. Fruit and Vegetable Consumption is Protective from Short Sleep and Poor Sleep Quality Among University Students from 28 Countries. *Nat. Sci. Sleep* **2020**, *12*, 627–633. [CrossRef] [PubMed]
27. Elsenburg, L.K.; Corpeleijn, E.; van Sluijs, E.M.; Atkin, A.J. Clustering and correlates of multiple health behaviours in 9–10 year old children. *PLoS ONE* **2014**, *9*, e99498. [CrossRef] [PubMed]
28. Dong, B.; Zou, Z.; Song, Y.; Hu, P.; Luo, D.; Wen, B.; Gao, D.; Wang, X.; Yang, Z.; Ma, Y.; et al. Adolescent Health and Healthy China 2030: A Review. *J. Adolesc. Health Off. Publ. Soc. Adolesc. Med.* **2020**, *67*, S24–S31. [CrossRef] [PubMed]
29. Shook, R.P.; Halpin, K.; Carlson, J.A.; Davis, A.; Dean, K.; Papa, A.; Sherman, A.K.; Noel-MacDonnell, J.R.; Summar, S.; Krueger, G.; et al. Adherence With Multiple National Healthy Lifestyle Recommendations in a Large Pediatric Center Electronic Health Record and Reduced Risk of Obesity. *Mayo Clin. Proc.* **2018**, *93*, 1247–1255. [CrossRef] [PubMed]
30. He, L.; Li, X.; Wang, W.; Wang, Y.; Qu, H.; Zhao, Y.; Lin, D. Clustering of multiple lifestyle behaviors among migrant, left-behind and local adolescents in China: A cross-sectional study. *BMC Public Health* **2021**, *21*, 542. [CrossRef]
31. Amaerjiang, N.; Li, M.; Xiao, H.; Zunong, J.; Li, Z.; Huang, D.; Vermund, S.H.; Pérez-Escamilla, R.; Jiang, X.; Hu, Y. Dehydration Status Aggravates Early Renal Impairment in Children: A Longitudinal Study. *Nutrients* **2022**, *14*, 335. [CrossRef]
32. Goraya, N.; Wesson, D.E. Management of the Metabolic Acidosis of Chronic Kidney Disease. *Adv. Chronic Kidney Dis.* **2017**, *24*, 298–304. [CrossRef]

 *nutrients*

Article

Leptin Levels of the Perinatal Period Shape Offspring's Weight Trajectories through the First Year of Age

Francesca Garofoli [1,*], Iolanda Mazzucchelli [2], Micol Angelini [1], Catherine Klersy [3], Virginia Valeria Ferretti [3], Barbara Gardella [4], Giulia Vittoria Carletti [4], Arsenio Spinillo [4], Chryssoula Tzialla [1] and Stefano Ghirardello [1]

1   Neonatal Unit and Neonatal Intensive Care Unit, Fondazione IRCCS Policlinico San Matteo, 27100 Pavia, Italy; m.angelini@smatteo.pv.it (M.A.); chryssoula.tzialla@unipv.it (C.T.); s.ghirardello@smatteo.pv.it (S.G.)
2   Unit of Rheumatology, Department of Internal Medicine and Therapeutics, Università di Pavia, 27100 Pavia, Italy; iolanda.mazzucchelli@unipv.it
3   Clinical Epidemiology and Biostatistics Unit, Fondazione IRCCS Policlinico San Matteo, 27100 Pavia, Italy; klersy@smatteo.pv.it (C.K.); v.ferretti@smatteo.pv.it (V.V.F.)
4   Unit of Obstetrics and Gynaecology, Fondazione IRCCS Policlinico San Matteo, 27100 Pavia, Italy; b.gardella@smatteo.pv.it (B.G.); g.carletti@smatteo.pv.it (G.V.C.); a.spinillo@smatteo.pv.it (A.S.)
*   Correspondence: lab.immunoneo@smatteo.pv.it; Tel.: +39-0382-502-865

Citation: Garofoli, F.; Mazzucchelli, I.; Angelini, M.; Klersy, C.; Ferretti, V.V.; Gardella, B.; Carletti, G.V.; Spinillo, A.; Tzialla, C.; Ghirardello, S. Leptin Levels of the Perinatal Period Shape Offspring's Weight Trajectories through the First Year of Age. *Nutrients* 2022, 14, 1451. https://doi.org/10.3390/nu14071451

Academic Editor: Zhiyong Zou

Received: 23 February 2022
Accepted: 29 March 2022
Published: 30 March 2022

Publisher's Note: MDPI stays neutral with regard to jurisdictional claims in published maps and institutional affiliations.

Copyright: © 2022 by the authors. Licensee MDPI, Basel, Switzerland. This article is an open access article distributed under the terms and conditions of the Creative Commons Attribution (CC BY) license (https://creativecommons.org/licenses/by/4.0/).

Abstract: Background: Leptin is a hormone regulating lifetime energy homeostasis and metabolism and its concentration is important starting from prenatal life. We aimed to investigate the association of perinatal leptin concentrations with growth trajectories during the first year of life. Methods: Prospective, longitudinal study, measuring leptin concentration in maternal plasma before delivery, cord blood (CB), and mature breast milk and correlating their impact on neonate's bodyweight from birth to 1 year of age, in 16 full-term (FT), 16 preterm (PT), and 13 intrauterine growth-restricted (IUGR) neonates. Results: Maternal leptin concentrations were highest in the PT group, followed by IUGR and FT, with no statistical differences among groups ($p = 0.213$). CB leptin concentrations were significantly higher in FT compared with PT and IUGR neonates (PT vs. FT; IUGR vs. FT: $p < 0.001$). Maternal milk leptin concentrations were low, with no difference among groups. Maternal leptin and milk concentrations were negatively associated with all the neonates' weight changes ($p = 0.017$ and $p = 0.006$), while the association with CB leptin was not significant ($p = 0.051$). Considering each subgroup individually, statistical analysis confirmed the previous results in PT and IUGR infants, with the highest value in the PT subgroup. In addition, this group's results negatively correlated with CB leptin ($p = 0.026$) and showed the largest % weight increase. Conclusions: Leptin might play a role in neonatal growth trajectories, characterized by an inverse correlation with maternal plasma and milk. PT infants showed the highest correlation with hormone levels, regardless of source, seeming the most affected group by leptin guidance. Low leptin levels appeared to contribute to critical neonates' ability to recover a correct body weight at 1 year. An eventual non-physiological "catch-up growth" should be monitored, and leptin perinatal levels may be an indicative tool. Further investigations are needed to strengthen the results.

Keywords: leptin; preterm neonate; intrauterine growth restriction (IUGR); growth trajectories; children health prognosis

1. Introduction

The adipose tissue is a dynamic endocrine organ that produces and secretes various adipokines and bioactive peptides, including leptin. Leptin is a hormone (16KD) synthesized by the obese (*ob*) gene, [1] the secreted amount is directly related to the mass of adipocytes. Leptin activates anorexigenic neuropeptide circuits and inhibits orexigenic routes after binding at hypothalamic receptors, decreasing food intake, and increasing energy expenditure.

Leptin stimulates lipid catabolism while constraining lipogenesis at the adipose cells level, through both central and peripheral energy homeostasis control and metabolism balance [2]. In addition, the physiological leptin concentration indicates that the body has sufficient energy stores, thus inhibiting appetite [3].

The role of leptin in the overall metabolism regulation starting at prenatal life has been previously suggested. In particular, the interplay between maternal and fetal leptin concentration might anticipate the aptitude to develop metabolic diseases in adulthood. Furthermore, maternal plasma, cord blood, and breast milk leptin concentrations from birth and during the first years of life seem good predictors of growth outline later in life [4–13].

Maternal leptin secretion rises during gestation due to the growing fetus and fat and body fluid increase. Maternal excessive weight gain during gestation is associated with gestational diabetes, pre-eclampsia, polycystic ovary syndrome, and preterm or growth-restricted newborns [14]. Thus, pregnancy is a critical period for programming children's metabolism in a manner that influences their long-term risk [13,14]. In addition, during early infancy, a rapid weight gain may result in an increased risk for obesity and non-communicable diseases (NCDs) in childhood and beyond [15–17].

Leptin is also produced by the mammary gland but can pass to breast milk from maternal serum, and the significate of maternal leptin has been previously investigated with contrasting results [18,19]. In recent years, maternal milk leptin has been evaluated for its potential role in the postnatal programming of healthy phenotype in adulthood [2].

This study aims to assess the contribution of maternal plasma, cord blood (CB), and breastmilk leptin concentrations to neonates' weight trajectory during the first year of life in different neonatal populations.

2. Methods

2.1. Study Design and Patients

This was a prospective longitudinal observational study. The Bioethical Committee of the Hospital approved the study and written informed consent was obtained from both parents before the neonate's enrollment. The results described herein have never been reported before, nevertheless this cohort was part of a population previously evaluated for cord blood and maternal and neonatal leptin from birth to 3 months of age, from the same authors [20]. This study pertains to fully breastfed infants only, evaluated during the first year of age. The inclusion criterion was breastfeeding until the end of the 3 months of life. We measured maternal plasma leptin one week before delivery, umbilical CB at birth and mature breast milk (one sample in the first two weeks from the delivery). Maternal anthropometric and clinical data and blood samples were collected before delivery; children's data were collected at birth and after 1 year of life.

Forty-five mother-infant dyads exclusively breastfeeding until the end of the third month of life could be enrolled. Dyads consisted of 16 full-term (FT), 16 preterm (PT), and 13 intra-uterine growth-restricted newborns (IUGR) and their mothers. FT babies included healthy newborns, gestational age (GA) $\geq$ 37 weeks, with adequate for gestational age (AGA) bodyweight. Infants born at GA 36 + 6 weeks of gestation or below were included in the PT group; IUGR newborns were born to mothers with altered flussimetric findings during ultrasound imaging and did not reach their genetically determined potential growth.

Exclusion criteria were infants with congenital malformations or infection, genetic syndromes, formula-fed, and infants large for gestational age.

2.2. Procedures

Leptin concentrations in maternal plasma and CB were determined as reported in the previous study [20]. Briefly, all plasma samples were processed simultaneously using Simple Plex™ Assays running on the Ella™ Automated immunoassay analyzer (ProteinSimple, Biotechne brand powered by R&D Systems, Inc., Minneapolis, MN, USA). (https://www.

rndsystems.com/products/simple-plex-human-leptin-cartridge_spckb-ps-000309, last accessed on 14 October 2018).

Maternal milk samples (30 mL) were obtained at the end of the first morning breast-feeding, by pumping manually for five minutes, and were stored at $-80\,^{\circ}$C until analyses.

To separate fat from the aqueous phase, all samples were thawed, vortexed, and centrifuged ($4000\times g$) for 30 min at 4 $^{\circ}$C, to separate fat from the aqueous phase. Leptin concentration in freshly skimmed milk samples were determined as for plasma samples. The Low limit of quantification (LLOQ) is 2.20 pg/mL; the Upper LOQ (ULOQ) is 13,810 pg/mL. The limit of detection is 1.71 pg/mL. The intra-assay precision (within an assay and each control was tested 16 times in one assay) is 93.2 ± 8.70 pg/mL for low QC and 4319 ± 143 for high quality control (QC). The inter-assay precision (between assays, replicates of each QC were tested in multiple assays performed by 3 technicians using 2 lots of reagent, QC was tested 16 times in one assay): 94.7 ± 9.70 pg/mL for low QC and 4305 ± 178 pg/mL for high QC.

2.3. Statistical Analysis

Quantitative variables were summarized as median and Interquartile range (IQR, 25th$-$75th percentiles). Qualitative variables were described as counts and percentage. The Kruskal–Wallis and Fisher exact tests were applied to compare quantitative and qualitative variables between the three groups of neonates (FT, PT and IUGR). For the analysis, we used a zero-skewness log-transformation of leptin concentrations. Generalized linear regression models, with Sandwich standard errors (to allow for intragroup correlation), were fitted to assess the association of the log-transformed leptin and the increase of offspring weight, while adjusting for neonate type and time. Using residuals derived from the model, the partial correlation of leptin and weight, was computed together with its 95% bootstrapped confidence interval (95% CI), overall and by neonate type. The corresponding scatterplots are presented. The interaction of leptin concentration and groups was tested to assess whether any effect modification by group was present.

A 2-sided p-value < 0.05 was considered statistically significant. The Bonferroni correction was used for post hoc comparisons. Data analysis was performed with the STATA statistical package (StataCorp. 2019. Stata Statistical Software: Release 17. StataCorp LLC: College Station, TX, USA).

2.4. Study Endpoints

To assess the association of leptin concentrations in maternal plasma, CB and maternal milk with growth trajectories up to one year of age in all the neonates, adjusting for confounders (time and neonate types).

To verify whether the associations depend on the specific subgroup of mother-neonate dyads: FT, PT, and IUGR.

3. Results

Maternal plasmatic leptin concentrations were highest in the PT group, followed by the IUGR and FT groups, without reaching statistical differences ($p = 0.213$). Conversely, CB leptin concentrations were highest in FT and lowest in PT and IUGR neonates (PT vs. FT and IUGR vs. FT $p < 0.001$, both). Maternal milk concentration showed large variability within the same group, without significant differences.

At 12 months of age, PT and IUGR change in weight was higher (absolute value), although not significantly, than FT neonates ($p = 0.272$), while percent (%) change in weight was statistically significant in PT compared to FT, similarly to IUGR infants compared to FT ($p < 0.001$, both) (Table 1, Supplementary Materials Table S1; Figure S1).

Table 1. Demographic data of mother-newborn couples, leptin concentrations, and weight increase in the three groups.

	FT (*n* = 16)	PT (*n* = 16)	IUGR (*n* = 13)	*p*-Value	Post Hoc Comparison *p*-Value—Bonferroni
Maternal age, years [#]	31 (30–32)	32 (29–36)	32 (30–37)	0.377	-
Cesarean section delivery [§]	1 (6.3%)	11 (68.8%)	10 (76.9%)	<0.001	PT vs. FT: 0.002 IUGR vs. FT: <0.001 IUGR vs. PT: >0.90
Maternal BMI before pregnancy *	23.5 (3.65) (range 16.5–29.8)	23.5 (3.98) (range 18.7–30.48)	23 (3.68) (range 19.0–31.2)	0.814	-
Maternal BMI at delivery *	28 (3.63) (range 20.8–33.2)	27.5 (4.0) (range 20.8–35.0)	26.0 (3.35) (range 220–33.6)	0.228	-
Male neonate [§]	8 (50%)	7 (43.8%)	5 (38.5%)	0.929	-
Gestational Age, weeks [#]	40.5 (39.4–41.0)	33.7 (29.7–34.6)	36.6 (32.1–37.3)	<0.001	PT vs. FT: <0.001 IUGR vs. FT: <0.001 IUGR vs. PT: 0.217
APGAR 1′	10 (9–10)	7.5 (5.5–9)	9 (6–9)	<0.001	PT vs. FT: <0.001 IUGR vs. FT: 0.027 IUGR vs. PT: 0.304
APGAR 5′	10 (10–10)	9 (8–10)	10 (7–10)	0.022	PT vs. FT: 0.003 IUGR vs. FT: 0.100 IUGR vs. PT: 0.402
Maternal plasmatic leptin concentration pg/mL [#]	44,473.5 (22,214.5–50,302.4)	75,575.2 (27,219.9–89,243.3)	71,768.5 (41,890.1–12,1378.1)	0.213	-
Maternal milk leptin concentration pg/mL [#]	620.6 (462.3–881.6)	622.0 (329.7–1235.0)	844.3 (444.9–1008.9)	0.782	-
CB leptin concentration pg/mL [#]	19,280.5 (11,800.8–32,894.8)	3958.1 (1982.9–11,545.6)	1588.8 (1223.1–6912.0)	<0.001	PT vs. FT: <0.001 IUGR vs. FT: <0.001 IUGR vs. PT: 0.632
Birth weight g [#]	3242.5 (3022.5–3585.0)	1802.5 (1367.5–2203.5)	1770.0 (1265.0–2265.0)	<0.001	PT vs. FT: <0.001 IUGR vs. FT: <0.001 IUGR vs. PT: >0.90
Change in weight g [#] (increase at 12 months)	6135 (5893–7110)	7062 (6275–8565)	6295 (5970–6955)	0.272	
% Change in weight (increase at 12 months)	191.7 (183.4–224.8)	374.9 (288.9–564.6)	341.6 (288.5–549.8)	<0.001	PT vs. FT: <0.001 IUGR vs. FT: <0.001 IUGR vs. PT: >0.90

* mean and (SD); [§] n and (%); [#] median and (IQR 25th–75th percentiles). Abbreviations, FT: full-term; PT: preterm; IUGR: intrauterine growth restriction; BMI: body mass index.

3.1. Association of Leptin Concentrations and Neonate's Weight Trajectories

Maternal plasma leptin concentration (log-transformed) was significantly associated with infants change in weight (p = 0.017), adjusting for time and neonate type (confounders) in the multivariable analysis; the partial correlation of maternal plasma leptin and change in weight was −0.27 (95% CI −0.48 to −0.05).

A marginally non-significant (p = 0.051) association of CB leptin concentration and weight change was observed. The partial correlation for confounders was −0.30 (95% CI −0.58 to 0.0).

Finally, maternal leptin milk concentration (log-transformed) was significantly associated with neonates' weight change (p = 0.006). The leptin-change in partial weight correlation was −0.33 (95% CI −0.51 to −0.15).

3.2. Association of Leptin Concentrations and Neonate's Subgroups Weight Trajectories

Figure 1 illustrates the association of change in weight and leptin for each compartment and each different type of neonates, together with the corresponding partial correlations. Neonatal characteristics did not significantly modify the relationship of weight changes and maternal plasma leptin concentrations in PT and IUGR compared to FT (p for interaction = 0.268); however, partial correlations ranged from −0.04 (FT) to −0.28 (PT) and to −0.52 (IUGR).

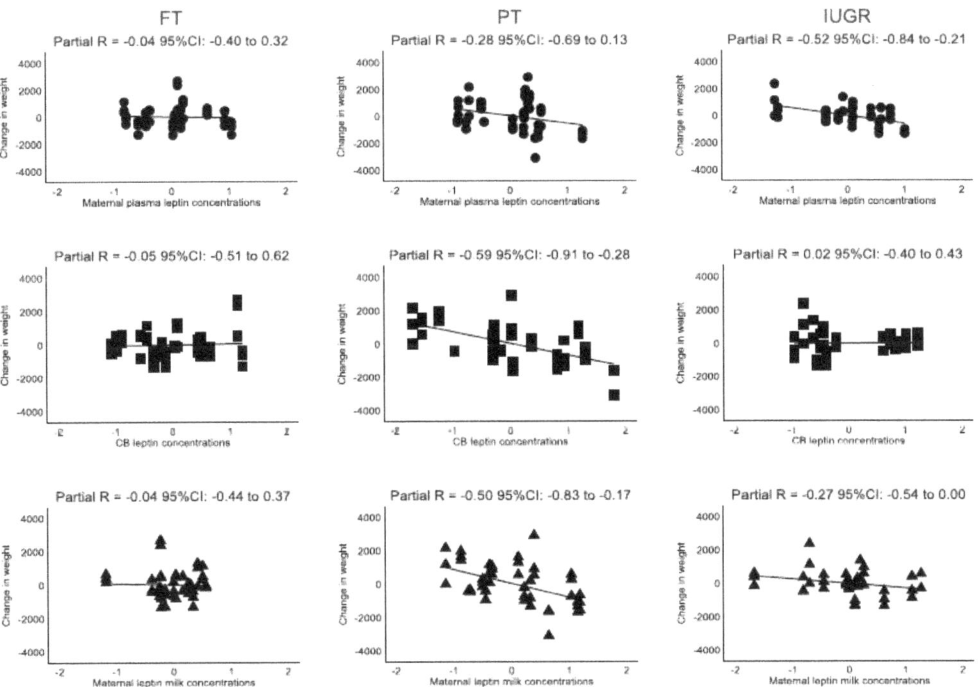

Figure 1. Partial correlations of leptin concentrations and change in weight by type of neonate. Residuals from the multivariable regression are plotted to account for time. Circles = maternal plasma leptin concentrations. Squares = CB leptin concentrations. Triangles = milk leptin concentrations.

In the case of CB leptin concentration, the effect modification by neonate type reached statistical significance (p = 0.026) and a significant partial correlation was elicited in the PT group (partial R −0.59 for PT; R −0.02 for IUGR; R −0.05 for FT). No significant modification by neonate type was found when considering the residual analysis from the multivariable regression plotting maternal milk and growth trajectory (*p* for interaction = 0.120). The inverse correlation was heterogeneous between the neonate's types and was highest in the PT group (partial R −0.50 for PT; R −0.27 for IUGR; R 0.04 for FT).

4. Discussion

We investigated the relationship of maternal plasma, cord blood, and milk leptin concentration with offspring's growth trajectories up to 1 year of age in three different mother-neonate dyads, FT, PT, and IUGR.

The multivariable analysis, adjusted for confounders, showed an inverse though weak correlation between leptin measured in the three different biological samples and growth trajectories up to one year of age. Maternal plasma leptin concentration was associated with infants change in weight (*p* = 0.017) and confirmed in the partial correlation when adjusting for time and neonate type (−0.27). Previous studies reported no association between maternal leptin concentration and birthweight/weight change [6,20]; or higher incidence of IUGR neonate in pre-eclamptic pregnancy, if maternal serum leptin levels were elevated [21]. Moreover, a significant rate of low-birth-weight infants in non-diabetic mothers with high leptin plasma concentration was reported [22,23]. Our results highlighted a marginally non-significant association between CB leptin concentration and weight change at one year of age for all neonates.

Conversely, the analysis of the specific sub-group demonstrated that PT neonates elicited a significant negative correlation. Other studies investigated the relationship between CB leptin concentration, birthweight, and growth pattern. Fonseca et al. [24] demonstrated a negative association between CB leptin with BMI and length change during the first six months of life of preterm infants, furthermore lower leptin levels correlated with greater catch-up growth. Equally, Karakosta et al. [25] evaluated the association of CB leptin and growth trajectories up to four years. Children with higher leptin concentration resulted in lower height, weight and BMI until early childhood; at the same time, a negative association was observed in small for gestation age infants and children with rapid infant growth during the first three months of life. Conversely, Chaoimh et al. [8] showed that cord blood leptin was positively associated with fat mass at birth, while higher concentrations were associated with slower weight gain in early infancy.

Since the preterm delivery leads to a premature separation from the maternal/placental leptin source, the infant is predisposed to postnatal hormone deficiency, enhanced by the lack of adipose tissue, which increases during the last three months of fetal life. Thus, leptin in the breastmilk could be a substitute source for neonates, particularly those born preterm [26]. In our study, leptin concentrations in maternal milk resulted in a similar small amount in each group, lower than the corresponding levels in maternal plasma and CB. Anyhow, maternal milk leptin was negatively associated with all the neonates' change in weight. In particular, the sub-group analysis showed the most significant values plotting mother's milk with PT neonate's growth trajectories.

Aydin et al. [27] confirmed that leptin levels in colostrum (2.01 ng/mL) and mature milk (2.04 ng/mL) were more than five-fold lower than the corresponding blood levels (11.54 ng/mL). Schuster et al. [12] measured leptin concentrations in maternal milk 22-fold lower than the maternal serum; the same study showed a negative association between breast milk leptin and the infant's weight increase during the first six months of life.

Dundar et al. [28] reported that maternal milk from small, large and adequate for gestational age infants had different leptin levels during the first months of life. Indeed, small for gestational age infants had a significant milk leptin reduction associated with higher weight gain. Andreas et al. highlighted no significant associations between milk leptin concentration and the timing of sample collection (measured at one week and three months post-partum) and hind and foremilk, demonstrating a constant concentration of the hormone through the first 3 months of lactation [29]. On the other hand, Kratzsch et al. [18] showed a variability of breast milk leptin concentrations depending, in particular, by maternal adipose tissue. In our study, maternal BMI, before pregnancy and at the delivery, resulted homogenous, without statistical differences among the three groups of mothers. In parallel, leptin concentrations in maternal milk resulted as well alike in the three groups and statistically associated with neonates' change in weight. Besides maternal influences, the role of breastmilk leptin concentration is still a matter of debate: a recent review highlighted contrasting findings concerning the potential action of milk leptin on modeling growth, either showing infants' association to weight changes or reporting no correlation [30].

Moreover, a recent study on 50 preterm infants found an association between higher maternal leptin intake and higher weight gain at 36 weeks of post menstrual age, suggesting a potential role for breastmilk leptin in earlier neonatal weight increase [31], which contrast with the known "appetite-inhibiting" role of the hormone [2].

Our study demonstrated for all group an inverse association between leptin of breast milk and neonatal growing trajectories, which was stronger in newborns born preterm. The outline of change in weight in the first year of life was significantly associated with low hormone levels: small and critical newborns struggle to rescue the appropriate size over time. Low leptin concentration contributes reaching the ideal body weight, avoiding the satiety perception, increasing feeding need. We found the most significant correlation between leptin concentration in each biological sample and weight gain in newborns born preterm.

Leptin concentrations were demonstrated to play a role in defensive mechanisms for survival, enhancing hunger, and expansion of energy stores needed for rapid growth. On the other side, monitoring for abnormal "catch-up growth" is mandatory. The association between excessive growth in infancy and overweight/obesity, insulin resistance, and elevated blood pressure in childhood or adulthood has been reported [32]. Therefore, measuring growth trajectory and leptin concentration is a valid tool for monitoring children's health. In particular, the period of early infancy known as "the first 1000 days of life", is a narrow but essential window for promoting future wellbeing [33,34].

During pregnancy, prevention measures can be adopted to promote healthy fetal growth and programming [17]. Similarly, in the post-natal period, it is possible to avoid future obesity and metabolic diseases, as an example, with prolonged breastfeeding that acts on an infant's microbiota and immune system. [35].

The low number of subjects included in our study requires caution in the interpretation of results, and we did not include other perinatal and postnatal factors that have a part in children growth. Contrariwise, the inverse correlation between leptin in PT infants and change in weight was significant throughout the first year of age, regardless of the source of the biological compartment.

We speculate that leptin plays a role in neonatal growth trajectories, characterized mainly by an inverse correlation with leptin concentration in maternal plasma and breast-milk. The PT infants elicited the highest negative association with each leptin source. Low hormone levels appeared to contribute to critical neonates' ability to recover a correct body-weight similar to full-term children at 1 year. An eventual non-physiological "catch-up growth" should be monitored, and leptin perinatal levels may be an indicative tool. Further investigations are needed to strengthen the results.

Supplementary Materials: The following are available online at https://www.mdpi.com/article/10.3390/nu14071451/s1, Table S1: Weight percentile at 12 month of age in the three different groups of children, Figure S1: Distribution of weight percentiles across types of neonate through 12 months of age (Fisher's test $p = 0.016$).

Author Contributions: Conceptualization, F.G., I.M. and C.T.; software, formal analysis, data curation, C.K., V.V.F. and M.A.; investigation, B.G. and G.V.C.; writing—original draft preparation, F.G.; writing—review and editing, I.M., B.G., A.S., C.T. and S.G.; supervision, A.S. and S.G. All authors have read and agreed to the published version of the manuscript.

Funding: This study is supported by grants from our Institute: Fondazione IRCCS Policlinico San Matteo, Piazzale Golgi 19, 27100 Pavia, Italy (Proj. 854; cod 08066315; 854-rcr2015 bis11). There is no further funding body supporting the design of the study; the collection, analysis, and interpretation of the data; writing the report; and the decision to submit the paper for publication of the manuscript.

Institutional Review Board Statement: The study complies with the Declaration of Helsinki. The Bioethical Committee of Fondazione IRCCS Policlinico S. Matteo, Pavia Italy, approved the study (prot Dir n.20140004843; proc.N.200140023494), and written informed consent was obtained from all parents before enrollment. Authors have full control of all primary data.

Informed Consent Statement: Bioethical Committee of Fondazione IRCCS Policlinico S. Matteo, Pavia Italy, approved the study (prot Dir n.20140004843; proc.N.200140023494), and written informed consent was obtained from both parents' patients before enrollment.

Data Availability Statement: Not applicable.

Conflicts of Interest: The authors declare no conflict of interest.

References

1. Zhang, Y.; Proenca, R.; Maffei, M.; Barone, M.; Leopold, L.; Friedman, J.M. Positional cloning of the mouse obese gene and its human homologue. *Nature* **1994**, *372*, 425–432. [CrossRef] [PubMed]
2. Palou, M.; Picó, C.; Palou, A. Leptin as a breast milk component for the prevention of obesity. *Nutr. Rev.* **2018**, *76*, 875–892. [CrossRef] [PubMed]
3. Ghadge, A.A.; Khaire, A.A. Leptin as a predictive marker for metabolic syndrome. *Cytokine* **2019**, *121*, 154735. [CrossRef] [PubMed]
4. Briffa, J.F.; McAinch, A.J.; Romano, T.; Wlodek, M.E.; Hryciw, D.H. Leptin in pregnancy and development: A contributor to adulthood disease? *Am. J. Physiol. Endocrinol. Metab.* **2015**, *308*, E335–E350. [CrossRef]
5. Weyermann, M.; Beermann, C.; Brenner, H.; Rothenbacher, D. Adiponectin and leptin in maternal serum, cord blood, and breast milk. *Clin. Chem.* **2006**, *52*, 2095–2102. [CrossRef]
6. Stefaniak, M.; Dmoch-Gajzlerska, E.; Mazurkiewicz, B.; Gajzlerska-Majewska, W. Maternal serum and cord blood leptin concentrations at delivery. *PLoS ONE* **2019**, *7*, e0224863. [CrossRef]
7. Mantzoros, C.S.; Rifas-Shiman, S.L.; Williams, C.J.; Fargnoli, J.L.; Kelesidis, T.; Gillman, M.W. Cord blood leptin and adiponectin as predictors of adiposity in children at 3 years of age: A prospective cohort study. *Pediatrics* **2009**, *123*, 682–689. [CrossRef]
8. Chaoimh, C.N.; Murray, D.M.; Kenny, L.C.; Irvine, A.D.; Hourihane, J.O.; Kiely, M. Cord blood leptin and gains in body weight and fat mass during infancy. *Eur. J. Endocrinol.* **2016**, *175*, 403–410. [CrossRef]
9. Karakosta, P.; Chatzi, L.; Plana, E.; Margioris, A.; Castanas, E.; Kogevinas, M. Leptin levels in cord blood and anthropometric measures at birth: A systematic review and meta-analysis. *Paediatr. Perinat. Epidemiol.* **2011**, *25*, 150–163. [CrossRef]
10. Bagias, C.; Sukumar, N.; Weldeselassie, Y.; Oyebode, O.; Saravanan, P. Cord Blood Adipocytokines and Body Composition in Early Childhood: A Systematic Review and Meta-Analysis. *Int. J. Environ. Res. Public Health.* **2021**, *18*, 1897. [CrossRef]
11. Ren, R.X.; Shen, Y. A meta-analysis of relationship between birth weight and cord blood leptin levels in newborns. *World J. Pediatr.* **2010**, *15*, 311–316. [CrossRef] [PubMed]
12. Schuster, S.; Hechler, C.; Gebauer, C.; Kiess, W.; Kratzsch, J. Leptin in maternal serum and breast milk: Association with infants' body weight gain in a longitudinal study over 6 months of lactation. *Pediatr. Res.* **2011**, *70*, 633–637. [CrossRef] [PubMed]
13. Schneider-Worthington, C.R.; Bahorski, J.S.; Fields, D.A.; Gower, B.A.; Fernández, J.R.; Chandler-Laney, P.C. Associations Among Maternal Adiposity, Insulin, and Adipokines in Circulation and Human Milk. *J. Hum. Lact.* **2021**, *37*, 714–722. [CrossRef] [PubMed]
14. Grieger, J.A.; Hutchesson, M.J.; Cooray, S.D.; Bahri Khomami, M.; Zaman, S.; Segan, L.; Teede, H.; Moran, L.J. A review of maternal overweight and obesity and its impact and cardiometabolic outcomes during pregnancy and postpartum. *Ther. Adv. Reprod. Health* **2021**, *15*, 2633494120986544. [CrossRef] [PubMed]
15. Zheng, M.; Lamb, K.E.; Grimes, C.; Laws, R.; Bolton, K.; Ong, K.K.; Campbellk, K. Rapid weight gain during infancy and subsequent adiposity: A systematic review and meta-analysis of evidence. *Obesity Rev.* **2018**, *19*, 321–332. [CrossRef]
16. Yajnik, C.S. Transmission of obesity-adiposity and related disorders from the mother to the baby. *Ann. Nutr. Metab.* **2014**, *64*, 8–17. [CrossRef]
17. Kapur, A. Links between maternal health and NCDs. *Best Pract. Res. Clin. Obstet. Gynaecol.* **2015**, *29*, 2–42. [CrossRef]
18. Kratzsch, J.; Bae, Y.J.; Kiess, W. Adipokines in human breast milk. *Best Pract. Res. Clin. Endocrinol. Metab.* **2018**, *32*, 27–38. [CrossRef]
19. Savino, F.; Sardo, A.; Rossi, L.; Benetti, S.; Savino, A.; Silvestro, L. Mother and Infant Body Mass Index, Breast Milk Leptin and Their Serum Leptin Values. *Nutrients* **2016**, *8*, 383. [CrossRef]
20. Garofoli, F.; Mazzucchelli, I.; Angelini, M.; Klersy, C.; Tinelli, C.; Carletti, G.V.; Calcaterra, V.; Gardella, B.; Tzialla, C. The dynamical interplay of perinatal leptin with birthweight and 3-month weight, in full-term, preterm, IUGR mother-infant dyads. *J. Matern. Fetal Neonatal Med.* **2020**, *8*, 1–7. [CrossRef]

21. De Knegt, V.E.; Hedley, P.L.; Kanters, J.K.; Thagaard, I.N.; Krebs, L.; Christiansen, M.; Lausten-Thomsen, U. The Role of Leptin in Fetal Growth during Pre-Eclampsia. *Int. J. Mol. Sci.* **2021**, *22*, 4569. [CrossRef] [PubMed]

22. Walsh, J.M.; Byrne, J.; Mahony, R.M.; Foley, M.E.; McAuliffe, F.M. Leptin, fetal growth and insulin resistance in non-diabetic pregnancies. *Early Hum. Dev.* **2014**, *90*, 271–274. [CrossRef] [PubMed]

23. Retnakaran, R.; Ye, C.; Hanley, A.J.; Connelly, P.W.; Sermer, M.; Zinman, B.; Hamilton, J.K. Effect of maternal weight, adipokines, glucose intolerance and lipids on infant birth weight among women without gestational diabetes mellitus. *Can. Med. Assoc. J.* **2012**, *184*, 1353–1360. [CrossRef] [PubMed]

24. Fonseca, V.M.; Sichieri, R.; Moreira, M.E.; Moura, A.S. Early postnatal growth in preterm infants and cord blood leptin. *J. Perinatol.* **2004**, *24*, 751–756. [CrossRef]

25. Karakosta, P.; Roumeliotaki, T.; Chalkiadaki, G.; Sarri, K.; Vassilaki, M.; Venihaki, M.; Malliaraki, N.; Kampa, M.; Castanas, E.; Kogevinas, M.; et al. Cord blood leptin levels in relation to child growth trajectories. *Metabolism* **2016**, *65*, 874–882. [CrossRef] [PubMed]

26. Steinbrekera, B.; Colaizy, T.T.; Vasilakos, L.K.; Johnson, K.J.; Santillan, D.A.; Haskell, S.E.; Roghair, R.D. Origins of neonatal leptin deficiency in preterm infants. *Pediatr. Res.* **2019**, *85*, 1016–1023. [CrossRef]

27. Aydin, S.; Ozkan, Y.; Erman, F.; Gurates, B.; Kilic, N.; Colak, R.; Gundogan, T.; Catak, Z.; Bozkurt, M.; Akin, O.; et al. Presence of obestatin in breast milk: Relationship among obestatin, ghrelin, and leptin in lactating women. *Nutrition* **2008**, *24*, 689–693. [CrossRef]

28. Dundar, N.O.; Anal, O.; Dundar, B.; Ozkan, H.; Caliskan, S.; Büyükgebiz, A. Longitudinal investigation of the relationship between breast milk leptin levels and growth in breast-fed infants. *J. Pediatr. Endocrinol. Metab.* **2005**, *2*, 181–187. [CrossRef]

29. Andreas, N.J.; Hyde, M.J.; Herbert, B.R.; Jeffries, S.; Santhakumaran, S.; Mandalia, S.; Holmes, E.; Modi, N. Impact of maternal BMI and sampling strategy on the concentration of leptin, insulin, ghrelin and resistin in breast milk across a single feed: A longitudinal cohort study. *BMJ Open* **2016**, *6*, e010778. [CrossRef]

30. Mazzocchi, A.; Giannì, M.L.; Morniroli, D.; Leone, L.; Roggero, P.; Agostoni, C.; De Cosmi, V.; Mosca, F. Hormones in Breast Milk and Effect on Infants' Growth: A Systematic Review. *Nutrients* **2019**, *11*, 1845. [CrossRef]

31. Joung, K.E.; Martin, C.R.; Cherkerzian, S.; Kellogg, M.; Belfort, M.B. Human Milk Hormone Intake in the First Month of Life and Physical Growth Outcomes in Preterm Infants. *J. Clin. Endocrinol. Metab.* **2021**, *106*, 1793–1803. [CrossRef] [PubMed]

32. Arisak, O.; Ichikawa, G.; Koyama, S.; Sairenchi, T. Childhood obesity: Rapid weight gain in early childhood and subsequent cardiometabolic risk. *Clin. Pediatr. Endocrinol.* **2020**, *29*, 135–142. [CrossRef] [PubMed]

33. Mameli, C.; Mazzantini, S.; Zuccotti, G.V. Nutrition in the First 1000 Days: The Origin of Childhood Obesity. *Int. J. Environ. Res. Public Health* **2016**, *13*, 838. [CrossRef] [PubMed]

34. Raspini, B.; Porri, D.; De Giuseppe, R.; Chieppa, M.; Liso, M.; Cerbo, R.M.; Civardi, E.; Garofoli, F.; Monti, M.C.; Vacca, M.; et al. Prenatal and postnatal determinants in shaping offspring's microbiome in the first 1000 days: Study protocol and preliminary results at one month of life. *Ital. J. Pediatr.* **2020**, *6*, 45. [CrossRef]

35. Van den Elsen, L.W.J.; Verhasselt, V. Human Milk Drives the Intimate Interplay Between Gut Immunity and Adipose Tissue for Healthy Growth. *Front. Immunol.* **2021**, *2*, 645415. [CrossRef]

Article

Most Commonly-Consumed Food Items by Food Group, and by Province, in China: Implications for Diet Quality Monitoring

Sheng Ma [1], Anna W. Herforth [2], Chris Vogliano [3] and Zhiyong Zou [1,*]

[1] National Health Commission Key Laboratory of Reproductive Health, Institute of Child and Adolescent Health, School of Public Health, Peking University, Beijing 100191, China; 2111210073@bjmu.edu.cn

[2] Department of Global Health and Population, Harvard T.H. Chan School of Public Health, Boston, MA 02115, USA; anna@annaherforth.net

[3] USAID Advancing Nutrition, Washington, DC 20532, USA; chris_vogliano@jsi.com

[*] Correspondence: harveyzou2002@bjmu.edu.cn

Abstract: Dietary quality is of great significance to human health at all country income levels. However, low-cost and simple methods for population-level assessment and monitoring of diet quality are scarce. Within these contexts, our study aimed to identify the sentinel foods nationally and by province of 29 food groups to adapt the diet quality questionnaire (DQQ) for China, and validate the effectiveness of the DQQ using data from the China Health and Nutrition Survey (CHNS). The DQQ is a rapid dietary assessment tool with qualitative and quantitative analysis to determine appropriate sentinel foods to represent each of 29 food groups. Dietary data of 13,076 participants aged 15 years or older were obtained from wave 2011 of CHNS, and each food and non-alcoholic beverage was grouped into 29 food groups of the DQQ. The data were analyzed to determine the most commonly consumed food items in each food group, nationally and in each province. Key informant interviews of 25 individuals familiar with diets in diverse provinces were also conducted to identify food items that may be more common in specific provinces. China's DQQ was finalized based on identification of sentinel foods from the key informant interviews, and initial national results of the quantitative data. Consumption of sentinel foods accounted for over 95% of people who consumed any food item in each food group, at national levels and in all provinces for almost all food groups, indicating the reliability of the sentinel food approach. Food-group consumption data can be obtained through DQQ to analyze dietary diversity as well as compliance with WHO global dietary guidance on healthy diets, providing a low-burden, food-group-based and simple method for China to evaluate diet quality at the whole population level.

Keywords: diet quality questionnaire; food groups; sentinel foods; dietary diversity

Citation: Ma, S.; Herforth, A.W.; Vogliano, C.; Zou, Z. Most Commonly-Consumed Food Items by Food Group, and by Province, in China: Implications for Diet Quality Monitoring. *Nutrients* **2022**, *14*, 1754. https://doi.org/10.3390/nu14091754

Academic Editor: Angeliki Papadaki

Received: 15 March 2022
Accepted: 21 April 2022
Published: 22 April 2022

Publisher's Note: MDPI stays neutral with regard to jurisdictional claims in published maps and institutional affiliations.

Copyright: © 2022 by the authors. Licensee MDPI, Basel, Switzerland. This article is an open access article distributed under the terms and conditions of the Creative Commons Attribution (CC BY) license (https://creativecommons.org/licenses/by/4.0/).

1. Introduction

Diet quality, including food quality and diversity, is as essential to human health as air is to human life. Poor diet quality is the main contributor to morbidity and mortality worldwide, with micronutrient deficiencies persisting and diet-related non-communicable diseases rising across the world, especially in low-income countries [1,2]. Dietary diversity, one of the best ways for people to meet the requirements for essential nutrients, has long been recognized as a key dimension of high-quality diets [3]. Furthermore, healthy diets must protect against diet-related non-communicable diseases (NCDs) [4]. Appeals to improve diet quality and nutrition call for more programmatic action to measure diet quality, promoting better nutritional status for people.

Many kinds of dietary quality indicators exist as important tools to evaluate the quality of the diet [5]. One major category is nutrient-based indicators, which require quantitative intake of food information and food-nutrient conversion tables to convert food weight into nutrient content, and further compare the nutrient adequacy rate, such as the Healthy Eating Index, the Healthy Diet Indicator, the Diet Quality Index, and the Chinese Healthy

Food Diversity (HFD) Index [6,7]. These indicators usually require substantial financial and technical support and are often designed for use in high-income countries/regions with life-cycle assessment databases, which are not available in low and middle-income countries/regions [5,8]. There is a need to be able to quickly and inexpensively monitor diet quality in populations worldwide. Therefore, another category is food/food group-based indicators, one of which that is widely used is Minimum Dietary Diversity for Women (MDD-W), for which only the consumption (not the quantity consumed) of a food group was needed [3]. However, the MDD-W only assesses 10 healthy food groups and has a weak correlation with non-communicable diseases (NCDs), which has brought many restrictions to its own application [4]. Dietary diversity scores, which calculate the number of food groups consumed during a given period, are validated indicators of micronutrient adequacy [9]. Currently, two standard dichotomous indicators correlated to achieving nutrient adequacy are used internationally: one is the minimum dietary diversity (MDD) ($\geq$5 of 8 food groups) [10], while the other is MDD-W ($\geq$5 of 10 food groups) for women of reproductive age (15–49 years) [3,11]. A second indicator, the Global Dietary Recommendations (GDR) score, has been developed as an indicator of diet patterns that protect against NCDs [4]. Based on WHO global recommendations [4], the GDR score consists of two subscales. The first is the GDR-Healthy score, including foods for healthy diets such as whole grains, legumes, dark green leafy vegetables, vitamin A-rich fruits, and nuts and seeds, while the other is the GDR-Limit score, including foods that should be limited to intake due to the high content of free sugar, salt, and total or saturated fat, such as grain-based sweets, deep-fried foods and packaged ultra-processed [12] salty snacks.

These indicators can be useful for helping to formulate intervention strategies and monitor intervention effects [3,13]. However, no consistent, reliable method for gathering food group consumption data had been standardized or adapted for individual countries with a wide variety of diets, including China. There is a need for a valid tool for food group consumption to reflect diet quality in China, promote the establishment of targets, and advocate and formulate policies aimed at improving diets and nutrition.

In the present study, we adapted the diet quality questionnaire (DQQ) for China and used data from the China Health and Nutrition Survey (CHNS) 2011 and qualitative interviews to determine the sentinel food items appropriate for China in each of 29 detailed food groups. Furthermore, we tested the validity of the questionnaire for each province.

The DQQ, a tool for collecting food group level data to reflect healthy dietary patterns at the whole population level [14], uses sentinel foods to capture food group consumption: since it is difficult to enumerate every single food currently available in each group in China, selecting foods known to be major contributors would be suitable. In addition, a previous study has already proven the reliability of the sentinel food method, reporting that using sentinel food data obtained similar results as those using all foods in dietary data collection [4]. This study is not only a part of international work on the validation of diet quality indicators that could be used globally, but also provides a new and more suitable method for China to investigate dietary diversity, helping to shape both national and regional programs and policies effectively.

2. Materials and Methods

2.1. Quantitative Dietary Data Source

The quantitative data in this study were obtained from wave 2011 of CHNS. The CHNS met the standards for the ethical treatment of participants and was approved by the Institutional Review Committee of the University of North Carolina at Chapel Hill, and the National Institute for Nutrition and Health, Chinese Center for Disease Control and Prevention (2015017). Written informed consent was obtained from all subjects before they participate the survey. Further details of the CHNS data have been previously published [15,16].

The 2011 wave of CHNS included nine provinces (Liaoning, Heilongjiang, Jiangsu, Shandong, Henan, Guangxi, and Guizhou) and three municipalities (Beijing, Shanghai, and Chongqing) that differ in economic development, geography, health indicators, and public resources [17]. Basic data cleaning was performed by the CHNS team before the datasets were uploaded online [18]. Using a multistage random cluster sampling method, 5923 families in 289 communities with over 15,000 participants in total were involved from urban and rural areas. A total of 648 participants with incomplete dietary information were excluded. Our study focused on respondents aged ≥15 years, resulting in the removal of 1927 individuals. Moreover, 74 respondents with implausible intakes (carbohydrate intake greater than 1500 g/d, calcium intake greater than 3000 mg/d, or sodium intake greater than 30 g/d) were further excluded. Finally, 13,076 subjects were included in the analysis. Specific exclusion and inclusion criteria are shown in Figure 1.

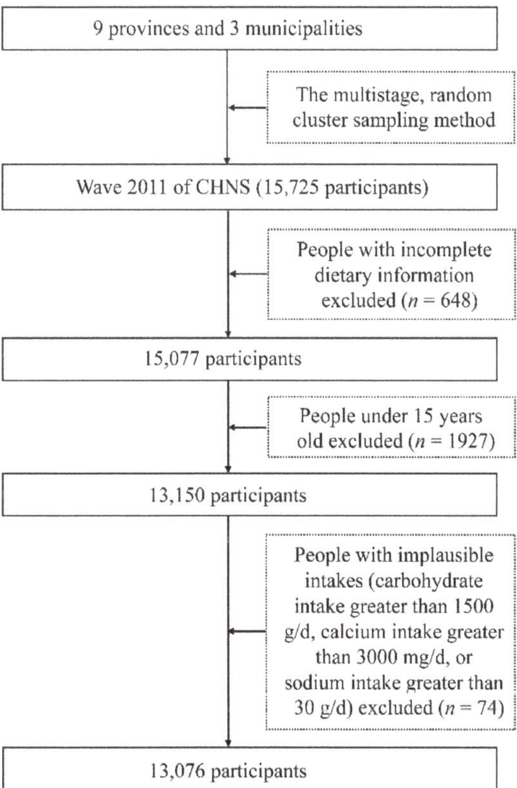

Figure 1. Exclusion and inclusion criteria of this study.

The basic sociodemographic characteristics and food consumption information of households were collected by trained interviewers from local Centers for Disease Control and Prevention through face-to-face interviews. Diet was assessed with a combination of 24-h recalls of individual dietary intake and a food inventory at the household level consisting of foods that they consumed at home or away from home in three consecutive days [19]. More detailed information on the dietary survey has been previously reported [20]. Since China's DQQ is designed to obtain food group intake data of the study population in the past 24 h, dietary intake data in the first day during the three consecutive days were used in this study to verify its validity.

2.2. Qualitative Data Collection

Key informant interviews of 25 people from diverse provinces were carried out by the research team. In each interview, key informants were asked to identify or confirm the food items they perceived to be most commonly consumed in each of the 29 DQQ food groups across China, inclusive of regions and seasons. Key informants ranged in occupation from health professionals, government officials, food industry professionals, consumers, expats, and students. These interviewees were familiar with the diets in their local regions, so the most commonly consumed food items in each food group were identified according to the data on how local interviewees name and refer to their local foods. Key informants were also asked to rank the popularity of each food, which influenced where each food item appears in the final DQQ tool question formulation. Each interview lasted 60–90 min and were conducted until consensus was reached. The China DQQ was drafted on the basis of these interviews, and accuracy was confirmed by using the national-level data analysis of the CHNS data.

2.3. Data Analysis

All foods and non-alcoholic beverages consumed by subjects were divided into 29 food groups of the DQQ as follows: (1) staple foods made from grains; (2) whole grains; (3) white root/tubers; (4) legumes; (5) vitamin A-rich orange vegetables; (6) dark green leafy vegetables; (7) other vegetables; (8) vitamin A-rich fruits; (9) citrus; (10) other fruits; (11) grain-based sweets; (12) other sweets; (13) eggs; (14) cheese; (15) yogurt; (16) processed meats; (17) unprocessed red meat (ruminant); (18) unprocessed red meat (nonruminant); (19) poultry; (20) fish and seafood; (21) nuts and seeds; (22) packaged ultra-processed salty snacks; (23) instant noodles; (24) deep fried foods; (25) fluid milk; (26) sweetened tea/coffee/milk drinks; (27) fruit juice; (28) sugar-sweetened beverages (SSBs) (sodas); (29) fast food.

To identify sentinel foods, all foods were ranked in descending order according to the consumption frequency contributions in each group nationally and by province, after which cumulative consumption frequency was calculated. Food items that captured consumption of 95% or more of the surveyed individuals, were identified nationally and by each province. Furthermore, the proportion of the population consuming the food group that was captured by the sentinel foods in each food group was calculated. For example, of all people who consumed any vitamin A-rich fruit, 96.5% were captured by the sentinel foods nationally. Data analyses were performed with the Statistical Package for Social Sciences (SPSS), version 25.0 for Windows.

3. Results

3.1. General Information

The sociodemographic characteristics and food consumption information of a total of 13,076 participants aged $\geq$15 years are described in Table 1. Nationally, 52.9% of the respondents were female and more than half (57.8%) were from rural villages. The mean age of all participants was 49.7 years with a standard deviation (SD) of 16.0 years old, while the average body mass index (BMI) was 22.8 ± 6.3 kg/m^2. The average intake of energy was 1929.1 ± 790.2 kcal nationally, with average intakes of carbohydrates, protein, and fat of 273.6 ± 130.8 g, 69.2 ± 34.9 g, and 66.1 ± 45.9 g, respectively. Characteristics for participants of each province are also shown in Table 1.

3.2. The DQQ for China

Sentinel foods identified through key informant interviews were used to draft the China-adapted DQQ. They were then checked against draft data analysis at the national level (but not at the province level). Table 2 shows the finalized DQQ for China.

Table 1. Characteristics of the study population.

Characteristics	National	Beijing	Liaoning	Heilongjiang	Shanghai	Jiangsu	Shandong	Henan	Hubei	Hunan	Guangxi	Guizhou	Chongqing
No. of participants	13,076	1098	978	960	1284	1143	1035	1099	983	1098	1344	1033	1021
Gender, %													
Male	47.1	47.5	47.0	47.7	46.9	46.8	47.0	46.8	46.2	47.0	48.0	47.4	46.2
Female	52.9	52.5	53.0	52.3	53.1	53.2	53.0	53.2	53.8	53.0	52.0	52.6	53.8
Location, %													
Urban	42.2	79.6	28.0	32.0	80.5	33.4	32.4	35.0	32.8	36.1	29.6	30.8	47.7
Rural	57.8	20.4	72.0	68.0	19.5	66.6	67.6	65.0	67.2	63.9	70.4	69.2	52.3
Age, yrs	49.7 ± 16.0	44.8 ± 14.9	52.4 ± 14.7	48.5 ± 14.0	48.7 ± 15.9	51.8 ± 16.7	52.1 ± 15.5	49.0 ± 16.4	51.3 ± 15.0	50.5 ± 15.9	48.1 ± 17.6	51.4 ± 17.1	49.4 ± 14.9
BMI, kg/m²	22.8 ± 6.3	24.7 ± 4.6	24.4 ± 5.6	24.1 ± 5.4	24.0 ± 3.9	23.0 ± 5.3	24.9 ± 4.6	20.7 ± 9.2	22.4 ± 6.3	23.1 ± 4.8	18.3 ± 8.9	22.0 ± 5.0	23.3 ± 4.6
Energy, kcal	1929.1 ± 790.2	1661.2 ± 713.2	1926.6 ± 712.7	2071.7 ± 760.8	1771.7 ± 652.6	2160.3 ± 828.5	1941.4 ± 821.9	1833.9 ± 799.3	2265.4 ± 1181.5	2159.8 ± 682.4	1814.1 ± 644.9	1961.0 ± 742.1	1661.6 ± 773.9
Carbohydrate, g	273.6 ± 130.8	231.4 ± 102.6	271.3 ± 112.2	302.4 ± 133.1	212.1 ± 94.8	290.5 ± 132.8	301.3 ± 141.9	309.0 ± 123.6	356.9 ± 196.5	266.8 ± 101.4	301.9 ± 121.9	259.2 ± 99.4	191.0 ± 94.5
Protein, g	69.2 ± 34.9	66.7 ± 35.2	65.6 ± 34.0	66.5 ± 31.5	78.5 ± 35.9	84.0 ± 41.9	70.6 ± 31.6	59.5 ± 26.8	79.6 ± 44.4	72.8 ± 33.3	68.9 ± 27.1	59.7 ± 28.5	54.9 ± 33.2
Fat, g	66.1 ± 45.9	56.6 ± 38.7	67.6 ± 40.1	71.7 ± 37.3	72.0 ± 37.8	77.7 ± 45.6	53.7 ± 36.6	45.8 ± 43.8	61.4 ± 46.1	92.2 ± 47.1	41.7 ± 28.4	79.8 ± 54.0	79.2 ± 60.5

Note: SD, standard deviation; BMI, body mass index.

Table 2. The diet quality questionnaire (DQQ) for China.

DIET QUALITY QUESTIONNAIRE: CHINA

Read:
Now I'd like to ask you some yes-or-no questions about foods and drinks that you consumed yesterday during the day or night, whether you had it at home or somewhere else.
First, I would like you to think about yesterday, from the time you woke up through the night. Think to yourself about the first thing you ate or drank after you woke up in the morning. Think about where you were when you had any food or drink in the middle of the day. Think about where you were when you had any evening meal, and any food or drink you may have had in the evening or late-night, and any other snacks or drinks you may have had between meals throughout the day or night.
I am interested in whether you had the food items I will mention even if they were combined with other foods.
Please listen to the list of foods and drinks, and if you ate or drank ANY ONE OF THEM, say yes.

(Do not read food group names)		Yesterday, did you eat any of the following foods?							(circle answer)
01 Staple foods made from grains	Global definition	rice	noodles	steamed bun	bread				YES or NO
	Suggested translation	米饭	面条	馒头	面包				
02 Whole grains	Global definition	corn	cornmeal	oats	millet	barley	brown rice	black rice	whole wheat bread	
	Suggested translation	玉米(鲜)	玉米面	燕麦片	小米	大麦/糌粑	糙米	黑米	全麦面包	
03 White roots/tubers	Global definition	potatoe	lotus root	starch noodles	yam	taro	turnip			YES or NO
	Suggested translation	马铃薯	藕(莲藕)	粉丝/粉条	山药	芋头	大头菜			
04 Legumes	Global definition	bean curd or tofu	bean curd sheet	soybean milk	soybeans	other dried beans				YES or NO
	Suggested translation	豆腐	豆腐皮	豆浆	黄豆	其他干豆类				

Table 2. *Cont.*

		Yesterday, did you eat any of the following vegetables?								
05 Vitamin A-rich orange vegetables	Global definition	carrots	pumpkin or butternut squash	sweet potatoes that are orange inside						YES or NO
	Suggested translation	胡萝卜	南瓜	红薯						
06.1 Dark green leafy vegetables	Global definition	Chinese cabbage	water spinach	Chinese spinach	rape	bok choy	sweet potato leaves	broccoli		YES or NO
	Suggested translation	大白菜	空心菜	菠菜	油菜	小白菜	番薯叶	西兰花		
06.2 Dark green leafy vegetables	Global definition	mustard leaves	chrysanthemum leaves	radish leaves	amaranth leaves	beet leaves	watercress			YES or NO
	Suggested translation	芥菜	茼蒿	萝卜叶	苋菜	甜菜叶	西洋菜			
07.1 Other vegetables	Global definition	cabbage	tomatoes	eggplant	loofah	green bean	local celery	cucumber		YES or NO
	Suggested translation	包菜/圆白菜	番茄	茄子	丝瓜	四季豆	芹菜	黄瓜		
07.2 Other vegetables	Global definition	mushrooms	lettuce	radish	cauliflower	seaweed	bamboo shoots	bell pepper	bean sprouts	YES or NO
	Suggested translation	蘑菇	生菜	萝卜	菜花	紫菜	笋	甜椒/柿子椒	豆芽	

		Yesterday, did you eat any of the following fruits?							
08 Vitamin A-rich fruits	Global definition	persimmon	cantaloupe	ripe mango	passion fruit	fresh or dried apricot	papaya		YES or NO
	Suggested translation	柿子	哈蜜瓜	芒果	百香果	杏或杏干	木瓜		
09 Citrus	Global definition	orange	tangerines	pomelo	grapefruit	kumquat			YES or NO
	Suggested translation	橙	橘子/柑橘	柚子	西柚	金桔			
10.1 Other fruits	Global definition	apple	pear	watermelon	banana	grapes	kiwi	dragonfruit	YES or NO
	Suggested translation	苹果	梨	西瓜	香蕉	葡萄	猕猴桃	火龙果	
10.2 Other fruits	Global definition	jujube	longan	wampee	lychee	pomegranate	cherry	peaches	YES or NO
	Suggested translation	枣子	龙眼	黄皮果	荔枝	石榴	樱桃	桃	

Table 2. *Cont.*

		Yesterday, did you eat any of the following sweets?								
11 Baked sweets	Global definition	cakes	cookies	sweet pastries	mooncake	rice dumplings	egg tarts			YES or NO
	Suggested translation	蛋糕	甜饼干	甜糕点	月饼	甜粽子/汤圆	蛋挞			
12 Other sweets	Global definition	candy	chocolates	jelly pudding	ice cream	popsicles				YES or NO
	Suggested translation	糖果	巧克力	果冻	冰淇淋	棒冰				
		Yesterday, did you eat any of the following foods of animal origin?								
13 Eggs	Global definition	chicken eggs	preserved duck eggs	quail eggs	pigeon eggs	goose eggs				YES or NO
	Suggested translation	鸡蛋	咸鸭蛋 (白), 松花蛋 (黑)	鹌鹑蛋	鸽子蛋	鹅蛋				
14 Cheese	Global definition	cheese								YES or NO
	Suggested translation	奶酪								
15 Yogurt	Global definition	yogurt								YES or NO
	Suggested translation	酸奶								
16 Processed meat	Global definition	sausages	bacon	ham	larou	luncheon meat	beef jerky	processed beef product	pork jerk	YES or NO
	Suggested translation	腊肠	培根	火腿	腊肉	午餐肉	牛肉干	酱牛肉	猪肉脯	
17 Unprocessed red meat (ruminant)	Global definition	beef	lamb sheep or goat	donkey	horse	organs from these animals				YES or NO
	Suggested translation	牛肉	羊肉	驴肉	马肉	内脏 (牛羊驴马)				
18 Unprocessed red meat (non-ruminant)	Global definition	pork	pork organs							YES or NO
	Suggested translation	猪肉	猪内脏							
19 Poultry	Global definition	chicken	duck	goose	pigeon	chicken gizzard				YES or NO
	Suggested translation	鸡	鸭	鹅	鸽子	鸡胗				
20 Fish & seafood	Global definition	fish	seafood							YES or NO
	Suggested translation	鱼	海鲜							

Table 2. *Cont.*

		Yesterday, did you eat any of the following other foods?								
21 Nuts & seeds	Global definition	sunflower seeds	pumpkin seeds	watermelon seeds	peanut	chestnut	walnuts	almonds	sesame paste	YES or NO
	Suggested translation	葵花子	南瓜子	西瓜子	花生	栗子	核桃	杏仁	芝麻酱	
22 Ultra-processed packaged salty snacks	Global definition	chips such as Lays, Pringles	Doritos	shrimp chips	macaroni crisp	spicy strip				YES or NO
	Suggested translation	薯片(乐事, 品客)	立体脆	虾条	通心脆	辣条				
23 Instant noodles	Global definition	instant noodles	instant rice noodles							YES or NO
	Suggested translation	方便面	速食米粉							
24 Deep fried foods	Global definition	French fries	fried bread stick	fried pancake	fried dough twist	fried glutinous rice ball	fried bean curd	chicken nuggets	deep fried meet	YES or NO
	Suggested translation	薯条	油条	油饼	麻花	炸糕	炸豆腐	鸡块	炸肉	
		Yesterday, did you have any of the following beverages?								
25 Fluid milk	Global definition	milk	milk powder							YES or NO
	Suggested translation	牛奶	奶粉							
26 Sweetened tea/coffee/milk drinks	Global definition	flavored milks	milk tea/bubble tea	nutri-Express	yakult	bottled tea beverage	coffee with sugar			YES or NO
	Suggested translation	果味奶	奶茶/珍珠奶茶	营养快线	酸乳饮料	瓶装茶饮料	咖啡加糖			
27 Fruit juice	Global definition	fruit juice	fruit juice beverage							YES or NO
	Suggested translation	果汁	果汁饮料							
28 SSBs (sodas)	Global definition	soft drinks such as Coca cola, Pepsi, Fanta, Sprite						sports drink	energy drink	YES or NO
	Suggested translation	软饮料, 如可口可乐, 百事可乐, 芬达, 雪碧						运动饮料	能量饮料	
		Yesterday, did you get food from any place like?								
29 Fast food	Global definition	KFC	McDonald's	Pizza Hut	Burger King	Subway	Dicos			YES or NO
	Suggested translation	肯德基	麦当劳	必胜客	汉堡王	赛百味	德克士			

Note: SSB, sugar-sweetened beverages.

3.3. Sentinel Foods Items and Consumption Proportions for Each Food Group, Nationally and by Province

The sentinel food items (represented by "X") in 29 food groups nationally and by province are shown in Table 3. The quantity and kind of sentinel foods differed by province, due to the diverse dietary habits or different local crops of distinct provinces. Taking the staple food made from grains (group 1) as an example, the sentinel food items in Liaoning Province were rice, steamed buns and bread, whereas those in Henan Province were noodles, steamed buns and bread. However, in almost every food group, different provinces contained at least one sentinel food item in common. For instance, carrots, pumpkin or butternut squash, and sweet potatoes that are orange inside were three national sentinel foods for vitamin A-rich orange vegetables (group 5), and 12 provinces or municipalities had the same or two of the three.

Moreover, in almost all cases the DQQ selected items represented >95% of people who consumed each food group nationally, and over 90% in each province (Figure 2; detailed data shown in Table S1). For example, in the legumes group (group 4), 98.1% of people consumed the sentinel foods (bean curd or tofu, bean curd sheet, soybean milk, soybeans, and other dried beans) nationally, while same sentinel foods captured >95% of people consuming any legumes in almost every province; only one province (Shanghai) would have required additional sentinel food (soy meat) in this category (shown in Tables 3 and S1). Therefore, although there were differences in most common food items between different provinces, the DQQ is valid for use in any province of China, in addition to use at national level.

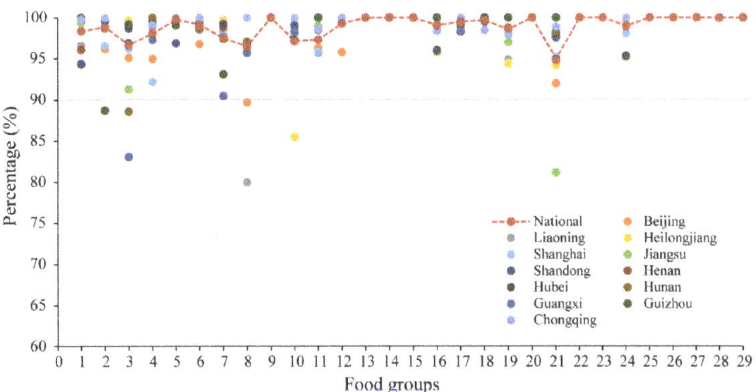

Figure 2. Percent consuming of DQQ sentinel foods in each food group compared to all items, nationally and by province (%).

Food group 1: staple foods made from grains; 2: whole grains; 3: white root/tubers; 4: legumes; 5: vitamin A-rich orange vegetables; 6: dark green leafy vegetables; 7: other vegetables; 8: vitamin A-rich fruits; 9: citrus; 10: other fruits; 11: grain-based sweets; 12: other sweets; 13: eggs; 14: cheese; 15: yogurt; 16: processed meats; 17: unprocessed red meat (ruminant); 18: unprocessed red meat (non-ruminant); 19: poultry; 20: fish and seafood; 21: nuts and seeds; 22: packaged ultra-processed salty snacks; 23: instant noodles; 24: deep fried foods; 25: fluid milk; 26: sweetened tea/coffee/milk drinks; 27: fruit juice; 28: sugar-sweetened beverages (SSBs) (sodas); 29: fast food.

Table 3. Sentinel foods for each food group nationally and by province.

Sentinel Foods	National	Beijing	Liaoning	Heilongjiang	Shanghai	Jiangsu	Shandong	Henan	Hubei	Hunan	Guangxi	Guizhou	Chongqing
Staple foods made from grains (group 1)													
rice	X	X	X	X	X	X	X		X	X	X	X	X
noodles	X	X					X	X	X	X			
steamed buns	X	X	X				X	X					
bread	X	X	X				X	X					
pancake ** in Beijing and Shandong		X					X						
Whole grain (group 2)													
millet	X	X	X	X	X	X	X	X	X	X	X		X
cornmeal	X	X	X	X		X	X	X	X		X	X	X
corn	X	X	X	X	X	X		X	X	X	X	X	X
whole wheat bread	X	X	X	X	X					X	X	X	X
oats	X	X			X	X				X			X
barley	X	X			X	X							
black rice	X												X
buckwheat * in Guizhou												X	
White root/tubers (group 3)													
potato	X	X	X	X	X	X	X	X	X	X	X	X	X
lotus root	X							X	X	X	X	X	X
starch noodles	X	X			X	X					X	X	X
turnip	X				X	X		X	X	X	X		X
yam	X	X			X		X				X	X	
taro	X				X	X	X		X	X			X
plantain ** in Jiangsu						X							
plantain * in Guangxi											X		
jicama * in Hunan										X			

Table 3. *Cont.*

Sentinel Foods	National	Beijing	Liaoning	Heilongjiang	Shanghai	Jiangsu	Shandong	Henan	Hubei	Hunan	Guangxi	Guizhou	Chongqing
Legumes (group 4)													
bean curd or tofu	X	X	X	X	X	X	X	X	X	X	X	X	X
bean curd sheet	X		X	X			X	X					X
soybean milk	X	X		X	X	X	X	X	X	X			X
soybeans	X	X			X	X	X	X	X	X	X	X	X
other dried beans	X	X	X	X	X	X	X	X	X	X		X	X
soy meat ** in Shanghai					X								
Vitamin A-rich orange vegetables (group 5)													
carrots	X	X	X	X	X	X	X	X	X	X	X	X	X
pumpkin or butternut squash	X	X	X	X		X	X	X	X	X	X	X	X
sweet potatoes that are orange inside	X	X	X		X	X	X	X		X			X
Dark green leafy vegetables (group 6)													
Chinese cabbage	X	X	X	X	X	X	X	X	X	X	X	X	X
water spinach	X					X			X		X		X
Chinses spinach	X	X	X	X	X	X	X	X	X	X			
rape	X	X		X	X	X	X	X	X				
bok choy	X	X	X		X	X		X	X	X	X	X	X
sweet potato leaves	X										X		X
broccoli	X	X			X	X							
mustard leaves	X		X							X			
chrysanthemum leaves	X					X							
radish leaves	X								X	X			
amaranth leaves	X									X			X

Table 3. *Cont.*

Sentinel Foods	National	Beijing	Liaoning	Heilongjiang	Shanghai	Jiangsu	Shandong	Henan	Hubei	Hunan	Guangxi	Guizhou	Chongqing
Other vegetables (group 7)													
cabbage	X	X	X	X	X	X	X	X	X	X	X	X	X
tomatoes	X	X	X	X	X	X	X	X	X	X	X	X	X
eggplant	X	X	X	X		X	X	X	X	X	X	X	X
loofah	X	X				X	X	X	X	X	X	X	X
cucumber	X	X	X	X			X	X	X	X	X	X	
local celery	X	X	X	X	X		X	X	X	X	X		
mushrooms	X	X	X	X	X	X					X	X	X
lettuce	X	X			X						X	X	X
radish	X		X		X	X	X	X	X		X		X
bean sprouts	X						X	X	X	X		X	X
cauliflower	X	X					X	X			X		
bamboo shoot	X				X	X					X		
bell pepper	X	X									X		
seaweed	X									X			
green beans	X	X	X	X					X	X	X	X	X
bitter melon ** in Guangxi and Guizhou											X	X	
zucchini ** in Guangxi											X		
chayote ** in Guizhou												X	
Vitamin A-rich fruits (group 8)													
persimmon	X	X	X	X	X	X	X	X	X	X		X	X
cantaloupe	X	X		X	X	X			X		X		
ripe mango	X			X							X		
papaya	X				X						X		
hawthorn berry * in Beijing and Liaoning		X	X										
Citrus (group 9)													
orange	X	X	X	X	X	X	X	X	X	X	X	X	X
tangerines	X	X	X			X				X	X		
pomelo	X	X	X		X	X			X	X			X
grapefruit	X	X											
kumquat	X							X					

Table 3. *Cont.*

Sentinel Foods	National	Beijing	Liaoning	Heilongjiang	Shanghai	Jiangsu	Shandong	Henan	Hubei	Hunan	Guangxi	Guizhou	Chongqing
Other fruits (group 10)													
apple	X	X	X	X	X	X	X	X	X	X	X	X	X
pear	X	X	X		X	X	X	X	X	X	X	X	X
watermelon	X			X		X					X	X	
banana	X		X	X	X	X	X	X	X	X	X		X
grapes	X	X	X	X		X	X		X	X	X	X	X
kiwi	X				X								X
dragonfruit	X	X			X								
jujube	X		X		X	X	X	X		X			X
longan	X									X	X		
wampee	X										X		
pomegranate	X									X			
lychee	X										X		
cherry	X											X	
peaches	X	X	X	X			X	X			X	X	
honeydew melon * in Heilongjiang				X									
Grain-based sweets (group 11)													
cakes	X	X	X		X	X	X	X	X	X	X	X	X
cookies	X	X	X	X	X	X	X	X			X	X	X
sweet pastries	X	X	X	X	X	X		X	X	X	X		
mooncake	X		X	X		X	X		X			X	X
rice dumplings	X											X	X
egg tart	X										X		
Other sweets (group 12)													
candy	X	X			X	X					X	X	X
chocolates	X	X			X	X		X			X		
popsicles	X	X	X	X		X					X	X	X
ice cream	X		X	X				X			X		
Jelly pudding	X		X		X	X	X	X			X		

Table 3. *Cont.*

Sentinel Foods	National	Beijing	Liaoning	Heilongjiang	Shanghai	Jiangsu	Shandong	Henan	Hubei	Hunan	Guangxi	Guizhou	Chongqing
Eggs (group 13)													
eggs	X	X	X	X	X	X	X	X	X	X	X	X	X
Cheese (group 14)													
cheese	X	X			X	X							
Yogurt (group 15)													
yogurt	X	X	X	X	X	X	X	X	X	X	X	X	X
Processed meats (group 16)													
ham	X	X	X	X	X	X	X	X	X	X	X		
bacon or larou	X				X	X			X	X	X	X	X
sausages	X	X	X	X	X	X	X		X	X	X	X	
beef jerky	X				X								
processed beef product	X	X				X	X						X
pork jerky	X		X		X	X	X	X	X		X	X	
Unprocessed red meat (ruminant) (group 17)													
beef	X	X	X	X	X	X	X	X	X	X	X	X	X
lamb sheep or goat	X	X	X		X		X	X		X		X	X
donkey	X			X			X						
organs from these animals	X				X	X				X	X		X
Unprocessed red meat (non-ruminant) (group 18)													
pork	X	X	X	X	X	X	X	X	X	X	X	X	X
pork organs	X					X		X					X
Poultry (group 19)													
chicken	X	X	X	X	X	X	X	X	X	X	X	X	X
duck	X	X			X	X	X		X	X	X	X	X
goose	X				X	X							
pigeon	X				X	X							
chicken gizzard	X										X		
chicken heart ** in Liaoning			X										
chicken liver ** in Heilongjiang				X									

Table 3. *Cont.*

Sentinel Foods	National	Beijing	Liaoning	Heilongjiang	Shanghai	Jiangsu	Shandong	Henan	Hubei	Hunan	Guangxi	Guizhou	Chongqing
Fish and seafood (group 20)													
fish	X	X	X	X	X	X	X	X	X	X	X	X	X
seafood	X	X	X	X	X	X	X	X			X	X	X
Nuts and seeds (group 21)													
peanuts	X	X	X	X	X	X	X	X	X	X	X	X	X
sunflower seeds	X	X	X	X	X		X	X		X		X	X
chestnuts	X	X	X		X	X	X		X	X		X	X
walnuts	X	X	X		X	X		X		X		X	X
sesame paste	X	X			X					X	X		X
almonds	X	X			X			X		X			
watermelon seeds	X				X		X		X				
pumpkin seeds	X				X							X	
seeds of Euryale ferox ** in Beijing		X											
seeds of Euryale ferox * in Jiangsu						X							
hazelnut ** in Liaoning			X										
pine nut ** in Heilongjiang				X									
pine nut * in Jiangsu						X							
lotus seeds ** in Guangxi											X		
Packaged ultra-processed salty snacks (group 22)													
chips	X	X	X		X				X			X	X
shrimp chips	X				X								
Instant noodles (group 23)													
instant noodles	X	X	X	X	X	X	X	X	X	X	X	X	X
Deep fried foods (group 24)													
fried bread stick	X	X	X	X	X	X	X	X	X	X	X	X	X
fried pancake	X	X	X	X	X	X	X	X	X	X		X	X
fried bean curd	X				X	X					X	X	
fried glutinous rice ball	X	X											X
chicken nugget	X							X					X

Table 3. *Cont.*

Sentinel Foods	National	Beijing	Liaoning	Heilongjiang	Shanghai	Jiangsu	Shandong	Henan	Hubei	Hunan	Guangxi	Guizhou	Chongqing
Fluid milk (group 25)													
milk	X	X	X	X	X	X	X	X	X	X	X	X	X
milk powder	X		X	X		X			X	X			
Sweet tea/coffee/milk drinks (group 26)													
flavored milk	X	X		X	X	X	X	X		X	X	X	X
Yakult	X	X			X	X	X	X	X	X	X		X
Nutri-Express	X				X						X		
bottled tea beverage	X											X	
Fruit juice (group 27)													
fruit juice	X	X	X	X	X	X	X	X	X	X	X	X	X
fruit juice beverage	X	X			X					X			
SSBs (sodas) (group 28)													
soft drink	X	X	X	X	X	X		X	X	X	X	X	X
sports drink	X	X			X			X					
energy drink	X									X			
Fast food (group 29)													
KFC	X	X	X		X	X	X						
burger	X	X			X		X			X			X
sandwich/sub	X				X	X							
pizza	X	X			X								

Note: "X" represents the sentinel food items of each food group nationally and by province; SSB = sugar-sweetened beverages; * means that the item is not included in the DQQ, and that <90.0% of people were captured by the DQQ sentinel foods in the corresponding province; ** means that the item is not included in the DQQ, and that between 90.0–95.0% of people were already captured by the DQQ sentinel foods in the corresponding province; The DQQ has included luncheon meat in group 16 (processed meat) even though it is not seen in the CHNS, since the people most likely to consume this food are outside of the sample (younger age groups); The DQQ has included horse in group 17 (uprocessed red meat (ruminant)) even though it is not seen in the CHNS, since the people most likely to consume this food are outside of the sample (ethnic minorities living in provinces not included in the survey).

4. Discussion

This present study validated the use of the DQQ in China, based on dietary intake data from CHNS 2011. It is the first study to report commonly-consumed food items in each of 29 food groups among Chinese residents by 12 provinces or municipalities, and to determine whether the DQQ sentinel food items are applicable for each province.

The DQQ for China is a low-burden tool to track trends in diet quality over time at low cost and without high technical expertise requirements or large-invested quantitative dietary intake surveys. If implemented widely, the diet quality monitoring data from the DQQ can give researchers and policymakers guidance to promote higher diet quality and heathier diet patterns in certain populations or in certain areas. Dietary data obtained from the DQQ could also be used to evaluate nutrient adequacy and trends of overnutrition, and may be useful in predicting the risk of diet-related NCDs due to its detailed food groups [21]. For China, a country with a long-standing food culture and a wide variety of food, DQQ would be helpful to measure the diet quality and diversity of Chinese people more accurately and more closely to the actual situation at national level. What's more, similar to a previous study which proved the reliability of sentinel food method [4], our results showed that sentinel foods the DQQ selected captured over 90% of people who consumed each food group in almost every province. Therefore, even if different dietary patterns exist, China DQQ can be effectively applied to each province to evaluate their trend of diet quality and to compare the differences between different provinces, so that decision makers could formulate intervention plans with a targeted approach.

This study has several strengths. First, to our knowledge, this was the first work that attempted to report commonly-consumed Chinese food items in 29 food groups nationally and by province, which could obtain more detailed information on dietary diversity at the population level. Second, the DQQ for China was validated to be effective and could be a low-burden indicator for diet quality measurement without substantial technical and financial support. Third, the development of DQQ for China was part of international collaborative research on diet quality which allowed for comparison among different population groups in different countries or regions. However, several limitations of this study should be noted. First, data collected by DQQ were effective at the population level and can be used to measure and track the diet quality of the population, but are not available for the dietary diagnosis of individuals. Second, DQQ is not designed to collect quantitative intake data, so quantitative surveys such as the CHNS will still be needed periodically to gather nutrient and energy intake information. Third, the application of diet quality indicators based on data from DQQ needs to be further verified in China. The GDR score has been validated in other countries [4]; work is underway to validate it in China. It is also possible to create a composite score which includes both the GDR score and Food Group Diversity Score (FGDS) to reflect nutrient adequacy and the global dietary recommendations, and further to capture the total diet quality [4]. These new scores require further validation, as they may complement or substitute for other indicators in use. Currently, the MDD-W (and its complement for the total population, the FGDS), is the most widely used diet-quality indicator based on food groups, but it did not perform well in reflecting total diet quality or meeting global dietary recommendations [4]. Previous studies reported that in various demographic groups in China, and in pregnant women in Bangladesh, MDD-W led to a high percentage of misclassification, indicating that its applications to other human groups should be well considered [17,22]. The development and application of DQQ may make improvements toward understanding diet quality at population level. Moreover, although this study used data from people over the age of 15, China's DQQ is developed for general population, including children and adolescents. The validity of China's DQQ would be validated in children and adolescents in the near future.

5. Conclusions

This study first described the most commonly consumed food items by 29 food groups nationally and by province and developed a low-burden, food group-based DQQ to evaluate diet quality at the population level. The DQQ could be used to evaluate multiple aspects of diet quality and risk factors for NCDs, and could be developed into a tool of low-burden, food group-based indicators, reflecting the dietary quality of the Chinese population.

Supplementary Materials: The following supporting information can be downloaded at: https://www.mdpi.com/article/10.3390/nu14091754/s1. Table S1: Percent consuming of DQQ sentinel foods in each food group compared to all items, nationally and by province (%).

Author Contributions: Conceptualization, Z.Z.; methodology, S.M., Z.Z., A.W.H. and C.V.; software, S.M.; validation, S.M., A.W.H. and C.V.; formal analysis, S.M.; investigation, A.W.H.; resources, A.W.H.; data curation, S.M.; writing—original draft preparation, S.M.; writing—review and editing, Z.Z., A.W.H. and C.V.; supervision, Z.Z.; project administration, Z.Z. All authors have read and agreed to the published version of the manuscript.

Funding: The study was funded by the National Natural Science Foundation of China (82073573 to Z.Z.) and the EU and BMZ through GIZ (Knowledge for Nutrition, K4N).

Institutional Review Board Statement: The study was approved by the Institutional Review Committee of the University of North Carolina at Chapel Hill, and the National Institute for Nutrition and Health, Chinese Center for Disease Control and Prevention.

Informed Consent Statement: Informed consent was obtained from all subjects.

Data Availability Statement: The dataset in the present study was open-accessed and freely obtained from the CHNS website with registration at https://www.cpc.unc.edu/projects/china/data/datasets/ (accessed on 22 March 2021).

Acknowledgments: This study used data from the China Health and Nutrition Survey (CHNS). We gratefully acknowledge the National Institute of Nutrition and Food Safety, China Centre for Disease Control and Prevention; the Carolina Population Centre, University of North Carolina at Chapel Hill; the National Institutes of Health (NIH; R01-HD30880, DK056350, and R01-HD38700); and the Fogarty International Centre, NIH, for their financial contribution towards the CHNS data collection and analysis files since 1989. We also thank all of the participants and the staff of CHNS involved in this study.

Conflicts of Interest: The authors declare no conflict of interest to disclose. The funders had no role in the design of the study, data collection, analyses, or interpretation of data, in the writing of the manuscript; or in the decision to publish the results.

References

1. GBD Risk Factor Collaborators. Health effects of dietary risks in 195 countries, 1990–2017: A systematic analysis for the Global Burden of Disease Study 2017. *Lancet* **2019**, *393*, 1958–1972. [CrossRef]
2. Popkin, B.M.; Corvalan, C.; Grummer-Strawn, L.M. Dynamics of the double burden of malnutrition and the changing nutrition reality. *Lancet* **2020**, *395*, 65–74. [CrossRef]
3. Food and Agriculture Organization of the United Nations (FAO). *Minimum Dietary Diversity for Woman*; FHI 360; Food and Agriculture Organization of the United Nations: Rome, Italy, 2016. [CrossRef]
4. Herforth, A.W.; Wiesmann, D.; Steele, E.M.; Andrade, G.; Monteiro, C.A. Introducing a Suite of Low-Burden Diet Quality Indicators That Reflect Healthy Diet Patterns at Population Level. *Curr. Dev. Nutr.* **2020**, *4*, nzaa168. [CrossRef] [PubMed]
5. Trijsburg, L.; Talsma, E.F.; De-Vries, J.H.M.; Kennedy, G.; Kuijsten, A.; Brouwer, I.D. Diet quality indices for research in low- and middle-income countries: A systematic review. *Nutr. Rev.* **2019**, *77*, 515–540. [CrossRef] [PubMed]
6. Gil, A.; De-Victoria, E.M.; Olza, J. Indicators for the evaluation of diet quality. *Nutr. Hosp.* **2015**, *31*, 128–144. [CrossRef] [PubMed]
7. Zhao, W.Z.; Zhang, J.; Zhao, A.; Wang, M.; Wu, W.; Tan, S.J.; Guo, M.F.; Zhang, Y.M. Using an introduced index to assess the association between food diversity and metabolic syndrome and its components in Chinese adults. *BMC Cardiovasc. Disord.* **2018**, *18*, 189. [CrossRef] [PubMed]
8. Jones, A.D.; Hoey, L.; Blesh, J.; Miller, L.; Green, A.; Shapiro, L.F. A Systematic Review of the Measurement of Sustainable Diets. *Adv. Nutr.* **2016**, *7*, 641–664. [CrossRef]
9. Diop, L.; Becquey, E.; Turowska, Z.; Huybregts, L.; Ruel, M.T.; Gelli, A. Standard Minimum Dietary Diversity Indicators for Women or Infants and Young Children Are Good Predictors of Adequate Micronutrient Intakes in 24–59-Month-Old Children and Their Nonpregnant Nonbreastfeeding Mothers in Rural Burkina Faso. *Nutr. J.* **2021**, *151*, 412–422. [CrossRef]

10. World Health Organization. *Global Nutrition Monitoring Framework: Operational Guidance for Tracking Progress in Meeting Targets for 2025*; World Health Organization: Geneva, Switzerland, 2017; ISBN 9789241513609.

11. Martin-Prével, Y.; Arimond, M.; Allemand, P.; Wiesmann, D.; Ballard, T.J.; Deitchler, M.; Dop, M.C.; Kennedy, G.; Lartey, A.; Lee, W.; et al. Development of a Dichotomous Indicator for Population-Level Assessment of Dietary Diversity in Women of Reproductive Age. *Curr. Dev. Nutr.* **2017**, *1*, cdn.117.001701. [CrossRef]

12. Monteiro, C.A.; Cannon, G.; Levy, R.B.; Moubarac, J.C.; Louzada, M.L.; Rauber, F.; Khandpur, N.; Cediel, G.; Neri, D.; Martinez-Steele, E.; et al. Ultra-processed foods: What they are and how to identify them. *Public Health Nutr.* **2019**, *22*, 936–941. [CrossRef]

13. Beckerman-Hsu, J.P.; Kim, R.; Sharma, S.; Subramanian, S.V. Dietary Variation among Children Meeting and Not Meeting Minimum Dietary Diversity: An Empirical Investigation of Food Group Consumption Patterns among 73,036 Children in India. *Nutr. J.* **2020**, *150*, 2818–2824. [CrossRef]

14. Herforth, A.; Steele, E.M.; Galixto, G.; Sattamini, I.; Olarte, D.; Ballard, T.; Monteiro, C. Development of a Diet Quality Questionnaire for Improved Measurement of Dietary Diversity and Other Diet Quality Indicators (P13-018-19). *Curr. Dev. Nutr.* **2019**, *3* (Suppl. S1), nzz036-P13. [CrossRef]

15. Zhang, B.; Zhai, F.Y.; Du, S.F.; Popkin, B.M. The China Health and Nutrition Survey, 1989–2011. *Obes. Rev.* **2014**, *15*, 2–7. [CrossRef]

16. Popkin, B.M.; Du, S.; Zhai, F.; Zhang, B. Cohort Profile: The China Health and Nutrition Survey—Monitoring and understanding socio-economic and health change in China, 1989–2011. *Int. J. Epidemiol.* **2010**, *39*, 1435–1440. [CrossRef] [PubMed]

17. Arimond, M.; Wiesmann, D.; Ramirez, S.R.; Levy, T.S.; Ma, S.; Zou, Z.Y.; Herforth, A.; Beal, T. *Food Group Diversity and Nutrient Adequacy: Dietary Diversity as a Proxy for Micronutrient Adequacy for Different Age and Sex Groups in Mexico and China*; Discussion Paper #9; Global Alliance for Improved Nutrition (GAIN): Geneva, Switzerland, 2021. [CrossRef]

18. Meng, L.P.; Wang, Y.; Li, T.; Loo-Bouwman, C.A.V.; Zhang, Y.M.; Szeto, I.M.Y. Dietary Diversity and Food Variety in Chinese Children Aged 3–17 Years: Are They Negatively Associated with Dietary Micronutrient Inadequacy? *Nutr. J.* **2018**, *10*, 1674. [CrossRef] [PubMed]

19. China Health and Nutrition Survey. Available online: https://www.cpc.unc.edu/projects/china/about/design/datacoll (accessed on 19 February 2022).

20. Wang, H.; Wang, D.; Ouyang, Y.; Huang, F.; Ding, G.; Zhang, B. Do Chinese Children Get Enough Micronutrients? *Nutr. J.* **2017**, *9*, 397. [CrossRef] [PubMed]

21. English, L.K.; Ard, J.D.; Bailey, R.L.; Bates, M.; Bazzano, L.A.; Boushey, C.J.; Brown, C.; Butera, G.; Callahan, E.H.; De-Jesus, J.; et al. Evaluation of Dietary Patterns and All-Cause Mortality: A Systematic Review. *JAMA Netw. Open* **2021**, *4*, e22277. [CrossRef] [PubMed]

22. Nguyen, P.H.; Huybregts, L.; Sanghvi, T.G.; Tran, L.M.; Frongillo, E.A.; Menon, P.; Ruel, M.T. Dietary Diversity Predicts the Adequacy of Micronutrient Intake in Pregnant Adolescent Girls and Women in Bangladesh, but Use of the 5-Group Cutoff Poorly Identifies Individuals with Inadequate Intake. *Nutr. J.* **2018**, *148*, 790–797. [CrossRef] [PubMed]

Article

Self-Reported Sedentary Behavior and Metabolic Syndrome among Children Aged 6–14 Years in Beijing, China

Ning Yin [1], Xiaohui Yu [2], Fei Wang [1], Yingjie Yu [2], Jing Wen [1], Dandan Guo [2], Yuanzhi Jian [1], Hong Li [2], Liyu Huang [2], Junbo Wang [1,3,*] and Yao Zhao [2,*]

[1] Department of Nutrition and Food Hygiene, School of Public Health, Peking University, Beijing 100191, China; yinning@bjmu.edu.cn (N.Y.); wangfei.duzhuo@163.com (F.W.); wenjing@bjmu.edu.cn (J.W.); 1610306107@pku.edu.cn (Y.J.)

[2] Beijing Center for Disease Prevention and Control, Beijing 100013, China; yxh770770@sina.com (X.Y.); yyj.jane.1982@163.com (Y.Y.); aldblm.tm@163.com (D.G.); benlih@sina.com (H.L.); huangly2006@163.com (L.H.)

[3] Beijing Key Laboratory of Toxicological Research and Risk Assessment for Food Safety, Peking University, Beijing 100191, China

* Correspondence: bmuwjbxy@bjmu.edu.cn (J.W.); yue1112@163.com (Y.Z.); Tel.: +86-10-8280-1575 (J.W.); +86-1368-158-3701 (Y.Z.)

Citation: Yin, N.; Yu, X.; Wang, F.; Yu, Y.; Wen, J.; Guo, D.; Jian, Y.; Li, H.; Huang, L.; Wang, J.; et al. Self-Reported Sedentary Behavior and Metabolic Syndrome among Children Aged 6–14 Years in Beijing, China. *Nutrients* **2022**, *14*, 1869. https://doi.org/10.3390/nu14091869

Academic Editor: Cristina Padez

Received: 14 April 2022
Accepted: 26 April 2022
Published: 29 April 2022

Publisher's Note: MDPI stays neutral with regard to jurisdictional claims in published maps and institutional affiliations.

Abstract: (1) Objective: This study aimed to examine the prevalence of metabolic syndrome (MetS) in children aged 6–14 years in Beijing, and to determine whether sedentary behavior is a risk factor. (2) Methods: Using a multistage stratified cluster random sampling method, 3460 students were selected for the Nutrition and Health Surveillance in Schoolchildren of Beijing (NHSSB). Data on children's sedentary behavior time and MetS indicators were collected using the questionnaires, physical measurements, and laboratory tests. MetS was defined according to the CHN2012 criteria, and logistic regression analysis was used to compare the effects of different sedentary time on MetS and its components. (3) Results: The overall prevalence of MetS among children aged 6–14 in Beijing was 2.4%, and boys, suburban children, and older age were associated with a higher prevalence (χ^2 values were 3.947, 9.982, and 27.463, respectively; $p < 0.05$). In boys, the prevalence rates of abdominal obesity, hyperglycemia, high triglycerides (TG), and low high-density lipoprotein cholesterol (HDL-C) were higher in the high-level sedentary behavior group than those in the low-level sedentary behavior group ($p < 0.05$); and in girls, the prevalence rates of high TG, low HDL-C, and MetS were higher in the high-level sedentary behavior group than those in the low-level sedentary behavior group ($p < 0.05$). After adjusting for confounding factors, the multivariate logistic regression results showed that compared with children with low-level sedentary behavior, the risks of abdominal obesity and low HDL-C were higher in boys with high-level sedentary behavior (odds ratio (OR) 1.51, 95% confidence interval (CI) 1.10–2.07, $p = 0.011$; OR 2.25, 95% CI 1.06–4.76, $p = 0.034$, respectively); while the risk of abdominal obesity was higher in girls with medium and high-level sedentary behavior (OR 1.52, 95% CI 1.01–2.27, $p = 0.043$; OR 1.59, 95% CI 1.04–2.43, $p = 0.032$, respectively). (4) Conclusions: Higher sedentary behavior time was related to the higher risk of MetS components among children aged 6–14 in Beijing. Reducing sedentary behavior may be an important method for preventing metabolic diseases.

Keywords: sedentary behavior; metabolic syndrome; children

Copyright: © 2022 by the authors. Licensee MDPI, Basel, Switzerland. This article is an open access article distributed under the terms and conditions of the Creative Commons Attribution (CC BY) license (https://creativecommons.org/licenses/by/4.0/).

1. Introduction

Metabolic syndrome (MetS) is a group of clinical syndromes that are closely related to lifestyle and characterized by a combination of obesity, hyperglycemia, hypertension, and dyslipidemia, including elevated triglyceride (TG) and low high-density lipoprotein cholesterol (HDL-C) [1,2].

In recent years, the prevalence of MetS in children and adolescents has gradually increased throughout the world [3–6]. Sina E et al. found that the prevalence of MetS was 5.5% in European children and adolescents aged 2–16 years [7]. According to the analysis by DeBoer et al., the prevalence of MetS by U.S. census division ranged from 4.6% to 13.6% among 4600 U.S. adolescents aged 12–19 years [8]. Ye et al. conducted a meta-analysis and found that the prevalence rates of MetS were between 1.8% and 2.6% among Chinese children and adolescents aged 6–20 years [9]. These data indicate that the prevalence of MetS in children and adolescents cannot be ignored. MetS or its components in childhood and adolescence may increase the risk of chronic diseases, such as cardiovascular disease (CVD) and type 2 diabetes mellitus (T2DM) in adulthood [10,11]. The increasing prevalence of MetS in children and adolescents may have implications for future global health burdens [12].

Sedentary behavior refers to the behavior in which energy consumption does not exceed 1.5 metabolic equivalents of task (METs) when sitting or in a reclining posture in an awake state in an education, family, community, and transportation environment [13–15]. It includes two types of sitting: sitting without looking at the screen and sitting in front of the screen; the time of the latter is called screen time. Song et al. analyzed the surveillance data of Chinese residents' nutrition and health status from 2010 to 2012 and found that the average daily sedentary behavior time of children aged 6–17 years was 2.92 h, and 85.8% of them had sedentary behaviors for more than two hours a day [16]. A survey conducted by Wu et al. in 12 provinces of China found that 16.2% and 41.5% of primary and middle school students, respectively, had screen time exceeding the recommended range of 2 h/day on study days and on weekends [17].

The "Guidelines on physical activity and sedentary behavior" issued by the World Health Organization (WHO) in 2020 suggests that, among children and adolescents, increased sedentary behavior is associated with the following adverse health outcomes: increased obesity, cardiovascular metabolism, and reduced sleep time; children and adolescents should limit the amount of time spent being sedentary [14]. At the same time, available studies have found that sedentary behavior is associated with an increased risk of CVD and MetS [15,18,19], and a systematic review by Tremblay et al. concluded that lower sedentary behavior, of any type, is associated with favorable health indicators [20].

China has experienced rapid social and economic changes in the past decade, consequently, lifestyle and health behaviors may have changed among Chinese children and adolescents. However, there are few descriptions of sedentary behavior in school-aged children in China, and few studies have analyzed the prevalence of MetS in school-aged children in Beijing. Therefore, we used data from the Nutrition and Health Surveillance in Schoolchildren of Beijing (NHSSB) in 2019 to describe the prevalence of MetS among children aged 6–14, analyze the relationship between sedentary behavior and MetS, and provide scientific evidence for its prevention and treatment.

2. Materials and Methods

2.1. Study Design and Participants

Beijing launched the Balanced Meal Campus Health Promotion Action Project in 2014–2020, which uses health education as the main means of guiding primary and secondary school students to develop healthy eating habits. To evaluate the changes in the nutritional health status of primary and middle school students during this period, the Beijing Center for Disease Prevention and Control conducted a baseline survey in 2015, a mid-term survey in 2017, and a third round of Nutrition and Health Surveillance in Schoolchildren of Beijing (NHSSB) in 2019. This surveillance adopted a multistage stratified cluster random sampling method (Figure 1).

This surveillance was approved by the Ethics Committee of the Beijing Center for Disease Prevention and Control (approval number: No. 14, 2019), and all participating students and their guardians signed the informed consent.

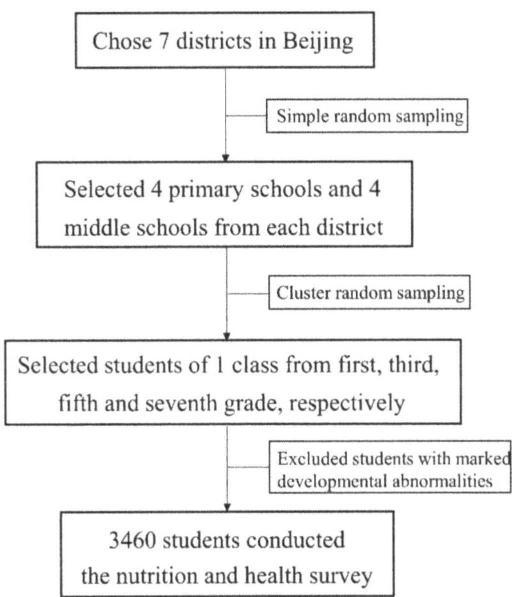

Figure 1. Flow diagram of the study participants.

2.2. Questionnaires Data Collection

The questionnaires used in this study were generated after expert argumentation based on the nutrition surveillance over the years. Questionnaires were designed for students and their parents separately. Students' nutrition, health-related knowledge, and eating behavior were obtained by the student questionnaire; basic family information, family behaviors, students' physical activity, and health information were obtained by the parental questionnaire. Questionnaire items on sedentary behaviors include (1) watching TV; (2) playing computers, tablets, mobile phones, and other electronic devices; (3) reading newspapers, novels, and other paper reading; (4) doing homework; (5) other sedentary activities. The sedentary behavior time of children on school days and weekends was measured separately. We only counted the sedentary behavior time of children after school considering that children have almost equal sedentary time in class.

2.3. Anthropometric Measurements

(1) Height and weight: The measurement of height and weight was carried out in accordance with the requirements of GB/T 26343-2010 "Technical standard for physical examination for students." Height was measured in centimeters (cm) to the nearest 0.1 cm; weight was measured in kilograms (kg) to the nearest 0.1 kg.

(2) Waist circumstance (WC): Using a tape measure, WC was measured horizontally at the midpoint between the inferior edge of the costal arch and the iliac crest in the mid-axillary line, at the end of a normal exhalation. The measurement was performed twice, with cm as the unit and accurate to 0.1 cm. The average of two repeated measurements was calculated for WC.

2.4. Blood Pressure Measurement

The measurement instrument was a digital sphygmomanometer (model HBP1300; Omron Healthcare (China) Co., Ltd., Liaoning, China). BP was obtained from the left arm using the appropriate cuff for each participant. The participant rested quietly for at least 5 min before the measurement and at the same time we ensured that the measurement environment was quiet and comfortable. BP was measured three times with a 1-min interval between repetitions; the average of three measures was calculated.

2.5. Laboratory Biochemical Examination

After 10–12 h overnight fasting, blood samples were collected by qualified medical physicians, centrifuged at 3000 rpm for 10 min within 30 min, collected the supernatant to aliquot, and then stored in a refrigerator at $-80\ °C$. Blood glucose, blood lipids, and other biochemical indicators were tested by an automatic biochemical analyzer (model 7600; Hitachi High-Tech (China) Co., Ltd., Beijing, China) and corresponding reagents (Wako Pure Chemical Industries, Ltd., Osaka, Japan). Glucose, TG, and HDL-C were analyzed by hexokinase method, free glycerol method, and direct method, respectively.

2.6. Calculation or Diagnostic Criteria

2.6.1. Sedentary Behavior

Sedentary behavior time was estimated by counting the minutes per day spent watching TV, computer, tablet computer, mobile phone, and other screens, reading newspapers, novels, and other paper reading time, and doing homework and other sedentary activities. Average sedentary behavior time could be calculated by the following formula:

$$\text{Sedentary behavior time (min/day)} = (\text{sedentary behavior time on school days} \times 5 + \text{sedentary behavior time on weekends} \times 2)/7$$

The sedentary behavior time was categorized in tertiles (low, medium, and high). There were 1090, 1114, and 1075 participants in the low-level, medium-level and high-level groups, respectively.

2.6.2. Metabolic Syndrome

MetS was defined according to the CHN2012 criteria [21]. Abdominal obesity is an essential component for the diagnosis of MetS. The waist-to-height ratio (WHtR) is used as the screening index [21–24], and the cut-off point of WHtR is 0.48 for boys and 0.46 for girls as the screening cut-off value for abdominal obesity [22]. At least two components of the following should be provided at the same time: (1) Hyperglycemia: a. Impaired fasting blood glucose (IFG): fasting blood glucose ≥ 5.6 mmol/L; b. or impaired glucose tolerance (IGT): oral glucose tolerance test (OGTT) 2 h blood glucose ≥ 7.8 mmol/L, but <11.1 mmol/L; c. or type 2 diabetes [21]. (2) Hypertension: using the rapid identification method, children aged ≥ 10 years with systolic blood pressure (SBP) ≥ 130 mmHg and/or diastolic blood pressure (DBP) ≥ 85 mmHg are identified to be hypertension; children aged 6–10 years with SBP ≥ 120 mmHg and/or DBP ≥ 80 mmHg are identified to be hypertension [22]. (3) Low HDL-C: HDL-C < 1.03 mmol/L or non-HDL-C ≥ 3.76 mmol/L. (4) High TG: TG ≥ 1.47 mmol/L [21].

2.7. Statistical Analysis

EpiData 3.0 software (The Epi Data Association, Odense, Denmark) was used for data entry, and all data received were double-checked. All analyses were performed with IBM SPSS version 26.0 software (IBM, Aromonk, NY, USA). Continuous variables were reported as mean $\pm$ standard deviation or the median and interquartile range (IQR), and categorical variables were reported as frequency (percentage). Chi-squared test was used to compare the constituent ratios between groups, *t*-test was used to compare the differences in height, weight, WC, glucose, and other indicators between the MetS and the non-MetS groups, nonparametric tests were used to compare differences in time to sedentary behavior between groups, and binary logistic regression was used to analyze the correlation between sedentary behavior and MetS and its components. A two-sided $p < 0.05$ was considered statistically significant.

3. Results

3.1. The Status Quo of MetS

The overall prevalence of MetS among students surveyed was 2.4%. The prevalence rates of abdominal obesity, hypertension, hyperglycemia, high TG, and low HDL-C were 23.0%, 4.0%, 7.9%, 7.8%, and 5.5%, respectively. Table 1 shows the distribution of participants' characteristics by MetS. Boys were more likely to have MetS than girls ($\chi^2 = 3.947$, $p = 0.047$) and suburbs were more prevalent in the MetS group than in the non-MetS group ($\chi^2 = 9.982$, $p < 0.01$). The prevalence rates of MetS among students in different age groups were also different ($\chi^2 = 27.463$, $p < 0.001$), and the prevalence was higher with increasing age. Children in the MetS group had longer sedentary behavior time ($Z = 3.409$, $p = 0.001$). The high-level sedentary behavior group had more MetS than the low-level sedentary behavior group ($\chi^2 = 9.363$, $p < 0.01$). All components of MetS differed between the two groups, and the MetS group showed worse profiles with higher WC, WHtR, systolic and diastolic blood pressure, fasting glucose, and triglyceride, and lower HDL-C ($p < 0.001$).

Table 1. Characteristics of participants by MetS #.

Group	Total ($n = 3392$)	Non-MetS ($n = 3311$)	MetS ($n = 81$)	*p*-Value
Gender				0.047 *
Boy	1724(50.8)	1674(50.6)	50(61.7)	
Girl	1668(49.2)	1637(49.4)	31(38.3)	
Age (years)				<0.001 ***
6~8	895(26.4)	886(26.8)	9(11.1)	
8~10	805(23.7)	796(24.0)	9(11.1)	
10~12	859(25.3)	831(25.1)	28(34.6)	
12~14	833(24.6)	798(24.1)	35(43.2)	
Residence				0.002 **
Urban	1328(39.2)	1310(39.6)	18(22.2)	
Suburbs	2064(60.8)	2001(60.4)	63(77.8)	
Height (cm)	140.95 ± 15.21	140.67 ± 15.13	152.33 ± 14.53	<0.001 ***
Weight (kg)	37.52 ± 14.12	36.96 ± 13.59	60.84 ± 15.86	<0.001 ***
Average sedentary time (min/day)	175.7(128.6228.6)	175.1(127.1227.9)	208.9(154.6251.8)	0.001 **
Sedentary behavior time group				0.009 **
Low level	1069(33.3)	1055(33.6)	14(18.4)	
Medium level	1091(33.9)	1064(33.9)	27(35.5)	
High level	1055(32.8)	1020(32.5)	35(46.1)	
Caregiver's education				0.125
Junior high school or below	477(14.1)	462(14.0)	15(18.5)	
High school/Technical school	640(18.9)	621(18.8)	19(23.5)	
College/Vocational college	703(20.7)	683(20.6)	20(24.7)	
Undergraduate or above	1571(46.3)	1544(46.6)	27(33.3)	
Per capita household income (CNY/year)				0.257
<20,000	307(9.1)	297(9.0)	10(12.3)	
20,000~39,999	489(14.4)	473(14.3)	16(19.8)	
40,000~69,999	721(21.3)	702(21.2)	19(23.5)	
≥70,000	1427(42.1)	1402(42.3)	25(30.9)	
Not clear	448(13.2)	437(13.2)	11(13.6)	

Table 1. *Cont.*

Group	Total (*n* = 3392)	Non-MetS (*n* = 3311)	MetS (*n* = 81)	*p*-Value
Leisure time MVPA (min/week)				0.826
≤60	1151(40.3)	1126(40.4)	25(39.7)	
61~120	710(24.9)	693(24.8)	17(27.0)	
121~240	611(21.4)	600(21.5)	11(17.5)	
>240	381(13.4)	371(13.3)	10(15.9)	
WC (cm)	62.40 ± 11.37	61.71 ± 10.45	82.60 ± 8.61	<0.001 ***
WHtR	0.44 ± 0.06	0.44 ± 0.06	0.54 ± 0.05	<0.001 ***
SBP (mmHg)	109.03 ± 10.14	108.62 ± 9.73	121.26 ± 11.16	<0.001 ***
DBP (mmHg)	64.32 ± 7.08	64.18 ± 7.01	68.12 ± 7.31	<0.001 ***
Fasting glucose (mmol/L)	5.03 ± 0.43	5.03 ± 0.42	5.36 ± 0.51	<0.001 ***
Serum TG (mmol/L)	0.78 ± 0.47	0.75 ± 0.41	1.83 ± 0.92	<0.001 ***
Serum HDL-C (mmol/L)	1.46 ± 0.29	1.47 ± 0.29	1.01 ± 0.18	<0.001 ***

#: Data were presented as means ± standard deviations, or median (P_{25}, P_{75}), or frequency values (percentages, %). Abbreviations: WC, waist circumference; WHtR, waist-to-height ratio; SBP, systolic blood pressure; DBP, diastolic blood pressure; TG, triglyceride; HDL-C, high-density lipoprotein cholesterol; MVPA, moderate-to-vigorous intensity physical activity. *: $p < 0.05$, **: $p < 0.01$, ***: $p < 0.001$.

3.2. The Status Quo of Sedentary Behavior

The median sedentary behavior time of the participants was 175.7 (128.6, 228.6) min, the mean was 182.9 ± 75.2 min, and 77.5% of participants had more than 2 h of sedentary behavior per day. There were no statistically significant differences between the sedentary behavior time of participants of different genders and regions (Z = 0.599, p = 0.446), but screen time of the boys was higher than that of the girls (Z = 4.034, p < 0.001), and screen time of suburban students was higher than that of urban students (Z = 7.697, p < 0.001). Older children had longer daily sedentary behavior time, and the difference in each age group was statistically significant (p < 0.001). Figure 2 shows the composition of sedentary behavior among children by gender, age, residence, and school days or weekends. The median sedentary behavior times on school days and weekends were 140.0 min and 240.0 min, respectively.

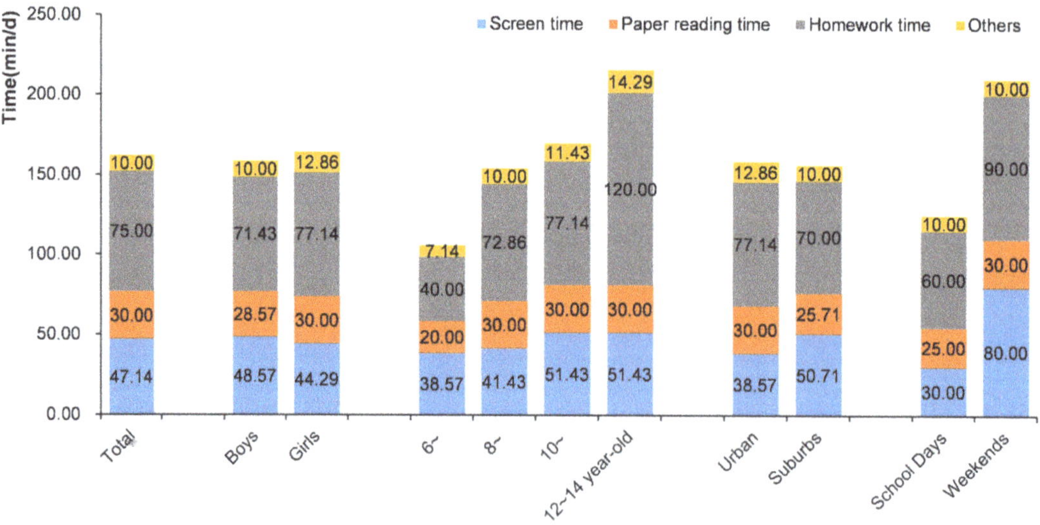

Figure 2. The composition of sedentary behavior time among children by gender, age, residence, and on school days or weekends.

3.3. Relationship between Sedentary Behavior and MetS

Table 2 shows the prevalence of MetS components in the different sedentary behavior time groups by gender. In boys, the prevalence rates of abdominal obesity, hyperglycemia, high TG, and low HDL-C were higher in the high-level sedentary behavior group than those in the low-level sedentary behavior group ($p < 0.05$); and in girls, the prevalence rates of high TG, low HDL-C, and MetS were higher in the high-level sedentary behavior group than those in the low-level sedentary behavior group ($p < 0.05$). Figure 3 shows the percentages of people with abnormal numbers of metabolism in different sedentary behavior time groups by gender. The high-level sedentary behavior group also had more metabolic abnormalities than the low-level sedentary behavior group in both boys and girls ($p < 0.001$).

Table 2. Prevalence of MetS components in different sedentary behavior time groups by gender #.

	Boys				Girls			
	Low Level	Medium Level	High Level	*p* Value	Low Level	Medium Level	High Level	*p* Value
Abdominal obesity	140(25.2)	168(30.3)	172(33.0)	0.018 *	68(13.2)	91(17.0)	90(16.9)	0.161
Hypertension	22(3.9)	25(4.4)	35(6.5)	0.101	13(2.5)	21(3.8)	17(3.2)	0.460
Hyperglycemia	38(6.7)	50(8.8)	61(11.4)	0.024 *	25(4.8)	33(6.0)	42(7.8)	0.123
High TG	30(5.3)	43(7.6)	50(9.3)	0.036 *	29(5.5)	50(9.1)	54(10.0)	0.020 *
Low HDL-C	16(2.8)	29(5.1)	38(7.1)	0.005 **	19(3.6)	37(6.7)	41(7.6)	0.017 *
MetS	10(1.8)	17(3.1)	19(3.6)	0.174	4(0.8)	10(1.9)	16(3.0)	0.031 *

#: Data were presented as frequency values (percentages, %). *p*-values were obtained by the chi-squared test, *: $p < 0.05$, **: $p < 0.01$. Abbreviations: TG, triglyceride; HDL-C, high-density lipoprotein cholesterol; MetS: metabolic syndrome.

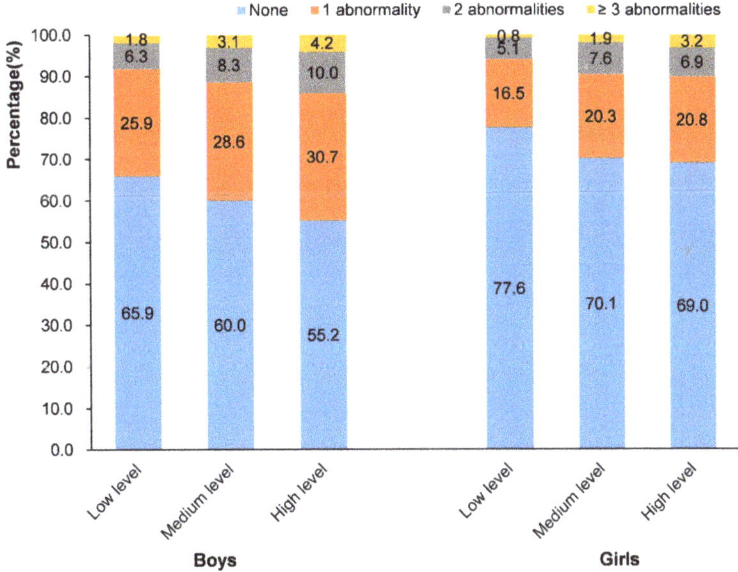

Figure 3. Percentages of people with abnormal numbers of metabolism in different sedentary behavior time groups by gender.

Single-factor logistic regression analysis of sedentary behavior time and MetS and its components revealed that compared with low-level sedentary behavior time, high-level sedentary behavior time had higher risks of abdominal obesity, hyperglycemia, high TG, low HDL-C, and MetS ($p < 0.05$). After adjusting for the confounding factors of age, gender, residence, caregiver's education, per capita household income, and leisure time moderate-to-vigorous intensity physical activity (MVPA), further analysis found that, compared with children with low-level sedentary behavior, the risks of abdominal obesity, high TG, and low HDL-C were higher in children with high-level sedentary behavior (odds ratio (OR) 1.49, 95% confidence interval (CI) 1.16–1.92, $p < 0.01$; OR 1.57, 95% CI 1.05–2.35, $p = 0.028$; OR 2.02, 95% CI 1.22–2.32, $p < 0.01$, respectively) (Table 3).

Table 3. Logistic regression analysis of sedentary behavior time, MetS, and its components of children aged 6–14 years (OR (95% CI of OR)).

Model	Sedentary Time	Abdominal Obesity	Hypertension	Hyperglycemia	High TG	Low HDL-C	MetS
Crude	Low level	1 (ref)	1 (ref)	1 (ref)	1 (ref)	1 (ref)	1 (ref)
	Medium level	1.29(1.05,1.59) *	1.30(0.83,2.03)	1.31(0.94,1.84)	1.59(1.14,2.23) **	1.90(1.25,2.89) **	1.91(1.00,3.67)
	High level	1.37(1.11,1.68) **	1.53(0.99,2.37)	1.73(1.25,2.39)**	1.87(1.34,2.61) ***	2.39(1.59,3.59) ***	2.59(1.38,4.83) **
Model 1	Low level	1 (ref)	1 (ref)	1 (ref)	1 (ref)	1 (ref)	1 (ref)
	Medium level	1.25(0.98,1.59)	1.05(0.61,1.82)	1.01(0.67,1.51)	1.44(0.98,2.14)	1.57(0.95,2.58)	1.39(0.65,2.96)
	High level	1.49(1.16,1.92) **	1.23(0.71,2.14)	1.26(0.84,1.89)	1.57(1.05,2.35) *	2.02(1.22,3.32) **	1.63(0.76,3.47)

Crude: Unadjusted confounding variables; Model 1: Adjusted for age, gender, residence, caregiver's education, per capita household income, and leisure time moderate-to-vigorous intensity physical activity; ref: reference; *: $p < 0.05$, **: $p < 0.01$, ***: $p < 0.001$.

A stratified analysis by gender found that compared with children with low-level sedentary behavior, the risks of abdominal obesity and low HDL-C were higher in boys with high-level sedentary behavior (OR 1.51, 95% CI 1.10–2.07, $p = 0.011$; OR 2.25, 95% CI 1.06–4.76, $p = 0.034$, respectively); the risk of abdominal obesity was higher in girls with medium and high-level sedentary behavior (OR 1.52, 95% CI 1.01–2.27, $p = 0.043$; OR 1.59, 95% CI 1.04–2.43, $p = 0.032$, respectively) (Table 4).

Table 4. Logistic regression analysis of sedentary behavior time, MetS, and its components of children aged 6–14 years by gender # (OR (95% CI of OR)).

Sedentary Time	Abdominal Obesity	Hypertension	Hyperglycemia	High TG	Low HDL-C	MetS
Boys						
Low level	1 (ref)	1 (ref)	1 (ref)	1 (ref)	1 (ref)	1 (ref)
Medium level	1.15(0.85,1.56)	0.67(0.32,1.41)	1.06(0.62,1.81)	1.35(0.77,2.36)	1.53(0.71,3.30)	1.18(0.45,3.06)
High level	1.51(1.10,2.07) *	0.91(0.44,1.86)	1.41(0.83,2.40)	1.53(0.87,2.70)	2.25(1.06,4.76) *	1.43(0.56,3.68)
Girls						
Low level	1 (ref)	1 (ref)	1 (ref)	1 (ref)	1 (ref)	1 (ref)
Medium level	1.52(1.01,2.27) *	1.87(0.80,4.37)	0.89(0.47,1.68)	1.56(0.89,2.71)	1.61(0.83,3.12)	2.08(0.58,7.42)
High level	1.59(1.04,2.43) *	1.89(0.77,4.64)	1.03(0.54,1.96)	1.61(0.91,2.86)	1.83(0.93,3.59)	2.25(0.61,8.26)

#: Adjusted for age, residence, caregiver's education, per capita household income, and leisure time moderate-to-vigorous intensity physical activity; *: $p < 0.05$.

4. Discussion

This study aimed to examine the prevalence of metabolic syndrome (MetS) in children aged 6–14 years in Beijing and determine whether sedentary behavior is a risk factor. Through the cross-sectional study design, the overall prevalence of MetS among children aged 6–14 in Beijing in 2019 was 2.4%, which was consistent with the results of the 2017 Guangzhou City and 2010 nationwide surveys among the same age group [25,26], which was lower than the 3.6% in Pudong New Area, Shanghai [27]. Boys were more likely to have MetS than girls, which was consistent with the results of most studies [25,26,28]. Possible reasons for this result are the physiological structure and genetic differences between boys and girls, and the prevalence of abdominal obesity among boys is much higher than that of girls. Some studies have shown that obesity is an important factor affecting the prevalence of MetS [2,29,30]. In addition, the screen time of the boys was higher than that of the girls. Carson et al. found that current evidence suggests that screen time may have a larger impact on health than overall sedentary time [19]. Suburbs were more prevalent in the MetS group than in the non-MetS group, which may also be related to the higher obesity rate and screen time among students in suburban areas than in urban areas in Beijing [31]. The prevalence rates of MetS among students in different age groups were also different. The overall trend was that prevalence goes up with increasing age. This trend was consistent with the results of previous studies [25,28]. The reason for this result may be that individual hormone levels and body fat distribution characteristics change significantly with age [32].

The study found that the median sedentary behavior time for children aged 6–14 in Beijing was 175.7 min, and the mean was 182.9 ± 75.2 min, which was similar to the average total sedentary behavior time in Guangzhou in 2017 of 3.06 ± 1.34 h [33], and 77.5% of students had more than 2 hours of sedentary behavior per day, which was lower than the 86.2% reported by Song et al. [16]. The median screen time was 47.14 min, and the average daily screen time exceeding 2 h for children accounted for 8.2%, slightly lower than the 9.0% in Guangzhou in 2017 [25] and higher than the 6.7% in Beijing in 2018 [34]. Changes in social development and lifestyles, as well as cultural differences in different regions, may be responsible for the different sedentary behavior time and screen time of children. In addition, inconsistencies in findings may reflect inconsistencies in data collection methods. Therefore, it is necessary to establish a uniform standard for the collection of sedentary time for children to facilitate comparisons between multiple regions. In this study, almost half of the sedentary behavior time came from homework, which is a problem that should be of concern. Moreover, the government of China released the Opinions on Further Reducing the Burden of Homework and Off-Campus Training for Compulsory Education Students ("Double Reduction" policy) on 24 July 2021; future studies focused on the effect of this policy on children's sedentary time are expected.

This study found that the prevalence of each MetS component was higher with longer sedentary time. In boys, the prevalence rates of abdominal obesity, hyperglycemia, high TG, and low HDL-C were higher in the high-level sedentary behavior group than those in the low-level sedentary behavior group; and in girls, the prevalence rates of high TG, low HDL-C, and MetS were higher in the high-level sedentary behavior group than those in the low-level sedentary behavior group. It suggested that the effects of sedentary behavior on MetS and its components might be different in different genders. Thus, gender should be considered in studies on metabolic outcomes. This study used waist-to-height ratio (WHtR) to screen for abdominal obesity. Several studies have shown that WHtR is more predictive of abdominal obesity in children and adolescents than body mass index [23,24,35]. It can be a useful substitute for body obesity when skinfold measurements are not available. Because height affects waist circumference, correcting WHtR can eliminate the effect of height and the potential impact on reference values related to age, gender, and race [36]. Research by Jorge Mota et al. found that boys with higher sedentary behaviors are more likely to suffer from central obesity [37]. Available evidence indicates that sedentary behavior is also independently associated with an increased risk of abnormal glucose metabolism, including decreased insulin sensitivity [38] and increased fasting insulin levels [39,40]. Hjorth et al. showed that a longer duration of accelerometer-derived sedentary time was significantly associated with lower high-density lipoprotein (HDL) cholesterol [41].

A longitudinal study by Greer et al. found that men with moderate (12–19 h/week) and high (>19 h/week) sedentary behaviors had 65% and 76% higher risks of MetS than men with low sedentary behaviors (<12 h/week), respectively [42]. Berg et al. used an accelerometer to measure sedentary behavior time, and further analysis found that an additional hour of static activity was associated with a 39% increase in the risk of MetS [43]. Edwardson et al. also conducted a meta-analysis and found that the longer the sedentary behavior, the greater the risk of MetS. Reducing sedentary behavior may play an important role in preventing MetS [44]. In this study, after adjusting for confounders, compared with children with low sedentary behavior time, the risk of MetS was not higher in children with high sedentary behavior time, but the risks of abdominal obesity and low HDL-C were higher in boys with high-level sedentary behavior, while the risk of abdominal obesity was higher in girls with medium and high-level sedentary behavior. A long sedentary behavior time has been suggested to be correlated with metabolic abnormalities in children. The mechanisms underlying the adverse effects of sedentary behavior on metabolic syndrome remain unclear. A possible reason for this is that prolonged sedentary behavior affects body weight and lipid metabolism [41,45].

When adjusting for confounding factors, leisure time MVPA, which might affect the prevalence of MetS, was not statistically significant, further verifying the difference between sedentary behavior and physical inactivity. A few studies have found that even if the time of MVPA meets the requirements of the recommended guidelines, excessive sedentary behavior still leads to metabolic abnormalities [42,45,46]. The amount of physical activity does not cancel out the negative effects of sedentary behavior, so it should be clear that sedentary behavior and physical inactivity are two different concepts that affect health independently.

However, this study has some limitations. First, this study was a cross-sectional investigation, which could not clarify the causal relationship between sedentary behavior and MetS. Therefore, a prospective study should be conducted. Second, the collection of sedentary behavior time depended on students' self-reports or parents' reports, which were prone to recall and reporting bias. In addition, sedentary traffic time refers to the behavior time in which energy consumption does not exceed 1.5 METs in the transportation environment. However, due to the lack of a feasible and effective measurement method for sedentary traffic time, it is not involved in this paper, which may lead to a lower total sedentary time and underestimate its influence. Therefore, objective devices such as accelerometers, inclinometers, and sports bracelets should be used to measure the level of sedentary behavior, or a reliable and effective subjective measurement method of sedentary behavior should be developed and used regularly. Third, the International Diabetes Federation (IDF) considers impaired fasting glucose (fasting blood glucose (FPG) $\geq$5.6 mmol/L) as an independent indicator of abnormal glucose metabolism. However, clinical studies have found that FPG is not stable enough for being easily influenced by factors such as emotion, stress, and diseases [22]. Therefore, OGTT or repeated measurement of FPG should be performed in future experiments.

5. Conclusions

Higher sedentary behavior time is related to the higher risk of MetS components among children aged 6–14 years in Beijing. Reducing sedentary behavior may be an important method for preventing metabolic diseases.

Author Contributions: Conceptualization, N.Y. and X.Y.; methodology, N.Y. and Y.Y.; software, N.Y.; validation, N.Y., F.W. and Y.Y.; formal analysis, N.Y.; investigation, X.Y., Y.Y., D.G., H.L. and L.H.; resources, J.W. (Junbo Wang) and Y.Z.; data curation, N.Y., J.W. (Jing Wen) and Y.J.; writing—original draft preparation, N.Y.; writing—review and editing, N.Y., Y.Y. and J.W. (Junbo Wang); project administration, J.W. (Junbo Wang) and Y.Z. All authors have read and agreed to the published version of the manuscript.

Funding: This research received no external funding.

Institutional Review Board Statement: This study was approved by the Ethics Committee of the Beijing Center for Disease Prevention and Control (approval number: No. 14, 2019).

Informed Consent Statement: Informed consent was obtained from all students and their guardians involved in the study.

Data Availability Statement: The data presented in this study are available on request from the corresponding author. The data are not publicly available due to privacy.

Acknowledgments: We are deeply thankful to all school personnel, students, and their parents of the selected schools for their cooperation in this project.

Conflicts of Interest: The authors declare no conflict of interest.

References

1. Zimmet, P.; Alberti, G.; Kaufman, F.; Tajima, N.; Silink, M.; Arslanian, S. The metabolic syndrome in children and adolescents. *Lancet* **2007**, *369*, 2059–2061. [CrossRef]
2. Mi, J. Childhood obesity and metabolic syndrome. *Chin. J. Child Health Care* **2007**, *3*, 221–223.

3.   Ford, E.S.; Li, C.; Zhao, G.; Pearson, W.S.; Mokdad, A.H. Prevalence of the metabolic syndrome among U.S. adolescents using the definition from the international diabetes federation. *Diabetes Care* **2008**, *31*, 587–589. [CrossRef] [PubMed]

4.   Li, Y.; Yang, X.; Zhai, F.; Kok, F.J.; Zhao, W.; Piao, J.; Zhang, J.; Cui, Z.; Ma, G. Prevalence of the metabolic syndrome in chinese adolescents. *Br. J. Nutr.* **2008**, *99*, 565–570. [CrossRef]

5.   Cook, S.; Weitzman, M.; Auinger, P.; Nguyen, M.; Dietz, W.H. Prevalence of a metabolic syndrome phenotype in adolescents: Findings from the third national health and nutrition examination survey, 1988–1994. *Arch. Pediatrics Adolesc. Med.* **2003**, *157*, 821–827. [CrossRef]

6.   Kim, H.S.; Demyen, M.F.; Mathew, J.; Kothari, N.; Feurdean, M.; Ahlawat, S.K. Obesity, metabolic syndrome, and cardiovascular risk in gluten-free followers without celiac disease in the united states: Results from the national health and nutrition examination survey 2009–2014. *Dig. Dis. Sci.* **2017**, *62*, 2440–2448. [CrossRef]

7.   Sina, E.; Buck, C.; Veidebaum, T.; Siani, A.; Reisch, L.; Pohlabeln, H.; Pala, V.; Moreno, L.A.; Molnar, D.; Lissner, L.; et al. Media use trajectories and risk of metabolic syndrome in European children and adolescents: The IDEFICS/I.Family cohort. *Int. J. Behav. Nutr. Phys. Act.* **2021**, *18*, 134. [CrossRef]

8.   DeBoer, M.D.; Filipp, S.L.; Gurka, M.J. Geographical variation in the prevalence of obesity and metabolic syndrome among us adolescents. *Pediatric Obes.* **2019**, *14*, e12483. [CrossRef]

9.   Ye, P.; Yan, Y.; Ding, W.; Dong, H.; Liu, Q.; Huang, G.; Mi, J. Prevalence of metabolic syndrome in Chinese children and adolescents: A meta-analysis. *Zhonghua Liuxingbingxue Zazhi* **2015**, *36*, 884–888.

10. Cornier, M.A.; Dabelea, D.; Hernandez, T.L.; Lindstrom, R.C.; Steig, A.J.; Stob, N.R.; van Pelt, R.E.; Wang, H.; Eckel, R.H. The metabolic syndrome. *Endocr. Rev.* **2008**, *29*, 777–822. [CrossRef]

11. De Boer, M.D. Assessing and managing the metabolic syndrome in children and adolescents. *Nutrients* **2019**, *11*, 1788. [CrossRef]

12. Kassi, E.; Pervanidou, P.; Kaltsas, G.; Chrousos, G. Metabolic syndrome: Definitions and controversies. *BMC Med.* **2011**, *9*, 48. [CrossRef] [PubMed]

13. Sedentary Behaviour Research Network. Letter to the editor: Standardized use of the terms "sedentary" and "sedentary behaviours". *Appl. Physiol. Nutr. Metab.* **2012**, *37*, 540–542. [CrossRef] [PubMed]

14. WHO. Who Guidelines on Physical Activity and Sedentary Behaviour. Available online: https://www.who.int/publications/i/item/9789240015128 (accessed on 19 April 2022).

15. Tremblay, M.S.; Aubert, S.; Barnes, J.D.; Saunders, T.J.; Carson, V.; Latimer-Cheung, A.E.; Chastin, S.F.M.; Altenburg, T.M.; Chinapaw, M.J.M. Sedentary behavior research network (sbrn)—Terminology consensus project process and outcome. *Int. J. Behav. Nutr. Phys. Act.* **2017**, *14*, 75. [CrossRef]

16. Song, C.; Guo, H.; Gong, W.; Zhang, Y.; Ding, C.; Feng, G.; Liu, A. Status of sedentary activities in the leisure time in chinese pupils in 2010–2012. *J. Hyg. Res.* **2017**, *46*, 705–708+721.

17. Wu, X.; Tao, S.; Zhang, S.; Zhang, Y.; Huang, K.; Tao, F. Analysis on risk factors of screen time among chinese primary and middle school students in 12 provinces. *Chin. J. Prev. Med.* **2016**, *50*, 508–513.

18. Mainous, A.G., 3rd; Tanner, R.J.; Rahmanian, K.P.; Jo, A.; Carek, P.J. Effect of sedentary lifestyle on cardiovascular disease risk among healthy adults with body mass indexes 18.5 to 29.9 kg/m^2. *Am. J. Cardiol.* **2019**, *123*, 764–768. [CrossRef]

19. Carson, V.; Hunter, S.; Kuzik, N.; Gray, C.E.; Poitras, V.J.; Chaput, J.P.; Saunders, T.J.; Katzmarzyk, P.T.; Okely, A.D.; Connor Gorber, S.; et al. Systematic review of sedentary behaviour and health indicators in school-aged children and youth: An update. *Appl. Physiol. Nutr. Metab.* **2016**, *41*, S240–S265. [CrossRef]

20. Tremblay, M.S.; le Blanc, A.G.; Kho, M.E.; Saunders, T.J.; Larouche, R.; Colley, R.C.; Goldfield, G.; Connor Gorber, S. Systematic review of sedentary behaviour and health indicators in school-aged children and youth. *Int. J. Behav. Nutr. Phys. Act.* **2011**, *8*, 98. [CrossRef]

21. Subspecialty Group of Endocrinologic, Hereditary and Metabolic Diseases, The Society of Pediatrics; Subspecialty Group of Cardiology, The Society of Pediatrics; Subspecialty Groups of Child Health Care, The Society of Pediatrics. Definition of metabolic syndrome in chinese children and adolescents and recommendations for prevention and treatment. *Chin. J. Pediatrics* **2012**, *50*, 420–422.

22. Liang, L.; Fu, J.; Du, J. The exploration and significance of the definition of metabolic syndrome in chinese children and adolescents. *Chin. J. Pediatrics* **2012**, *2012*, 401–404.

23. Lee, C.M.; Huxley, R.R.; Wildman, R.P.; Woodward, M. Indices of abdominal obesity are better discriminators of cardiovascular risk factors than bmi: A meta-analysis. *J. Clin. Epidemiol.* **2008**, *61*, 646–653. [CrossRef] [PubMed]

24. Guidelines for Prevention and Control of Childhood Obesity Research Committee. *Guidelines for Prevention and Control of Childhood Obesity*; People's Medical Publishing House Co., Ltd.: Beijing, China, 2021.

25. Wang, H.; Gui, Z.; Zhang, J.; Cai, L.; Tan, W.; Chen, Y. Association between screen time and metabolic syndrome among 6–13 years old children in guangzhou. *Chin. J. Sch. Health* **2019**, *40*, 1780–1783.

26. Chinese Work Group of Pediatric Metabolic Syndrome. Prevalence of metabolic syndrome of children and adolescent students in chinese six cities. *Chin. J. Pediatrics* **2013**, *6*, 409–413.

27. Bai, P.; Shen, H.; Yang, C.; Fu, L.; Wang, H.; Shi, Y. Prevalence of metabolic syndrome among schoolchildren in pudong new area of shanghai. *J. Environ. Occup. Med.* **2015**, *32*, 852–855.

28. Wittcopp, C.; Conroy, R. Metabolic syndrome in children and adolescents. *Pediatrics Rev.* **2016**, *37*, 193–202. [CrossRef]

29. Liang, J.; Chen, Y.; Yang, W.; Cao, M.; Dai, M.; Mai, J.; Ma, J.; Jing, J. Relationship between overweight and obesity and metabolic syndrome among children and adolescents in guangzhou. *Chin. J. Sch. Health* **2015**, *36*, 1851–1854.
30. Chen, T.; Ji, C. Relationship between waist circumference and body mass index and metabolic syndrome related traits among middle school students in beijing. *J. Peking Univ.* **2012**, *44*, 355–358.
31. Beijing Municipal Government. *Beijing 2018 Annual Health and Population Health Status Report*; People's Medical Publishing House Co., Ltd.: Beijing, China, 2019; p. 9.
32. Wang, J.; Zhu, Y.; Cai, L.; Jing, J.; Chen, Y.; Mai, J.; Ma, L.; Ma, Y.; Ma, J. Metabolic syndrome and its associated early-life factors in children and adolescents: A cross-sectional study in guangzhou, china. *Public Health Nutr.* **2016**, *19*, 1147–1154. [CrossRef]
33. Chen, Y.; Chen, Y.; Gui, Z.; Bao, W.; Zhang, J.; Tan, K.; Zhang, S.; Cha, L. Association of sedentary behaviors with visual acuity among primary school students: A cohort study. *Chin. J. Sch. Health* **2021**, *42*, 1144–1147.
34. An, M.; Chen, T.; Ma, J. Association between screen time and overweight and obesity among middle and primary school students in fangshan district of beijing city. *Chin. J. Sch. Health* **2018**, *39*, 506–508.
35. Brambilla, P.; Bedogni, G.; Heo, M.; Pietrobelli, A. Waist circumference-to-height ratio predicts adiposity better than body mass index in children and adolescents. *Int. J. Obes.* **2013**, *37*, 943–946. [CrossRef] [PubMed]
36. Navti, L.K.; Samani-Radia, D.; David McCarthy, H. Children's body fatness and prevalence of obesity in relation to height for age. *Ann. Hum. Biol.* **2014**, *41*, 84–90. [CrossRef] [PubMed]
37. Mota, J.; Silva Dos Santos, S.; Santos, A.; Seabra, A.; Vale, S. Association between sedentary behavior time and waist-to-height ratio in preschool children. *Am. J. Hum. Biol. Off. J. Hum. Biol. Counc.* **2016**, *28*, 746–748. [CrossRef]
38. Lahjibi, E.; Heude, B.; Dekker, J.M.; Højlund, K.; Laville, M.; Nolan, J.; Oppert, J.M.; Balkau, B. Impact of objectively measured sedentary behaviour on changes in insulin resistance and secretion over 3 years in the risc study: Interaction with weight gain. *Diabetes Metab.* **2013**, *39*, 217–225. [CrossRef]
39. Ford, E.S.; Li, C.; Zhao, G.; Pearson, W.S.; Tsai, J.; Churilla, J.R. Sedentary behavior, physical activity, and concentrations of insulin among us adults. *Metab. Clin. Exp.* **2010**, *59*, 1268–1275. [CrossRef]
40. Helmerhorst, H.J.; Wijndaele, K.; Brage, S.; Wareham, N.J.; Ekelund, U. Objectively measured sedentary time may predict insulin resistance independent of moderate- and vigorous-intensity physical activity. *Diabetes* **2009**, *58*, 1776–1779. [CrossRef]
41. Hjorth, M.F.; Chaput, J.P.; Damsgaard, C.T.; Dalskov, S.M.; Andersen, R.; Astrup, A.; Michaelsen, K.F.; Tetens, I.; Ritz, C.; Sjödin, A. Low physical activity level and short sleep duration are associated with an increased cardio-metabolic risk profile: A longitudinal study in 8–11 year old danish children. *PLoS ONE* **2014**, *9*, e104677. [CrossRef]
42. Greer, A.E.; Sui, X.; Maslow, A.L.; Greer, B.K.; Blair, S.N. The effects of sedentary behavior on metabolic syndrome independent of physical activity and cardiorespiratory fitness. *J. Phys. Act. Health* **2015**, *12*, 68–73. [CrossRef]
43. van der Berg, J.D.; Stehouwer, C.D.; Bosma, H.; van der Velde, J.H.; Willems, P.J.; Savelberg, H.H.; Schram, M.T.; Sep, S.J.; van der Kallen, C.J.; Henry, R.M.; et al. Associations of total amount and patterns of sedentary behaviour with type 2 diabetes and the metabolic syndrome: The maastricht study. *Diabetologia* **2016**, *59*, 709–718. [CrossRef]
44. Edwardson, C.L.; Gorely, T.; Davies, M.J.; Gray, L.J.; Khunti, K.; Wilmot, E.G.; Yates, T.; Biddle, S.J. Association of sedentary behaviour with metabolic syndrome: A meta-analysis. *PLoS ONE* **2012**, *7*, e34916. [CrossRef] [PubMed]
45. Ding, J.; Ma, H.; Wang, Y. Sedentary behavior and cardiovascular health risk. *Chin. Heart J.* **2021**, *33*, 330–334. [CrossRef]
46. Zhang, B.; Sun, J. Advance and enlightenment of foreign studies on sedentary behavior. *J. Environ. Occup. Med.* **2017**, *34*, 740–744, 748.

Article

Effects of a Formula with scGOS/lcFOS (9:1) and Glycomacropeptide (GMP) Supplementation on the Gut Microbiota of Very Preterm Infants

Xue Yu [1,†], Yan Xing [2,†], Hui Liu [2], Yanmei Chang [2], Yanxia You [2], Yuqi Dou [1], Bin Liu [3], Qi Wang [4], Defu Ma [1,*], Lijun Chen [3,*] and Xiaomei Tong [2,*]

[1] School of Public Health, Peking University Health Science Center, Beijing 100191, China; 1911210153@pku.edu.cn (X.Y.); dyqhoney@126.com (Y.D.)
[2] Department of Pediatrics, Peking University Third Hospital, Beijing 100191, China; yxsxz@outlook.com (Y.X.); adelebjmu@126.com (H.L.); changyanmei1226@163.com (Y.C.); 13911809214@163.com (Y.Y.)
[3] National Engineering Center of Dairy for Maternal and Child Health, Beijing Sanyuan Foods Co., Ltd., Beijing 100163, China; liubin@sanyuan.com.cn
[4] Cuiying Biomedical Research Center, Lanzhou University Second Hospital, Lanzhou 730030, China; ery_wangqery@lzu.edu.cn
[*] Correspondence: madefu@bjmu.edu.cn (D.M.); chenlijun@sanyuan.com.cn (L.C.); tongxm2007@126.com (X.T.)
[†] These authors contributed equally to this work.

Citation: Yu, X.; Xing, Y.; Liu, H.; Chang, Y.; You, Y.; Dou, Y.; Liu, B.; Wang, Q.; Ma, D.; Chen, L.; et al. Effects of a Formula with scGOS/lcFOS (9:1) and Glycomacropeptide (GMP) Supplementation on the Gut Microbiota of Very Preterm Infants. *Nutrients* **2022**, *14*, 1901. https://doi.org/10.3390/nu14091901

Academic Editor: Elvira Verduci

Received: 22 March 2022
Accepted: 27 April 2022
Published: 1 May 2022

Publisher's Note: MDPI stays neutral with regard to jurisdictional claims in published maps and institutional affiliations.

Copyright: © 2022 by the authors. Licensee MDPI, Basel, Switzerland. This article is an open access article distributed under the terms and conditions of the Creative Commons Attribution (CC BY) license (https://creativecommons.org/licenses/by/4.0/).

Abstract: Microbial colonization of very preterm (VPT) infants is detrimentally affected by the complex interplay of physiological, dietary, medical, and environmental factors. The aim of this study was to evaluate the effects of an infant formula containing the specific prebiotic mixture of scGOS/lcFOS (9:1) and glycomacropeptide (GMP) on the composition and function of VPT infants' gut microbiota. Metagenomic analysis was performed on the gut microbiota of VPT infants sampled at four time points: 24 h before the trial and 7, 14, and 28 days after the trial. Functional profiling was aggregated into gut and brain modules (GBMs) and gut metabolic modules (GMMs) based on the Kyoto Encyclopedia of Genes and Genomes (KEGG) pathways. *Enterococcus faecium*, *Escherichia coli*, *Klebsiella aerogenes*, and *Klebsiella pneumoniae* were dominant species in both the test group and the control group. After the 4-week intervention, the abundance of *Bifidobacterium* in the test group was significantly increased. We found two GBMs (quinolinic acid synthesis and kynurenine degradation) and four GMMs (glutamine degradation, glyoxylate bypass, dissimilatory nitrate reduction, and preparatory phase of glycolysis) were significantly enriched in the test group, respectively. The results of this study suggested that formula enriched with scGOS/lcFOS (9:1) and GPM is beneficial to the intestinal microccology of VPT infants.

Keywords: preterm; prebiotics; glycomacropeptide; metagenomics

1. Introduction

Globally, the preterm rate is rising and is the major cause of infant mortality. Preterm infants have immature and fragile organs, which leads them to be more susceptible to a series of health problems, especially when they are born very preterm (VPT) (born 28 to <32 weeks of gestation) [1]. Despite recent advances in neonatal care, VPT infants remain at high risk of necrotizing enterocolitis, respiratory problems, neonatal jaundice, and neurodevelopmental impairment [2].

Due to various health challenges, VPT infants are usually hospitalized in the neonatal intensive care unit (NICU) for an extended period of time, being put on artificial respiration and fed artificially or parenterally. Moreover, antibiotics and other medications are widely used during the treatment of these infants. All these factors may interfere with the natural pattern of microbiota acquisition and development, resulting in an aberrant establishment or deviation of the composition of the gut microbiota [3,4]. The microbiota of healthy term infants is dominated by *Bifidobacterium* and *Bacteroides*; however, these bacteria only exist in low abundance in premature infants [5]. In contrast, premature infants show low diversity and increased colonization of potentially pathogenic bacteria from Gram-negative *Enterobacteriaceae* of *Proteobacteria* [6]. These alterations may dramatically affect short- and long-term health [7].

Nutrition is a key factor in shaping the composition and function of the early microbiota [8]. Breastmilk is considered the gold standard of infant nutrition and is recommended for preterm infants [9]. However, preterm infants (especially <32 weeks of gestation) often face the dilemma of insufficient breast milk. In this condition, infant formula provides a healthy alternative that attempts to mimic the nutritional content of breast milk [10]. The most common infant formula is based on bovine milk, which has a very different nutritional composition than human breast milk. Human milk oligosaccharides (HMOs) not only have a "prebiotic" effect [11] but are also rich in sialic acid found in brain gangliosides [12], the content of which is significantly lower in formula than in breast milk [13]. Sialic acids are a class of alpha-keto acid sugars with a nine-carbon backbone [14]. A study on piglets confirmed that sialic acid derived from casein glycomacropeptide could improve the learning and memory abilities of piglets during early development [15].

Studies have found that the addition of prebiotics in term infant formula promotes the development of a neonatal gut microbiota resembling that of breast-fed infants [16]. Furthermore, feeding infant formula containing a GOS/FOS mixture has a positive effect on bifidobacterial abundance [17]. Glycomacropeptide (GMP) is a glycopeptide rich in sialic acid that can also promote the growth of beneficial bacteria and bind to pathogenic bacteria [18]. It has numerous biological effects on gut health, including preventing pathogen adhesion, decreasing intestinal barrier dysfunction, limiting lipopolysaccharide production, and attenuating inflammation [19].

However, most of the previous studies only focused on individual prebiotic-associated microorganisms and did not elucidate the whole gut microbiota of subjects; some studies reported gut microbial composition at the genus level without a detailed description of the species-specific effects of prebiotics. In the present study, we hypothesized that preterm formula supplemented with a prebiotic mixture of scGOS/lcFOS and GMP would facilitate the establishment of normal gut microbiota in VPT infants and confer health benefits. The aim of our study was to investigate the effects of a preterm infant formula enriched with scGOS/lcFOS and GMP on the composition and function of healthy VPT infants' gut microbiota, using metagenomic data collected at four time points. The effects on infant growth and stool characteristics were further aims of this study.

2. Materials and Methods

2.1. Study Design and Subjects

This study is a prospective, non-randomized, controlled trial conducted between October 2019 and November 2020 in the NICU of Peking University Third Hospital. The study was approved by the Peking University Third Hospital Medical Science Research Ethics Committee (No. S2016159). It was registered in the Chinese Clinical Trials Register (registration number ChiCTR2100051988). Written informed consent was obtained from parents before infant enrollment in the study.

Healthy VPT infants born at a gestational age between 28 and 32 weeks and whose mothers could not offer sufficient breastmilk or elected not to breast-feed were recruited as the study subjects. Exclusion criteria included severe gastrointestinal dysfunction, congenital malformations, genetic diseases, or any other disease requiring surgery. The infants were assigned to receive either a standard formula (control group) or an experimental formula (test group) according to the preference of parents.

Basic clinical information about the parents and their offspring was obtained through the medical records, including the mother's age, mode of delivery, gestational age, infant sex, Apgar score, antenatal and postpartum antibiotic therapy, birth weight, length, and head circumference.

Due to considerations of feeding tolerance and safety, the trial was started when the infants' enteral feeding volume reached 80 mL/kg/d. Prior to this, both groups were fed the same standard formula for preterm infants. When the prescribed enteral feeding amount was reached, the test group was fed the experimental formula, and the control group continued to feed the standard formula. The control and experimental formulas were comparable in nutritional composition, except that the experimental formula contained a prebiotic mixture providing 0.65 g scGOS/lcFOS (9:1) and casein GMP providing 40 mg sialic acid/100 mL. Both formulas were bovine milk-based, containing energy 316 kJ/100 mL, protein 0.68 g/100 kJ, and lipid 1.30 g/100 kJ. The intervention was carried out when the infant was hospitalized in the NICU and lasted for 28 days. During this period, they underwent a daily clinical follow-up, including feeding volume, stool consistency, and stool frequency.

2.2. Stool Collection

Fresh stools were collected by a trained nurse from the infants' diapers within 24 h before, 7 days after, 14 days after, and 28 days after the intervention using a collection tube containing 8 mL of DNA stabilizer reagent (PSP® Spin Stool DNA Plus Kit, STRATEC Biomedical AG, Birkenfeld, Germany). The samples were kept frozen at −80 °C until DNA extraction.

2.3. Fecal DNA Extraction and Metagenomic Sequencing

The metagenomic DNA was extracted using MagPure Stool DNA kf kit (MP, Guangzhou, China), following the manufacturer's instructions, which included a step of mechanical cell disruption by bead beating.

Metagenomic sequencing was performed by China National GeneBank (Shenzhen, China), following the protocol published previously [20]. Briefly, DNA was fragmented and barcoded, then subjected to amplification to produce DNA nanoballs. High-throughput sequencing was performed on BGISEQ-500. Adaptor and low-quality reads were removed, and human DNA reads were filtered out.

2.4. Taxonomic and Functional Profiling

MetaPhlAn 3.0 was used to estimate the relative abundance of taxonomic profiles. Putative amino acid sequences were translated from the gene catalog [21] and aligned against the proteins or domains in the Kyoto Encyclopedia of Genes and Genomes (KEGG) databases (release 79.0, with animal and plant genes removed) using BLASTP (v2.26, default parameter, except -m 8 -e 1e-5 -F -a 6 -b 50). Each protein was assigned to a KEGG orthologous (KO) group on the basis of the highest-scoring annotated hit(s) containing at least one segment pair scoring over 60 bits. The relative abundance profile of KOs was determined by summing the relative abundance of genes from each KO using the mapped reads per sample [21]. The abundance of each gut metabolic module (GMM) (-a 2 -d GMM.v1.07.txt -s average) and gut neuroactive module (GBM) (default parameter) were calculated as shown in the former article [22,23].

2.5. Statistical Analysis

All statistical analyses were performed using R statistical software version 4.0.3. Baseline characteristics and clinical results of the study participants were presented as mean ± SD for continuous variables and frequencies for categorical variables.

Alpha-diversity (Shannon index) was calculated using the Vegan package and compared by using Wilcoxon rank-sum test. Phylogenetic measures of beta-diversity based on the genus level abundance profile were also calculated by using the Vegan package, and PCoA plot based on Bray–Curtis distances were performed using the ggplot2 package. The top two principal coordinates (PC1 and PC2, representing the maximum amount of variation present in the dataset) were compared in each group. In order to investigate the specific differences in the gut microbiome composition and function between the test and control groups, Wilcoxon rank-sum test was used, and p-values were corrected for multiple testing using the Benjamin and Hochberg method. In order to compare the microbiota community structure across four time points, Kruskal–Wallis test was used and pairwise test for multiple comparisons was performed using Bonferroni's p-adjustment method. Statistical significance was assumed at $p < 0.05$.

2.6. Data Availability

The datasets used to analyze for this study can be found in the China National Genebank (CNGB) with program ID CNP0001742.

3. Results

A total of 155 samples from day 0 (24 h before intervention), 7 days, 14 days, and 28 days after intervention were obtained from 72 infants (Figure 1). One participant in the test group dropped out because of sepsis, and one in the control group dropped out because of necrotizing enterocolitis, and they were not included in the analysis. In the test group, an average of 2.3 stool samples were collected per participant, and in the control group, an average of 2 stool samples were collected per participant. The characteristics of the participants are shown in Table 1. Thirty-seven infants were allocated to the test group and thirty-five to the control group in the present research. No significant differences were found in baseline characteristics between the two study groups except that the test group was more mature ($p = 0.040$) and had a shorter hospitality stay than the control group ($p = 0.039$). However, there was no significant difference in the corrected gestational age at the beginning of the intervention.

Table 1. Characteristics of the study participants.

	Test Group (n = 37)	Control Group (n = 35)	p
Male sex (n [%])	14 (37.8)	20 (58.8)	0.077
Siblings at birth (n [% yes])	11 (29.7)	13 (37.1)	0.505
Cesarean delivery (n [%])	23 (62.2)	22 (64.7)	0.824
Maternal antibiotics at delivery (n [% yes])	11 (29.7)	10 (30.3)	0.958
Infant antibiotics after inclusion (n [% yes])	12 (32.4)	14 (40.0)	0.504
Maternal age (year)	32.2 ± 4.8	32.8 ± 4.7	0.589
Birth weight (g)	1329.2 ± 268.1	1265.1 ± 257.1	0.305
Birth length (cm)	38.3 ± 3.3	37.7 ± 2.5	0.384
Head circumference (cm)	27.6 ± 1.7	26.9 ± 1.7	0.060
1-min Apgar score	8.6 ± 2.0	8.2 ± 2.1	0.356
5-min Apgar score	9.4 ± 1.1	9.0 ± 1.2	0.227
10-min Apgar score	9.5 ± 0.8	9.4 ± 0.8	0.562
Gestational age at birth (week)	30.4 ± 1.8	29.5 ± 1.6	0.040
Corrected age at the beginning of intervention (week)	32.4 ± 1.3	31.8 ± 1.5	0.095
Hospitalization (day)	41.4 ± 15.2	49.4 ± 16.9	0.039

Continuous variables are represented as the mean ± SD.

PCoA based on Bray–Curtis dissimilarity was used to examine the microbial community structure across different groups and stages (Figure 2a). The distribution of samples by groups and stages is shown along the first and second axes of the PCoA plot. Along the first axis, the value of PC1 in the control group showed an increasing trend over time, and there was a significant difference in microbiota community structure between stages three and four. A similar trend was found in the test group, but no significant difference was observed between the four stages. Along the second axis, the microbiota community structure of the test group showed significant differences between stages one and two, as well as stages one and three. There was no such difference found in the control group.

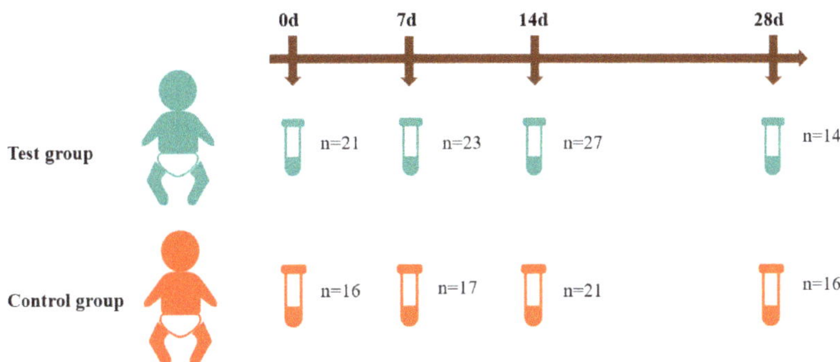

Figure 1. The study cohort. A total of 72 very preterm infants were included in this study, among which 37 infants were given a test formula containing prebiotics and sialic acid. Fecal samples were collected 24 h before, 7 days after, 14 days after, and 28 days after the intervention.

We compared the number of genera between the test group and the control group over the sampled time points (Figure 2b). We found that the genus number of the control group showed an increasing trend over time while the test group remained constant. The number of genera in the control group at stage four was significantly higher than that at stage one ($p = 0.0036$) and stage three ($p = 0.046$). After 28 days of intervention, the genus number of the control group was significantly higher than that of the test group, indicating that the test formula had an effect on the richness of microbiota in the VPT infant gut ($p = 0.028$).

The alpha diversity of the infant gut microbiota was measured by the Shannon index. There was no significant change in the microbial diversity within either group over the intervention period. When comparing between feeding groups, the test group was significantly lower than the control group at 7 d ($p = 0.009$) and 28 d ($p = 0.031$) (Figure 2c).

The main phyla in the feces in both the test group and the control group were Proteobacteria, followed by Firmicutes and Actinobacteria. Consistent with the results of previous studies, we detected that *Enterobacteriaceae*-related genera such as *Klebsiella*, *Escherichia*, and *Enterobacter*, and genera of *Enterococcus*, *Clostridium*, and *Bifidobacterium* predominated the intestinal microbiome of VPT infants (Figure 3a). The relative abundance of the gut microbiota in both the test and control groups fluctuated during the intervention period.

We found that *Enterococcus faecium*, *Escherichia coli*, *Klebsiella aerogenes*, and *Klebsiella pneumoniae* were dominant species in both groups (Figure 3b). The test group was characterized by a higher relative abundance of *Bifidobacterium* on day 28 (adjusted $p = 0.023$, Figure 3c). Four *Bifidobacterium* species were detected: *B. longum*, *B. animalis*, *B. breve*, and *B. pseudocatenulatum*. Among them, *B. longum* was most affected by the administered prebiotics and showed the greatest increase in the test group (Figure 3c).

Using gut and brain modules (GBMs) and gut metabolic modules (GMMs), we evaluated the functional capacity development of gut microbiota. After 28 days of intervention, two significantly changed GBMs between two groups were observed. Quinolinic acid synthesis and kynurenine degradation were enriched in the test group (Figure 4a,b). Four significantly changed GMMs between the two groups were observed. Glutamine degradation, glyoxylate bypass, dissimilatory nitrate reduction, and preparatory phase of glycolysis (Figure 4c–f) were enriched in the test group compared with the control group.

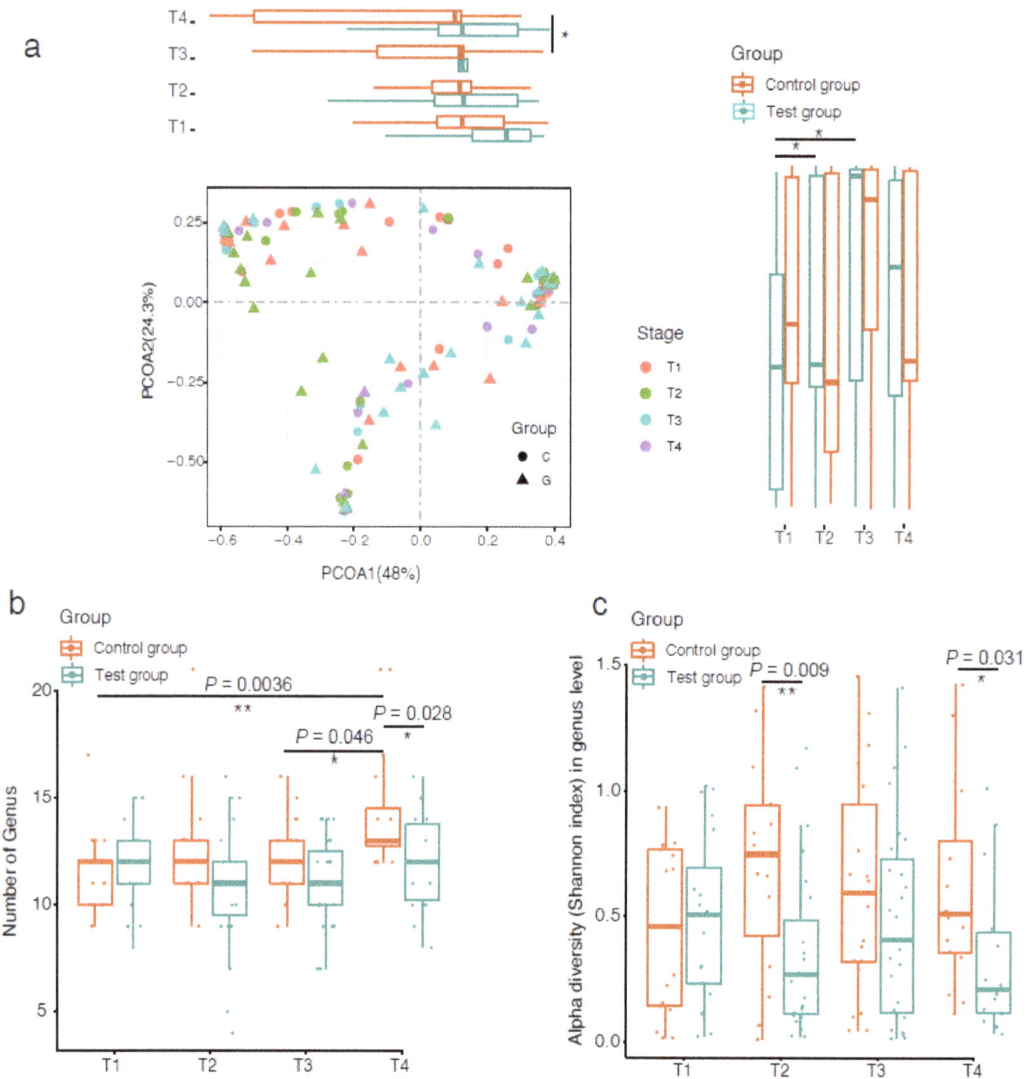

Figure 2. The effect of formula supplemented with prebiotics and sialic acid on microbial diversity in the VPT infant gut. PCoA of Bray–Curtis distances based on the profile of genera, * indicates difference at *p* value < 0.05, ** indicates difference at *p* value < 0.01 (**a**); comparison of genus numbers between the test group and control group (**b**); alpha diversity of gut microbiota measured by the Shannon index (**c**).

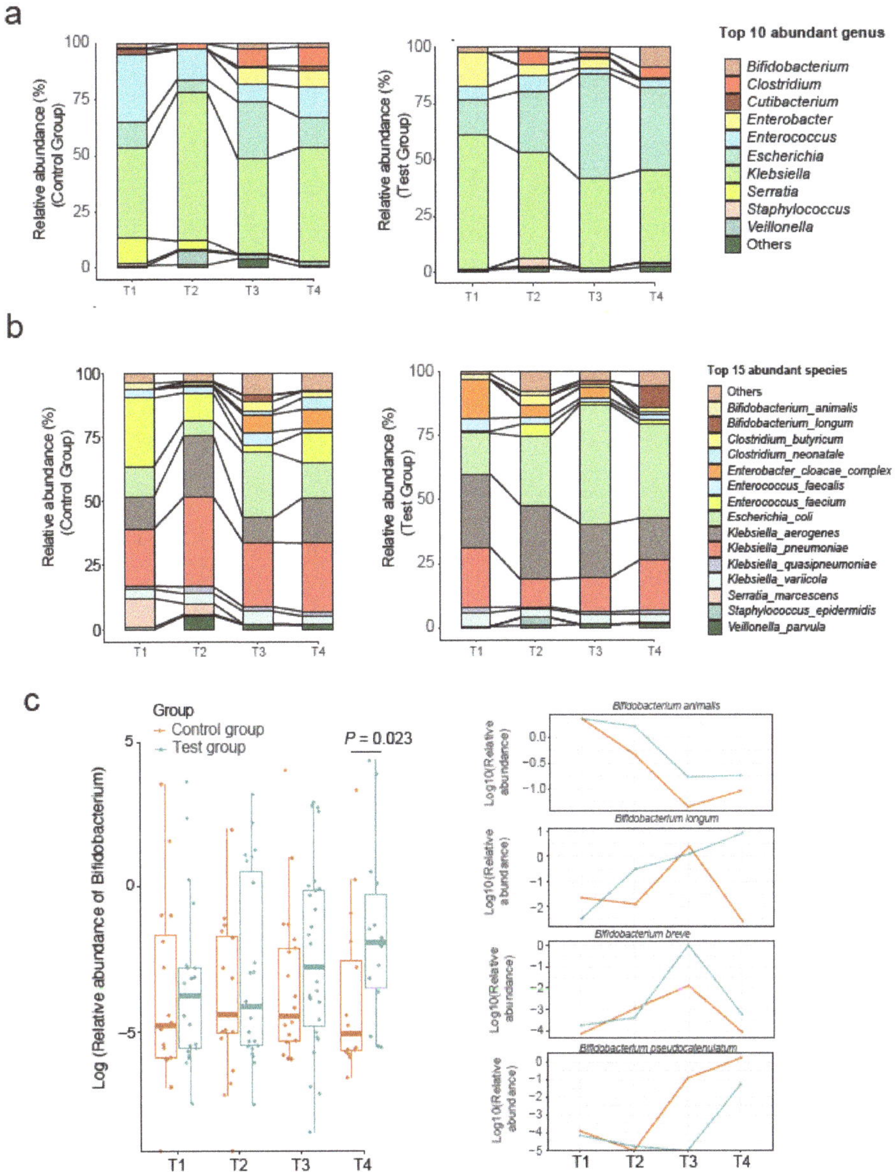

Figure 3. Relative abundance of gut microbiota at the genus level (**a**) and species level (**b**). Comparison of the relative abundance of *Bifidobacterium* between the test group and the control group (**c**).

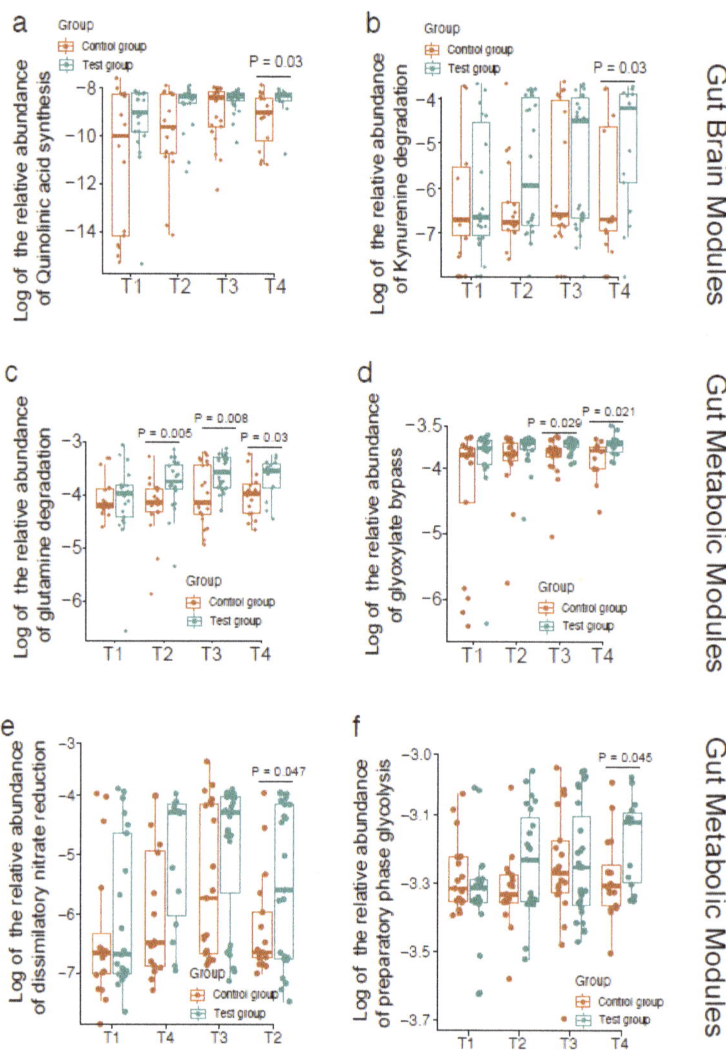

Figure 4. Comparison of gut and brain modules (GBMs) and gut metabolic modules (GMMs) between the test group and the control group. Quinolinic acid synthesis (**a**), kynurenine degradation (**b**), glutamine degradation (**c**), glyoxylate bypass (**d**), dissimilatory nitrate reduction (**e**) and preparatory phase of glycolysis (**f**).

Infants in both groups consumed an increasing amount of formula during the study period. No significant difference was observed between the two groups in daily intake of formula, weight, length, and head circumference. No significant differences were shown at any timepoint for stool consistency and stool frequency between the test and control groups (Table 2).

Table 2. Clinical results at 4 time points.

	T1	T2	T3	T4
Infant feeding volume (mL/day)				
Test	175.1 ± 63.5	214.7 ± 69.5	258.4 ± 55.1	297.5 ± 38.0
Control	171.9 ± 78.3	199.0 ± 64.2	249.7 ± 59.7	293.9 ± 36.8
p value	0.136	0.719	0.217	0.314
Weight (g)				
Test	1420.9 ± 199.9	1459.2 ± 186.4	1634.6 ± 199.5	1900.0 ± 197.2
Control	1393.1 ± 210.1	1456.8 ± 222.8	1599.1 ± 235.3	1900.0 ± 129.2
p value	0.582	0.966	0.583	0.699
Length (cm)				
Test	39.8 ± 2.9	40.3 ± 2.1	41.3 ± 2.3	44.7 ± 2.5
Control	38.8 ± 2.7	40.0 ± 2.6	41.1 ± 2.7	44.4 ± 2.8
p value	0.231	0.736	0.814	0.794
Head circumference (cm)				
Test	28.0 ± 2.0	28.3 ± 2.0	29.1 ± 2.1	31.2 ± 3.1
Control	27.5 ± 1.8	28.2 ± 2.3	28.6 ± 2.3	29.7 ± 2.5
p value	0.323	0.861	0.488	0.187
Stool consistency (Bristol stool score)				
Test	3.9 ± 0.5	4.0 ± 0.0	3.9 ± 0.4	4.0 ± 0.0
Control	3.8 ± 0.6	3.9 ± 0.3	3.9 ± 0.3	3.9 ± 0.4
p value	0.359	0.103	0.865	0.253
Stool frequency (times/day)				
Test	1.7 ± 1.5	1.7 ± 1.3	2.2 ± 1.5	2.3 ± 1.1
Control	1.9 ± 1.2	2.1 ± 1.4	2.2 ± 1.3	1.7 ± 1.3
p value	0.643	0.268	0.908	0.778

4. Discussion

To the authors' knowledge, the present study was the first study using metagenomic shotgun sequencing to profile the gastrointestinal microbiome of VPT infants. We found that after the 4-week intervention of the specific prebiotic mixture of scGOS/lcFOS (9:1) and GPM, the abundance of *Bifidobacterium* in the test group was significantly increased. Gut and brain modules (GBMs) and gut metabolic modules (GMMs) showed differences in neuroactive compound production and energy source utilization between the test and control groups.

It is acknowledged that gestational age is an important factor in the establishment of the infant gut microbiota [24]. VPT infants have a delayed progression to a *Bifidobacterium*-dominated microbiota compared to term infants [6]. Consistent with previous studies [25,26], we observed that *Klebsiella*, *Escherichia*, and *Enterococcus* predominated the intestinal microbiome of VPT infants. Although the gut microbiota of VPT infants is highly dynamic, the composition and longitudinal progression trend of the gut microbiota in the two groups were similar because they were both fed with formula.

In our study, we found a significant increase in the abundance of *Bifidobacterium* in the test group, with a proportion similar to that reported in breast-fed counterparts [27]. Both scGOS/lcFOS and GPM have been reported to promote the growth of *Bifidobacterium*. A dose-related bifidogenic effect of GOS/FOS supplementation has previously been shown in formula-fed term infants, ranging between 0.4–0.8 g/dL [28], which was comparable to our study. On the other hand, Korpela et al. studied probiotic and galactooligosaccharides (GOS) supplementation in breast-fed and formula-fed neonates; they found the bifidobacterial community in the formula-fed infants have a weaker response to the supplement compared to the breast-fed infants. Only breast-fed infants showed the expected increase in bifidobacteria and reduction in *Proteobacteria* and *Clostridia* [29]. GMP supplementation promoted the growth of *Bifidobacterium* in the feces of rats with atopic dermatitis and increased the contents of acetic acid and butyric acid [30]. The administration of GMP to 6-week-old infants for 6 months augmented the levels of *Bifidobacterium* compared with baseline [31]. At the species level, we found 4 *Bifidobacterium* species in the VPT infant gut: *B. longum, B. animalis, B. breve,* and *B. pseudocatenulatum.* The former two were relatively abundant in the VPT infant gut, and *B. longum* showed the greatest response to the intervention. This finding was in correspondence with a previous study in bacterial cultures, which showed that *B. longum* strains had the ability to utilize both GOS and FOS and used a major extent for growth [32].

Moreover, the premature gut microbiota is different not only in composition but also in functionality. The main short-chain fatty acids (SCFAs) produced by the intestinal microbiota were found at lower levels in fecal samples from preterm and VLBW infants than in the feces of full-term infants [33]. In this study, we found that the intervention also improved the function of the microbial community. Glutamine is the amino group donor for many cellular biosynthetic reactions and serves as a storage reservoir of ammonia, playing a central role in nitrogen metabolism [34]. The glutamine degradation enriched in the test group might indicate the enhanced ability of microorganisms to utilize nitrogen sources. Glyoxylate bypass is also known as the glyoxylate shunt, in which acetyl-CoA is converted to succinate for the synthesis of carbohydrates. The glyoxylate shunt acts as a microbial survival pathway [35,36] and is essential for the production of bacterial acetate and fatty acid metabolism [37]. Dissimilatory nitrate reduction is a pathway related to bacteria's respiration. Nitrate is one of the alternative electron acceptors allowing bacteria to respire in the absence of oxygen [38]. The glycolytic pathway is a major metabolic pathway for microbial fermentation, and the preparatory phase forms a key intermediate of the pathway. In the preparatory phase of glycolysis, one glucose molecule is converted to two glyceraldehyde-3- phosphate [39]. The kynurenine pathway is responsible for 90% of tryptophan metabolism, and the downstream metabolites kynurenic acid and quinolinic acid of this pathway have recently been identified as relevant for the nervous system, as they exert neuroprotective and excitotoxic effects, respectively, through their interaction with N-methyl-D-aspartate (NMDA) receptors [40]. Recently, it became evident that intestinal bacteria can affect brain function and behavior through signaling pathways of the microbiome gut–brain axis [41].

Since previous studies have reported that the preterm infant gut cluster is independent of sex or delivery mode [42], we did not subgroup these factors in the subject recruitment and analysis. Antibiotics are another factor that has an important influence on the intestinal microbiota. In this study, all infants received one course of antibiotics, with a third receiving at least one additional course. Thus, despite the potential confounding effects of antibiotic use, this study represents a typical characterization of the bacterial community of VPT infants under clinical supervision.

The present study has several limitations. Firstly, although the study was conducted out of humanitarian interest and left to parents to decide which preterm formula to feed their infants, this non-randomized study design may give rise to imbalances and biased estimates of treatment effects [43]. A relatively small sample size may limit the power of statistical analysis. However, our study found some significant differences in the composition and function of the gut microbiota between the test and control groups, which were consonant with previous findings. The loss of our study samples was mainly attributed to (i) extracting DNA from VPT infant stool samples being challenging and (ii) subjects being unable to adhere to our longitudinal intervention and collection schedule when discharged from the NICU. It is of great importance that safety is taken into account when trials are performed in VPT infants because those infants are at increased risk for infections, so the trials are conducted under close supervision in the NICU. This also resulted in a short trial period, which was insufficient to observe the trajectory trends of the intestinal microbiome over a long period of time.

5. Conclusions

In summary, our results show that formula enriched with scGOS/lcFOS (9:1) and GPM is safe for VPT infants, and it promotes the growth of *Bifidobacterium*. VPT infants fed the experimental formula have a microbiota more active in neuroactive compound production and energy source utilization, which might benefit their health. However, future randomized controlled studies with larger samples are warranted to further confirm these findings.

Author Contributions: X.Y.: investigation, methodology, formal analysis, writing—original draft preparation; H.L.: investigation, visualization; Y.C.: investigation, data curation; Y.Y.: investigation; Y.D.: investigation; B.L.: resources; Q.W.: software; D.M.: conceptualization, supervision, writing—review and editing; L.C.: project administration; X.T.: funding acquisition; Y.X.: methodology, project administration. All authors have read and agreed to the published version of the manuscript.

Funding: This research was funded by Natural Science Foundation of Beijing, China (NO. S160004), National Key Research and Development Program of China (2021YFC2700700), and Peking University Third Hospital Research Fund for outstanding overseas returnees (No. BYSYLXHG2019005).

Institutional Review Board Statement: The study was conducted according to the guidelines of the Declaration of Helsinki, and approved by Peking University Third Hospital Medical Science Research Ethics Committee (protocol code S2016159 and date of approval: 28 September 2017).

Informed Consent Statement: Informed consent was obtained from all Participants involved in the study.

Data Availability Statement: The datasets used to analyze for this study can be found in the China National Genebank (CNGB) with program ID CNP0001742.

Acknowledgments: We gratefully acknowledge Shengping Han from China National GeneBank for his invaluable assistance with the project.

Conflicts of Interest: The authors declare no conflict of interest. The funders had no role in the design of the study; in the collection, analyses, or interpretation of data; in the writing of the manuscript, or in the decision to publish the results.

References

1. Vogel, J.P.; Chawanpaiboon, S.; Moller, A.B.; Watananirun, K.; Bonet, M.; Lumbiganon, P. The global epidemiology of preterm birth. *Best Pr. Res. Clin. Obs. Gynaecol.* **2018**, *52*, 3–12. [CrossRef] [PubMed]
2. Platt, M.J. Outcomes in preterm infants. *Public Health* **2014**, *128*, 399–403. [CrossRef] [PubMed]
3. Aguilar-Lopez, M.; Dinsmoor, A.M.; Ho, T.T.B.; Donovan, S.M. A systematic review of the factors influencing microbial colonization of the preterm infant gut. *Gut Microbes* **2021**, *13*, 1–33. [CrossRef]
4. Groer, M.W.; Gregory, K.E.; Louis-Jacques, A.; Thibeau, S.; Walker, W.A. The very low birth weight infant microbiome and childhood health. *Birth Defects Res. C Embryo Today* **2015**, *105*, 252–264. [CrossRef] [PubMed]
5. Matamoros, S.; Gras-Leguen, C.; Le Vacon, F.; Potel, G.; de La Cochetiere, M.F. Development of intestinal microbiota in infants and its impact on health. *Trends Microbiol.* **2013**, *21*, 167–173. [CrossRef] [PubMed]
6. Buffet-Bataillon, S.; Bellanger, A.; Boudry, G.; Gangneux, J.P.; Yverneau, M.; Beuchée, A.; Blat, S.; Le Huërou-Luron, I. New Insights Into Microbiota Modulation-Based Nutritional Interventions for Neurodevelopmental Outcomes in Preterm Infants. *Front. Microbiol.* **2021**, *12*, 676622. [CrossRef] [PubMed]
7. Gasparrini, A.J.; Crofts, T.S.; Gibson, M.K.; Tarr, P.I.; Warner, B.B.; Dantas, G. Antibiotic perturbation of the preterm infant gut microbiome and resistome. *Gut Microbes* **2016**, *7*, 443–449. [CrossRef]
8. Tamburini, S.; Shen, N.; Wu, H.C.; Clemente, J.C. The microbiome in early life: Implications for health outcomes. *Nat. Med.* **2016**, *22*, 713–722. [CrossRef]
9. Harding, J.E.; Cormack, B.E.; Alexander, T.; Alsweiler, J.M.; Bloomfield, F.H. Advances in nutrition of the newborn infant. *Lancet* **2017**, *389*, 1660–1668. [CrossRef]
10. Boquien, C.Y. Human Milk: An Ideal Food for Nutrition of Preterm Newborn. *Front. Pediatr.* **2018**, *6*, 295. [CrossRef]
11. Bode, L. Human milk oligosaccharides: Every baby needs a sugar mama. *Glycobiology* **2012**, *22*, 1147–1162. [CrossRef] [PubMed]
12. Wang, B.; McVeagh, P.; Petocz, P.; Brand-Miller, J. Brain ganglioside and glycoprotein sialic acid in breastfed compared with formula-fed infants. *Am. J. Clin. Nutr.* **2003**, *78*, 1024–1029. [CrossRef] [PubMed]
13. Wicinski, M.; Sawicka, E.; Gebalski, J.; Kubiak, K.; Malinowski, B. Human Milk Oligosaccharides: Health Benefits, Potential Applications in Infant Formulas, and Pharmacology. *Nutrients* **2020**, *12*, 266. [CrossRef]

14. Schauer, R.; Kamerling, J.P. Exploration of the Sialic Acid World. *Adv. Carbohydr. Chem. Biochem.* **2018**, *75*, 1–213. [CrossRef] [PubMed]

15. Wang, B.; Yu, B.; Karim, M.; Hu, H.; Sun, Y.; McGreevy, P.; Petocz, P.; Held, S.; Brand-Miller, J. Dietary sialic acid supplementation improves learning and memory in piglets. *Am. J. Clin. Nutr.* **2007**, *85*, 561–569. [CrossRef] [PubMed]

16. Vandenplas, Y.; De Greef, E.; Veereman, G. Prebiotics in infant formula. *Gut Microbes* **2014**, *5*, 681–687. [CrossRef]

17. Knol, J.; Scholtens, P.; Kafka, C.; Steenbakkers, J.; Gro, S.; Helm, K.; Klarczyk, M.; Schöpfer, H.; Böckler, H.M.; Wells, J. Colon microflora in infants fed formula with galacto- and fructo-oligosaccharides: More like breast-fed infants. *J. Pediatr. Gastroenterol. Nutr.* **2005**, *40*, 36–42. [CrossRef]

18. Nakajima, K.; Tamura, N.; Kobayashi-Hattori, K.; Yoshida, T.; Hara-Kudo, Y.; Ikedo, M.; Sugita-Konishi, Y.; Hattori, M. Prevention of intestinal infection by glycomacropeptide. *Biosci. Biotechnol. Biochem.* **2005**, *69*, 2294–2301. [CrossRef]

19. Foisy Sauvé, M.; Spahis, S.; Delvin, E.; Levy, E. Glycomacropeptide: A Bioactive Milk Derivative to Alleviate Metabolic Syndrome Outcomes. *Antioxid. Redox Signal.* **2021**, *34*, 201–222. [CrossRef]

20. Fang, C.; Zhong, H.; Lin, Y.; Chen, B.; Han, M.; Ren, H.; Lu, H.; Luber, J.M.; Xia, M.; Li, W.; et al. Assessment of the cPAS-based BGISEQ-500 platform for metagenomic sequencing. *Gigascience* **2018**, *7*, 1–8. [CrossRef]

21. Xie, H.; Guo, R.; Zhong, H.; Feng, Q.; Lan, Z.; Qin, B.; Ward, K.J.; Jackson, M.A.; Xia, Y.; Chen, X.; et al. Shotgun Metagenomics of 250 Adult Twins Reveals Genetic and Environmental Impacts on the Gut Microbiome. *Cell Syst.* **2016**, *3*, 572–584. [CrossRef] [PubMed]

22. Vieira-Silva, S.; Falony, G.; Darzi, Y.; Lima-Mendez, G.; Garcia Yunta, R.; Okuda, S.; Vandeputte, D.; Valles-Colomer, M.; Hildebrand, F.; Chaffron, S.; et al. Species-function relationships shape ecological properties of the human gut microbiome. *Nat. Microbiol.* **2016**, *1*, 16088. [CrossRef]

23. Valles-Colomer, M.; Falony, G.; Darzi, Y.; Tigchelaar, E.F.; Wang, J.; Tito, R.Y.; Schiweck, C.; Kurilshikov, A.; Joossens, M.; Wijmenga, C.; et al. The neuroactive potential of the human gut microbiota in quality of life and depression. *Nat. Microbiol.* **2019**, *4*, 623–632. [CrossRef]

24. Milani, C.; Duranti, S.; Bottacini, F.; Casey, E.; Turroni, F.; Mahony, J.; Belzer, C.; Delgado Palacio, S.; Arboleya Montes, S.; Mancabelli, L.; et al. The First Microbial Colonizers of the Human Gut: Composition, Activities, and Health Implications of the Infant Gut Microbiota. *Microbiol. Mol. Biol. Rev.* **2017**, *81*, e00036-17. [CrossRef] [PubMed]

25. Gibson, M.K.; Wang, B.; Ahmadi, S.; Burnham, C.A.; Tarr, P.I.; Warner, B.B.; Dantas, G. Developmental dynamics of the preterm infant gut microbiota and antibiotic resistome. *Nat. Microbiol.* **2016**, *1*, 16024. [CrossRef] [PubMed]

26. Hui, Y.; Smith, B.; Mortensen, M.S.; Krych, L.; Sørensen, S.J.; Greisen, G.; Krogfelt, K.A.; Nielsen, D.S. The effect of early probiotic exposure on the preterm infant gut microbiome development. *Gut Microbes* **2021**, *13*, 1951113. [CrossRef] [PubMed]

27. Ford, S.L.; Lohmann, P.; Preidis, G.A.; Gordon, P.S.; O'Donnell, A.; Hagan, J.; Venkatachalam, A.; Balderas, M.; Luna, R.A.; Hair, A.B. Improved feeding tolerance and growth are linked to increased gut microbial community diversity in very-low-birth-weight infants fed mother's own milk compared with donor breast milk. *Am. J. Clin. Nutr.* **2019**, *109*, 1088–1097. [CrossRef] [PubMed]

28. Moro, G.; Minoli, I.; Mosca, M.; Fanaro, S.; Jelinek, J.; Stahl, B.; Boehm, G. Dosage-related bifidogenic effects of galacto- and fructooligosaccharides in formula-fed term infants. *J. Pediatr. Gastroenterol. Nutr.* **2002**, *34*, 291–295. [CrossRef]

29. Korpela, K.; Salonen, A.; Vepsäläinen, O.; Suomalainen, M.; Kolmeder, C.; Varjosalo, M.; Miettinen, S.; Kukkonen, K.; Savilahti, E.; Kuitunen, M.; et al. Probiotic supplementation restores normal microbiota composition and function in antibiotic-treated and in caesarean-born infants. *Microbiome* **2018**, *6*, 182. [CrossRef]

30. Jiménez, M.; Muñoz, F.C.; Cervantes-García, D.; Cervantes, M.M.; Hernández-Mercado, A.; Barrón-García, B.; Moreno Hernández-Duque, J.L.; Rodríguez-Carlos, A.; Rivas-Santiago, B.; Salinas, E. Protective Effect of Glycomacropeptide on the Atopic Dermatitis-Like Dysfunctional Skin Barrier in Rats. *J. Med. Food* **2020**, *23*, 1216–1224. [CrossRef]

31. Brück, W.M.; Redgrave, M.; Tuohy, K.M.; Lönnerdal, B.; Graverholt, G.; Hernell, O.; Gibson, G.R. Effects of bovine alpha-lactalbumin and casein glycomacropeptide-enriched infant formulae on faecal microbiota in healthy term infants. *J. Pediatr. Gastroenterol. Nutr.* **2006**, *43*, 673–679. [CrossRef] [PubMed]

32. Sims, I.M.; Tannock, G.W. Galacto- and Fructo-oligosaccharides Utilized for Growth by Cocultures of Bifidobacterial Species Characteristic of the Infant Gut. *Appl. Environ. Microbiol.* **2020**, *86*, e00214-20. [CrossRef] [PubMed]

33. Arboleya, S.; Sánchez, B.; Solís, G.; Fernández, N.; Suárez, M.; Hernández-Barranco, A.M.; Milani, C.; Margolles, A.; de Los Reyes-Gavilán, C.G.; Ventura, M.; et al. Impact of Prematurity and Perinatal Antibiotics on the Developing Intestinal Microbiota: A Functional Inference Study. *Int. J. Mol. Sci.* **2016**, *17*, 649. [CrossRef] [PubMed]

34. Soberón, M.; González, A. Physiological role of glutaminase activity in Saccharomyces cerevisiae. *J. Gen. Microbiol.* **1987**, *133*, 1–8. [CrossRef]

35. Belizário, J.E.; Faintuch, J.; Garay-Malpartida, M. Gut Microbiome Dysbiosis and Immunometabolism: New Frontiers for Treatment of Metabolic Diseases. *Mediat. Inflamm.* **2018**, *2018*, 2037838. [CrossRef]

36. Németh, B.; Doczi, J.; Csete, D.; Kacso, G.; Ravasz, D.; Adams, D.; Kiss, G.; Nagy, A.M.; Horvath, G.; Tretter, L.; et al. Abolition of mitochondrial substrate-level phosphorylation by itaconic acid produced by LPS-induced Irg1 expression in cells of murine macrophage lineage. *FASEB J.* **2016**, *30*, 286–300. [CrossRef]

37. Ahn, S.; Jung, J.; Jang, I.A.; Madsen, E.L.; Park, W. Role of Glyoxylate Shunt in Oxidative Stress Response. *J. Biol. Chem.* **2016**, *291*, 11928–11938. [CrossRef]

38. Allison, C.; Macfarlane, G.T. Dissimilatory nitrate reduction by Propionibacterium acnes. *Appl. Environ. Microbiol.* **1989**, *55*, 2899–2903. [CrossRef]
39. Barnett, J.A. A history of research on yeasts 5: The fermentation pathway. *Yeast* **2003**, *20*, 509–543. [CrossRef]
40. Cohen Kadosh, K.; Muhardi, L.; Parikh, P.; Basso, M.; Jan Mohamed, H.J.; Prawitasari, T.; Samuel, F.; Ma, G.; Geurts, J.M. Nutritional Support of Neurodevelopment and Cognitive Function in Infants and Young Children-An Update and Novel Insights. *Nutrients* **2021**, *13*, 199. [CrossRef]
41. Cryan, J.F.; O'Riordan, K.J.; Cowan, C.S.M.; Sandhu, K.V.; Bastiaanssen, T.F.S.; Boehme, M.; Codagnone, M.G.; Cussotto, S.; Fulling, C.; Golubeva, A.V.; et al. The Microbiota-Gut-Brain Axis. *Physiol. Rev.* **2019**, *99*, 1877–2013. [CrossRef] [PubMed]
42. Rao, C.; Coyte, K.Z.; Bainter, W.; Geha, R.S.; Martin, C.R.; Rakoff-Nahoum, S. Multi-kingdom ecological drivers of microbiota assembly in preterm infants. *Nature* **2021**, *591*, 633–638. [CrossRef] [PubMed]
43. Reeves, B.C. Principles of research: Limitations of non-randomized studies. *Surgery* **2008**, *26*, 120–124.

Article

Associations between Breastfeeding Duration and Obesity Phenotypes and the Offsetting Effect of a Healthy Lifestyle

Jiajia Dang [1,2], Ting Chen [1,2], Ning Ma [1,2], Yunfei Liu [1,2], Panliang Zhong [1,2], Di Shi [1,2], Yanhui Dong [1,2], Zhiyong Zou [1,2], Yinghua Ma [1,2], Yi Song [1,2,*] and Jun Ma [1,2,*]

[1] Institute of Child and Adolescent Health, School of Public Health, Peking University, Beijing 100191, China; dangjj@bjmu.edu.cn (J.D.); chentbjmu@163.com (T.C.); mnmaning@bjmu.edu.cn (N.M.); liuyunfei_pku@163.com (Y.L.); 1610306219@pku.edu.cn (P.Z.); 2111210123@bjmu.edu.cn (D.S.); dongyanhui@bjmu.edu.cn (Y.D.); harveyzou2002@bjmu.edu.cn (Z.Z.); yinghuama@bjmu.edu.cn (Y.M.)

[2] National Health Commission Key Laboratory of Reproductive Health, Peking University, Beijing 100191, China

* Correspondence: songyi@bjmu.edu.cn (Y.S.); majunt@bjmu.edu.cn (J.M.); Tel.: +86-10-82801624 (Y.S.); Fax: +86-10-82801178 (Y.S.)

Citation: Dang, J.; Chen, T.; Ma, N.; Liu, Y.; Zhong, P.; Shi, D.; Dong, Y.; Zou, Z.; Ma, Y.; Song, Y.; et al. Associations between Breastfeeding Duration and Obesity Phenotypes and the Offsetting Effect of a Healthy Lifestyle. *Nutrients* 2022, *14*, 1999. https://doi.org/10.3390/nu14101999

Academic Editor: Ekhard E. Ziegler

Received: 2 April 2022
Accepted: 6 May 2022
Published: 10 May 2022

Publisher's Note: MDPI stays neutral with regard to jurisdictional claims in published maps and institutional affiliations.

Copyright: © 2022 by the authors. Licensee MDPI, Basel, Switzerland. This article is an open access article distributed under the terms and conditions of the Creative Commons Attribution (CC BY) license (https://creativecommons.org/licenses/by/4.0/).

Abstract: *Background*: Additional metabolic indicators ought to be combined as outcome variables when exploring the impact of breastfeeding on obesity risk. Given the role of a healthy lifestyle in reducing obesity, we aimed to assess the effect of breastfeeding duration on different obesity phenotypes according to metabolic status in children and adolescents, and to explore the offsetting effect of healthy lifestyle factors on the associations between breastfeeding duration and obesity phenotypes. *Methods*: A total of 8208 eligible children and adolescents aged 7–18 years were recruited from a Chinese national cross-sectional study conducted in 2013. Anthropometric indicators were measured in the survey sites, metabolic indicators were tested from fasting blood samples, and breastfeeding duration and sociodemographic factors were collected by questionnaires. According to anthropometric and metabolic indicators, obesity phenotypes were divided into metabolic healthy normal weight (MHNW), metabolic unhealthy normal weight (MUNW), metabolic healthy obesity (MHO), and metabolic unhealthy obesity (MUO). Four common obesity risk factors (dietary consumption, physical activity, screen time, and sleep duration) were used to construct a healthy lifestyle score. Scores on the lifestyle index ranged from 0 to 4 and were further divided into unfavorable lifestyles (zero or one healthy lifestyle factor), intermediate lifestyles (two healthy lifestyle factors), and favorable lifestyle (three or four healthy lifestyle factors). Multinomial logistic regression was used to estimate the odds ratio (OR) and 95% confidence interval (95% CI) for the associations between breastfeeding duration and obesity phenotypes. Furthermore, the interaction terms of breastfeeding duration and each healthy lifestyle category were tested to explore the offsetting effect of lifestyle factors. *Results*: The prevalence of obesity among Chinese children and adolescents aged 7–18 years was 11.0%. Among the children and adolescents with obesity, the prevalence of MHO and MUO was 41.0% and 59.0%, respectively. Compared to the children and adolescents who were breastfed for 6–11 months, prolonged breastfeeding ($\geq$12 months) increased the risks of MUNW (OR = 1.35, 95% CI: 1.19–1.52), MHO (OR = 1.61, 95% CI: 1.27–2.05), and MUO (OR = 1.46, 95% CI: 1.20–1.76). When stratified by healthy lifestyle category, there was a typical dose–response relationship between duration of breastfeeding over 12 months and MUNW, MHO, and MUO, with an increased risk of a favorable lifestyle moved to an unfavorable lifestyle. *Conclusions*: Prolonged breastfeeding ($\geq$12 months) may be associated with increased risks of MUNW, MHO, and MUO, and the benefits of breastfeeding among children and adolescents may begin to wane around the age of 12 months. The increased risks may be largely offset by a favorable lifestyle.

Keywords: breastfeeding; obesity phenotypes; healthy lifestyle; children and adolescents

---

1. Introduction

Recently, childhood obesity has been among the looming global public health issues, which could cause metabolic abnormalities, type 2 diabetes, cardiovascular diseases, and tumors in adulthood, thus representing a serious medical burden [1–3]. China has the largest number of children and adolescents with obesity in the world [4], where the prevalence of overweight and obesity among children and adolescents aged 6–17 years has exceeded 19% [5]. Body mass index (BMI) is a combination of weight and height, thus representing a typical way to identify obesity. However, it has been shown that individuals with the same BMI may have different status of metabolic components, including blood glucose, blood pressure, and blood lipid levels [6]. Therefore, the definition of obesity by BMI alone does not accurately reflect obesity-related metabolic status and susceptibility to metabolic diseases. There are two obesity phenotypes based on BMI combined with metabolic components. People with obesity and normal metabolic characteristics are considered as metabolic healthy obesity (MHO), while people with obesity and metabolically unhealthy status are defined as metabolic unhealthy obesity (MUO) [7,8]. MUO is associated with a higher risk of cardiovascular disease and mortality compared to MHO [9]. Given that different obesity phenotypes have varying disease risks, identifying the modifiable risks and protective factors of obesity phenotypes may contribute to future precise stratified management of obesity intervention.

Among modifiable risk factors for childhood obesity in the first 1000 days of life, breastfeeding was considered as an effective protective factor to reduce the risk of childhood obesity due to the bioactive compounds of breast milk [10–15]. However, breastfeeding as a measure to prevent overweight and obesity in children and adolescents has produced conflicting results. Some systematic reviews, most of which were conducted in European countries [16–18], as well as several studies from the United States [19–21] and Brazil [22], reported that breastfeeding duration was inversely related to the risk of obesity in children and adolescents. However, a large randomized controlled trial that promoted longer breastfeeding duration in Belarus did not show significant or meaningful changes in obesity, blood pressure, or cardiometabolic risk factors in children and adolescents [23–25]. Another cohort study in Japan had similar findings [26]. In addition, cohort studies in Finland and Sweden found a positive association between breastfeeding duration and obesity [27,28]. According to these findings, the association between breastfeeding duration and childhood obesity may vary by study design, ethnicities, confounding factors, and the definition of obesity itself, thus necessitating further exploration [29,30].

Previous studies that measured only BMI in children may have underestimated the true impact of breastfeeding on obesity risk [31], and more metabolic indicators need to be combined as outcome variables such as MUO with higher health damage to assess the effect of breastfeeding. More studies urgently ought to explore the associations between breastfeeding duration and obesity phenotypes. Studies have found that breastfeeding was associated with healthier eating behaviors in later childhood, such as higher consumption of vegetables and fruits and lower consumption of sugary drinks and ultra-processed foods [32–34]. Additionally, a large number of studies showed that a favorable lifestyle such as healthy diet, adequate sleeping duration, and physical activity established in later life would significantly decrease the risk of childhood obesity [35–37]. However, it remains unclear whether adherence to a healthy lifestyle could influence the associations between breastfeeding duration and obesity or obesity phenotypes.

We hypothesized that breastfeeding duration was related to obesity phenotypes, and that these associations might be affected by a healthy lifestyle. In this study, we aimed to assess the effect of breastfeeding duration on different obesity phenotypes and to investigate whether a healthy lifestyle would modify these associations among children and adolescents aged 7–18 years using the data from a Chinese national cross-sectional study in 2013.

2. Materials and Methods

2.1. Study Design and Population

Data were collected and maintained by a national cross-sectional study conducted in September 2013. Children and adolescents aged 7–18 years from seven provinces, namely, Tianjin, Shanghai, Chongqing, Liaoning, Hunan, Ningxia, and Guangdong, were selected by a multistage cluster random sampling method. The seven provinces in this study came from different economic levels and different geographical regions of China. The detailed description of the study design was reported in a previous study [38]. In brief, 3–4 districts were randomly selected from each province, with 12–16 schools randomly selected from each district, and 2–3 classes per grade randomly selected from each school. A total of 15,733 children and adolescents aged 7–18 years with a blood sample were included (Figure 1). In our study, children and adolescents with normal weight and obesity were extracted, and cases with missing data about systolic blood pressure (SBP), diastolic blood pressure (DBP), and breastfeeding were excluded. The remaining 8208 children and adolescents were included in final analysis.

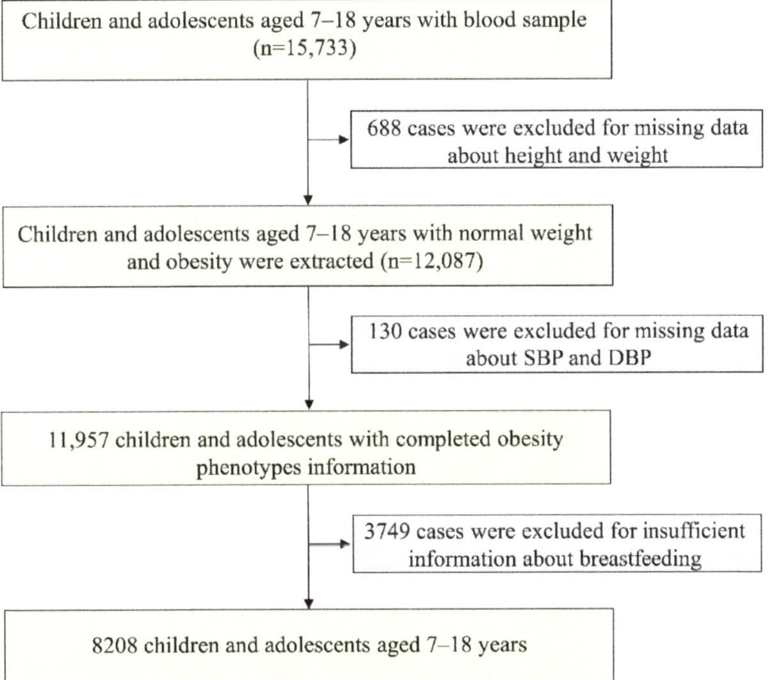

Figure 1. Flow chart of data. SBP, systolic blood pressure; DBP, diastolic blood pressure.

2.2. Data Collection and Questionnaire Survey

2.2.1. Anthropometric Measurements

All anthropometric measurements were conducted using standardized instruments and procedures. Height (cm) was measured to the nearest 0.1 cm by a portable stadiometer (model TZG, China) where participants need to stand straight with light clothes and without shoes. Body weight (kg) was assessed by a lever-type weight scale (model RGT-140, China) to the nearest 0.1 kg. Body mass index (BMI) was then calculated as weight (kg) divided by the square of the height (m). Children and adolescents were required to sit quietly for at least 5 min prior to the first measurement of SBP (mmHg) and DBP (mmHg) using a mercury sphygmomanometer (model XJ1ID, Shanghai Medical Instruments Co.,

Ltd., Shanghai, China). Each indicator was measured twice, and the mean of the two measurements was calculated for the final analysis. Specifically, a 5 min break was allowed between the two measurements of blood pressure.

2.2.2. Blood Sample Collection and Detection

Venous blood was collected by venipuncture after fasting for 12 h. After centrifuging at 3000 rpm for 10 min, serum was collected and transported to the experimental center at low temperature ($-80\,^\circ$C). Blood biochemical analyses were performed by a biomedical analysis company certified by Peking University [39]. Fasting plasma glucose (FPG), triglyceride (TG), and high-density lipoprotein cholesterol (HDL-C) were analyzed using the hexokinase method, enzymatic method, and clearance method, respectively.

2.2.3. Questionnaire Survey

A self-administered questionnaire of children and adolescents was used to obtain information on basic characteristics (e.g., age, sex, residence, and single-child status), dietary consumption (meat, sugar-sweetened beverage, fruit, and vegetable consumption), physical activity, screen time, and sleep duration. The questionnaires were revised in the early stage of our project and validated by experts, before being deemed feasible and acceptable for children and parents. Some questionnaires (3%) from the same participants were re-examined within a week. All questionnaires were also checked for logicality and integrity. Children aged 7–9 years were administered the questionnaire with the assistance of their parents. We gathered the frequency (day) and amount (servings) of each food over the course of a week. The average daily intake of a single food was calculated using the following equation: (day $\times$ quantity in each of those days)/7 [40].

A self-administered parent questionnaire was used to obtain information on breastfeeding duration, birth weight, delivery time, and delivery model for children and adolescents, education level and tobacco and alcohol consumption for parents, maternal age at delivery, and family household income. Family history of diseases was considered as having a family history of obesity, hypertension, diabetes mellitus, or cerebrovascular disease when either paternal or maternal diagnosis was self-reported.

2.3. Definition and Categorization of Indicators

2.3.1. Obesity Phenotypes

Obesity phenotypes included metabolic healthy normal weight (MHNW), metabolic unhealthy normal weight (MUNW), MHO, and MUO according to expert consensus on the definition of metabolically healthy obesity and screening metabolically healthy obesity in Chinese children [41]. Obesity phenotypes were evaluated on the basis of (1) BMI (normal weight was determined as BMI < 85th percentile and obesity was determined as BMI $\geq$ 95th percentile for sex- and age-specific group [42]) and (2) metabolic abnormalities ((a) hypertension: SBP and/or DBP $\geq$ 95th percentile for sex- and age-specific group; (b) elevated fasting glucose: FPG $\geq$ 5.6 mmol/L; (c) high TG: TG $\geq$ 1.70 mmol/L; (d) low HDL-C: HDL-C < 1.03 mmol/L). MHNW was defined as normal weight without metabolic abnormalities. MUNW was defined as normal weight with 1–4 metabolic abnormalities. MHO was defined as obesity without metabolic abnormalities. MUO was defined as obesity with 1–4 metabolic abnormalities.

2.3.2. Breastfeeding Duration

Children and adolescents were categorized into three subgroups of 0–5 month(s), 6–11 months, and $\geq$12 months according to breastfeeding duration (in month). Breastfeeding was not limited to exclusive breastfeeding.

2.3.3. Healthy Lifestyle

Four common obesity risk factors (dietary consumption, physical activity, screen time, and sleep duration) were used to construct a healthy lifestyle score [43,44]. A healthy diet was based on four regularly consumed foods (meat, sugar-sweetened beverage, fruit, and vegetable consumption) linked to childhood obesity [45,46]. In this study, dietary optimum components were defined as daily intake of fruits of $\geq$3 servings (one serving is about 100 g), vegetables of $\geq$4 servings (one serving is about 100 g), meat products of 2–3 servings (one serving is about 50 g), and weekly intake of sugar-sweetened beverage of <1 serving (one serving is about 250 mL) according to the dietary guidelines for school-age children in China (2016) [47]. Children and adolescents were considered to have a healthy diet if they consumed at least two of the four common foods in the required servings. Physical activity was categorized by the cutoff of moderate to vigorous physical activity for 1 h per day [48]. Screen time was defined as <2 h of screen time per day and $\geq$2 h per day. Sleep duration was defined as sufficient sleep and insufficient sleep according to the cutoff of 9 h for children and adolescents [49]. Children and adolescents scored one point for each of the four defined health behaviors. Scores on the lifestyle index ranged from 0–4, and they were further divided into unfavorable lifestyles (zero or one healthy lifestyle factor), intermediate lifestyles (two healthy lifestyle factors), and favorable lifestyles (three or four healthy lifestyle factors) [37].

2.3.4. Confounding Variables

In this study, age, sex, residence, single-child status, delivery model, delivery date, birth weight, family history of diseases, parental education level, parental tobacco and alcohol consumption, maternal age at delivery, and family household income were considered as confounding variables. Children and adolescents were categorized according to their birth weight into low birth weight (LBW, birth weight < 2500 g), normal birth weight (NBW, birth weight: 2500–3999 g), and high birth weight (HBW, birth weight $\geq$ 4000 g).

2.4. Statistics Analysis

Quantitative variables were shown as mean (standard deviation) or median (interquartile range) according to the normality of distribution, and categorical variables were shown as number (percentage). Nonparametric test or Bonferroni multiple comparison method and chi-squared test were used to compare difference of quantitative data and categorical data, respectively, between different breastfeeding duration groups. A multinomial logistic regression model was used to estimate the odds ratio (OR) and 95% confidence interval (95% CI) for the associations between breastfeeding duration and obesity phenotypes, and MHNW was considered as the reference group. The interaction terms of breastfeeding duration and healthy lifestyle category were also tested. When interactions were significant, stratified analyses were performed. The final model was adjusted for age, sex, residence, single-child status, delivery model, delivery time, birth weight, family history of diseases, parental education level, parental tobacco and alcohol consumption, maternal age at delivery, and family household income. All statistical analyses were performed with IBM SPSS Statistics version 26.0 and GraphPad Prism 9. A two-tailed *p*-value <0.05 was considered statistically significant.

3. Results

3.1. Characteristics of Participants

In the total population, the prevalence of obesity was 11.0%, and the prevalence of MHO and MUO was 4.4% and 6.6%, respectively. Table 1 shows the characteristics of children and adolescents, as well as their parents, by breastfeeding duration groups. A total of 8208 eligible children and adolescents, 2373 in the breastfeeding duration for 0–5 months group, 3184 in the breastfeeding duration for 6–11 months group, and 2651 in the breastfeeding duration for $\geq$12 months group, were included in the final analysis. Among the children and adolescents with obesity, the prevalence of MHO and MUO was 41.0% and 59.0% respectively.

Table 1. Demographic characteristics of eligible children and adolescents and their parents, stratified by breastfeeding duration (n = 8208).

		Breastfeeding Duration (Months)			
	Total	0–5 (n = 2373)	6–11 (n = 3184)	≥12 (n = 2651)	*p*-Value
Characteristics of children and adolescents					
Age *	11.2 (4.0)	11.1 (6.0)	11.2 (6.0)	11.3 (4.0)	0.016
Sex					0.013
Boys	3914 (47.7)	1180 (49.7) [#]	1458 (45.8)	1276 (48.1)	
Girls	4294 (52.3)	1193 (50.3) [#]	1726 (54.2)	1375 (51.9)	
Residence					<0.001
Urban	4623 (56.3)	1533 (64.6) [#]	1844 (57.9)	1246 (47.0) [#]	
Rural	3585 (43.7)	840 (35.4) [#]	1340 (42.1)	1405 (53.0) [#]	
Obesity phenotypes					
Normal weight					<0.001
MHNW	4776 (68.1)	1472 (72.8) [#]	1927 (69.5)	1377 (62.0) [#]	
MUNW	2242 (31.9)	551 (27.2) [#]	847 (30.5)	844 (38.0) [#]	
Obesity					0.002
MHO	488 (41.0)	171 (48.9) [#]	153 (37.3)	164 (38.1)	
MUO	702 (59.0)	179 (51.1) [#]	257 (62.7)	266 (61.9)	
Healthy lifestyle category					0.001
Favorable	1213 (14.8)	367 (15.5)	487 (15.3)	359 (13.5) [#]	
Intermediate	2838 (34.6)	825 (34.8)	1150 (36.1)	863 (32.6) [#]	
Unfavorable	4157 (50.6)	1181 (49.8)	1547 (48.6)	1429 (53.9)	
Single-child status					<0.001
Single children	5493 (66.9)	1733 (73.0) [#]	2188 (68.7)	1572 (59.3) [#]	
Non-single children	2715 (33.1)	640 (27.0) [#]	996 (31.3)	1079 (40.7) [#]	
Delivery time					<0.001
Normal	3802 (46.3)	1083 (45.6) [#]	1340 (42.1)	1379 (52.0) [#]	
Premature delivery	2435 (29.7)	803 (33.8)	993 (31.2)	639 (24.1) [#]	
Delayed delivery	1971 (24.0)	487 (20.5) [#]	851 (26.7)	633 (23.9) [#]	
Delivery model					<0.001
Caesarean	3428 (41.8)	1192 (50.2) [#]	1288 (40.5)	948 (35.8) [#]	
Eutocia	4780 (58.2)	1181 (49.8) [#]	1896 (59.5)	1703 (64.2) [#]	
Birth weight					<0.001
NBW	6881 (83.8)	1954 (82.3)	2665 (83.7)	2262 (85.3)	
LBW	569 (6.9)	213 (9.0)	233 (7.3)	123 (4.6) [#]	
HBW	758 (9.2)	206 (8.7)	286 (9.0)	266 (10.0)	
Family history of diseases [$]					0.022
Yes	1254 (15.3)	403 (17.0) [#]	459 (14.4)	392 (14.8)	
No	6954 (84.7)	1970 (83.0) [#]	2725 (85.6)	2259 (85.2)	
Parental characteristics					
Paternal education level					<0.001
Primary or below	581 (7.1)	133 (5.6)	169 (5.3)	279 (10.5) [#]	
Secondary or equivalent	5338 (65.0)	1381 (58.2) [#]	2010 (63.1)	1947 (73.4) [#]	
Junior college or above	2289 (27.9)	859 (36.2) [#]	1005 (31.6)	425 (16.0) [#]	
Maternal education level					<0.001
Primary or below	769 (9.4)	159 (6.7) [#]	273 (8.6)	337 (12.7) [#]	
Secondary or equivalent	5221 (63.6)	1351 (56.9) [#]	1956 (61.4)	1914 (72.2) [#]	
Junior college or above	2218 (27.0)	863 (36.4) [#]	955 (30.0)	400 (15.1) [#]	
Paternal tobacco consumption					<0.001
Yes	4327 (52.7)	1220 (51.4)	1612 (50.6)	1495 (56.4) [#]	
No	3881 (47.3)	1153 (48.6)	1572 (49.4)	1156 (43.6) [#]	
Maternal tobacco consumption					0.001
Yes	97 (1.2)	38 (1.6) [#]	20 (0.6)	39 (1.5) [#]	
No	8111 (98.8)	2335 (98.4) [#]	3164 (99.4)	2612 (98.5) [#]	
Paternal alcohol consumption					0.120
Yes	2560 (31.2)	732 (30.8)	962 (30.2)	866 (32.7)	
No	5648 (68.8)	1641 (69.2)	2222 (69.8)	1785 (67.3)	
Maternal alcohol consumption					0.453
Yes	152 (1.9)	48 (2.0)	62 (1.9)	42 (1.6)	
No	8056 (98.1)	2325 (98.0)	3122 (98.1)	2609 (98.4)	
Maternal age at delivery *	26.4 (4.8)	27.0 (4.8) [#]	26.1 (4.7)	26.1 (4.8)	<0.001
Monthly household income					0.835
<5000 CNY	6803 (82.9)	1958 (82.5)	2641 (82.9)	2204 (83.1)	
≥5000 CNY	1405 (17.1)	415 (17.5)	543 (17.1)	447 (16.9)	

Note: Categorical variables are expressed by frequency value (percentage, %). * Quantitative variables are shown as median (interquartile range) because the data did not follow a normality of distribution. [$] Family history of diseases includes obesity, hypertension, diabetes mellitus, and cerebrovascular disease. [#] Significant difference compared with the breastfeeding duration for 6–11 months group. MHNW, metabolic healthy normal weight; MUNW, metabolic unhealthy normal weight; MHO, metabolic healthy obesity; MUO, metabolic unhealthy obesity; NBW, normal birth weight; LBW, low birth weight; LBW, high birth weight.

Compared to the breastfeeding duration for 6–11 months group, children and adolescents who breastfed for ≥12 months had a lower proportion of MHNW (62.0% vs. 69.5%) and a higher proportion of MUNW (38.0% vs. 27.2%); children and adolescents who breastfed for 0–5 months had a higher proportion of MHO (48.9% vs. 37.3%) and a lower proportion of MUO (51.1% vs. 62.7%). A lower proportion of children and adolescents with breastfeeding duration for ≥12 months engaged in a favorable (15.3% vs. 13.5%) and intermediate lifestyle (36.1% vs. 32.6%). Parents in the breastfeeding duration for 0–5 months group had the highest education level (36.2% and 36.4% for junior college or above); however, those in the breastfeeding duration for ≥12 months group had the lowest education level (16.0% and 15.1% for junior college or above). No significant difference was found in monthly household income between different breastfeeding duration groups.

3.2. Associations between Breastfeeding Duration and Obesity Phenotypes

Figure 2 and Supplementary Table S1 show the OR values for the associations between breastfeeding duration and obesity phenotypes. Compared to the children and adolescents who were breastfed for 6–11 months, participants who were in the longer breastfeeding duration group (≥12 months) had higher risks of MUNW (OR = 1.35, 95% CI: 1.19–1.52), MHO (OR = 1.61, 95% CI: 1.27–2.05), and MUO (OR = 1.46, 95% CI: 1.20–1.76) after adjusting for age, sex, residence, single-child status, delivery model, delivery date, birth weight, family history of diseases, parental education level, parental tobacco and alcohol consumption, maternal age at delivery, and family household income. In Model 1 (crude model, see Supplementary Table S1), participants who were in the breastfeeding duration for 0–5 months group had a higher risk of MHO (OR = 1.46, 95% CI: 1.16–1.84); however, after adjusting for confounders, these associations turned out to be nonsignificant (Model 3 in Supplementary Table S1).

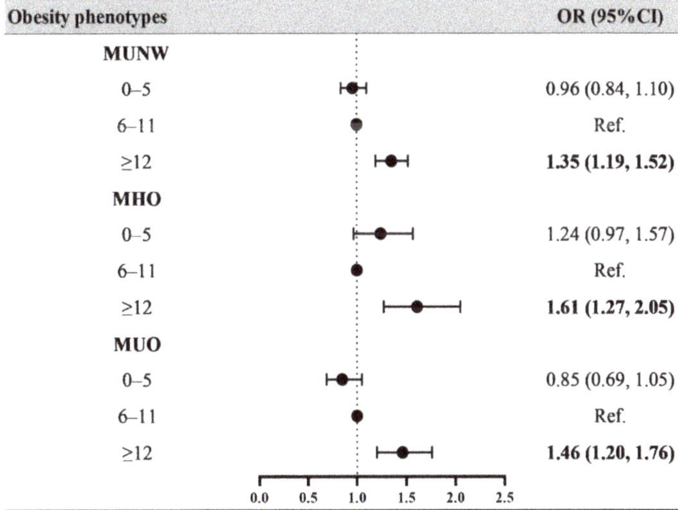

Figure 2. Associations between breastfeeding duration and obesity phenotypes. Obesity phenotypes include metabolic healthy normal weight (MHNW, reference group), metabolic unhealthy normal weight (MUNW), metabolic healthy obesity (MHO), and metabolic unhealthy obesity (MUO). Bold values of OR (95% CI) are statistically significant ($p < 0.05$). Adjusted for age, sex, residence, single-child status, delivery model, delivery date, family history of diseases (obesity, hypertension, diabetes mellitus and cerebrovascular disease), parental education level, parental tobacco and alcohol consumption, maternal age at delivery, and family household income.

3.3. The Offsetting Effect of Healthy Lifestyle

When breastfeeding duration and healthy lifestyle category were evaluated as interaction variables for obesity phenotypes, it was discovered that healthy lifestyle category and breastfeeding duration for ≥12 months had a significant interaction effect on MUNW, MHO, and MUO (Supplementary Table S2). Figure 3 shows the adjusted OR values for associations between breastfeeding duration and obesity phenotypes stratified by healthy lifestyle category. Although the OR values in certain groups were not statistically significant, there was a typical dose–response relationship between breastfeeding duration for more than 12 months and obesity phenotypes, with increased risks of MUNW (favorable: OR = 1.08, 95% CI: 0.77–1.51; intermediate: OR = 1.35, 95% CI: 1.09–1.68; unfavorable: OR = 1.40, 95% CI: 1.18–1.66), MHO (favorable: OR = 1.25, 95% CI: 0.66–2.37; intermediate: OR = 1.81, 95% CI: 1.17–2.81; unfavorable: OR = 1.61, 95% CI: 1.15–2.23), and MUO (favorable: OR = 0.93, 95% CI: 0.53–1.61; intermediate: OR = 1.41, 95% CI: 1.03–1.94; unfavorable: OR = 1.68, 95% CI: 1.27–2.21) when a favorable lifestyle moved to an unfavorable lifestyle.

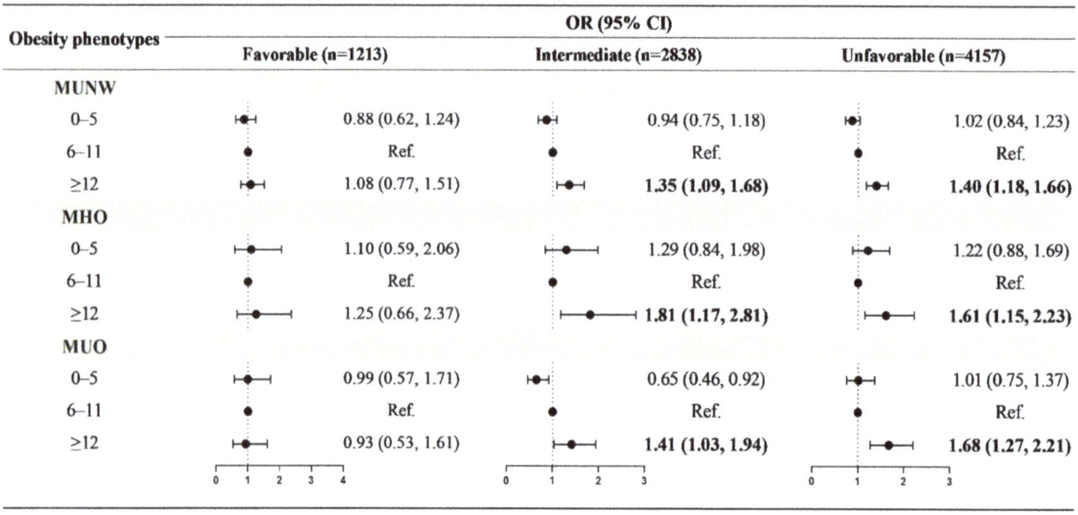

Obesity phenotypes	OR (95% CI)		
	Favorable (n=1213)	Intermediate (n=2838)	Unfavorable (n=4157)
MUNW			
0–5	0.88 (0.62, 1.24)	0.94 (0.75, 1.18)	1.02 (0.84, 1.23)
6–11	Ref.	Ref.	Ref.
≥12	1.08 (0.77, 1.51)	**1.35 (1.09, 1.68)**	**1.40 (1.18, 1.66)**
MHO			
0–5	1.10 (0.59, 2.06)	1.29 (0.84, 1.98)	1.22 (0.88, 1.69)
6–11	Ref.	Ref.	Ref.
≥12	1.25 (0.66, 2.37)	**1.81 (1.17, 2.81)**	**1.61 (1.15, 2.23)**
MUO			
0–5	0.99 (0.57, 1.71)	0.65 (0.46, 0.92)	1.01 (0.75, 1.37)
6–11	Ref.	Ref.	Ref.
≥12	0.93 (0.53, 1.61)	**1.41 (1.03, 1.94)**	**1.68 (1.27, 2.21)**

Figure 3. Associations between breastfeeding duration and obesity phenotypes in different healthy lifestyle groups. Obesity phenotypes include metabolic healthy normal weight (MHNW, reference group), metabolic unhealthy normal weight (MUNW), metabolic healthy obesity (MHO), and metabolic unhealthy obesity (MUO). Adjusted for age, sex, residence, single-child status, delivery model, delivery date, birth weight, family history of diseases (obesity, hypertension, diabetes mellitus and cerebrovascular disease), parental education level, parental tobacco and alcohol consumption, maternal age at delivery, and family household income. Bold values of OR (95% CI) are statistically significant ($p < 0.05$).

4. Discussion

This study is the first to report the national prevalence of different obesity phenotypes in China. The prevalence of obesity among Chinese children and adolescents aged 7–18 years was 11.0%. Among the children and adolescents with obesity, the prevalence of MHO and MUO was 41.0% and 59.0%, respectively. On the basis of the different obesity phenotype outcomes, we found that prolonged breastfeeding (≥12 months) was associated with increased risks of MUNW, MHO, and MUO, while a favorable lifestyle may offset the negative effects to some extent.

Consistent with the findings of existing studies [50,51], our results support the conclusion that prolonged breastfeeding is detrimental. Adults of a birth cohort study in Finland who breastfed for 8 months or more reported an 8.8% increase in overweight compared to those who breastfed for a shorter period of time [51]. A prospective cohort study in Sweden found that obesity prevalence was slightly higher for children aged 4 years with breastfeeding for 12 months than those who were breastfed for 9 months [30]. We speculated that the benefits of breastfeeding might begin to wane around the age of 12 months. The physiological mechanism via which prolonged breastfeeding might increase the risk of obesity is chronic exposure to maternal hormones present in breast milk, which could theoretically alter infant lipid metabolism and increase body fat composition later in life [51]. For example, a cohort study in the United Kingdom showed that prolonged breastfeeding was associated with higher cholesterol concentrations in later life [31], while mothers who breastfed for longer than 12 months had increased free thyroxine concentrations [52], which were associated with the regulation of circulating lipoprotein concentrations [53]. In addition, there were U-shaped associations between breastfeeding duration and body fat percentage and blood pressure [31,51]. Infants in poorer families tend to be breastfed for longer [51,54]. For children older than 12 months with a better supply of complementary food in high-income countries, additional breast milk may increase fat intake since the primary nutrient in breast milk after 12 months for infant is fat [55]. These facts may be significant reasons for the higher risk of MHO and MUO in longer periods ($\geq$12 months) of breastfeeding.

A previous study from the same database found that prolonged breastfeeding ($\geq$12 months) was associated with low lipid level and reduced risk of abnormal blood lipids; however, the beneficial effects were mainly reflected in total cholesterol (TC) [56]. Although the OR values of HDL-C and TG were not statistically significant, the HDL-C levels in children and adolescents who were breastfed for more than 12 months were lower than those in the non-breastfeeding group ($p < 0.01$). Only HDL-C and TG overlapped with the metabolic indicators of obesity phenotypes in the present study. Therefore, the results of the two studies are not contradictory, but are presented from different perspectives due to the role of breastfeeding duration in different diseases. Actually, this is the first study in the world to examine the associations between different obesity phenotypes and breastfeeding duration. This study provides a standardized outcome and detailed extensive subgroup analysis, serving as a reference and a more specific association for future investigation.

Breastfeeding for shorter periods of time might be hazardous due to the negative effects of alternate feeding [51]. Nevertheless, in our study, we did not find a role for short-term breastfeeding. The present study discovered that more educated parents established a shorter breastfeeding duration. Highly educated parents may turn to other alternatives such as more intensive care and promoting a healthier lifestyle for their children in later life to prevent obesity. However, the present study did not collect specific information on artificial feeding and food availability. Formula-fed children had higher protein intakes than breastfed children, relative to exclusive breastfeeding [57]. Nevertheless, higher protein intake is thought to increase insulin and insulin-like growth factor-1 (IGF-1) secretion in early childhood, which is associated with early obesity rebound [57]. Breastfed children could automatically control their food intake according to their own requirements, while formula-fed babies were passive. In addition, introducing complementary foods as early as 4 months of age or before increased the risk of overweight in childhood [58]. Thus, the findings may only be applicable to areas with a diverse food market, such as China.

The World Health Organization (WHO) recommends exclusive breastfeeding for 6 months and continued breastfeeding until 2 years or more [59], which is a very ideal goal to reach [60,61], especially in those countries and regions with limited drinking water and food security and diversity [62]. In fact, many Western countries, including 65% of European members and the United States, choose not to fully follow this recommendation [63]. The American Academy of Pediatrics recommends exclusive breastfeeding for 6 months and continued breastfeeding until 1 year or more [64], while guidelines from the European Society of Pediatric Gastroenterology, Hepatology, and Nutrition state that complementary foods could be introduced between the 17th and 26th weeks of life [65]. Moreover, Grummer-Strawn and Rollins emphasized the difficulty of conclusions from breastfeeding research due to the different conclusions of many studies, methodological weaknesses, and the varied mechanisms of breastfeeding affecting health [66]. Despite a systematic review which demonstrated that breastfeeding was linked to a 12% lower incidence of obesity when compared to never breastfeeding [67], this association might no longer be present after controlling for maternal BMI, smoking, and socioeconomic variables, as argued by several well-designed studies [23,27]. Given that varied settings with different social atmospheres, food supplies, and individual characteristics, the recommended breastfeeding duration from WHO might not be suitable for every country, which also need to establish individual policies according to their specific conditions.

Research has shown that, further away from infancy, there is a greater impact of other lifestyle factors (such as diet, exercise) on obesity [21]. When children adhered to a favorable lifestyle in later life, they could have a 50% lower risk of obesity than those who adhered to an unfavorable lifestyle [37]. These findings suggest that the increased risk of childhood obesity induced by prolonged breastfeeding duration might be largely offset by a favorable lifestyle later in life.

Despite these promising results, several limitations remained. Firstly, the associations observed in this study were from a cross-sectional study, leading to weak causal inference. Secondly, due to excessive sample deletion, there may have been some selection bias in the samples of this study. Thirdly, the reliability and validity information of the questionnaire was not provided in this study, and the validity of the questionnaire was not verified. In particular, breastfeeding was not collected according to the scale; thus, there might have been information bias. Fourthly, the information about breastfeeding was retrospectively investigated, and there was a certain recall bias, leading to overestimation or underestimation of effect values. In addition, the breastfeeding group and healthy lifestyle category were artificially divided; thus, misclassification bias existed. Fifthly, the role of formula milk powder and complementary foods in infant feeding was not excluded in this study. Additionally, despite the analysis of many variables that could have led to bias, residual confounding existed. Lastly, this study was not a large-scale national survey, and more regional participation is needed in the future.

5. Conclusions

Prolonged breastfeeding ($\geq$12 months) may be associated with increased risks of MUNW, MHO, and MUO, and the benefits of breastfeeding might begin to wane around the age of 12 months among children and adolescents. The increased risks may be largely offset by a favorable lifestyle such as healthy diet, limited screen time, adequate physical activity, and sufficient sleep duration. Breastfeeding duration for no more than 12 months and the establishment of a healthy lifestyle are recommended to prevent different obesity phenotypes for children and adolescents in China. Breastfeeding duration until 2 years or more, as recommended by the WHO, may be amended on the basis of the actual situation in different countries, such as the addition of children's complementary foods, and policies in the light of evidence-based studies should be formulated.

Supplementary Materials: The following supporting information can be downloaded at https://www.mdpi.com/article/10.3390/nu14101999/s1: Table S1. Associations between breastfeeding duration and obesity phenotypes; Table S2. Risk of obesity phenotypes according to healthy lifestyle category and breastfeeding duration.

Author Contributions: Conceptualization, J.D. and Y.S.; data curation, T.C., N.M., Y.L., P.Z. and D.S.; formal analysis, J.D.; funding acquisition, Y.S.; investigation, Z.Z., Y.M. and J.M.; methodology, J.D. and T.C.; project administration, Y.D., Z.Z., Y.M., Y.S. and J.M.; resources, Y.D., Z.Z., Y.M., Y.S. and J.M.; software, J.D., Y.L. and P.Z.; supervision, Y.M., Y.S. and J.M.; validation, Y.D., Z.Z. and Y.S.; visualization, N.M. and D.S.; writing—original draft, J.D. and T.C.; writing—review and editing, T.C., N.M., Y.L., P.Z., D.S., Y.D., Z.Z., Y.M., J.M. and Y.S. All authors have read and agreed to the published version of the manuscript.

Funding: The present research was supported by funding from the Natural Science Foundation of Beijing (Grant No. 7222247 to Y.S.)

Institutional Review Board Statement: This study was conducted according to the guidelines laid down in the Declaration of Helsinki, and all procedures involving students were approved by the Ethics Committee of Peking University (NO. IRB0000105213034).

Informed Consent Statement: All participants and their parents voluntarily signed informed consent forms.

Data Availability Statement: The data supporting the conclusions of this article can be made available from the corresponding author upon request.

Acknowledgments: The authors gratefully acknowledge all the schoolteachers and doctors for their assistance in data collection and all students and their guardians for participating in the study. We also would like to thank Patrick W.C. Lau from Hong Kong Baptist University for his contribution to the English editing.

Conflicts of Interest: The authors declare no conflict of interest.

References

1. Mokdad, A.H.; Ford, E.S.; Bowman, B.A.; Dietz, W.H.; Vinicor, F.; Bales, V.S.; Marks, J.S. Prevalence of obesity, diabetes, and obesity-related health risk factors, 2001. *JAMA* **2003**, *289*, 76–79. [CrossRef] [PubMed]
2. Canoy, D.; Boekholdt, S.M.; Wareham, N.; Luben, R.; Welch, A.; Bingham, S.; Buchan, I.; Day, N.; Khaw, K. Body fat distribution and risk of coronary heart disease in men and women in the European Prospective Investigation Into Cancer and Nutrition in Norfolk cohort: A population-based prospective study. *Circulation* **2007**, *116*, 2933–2943. [CrossRef] [PubMed]
3. Bijlsma, J.W.; Berenbaum, F.; Lafeber, F.P. Osteoarthritis: An update with relevance for clinical practice. *Lancet* **2011**, *377*, 2115–2126. [CrossRef]
4. Di Cesare, M.; Sorić, M.; Bovet, P.; Miranda, J.J.; Bhutta, Z.; Stevens, G.A.; Laxmaiah, A.; Kengne, A.-P.; Bentham, J. The epidemiological burden of obesity in childhood: A worldwide epidemic requiring urgent action. *BMC Med.* **2019**, *17*, 212. [CrossRef]
5. Pan, X.F.; Wang, L.; Pan, A. Epidemiology and determinants of obesity in China. *Lancet Diabetes Endocrinol.* **2021**, *9*, 373–392. [CrossRef]
6. Kramer, C.K.; Zinman, B.; Retnakaran, R. Are metabolically healthy overweight and obesity benign conditions? A systematic review and meta-analysis. *Ann. Intern. Med.* **2013**, *159*, 758–769. [CrossRef]
7. Eftekharzadeh, A.; Asghari, G.; Serahati, S.; Hosseinpanah, F.; Azizi, A.; Barzin, M.; Mirmiran, P.; Azizi, F. Predictors of incident obesity phenotype in nonobese healthy adults. *Eur. J. Clin. Investig.* **2017**, *47*, 357–365. [CrossRef]
8. Mirzaei, B.; Abdi, H.; Serahati, S.; Barzin, M.; Niroomand, M.; Azizi, F.; Hosseinpanah, F. Cardiovascular risk in different obesity phenotypes over a decade follow-up: Tehran Lipid and Glucose Study. *Atherosclerosis* **2017**, *258*, 65–71. [CrossRef]
9. Rhee, E.-J.; Lee, M.K.; Kim, J.D.; Jeon, W.S.; Bae, J.C.; Park, S.E.; Park, C.-Y.; Oh, K.-W.; Park, S.-W.; Lee, W.-Y. Metabolic health is a more important determinant for diabetes development than simple obesity: A 4-year retrospective longitudinal study. *PLoS ONE* **2014**, *9*, e98369. [CrossRef]
10. Marseglia, L.; Manti, S.; D'Angelo, G.; Cuppari, C.; Salpietro, V.; Filippelli, M.; Trovato, A.; Gitto, E.; Salpietro, C.; Arrigo, T. Obesity and breastfeeding: The strength of association. *Women Birth* **2015**, *28*, 81–86. [CrossRef]
11. Cope, M.B.; Allison, D.B. Critical review of the World Health Organization's (WHO) 2007 report on 'evidence of the long-term effects of breastfeeding: Systematic reviews and meta-analysis' with respect to obesity. *Obes. Rev.* **2008**, *9*, 594–605. [CrossRef] [PubMed]
12. Mosca, F.; Giannì, M.L. Human milk: Composition and health benefits. *Pediatr. Med. Chir.* **2017**, *39*, 155. [CrossRef] [PubMed]
13. Pietrobelli, A.; Agosti, M. Nutrition in the First 1000 Days: Ten Practices to Minimize Obesity Emerging from Published Science. *Int. J. Environ. Res. Public Health* **2017**, *14*, 1491. [CrossRef] [PubMed]
14. Horta, B.L.; Loret de Mola, C.; Victora, C.G. Long-term consequences of breastfeeding on cholesterol, obesity, systolic blood pressure and type 2 diabetes: A systematic review and meta-analysis. *Acta Paediatr.* **2015**, *104*, 30–37. [CrossRef]
15. Victora, C.G.; Bahl, R.; Barros, A.J.D.; Franca, G.V.A.; Horton, S.; Krasevec, J.; Murch, S.; Sankar, M.J.; Walker, N.; Rollins, N.C.; et al. Breastfeeding in the 21st century: Epidemiology, mechanisms, and lifelong effect. *Lancet* **2016**, *387*, 475–490. [CrossRef]
16. Arenz, S.; Rückerl, R.; Koletzko, B.; von Kries, R. Breast-feeding and childhood obesity—A systematic review. *Int. J. Obes.* **2004**, *28*, 1247–1256. [CrossRef]
17. Harder, T.; Bergmann, R.; Kallischnigg, G.; Plagemann, A. Duration of breastfeeding and risk of overweight: A meta-analysis. *Am. J. Epidemiol.* **2005**, *162*, 397–403. [CrossRef]
18. Spatz, D.L. Preventing obesity starts with breastfeeding. *J. Perinat. Neonatal Nutr.* **2014**, *28*, 41–50. [CrossRef]
19. Hawkins, S.S.; Baum, C.F.; Rifas-Shiman, S.L.; Oken, E.; Taveras, E.M. Examining Associations between Perinatal and Postnatal Risk Factors for Childhood Obesity Using Sibling Comparisons. *Child. Obes.* **2019**, *15*, 254–261. [CrossRef]

20. Colen, C.G.; Ramey, D.M. Is breast truly best? Estimating the effects of breastfeeding on long-term child health and wellbeing in the United States using sibling comparisons. *Soc. Sci. Med.* **2014**, *109*, 55–65. [CrossRef]

21. Metzger, M.W.; McDade, T.W. Breastfeeding as obesity prevention in the United States: A sibling difference model. *Am. J. Hum. Biol.* **2010**, *22*, 291–296. [CrossRef] [PubMed]

22. Contarato, A.A.; Rocha, E.D.; Czarnobay, S.A.; Mastroeni, S.S.; Veugelers, P.J.; Mastroeni, M.F. Independent effect of type of breastfeeding on overweight and obesity in children aged 12–24 months. *Cad. Saude Publ.* **2016**, *32*, e00119015.

23. Kramer, M.S.; Matush, L.; Vanilovich, I.; Platt, R.W.; Bogdanovich, N.; Sevkovskaya, Z.; Dzikovich, I.; Shishko, G.; Collet, J.; Martin, R.M.; et al. Effects of prolonged and exclusive breastfeeding on child height, weight, adiposity, and blood pressure at age 6.5 y: Evidence from a large randomized trial. *Am. J. Clin. Nutr.* **2007**, *86*, 1717–1721. [CrossRef] [PubMed]

24. Martin, R.M.; Patel, R.; Kramer, M.S.; Guthrie, L.; Vilchuck, K.; Bogdanovich, N.; Sergeichick, N.; Gusina, N.; Foo, Y.; Palmer, T.; et al. Effects of promoting longer-term and exclusive breastfeeding on adiposity and insulin-like growth factor-I at age 11.5 years: A randomized trial. *JAMA* **2013**, *309*, 1005–1013. [CrossRef] [PubMed]

25. Martin, R.M.; Patel, R.; Kramer, M.S.; Vilchuck, K.; Bogdanovich, N.; Sergeichick, N.; Gusina, N.; Foo, Y.; Palmer, T.; Thompson, J.; et al. Effects of promoting longer-term and exclusive breastfeeding on cardiometabolic risk factors at age 11.5 years: A cluster-randomized, controlled trial. *Circulation* **2014**, *129*, 321–329. [CrossRef] [PubMed]

26. Sata, M.; Yamagishi, K.; Sairenchi, T.; Irie, F.; Sunou, K.; Watanabe, H.; Iso, H.; Ota, H. Breastfeeding in Infancy in Relation to Subsequent Physical Size: A 20-year Follow-up of the Ibaraki Children's Cohort Study (IBACHIL). *J. Epidemiol.* **2021**, *advance online publication*.

27. Wallby, T.; Lagerberg, D.; Magnusson, M. Relationship Between Breastfeeding and Early Childhood Obesity: Results of a Prospective Longitudinal Study from Birth to 4 Years. *Breastfeed Med.* **2017**, *12*, 48–53. [CrossRef]

28. Fall, C.H.; Borja, J.B.; Osmond, C.; Richter, L.; Bhargava, S.K.; Martorell, R.; Stein, A.D.; Barros, F.C.; Victora, C.G.; the COHORTS Group. Infant-feeding patterns and cardiovascular risk factors in young adulthood: Data from five cohorts in low- and middle-income countries. *Int. J. Epidemiol.* **2011**, *40*, 47–62. [CrossRef]

29. Taveras, E.M.; Gillman, M.W.; Kleinman, K.; Rich-Edwards, J.W.; Rifas-Shiman, S.L. Racial/ethnic differences in early-life risk factors for childhood obesity. *Pediatrics* **2010**, *125*, 686–695. [CrossRef]

30. Kelly, Y.J.; Watt, R.G.; Nazroo, J.Y. Racial/ethnic differences in breastfeeding initiation and continuation in the United Kingdom and comparison with findings in the United States. *Pediatrics* **2006**, *118*, e1428–e1435. [CrossRef]

31. Ma, J.; Qiao, Y.; Zhao, P.; Li, W.; Katzmarzyk, P.T.; Chaput, J.; Fogelholm, M.; Kuriyan, R.; Lambert, E.; Maher, C.; et al. Breastfeeding and childhood obesity: A 12-country study. *Matern. Child Nutr.* **2020**, *16*, e12984. [CrossRef]

32. Hamulka, J.; Zielinska, M.A.; Jeruszka-Bielak, M.; Górnicka, M.; Głąbska, D.; Guzek, D.; Hoffmann, M.; Gutkowska, K. Analysis of Association between Breastfeeding and Vegetable or Fruit Intake in Later Childhood in a Population-Based Observational Study. *Int. J. Environ. Res. Public Health* **2020**, *17*, 3755. [CrossRef] [PubMed]

33. Perrine, C.G.; Galuska, D.A.; Thompson, F.E.; Scanlon, K.S. Breastfeeding duration is associated with child diet at 6 years. *Pediatrics* **2014**, *134* (Suppl. 1), S50–S55. [CrossRef] [PubMed]

34. Fonseca, P.C.D.A.; Ribeiro, S.A.V.; Andreoli, C.S.; de Carvalho, C.A.; Pessoa, M.C.; de Novaes, J.F.; Priore, S.E.; Franceschini, S.D.C.C. Association of exclusive breastfeeding duration with consumption of ultra-processed foods, fruit and vegetables in Brazilian children. *Eur. J. Nutr.* **2019**, *58*, 2887–2894. [CrossRef] [PubMed]

35. Kang, M.; Yoo, J.E.; Kim, K.; Choi, S.; Park, S.M. Associations between birth weight, obesity, fat mass and lean mass in Korean adolescents: The Fifth Korea National Health and Nutrition Examination Survey. *BMJ Open* **2018**, *8*, e018039. [CrossRef]

36. Vasylyeva, T.L.; Barche, A.; Chennasamudram, S.P.; Sheehan, C.; Singh, R.; Okogbo, M.E. Obesity in prematurely born children and adolescents: Follow up in pediatric clinic. *Nutr. J.* **2013**, *12*, 150. [CrossRef]

37. Wang, Z.H.; Zou, Z.Y.; Dong, Y.H.; Xu, R.B.; Yang, Y.D.; Ma, J. A Healthy Lifestyle Offsets the Increased Risk of Childhood Obesity Caused by High Birth Weight: Results from a Large-Scale Cross-Sectional Study. *Front. Nutr.* **2021**, *8*, 736900. [CrossRef]

38. Dong, Y.H.; Zou, Z.Y.; Yang, Z.P.; Wang, Z.H.; Jing, J.; Luo, J.Y.; Zhang, X.; Luo, C.Y.; Wang, H.; Zhao, H.P.; et al. Association between high birth weight and hypertension in children and adolescents: A cross-sectional study in China. *J. Hum. Hypertens.* **2017**, *31*, 737–743. [CrossRef]

39. Chen, Y.; Ma, L.; Ma, Y.; Wang, H.; Luo, J.; Zhang, X.; Luo, C.; Wang, H.; Zhao, H.; Pan, D.; et al. A national school-based health lifestyles interventions among Chinese children and adolescents against obesity: Rationale, design and methodology of a randomized controlled trial in China. *BMC Public Health* **2015**, *15*, 210. [CrossRef]

40. Rehm, C.D.; Peñalvo, J.L.; Afshin, A.; Mozaffarian, D. Dietary Intake Among US Adults, 1999–2012. *JAMA* **2016**, *315*, 2542–2553. [CrossRef]

41. Association for Maternal Child Health Study; Expert Committee on Obesity Controlling for Women Children; Expert Committee on Definition of Metabolically Healthy Obesity; Screening Metabolically Unhealthy Obesity in Chinese Children. The expert consensus on definition of metabolically healthy obesity and screening metabolically healthy obesity in Chinese children. *Matern. Child Health China* **2019**, *30*, 1487–1490.

42. National Health Commission of the People's Republic of China. *Screening for Overweight and Obesity among School-Age Children and Adolescents*; National Health Commission of the People's Republic of China: Beijing, China, 2018.

43. Liberali, R.; Kupek, E.; Assis, M.A.A. Dietary Patterns and Childhood Obesity Risk: A Systematic Review. *Child Obes.* **2020**, *16*, 70–85. [CrossRef]

44. Salvo, D.; Ranjit, N.; Nielsen, A.; Akhavan, N.; van den Berg, A. Characterizing Micro-scale Disparities in Childhood Obesity: Examining the Influence of Multilevel Factors on 4-Year Changes in BMI, Healthy Eating, and Physical Activity, Among a Cohort of Children Residing in Disadvantaged Urban Enclaves. *Front. Public Health* **2019**, *7*, 301. [CrossRef] [PubMed]

45. Essman, M.; Popkin, B.M.; Corvalán, C.; Reyes, M.; Taillie, L.S. Sugar-Sweetened Beverage Intake among Chilean Preschoolers and Adolescents in 2016: A Cross-Sectional Analysis. *Nutrients* **2018**, *10*, 1767. [CrossRef]

46. Noor, S.; Dehghan, M.; Lear, S.A.; Swaminathan, S.; Ibrahim, Q.; Rangarajan, S.; Punthakee, Z. Relationship between diet and acculturation among South Asian children living in Canada. *Appetite* **2020**, *147*, 104524. [CrossRef] [PubMed]

47. Chinese Nutrition Society. *Dietary Guidelines for School-Age Children*; People 's Medical Publishing House: Beijing, China, 2016.

48. Fan, M.; Lyu, J.; He, P. Chinese guidelines for data processing and analysis concerning the International Physical Activity Questionnaire. *Zhonghua Liu Xing Bing Xue Za Zhi* **2014**, *35*, 961–964.

49. Hirshkowitz, M.; Whiton, K.; Albert, S.M.; Alessi, C.; Bruni, O.; DonCarlos, L.; Hazen, N.; Herman, J.; Adams Hillard, P.J.; Katz, E.S.; et al. National Sleep Foundation's updated sleep duration recommendations: Final report. *Sleep Health* **2015**, *1*, 233–243. [CrossRef]

50. Martin, R.M.; Kramer, M.S.; Patel, R.; Rifas-Shiman, S.L.; Thompson, J.; Yang, S.; Vilchuck, K.; Bogdanovich, N.; Hameza, M.; Tilling, K.; et al. Effects of Promoting Long-term, Exclusive Breastfeeding on Adolescent Adiposity, Blood Pressure, and Growth Trajectories: A Secondary Analysis of a Randomized Clinical Trial. *JAMA Pediatrics* **2017**, *171*, e170698. [CrossRef]

51. O'Tierney, P.F.; Barker, D.J.; Osmond, C.; Kajantie, E.; Eriksson, J.G. Duration of breast-feeding and adiposity in adult life. *J. Nutr.* **2009**, *139*, 422S–425S. [CrossRef]

52. Phillips, D.I.; Barker, D.J.; Osmond, C. Infant feeding, fetal growth and adult thyroid function. *Acta Endocrinol.* **1993**, *129*, 134–138. [CrossRef]

53. Heimberg, M.; Olubadewo, J.O.; Wilcox, H.G. Plasma lipoproteins and regulation of hepatic metabolism of fatty acids in altered thyroid states. *Endocr. Rev.* **1985**, *6*, 590–607. [CrossRef]

54. Fall, C.H.; Barker, D.J.; Osmond, C.; Winter, P.D.; Clark, P.M.; Hales, C.N. Relation of infant feeding to adult serum cholesterol concentration and death from ischaemic heart disease. *BMJ* **1992**, *304*, 801–805. [CrossRef] [PubMed]

55. Agostoni, C.; Decsi, T.; Fewtrell, M.; Goulet, O.; Kolacek, S.; Koletzko, B.; Michaelsen, K.F.; van Goudoever, J. Complementary feeding: A commentary by the ESPGHAN Committee on Nutrition. *J. Pediatr. Gastroenterol. Nutr.* **2008**, *46*, 99–110. [CrossRef] [PubMed]

56. Li, Y.; Gao, D.; Chen, L.; Ma, T.; Ma, Y.; Chen, M.; Dong, B.; Dong, Y.; Ma, J.; Arnold, L. The Association between Breastfeeding Duration and Lipid Profile among Children and Adolescents. *Nutrients* **2021**, *13*, 2728. [CrossRef] [PubMed]

57. Escribano, J.; The European Childhood Obesity Trial Study Group; Luque, V.; Ferré, N.; Mendez-Riera, G.; Koletzko, B.; Grote, V.; Demmelmair, H.; Bluck, L.; Wright, A.; et al. Effect of protein intake and weight gain velocity on body fat mass at 6 months of age: The EU Childhood Obesity Programme. *Int. J. Obes.* **2012**, *36*, 548–553. [CrossRef]

58. Pearce, J.; Taylor, M.A.; Langley-Evans, S.C. Timing of the introduction of complementary feeding and risk of childhood obesity: A systematic review. *Int. J. Obes.* **2013**, *37*, 1295–1306. [CrossRef]

59. Sobti, J.; Mathur, G.P.; Gupta, A. WHO's proposed global strategy for infant and young child feeding: A viewpoint. *J. Indian Med. Assoc.* **2002**, *100*, 502–504.

60. Forsyth, J.S. Policy and pragmatism in breast feeding. *Arch. Dis. Child.* **2011**, *96*, 909–910. [CrossRef]

61. Fewtrell, M.; Wilson, D.C.; Booth, I.; Lucas, A. Six months of exclusive breast feeding: How good is the evidence? *BMJ* **2010**, *342*, c5955. [CrossRef]

62. Nieuwoudt, S.; Daniels, L.; Sayed, N. *Food and Nutrition Security of Infants and Young Children: Breastfeeding and Complementary Feeding*; Children's Institute, University of Cape Town: Cape Town, South Africa, 2021.

63. OECD Directorate for Employment LaSASPD. OECD Family Database. CO1.5 Breastfeeding Rates. 2009. Available online: www.oecd.org/dataoecd/30/56/43136964.pdf (accessed on 13 March 2022).

64. American Academy of Pediatrics. Breastfeeding and the use of human milk. *Pediatrics* **2012**, *129*, e827–e841. [CrossRef]

65. ESPGHAN Committee on Nutrition; Agostoni, C.; Braegger, C.; Decsi, T.; Kolacek, S.; Koletzko, B.; Michaelsen, K.F.; Mihatsch, W.; Moreno, L.A.; Puntis, J.; et al. Breast-feeding: A commentary by the ESPGHAN Committee on Nutrition. *J. Pediatr. Gastroenterol. Nutr.* **2009**, *49*, 112–125. [CrossRef]

66. Grummer-Strawn, L.M.; Rollins, N. Summarising the health effects of breastfeeding. *Acta Paediatr.* **2015**, *104*, 1–2. [CrossRef] [PubMed]

67. Bernardo, H.; Cesar, V.; World Health Organization. *Long-Term Effects of Breastfeeding: A Systematic Review*; World Health Organization: Geneva, Switzerland, 2013.

Article

Health Safety Assessment of Ready-to-Eat Products Consumed by Children Aged 0.5–3 Years on the Polish Market

Anita Żmudzińska *, Anna Puścion-Jakubik, Joanna Bielecka, Monika Grabia, Jolanta Soroczyńska, Konrad Mielcarek and Katarzyna Socha

Department of Bromatology, Faculty of Pharmacy with the Division of Laboratory Medicine, Medical University of Białystok, Mickiewicza 2D Street, 15-222 Białystok, Poland; anna.puscion-jakubik@umb.edu.pl (A.P.-J.); joanna.bielecka@umb.edu.pl (J.B.); monika.grabia@umb.edu.pl (M.G.); jolanta.soroczynska@umb.edu.pl (J.S.); konrad.mielcarek@umb.edu.pl (K.M.); katarzyna.socha@umb.edu.pl (K.S.)
* Correspondence: anita.zmudzinska@umb.edu.pl; Tel.: +48-85-748-5469

Citation: Żmudzińska, A.; Puścion-Jakubik, A.; Bielecka, J.; Grabia, M.; Soroczyńska, J.; Mielcarek, K.; Socha, K. Health Safety Assessment of Ready-to-Eat Products Consumed by Children Aged 0.5–3 Years on the Polish Market. *Nutrients* **2022**, *14*, 2325. https://doi.org/10.3390/nu14112325

Academic Editor: Zhiyong Zou

Received: 5 May 2022
Accepted: 30 May 2022
Published: 1 June 2022

Publisher's Note: MDPI stays neutral with regard to jurisdictional claims in published maps and institutional affiliations.

Copyright: © 2022 by the authors. Licensee MDPI, Basel, Switzerland. This article is an open access article distributed under the terms and conditions of the Creative Commons Attribution (CC BY) license (https://creativecommons.org/licenses/by/4.0/).

Abstract: Toxic elements have a negative impact on health, especially among infants and young children. Even low levels of exposure can impair the normal growth and development of children. In young children, all organs and metabolic processes are insufficiently developed, making them particularly vulnerable to the effects of toxic elements. The aim of this study is to estimate the concentration of toxic elements in products consumed by infants and young children. The health risk of young children due to consumption of ready-made products potentially contaminated with As (arsenic), Cd (cadmium), Hg (mercury), and Pb (lead) was also assessed. A total of 397 samples (dinners, porridges, mousses, snacks "for the handle", baby drinks, dairy) were analyzed for the content of toxic elements. Inductively coupled plasma mass spectrometry (ICP-MS) was used to assess As, Cd, and Pb concentration. The determination of Hg was performed by atomic absorption spectrometry (AAS). In order to estimate children's exposure to toxic elements, the content of indicators was also assessed: estimated daily intake (EDI), estimated weekly intake (EWI), provisional tolerable weekly intake (PTWI), provisional tolerable monthly intake (PTMI), the benchmark dose lower confidence limit (BMDL), target hazard quotient (THQ), hazard index (HI), and cancer risk (CR). The average content of As, Cd, Hg, and Pb for all ready-made products for children is: 1.411 ± 0.248 µg/kg, 2.077 ± 0.154 µg/kg, 3.161 ± 0.159 µg/kg, and 9.265 ± 0.443 µg/kg, respectively. The highest content As was found in wafer/crisps (84.71 µg/kg); in the case of Cd, dinners with fish (20.15 µg/kg); for Hg, dinners with poultry (37.25 µg/kg); and for Pb, fruit mousse (138.99 µg/kg). The results showed that 4.53% of the samples attempted to exceed Pb, and 1.5% exceeded levels of Hg. The highest value of THQ was made in the case of drinks, for Cd and Pb in mousses for children, and Hg for dairy products. The THQ, BMDL, and PTWI ratios were not exceeded. The analyzed ready-to-eat products for children aged 0.5–3 years may contain toxic elements, but most of them appear to be harmless to health.

Keywords: baby food; food contaminant; toxic elements; food exposure; children safety; children's health; arsenic; cadmium; mercury; lead

1. Introduction

Toxic elements are found in the human environment, and, as such, they can have a negative effect on health. Their consumption is especially dangerous in the early years of life, as this can affect children's development, even at low exposure levels. Among infants and young children, changes in the structure and function of the main organs occur very quickly, making them a group more susceptible to the negative effects of toxic elements [1]. Infants and young children are characterized by a higher resting metabolism and a higher food consumption per kilogram of body weight. Additionally, changes in body composition can affect the absorption, distribution, and storage of toxic elements in the organs of a growing organism. The digestive and endocrine systems are not fully developed at the time of birth, and vary in their maturation rates. Renal filtration among young children is only 30–40% developed. The differences in the development of the organism affect the degree of exposure and the distribution of pollutants; thus, the effects of exposure may be much more severe than in adults [1].

The inorganic form of arsenic (As) is listed by the IARC (Group 1) as a carcinogen [2]. Exposure to As has a toxic effect on diseases of the nervous system, children's nervous development, respiratory system, and skin. The typical sources of As in food are rice processing products, dairy products, and products intended for infants [3].

Cadmium (Cd) is also on the IARC list of carcinogenic pollutants (Group 1) [2]. As with Pb, Cd can be neurotoxic in children at doses lower than the TWI (tolerable weekly intake). Exposure to Cd disrupts osteoblast metabolism, collagen production, and also increases the urinary excretion of calcium and phosphorus. In children, it can disturb growth and damage bones [4]. Cd is present in grains, rice, fish and seafood, and vegetables [5].

Mercury (Hg) is not classified in the IARC list for carcinogenicity (Group 3) [2]. Hg enters the body through food, contaminated water, cosmetics, soil, and other Hg-containing products. The toxicity of this element depends on the time of exposure, dose, and type of Hg compound. Methylmercury (MeHg) is a carbon-bound form of Hg that is highly neurotoxic in infants and young children [6]. MeHg may cause central nervous system dysfunction in children. Exposure to Hg can cause neuromotor dysfunction, mental retardation, and cerebral palsy [7]. In adults, the half-life of Hg can be up to 90 days, and in children this time may be prolonged [8]. The main source of Hg is fish and seafood, with the highest amounts being among predatory fish [6].

Lead (Pb) is recognized as "possibly carcinogenic to humans (Group 2B)" in the International Agency of Research on Cancer (IARC) list of carcinogenic contaminants [2]. Its neurotoxic effect was observed at the lowest tested level, therefore a safe threshold for exposure to Pb cannot be assumed [9]. Exposure to Pb can impair the development of young children by damaging the nervous system. Long-term exposure results in difficulties with concentration and attention. and a lower IQ [10]. The higher the Pb concentration in the blood, the more nervous system disorders develop in children [11]. Pb can be found in meat, fish and seafood, grain products, dairy products, and fruit and vegetables [12].

In the European Union (EU), products for infants and young children are subject to specific standards regarding the composition, nutritional value, and contamination of food. The regulation requires food producers to evaluate toxic elements in products intended for infants and young children. According to the regulation of the European Commission, the permissible concentrations of toxic elements in children's products are more restrictive compared to conventional food [13].

In Poland and in Europe, there are no extensive publications assessing the content of toxic elements in children's products. Studies from outside of Europe show that the content of As and Cd are exceeded in 75% and 14% of samples tested; therefore, there is a need to evaluate food for this age group [14,15].

The aim of the study was to assess the safety of ready-to-eat products for children aged 0.5–3 years in terms of the content of toxic elements. In addition, the health risk of infants and young children as a result of exposure to toxic elements in ready-to-eat food was assessed and compared with the applicable standards.

2. Materials and Methods

2.1. Sample Collection

The research material consisted of 397 samples. Table 1 shows the number of samples with product categories.

Table 1. The number of samples with product categories.

Type of Products	*n*
BABY DINNERS	102
-with poultry	24
-with beef	16
-with pork	13
-with fish	18
-with rabbit	11
-vegetarian	20
PORRIDGE	50
-with milk	8
-with milk and fruit	15
-cereal gluten	12
-gluten-free cereal	15
FRUIT AND VEGETABLE MOUSSES	58
-fruit and vegetables	9
-fruit	33
-fruit and cereal	6
-fruit and dairy	6
-vegetables	4
BABY DRINKS	64
-fruit drinks and water	22
-fruit juices	42
SNACKS "FOR THE HAND"	62
-waffle/crisps	30
-biscuits/cookies	17
-fruit bars	15
DAIRY	60
-yellow cheese	28
-yogurt	32
TOTAL	397

The aim was to collect as many different categories as possible to obtain representative groups. The selection of samples reflects the range of products on the Polish market. The samples were purchased in brick-and-mortar stores (hypermarkets, discount stores, and children's food stores located in Białystok) in north-eastern Poland, as well as in Polish online stores between December 2020 and September 2021. We have selected the most accessible and most consumed products by children aged 0.5–3 years old. The products were sourced from leading baby food producers, such as Nestle, Nutricia, Hipp, Humana, Holle, and Helpa. Most of the analyzed products can be consumed by children from 6 months of age, while products such as biscuits, cookies, crips and fruit bars should be consumed after 12 months of age. Conventional products for the majority of consumers were also analyzed, such as cheese, milk desserts, and yoghurts, since they are often eaten by children.

2.2. Sample Digestion

All products were homogenized in a grinder or ground in a mortar, weighed (sample weight 0.25–0.35 g, with accuracy 0.001 g), and then placed in mineralization vessels made of polytetrafluoroethylene. The next step was the addition of 4 mL of concentrated 69% HNO_3 (Berghof, Speedwave, Eningen, Germany). The reagent was added according to the recommendations of the producer. Closed-loop microwave mineralization took place in four phases: first phase with a duration of 10 min, at a temperature of 170 °C, pressure 20 atm, power 90%; second phase with a duration of 10 min, at a temperature of 190 °C, pressure 30 atm, power 90%; third phase with a duration of 40 min, temperature 210 °C, pressure 40 atm, power 90%; and fourth phase with a cooling duration of 18 min, temperature 50 °C, pressure 40 atm, power 0%. After mineralization, the samples were quantitatively transferred to polypropylene vessels and diluted 10 times.

2.3. Analysis of Toxic Elements

In accordance with the regulations of the European Commission [16,17], individual types of food products have a daily limit of the consumption of toxic elements, which is the level at which they are safe for health.

2.3.1. As, Cd, Pb

Inductively coupled plasma mass spectrometry (ICP-MS NexION 300D, PerkinElmer, Waltham, MA, USA) with a kinetic discrimination chamber (KED) was used for As analysis, while Cd and Pb analysis was performed in standard mode. To correct for polyatomic interference in this configuration, kinetic energy discrimination and collisions were applied. The limit of detection (LOD) was determined by producing 10 independent blank measurements. The LOD values were as follows: for As: 0.019 µg/kg; for Cd: 0.017 µg/kg; for Pb: 0.16 µg/kg, taking LOD as a three-fold standard deviation from the mean value of the sample concentration.

2.3.2. Hg

Hg analysis was determined without the initial mineralization process. Determination of Hg was performed by atomic absorption spectrometry (AAS) using the amalgamation technique (AMA-254, Leco Corp., Altec Ltd., Prague, Czech Republic). The samples were weighed (sample weight 0.018–0.023 g), placed in a nickel cuvette, and analyzed. In the first step, the sample was dried and ashed with oxygen at 600 °C. Hg vapors were collected by the amalgamator. In the last step, the element was released from the amalgamator and measured by atomic absorption spectrometry at 254 nm. The LOD for each sample was 0.003 ng/kg.

2.3.3. Certified Reference Materials

Before starting the analyses, the certified reference material (CRM) was marked to verify the accuracy of the method. In the case of baby dinners, it was Simulated diet D (Swedish National Food Administration, Livsmedelsverket, Uppsala, Sweden); for porridges and snacks—corn flour (Institute of Nuclear Chemistry and Technology, Warsaw, Poland); for dairy products—skim milk powder (Community Bureau of Reference BCR). A total of six independent samples were performed for each CRM. Table 2 shows the results of the quality control of the certified reference materials.

Table 2. The results of the quality control of the certified reference materials.

Element	Declared Concentration in CRM (µg/kg)	Recovery (%)	Precision (%)
CORN FLOUR (INCT-CF-3)			
As	10 *	101.3	4.5
Cd	7 *	99.1	4.1
Hg	1.5 *	101.7	4.6
Pb	52 *	98.8	2.3
SKIM MILK POWDER (CRM 63R)			
Hg	0.19 ± 0.02	98.3	3.8
Pb	18.5 ± 2.7	100.9	4.3
SIMULATED DIET D			
As	<50 *	-	3.9
Cd	478 ± 26	98.4	2.4
Hg	52 *	102.0	3.2
Pb	218 ± 13	99.6	2.2

* Informational value.

2.4. Risk Assessment

To estimate the exposure to toxic elements in a short time, indicators such as estimated daily intake (EDI), estimated weekly intake (EWI), provisional tolerable weekly intake (PTWI), provisional tolerable monthly intake (PTMI) were used. Long-term exposure to toxic elements was estimated using the following indicators: the benchmark dose lower confidence limit (BMDL), target hazard quotient (THQ), hazard index (HI), and cancer risk (CR) [18].

Currently, the application of the PTWI index for Pb and As, previously recommended by EFSA in 2002, has been withdrawn. A more appropriate indicator is the BMDL, which is used to compare the coverage of the benchmark dose lower confidence limit [18].

In order to determine the dose of the oral exposure of toxic elements, the EDI values were calculated in accordance with the equation:

$$EDI = C \times Cons,$$

where C is the average concentration of a given toxic element in products for children (mg/kg), and Cons is the average daily consumption (kg) of product (children's dinners, porridges, mousses, drinks, snacks, dairy products) by young children aged 0.5–3 years (average: 1 year and 9 months) in Poland [19]. Standard portions consumed by children and declared by the producer were adopted. Based on the EDI value, the average weekly consumption (EWI) was calculated by multiplying the given result by 7 (7 days).

The PTWI value was also calculated according to the equation:

$$PTWI = EDI \times 7/BW,$$

where BW is the average body weight (the adopted average body weight for this age group is 10.5 kg). The PTWI value for Cd is 7 µg/kg BW/day, and for Hg, it is 1.6 µg/kg BW/day [16,20].

The BMDL index was also determined according to the equation:

$$BMDL = EDI/BW$$

BMDL values for As are 3 µg/kg/BW/day, with a benchmark dose lower confidence of 0.5% (BMDL0.5) according to FAO/WHO standards. For Pb, it is 0.5–6 µg/kg BW/day. PTWI values which are above the levels set by the standards may pose a health hazard. The results are presented as a percentage of the reference value [18].

The THQ index is also determined to produce adverse health effects due to the properties of the toxic elements, according to the equation:

$$THQ = (Fr \times D \times Cons \times C)/(RfD \times BW \times T) \times 10^{-3},$$

where Fr is the frequency of exposure per year (365 days), D is the exposure time (70 years), Cons is the average daily consumption of the product [g], C is the average concentration of a given toxic element in the product, and RfD is the oral reference dose, which is respectively 0.3 μg/kg BW/day for As and Hg, and 1 μg/kg BW/day for Cd and Pb, according to the guidelines of the United States Environmental Protection Agency (US EPA) [17]. If the THQ value is >1, it indicates a potential non-carcinogenic risk; if THQ is <1, it indicates a low non-carcinogenic risk [21].

Cancer risk (CR) is used to determine the carcinogenic risk for each toxic square, calculated according to the equation:

$$CR = (Fr \times D \times EDI \times SF)/T \times 10^{-3},$$

where SF stands for the slope factor of cancer. According to US EPA standards, the SF value is 1.5 mg/kg/day for As, 6.3 mg/kg/day for Cd, and 0.0085 mg/kg/day for Pb. If the CR index is above 10^{-4}, it refers to an increased carcinogenic risk. The THQ and CR values for Hg were not calculated since, according to the US EPA position, it is not considered a carcinogenic element [21].

2.5. Statistical Analysis

The obtained results were analyzed using the Statistica software (TIBCO Software Inc., Palo Alto, CA, USA). The Shapiro–Wilk test was performed, and the distribution was found to be non-normal. Non-parametric Mann–Whitney U tests and the Kruskal–Wallis ANOVA test were used to analyze the concentrations of toxic elements in various product groups. The results were summarized using the median and quartiles. To compare the results of our own research with other authors, the tables contain the mean, along with the standard deviation, as well as the maximum and minimum values. In order to assess significance, the values $p < 0.05$, $p < 0.01$, and $p < 0.001$ were adopted.

3. Results

3.1. Content of As, Cd, Hg, and Pb

The average content of As in all samples was 1.41 ± 0.25 μg/kg, while in 69 samples, no As was recorded. These were mainly products without the addition of cereals (fruit and vegetable mousses, drinks and dairy products), and the highest value was determined in a fruit bar with the addition of rice (84.71 μg/kg). Among all of the categories, the lowest As content was found in fruit and cereal mousses (0.16 ± 0.09 μg/kg) and dairy products (0.10 ± 0.16 μg/kg). In turn, the highest As value was recorded in porridges (2.30 ± 0.33 μg/kg), and, in particular, in gluten-free porridges (4.31 ± 0.72 μg/kg). High concentrations were also obtained in snacks "for the hand" (2.92 ± 10.77 μg/kg), and especially in wafers/chips (4.88 ± 15.34 μg/kg). In accordance with the level of the European Commission [22], the maximum allowable level of As in food is 100 μg/kg (in cereals, fruit and vegetables, and meat). There were 69 samples for As below the limit of detection. The median As content results with quartile 1 (Q1) and quartile 3 (Q3) in the analyzed product groups are presented in Figure 1A.

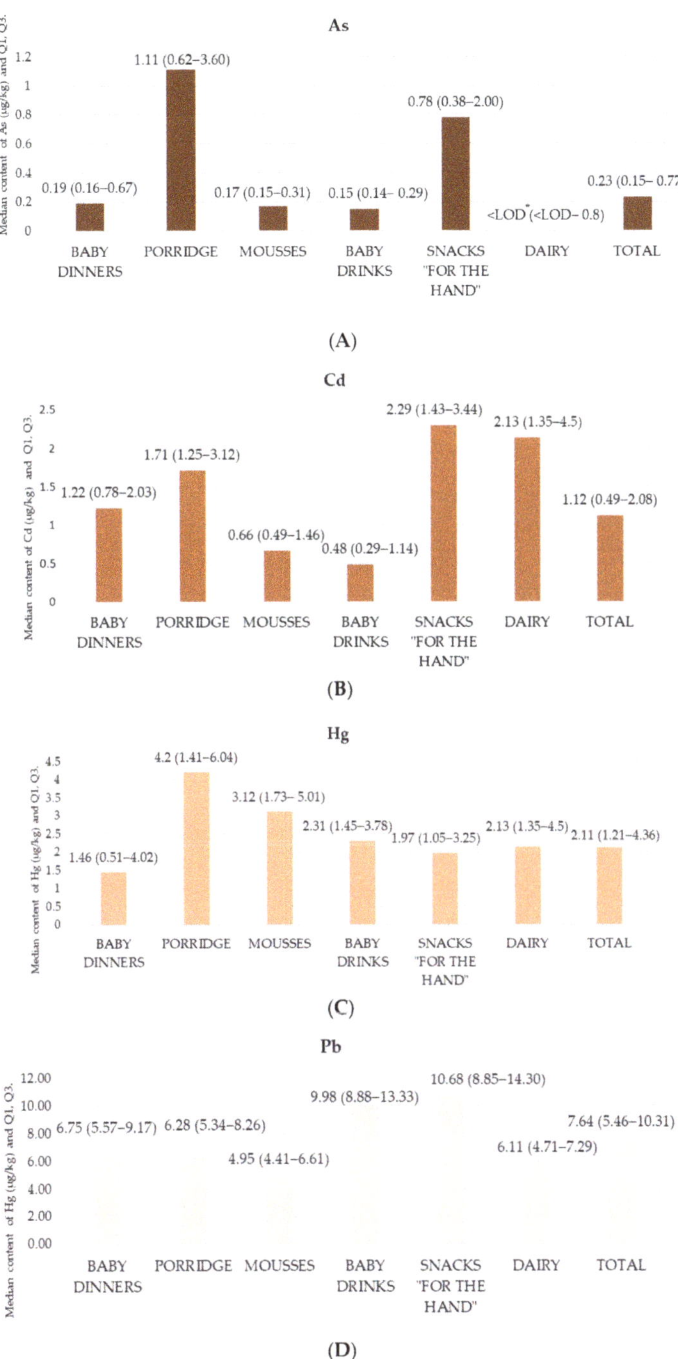

Figure 1. Median As (**A**), Cd (**B**), Hg (**C**) and Pb (**D**) content with quartile 1 (Q1) and quartile 3 (Q3) in the analyzed product groups. * LOD—limit of detection.

The mean concentration of Cd in all the tested samples was 2.08 ± 0.15 µg/kg. The lowest concentration of Cd in products was 0.14 µg/kg, recorded in juices for children (apple, grape, chokeberry, raspberry). The highest Cd concentration was 20.15 µg/kg, and was in a ready-to-eat dinner for children after 6 months of age, consisting mainly of vegetables and salmon. Taking into account the product categories, the lowest Cd values were recorded in dairy products (0.95 ± 2.56 µg/kg) and especially in the yoghurt subgroup (0.45 ± 0.78 µg/kg), mousses (1.39 ± 2.31 µg/kg), and also in fruit juices (0.66 ± 2.01 µg/kg). The highest levels of Cd were obtained in drinks for children (3.39 ± 1.30 µg/kg), especially in fruit drinks and water (4.17 ± 5.98 µg/kg) and snacks "for the hand" (3.09 ± 1.69 µg/kg), especially in the waffle/crisps subcategory (3.51 ± 3.38 µg/kg). According to the European Commission Regulation (EU) 2021/1323 of 10 August 2021 amending Regulation (EC) No 1881/2006 as regards maximum levels of Cd in certain foodstuffs, the maximum level of Cd is 100 µg/kg for cereal, 40 µg/kg for baby food, 20 µg/kg for fruits and vegetables, 20 µg/kg for juices, and 50 µg/kg for meat and fish [17]. The daily limit for Cd was not exceeded; therefore, it has not been included in the Table 3. There were 47 samples for Cd below the limit of detection. The median Cd content results with quartile 1 (Q1) and quartile 3 (Q3) in the analyzed product groups are presented in Figure 1B.

Table 3. Maximum allowable content of As, Cd, Hg, and Pb in food and the number of baby food samples that exceed the maximum allowable content [16,17,22].

Baby Food Samples	Maximum Allowable Content (mg/kg Product Weight)				Exceeding Maximum Allowable Content							
					As		Cd		Hg		Pb	
	As	Cd	Hg	Pb	*n*	%	*n*	%	*n*	%	*n*	%
Dinners (*n* = 102)	-	0.04	0.01	0.02	-	-	0	0	3	0.75	1	0.25
Porridges (*n* = 50)	0.5	0.04	0.01	0.02	0	0	0	0	0	0.00	2	0.50
Mousses (*n* = 58)	0.5	0.02	0.01	0.01	0	0	0	0	2	0.50	1	0.25
Drinks (*n* = 64)	-	0.02	0.02	0.02	-	-	0	0	1	0.25	7	1.76
Baby snacks (*n* = 62)	-	0.04	0.01	0.02	-	-	0	0	0	0.00	6	1.50
Dairy (*n* = 60)	-	-	0.01	0.02	-	-	-	-	0	0.00	1	0.25
TOTAL (*n* = 397)					-	-	-	-	6	1.50	18	4.53

n—number of samples, %—percentage of exceeded samples.

The average Hg level in all samples was 3.16 ± 0.16 µg/kg, The lowest value was recorded in the fruit and cereal bar (the main ingredients of which are grape juice, oatmeal, apple juice) at 0.005 µg/kg, while the highest Hg value was 37.25 µg/kg in a lunch based on vegetables, rice, and hake. Among all categories, the lowest amounts of Hg were found in snacks "for the hand" (2.36 ± 1.69 µg/kg) and in the subcategory of vegetable mousses (2.26 ± 0.83 µg/kg). The highest amounts of Hg were recorded in porridges (4.20 ± 0.35 µg/kg), in particular in gluten-free (4.89 ± 0.69 µg/kg) and gluten (4.86 ± 0.60 µg/kg) porridges; in the subcategories, the highest levels were recorded in fish dinners (as much as 9.213 ± 0.571 µg/kg) and fruit and vegetable mousses (5.748 ± 1.118 µg/kg). In accordance with European Commission Regulation (EU) 2018/73 of 16 January 2018 amending Annexes II and III to Regulation (EC) No 396/2005 of the European Parliament and of the Council, regarding the maximum residue levels for Hg compounds in or on certain products, the maximum allowable level of Hg is: 5 µg/kg for fish, 10 µg/kg for cereal, baby food, fruit and vegetables, meat, and milk, and 20 µg/kg for juices [16]. Table 3 shows the number of samples that exceeded the maximum allowable limits of Hg and Pb. The limit for Hg was exceeded in 6 samples, which constitutes 1.5% of all samples. The highest amount of the exceeded Hg limit was recorded in lunches (three samples). The daily limit was not exceeded in porridges, snacks for children, and dairy products. There were the same samples for Hg below the limit of detection. The median Hg content results with quartile 1 (Q1) and quartile 3 (Q3) in the analyzed product groups are presented in Figure 1C.

The mean concentration of Pb in all samples was 9.27 ± 0.44 µg/kg, with the lowest recorded concentration being 0.46 µg/kg in children's juice (based on apple, grape, aronia, raspberry). The highest Pb value was 138.99 µg/kg, found in a mousse (based on apple, cottage cheese, and grape juice). Among all categories, the lowest Pb level was recorded in drinks for children (1.14 ± 0.98 µg/kg). The highest concentrations of Pb were found in snacks "for the hand" (12.80 ± 7.56 µg/kg), porridges (8.09 ± 0.70 µg/kg), mousses for children (7.97 ± 0.02 µg/kg), especially fruit-based mousses (9.40 ± 4.06 µg/kg), and in children's dinners (7.55 ± 0.41 µg/kg). According to the regulations of the European Commission, the maximum permissible level of Pb is: 10 µg/kg for meat, 20 µg/kg for cereal, baby food and milk, 30 µg/kg for fish, 50 µg/kg for juices, and 100 µg/kg for fruit and vegetables [20]. The median Pb content results with quartile 1 (Q1) and quartile 3 (Q3) in the analyzed product groups are presented in Figure 1D. There were 10 samples for Pb below the limit of detection.

Detailed content of toxic elements in individual groups and subgroups is presented in Table S1 (Supplementary Materials).

There were 18 exceedances in the content of the daily Pb limit (4.53%). The highest Pb content was recorded in drinks (seven samples) and baby snacks (six samples). The lowest Pb exceedances were recorded in dairy products (one sample), mousses (one sample), and dinners (one sample).

Table 4 presents statistically significant differences in the content of toxic elements in the analyzed products. The most statistically significant differences were noted in the As content. The As content in lunches differed statistically significantly in drinks ($p < 0.01$). Differences were also noted in porridges and drinks ($p < 0.01$), mousses and drinks ($p < 0.01$), baby snacks and drinks ($p < 0.01$), dairy and drinks ($p < 0.01$), and groups: poultry dinners and fish dinners ($p < 0.001$), beef dinners and fish dinners ($p < 0.001$), beef dinners and fish dinners ($p < 0.001$), pork dinners and fish dinners ($p < 0.001$), rabbit dinners and fish dinners ($p < 0.001$), vegetarian dinners and fish dinners ($p < 0.001$), and milk porridge and gluten-free cereal ($p < 0.001$).

Table 4. Significant statistical differences in the analyzed product groups.

Toxic Elements	Analyzed Food Group	*p*-Value
	Dinners—drinks	<0.01
	Porridges—drinks	<0.01
	Mousses—drinks	<0.01
	Baby snacks—drinks	<0.01
	Dairy—drinks	<0.01
As	Poultry dinners—fish dinners	<0.001
	Beef dinners—fish dinners	<0.001
	Pork dinners—fish dinners	<0.001
	Rabbit dinners—fish dinners	<0.001
	Vegetarian dinners—fish dinners	<0.001
	Milk porridge—gluten-free cereal	<0.001
Hg	Mousses—dairy	<0.05
	Pork dinners—vegetarian dinners	<0.05
Pb	Fruit drinks and water—fruit juices	<0.01

There were also statistically significant differences in the Hg content in mousses and dairy ($p < 0.05$), as well as in the pork dinners and vegetarian dinners subgroup ($p < 0.05$). For the Pb content, one difference was found between fruit drinks and water–fruit juices ($p < 0.01$). In this study, no statistically significant differences in the Cd content were found.

3.2. Risk Assessment

The health indicators of all analyzed products are summarized in Table 5. In the case of PTWI and BMDL indicators, no exceedances were found in the tested assortment.

Table 5. Average value of the ratios: EDI, EWI, PTWI, and BMDL in all analyzed groups.

Type of Products	Elements	EDI (µg/Day)	EWI (µg/Week)	PTWI (%PTWI) (µg/kg/BW/Week)	BMDL (%BMDL) (µg/kg/BW/Day)
Baby dinners (*n* = 102)	As	0.419 ± 0.909	2.938 ± 6.368	NA	0.042 ± 0.091 (1.40%)
	Cd	0.515 ± 0.744	3.607 ± 5.205	0.362 ± 0.523 (5.17%)	NA
	Hg	0.544 ± 0.579	3.804 ± 4.049	0.382 ± 0.407 (23.89%)	NA
	Pb	1.651 ± 0.922	11.563 ± 6.458	NA	0.166 ± 0.092 (33.2%)
Porridges (*n* = 50)	As	0.051 ± 0.054	0.362 ± 0.380	NA	0.005 ± 0.005 (0.17%)
	Cd	0.064 ± 0.065	0.449 ± 0.459	0.045 ± 0.046 (0.64%)	NA
	Hg	0.097 ± 0.057	0.681 ± 0.402	0.068 ± 0.040 (4.28%)	NA
	Pb	0.187 ± 0.114	1.313 ± 0.804	NA	0.018 ± 0.011 (3.77%)
Mousses (*n* = 58)	As	0.021 ± 0.017	0.148 ± 0.119	NA	0.002 ± 0.001 (0.07%)
	Cd	0.159 ± 0.226	1.0113 ± 1.583	0.111 ± 0.159 (1.59%)	NA
	Hg	0.381 ± 0.239	2.669 ± 1.677	0.268 ± 0.168 (16.76%)	NA
	Pb	0.892 ± 2.008	6.245 ± 14.056	NA	0.08 ± 0.02 (17.93%)
Baby drinks (*n* = 64)	As	0.037 ± 0.086	0.264 ± 0.607	NA	0.003 ± 0.008 (0.12%)
	Cd	0.078 ± 0.165	0.548 ± 1.551	0.055 ± 0.116 (0.78%)	NA
	Hg	0.122 ± 0.080	0.858 ± 0.560	0.086 ± 0.056 (5.39%)	NA
	Pb	0.553 ± 0.320	3.874 ± 2.242	NA	0.055 ± 0.032 (11.12%)
Snacks (*n* = 62)	As	0.242 ± 0.870	1.697 ± 6.091	NA	0.024 ± 0.087 (0.81%)
	Cd	0.252 ± 0.226	1.767 ± 1.585	0.177 ± 0.159 (2.53%)	NA
	Hg	0.192 ± 0.136	1.348 ± 0.958	0.135 ± 0.096 (8.46%)	NA
	Pb	1.053 ± 0.610	7.375 ± 4.276	NA	0.105 ± 0.061 (21.17%)
Dairy (*n* = 60)	As	0.007 ± 0.034	0.054 ± 0.233	NA	0.000 ± 0.003 (0.02%)
	Cd	0.063 ± 0.081	0.436 ± 0.563	0.044 ± 0.056 (0.63%)	NA
	Hg	0.391 ± 0.668	2.726 ± 4.641	0.275 ± 0.470 (17.21%)	NA
	Pb	1.168 ± 1.491	8.047 ± 10.397	NA	0.117 ± 0.149 (23.47%)
TOTAL (*n* = 397)	As	0.162 ± 0.598	1.139 ± 4.196	NA	0.016 ± 0.060 (0.54%)
	Cd	0.224 ± 0.439	1.571 ± 3.076	0.157 ± 0.309 (2.25%)	NA
	Hg	0.314 ± 0.438	2.204 ± 3.070	0.221 ± 0.308 (13.84%)	NA
	Pb	1.006 ± 1.192	7.046 ± 8.348	NA	0.101 ± 0.119 (20.23%)

BMDL—the benchmark dose lower confidence limit, EDI—estimated daily intake, EWI—estimated weekly intake, NA—not applicable, PTMI—provisional tolerable monthly intake, PTWI—provisional tolerable weekly intake.

Baby dinners have the highest EDI and EWI values for As and Cd (EDI_{As} = 0.419 ± 0.909 mg/day, EWI_{As} = 2.938 ± 6.368 mg/week, EDI_{Cd} = 0.515 ± 0.744, EWI_{Cd} = 3.607 ± 5.205). The estimated daily consumption was the lowest in dairy products in the case of As (EDI = 0.007 ± 0.034 mg/day, EWI = 0.054 ± 0.233 mg/week), while in the case of Cd, this was in dairy products (EDI = 0.063 ± 0.081 mg/day, EWI = 0.436 ± 0.563 mg/week) and porridges (EDI = 0.064 ± 0.065 mg/day, EWI = 0.449 ± 0.459 mg/week). Baby dinners showed the highest value of EDI and EWI in the case of Pb (EDI = 1.651 ± 0.922 mg/day, EWI = 11.563 ± 6.458 mg/week), while in the case of Hg, the highest value was found in mousses (EDI = 0.381 ± 0.239 mg/day, EWI = 2.669 ± 1.677 mg/week). The lowest EDI value for Hg and Pb was determined in porridges (EDI_{Hg} = 0.097 ± 0.057 mg/day, EWI_{Hg} = 0.681 ± 0.402 mg/week, EDI_{Pb} = 0.187 ± 0.114 mg/day, EWI_{Pb} = 1.313 ± 0.804 mg/week). In the case of the PTWI and BMDL indicators, the specified limits were not exceeded (Table 5). The PTWI in all ready-to-eat samples for Cd was 0.157 ± 0.308 µg/kg, which was 2.25% of the norm; for Hg, it was 0.221 ± 0.308 µg/kg, which was 13.84% of the norm. The $BMDL_{As}$

was 0.016 ± 0.06 µg/kg, which was 0.54% of the norm, while in the case of Pb, it was 0.101 ± 0.119 µg/kg (20.23% of the norm).

The estimated THQ and CR ratios for the analyzed toxic elements are presented in Table 6. The highest values of the THQ index for all elements were recorded for Hg in dairy products ($1.14 \times 10^{-7} \pm 6.18 \times 10^{-7}$). In the case of As, the highest value of the index was $1.41 \times 10^{-7} \pm 3.04 \times 10^{-7}$, and these levels were found in baby dinners. In the case of Cd, the highest THQ was recorded in porridges for children, and THQ was $1.31 \times 10^{-8} \pm 2.01 \times 10^{-8}$. In the case of Hg, it was $1.01 \times 10^{-7} \pm 1.19 \times 10^{-7}$, found in dinners for children. In the case of Pb, it was porridges $1.01 \times 10^{-7} \pm 1.19 \times 10^{-7}$. The norms of THQ values were not exceeded in the tested assortment, therefore there is no increased risk of non-carcinogens effect. The CR index was not exceeded, so the research shows that the risk of developing cancer as a result of consuming the analyzed products is low.

Table 6. Estimated values of THQ and CR indices of toxic elements (As, Cd, Hg, and Pb) in the analyzed food products.

Type of Products	THQ X ± SD (Min–Max)			
	As	Cd	Hg	Pb
Baby dinners ($n = 102$)	$1.41 \times 10^{-7} \pm 3.04 \times 10^{-7}$ (9.61×10^{-7}–1.26×10^{-6})	$2.26 \times 10^{-8} \pm 4.41 \times 10^{-8}$ (1.75×10^{-9}–1.56×10^{-6})	$1.01 \times 10^{-7} \pm 1.19 \times 10^{-7}$ (1.36×10^{-10}–8.56×10^{-7})	$1.06 \times 10^{-7} \pm 1.46 \times 10^{-7}$ (1.75×10^{-9}–1.56×10^{-6})
Porridges ($n = 50$)	$1.82 \times 10^{-8} \pm 1.82 \times 10^{-8}$ (1.23×10^{-8}–2.44×10^{-6})	$1.31 \times 10^{-8} \pm 2.01 \times 10^{-8}$ (1.82×10^{-8}–2.31×10^{-6})	$1.06 \times 10^{-7} \pm 1.46 \times 10^{-7}$ (7.66×10^{-10}–8.65×10^{-7})	$1.01 \times 10^{-7} \pm 1.19 \times 10^{-7}$ (1.88×10^{-8}–8.65×10^{-7})
Mousses ($n = 58$)	$7.11 \times 10^{-9} \pm 5.70 \times 10^{-9}$ (1.31×10^{-8}–2.79×10^{-6})	$2.13 \times 10^{-8} \pm 2.44 \times 10^{-8}$ (1.14×10^{-11}–8.78×10^{-7})	$1.06 \times 10^{-7} \pm 1.19 \times 10^{-7}$ (1.36×10^{-10}–8.65×10^{-7})	c$3.21 \times 10^{-7} \pm 2.88 \times 10^{-7}$ (7.03×10^{-9}–8.55×10^{-6})
Baby drinks ($n = 64$)	$1.29 \times 10^{-8} \pm 2.92 \times 10^{-8}$ (1.31×10^{-8}–1.47×10^{-6})	$3.06 \times 10^{-8} \pm 4.40 \times 10^{-8}$ (8.54×10^{-11}–4.46×10^{-7})	$1.98 \times 10^{-7} \pm 1.70 \times 10^{-7}$ (2.06×10^{-10}–4.50×10^{-6})	$2.98 \times 10^{-7} \pm 3.17 \times 10^{-7}$ (1.05×10^{-9}–1.26×10^{-6})
Snacks ($n = 62$)	$8.12 \times 10^{-8} \pm 2.91 \times 10^{-7}$ (5.95×10^{-9}–2.31×10^{-6})	$4.99 \times 10^{-8} \pm 1.10 \times 10^{-8}$ (8.52×10^{-11}–9.89×10^{-7})	$1.76 \times 10^{-7} \pm 4.46 \times 10^{-7}$ (6.11×10^{-11}–1.05×10^{-7})	$4.51 \times 10^{-7} \pm 1.19 \times 10^{-7}$ (1.46×10^{-10}–1.16×10^{-6})
Dairy ($n = 60$)	$2.66 \times 10^{-9} \pm 1.127 \times 10^{-8}$ (28.84×10^{-10}–2.01×10^{-7})	$2.63 \times 10^{-9} \pm 2.33 \times 10^{-8}$ (1.47×10^{-11}–1.46×10^{-7})	$1.14 \times 10^{-7} \pm 6.18 \times 10^{-7}$ (1.99×10^{-10}–8.21×10^{-7})	$1.91 \times 10^{-7} \pm 1.19 \times 10^{-7}$ (1.75×10^{-9}–1.56×10^{-7})
TOTAL ($n = 397$)	$5.44 \times 10^{-8} \pm 2.01 \times 10^{-8}$ (28.84×10^{-10}–1.26×10^{-6})	$2.61 \times 10^{-8} \pm 4.42 \times 10^{-8}$ (1.47×10^{-11}–1.56×10^{-6})	$1.65 \times 10^{-7} \pm 1.47 \times 10^{-7}$ (6.11×10^{-11}–4.50×10^{-6})	$2.82 \times 10^{-7} \pm 2.21 \times 10^{-7}$ (1.46×10^{-10}–1.16×10^{-6})
Mean CR	$2.44 \times 10^{-7} \pm 8.97 \times 10^{-7}$ (2.07×10^{-7} 1.03×10^{-5})	$1.41 \times 10^{-6} \pm 2.77 \times 10^{-6}$ (5.34×10^{-6} 2.79×10^{-5})	NA	$8.56 \times 10^{-9} \pm 1.01 \times 10^{-8}$ (1.48×10^{-10}–1.32×10^{-7})

CR—cancer risk, Max—maximum, Min—minimum, NA—not applicable, SD—standard deviation, THQ—target hazard quotient, X—mean.

The analyzed products were also divided according to the intended use of the given products in terms of age category: for children 6–12 months, for children under 12 months, and products without an age declaration. Table 7 presents the characteristics of age groups in terms of mean, median, and other statistical parameters. The age declaration group includes dinners, porridges, and mousses. The remaining product groups (drinks, snacks, and dairy products for children) are included in the category without an age declaration.

Table 7. The average content and median of toxic elements in the studied groups and significant statistical differences, taking into account the intended use of products for age groups.

Type of Products	Elements	n	X ± SD (µg/kg)	Min–Max (µg/kg)	Me (µg/kg)	Q1–Q3 (µg/kg)	Toxic Elements	Analyzed Food Group	*p*-Value
			Average Content of Analyzed Parameters				**Statistical Differences**		
for children 6–12 months (a)	As	107	2.01 ± 3.67	0.00–16.97	0.44	0.17–1.72	As	a/b	<0.01
	Cd		2.22 ± 3.14	0.31–20.15	1.33	0.68–2.02		b/c	<0.001
	Hg		3.20 ± 2.86	0.01–9.88	2.05	0.83–5.28			
	Pb		7.82 ± 4.76	0.52–32.28	6.70	5.25–9.89			
for children under 12 months (b)	As	104	1.01 ± 2.54	0.00–14.63	0.19	0.16–0.50	Cd	a/c	<0.001
	Cd		2.17 ± 2.78	0.16–13.34	1.15	0.64–2.20		b/c	<0.01
	Hg		3.09 ± 2.22	0.17–9.12	2.53	1.240–4.47			
	Pb		7.74 ± 13.55	0.77–134.00	5.81	4.72–7.45			
without an age declaration (c)	As	186	1.35 ± 6.42	0.00–84.71	0.20	0.00–0.66	Pb	a/b	<0.05
	Cd		1.99 ± 3.23	0.14–18.54	0.63	0.27–2.24		a/c	<0.001
	Hg		2.68 ± 1.86	0.00–9.12	2.11	1.27–3.74		b/c	<0.001
	Pb		10.95 ± 7.02	0.460–48.18	9.38	7.18–12.09			

a—products for children 6–12 months, b—products for children under 12 months, c—products without an age declaration, Max—maximum, Me—median, Min—minimum, SD—standard deviation, Q1—quartile 1, Q3—quartile 3, X—mean.

The highest average contents of As, Cd, and Hg were recorded in the group of products intended for children aged 6–12 months (As: 2.01 ± 3.67 µg/kg, Cd: 2.22 ± 3.14 µg/kg, Hg: 3.20 ± 2.86 µg/kg). In the case of Pb, it was in the group of products without age declaration (10.95 ± 7.02 µg/kg). The highest mean content of As was recorded in the group of products for children 6–12 months (2.01 ± 3.67 µg/kg), and the lowest for children under 12 months (1.01 ± 2.54 µg/kg). The highest mean Cd content was observed in the group of products for children aged 6–12 months (2.22 ± 3.14 µg/kg), and the lowest in the group of products without declaration (1.98 ± 3.23 µg/kg), although these differences were small (the statistical significance of the analyzed groups was presented in Table 7). The highest mean Hg content was recorded in the group of products intended for children 6–12 months (3.20 ± 2.86 µg/kg), and the lowest mean in the group of products without declaration (2.68 ± 1.86 µg/kg), while there were not statistically significant groups (Table 7). In the case of Pb, the highest mean was recorded in the group of products without age declaration (10.95 ± 7.02 µg/kg), and in the remaining groups the mean was very similar (in the group of products for children 6–12 months: 7.82 ± 4.76 µg/kg, in the group above 12 months: 7.74 ± 13.55 µg/kg).

The most statistically significant differences were recorded in the case of Pb. The Pb content differed statistically significantly in the group of products intended for children aged 6–12 months and over 12 months ($p < 0.01$); in the group of 6–12 months and in the first group without age declaration ($p < 0.001$); in the group of products above 12 months month and the product group without age declaration ($p < 0.001$). In the case of As, statistical significance was noted in the groups of products for children aged 6–12 months and over 12 months of age ($p < 0.01$); and in the group of 6–12 months in the group without age declaration ($p < 0.001$). The content of Cd differed statistically significantly in the group of products intended for children 6–12 months, the group without age declaration ($p < 0.001$); and in the group over 12 months of age, and the group without age declaration ($p < 0.01$).

4. Discussion

In our study, we found that toxic elements are present in ready-to-eat baby foods, and that some foods contain Hg and Pb levels of concern.

Our research found that the permissible daily limit was not exceeded for As. In a Spanish study, As was detected in three samples, one of which exceeded the acceptable toxicological standards [23]. The study by Rothenberg et al., (2017) examined the content of As in rice products for children. These products had a higher content of the tested elements compared to wheat and oat products, but they did not exceed the permissible standard [24]. A similar study by Gu et al., (2020) assessed the As content in products for infants and young children. Despite the fact that total As was found in 88% of the samples, it did not exceed the total permissible content, while 75% of the samples exceeded the standard for inorganic As for infants and young children [14]. In turn, Ljung et al., (2011) assessed the content of toxic elements in, among others, products based on milk, oats, spelt, and whole grain rice. The researcher showed that rice-based products were the largest source of As (up to 30 µg/kg) [25]. A Nigerian study (2020) showed that the tolerable daily intake of As was exceeded in baby food. The mean daily consumption of As was 437.1 µg/kg (771% PTDI). Abnormalities for As have been observed in products containing rice [26]. Rice is the most abundant product containing As. In Poland, rice is consumed rarely, and among children under 3 years of age, the average consumption of rice is 17.9 g/day The range of rice-based products on the Polish market is relatively small, and they are not so widely consumed by children, so should not pose a threat to children's health.

In our own study, the maximum allowable content was not exceeded for Cd. Gardener et al., (2019) showed that Cd was found in 57% of samples. Out of 91 trials of baby food, 14% exceeded the acceptable limits for Cd consumption [15]. The study by Kim et al., (2014) assessed the exposure to toxic elements in the daily food ration. The mean concentration of Cd in the diet was 0.38 ± 0.20 µg/kg BW/day, respectively. In the case of Cd, the reference value was exceeded (42%) [27]. De Castro et al., (2010) Cd concentrations were higher than in our study, but did not exceed the acceptable limits (109 µg/kg) [28]. In the study (2019), the mean daily exposure to Cd was 0.43 µg/kg BW/day [29]. In our study, the average daily consumption for Cd was lower (0.224 ± 0.439 µg/kg). In the study by Winiarska-Mleczan (2012), the average consumption in one meal (ready-made dessert for children) covers 2% of the PTDI of Cd, while the average consumption of a bottle of drink for children is 2% of the PTDI of Cd [30]. In a study, Chen et al., (2021) showed that, among all age groups, children up to 5 years old showed the highest concentrations of Cd and the highest risk of health risk. The total Cd exposure index for children aged 0.5–5 years was 0.00325, and the Cd level was 3.9 times the PTMI norm on Cd, suggesting that children are at an unacceptable level of health risk. [31]. This is a study carried out in China, where there are different eating habits than in Europe (e.g., higher consumption of e.g., rice), which may be the reason for such a high rate. An Egyptian study analyzed the content of Cd in dairy products consumed by children. The estimated daily intake for Cd was 0.33 µg/kg bw/day, which is 39.8% of the TDI [32]. In comparison, in our study, the EDI for Cd in dairy products was higher and amounts to 0.43 ± 0.56 µg/kg. In our study, the highest Cd value was found for the salmon-based children's dinner (20.15 µg/kg), which may suggest greater caution in terms of children's consumption of fish. Among different groups, the highest value of Cd was recorded in drinks for children. The reason for high concentrations of Cd may be the high presence of Cd in the water due to the affinity for its accumulation [12]. The presence of Cd in children's products should be of concern, since Cd is neurotoxic in amounts lower than the reference doses, which means that any dose may have adverse health effects and impair children's development.

In our study, the maximum allowable content was exceeded for Hg in 1.5% of samples analyzed. In the study by Guerin et al., (2018), the Hg content of 291 products for infants and young children, as well as conventional foods consumed by this group, on the French market was determined. This element was detected in 7.6% of all samples, while none of the products exceeded the permitted Hg limit [33]. Higher Hg concentrations were also found in rice products, as compared to oat products [24]. In our study, the product with the highest Hg content was based on rice and hake. In the study of Spungen et al., (2019), the mean daily exposure to Hg was 0.12 µg/kg bw/day in children aged 1–3 years old [29]. In our study, the estimated daily intake was higher (0.31 ± 0.44 µg/kg). The study by Martins et al., (2013) assessed the exposure of infants and young children to Hg derived from processed cereal-based foods and infant food (vegetable-, meat-, fish-, and fruit-based meals). It has been shown that the average consumption of the analyzed products is from 0.2 to 0.4% of the PTWI of Hg [34]. In our study, the coverage of PTWI of Hg is significantly higher (9.6%), but it should not be a cause for concern. The study by Igwaze et al., (2020) showed that mean daily consumption of Hg was 23.7 µg/kg (41.8% of PTDI). The mean daily Hg intake was below the PTDI value [26]. In a Spanish study evaluating infant dietary Hg exposure, the EDI for Hg was shown to be significantly higher, at 1.5 µg/kg BW/day (EWI = 11 BW/day). In our study, the EDI for Hg was 0.31 ± 0.43 µg/kg BW/day (EWI = 2.20 ± 0.07 µg/kg) [35]. Moreover, it is particularly interesting for public health that the tested samples showed exceeded Hg concentrations in six samples, including three categorized as children's lunches, two of which contained fish. This should be of concern, especially in view of the potential risks posed by frequent consumption of fish-based products. Hg damages the central nervous system of children, impairs the immune function, and leads to dysfunction of the children's bloodstream.

In our study, the maximum allowable content was exceeded for Pb in 4.5% of samples. The Gardener et al., (2019) study assessed the content of Pb in products consumed by infants and young children (infant food, cereals, drinks, pouches). Pb was found in 37% of samples [15]. In the study by Škrbić et al., (2016) which assessed the content of toxic elements in 90 products for children (porridge fruit, porridge vegetable, meat and fish, porridge, corn and rice porridge, and yogurt-based products), Pb was detected in two products [23]. Ljung et al., (2011) showed that Pb was found in rice-based products (13 µg/kg) [25]. In the study by Kim et al., (2014), the mean concentration of Pb was 0.47 ± 0.14 µg/kg BW/day and 35% of samples exceeded the maximum allowable content limit [27]. In a study by Rasic Misic et al., (2022), Pb was detected in 31% of baby and toddler juice and chafing samples. The mean concentrations of Pb were 12.0 µg/kg in fruit-based food, 29.0 µg/kg in vegetable-based food, and 34.50 µg/kg in meat-based food. In all juices, the Pb value was higher than the maximum concentration in food for young children (0.05 mg/kg). In our research, similar conclusions were noted, since the most exceedances of the permissible daily norm were recorded in beverages—1.76% of all products (7 samples). In our study, mean Pb concentrations were also high in fruit-based mousses (9.40 ± 4.06 µg/kg). Higher Pb concentrations in fruit and vegetable products may come from contaminated soil or production processes [36]. The median of Pb in baby products from Brazil is significantly higher than in our research. It was shown that the median Pb was 33 µg/kg; therefore, the FAO/WHO standards were exceeded in 63% of Pb [28]. In the French study assessing the Pb content in food for infants and children, Pb was found in most of the analyzed samples (mean 2.4 µg/kg). The highest concentrations of Pb were found in the group of biscuits and bars (9.6 µg/kg), croissants (8.2 µg/kg), and cake and bread (5.5 µg/kg). Among all of the products, the highest content was found in chocolate biscuits (26.2 µg/kg), chocolate flakes for babies (19.7 µg/kg), and spinach (16.1 µg/kg). The products containing chocolate were characterized by significantly higher Pb concentrations [37]. In another Polish study (2012), the average consumption of desserts for children covers 2.2% of the PTDI of Pb, while the average consumption of a bottle of a drink for children is 13.6% of the PTDI of Pb [33]. In a Japanese study (2019), the weekly Pb exposure for children aged 1–3.5 years was 3.28 ± 0.26 µg/kg BW/week. Pb concentration was higher in children than in adults, but it did not exceed the acceptable standard [38]. In another study, the content of Pb was 1.27 µg/kg BW/day, which is 35.3% of the TDI. In our study, the EDI values for Pb are comparable (1.006 ± 1.192 µg/kg) [37]. In 18 samples, Pb was above the threshold values, and the highest amounts were found in drinks (7) and snacks (6). Pb contamination can come from fruits and grains, which are the main ingredient of these products. In the study by Kim et al, Pb was also present in high concentrations in fruits [27]. The source of Pb can be soil and water from highly industrialized areas. Fruit and grains are one of the main components of the diet of infants and young children, so they should not be abandoned, but it might be worthwhile to encourage children to be more varied in their diet to minimize the risk of possible Pb exposure. Exposure to this element disturbs the development of the central nervous system, and causes problems with concentration; thus, it is important to avoid chronic exposure.

Our results indicate that baby food may be safe, and the estimated health risk indicators did not show an increased health risk in terms of exposure of toxic elements in ready-made products for children. However, it is worrying that some products have exceeded the daily limit for Pb (18 samples) and Hg (6 samples). In Poland, around 60% of parents use convenience food, which should cause concern, especially in children, who often base their nutrition on ready-to-eat foods [39].

According to our results, it is worth paying attention to ready-to-eat products containing fish. In the study, children's meals containing salmon had the highest levels of Cd (20.15 µg/kg), and those containing hake had the highest levels of Hg (37.25 µg/kg). Among the assortment of lunches intended for children over 1 year of age was a ready-made tuna dinner that should not be eaten at all in this age group. Some fish products are especially recommended for children (salmon, cod, mackerel, eel), while, due to the large accumulation of pollutants harmful to health, perch, shark, swordfish, and tuna are not recommended. The consumption of fish by children brings many health benefits, so we should not exclude them in the diet of children, while parents should pay attention to the selection of fish species that do not pose a health risk [40].

In our study, cereal products (especially those containing rice and rice products) contained high amounts of As (rice bar 84.71 µg/kg, baby porridge group, waffles and crips group). Cereals for children and snacks were characterized by one of the higher concentrations of toxic elements (in the case of As, gluten-free and gluten-free cereals; in the case of Cd, waffles/crips; in the case of Hg, porridges; in the case of Pb, snacks "for the hand").

This allows us to conclude that cereal products easily accumulate toxic elements and, if consumed in large amounts, may pose a health hazard to infants and young children.

An unexpected observation is the high contribution of Cd in the group of beverages for children (4.17 ± 5.98 µg/kg), and the exceedance of the daily limit of Pb content in beverages (7 samples, 1.76% of all trials). This may result from contamination of the water used for the production of beverages or flavored waters. In Poland, there is an obligation to regularly test water for, e.g., toxic elements by provincial sanitary and epidemiological stations. This water has the correct parameters, while the problem may be an outdated sewage system located, for example, close to a factory, from which harmful substances, including toxic elements, may be released. Table 8 shows the content of toxic elements in baby food compared to the results of other authors.

Table 8. The content of toxic elements in baby food shown in studies by other authors.

Type of Products	As (µg/kg)	Cd (µg/kg)	Hg (µg/kg)	Pb (µg/kg)	Reference
Baby food *	NA	Me = 2.8 Max = 103.9	NA	Max = 183.6	[15]
Baby desserts Baby juices Baby dinners	NA NA NA	Me = 0.71 Max = 14.90 Me = 0.80 Max = 1.34 Me = 8.31 Max = 28.2	NA NA NA	Me = 2.14 Max = 5.11 Me = 20.6 Max = 41.0	[30]
Baby food	<LOD Max = 0.89	<LOD Max < LOD	NA NA	<LOD <LOD	[23]
Baby dinners	NA	NA	Me = 0.80 Max = 7.40	NA	[33]
Cereals food	NA	NA	Me = 0.58 Max = 1.0	NA	

Table 8. *Cont.*

Type of Products	As (µg/kg)	Cd (µg/kg)	Hg (µg/kg)	Pb (µg/kg)	Reference
Rise baby food	Me = 0.088 Max = 0.150	NA NA	Me = 0.062 Max = 0.3	NA NA	[24]
Rise baby food	NA	NA	X = 0.94 ± 0.47	NA	[14]
Milk-based food Spelt based food Oat-based food Rice-based food	X = 0.23 ± 0.05 X = 3.8 ± 0.05 X = 2.4 ± 0.19 X = 1.7 ± 0.04	X = 0.23 ± 0.05 X = 2.4 ± 0.11 X = 3.3 ± 0.14 X = 33 ± 0.56	NA NA NA NA	X = 1.2 ± 0.19 X = 1.8 ± 0.12 X = 3.1 ± 0.23 X = 1.2 ± 0.12	[25]
Baby food	NA	X = 0.38 ± 0.2	X = 0.22 ± 0.08	X = 0.47 ± 0.14	[27]
Cereal-based food Baby food	NA NA	NA NA	Me = 0.50 Me = 0.40	NA NA	[34]
Baby food	NA	Me = 33.0	NA	Me = 109	[28]
Milk-based food	NA	X = 0.05 ± 0.005	NA	X = 0.21 ± 0.02	[32]
Milk-based food	NA	NA	Me = 0.03	NA	[41]
Fruit-based food Vegetable-based food Meat based food	<LOD	<LOD	NA	X = 12.0 X = 29.0 X = 34.50	[36]
Cereal-based food	X = 0.68 ± 0.67	NA	NA	NA	[26]
Baby food Milk-based food Cereal-based food Baby mouses Baby drinks	NA	NA	NA	X = 3.4 ± 2.01 X = 1.11 ± 0.21 X = 1.40 ± 1.95 X = 2.15 ± 2.08 X = 3.72 ± 2.31	[37]

* Baby food—vegetable, meat, fish, fruit-based samples, LOD—limits of detection, Max—maximum, Me—median, Min—minimum, NA—not applicable, X—mean.

5. Conclusions

The range of ready-to-eat products for children aged 0.5–3 years is a source of toxic elements, but most of the assessed products shown do not pose a health risk. In our study, the estimated health risk indicators did not show an increased health risk in terms of exposure to toxic elements in ready-made products for children. In some samples, Hg and Pb concentrations were exceeded, which may suggest that these products are not completely safe for children. There is a need for monitoring in ready-to-eat products intended for children aged 0.5–3 years. Particular attention should be paid to children's consumption of fish-based products, cereal products, snacks, and flavored drinks and waters. These product groups should be subject to special control with regard to the health safety of children already at the production stage.

Supplementary Materials: The following supporting information can be downloaded at: https://www.mdpi.com/article/10.3390/nu14112325/s1, Table S1: The content of toxic elements in tested products.

Author Contributions: Conceptualization, K.S. and A.P.-J.; methodology, J.B., J.S., M.G. and K.M.; software, A.Ż.; formal analysis, A.Ż., K.S. and A.P.-J.; investigation, A.Ż.; resources, A.Ż.; data curation, A.Ż. and A.P.-J.; writing—original draft preparation, A.Ż.; writing—review and editing, K.S. and A.P.-J.; visualization, A.Ż. and A.P.-J.; supervision, K.S. All authors have read and agreed to the published version of the manuscript.

Funding: This research was funded by Medical University of Bialystok, funding number: SUB/2/DN/21/001/2216 and SUB/2/DN/22/004/2216.

Institutional Review Board Statement: Not applicable.

Informed Consent Statement: Not applicable.

Data Availability Statement: Detailed data are available from the authors.

Conflicts of Interest: The authors declare no conflict of interest. The funders had no role in the design of the study; in the collection, analyses, or interpretation of data; in the writing of the manuscript; or in the decision to publish the results.

References

1. Patriarca, M.; Menditto, A.; Rossi, B.; Lyon, T.D.B.; Fell, G.S. Environmental exposure to metals of newborns, infants and young children. *Microchem. J.* **2000**, *67*, 351–361. [CrossRef]
2. IARC Working Group on the Evaluation of Carcinogenic Risks to Humans. Arsenic, Metals, Fibres, and Dusts. *IARC Monogr. Eval. Carcinog. Risks Hum.* **2012**, *100*, 11–465.
3. Gundert-Remy, U.; Damm, G.; Foth, H. High exposure to inorganic arsenic by food: The need for risk reduction. *Arch. Toxicol.* **2015**, *89*, 2219–2227. [CrossRef] [PubMed]
4. Qing, Y.; Yang, J.; Chen, Y.; Shi, C.; Zhang, Q.; Ning, Z.; Yu, Y.; Li, Y. Urinary cadmium in relation to bone damage: Cadmium exposure threshold dose and health-based guidance value estimation. *Ecotoxicol. Environ. Saf.* **2021**, *226*, 112824. [CrossRef] [PubMed]
5. Iwegbue, C.M.A.; Onyonyewoma, U.A.; Bassey, F.I.; Nwajei, G.E.; Bice, S.M. Concentrations and Health Risk of Polycyclic Aromatic Hydrocarbons in Some Brands of Biscuits in the Nigerian Market. *Hum. Ecol. Risk Assess. Int. J.* **2015**, *21*, 338–357. [CrossRef]
6. Llop, S.; Murcia, M.; Aguinagalde, X.; Vioque, J.; Rebagliato, M.; Cases, A.; Iñiguez, C.; Lopez-Espinosa, M.J.; Amurrio, A.; Navarrete-Muñoz, E.M.; et al. Exposure to mercury among Spanish preschool children: Trend from birth to age four. *Environ. Res.* **2014**, *132*, 83–92. [CrossRef]
7. Young, E.C.; Davidson, P.W.; Wilding, G.; Myers, G.J.; Shamlaye, C.; Cox, C.; de Broeck, J.; Bennett, C.M.; Reeves, J.S. Association between prenatal dietary methyl mercury exposure and developmental outcomes on acquisition of articulatory-phonologic skills in children in the Republic of Seychelles. *Neurotoxicology* **2020**, *81*, 353–357. [CrossRef]
8. Dack, K.; Fell, M.; Taylor, C.M.; Havdahl, A.; Lewis, S.J. Mercury and Prenatal Growth: A Systematic Review. *Int. J. Environ. Res. Public Health* **2021**, *18*, 7140. [CrossRef]
9. Etchevers, A.; Bretin, P.; Lecoffre, C.; Bidondo, M.L.; Le Strat, Y.; Glorennec, P.; Le Tertre, A. Blood lead levels and risk factors in young children in France, 2008–2009. *Int. J. Hyg. Environ. Health* **2014**, *217*, 528–537. [CrossRef]
10. Bodeau-Livinec, F.; Glorennec, P.; Cot, M.; Dumas, P.; Durand, S.; Massougbodji, A.; Ayotte, P.; Le Bot, B. Elevated Blood Lead Levels in Infants and Mothers in Benin and Potential Sources of Exposure. *Int. J. Environ. Res. Public Health* **2016**, *13*, 316. [CrossRef]
11. Téllez-Rojo, M.; Bellinger, D.; Arroyo-Quiroz, C.; Lamadrid-Figueroa, H.; Mercado-Garcia, A.; Schnaas-Arrieta, L.; Wright, R.; Hernandez-Avila, M.; Hu, H. Longitudinal associations between blood lead concentrations lower than 10 microg/dL and neurobehavioral development in environmentally exposed children in Mexico City. *Pediatrics* **2006**, *118*, 323–330. [CrossRef] [PubMed]
12. Satarug, S.; Gobe, G.C.; Vesey, D.A.; Phelps, K.R. Cadmium and Lead Exposure, Nephrotoxicity, and Mortality. *Toxics* **2020**, *8*, 86. [CrossRef] [PubMed]
13. Commission Regulation (EU) of the European Parliament and of the Council 852/2004 of 29 April 2004 on the Hygiene of Foodstuffs. Available online: https://eur-lex.europa.eu/legal-content/EN/TXT/?uri=celex%3A32004R0852 (accessed on 31 May 2022).
14. Gu, Z.; de Silva, S.; Reichman, S.M. Arsenic Concentrations and Dietary Exposure in Rice-Based Infant Food in Australia. *Int. J. Environ. Res. Public Health* **2020**, *17*, 415. [CrossRef] [PubMed]
15. Gardener, H.; Bowen, J.; Callan, S.P. Lead and cadmium contamination in a large sample of United States infant formulas and baby foods. *Sci. Total Environ.* **2019**, *651*, 822–827. [CrossRef]
16. Commission Regulation (EU) 2018/73 of 16 January 2018 Amending Annexes II and III to Regulation (EC) No 396/2005 of the European Parliament and of the Council as Regards Maximum Residue Levels for Mercury Compounds in or on Certain Products. Available online: https://eur-lex.europa.eu/legal-content/EN/TXT/HTML/?uri=CELEX:32018R0073&from=PL (accessed on 30 January 2022).
17. Commission Regulation (EU) 2021/1323 of 10 August 2021 Amending Regulation (EC) No 1881/2006 as Regards Maximum Levels of Cadmium in Certain Foodstuffs. Available online: https://eur-lex.europa.eu/eli/reg/2021/1323/oj (accessed on 31 May 2022).
18. World Health Organization; Food and Agriculture Organization of the United Nations & Joint FAO/WHO Expert Committee on Food Additives. Evaluation of certain food additives and contaminants: Eightieth report of the Joint FAO/WHO Expert Committee on Food Additives. In *Proceedings of the 80th Meeting, Rome, Italy, 2016*; World Health Organization: Geneva, Switzerland, 2016.
19. Weker, H.; Barańska, M.; Riahi, A.; Strucińska, M.; Więch, M.; Rowicka, G.; Dyląg, H.; Klemarczyk, W.; Bzikowska, A.; Socha, P. Nutrition of infants and young children in Poland—Pitnuts 2016. *Dev. Period. Med.* **2017**, *21*, 13–28.

20. Commission Regulation (EC) No 1881/2006 of 19 December 2006 Setting Maximum Levels for Certain Contaminants in Foodstuffs. Available online: https://eur-lex.europa.eu/LexUriServ/LexUriServ.do?uri=OJ:L:2006:364:0005:0024:EN:PDF (accessed on 30 January 2022).

21. Environmental Protection Agency. *A Review of the Reference Dose and Reference Concentration Processes*; Risk Assessment Forum United States Environmental Protection Agency, United States Environmental Protection Agency: Washington, DC, USA, 2002.

22. *GB 2762-2012*; Polish National Food Safety Standard, Maximum Contamination Levels in Foodstuffs. National Commission for Health and Family Planning. General Veterinary Inspectorate: Warsaw, Poland, 2014.

23. Škrbić, B.; Živančev, J.; Jovanović, G.; Farre, M. Essential and toxic elements in commercial baby food on the Spanish and Serbian market. *Food Addit. Contam. Part B Surveill.* **2017**, *10*, 27–38. [CrossRef]

24. Rothenberg, S.E.; Jackson, B.P.; Carly McCalla, G.; Donohue, A.; Emmons, A.M. Co-exposure to methylmercury and inorganic arsenic in baby rice cereals and rice-containing teething biscuits. *Environ. Res.* **2017**, *159*, 639–647. [CrossRef]

25. Ljung, K.; Palm, B.; Grandér, M.; Vahter, M. High concentrations of essential and toxic elements in infant formula and infant foods—A matter of concern. *Food Chem.* **2011**, *127*, 943–951. [CrossRef]

26. Igweze, Z.N.; Ekhator, O.C.; Nwaogazie, I.; Orisakwe, O.E. Public Health and Paediatric Risk Assessment of Aluminium, Arsenic and Mercury in Infant Formulas Marketed in Nigeria. *Sultan Qaboos Univ. Med. J.* **2020**, *20*, e63–e70. [CrossRef]

27. Kim, D.W.; Woo, H.D.; Joo, J.; Park, K.S.; Oh, S.Y.; Kwon, H.J.; Park, J.D.; Hong, Y.S.; Sohn, S.J.; Yoon, H.J.; et al. Estimated long-term dietary exposure to lead, cadmium, and mercury in young Korean children. *Eur. J. Clin. Nutr.* **2014**, *68*, 1322–1326. [CrossRef]

28. De Castro, C.S.; Arruda, A.F.; Da Cunha, L.R.; De Souza, J.R.; Braga, J.W.; Dórea, J.G. Toxic metals (Pb and Cd) and their respective antagonists (Ca and Zn) in infant formulas and milk marketed in Brasilia, Brazil. *Int. J. Environ. Res. Public Health* **2010**, *7*, 4062–4077. [CrossRef]

29. Spungen, J.H. Children's exposures to lead and cadmium: FDA total diet study 2014-16. *Food Addit. Contam.* **2019**, *36*, 893–903. [CrossRef] [PubMed]

30. Winiarska-Mieczan, A.; Kiczorowska, B. Determining the content of lead and cadmium in infant food from the Polish market. *Int. J. Food Sci. Nutr.* **2012**, *63*, 708–712. [CrossRef] [PubMed]

31. Chen, Y.; Qu, J.; Sun, S.; Shi, Q.; Feng, H.; Zhang, Y.; Cao, S. Health risk assessment of total exposure from cadmium in South China. *Chemosphere* **2021**, *269*, 128673. [CrossRef] [PubMed]

32. Meshref, A.M.S.; Moselhy, W.A.; Hassan, N.E.H.Y. Heavy metals and trace elements levels in milk and milk products. *J. Food Meas. Charact.* **2014**, *8*, 381–388. [CrossRef]

33. Guérin, T.; Chekri, R.; Chafey, C.; Testu, C.; Hulin, M.; Noël, L. Mercury in foods from the first French total diet study on infants and toddlers. *Food Chem.* **2018**, *239*, 920–925. [CrossRef]

34. Martins, C.; Vasco, E.; Paixão, E.; Alvito, P. Total mercury in infant food, occurrence and exposure assessment in Portugal. *Food Addit. Contam. Part B Surveill.* **2013**, *6*, 151–157. [CrossRef]

35. Junqué, E.; Garí, M.; Arce, A.; Torrent, M.; Sunyer, J.; Grimalt, J.O. Integrated assessment of infant exposure to persistent organic pollutants and mercury via dietary intake in a central western Mediterranean site (Menorca Island). *Environ. Res.* **2017**, *156*, 714–724. [CrossRef]

36. Misic, I.D.R.; Tosic, S.B.; Pavlovic, A.N.; Pecev-Marinkovic, E.T.; Mrmosanin, J.M.; Mitic, S.S.; Stojanovic, G.S. Trace element content in commercial complementary food formulated for infants and toddlers: Health risk assessment. *Food Chem.* **2022**, *378*, 132113. [CrossRef]

37. Guérin, T.; Le Calvez, E.; Zinck, J.; Bemrah, N.; Sirot, V.; Leblanc, J.C.; Chekri, R.; Hulin, M.; Noël, L. Levels of lead in foods from the first French total diet study on infants and toddlers. *Food Chem.* **2017**, *237*, 849–856. [CrossRef]

38. Ohtsu, M.; Mise, N.; Ikegami, A.; Mizuno, A.; Kobayashi, Y.; Nakagi, Y.; Nohara, K.; Yoshida, T.; Kayama, F. Oral exposure to lead for Japanese children and pregnant women, estimated using duplicate food portions and house dust analyses. *Environ. Health Prev. Med.* **2019**, *24*, 72. [CrossRef] [PubMed]

39. Weker, H.; Barańska, M.; Dyląg, H.; Riahi, A.; Więch, M.; Strucińska, M.; Kurpińska, P.; Rowicka, G.; Klemarczyk, W. Analysis of nutrition of children aged 13–36 months in Poland: A nation-wide study. *Med. Wieku Rozwoj* **2011**, *15*, 224–231.

40. Mozaffarian, D.; Rimm, E.B. Fish intake, contaminants, and human health: Evaluating the risks and the benefits. *JAMA* **2006**, *296*, 1885–1899. [CrossRef] [PubMed]

41. Dabeka, R.W.; McKenzie, A.D. Survey of total mercury in infant formulae and oral electrolytes sold in Canada. *Food Addit. Contam. Part B Surveill.* **2012**, *5*, 65–69. [CrossRef] [PubMed]

Article

The Mediating Effect of Inflammation between the Dietary and Health-Related Behaviors and Metabolic Syndrome in Adolescence

Ui-Jeong Kim [1], Eun-Jeong Choi [2], Hyunjin Park [1], Hye-Ah Lee [3], Bomi Park [4], Haesoon Kim [5], Youngsun Hong [6], Seungyoun Jung [7] and Hyesook Park [1,*]

[1] Department of Preventive Medicine, Graduate Program in System Health Science and Engineering, Ewha Womans University, Seoul 07804, Korea; its0912@naver.com (U.-J.K.); hyunjin9191@ewhain.net (H.P.)

[2] Department of Preventive Medicine, College of Medicine, Ewha Womans University, Seoul 07804, Korea; choiej_89@naver.com

[3] Clinical Trial Center, Mokdong Hospital, Ewha Womans University, Seoul 07985, Korea; khyeah@ewha.ac.kr

[4] Department of Preventive Medicine, College of Medicine, Chung-Ang University, Seoul 06974, Korea; bomi.s.park@gmail.com

[5] Department of Pediatrics, College of Medicine, Ewha Womans University, Seoul 07804, Korea; hyesk@ewha.ac.kr

[6] Department of Internal Medicine, College of Medicine, Ewha Womans University, Seoul 07804, Korea; imhys@ewha.ac.kr

[7] Department of Nutritional Science and Food Management, Graduate Program in System Health Science and Engineering, Ewha Womans University, Seoul 03760, Korea; sjung131@ewha.ac.kr

* Correspondence: hpark@ewha.ac.kr

Citation: Kim, U.-J.; Choi, E.-J.; Park, H.; Lee, H.-A.; Park, B.; Kim, H.; Hong, Y.; Jung, S.; Park, H. The Mediating Effect of Inflammation between the Dietary and Health-Related Behaviors and Metabolic Syndrome in Adolescence. *Nutrients* 2022, 14, 2339. https://doi.org/10.3390/nu14112339

Academic Editor: Zhiyong Zou

Received: 9 May 2022
Accepted: 1 June 2022
Published: 2 June 2022

Publisher's Note: MDPI stays neutral with regard to jurisdictional claims in published maps and institutional affiliations.

Copyright: © 2022 by the authors. Licensee MDPI, Basel, Switzerland. This article is an open access article distributed under the terms and conditions of the Creative Commons Attribution (CC BY) license (https://creativecommons.org/licenses/by/4.0/).

Abstract: Chronic diseases develop via complex pathways, depending on the degree of exposure to risk factors from early in life and childhood onward. Metabolic syndrome has multiple risk factors, including genetic factors, inappropriate diet, and insufficient physical activity. This study classified health-related behavior classes in childhood and adolescents and analyzed the direct and indirect effects of each class on the metabolic risk in inflammation-mediated pathways. We identified the health-related lifestyle classes based on health-related behavior indicators in subjects aged 3–15 years who participated in the Ewha Birth and Growth Cohort Study by using a latent class analysis. A mediation analysis was performed to access the direct and indirect effects of each class on the continuous metabolic syndrome score (cMetS), with the inflammatory index used as a mediating factor. Subjects were classified into inactive and positive lifestyle classes according to their characteristics. In the inactive lifestyle class, interleukin (IL)-6 and cMetS had a significant association. The study confirmed that IL-6 exerts a significant indirect effect between inactive lifestyle and cMetS. This result supports previous studies. Since the health behaviors of children and adolescents can affect the likelihood of subsequent metabolic syndrome, appropriate health behavior interventions for this period are needed.

Keywords: latent class analysis; mediation analysis; metabolic syndrome; health-related behavior; cohort

1. Introduction

Metabolic syndrome is diagnosed when a patient has three or more of the following: abdominal obesity, high blood pressure, high fasting glucose level, hypertriglycemia, or low high-density lipoprotein cholesterol (HDL-c) levels [1]. It increases the risks of death from cardiovascular disease, type 2 diabetes, and stroke [2]. The worldwide prevalence of metabolic syndrome is 20–25% in adults [3,4] and 19.2% in children [5]. In Korea, one in five adults (22.9%) had metabolic syndrome in 2018 [6].

Chronic disease develops via complex pathways, depending on the degree of exposure to risk factors from early in life and childhood onward. The World Health Organization

(WHO) states that a "life-course approach" should be taken to prevent chronic diseases [7]. Metabolic syndrome has multiple risk factors, including genetic factors, inappropriate dietary habits, and insufficient physical activity [8,9]. Risk factors related to lifestyle habits accumulated over the time, and many studies have been reported on the relationship between lifestyle and health. A recent study reported a change in life expectancy with the cessation of smoking, consumption of a healthy diet, performance of moderate-intensity exercise for at least 30 min a day, and maintenance of a normal body weight. Cancer patients who implemented healthy habits survived for an average of 22.9 years, which was more than twice as long as those who did not (11 years) [10]. Moreover, some studies reported that low-level chronic inflammation may cause metabolic syndrome [11].

Adolescence is a transitional period from puberty to adulthood and is an important stage in the development and maintenance of health-related behaviors [12]. It has been reported that health-related behaviors and lifestyles learned in early childhood and adolescence are difficult to change, so it can be said that chronic diseases are caused by health behaviors acquired during this period [13]. However, the majority of adolescents are reported to have a lower adherence to healthy lifestyles even when they are critically ill [14]. This low prevalence of healthy lifestyles during adolescence poses a significant public health challenge as it increases future chronic disease risk later in adulthood and lowers the quality of life during adolescence.

Therefore, adolescence can be regarded as an important period for preventive management in which chronic diseases can be prevented through interventions for the formation of healthy habits [15]. This study classified health-related behavior classes in childhood and adolescents and analyzed the direct and indirect effects of each class on metabolic risk in inflammation-mediated pathways.

2. Materials and Methods

2.1. Participants

This study was conducted on the prospective cohort of the Ewha Birth and Growth Study, which enrolled mothers at 24–28 weeks of pregnancy from 2001 to 2006 at the Department of Obstetrics and Gynecology, Ewha Womans University Mokdong Hospital (Seoul, Korea). Since 2005, 940 children of these women who agreed to participate in the study have been followed annually [16]. At each follow-up examinations, blood and urine samples collection after fasting for at least 8 h, anthropometric measurements, and questionnaires were performed. This study examined 249 subjects aged 13–15 years who participated in at least two follow-up examinations and did not have missing metabolic index data. All participants provided informed consent and the study protocol was approved by the Institutional Review Board of Ewha Womans University Seoul Hospital (IRB number: SEUMC 2020-07-016).

2.2. Exposure

Physical activity, sedentary behavior, and dietary habits are important modifiable determinants in adolescents. The health-related behavior classes were based on health-related behavior indicators: subjective health status, physical activity, dietary inflammatory index, and secondhand smoke exposure. These indicators were measured repeatedly at the ages of 3–15 using questionnaires and included in the analysis if observed at least twice.

To measure the subjective health, physical activity, and sedentary lifestyle of children aged from 3 to 15 years during the study follow-up, we used modified questionnaires with reference to the Korea National Health and Nutrition Examination Survey. Subjective health status was determined based on self-reported health status (very good, good, average, bad, or very bad). The performance of vigorous physical activity was determined if participants reported more than three sessions of at least 20 min activities on each occasion that made them out of breath or sweaty over the past week. The daily leisure sedentary lifestyle was estimated based on the time spent on activities, such as watching TV, surfing the internet, and playing games, over the past week. Then, participants were determined as "inactive" if they reported more than three times of daily sedentary leisure greater than two hours over the past week. Shivappa et al. [17] developed a dietary inflammatory index (DII) through a systematic literature review. The index is used to estimate dietary inflammation based on the intakes of 36 nutrients and 9 foods. Calculation of the DII is based on dietary intake that provided a robust estimate of a mean and standard deviation in a world database for each parameter. Dietary data for each participant expressed an individual's exposure as a Z-score. We get the "standard mean" by subtracting from the actual reported parameter value and divided by its standard deviation. This Z-score was converted to a percentile score to minimize the effect of right-skewing and centered values on 0 and bounded them between -1 (reflect the anti-inflammatory potential of the diet) and $+1$ (reflect proinflammatory potential of the diet), each percentile score was doubled and then "1" was subtracted. Then, this value was respectively multiplied by the inflammatory effect score of each food to obtain the specific DII score of each parameter. All specific DII scores were then summed for each participant in the study [17]. We investigated 24 food groups using a food frequency questionnaire at the age of 13–15 years and we assessed them based on the method from Shivappa et al. [17]. The urine level of cotinine, a nicotine metabolite, was used as a measure of secondhand smoke. Cotinine was measured using a high-performance liquid chromatography–triple tandem mass detector, HPLC–MS/MS (Agilent 6490b, Agilent, Santa Clara, CA, USA). Urine samples were collected at each follow-up and stored at $-80\ °C$. The limit of detection (LOD) was 0.141 µg/L; any measured value lower than LOD was replaced with $LOD/\sqrt{2}$.

2.3. Mediator

The inflammatory markers high-sensitivity C-reactive protein (hs-CRP) and inter-leukin (IL)-6 in the adolescents were considered as mediator. hs-CRP was measured using a particle-enhanced immune turbidimetric assay (Cobas 8000 C702 analyzer; Roche, Mannheim, Germany) and IL-6 was measured using a VersaMax microplate reader (Molecular Devices, Sunnyvale, CA, USA). Since the hs-CRP and IL-6 data were not normally distributed, they were converted into log values. The LOD for hs-CRP was 0.15 mg/dL and any measured value lower than LOD was replaced with $LOD/\sqrt{2}$. The coefficients of variation (CV) of hs-CRP and IL-6 were less than 10% in all measurements.

2.4. Outcome

Since the components of metabolic syndrome in adolescence have not been defined [18], continuous metabolic syndrome scores (cMetS) at the age of 13–15 years was calculated using body mass index (BMI), mean arterial pressure (MAP), fasting blood glucose, triglyceride (TG), and high-density lipoprotein-cholesterol (HDL-c) to determine the risk of MetS in adolescence. BMI was used as an index of obesity and calculated as weight (kg) divided by height squared (m^2). MAP is an index of blood pressure that has a smaller standard deviation than the systolic blood pressure (SBP) and diastolic blood pressure (DBP) and it was calculated as DBP + [(SBP − DBP)/3]. Blood chemistry tests were performed for fasting blood glucose, TG, and HDL-c levels. Each metabolic indicator was standardized using the Z-score method with consideration of sex. To calculated cMetS, Z-scores of each metabolic indicator were summed. Since HDL-c has a reversal causality with cMetS, its Z-score was multiplied by -1 [19,20].

2.5. Covariates

Sex, age, and parents' monthly household income were considered as covariates. The parents' monthly household income was divided into three categories (less than 3 million won, 3–5 million won, and 5 million won or more). The association between these factors and health-related behaviors has been demonstrated previously [21,22].

2.6. Statistical Analysis

For the descriptive statistics, the mean and standard deviation were calculated for normally distributed continuous variables and nonnormally distributed continuous variables are shown as median and interquartile range (IQR). Categorical variables are presented as frequencies with percentages.

We constructed a trajectory model in ages 3 to 15 years to assess the changes in participant's health-related behaviors using PROC TRAJ in the SAS program (SAS Institute, Cary, NC, USA). We selected the appropriate model based on the group distribution and the Bayesian information criterion (BIC) [23].

Based on the changes in behavior patterns, a latent class analysis (LCA) was performed to identify health-related behavior classes. LCA can distinguish different classes. The main assumption of LCA is that latent class and observational variables are categorical, and each observed variable is independent and conditional on the class. The latent class model estimates the probability of a given individual belonging to each class [24]. Typically, the number of classes in the model produced by LCA is increased one class at a time. Several model-fit indices are used, including the Akaike information criterion (AIC) and BIC. Since the BIC is a commonly used criterion [25], it was prioritized here.

The fitness indexes of the LCA models of patterns of health-related behavior changes were compared. Two concise statistical LCA models with best fits measured by a lower BIC were selected (BIC = 69.99, aBIC = 35.12) (Supplementary Materials Table S1).

To confirm each health-related behavior class and its effect on cMetS, an association analysis was performed, and a mediation analysis was performed considering inflammatory markers as a mediating factor. A mediation analysis is a statistical method for determining causality between input variables and the outcome variable. Causal variables can have direct or indirect (via a mediator) effects on the outcome variable. This study used the Process-Macro proposed by Hayes [26].

All statistical analyses were performed using SAS software (version 9.4; SAS Institute, Cary, NC, USA). In all analyses, p values < 0.05 in two-tailed tests were considered statistically significant.

3. Results

Table 1 summarizes the characteristics of the subjects. There were 122 males (49.0%) and 127 females (51.0%), with an average age of 13.28 ± 0.59 years. Of the subjects, 58.9% had a "good" subjective health status, and there was no sex difference. Regarding vigorous physical activity, once or twice a week was the most dominant class (38.5%), while that for sedentary leisure time was ≥2 h a day (42.7%). Of the boys and girls, 35.3% and 24.0%, respectively, performed vigorous physical activity at least 3–4 times a week, while 15.6% and 31.2%, respectively, rarely participated in vigorous physical activity. Physical activity levels differed significantly between the sexes, while sedentary leisure time did not. The dietary inflammatory index was positive for girls (0.24 ± 1.95), but not for boys (−0.25 ± 1.72). There was no gender difference in the monthly household income.

Figure 1 shows the changes in health-related behavioral indicators. Regarding subjective health status, the respondents reported that it was getting better (32.7%) and worse (67.3%), respectively. The respondents reported an increase (44.1%) and slight decrease (55.9%), respectively, regarding their vigorous physical activity, while their sedentary leisure levels remained high (27.2%) and low (72.8%), respectively. For secondhand smoke exposure in childhood, high (4.7%) and low (95.3%) exposure groups were distinguished.

Table 1. Characteristics of study subjects aged 13–15 years.

	Total (*n* = 249)	Boys (*n* = 122)	Girls (*n* = 127)	*p*-Value
Age (years)	13.28 ± 0.59	13.24 ± 0.53	13.31 ± 0.64	0.301
Subjective health status *	183 (58.9%)	96 (39.0%)	87 (35.4%)	0.223
Vigorous physical activity (more than 20 min)				
Never	58 (23.3%)	19 (15.6%)	39 (31.2%)	0.004
1–2 times/week	95 (38.5%)	45 (36.9%)	50 (40.0%)	
3–4 times/week	73 (29.6%)	43 (35.3%)	30 (24.0%)	
≥5 times/week	21 (8.5%)	15 (12.3%)	6 (4.8%)	
Sedentary lifestyle				
Never	4 (1.63%)	1 (0.8%)	3 (2.4%)	0.988
Less than 1 h/day	64 (26.0%)	31 (26.4%)	33 (26.4%)	
1–2 h/day	73 (29.7%)	36 (29.8%)	37 (29.6%)	
More than 2 h/day	105 (42.7%)	53 (43.8%)	52 (41.6%)	
Dietary Inflammation Index	0.00 ± 1.85	−0.25 ± 1.72	0.24 ± 1.95	0.037
hs-CRP (mg/dL)	0.16 (0.11, 0.38)	0.20 (0.11, 0.48)	0.11 (0.11, 0.33)	0.013
IL-6 (pg/mL)	2.74 (2.02, 3.65)	2.72 (2.01, 3.48)	2.76 (2.08, 3.76)	0.788
cMetS	0.00 ± 3.03	0.00 ± 3.15	0.00 ± 2.92	0.425
Monthly household income, KRW				
<KRW 3 million	16 (6.6%)	7 (5.8%)	9 (7.3%)	0.739
KRW 3–5 million	68 (27.9%)	34 (28.1%)	34 (27.6%)	
≥KRW 5 million	160 (65.6%)	80 (66.1%)	80 (65.0%)	

* Subjective health good status. Values are presented as mean ± SD (standard deviation) or median (interquartile range) or *n* (%). hs-CRP, high-sensitivity C-reactive protein; IL-6, Interleukin-6; cMetS, continuous metabolic syndrome risk score; KRW, Korean won.

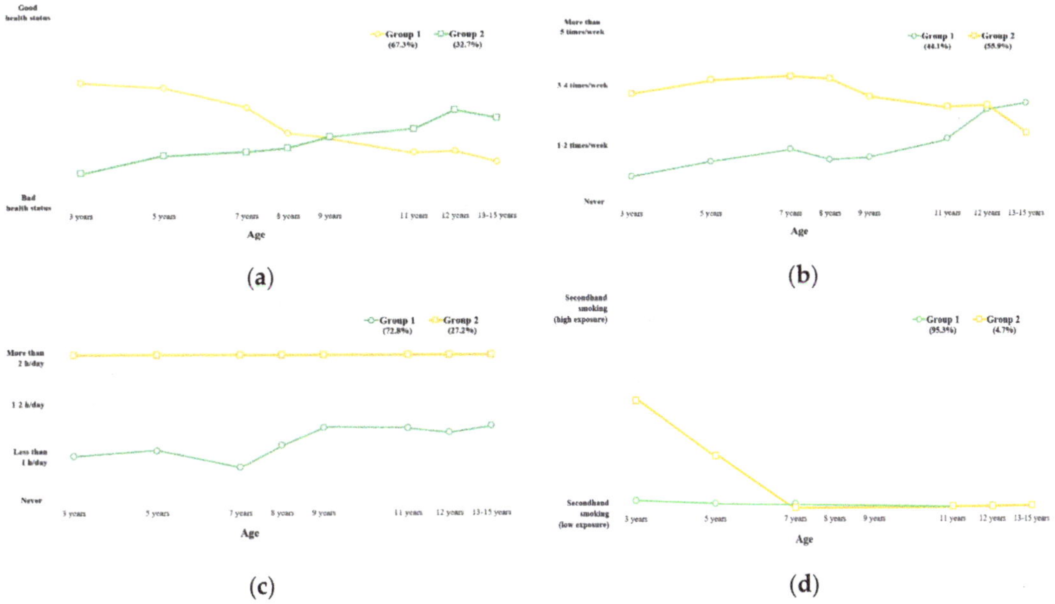

Figure 1. Trajectories of health-related behaviors from 3 to 15 years of age: (**a**, Subjective health status) 1: getting worse health status, 2: getting better health status; (**b**, Vigorous physical activity) 1: increasing physical activity group, 2: slight decreasing physical activity group; (**c**, Sedentary lifestyle) 1: low sedentary lifestyle group, 2: high sedentary lifestyle group; (**d**, Cotinine) 1: low level of cotinine as secondhand smoking in childhood group, 2: high level of cotinine as secondhand smoking in childhood.

Figure 2 shows the conditional probabilities of health-related behaviors. In the figure, the closer each specific health-related behavior factor is to 1.00, the higher the probability that the subjects in each group have a healthy behavior for each specific factor. Class 1 (inactive lifestyle class, $n = 102$) subjects were less likely to engage in vigorous physical activity, and more likely to engage in sedentary leisure activities. Class 2 (positive lifestyle class, $n = 127$) subjects tended to have an overall healthy pattern of behavior. There was no significant difference between the classes in subjective health status, likelihood of having an inflammatory diet, or exposure to secondhand smoke in childhood and childhood.

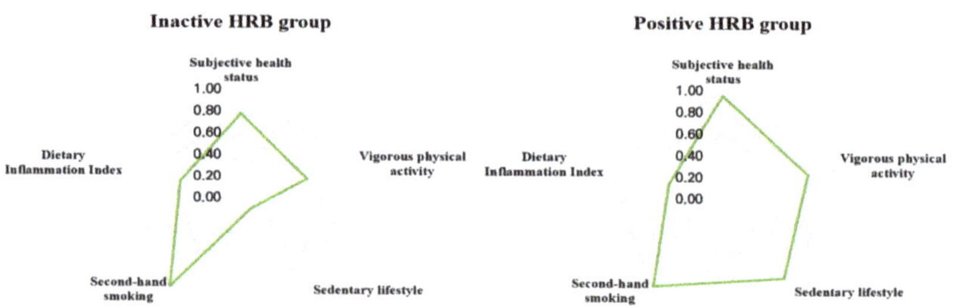

Figure 2. Health-related behavior patterns for each group. HRB, health-related behavior.

Table 2 shows the results from the models using a linear regression analysis. In the crude model, the inactive health-related behavior class was significantly associated with greater levels of IL-6 ($\beta = 0.187$, SE (standard error) = 0.069, $p = 0.007$) and cMetS ($\beta = 0.921$, SE = 0.392, $p = 0.020$), whereas the positive association between the inactive health-related behavior class and hs-CRP ($\beta = 0.228$, SE = 0.124) were marginally significant ($p = 0.067$). After adjusting for sex, age, and monthly household income, the positive associations remained significant with IL-6 ($\beta = 0.168$, SE = 0.072, $p = 0.02$) and marginally significant with cMetS ($\beta = 0.751$, SE = 0.405, $p = 0.065$). However, hs-CRP was not statistically significantly associated with inactive health-related behavior class.

Table 2. Multiple linear regression analysis for the association between latent group for health-related behaviors and metabolic risk factors.

		hs-CRP			IL-6			cMetS		
		β	SE	*p*-Value	β	SE	*p*-Value	β	SE	*p*-Value
Crude model	Inactive HRB group Positive HRB Group	0.228	0.124	0.067	0.187	0.069	0.007	0.921	0.392	0.02
Adjusted model	Inactive HRB group Positive HRB group	0.202	0.129	0.118	0.168	0.072	0.02	0.751	0.405	0.065

Reference group: positive HRB group. Adjusted model was adjusted by sex, age, and monthly household income. HRB, health-related behavior; SE: standard error.

Mediation analysis was performed to access the direct and indirect effects of each class on the metabolic risk in inflammation-mediated pathways (Figure 3). When hs-CRP was considered as a mediator, we observed a nonsignificant direct effect of the inactive health-related behavior class on cMetS ($\beta = 0.482$, 95% CI (confidence interval) -0.268 to 1.231). Similarly, there was a suggestion of an indirect effect of the inactive health-related behavior class on cMetS via hs-CRP, but the results did not reach statistical significance ($\beta = 0.238$, 95% CI -0.065 to 0.580) (Figure 3A). When IL-6 was considered as a mediator, we observed a nonsignificant direct effect of the inactive health-related behavior class on cMetS ($\beta = 0.477$, 95% CI -0.320 to 1.273). However, we observed a significant indirect effect of the inactive health-related behavior class on cMetS via the IL-6 pathway ($\beta = 0.220$, 95% CI 0.040 to 0.456). When analyzing the specific IL-6 pathway, an inactive health-related behavior was significantly positively associated with IL-6 ($\beta = 0.170$, $p < 0.05$) and IL-6 was significantly associated with cMetS (β = 1.294, *p* < 0.01). The entire path was "inactive health-related behavior class" –"IL-6" –"cMetS" (Figure 3B).

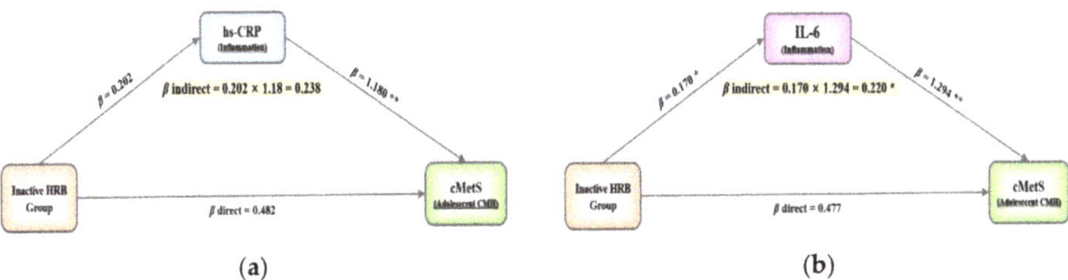

Figure 3. The effect of mediators (M) on the association between health-related behaviors group (X) and continuous metabolic syndrome score (Y): (**a**, hs-CRP) considering hs-CRP as mediators; (**b**, IL-6) considering IL-6 as mediators. * *p* < 0.05, ** *p* < 0.01. HRB, health-related behavior; hs-CRP, high-sensitivity C-reactive protein; IL-6, interleukin-6; cMetS, continuous metabolic syndrome score; CMH, Cardiometabolic health. All models were adjusted by sex, age, and monthly household income.

4. Discussion

This study distinguished between inactive lifestyle class and positive lifestyle class based on repeated measures of health-related behaviors. The IL-6 level of the inactive lifestyle class was 0.168 higher than that of the positive lifestyle class, while cMetS was 0.751 higher. Moreover, IL-6 had a significant indirect effect on the relationship between the inactive lifestyle class and cMetS in the mediated pathway (β = 0.220, 95% CI 0.040–0.456).

An association between an inactive lifestyle and metabolic syndrome has been reported previously, and the association between negative health behaviors and metabolic syndrome is also well known. In a study examining the effects of health-related behavioral changes on the prevalence of metabolic syndrome in adults over 40 years of age, the prevalence increased by 9% in a persistent heavy drinking compared to persistent moderate drinking group. In addition, the risk of MetS decreased by 30.3% in the "continuous physical activity group" compared to "continuous passive activity group" [27].

Many studies have reported an association between physical activity and metabolic syndrome. In this study, the physical activity indicators were daily leisure time and the frequency of vigorous physical activity; the cMetS was high in the low-physical activity group. A Portuguese study found that a sedentary lifestyle, excessive caloric intake, and obesity tended to increase the incidences of diabetes and metabolic syndrome [28]. Meanwhile, a Swedish population-based cohort study reported that the odds ratio of metabolic disease among those who cited "watching TV" and "low physical activity" in the context of their daily leisure time during adolescence was 2.14 times higher than that of those who were physically active [29].

A study of the eating patterns of 14-year-olds in Western Australia reported that the risk of metabolic disease was about 2.5 times higher in those frequently consuming a Western diet [30]. In a 2018 meta-analysis of cross-sectional and cohort studies, a high dietary fiber intake was inversely proportional to metabolic syndrome in the general population [31,32]. A cohort study reported a significant association of the Dietary Inflammatory Index with metabolic syndrome and its components, even after adjusting for various potential confounders [33]. However, in this study, we found no significant difference in the DII between the inactive and positive lifestyle groups. Given that individuals who participate in cohorts tend to be more health conscious, this result may reflect an overall high quality of diet of our study participants with less variation in dietary inflammation potential. Indeed, the distribution of DII in our study was relatively narrow with a low mean value.

An association between bad eating habits and metabolic syndrome has been reported. The Korean Nurses' Health Study, which analyzed the risks posed by the unhealthy eating habits associated with the nursing profession, found that consuming ≥50% of one's daily calories after 7 pm, frequent consumption of carbonated drinks, and irregular or short meals were associated with metabolic syndrome [34].

In the mediated pathway analysis conducted in this study, a significant relationship of IL-6, but not hs-CRP, with the cMetS was observed. The Mater-University of Queensland Study of Pregnancy cohort study demonstrated that negative health behaviors, such as smoking, partially contributed to inflammation [35]. Low-level chronic inflammation may be associated with metabolic dysfunction, which can lead to metabolic syndrome [36]. Although the latter study examined adults, the results were similar to this study, in suggesting that inflammation contributes to the long-term risk of disease.

LCA is a popular method for identifying classes, which uses a maximum likelihood estimation to distinguish internally homogeneous and externally heterogeneous subgroups. The relationships of these subgroups with risk and protective factors can then be investigated [37]. Another advantage of LCA is that it is a model-based approach and is thus more flexible than conventional clustering techniques; it determines the probability that a given individual belongs to a specific class, such that patient selection criteria are less arbitrary [24,38].

This study had several limitations. First, since the subjects were recruited from a single hospital, it is difficult to generalize the results. Second, since inflammatory indicators and metabolic health were both investigated in adolescence, there may have been a bias toward an inverse correlation, i.e., measurement error. Third, since the questionnaire used was self-written, there may have been a reporting bias. Lastly, our study population had relatively high levels of vigorous physical activity (48.7% in boys and 28.5% in girls). Nonetheless, these levels of physical activity were similarly observed in the national data (45.8% in boys and 23.7% in girls from the Korea Youth Risk Behavior Web-based Survey [39]), supporting the representativeness of our data when applied to the Korean population.

Despite its limitations, this study identified latent classes based on changes in health-related behaviors and confirmed a significant indirect effect of inflammatory indicators on the relationships of the changes in health-related behaviors with metabolic syndrome. These results support previous studies. Since the health behaviors of children and adolescents can affect the likelihood of subsequent metabolic syndrome, appropriate health behavior interventions for this period are needed.

5. Conclusion

In summary, we classified health-related behavior classes in childhood and adolescents, and analyzed the direct and indirect effects of each class on metabolic risk in inflammation-mediated pathway. The IL-6 level and cMetS of the inactive lifestyle class was higher than that of the positive life-style class. Also, IL-6 had a significant indirect effect on the relationship between the inactive lifestyle class and cMetS in the mediated pathway. This study is meaningful in that it has confirmed a significant effect on the relationship between health-related behavior and cMets in the indirect path considering inflammation indicator and supports the results of previous studies.

Supplementary Materials: The following are available online at https://www.mdpi.com/article/10.3390/nu14112339/s1, Table S1: The fit index for each latent class model.

Author Contributions: Conceptualization, U.-J.K. and H.P. (Hyesook Park); methodology, H.-A.L., B.P. and H.P. (Hyesook Park); validation, H.K., Y.H., and S.J.; formal analysis, U.-J.K. and E.-J.C.; writing—original draft preparation, U.-J.K. and H.P. (Hyesook Park); writing—review and editing, U.-J.K., E.-J.C., H.P. (Hyunjin Park), H.-A.L., B.P., H.K., Y.H., S.J. and H.P. (Hyesook Park); visualization, U.-J.K. and E.-J.C.; supervision, H.P. (Hyesook Park). All authors have read and agreed to the published version of the manuscript.

Funding: This research was supported by the Basic Science Research Program through the National Research Foundation of Korea (NRF), funded by the Ministry of Science and ICT (NRF-2020R1F1A1062227).

Institutional Review Board Statement: The study was conducted in accordance with the Declaration of Helsinki, and approved by the Institutional Review Board of Ewha Womans University Seoul Hospital (IRB number: SEUMC 2020-07-016 and approved on 16 July 2020).

Informed Consent Statement: Informed consent was obtained from all subjects involved in the study.

Data Availability Statement: The cohort data are not freely available, but the Ewha Birth and Growth Study team welcomes collaborations with other researchers. For further information, contact Hyesook Park (the corresponding author).

Conflicts of Interest: The authors declare no conflict of interest.

References

1. Alberti, K.G.; Eckel, R.H.; Grundy, S.M.; Zimmet, P.Z.; Cleeman, J.I.; Donato, K.A.; Fruchart, J.C.; James, W.P.; Loria, C.M.; Smith, S.C., Jr. Harmonizing the metabolic syndrome: A joint interim statement of the International Diabetes Federation Task Force on Epidemiology and Prevention; National Heart, Lung, and Blood Institute; American Heart Association; World Heart Federation; International Atherosclerosis Society; and International Association for the Study of Obesity. *Circulation* **2009**, *120*, 1640–1645. [CrossRef] [PubMed]
2. Saklayen, M.G. The Global Epidemic of the Metabolic Syndrome. *Curr. Hypertens. Rep.* **2018**, *20*, 12. [CrossRef] [PubMed]
3. Ranasinghe, P.; Mathangasinghe, Y.; Jayawardena, R.; Hills, A.P.; Misra, A. Prevalence and trends of metabolic syndrome among adults in the asia-pacific region: A systematic review. *BMC Public Health* **2017**, *17*, 101. [CrossRef] [PubMed]
4. Belete, R.; Ataro, Z.; Abdu, A.; Sheleme, M. Global prevalence of metabolic syndrome among patients with type I diabetes mellitus: A systematic review and meta-analysis. *Diabetol. Metab. Syndr.* **2021**, *13*, 25. [CrossRef] [PubMed]
5. Friend, A.; Craig, L.; Turner, S. The prevalence of metabolic syndrome in children: A systematic review of the literature. *Metab. Syndr. Relat. Disord.* **2013**, *11*, 71–80. [CrossRef] [PubMed]
6. Huh, J.H.; Kang, D.R.; Kim, J.Y.; Koh, K.K. Metabolic syndrome fact sheet 2021: Executive report. *CardioMetabolic Syndr. J.* **2021**, *1*, 125–134. [CrossRef]
7. Baird, J.; Jacob, C.; Barker, M.; Fall, C.H.; Hanson, M.; Harvey, N.C.; Inskip, H.M.; Kumaran, K.; Cooper, C. Developmental Origins of Health and Disease: A Lifecourse Approach to the Prevention of Non-Communicable Diseases. *Healthcare* **2017**, *5*, 14. [CrossRef]
8. Al-Hamad, D.; Raman, V. Metabolic syndrome in children and adolescents. *Transl. Pediatr.* **2017**, *6*, 397–407. [CrossRef]
9. Brage, S.; Wedderkopp, N.; Ekelund, U.; Franks, P.W.; Wareham, N.J.; Andersen, L.B.; Froberg, K. Features of the metabolic syndrome are associated with objectively measured physical activity and fitness in Danish children: The European Youth Heart Study (EYHS). *Diabetes Care* **2004**, *27*, 2141–2148. [CrossRef]
10. Li, Y.; Schoufour, J.; Wang, D.D.; Dhana, K.; Pan, A.; Liu, X.; Song, M.; Liu, G.; Shin, H.J.; Sun, Q.; et al. Healthy lifestyle and life expectancy free of cancer, cardiovascular disease, and type 2 diabetes: Prospective cohort study. *bmj* **2020**, *368*, l6669. [CrossRef]
11. Yudkin, J.S.; Kumari, M.; Humphries, S.E.; Mohamed-Ali, V. Inflammation, obesity, stress and coronary heart disease: Is interleukin-6 the link? *Atherosclerosis* **2000**, *148*, 209–214. [CrossRef]
12. Miranda, V.P.N.; Coimbra, D.R.; Bastos, R.R.; Miranda Júnior, M.V.; Amorim, P. Use of latent class analysis as a method of assessing the physical activity level, sedentary behavior and nutritional habit in the adolescents' lifestyle: A scoping review. *PLoS ONE* **2021**, *16*, e0256069. [CrossRef]
13. Grøntved, A.; Ried-Larsen, M.; Møller, N.C.; Kristensen, P.L.; Wedderkopp, N.; Froberg, K.; Hu, F.B.; Ekelund, U.; Andersen, L.B. Youth screen-time behaviour is associated with cardiovascular risk in young adulthood: The European Youth Heart Study. *Eur. J. Prev. Cardiol.* **2014**, *21*, 49–56. [CrossRef]
14. Mozzillo, E.; Zito, E.; Maffeis, C.; De Nitto, E.; Maltoni, G.; Marigliano, M.; Zucchini, S.; Franzese, A.; Valerio, G. Unhealthy lifestyle habits and diabetes-specific health-related quality of life in youths with type 1 diabetes. *Acta Diabetol.* **2017**, *54*, 1073–1080. [CrossRef]
15. Morrison, J.A.; Friedman, L.A.; Wang, P.; Glueck, C.J. Metabolic syndrome in childhood predicts adult metabolic syndrome and type 2 diabetes mellitus 25 to 30 years later. *J. Pediatr.* **2008**, *152*, 201–206. [CrossRef]
16. Lee, H.A.; Park, B.; Min, J.; Choi, E.J.; Kim, U.J.; Park, H.J.; Park, E.A.; Cho, S.J.; Kim, H.S.; Lee, H.; et al. Cohort profile: The Ewha Birth and Growth Study. *Epidemiol. Health* **2021**, *43*, e2021016. [CrossRef]
17. Shivappa, N.; Steck, S.E.; Hurley, T.G.; Hussey, J.R.; Hébert, J.R. Designing and developing a literature-derived, population-based dietary inflammatory index. *Public Health Nutr.* **2014**, *17*, 1689–1696. [CrossRef]
18. Eisenmann, J.C. On the use of a continuous metabolic syndrome score in pediatric research. *Cardiovasc. Diabetol.* **2008**, *7*, 17. [CrossRef]
19. Shafiee, G.; Kelishadi, R.; Heshmat, R.; Qorbani, M.; Motlagh, M.E.; Aminaee, T.; Ardalan, G.; Taslimi, M.; Poursafa, P.; Larijani, B. First report on the validity of a continuous Metabolic syndrome score as an indicator for Metabolic syndrome in a national sample of paediatric population—The CASPIAN-III study. *Endokrynol. Pol.* **2013**, *64*, 278–284. [CrossRef]

20. Heshmat, R.; Heidari, M.; Ejtahed, H.S.; Motlagh, M.E.; Mahdavi-Gorab, A.; Ziaodini, H.; Taheri, M.; Shafiee, G.; Beshtar, S.; Qorbani, M.; et al. Validity of a continuous metabolic syndrome score as an index for modeling metabolic syndrome in children and adolescents: The CASPIAN-V study. *Diabetol. Metab. Syndr.* **2017**, *9*, 89. [CrossRef]

21. Meader, N.; King, K.; Moe-Byrne, T.; Wright, K.; Graham, H.; Petticrew, M.; Power, C.; White, M.; Sowden, A.J. A systematic review on the clustering and co-occurrence of multiple risk behaviours. *BMC Public Health* **2016**, *16*, 657. [CrossRef]

22. McAloney, K.; Graham, H.; Law, C.; Platt, L. A scoping review of statistical approaches to the analysis of multiple health-related behaviours. *Prev. Med.* **2013**, *56*, 365–371. [CrossRef]

23. Jones, B.L.; Nagin, D.S. Advances in group-based trajectory modeling and an SAS procedure for estimating them. *Sociol. Methods Res.* **2007**, *35*, 542–571. [CrossRef]

24. Hagenaars, J.A.; McCutcheon, A.L. *Applied Latent Class Analysis*; Cambridge University Press: Cambridge, UK, 2002.

25. Tein, J.Y.; Coxe, S.; Cham, H. Statistical Power to Detect the Correct Number of Classes in Latent Profile Analysis. *Struct. Equ. Modeling Multidiscip. J.* **2013**, *20*, 640–657. [CrossRef]

26. Hayes, A.F. *Introduction to Mediation, Moderation, and Conditional Process Analysis: A Regression-Based Approach*; Guilford Publications: New York, NY, USA, 2017.

27. Yim, E.; Lee, K.; Park, I.; Lee, S. The Prevalence of Metabolic Syndrome and Health-Related Behavior Changes: The Korea National Health Examination Survey. *Healthcare* **2020**, *8*, 134. [CrossRef]

28. Santos, A.C.; Ebrahim, S.; Barros, H. Alcohol intake, smoking, sleeping hours, physical activity and the metabolic syndrome. *Prev. Med.* **2007**, *44*, 328–334. [CrossRef]

29. Wennberg, P.; Gustafsson, P.E.; Dunstan, D.W.; Wennberg, M.; Hammarström, A. Television viewing and low leisure-time physical activity in adolescence independently predict the metabolic syndrome in mid-adulthood. *Diabetes Care* **2013**, *36*, 2090–2097. [CrossRef]

30. Ambrosini, G.L.; Huang, R.C.; Mori, T.A.; Hands, B.P.; O'Sullivan, T.A.; de Klerk, N.H.; Beilin, L.J.; Oddy, W.H. Dietary patterns and markers for the metabolic syndrome in Australian adolescents. *Nutr. Metab. Cardiovasc. Dis.* **2010**, *20*, 274–283. [CrossRef] [PubMed]

31. Chen, J.P.; Chen, G.C.; Wang, X.P.; Qin, L.; Bai, Y. Dietary Fiber and Metabolic Syndrome: A Meta-Analysis and Review of Related Mechanisms. *Nutrients* **2017**, *10*, 24. [CrossRef] [PubMed]

32. Wei, B.; Liu, Y.; Lin, X.; Fang, Y.; Cui, J.; Wan, J. Dietary fiber intake and risk of metabolic syndrome: A meta-analysis of observational studies. *Clin. Nutr.* **2018**, *37*, 1935–1942. [CrossRef] [PubMed]

33. Abdollahzad, H.; Pasdar, Y.; Nachvak, S.M.; Rezaeian, S.; Saber, A.; Nazari, R. The Relationship Between the Dietary Inflammatory Index and Metabolic Syndrome in Ravansar Cohort Study. *Diabetes Metab. Syndr. Obes.* **2020**, *13*, 477–487. [CrossRef]

34. Jung, H.; Dan, H.; Pang, Y.; Kim, B.; Jeong, H.; Lee, J.E.; Kim, O. Association between Dietary Habits, Shift Work, and the Metabolic Syndrome: The Korea Nurses' Health Study. *Int. J. Environ. Res. Public Health* **2020**, *17*, 7697. [CrossRef]

35. Raposa, E.B.; Bower, J.E.; Hammen, C.L.; Najman, J.M.; Brennan, P.A. A developmental pathway from early life stress to inflammation: The role of negative health behaviors. *Psychol. Sci.* **2014**, *25*, 1268–1274. [CrossRef]

36. Yudkin, J.S.; Juhan-Vague, I.; Hawe, E.; Humphries, S.E.; di Minno, G.; Margaglione, M.; Tremoli, E.; Kooistra, T.; Morange, P.E.; Lundman, P.; et al. Low-grade inflammation may play a role in the etiology of the metabolic syndrome in patients with coronary heart disease: The HIFMECH study. *Metabolism* **2004**, *53*, 852–857. [CrossRef]

37. Berlin, K.S.; Williams, N.A.; Parra, G.R. An introduction to latent variable mixture modeling (part 1): Overview and cross-sectional latent class and latent profile analyses. *J. Pediatr. Psychol.* **2014**, *39*, 174–187. [CrossRef]

38. Miettunen, J.; Nordström, T.; Kaakinen, M.; Ahmed, A.O. Latent variable mixture modeling in psychiatric research–a review and application. *Psychol. Med.* **2016**, *46*, 457–467. [CrossRef]

39. *The 17th (2021) Korea Youth Risk Behavior Web-Based Survey Statistics Report*; Ministry of Education, Ministry of Health & Welfare, Korea Disease Control and Prevention Agency: Cheongju, Korea, 2021.

Article

Investigation of the Associations between Diet Quality and Health-Related Quality of Life in a Sample of Swedish Adolescents

Callum Regan [1,*], Hedda Walltott [1], Karin Kjellenberg [1], Gisela Nyberg [1,2] and Björg Helgadóttir [1,3]

[1] The Swedish School of Sport and Health Sciences (GIH), 114 86 Stockholm, Sweden; hedda.walltott@gih.se (H.W.); karin.kjellenberg@gih.se (K.K.); gisela.nyberg@gih.se (G.N.); bjorg.helgadottir@ki.se (B.H.)
[2] Department of Global Public Health, Karolinska Institutet, 171 77 Stockholm, Sweden
[3] Division of Insurance Medicine, Department of Clinical Neuroscience, Karolinska Institutet, 171 77 Stockholm, Sweden
* Correspondence: callum.regan@gih.se

Abstract: Most adolescents do not consume a high-quality diet, while self-reported mental health problems within this group are increasing. This study aimed to investigate the association between diet quality and health-related quality of life, and to explore the differences in diet quality and health-related quality of life between gender and parental education status. In this cross-sectional study, a detailed web-based recall method was implemented to determine dietary intake, which was analysed using the newly developed Swedish Healthy Eating Index for Adolescents 2015 (SHEIA15) and the Riksmaten Adolescents Diet Diversity Score (RADDS), to determine diet quality. The KIDSCREEN-10 questionnaire was used to measure health-related quality of life, and parental education was self-reported through questionnaires. Parental education was divided into two groups: $\leq$12 years or >12 years. The study included 1139 adolescents from grade 7 (13–14 years old), 51% were girls. The results showed that girls had higher scores for healthy eating and diet diversity but lower scores for health-related quality of life. A positive association was found between diet diversity and health-related quality of life (Adj R^2 = 0.072, p = 0.001), between vegetable/fruit consumption and health-related quality of life (Adj R^2 = 0.071, p = 0.002), and between healthy eating and diet diversity (Adj R^2 = 0.214, p < 0.001). No association was found between healthy eating and health-related quality of life for all participants. The mean scores for healthy eating and diet diversity were significantly higher in the higher education parental group. In conclusion, higher diet diversity and increased fruit and vegetable consumption could be a strategy to improve health-related quality of life among adolescents. There is a need to promote better diet quality, especially in households of low parental education. In addition, there is a further need to investigate the potential benefits of improved diet quality on mental health and overall well-being.

Keywords: diet quality; health-related quality of life; Swedish Healthy Eating Index for Adolescents 2015 (SHEIA15); Riksmaten Adolescents Diet Diversity Score (RADDS)

Citation: Regan, C.; Walltott, H.; Kjellenberg, K.; Nyberg, G.; Helgadóttir, B. Investigation of the Associations between Diet Quality and Health-Related Quality of Life in a Sample of Swedish Adolescents. *Nutrients* **2022**, *14*, 2489. https://doi.org/10.3390/nu14122489

Academic Editor: Zhiyong Zou

Received: 25 May 2022
Accepted: 14 June 2022
Published: 15 June 2022

Publisher's Note: MDPI stays neutral with regard to jurisdictional claims in published maps and institutional affiliations.

Copyright: © 2022 by the authors. Licensee MDPI, Basel, Switzerland. This article is an open access article distributed under the terms and conditions of the Creative Commons Attribution (CC BY) license (https://creativecommons.org/licenses/by/4.0/).

1. Introduction

Adolescents across Europe often are not meeting the national dietary guidelines [1], thus, not consuming a high-quality diet. Concurrently, the number of adolescents reporting poor mental health is increasing [1]. Mental health is complex, with a variety of environmental and personal factors contributing to poor mental health status, including diet quality [2]. A high-quality diet includes: the consumption of nutrient-dense foods such as fruits, vegetables, and wholegrains, while limiting intake of red and processed meat, salt, and added sugar [3]. Existing evidence from systematic reviews has shown a positive association between diet quality and mental health in adolescents [4,5]. Additionally, a recent systematic review also found a positive association between diet quality and health-related quality of life in adolescents [2]. However, very few of the studies included in the abovementioned systematic reviews [2,4,5] have incorporated diet diversity within the scope of diet quality. Therefore, studies that specifically investigate the association between diet diversity and health-related quality of life are needed. This current study investigates this association using a newly designed and population-specific diet diversity index.

Adopting dietary guidelines is a key strategy for improving diet quality in adolescents [6]. The World Health Organization (WHO) reported an estimated number of two out of three adolescents in the European region and Canada that do not consume enough nutrient-rich foods [1]. Furthermore, results from the national Swedish survey Riksmaten Adolescents from 2016 to 2017 found that the average consumption of fruits and vegetables among Swedish adolescents was 50% less than the recommended 500 g daily amount [7]. The same report also found that participants consumed more red meat and charcuteries than recommended, and that 17% of the overall intake came from candy, cookies, and sugar-sweetened beverages [7]. Furthermore, in a population-based survey of Swedish adolescents, girls were found to consume a healthier diet than boys and 13–17-year-old adolescents were shown to adhere to unhealthier dietary patterns than 7–12-year-old adolescents [8].

To assess diet quality in adolescents, several approaches can be utilised, such as: adherence to a Mediterranean diet [9] or measuring unhealthy dietary patterns [5]. Another approach is to use dietary indices that are in accordance with dietary guidelines and recommendations [10]. The Swedish Healthy Eating Index for Adolescents (SHEIA15) and the Riksmaten Adolescents Diet Diversity Score (RADDS) were developed to capture healthy eating and diet diversity, respectively, as a measure of diet quality in Swedish adolescents [10] and were utilised in this study.

Investigating the association between diet quality and mental health in adolescents is important due to the global prevalence of the latter, accounting for 16% of disease burden amongst 10–19-year-old adolescents [11]. Globally, around 10% of children and adolescents experience mental health challenges [12], and a recent cross-national study found that 22% of adolescents (n = 6245) across eight European countries reported at least one mental health disorder [13]. Girls tend to report more life satisfaction complaints, for example, feeling low and nervous and are more likely to report multiple health complaints as compared with boys [14]. Furthermore, the results from the Health Behaviour in School-Aged Children (2017/2018) showed that the percentage of Swedish adolescents aged 11 to 15 years old who felt nervous or depressed had doubled since the 1980s and these health complaints were more prevalent in girls than boys [15]. Moreover, this study found that Swedish children were one of the highest groups to report these health complaints as compared with children from other European countries [15].

One way to assess mental health in adolescents is to measure health-related quality of life, which is a subset of mental health. Health-related quality of life measures the perceived physical, emotional, mental, and functional well-being of an individual [16]. It is crucial to measure domains of health in adolescents to prevent more serious long-term conditions [17]. Lower health-related quality of life has been associated with high mental health symptoms in adolescents and therefore health-related quality of life should be explored in this population [18].

The association between diet quality and health-related quality of life or other aspects of mental health in adolescents has been investigated. A previous cross-sectional analysis showed that a healthy diet, where higher scores indicated that more healthy dietary practices were being followed, was associated with higher health-related quality of life scores [19], and a past prospective cohort, measuring diet quality in a similar way, showed that improved diet quality was associated with positive mental health outcomes [20]. Additionally, a systematic review reported a dose-response relationship between health-related quality of life and diet quality, where low diet quality was associated with lower health-related quality of Life [2]. A possible explanation for the association between a low-quality diet and low health-related quality of life, or poor mental health is that diets relatively low in nutrient density can lead to nutrient deficiencies, which, in turn, are associated with mental health problems [4]. However, the possibility of reverse causality cannot be ignored [5]. Potential confounders also need to be considered when investigating the association between diet quality and health-related quality of life, such as parental education, which can have a significant influence on overall adolescent health-related quality of life [21].

Parental education is a major risk factor in developing mental health problems in adolescents [21], and can influence diet quality [22], hence, the importance to incorporate this factor when investigating the association between diet quality and health-related quality of life. Swedish adolescents with parents who have higher education have been found to have healthier dietary patterns [8], with one recent cross-sectional study finding high scores for healthy eating and diet diversity to be associated with higher education households [10]. It is proposed that parents with more educational experience are more likely to make healthier food choices for their family attributed to better nutritional knowledge [22].

Adequate nutrition is pivotal for growth and the increasing physiological needs of adolescents [20]. Given the relatively high prevalence of self-reported mental health problems in Swedish adolescents [14], it is important to investigate the association between diet quality and health-related quality of life in this population. It is also pivotal to explore different measures of diet quality and to examine the influence that following national dietary guidelines has on health-related quality of life. Healthy eating and diet diversity have both been proposed to measure diet quality and have been investigated separately with health-related quality of life. Thus, the aim of the current study was to investigate the association between healthy eating and diet diversity (diet quality) with health-related quality of life (a measure of mental health) in a sample of 13–14-year-old adolescents in Sweden. Additionally, potential differences in scores for healthy eating, diet diversity, and health-related quality of life between gender and parental education status were explored.

To the best of our knowledge, this is the first study to investigate the association between diet quality and health-related quality of life in Swedish adolescents. Relatively few studies have investigated this association in healthy adolescent populations [5]. Moreover, this study incorporates diet diversity into diet quality which is unique and is the first to utilise the newly developed diet quality indices SHEIA15 and RADDS that follow Swedish national dietary guidelines [10] in relation to health-related quality of life.

2. Materials and Methods

2.1. Study Design and Study Population

This study utilised a cross-sectional method. This study included a final sample size of 1139 adolescents from grade 7 (13–14-year-old girls/boys) from 34 different schools (Figure 1) within 2–3 h travel time from Stockholm.

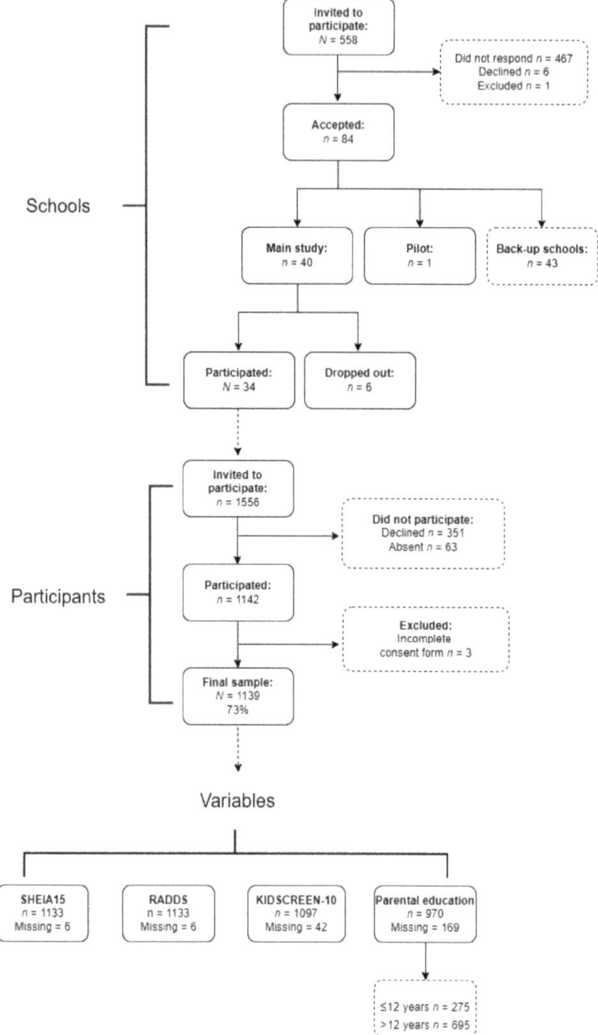

Figure 1. Flowchart showing recruitment of schools and participants and *N* values for variables.

2.2. Recruitment

All schools within a two–three-hour radius of the Swedish School of Sport and Health Sciences (GIH) were invited to participate in the study (*n* = 558) (Figure 1). Schools with less than 15 students in grade 7 and schools with a sports profile were excluded. The principal of each school received an invitation to participate, on 11 March 2019. The required school sample size was achieved after the first invitation; 40 schools were selected to be included, which varied in geographical location, size, and parental education status (Figure 1). One of the schools was involved in the pilot study, conducted on 3 September 2019, and six schools dropped out after initial recruitment, most likely due to time constraints (Figure 1).

Participants and guardians received an invitation letter and a consent form either through post or given directly by the teachers. Written consent was provided by both the participants and guardians before measurements could take place. The numbers of participants who declined to participate or were excluded from the study are shown in

Figure 1. Only participants fluent or with a good comprehension of the Swedish language were included. The participants went to the GIH for half a day between September and December 2019, for data to be collected and for measurements, including health-related quality of life, dietary intake, body weight, and height. A week after participants visited the GIH, parental questionnaires were sent out electronically to the parents.

2.3. Assessment of Dietary Intake

Dietary intake was self-reported with a validated web-based method called Riks-matenFlexDiet (developed by the Swedish Food Agency, Uppsala, Sweden for Riksmaten Adolescents 2016/17); using a repeated 24 h multiple pass recall method [23], which has been validated against 24 h recall interview methods and has been shown to be reliable in recording total energy, fruit/vegetable, and wholegrain intakes [24]. Participants were given instructions and demonstrations on how to record their diets. Three dietary recording days were completed, two weekdays and one weekend day. All days were recorded retrospectively, however, day two was excluded from analyses as this was the day of the school visit and researchers at the GIH may have influenced the participants' food choices. Thus, dietary intake was assessed using an average of two diet recording days (*n* = 1036) or just one, if only one day was completed (*n* = 97). Six participants had no dietary recording days. Participants documented their dietary intake using a food list containing 778 foods (Swedish Food Agency Food Composition Database, version Riksmaten Adolescents 2016/17), dishes, and beverages which was adapted for adolescents from Riksmaten Adolescents 2016/17 [10]. Many of these dishes and food items had corresponding photographic images which gave an indication of portion size to the participants.

2.4. Plausible Energy Reporters

Participants with dietary intake records for two days and with information on energy expenditure (*n* = 1008) were classified into under, plausible, or over-energy reporters from a calculated ratio of energy intake (EI) and total energy expenditure (TEE). Confidence intervals (CIs) around the mean of EI/TEE were calculated based on that reported by [7]. Those with a ratio of less than 0.5114 were classified as under-reporters and those with a ratio of more than 1.1286 were classified as over-reporters. TEE was based on prediction equations using counts per minute from accelerometer data and measured body weight (kg) of participants, as described in [25].

2.5. Dietary Indices to Measure Diet Quality from Dietary Intake
2.5.1. Swedish Healthy Eating Index for Adolescents 2015 (SHEIA15)

To assess healthy eating, the SHEIA15 index was used which was developed by [10]. The SHEIA15 index was based on the 2015 Swedish food-based dietary guidelines, referred to as "Find your way" [26] and the Nordic nutrient recommendations [3]. Components of SHEIA15, include vegetables and fruit, fibre, fish and shellfish, red and processed meat, saturated fatty acids (SFAs), mono-unsaturated fatty acids (MUFAs), poly-unsaturated fatty acids (PUFAs), added sugar, and wholegrains [10]. Composing SHEIA15 scores involved grouping food from the subcomponents, for example, red meat together, from products and composite dishes consumed by the participants. New variables were computed to sum together total intake of that food group for each participant and was done for all components of SHEIA15. The SHEIA15 scores were calculated using a ratio of actual intake and recommended intake and adapted from [27]. This could be intake in grams or energy percent values, for instance, total intakes of SFAs (C4:0–C20:0), MUFAs (C16:1–C18:1), and PUFAs (C18:2–C21:6) in grams were converted to kcal per gram and divided by total energy intake (kcal) to acquire daily energy percent values. This was also done for added sugar. The criterion for construction of SHEIA15 components can be found in [10].

Scores were given for each SHEIA15 component and scores above one were rounded to one, for example, a participant consuming 600 g of vegetables/fruits gained a score – 1.2 (600/500, with 500 g = the recommended intake value) which would then be rounded to 1.0. Mean scores for all nine SHEIA15 components from day one and day three were added together to obtain a final SHEIA15 score. Scores ranged between 0 and 9 for each participant, with higher scores indicating better healthy eating.

2.5.2. The Riksmaten Adolescents Diet Diversity Score (RADDS)

To assess diet diversity, the RADDS index was used, which was developed by [10]. The RADDS index is based on food groups from "Find your way" [26] and accounts for the participant consuming a particular food group in the dietary recording period but does not consider amounts or frequencies of food consumption. There are 17 components of RADDS, including cabbage, root vegetables, fruit, red meat, poultry, oily fish, and milk (Figure 2). Composing the RADDS scores involved grouping intakes of relevant food groups together from products and composite dishes that the participants had consumed. Composing some of the RADDS components was more complex and contained more calculations than that of SHEIA15, because the intake of many RADDS components from composite dishes could not be specified in RiksmatenFlexDiet, for example, a participant could not specify that it was wholegrain pasta within composite dishes [10]. Therefore, only consumption of specific wholegrain products could be assessed; participants did not score any points if they consumed desserts with fruits, juice, and dried fruit as these were not within the dietary guidelines. The criterion for RADDS components can be found in [10].

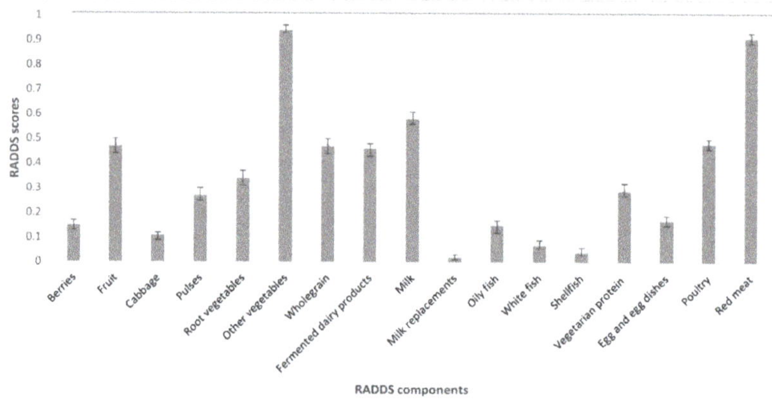

Figure 2. Mean scores for the Riksmaten Adolescents Diet Diversity Score (RADDS) scores for each component as compared with the RADDS reference value. Error bars = 95% confidence intervals, reference value = 1 for acquiring a RADDS score of 0.

Participants gained a point for eating a mean intake of ≥ 5 g for each individual RADDS component over the 2 dietary recording days. Mean scores for all 17 RADDS components from day one and day three were added together to obtain a final RADDS score. The scores ranged between 0 and 17 for each participant, with higher scores indicating better diet diversity.

2.6. Assessment of Health-Related Quality of Life

Health-related quality of life was measured using the KIDSCREEN-10 (KS-10) index and was completed by participants at the GIH where they were given instructions on how to fill in the questionnaire. KS-10 is a general 10 item questionnaire shortened from the Kidscreen-52 and -27 versions aimed at children and young adults aged 8–18 years and provides one global score for health-related quality of life [28]. The KS-10 questionnaire provides a global score from items in the longer KS versions. The dimensions represented in the longer versions include physical well-being, psychological well-being, moods and emotions, self-perception, autonomy, parent relations and home life, social support and peers, school environment, social acceptance (bullying), and financial resources [28]. The KS-10 questionnaire has been found to have a 0.82 Cronbach alpha reliability score and a strong correlation with general factor scores with the longer Kidscreen-52 version [29]. Responses are measured using a five-point Likert scale, with answers ranging from "not at all" to "extremely" or from "never" to "always". Responses were coded so that higher values were an indicator of better health-related quality of life, with scores ranging from 10 to 50 [28]. Selected questions include: Have you felt full of energy? Have you felt sad? Have you got on well at school? Have your parent(s) treated you fairly? [29]. Scores for KS-10 Items 3 and 4 were reversed, as suggested in the documentation for KS-10 [28]. The items were added up (Items 1–10) to form a scale from 10 to 50 and calculated in accordance with the Kidscreen manual [28].

2.7. Assessing Parental Education

Questions regarding education status were answered by the parent(s) or legal guardian(s) and their partners ($n = 970$) via parental questionnaires. Parental education status was divided by length of education experience and grouped into ≤12 years or >12 years. The parent or legal guardian with the highest level of education status in the household was used for the classification into the parental education groups. Parental education was used as a proxy for socioeconomic status (SES) and represented a SES factor.

2.8. Confounders

In addition to parental education, gender, body mass index (BMI), and home country were used in the regression analysis when investigating associations between SHEIA15, RADDS, and KS-10. Self-identified gender was answered by the participant and weight and height was measured through standardised procedures at the GIH for BMI calculations. BMI scores were calculated using the International Obesity Task Force age-standardised cutoffs and categorised into underweight, normal weight, overweight and obese [30]. Questions regarding home country of the adolescents were self-reported by parents/legal guardians ($n = 1129$). Parental country of birth was also reported by parents/legal guardians ($n = 1117$), and this variable was only used for descriptive purposes (Table 1).

Table 1. Descriptives of participants divided by gender.

	Total **N = 1139**	Girls **n = 580**	Boys **n = 558**			
	Mean (SD)	Mean (SD)	Mean (SD)	*p*-Value [c]	*t*-Value	DF
Age	13.4 (0.3)	13.4 (0.3)	13.4 (0.4)	0.148	−1.45	1137
SHEIA15 [a]	5.3 (1.1)	5.4 (1.1)	5.2 (1.1)	**0.008**	2.64	1131
RADDS [a]	5.9 (2.1)	6.1 (2.0)	5.8 (2.2)	**0.011**	2.56	1131
KS-10 [b]	39.6 (5.4)	38.3 (5.2)	41 (5.3)	**<0.001**	−8.30	1095
	N (%)	n (%)	n (%)	*p*-Value [d]		
BMI categories				0.181		
Underweight	89 (7.8)	38 (6.6)	51 (9.2)			
Normal weight	815 (71.8)	430 (74.1)	384 (69.3)			
Overweight	179 (15.8)	89 (15.3)	90 (16.2)			
Obese	52 (4.6)	23 (4.0)	29 (5.2)			
Parental education						
≤12 years	275 (28.4)	144 (29)	131 (27.8)	0.674		
>12 years	695 (71.6)	353 (71)	341 (72.2)			

Table 1. *Cont.*

	Total **N = 1139**	Girls **n = 580**	Boys **n = 558**	
Parental country of birth				
Swedish born and at least one Swedish parent	800 (71.6)	414 (72.5)	386 (70.8)	0.534
Born outside of Sweden or both parents born outside of Sweden	317 (28.4)	457 (27.5)	159 (29.2)	
Home country				
Sweden	967 (85.7)	490 (84.9)	476 (86.4)	0.758
EU (inc. Nordic countries)	46 (4.1)	24 (4.2)	22 (4.0)	
Outside EU	116 (10.3)	63 (10.9)	53 (9.6)	

Independent *t*-test comparing mean values of variables between gender and chi-squared frequency distribution. Abbreviations: Swedish Healthy Eating Index for Adolescents 2015 (SHEIA15), The Riksmaten Adolescents Diet Diversity Score (RADDS), KIDSCREEN-10 (KS-10), body mass index (BMI), standard deviation (SD), and degrees of freedom (DF). [a] $n = 1133$; [b] $n = 1097$; [c] *p*-value for independent *t*-tests; [d] *p*-value for Pearson's χ2. Significant values shown in bold.

2.9. Statistical Methods

The statistical analytical plan was prespecified before beginning statistical testing. The IBM Statistics SPSS 27 software was used for all data cleaning and analyses. Continuous dependent variables were all checked for normality through histogram, skewness, and kurtosis analysis. Descriptive statistics are described using the means and standard deviations for continuous variable whilst proportions of distribution are used for categorical variables (Table 1). Independent *t*-tests were used to compare the difference of age, SHEIA15, RADDS, and KS-10 between girls and boys (Table 1)Pearson's χ2 was used to investigate distribution of proportions between categorical variables, for example, parental education and gender (Table 1). Correlation matrixes using Pearson's R values were completed for the 3 main variables: SHEIA15, RADDS and KS-10, for all participants (Table S1, Supplementary Material).

Multiple linear regression models were constructed to investigate the associations among SHEIA15, RADDS, and KS-10 (Tables 2 and 3), using confounders mentioned in Section 2.8. Models 1 and 2 investigated the association between SHEIA15 and KS-10 with different confounders, whilst Models 3 and 4 investigated the association between RADDS and KS-10 with different confounders. Additionally, a multi-level mixed linear regression model was applied, with two levels: the school level and individual/student level for comparing the association between SHEIA15 and KS-10, as well as between RADDS and KS-10 (results not shown). A random intercept was modelled for each school. Assumptions for regression were checked for all regression analyses, including normality, normal PP plots of residuals, collinearity, and Durbin–Watson test. Morewover, ndependent *t*-tests were used to compare mean scores of SHEIA15, RADDS, and KS-10 between parent education groups (Table 4). A *p*-value <0.05 was used to detect statistical significance for all tests; unstandardised β coefficient 95% and CI were used for regression analyses.

Table 2. Associations between SHEIA and KS-10 and between RADDS and KS-10.

	KS-10					
	Total		Girl		Boy	
Main Exposure: **SHEIA**	Unstandardised β Coefficient	*p*-Value	Unstandardised β Coefficient (95% CI)	*p*-Value	Unstandardised β Coefficient (95% CI)	*p*-Value
Model 1 SHEIA15	0.16 (−0.11, 0.44)	0.254	0.09 (−0.29, 0.47)	0.637	0.28 (−0.12, 0.68)	0.171

Table 2. *Cont.*

| | KS-10 | | | | | |
	Total		Girl		Boy	
Gender						
Girl	REF	**<0.001**	N/A	**N/A**	**N/A**	**N/A**
Boy	2.60 (1.98, 3.22)					
Model 2 SHEIA15	0.16 (−0.14, 0.46)	0.291	0.07 (−0.35, 0.48)	0.755	0.32 (−0.12, 0.75)	0.159
Gender						
Girl	REF	**<0.001**	N/A	N/A	N/A	N/A
Boy	2.52 (1.85, 3.19)					
BMI	−0.11 (−0.20, −0.01)	**0.024**	−0.08 (−0.22, 0.05)	0.230	−0.15 (−0.28, −0.01)	**0.034**
Home country						
Sweden	REF					
EU	−1.06 (−2.76, 0.63)	0.219	−0.74 (−3.07, 1.60)	0.535	−1.54 (−4.00, 0.92)	0.223
Outside EU	−0.79 (−1.91, 0.32)	0.163	0.17 (−1.35, 1.69)	0.822	−2.08 (−3.74, −0.42)	**0.014**
Parental education						
≤12 years	REF					
	−0.15 (−0.92, 0.62)	0.700	0.65 (−0.41, 1.71)	0.228	−1.06 (−2.18, 0.06)	0.064
Main exposure: RADDS						
Model 3 RADDS	0.28 (0.14, 0.43)	**<0.001**	0.29 (0.08, 0.50)	**0.008**	0.28 (0.07, 0.48)	**0.008**
Gender						
Girl	REF	**<0.001**	N/A	N/A	N/A	N/A
Boy	2.72 (2.10, 3.34)					
Model 4 RADDS	0.26 (0.11, 0.42)	**0.001**	0.25 (0.02, 0.48)	**0.035**	0.25 (0.03, 0.47)	**0.024**
Gender						
Girl	REF	**<0.001**	N/A	N/A	N/A	N/A
Boy	2.65 (1.98, 3.32)					
BMI	−0.09 (−0.19, 0.00)	0.059	−0.07 (−0.20, 0.07)	0.335	−0.14 (−0.27, −0.00)	**0.047**
Home country						
Sweden	REF					
EU	−1.00 (−2.69, 0.69)	0.244	−0.68 (−3.00, 1.64)	0.565	−1.48 (0.97, 0.99)	0.235
Outside EU	−0.70 (−1.81, 0.41)	0.219	0.19 (−1.32, 1.70)	0.803	−1.90 (−3.56, −0.24)	**0.025**
Parental education						
≤12 years	REF	0.529	0.47 (−0.59, 1−53)	0.384	−1.04 (−2.14, 0.06)	0.064
	−0.24 (−1.00, 0.52)					

Linear regression models using KS-10 as a continuous dependent variable, with SHEIA15 and RADDS as independent variables. All models adjusted for confounders and stratified by gender; significant values shown in bold. Abbreviations: Swedish Healthy Eating Index for Adolescents 2015 (SHEIA15), body mass index (BMI), Riksmaten Adolescents Diet Diversity Score (RADDS), confidence intervals (CI), reference value (REF), not applicable (N/A), and KIDSCREEN-10 (KS-10). Significant values shown in bold.

Table 3. Association between SHEIA and RADDS.

| | RADDS | |
	Unstandardised β Coefficient	*p*-Value
Main exposure: SHEIA	Total	Total

Table 3. *Cont.*

	RADDS	
	Unstandardised β Coefficient	*p*-**Value**
Model 1 SHEIA15	0.83 (0.73, 0.93)	**<0.001**
Gender Girl	REF	0.122
Boy	−0.18 (−0.40, 0.15)	
Model 2 SHEIA15	0.80 (0.69, 0.91)	**<0.001**
Gender Girl	REF	0.068
Boy	−0.22 (−0.46, 0.02)	
BMI	−0.06 (−0.09, −0.02)	**0.001**
Home country Sweden	REF	
EU	−0.24 (−0.85, −0.37)	0.433
Outside EU	−0.37 (−0.82, 0.00)	0.073
Parental education ≤12 years	REF	
>12 years	0.34 (0.07, 0.62)	**0.015**

Linear regression model using RADDS as a continuous dependent variable with SHEIA15 as an independent variable. All models adjusted for confounders; significant values shown in bold. Abbreviations: Swedish Healthy Eating Index for Adolescents 2015 (SHEIA15), body mass index (BMI), confidence interval (CI), reference value (REF), and The Riksmaten Adolescents Diet Diversity Score (RADDS). Significant values shown in bold.

Table 4. Mean scores for RADDS, SHEIA15, and KS-10 across parental education groups.

	Parental Education														
	≤12 Years, Mean (SD)			>12 Years, Mean (SD)			*p*-Value			*t*-Value			DF		
	Total	Girl	Boy	Total	Girl	Boy	Total	Girl	Boy	Total	Girl	Boy	Total	Girl	Boy
RADDS	5.5 (1.9)	5.6 (1.9)	5.4 (2.0)	6.3 (2.1)	6.5 (2.0)	6.1 (2.2)	**<0.001**	**<0.001**	**0.004**	−5.17	−4.56	−2.90	965	496	468
SHEIA15	5.0 (1.1)	5.2 (1.1)	4.9 (1.1)	5.5 (1.1)	5.5 (1.1)	5.4 (1.1)	**<0.001**	**0.001**	**<0.001**	−5.31	−3.29	−4.33	965	496	468
KS-10	9.4 (5.8)	37.7 (5.6)	41.3 (5.5)	39.6 (5.2)	38.5 (5.1)	40.8 (5.1)	0.668 [a]	0.147	0.349	−0.45	−1.45	0.94	942	486	455

Independent *t*-tests, significant values shown in bold. Abbreviations: Riksmaten Adolescents Diet Diversity Score (RADDS), Swedish Healthy Eating Index for Adolescents 2015 (SHEIA15), KIDSCREEN-10 (KS-10), and standard deviation (SD), independent *t*-tests, [a] Equal variances not assumed. Significant values shown in bold.

3. Results

Participants were evenly divided by gender: girls (51%) and boys (49%), with mean ages of 13 ± 0.3 and 13 ± 0.4 years, respectively (Table 1). In addition, 71.6% of the participants had at least one parent who had >12 years of education and 85.7% of the participants stated that Sweden was the country they were born in (Table 1). A total of 1129 participants had valid dietary recalls on day one and 1040 participants for day three. The SHEIA15 (Swedish Healthy Eating Index for Adolescents 2015) scores (for healthy eating) ranged from 2.29 to 8.49, the RADDS (Riksmaten Adolescents Diet Diversity Score) scores (for diet diversity) ranged from 0 to 13, and the KS-10 scores (health-related quality of life) ranged from 18–50, with higher scores indicating better outcomes for all three variables. There were no significant Pearson's χ2 associations between gender and the following variables: BMI (body mass index) status, parental education, country of birth, and home country (Table 1). The mean values for both SHEIA15 and RADDS scores were significantly higher in girls as compared with boys (*p* < 0.008 and *p* = 0.011, respectively), whilst the mean value for KS-10 was significantly higher in boys as compared with girls (*p* < 0.001) (Table 1).

3.1. Associations among SHEIA15, RADDS, and KS-10

In a multiple linear regression analysis between SHEIA15 and KS-10, no significant association was found for all participants or when stratifying for gender (Models 1 and 2, Table 2). Nonetheless a significant positive association was found between the vegetable/fruit SHEIA15 component and KS-10 for all participants when adjusting for gender, BMI, home country, and parental education (Adj R^2 = 0.069, p = 0.002, unstandardised β = 1.95) and the vegetable/fruit SHEIA15 component was a significant variable in the regression model when stratifying for gender (p = 0.027 and p = 0.04, for girls and boys, respectively). Moreover, multiple regression analysis showed a significant positive association between RADDS and KS-10 for all participants when adjusting for gender, BMI, home country, and parental education (Model 4, Table 2) (Adj R^2 = 0.072, p = 0.001), and RADDS was a significant variable in the regression model when stratifying for gender (p = 0.035 and p = 0.024 for girls and boys respectively) (Model 4, Table 2). Additionally, applying the multi-level mixed linear regression model, accounting for potential school clustering effect, did not change the non-significant association between SHEIA15 and KS-10 and the association between RADDS and KS-10 remained positive and significant (p = 0.001 for RADDS in multi-level mixed model) (results not shown). Furthermore, multiple linear regression analyses showed a positive association between SHEIA15 and RADDS for all participants when adjusting for confounders (Model 2, Table 3) (Adj R^2 = 0.214, p < 0.001).

3.2. SHEIA15, RADDS, and KS-10 Scores in Relation to Parental Education

The mean RADDS and SHEIA15 scores for all participants as well as for girls and boys were found to be significantly higher in the >12 years parent education group as compared with the ≤12 years (p < 0.05) (Table 4), whereas the mean KS-10 scores were not significantly different between parental education groups (Table 4).

3.3. Intake and Scores of SHEIA15 and RADDS in Relation to Mean Dietary Intakes

3.3.1. SHEIA15

For all participants, the mean consumption of fibre, PUFAs, vegetables/fruits, and fish/shellfish fell below their corresponding recommended values (Table S2), whereas the mean intake of red and processed meat, SFAs, and added sugars were above recommended intakes at 500 g per week, i.e., 13.2 E% and 10.7 E%, respectively (Table S2). The only significant difference in mean consumption of SHEAI15 components between girls and boys was that boys consumed a significantly higher amount of red and processed meat as compared with girls (98.7 ± 4 g/day as compared with 86.9 ± 3.1 g/day, p = 0.019). Only the consumption of MUFAs were in line with recommended values for all participants (Table S2).

3.3.2. RADDS

For all participants, the overall mean intakes for the 17 RADDS subgroups ranged between 1.29 and 197 g/d, with shellfish consumption receiving the lowest intake and milk the highest intake (results not shown). The RADDS mean scores ranged between 0.02 and 0.94 and the mean scores for RADDS components against a reference value = 1 (Figure 2). The "milk replacements" component acquired the lowest RADDS score and "other vegetables" component received the highest RADDS score (Figure 2). The most common vegetables contributing to the "other vegetables" score included: cucumber, lettuce, green salad, tomato, and avocado.

3.4. Plausible Energy Reporters

Among the participants, 68.5% of participants were found to be plausible energy reporters, whilst 15.8% of participants were classified as under-reporters, and 15.8% of participants were classified as over-reporters (n = 1008). Multiple linear regression analyses were completed again for all plausible energy reporters. The main findings included. A significant association between RADDS and KS-10 for all plausible energy reporters with all confounders (p = 0.00). A significant association was present between SHEIA15 and RADDS (p = 0.00) and remained significant when confounders were added (p = 0.00). No significant association was present between SHEIA15 and KS-10 (p = 0.620).

4. Discussion

This study investigated associations among healthy eating, the Swedish Healthy Eating Index for Adolescents 2015 (SHEIA15), diet diversity, the Riksmaten Adolescents Diet Diversity Score (RADDS), and health-related quality of life using KS-10, in a sample of predominantly Swedish adolescents. Additionally, the comparison in scores of healthy eating, diet diversity, and health-related quality of life among parental education groups was explored. This study is unique because no other study, to the best of our knowledge, has investigated associations among these factors in this group of adolescents. Additionally, diet diversity was incorporated into diet quality which is seldom, and an association between diet diversity and health-related quality of life was found which is a novel finding.

Key findings in this study include: a significant positive association between diet diversity and health-related quality of life, a significant positive association between the fruit and vegetable component of healthy eating and health-related quality of life, a significant positive association between healthy eating and diet diversity, no significant association between healthy eating and health-related quality of life, and mean scores for healthy eating and diet diversity were significantly higher in parental education group >12 years as compared with ≤12 years.

Higher diet diversity (a component of diet quality [10]) and its association with higher health-related quality of life remained to be significant when accounting for schools, highlighting the robustness of this relationship. Although previous studies have found aspects of high quality diets to be associated with better health-related quality of life in adolescents [2,5,19], they have not found diet diversity to be positively associated with health-related quality of life. Very few studies have incorporated diet diversity into diet quality when investigating the association between diet quality and health-related quality of life, and yet this could be an important factor contributing to overall diet quality. Therefore, it is important to include measures of diet diversity in similar future investigations and to test whether the current findings could be generalizable to adolescents residing in other countries, although, it is acknowledged that other countries may need to implement their own diet quality indices tailored to specific food dietary patterns and cultural habits, for example, the Finnish Children Healthy Eating Index [31]. Since diet quality is measured in different ways, this impedes comparisons between studies investigating the relationship between diet quality and health-related quality of life; with many dietary recording methods subject to large recall errors and social desirability bias [5]. More consistent methods, for example, based on national dietary guidelines to analyse diet quality should be utilised in future studies to allow more valid comparisons to be made.

The significant positive association between healthy eating and diet diversity, as well as girls having higher scores for both, is consistent with findings from [10] and in line with findings that have shown that girls tend to make healthier food choices than boys [8,32,33]. The lower health-related quality of life in girls as compared with boys is comparable to self-reports of greater psychosomatic complaints in girls than boys [14]. However, the health-related quality of life for girls and boys on average were lower in the current sample than the recently developed aged-based norms for 11–13-year-old adolescents, where a score >42.52 was considered to be good health-related quality of life [34]. This is in line with results finding Swedish adolescents to have more mental health-related problems than other European countries [14,15]. It is difficult to know whether these differences represent a clinically meaningful difference due to the potential overestimation in classification accuracy used for the age-based norms [34]. Nonetheless, health-related quality of life (measured by mean international KS-10 t-scores, results not shown) for all participants was found to be similar as compared with a similarly aged but much larger cohort of 3283 Swedish adolescents, therefore, the current sample appears to be relatively representative of the Swedish adolescent population [29].

Higher scores for healthy eating and diet diversity in the higher education parental group is consistent with findings in [10] and comparable to findings from [8], where Swedish adolescents with parents of higher education had healthier food habits. Clearly, there are multiple points of influence by parents on household dietary intake, since parents generally act as role models, they are responsible for what is purchased in the home, and parents may influence healthier dietary habits because they themselves follow healthy dietary patterns [35].

While parental education was an important determinant of healthy eating in the current study, it may be the case that less well-educated parents are aware of dietary recommendations but find healthy foods more expensive and less accessible [36]. Therefore, advocating healthy eating practices may not only be linked to education but to overall affluence and access. Healthier food options and consuming a more varied diet are more expensive as compared with high energy-dense and nutrient-poor foods; implying that high food costs are a potential barrier for low-income households to consume a healthier diet [36]. Evidence has shown that children with parents of low educational background consume a less healthy and cheaper diet [36]. This could be an explanation for the results found in this study, however, we did not measure income, and thus, we cannot draw any definite conclusions on this matter. Nevertheless, foods that constitute a high-quality diet should be made more accessible and affordable for less affluent households.

The absence of a significant association between healthy eating and health-related quality of life may be partially due to the proportion of under-reporting or because many adolescents are not consuming a healthy diet, and even if so, this may not have been captured in the days of diet recording. However, the results from this study indicate a relationship between consuming more fruit and vegetables and higher health-related quality of life. This is comparable to results from an intervention which found that those consuming a high fruit/vegetable diet as compared with other intervention and control groups also had improvements in psychological well-being [37] and to a recent study that found those who had a higher frequency of fruit and vegetable consumption had higher health-related quality of life [33]. Furthermore, a cross-sectional analysis study in 12 different countries ($N = 5759$) found healthier food choices to be associated with better health-related quality of life, measured by KS-10 in children aged 9–11 [38]. Thus, specific food components of the diet could be important for better health-related quality of life and overall well-being, and therefore, should be investigated in future studies, rather than grouping many components into one.

The mean consumption of fruit and vegetables was more than 50% (208 g) below the recommended intake value and comparable to previous results that have shown that Swedish adolescents did not achieve the recommended intakes of fruit and vegetables [7,8,10]. Additionally, 14 diet diversity components acquired a mean score less than 0.5, which was comparable to 11 diet diversity scores (with the same diet diversity components used) receiving a score less than 0.5 in an adolescent population [10]. A relatively low proportion of under-reporters was found in this sample, indicating that these results are relatively accurate. The proportion of plausible energy reporters is higher as compared with findings in [10] and as compared with a literature review finding approximately 50% of adolescents to be plausible energy reporters from 15 different studies [39]. Based on the results from this study and previous research that showed most adolescents are not meeting recommended nutritional intakes [1], it can be concluded that awareness regarding the impact that diet quality has on aspects of mental health, needs to be promoted to help to increase adherence to dietary recommendations in this population. However, it is important to note that some of the national dietary recommendations and food guidelines may be set too high for some adolescents and not achievable for many on a daily basis, for example, 500 g of fruit/vegetables per day. Revision of the recommendations for adolescents may be worth considering to make them more realistic.

4.1. Limitations and Strengths

A limitation of the study is that the RiksmatenFlexDiet dietary recording method is subjective and, similar to all other self-report measures, it is prone to recall error and social desirability bias. It also calculates the average amount consumed from each of the food groups that constitute meals and not exact amounts; however, it can calculate ingredients from composite dishes [10], which is a strength. Another strength of the RiksmatenFlexDiet is that it provides photographic images and detailed descriptions of foods to aid identification of portion sizes. Additionally, mean healthy eating and diet diversity scores from dietary intake were calculated from just one or two days, and thus, perhaps did not capture food groups that were consumed less frequently, for example, fish and shellfish. Thus, increasing the days of dietary recording would be able to capture the intake of these foods, diet diversity, and healthy eating to a larger extent, and thus, a more representative picture of habitual dietary habits. This could make better use of the indices and could lead to a higher range of results on the spectrum of healthy eating and diet diversity.

There are limitations and strengths associated with using these dietary indices, since they are reliant on dietary recommendations and guidelines [3,26] within the Nordic countries including Sweden. These indices are population specific and are in line with national dietary guidelines; if they are met, they have been proposed to have important implications for health outcomes and to improve overall diet quality [3]. However, they fail to capture some foods in a potential diet. The diet diversity index, although able to capture dietary variety, does not consider the amount consumed of each component after 5 g, which is a relatively small amount. If, for example, a participant consumed a mean value of 100 g of cabbage and another 5 g of cabbage, both amounts would receive one point for the cabbage diet diversity score. Furthermore, the diet diversity index does not consider desserts with fruits, juices, and dried fruits (as they are not included in the guidelines), as well as wholegrain and egg consumption in composite dishes [10], subsequently introducing some measurement error of these food groups. It is also important to note that just because a participant scored low on either or both indices does not mean that their overall diet was unhealthy or lacking in diversity. For instance, a vegetarian/vegan would not score high in the fish and shellfish SHEIA15 component.

The use of KS-10 to measure health-related quality of life is a strength given its cross-cultural relevance and that it is convenient to use for studies with large sample sizes. KS-10 has been found to be a reliable measure for health-related quality of life in adolescents aged 8–18 years across 13 different EU countries ($N = 22,830$), including Swedish adolescents and validated against other health-related quality of life methods [29]. Moreover, using a short 10-item questionnaire may have been less burdensome as compared with the 27- or 52-item KS-10 questionnaires, and thus, using the shorter version could have provided more reliable answers regarding the participants' health-related quality of life. Nonetheless, the KS-10 provides a global score from dimensions in the longer KS versions, and thus, significant interactions related to specific dimensions may be overlooked. A further limitation when measuring health-related quality of life is that this was only measured on one day, which may not represent participants' overall health-related quality of life. The current analysis was of cross-sectional design, capturing only a snapshot in time, and thus, cannot infer any causations from the associations found between different exposure and outcome variables. Nevertheless, we used a relatively large sample size for this population, one sufficiently large enough to detect significance between subgroups, for example, boys and girls. Furthermore, it would have been a strength to incorporate information about parental income as a SES-F. This could have helped to explain the associations among healthy eating, diet diversity, and health-related quality of life; those with greater economy are more likely to acquire healthy diets [40].

4.2. Future Perspectives

Considering that most mental health problems occur in adolescence through to adulthood [20], the period of adolescence presents an ideal time for interventions to promote healthy lifestyle changes. Improving the variety and quality of foods provided to adolescents could, in turn, lead to healthier lifestyles and improvements in overall health-related quality of life, which could reduce potential future mental health problems. Thus, studies investigating the relationship between diet quality and health-related quality of life should focus on adolescent populations.

There is a need to conduct more intervention studies, such as the study carried out by Conner et al. (2017) [37] to establish causation between exposure and outcome variables and to assess if diet causes changes in health-related quality of life or vice versa. For instance, comparing differences in health-related quality of life between a high-quality diet intervention group and baseline control in a randomised controlled trial that recruits individuals already consuming unhealthy diets, could be an insightful analysis. In two recent systematic reviews only, correlational studies were identified [2,5], showing that there is a lack of intervention studies to test the causal relationship between diet and health-related quality of life.

The school lunch scheme in Sweden plays a significant role in the quality of adolescents' diets and Sweden is among very few countries to provide school lunches free of charge [41]. The school offers hot meals, salad bars, milk, butter, and water and has been found to be more nutritious than meals provided at home [42]. School lunches may even-out diet-related inequalities associated with education, and it was found that there were not many significant differences in energy and nutrient intake in school lunches between high and low parental education groups [42]. Given this, the disparity in diet quality among different classes of parental education, may be due to food provided at home. However, it is important to bear in mind that financial constraints are likely to have an influence on diet quality at home. Therefore, a deeper investigation into the differences between diet quality at school and home in different sociodemographic areas within Sweden is recommended to identify the SES factors, for example, parental income, that have the most influence on diet quality and investigate associations among different SES factors.

In essence, future studies should incorporate multiple SES factors in analyses, for example, schools from different socioeconomic areas, schools in rural and urban areas, and parental income/wealth, since they can have major influences on diet quality [41] and on adolescents' health-related quality of life [21]. Moreover, sample sizes should aim to represent the general population in terms of home country, country of origin, and refugees that may not speak the native language, to improve the generalisability of findings.

5. Conclusions

This study replicates and extends the evidence base regarding the association between diet quality and health-related quality of life in adolescents. Higher diet diversity was shown to be associated with better health-related quality of life, an insightful and unique finding in this area of research. Moreover, higher consumption of fruit and vegetables was shown to be positively associated with health-related quality of life, indicating the importance of fruit and vegetable consumption for well-being. Therefore, encouraging and promoting a varied diet and consumption of fruit and vegetables could be advocated to improve aspects of mental health in adolescents. Additionally, less healthy, and less diverse diets were found in households with parents who had experienced fewer years of formal education. However, the importance of parental income should not be overlooked, as economical constraints may be of more significance than education in obtaining healthy diets. Buying energy-dense and nutrient-poor foods and acquiring diets of lower quality have been associated with individuals who have fewer economic capabilities [40]. Thus, future studies would benefit from incorporating a variety of SES factors, especially parental income. Intervention study designs to establish causation and the incorporation of diet diversity into diet quality, when investigating the associations between diet quality and health-related quality of life, are also recommended.

Supplementary Materials: The following supporting information can be downloaded at: https://www.mdpi.com/article/10.3390/nu14122489/s1, Table S1: Correlation matrix between SHEIA15, RADDS and KS-10; Table S2: Mean intake for SHEIA15 components and associated reference values based on national recommendations and guidelines.

Author Contributions: Conceptualization, C.R., H.W., K.K., G.N. and B.H.; methodology, C.R., H.W., K.K., G.N. and B.H.; software, C.R. and H.W.; validation, C.R., H.W., K.K., G.N. and B.H.; C.R. and H.W.; investigation, C.R. and H.W.; resources, C.R., H.W., K.K., G.N. and B.H.; data curation, C.R., H.W., K.K., G.N. and B.H.; writing—original draft preparation, C.R. and H.W.; writing—review and editing, C.R., H.W., K.K., G.N. and B.H.; visualization, C.R., H.W., K.K., G.N. and B.H.; supervision, K.K. and B.H.; project administration, K.K., G.N. and B.H.; funding acquisition, G.N. All authors have read and agreed to the published version of the manuscript.

Funding: This study is part of the "Physical activity for healthy brain functions in school youth" project, which was funded by the Knowledge Foundation (grant 20180040) and conducted in collaboration with Coop, IKEA, Skanska, Skandia, Stockholm Consumer Cooperative Society, and the Swedish Crown Princess Couple's Foundation/Generation Pep.

Institutional Review Board Statement: The study was ethically approved by the Swedish Ethical Review Authority (Dnr 2019-03579) and conducted in accordance with the Declaration of Helsinki.

Informed Consent Statement: Written consent was received from both the students and the parents.

Data Availability Statement: The datasets are not available for download to protect the confidentiality of the participants. The data are held at The Swedish School of Sport and Health Sciences.

Acknowledgments: We would like to extend our deepest gratitude to all the students, parents, and school staff who participated in the study and the research assistants that took part in the data collection. We would also like to thank friends and family for their moral and emotional support of our efforts. Finally, a big thank you to Ulrika Sofia Jakobsson who helped spark creativity and finesse.

Conflicts of Interest: The authors declare no conflict of interest. The sponsors had no role in the design, execution, interpretation, writing, or publishing of the study.

Abbreviations

Abbrevation	Key Terms
SHEIA15	Swedish Healthy Eating Index for Adolescents (a measure of healthy eating)
RADDS	Riksmaten Adolescents Diet Diversity Score (a measure of diet diversity)
KS-10	Kidscreen-10 (a measure of health-related quality of life)
SES factors	Socioeconomic status factors

References

1. WHO Report on Health Behaviours of 11–15-Year-Olds in Europe Reveals More Adolescents Are Reporting Mental Health Concerns. Available online: https://www.euro.who.int/en/media-centre/sections/press-releases/2020/who-report-on-health-behaviours-of-1115-year-olds-in-europe-reveals-more-adolescents-are-reporting-mental-health-concerns (accessed on 8 April 2021).
2. Wu, X.Y.; Zhuang, L.H.; Li, W.; Guo, H.W.; Zhang, J.H.; Zhao, Y.K.; Hu, J.W.; Gao, Q.Q.; Luo, S.; Ohinmaa, A.; et al. The influence of diet quality and dietary behavior on health-related quality of life in the general population of children and adolescents: A systematic review and meta-analysis. *Qual. Life Res.* **2019**, *28*, 1989–2015. [CrossRef] [PubMed]
3. Nordic Council of Ministers. Nordic Nutrition Recommendations 2012. In *Integrating Nutrition and Physical Activity*, 5th ed.; Nordic Council of Ministers: Copenhagen, Denmark, 2012; p. 629. Available online: www.norden.org (accessed on 21 January 2021).
4. O'Neil, A.; Quirk, S.E.; Housden, S.; Brennan, S.L.; Williams, L.J.; Pasco, J.A.; Berk, M.; Jacka, F.N. Relationship Between Diet and Mental Health in Children and Adolescents: A Systematic Review. *Am. J. Public Health* **2014**, *104*, E31–E42. [CrossRef]
5. Khalid, S.; Williams, C.M.; Reynolds, S.A. Is there an association between diet and depression in children and adolescents? A systematic review. *Br. J. Nutr.* **2016**, *116*, 2097–2108. [CrossRef] [PubMed]
6. Dalwood, P.; Marshall, S.; Burrows, T.L.; McIntosh, A.; Collins, C.E. Diet quality indices and their associations with health-related outcomes in children and adolescents: An updated systematic review. *Nutr. J.* **2020**, *19*, 43. [CrossRef] [PubMed]

7. Lemming, E.W.; Moraeus, L.; Sipinen, J.P.; Lindroos, A.K. Riksmaten Ungdom 2016–2017, Så äter Ungdomar i Sverige. Available online: https://www.livsmedelsverket.se/globalassets/publikationsdatabas/rapporter/2018/2018-nr-14-riksmatenungdom-huvudrapport_del-1-livsmedelskonsumtion.pdf (accessed on 17 March 2021).

8. Rosell, M.; Carlander, A.; Cassel, S.; Henriksson, P.; Hook, M.J.S.; Lof, M. Generation Pep Study: A population-based survey on diet and physical activity in 12,000 Swedish children and adolescents. *Acta Paediatr.* **2021**, *110*, 2597–2606. [CrossRef]

9. Muros, J.J.; Perez, F.S.; Ortega, F.Z.; Sanchez, V.M.G.; Knox, E. The association between healthy lifestyle behaviors and health-related quality of life among adolescents. *J. Pediatr.* **2017**, *93*, 406–412. [CrossRef]

10. Moraeus, L.; Lindroos, A.K.; Lemming, E.W.; Mattisson, I. Diet diversity score and healthy eating index in relation to diet quality and socio-demographic factors: Results from a cross-sectional national dietary survey of Swedish adolescents. *Public Health Nutr.* **2020**, *23*, 1754–1765. [CrossRef]

11. Adolescent Mental Health in the WHO European Region. Available online: https://www.euro.who.int/__data/assets/pdf_file/0005/383891/adolescent-mh-fs-eng.pdf (accessed on 27 January 2021).

12. Improving the Mental and Brain Health of Children and Adolescents. Available online: https://www.who.int/activities/Improving-the-mental-and-brain-health-of-children-and-adolescents (accessed on 8 April 2021).

13. Husky, M.M.; Boyd, A.; Bitfoi, A.; Carta, M.G.; Chan-Chee, C.; Goelitz, D.; Koc, C.; Lesinskiene, S.; Mihova, Z.; Otten, R.; et al. Self-reported mental health in children ages 6-12 years across eight European countries. *Eur. Child Adolesc. Psychiatry* **2018**, *27*, 785–795. [CrossRef]

14. Inchley, J.; Currie, D.; Budisavljevic, S.; Torsheim, T.; Jaastad, A.; Cosma, A.; Kelly, C.; Arnarsson, Á.M.; Samdal, O. *Spotlight on Adolescent Health and Well-Being. Findings from the 2017/2018 Health Behaviour in School-Aged Children (HBSC) Survey in Europe and Canada*; WHO Regional Office for Europe: Copenhagen, Denmark, 2020.

15. Dalman, C.; Bremberg, S.; Åhlén, J.; Ohlis, A.; Agardh, E.; Wicks, S.; Lundin, A. *Mental Well-Being, Mental Distress and Mental Disorders among Children and Young Adults*; Forte: Stockholm, Sweden, 2021.

16. Gill, T.M.; Feinstein, A.R. A critical-appraisal of the quality of quality-of-life measurements. *Am. Med. Assoc.* **1994**, *272*, 619–626. [CrossRef]

17. Kieling, C.; Baker-Henningham, H.; Belfer, M.; Conti, G.; Ertem, I.; Omigbodun, O.; Rohde, L.A.; Srinath, S.; Ulkuer, N.; Rahman, A. Global Mental Health 2 Child and adolescent mental health worldwide: Evidence for action. *Lancet* **2011**, *378*, 1515–1525. [CrossRef]

18. O'Loughlin, R.; Hiscock, H.; Pan, T.; Devlin, N.; Dalziel, K. The relationship between physical and mental health multimorbidity and children's health-related quality of life. *Qual. Life Res.* **2022**, *31*, 2119–2131. [CrossRef] [PubMed]

19. Bolton, K.A.; Jacka, F.; Allender, S.; Kremer, P.; Gibbs, L.; Waters, E.; de Silva, A. The association between self-reported diet quality and health-related quality of life in rural and urban Australian adolescents. *Aust. J. Rural Health* **2016**, *24*, 317–325. [CrossRef] [PubMed]

20. Jacka, F.N.; Kremer, P.J.; Berk, M.; de Silva-Sanigorski, A.M.; Moodie, M.; Leslie, E.R.; Pasco, J.A.; Swinburn, B.A. A Prospective Study of Diet Quality and Mental Health in Adolescents. *PLoS ONE* **2011**, *6*, e24805. [CrossRef] [PubMed]

21. von Rueden, U.; Gosch, A.; Rajmil, L.; Bisegger, C.; Ravens-Sieberer, U.; The European KIDSCREEN Group. Socioeconomic determinants of health related quality of life in childhood and adolescence: Results from a European study. *J. Epidemiol. Community Health* **2006**, *60*, 130–135. [CrossRef]

22. Beghin, L.; Dauchet, L.; De Vriendt, T.; Cuenca-Garcia, M.; Manios, Y.; Toti, E.; Plada, M.; Widhalm, K.; Repasy, J.; Huybrechts, I.; et al. Influence of parental socio-economic status on diet quality of European adolescents: Results from the HELENA study. *Br. J. Nutr.* **2014**, *111*, 1303–1312. [CrossRef]

23. Moraeus, L.; Lemming, E.W.; Hursti, U.K.K.; Arnemo, M.; Sipinen, J.P.; Lindroos, A.K. Riksmaten Adolescents 2016-17: A national dietary survey in Sweden-design, methods, and participation. *Food Nutr. Res.* **2018**, *62*, 1381. [CrossRef]

24. Lindroos, A.K.; Sipinen, J.P.; Axelsson, C.; Nyberg, G.; Landberg, R.; Leanderson, P.; Amemo, M.; Lemming, E.W. Use of a Web-Based Dietary Assessment Tool (RiksmatenFlex) in Swedish Adolescents: Comparison and Validation Study. *J. Med. Internet Res.* **2019**, *21*, 12572. [CrossRef]

25. Ekelund, U.; Sjostrom, M.; Yngve, A.; Poortvliet, E.; Nilsson, A.; Froberg, K.; Wedderkopp, N.; Westerterp, K. Physical activity assessed by activity monitor and doubly labeled water in children. *Med. Sci. Sports Exerc.* **2001**, *33*, 275–281. [CrossRef]

26. Livsmedelsverket: Swedish Food Agency Find Your Way to Eat Greener, Not Too Much and Be Active. Available online: https://www.livsmedelsverket.se/globalassets/publikationsdatabas/andra-sprak/kostraden/kostrad-eng.pdf (accessed on 15 March 2021).

27. Knudsen, V.K.; Fagt, S.; Trolle, E.; Matthiessen, J.; Groth, M.V.; Biltoft-Jensen, A.; Sorensen, M.R.; Pedersen, A.N. Evaluation of dietary intake in Danish adults by means of an index based on food-based dietary guidelines. *Food Nutr. Res.* **2012**, *56*, 17129. [CrossRef]

28. Ravens-Sieberer, U.; Gosch, A.; Erhart, M.; Von Rueden, U.; Nickel, J.; Maria Kurth, B.; Duer, W.; Fuerth, K.; Czemy, L.; Auquier, P.; et al. *The KIDSCREEN Questionnaires Quality of Life Questionnaires for Children and Adolescents-Handbook*; Pabst Science Publishers: Lengerich, Germany, 2006.

29. Ravens-Sieberer, U.; Erhart, M.; Rajmil, L.; Herdman, M.; Auquier, P.; Bruil, J.; Power, M.; Duer, W.; Abel, T.; Czemy, L.; et al. Reliability, construct and criterion validity of the KIDSCREEN-10 score: A short measure for children and adolescents' well being and health-related quality of life. *Qual. Life Res.* **2010**, *19*, 1487–1500. [CrossRef]

30. Cole, T.J.; Lobstein, T. Extended international (IOTF) body mass index cut-offs for thinness, overweight and obesity. *Pediatr. Obes.* **2012**, *7*, 284–294. [CrossRef] [PubMed]
31. Cooke, L.J.; Wardle, J. Age and gender differences in children's food preferences. *Br. J. Nutr.* **2005**, *93*, 741–746. [CrossRef] [PubMed]
32. Davison, J.; Stewart-Knox, B.; Connolly, P.; Lloyd, K.; Dunne, L.; Bunting, B. Exploring the association between mental wellbeing, health-related quality of life, family affluence and food choice in adolescents. *Appetite* **2021**, *158*, 105020. [CrossRef] [PubMed]
33. Hirschfeld, G.; von Brachel, R.; Thiele, C. Screening for health-related quality of life in children and adolescents: Optimal cut points for the KIDSCREEN-10 for epidemiological studies. *Qual. Life Res.* **2020**, *29*, 529–536. [CrossRef]
34. Kyttala, P.; Erkkola, M.; Lehtinen-Jacks, S.; Ovaskainen, M.L.; Uusitalo, L.; Veijola, R.; Simell, O.; Knip, M.; Virtanen, S.M. Finnish Children Healthy Eating Index (FCHEI) and its associations with family and child characteristics in pre-school children. *Public Health Nutr.* **2014**, *17*, 2519–2527. [CrossRef]
35. Scaglioni, S.; De Cosmi, V.; Ciappolino, V.; Parazzini, F.; Brambilla, P.; Agostoni, C. Factors Influencing Children's Eating Behaviours. *Nutrients* **2018**, *10*, 706. [CrossRef]
36. Rydén, P.J.; Hagfors, L. Diet cost, diet quality and socio-economic position: How are they related and what contributes to differences in diet costs? *Public Health Nutr.* **2011**, *14*, 1680–1692. [CrossRef]
37. Conner, T.S.; Brookie, K.L.; Carr, A.C.; Mainvil, L.A.; Vissers, M.C.M. Let them eat fruit! The effect of fruit and vegetable consumption on psychological well-being in young adults: A randomized controlled trial. *PLoS ONE* **2017**, *12*, e0171206. [CrossRef]
38. Dumuid, D.; Olds, T.; Lewis, L.K.; Martin-Fernandez, J.A.; Katzmarzyk, P.T.; Barreira, T.; Broyles, S.T.; Chaput, J.P.; Fogelholm, M.; Hu, G.; et al. Health-Related Quality of Life and Lifestyle Behavior Clusters in School-Aged Children from 12 Countries. *J. Pediatr.* **2017**, *183*, 178–183. [CrossRef]
39. Forrestal, S.G. Energy intake misreporting among children and adolescents: A literature review. *Matern. Child Nutr.* **2011**, *7*, 112–127. [CrossRef]
40. Darmon, N.; Drewnowski, A. Does social class predict diet quality? *Am. J. Clin. Nutr.* **2008**, *87*, 1107–1117. [CrossRef] [PubMed]
41. Livsmedelsverket: Swedish Food Agency. School Lunches. Available online: https://www.livsmedelsverket.se/en/food-habits-health-and-environment/maltider-i-vard-skola-och-omsorg/skola (accessed on 3 June 2022).
42. Colombo, P.E.; Patterson, E.; Elinder, L.S.; Lindroos, A.K. The importance of school lunches to the overall dietary intake of children in Sweden: A nationally representative study. *Public Health Nutr.* **2020**, *23*, 1705–1715. [CrossRef] [PubMed]

Article

Mealtime Regularity Is Associated with Dietary Balance among Preschool Children in Japan—A Study of Lifestyle Changes during the COVID-19 Pandemic

Yuki Tada [1,*], **Yukari Ueda** [2], **Kemal Sasaki** [3], **Shiro Sugiura** [4], **Mieko Suzuki** [5], **Hiromi Funayama** [6], **Yuka Akiyama** [7], **Mayu Haraikawa** [8] and **Kumi Eto** [9]

[1] Department of Nutritional Science, Faculty of Applied Bioscience, Tokyo University of Agriculture, Sakuragaoka 1-1-1, Setagaya 156-8502, Tokyo, Japan
[2] Department of Health and Nutrition, Faculty of Health and Nutrition, Osaka Shoin Women's University, Hishiyanishi 4-2-26, Higashiosaka 577-8550, Osaka, Japan; ueda.yukari@osaka-shoin.ac.jp
[3] Department of Food and Health Sciences, Jissen Women's University, Osakaue 4-1-1, Hino 191-8510, Tokyo, Japan; kemal-s@umin.org
[4] Maternal and Child Health Center, Aichi Children's Health and Medical Center, Morioka 7-426, Obu 474-8710, Aichi, Japan; bee_jayz@hotmail.com
[5] Department of Early Childhood Care and Development, Tamagawa University, Tamagawagakuen 6-1-1, Machida 194-8610, Tokyo, Japan; mieko@edu.tamagawa.ac.jp
[6] Department of Pediatric Dentistry, Tsurumi University School of Dental Medicine, Tsurumi 2-1-3, Tsurumi, Yokohama 230-8501, Kanagawa, Japan; funayama-h@tsurumi-u.ac.jp
[7] Department of Health Sciences, University of Yamanashi, Shimokato 1110, Chuo 409-3898, Yamanashi, Japan; yukaa@yamanashi.ac.jp
[8] Department of Child Studies, Seitoku University, Iwase 550, Matsudo 271-8555, Chiba, Japan; mharai@wa.seitoku.ac.jp
[9] School of Nutrition Sciences, Kagawa Nutrition University, Chiyoda 3-9-21, Sakado 350-0288, Saitama, Japan; ekumi@eiyo.ac.jp
* Correspondence: y3tada@nodai.ac.jp; Tel.: +81-(0)3-5477-2670

Citation: Tada, Y.; Ueda, Y.; Sasaki, K.; Sugiura, S.; Suzuki, M.; Funayama, H.; Akiyama, Y.; Haraikawa, M.; Eto, K. Mealtime Regularity Is Associated with Dietary Balance among Preschool Children in Japan—A Study of Lifestyle Changes during the COVID-19 Pandemic. *Nutrients* 2022, 14, 2979. https://doi.org/10.3390/nu14142979

Academic Editor: Zhiyong Zou

Received: 26 June 2022
Accepted: 16 July 2022
Published: 21 July 2022

Publisher's Note: MDPI stays neutral with regard to jurisdictional claims in published maps and institutional affiliations.

Copyright: © 2022 by the authors. Licensee MDPI, Basel, Switzerland. This article is an open access article distributed under the terms and conditions of the Creative Commons Attribution (CC BY) license (https://creativecommons.org/licenses/by/4.0/).

Abstract: The novel coronavirus-19 (COVID-19) pandemic has considerably impacted children's lives. The aim of this study was to determine whether the pandemic affected mealtime regularity among preschool children and whether maintaining regular mealtimes or changes in mealtime regularity during the pandemic were related to dietary balance, including chronological relationships. This online cross-sectional survey involving individuals registered with a company that provides meals to children aged 2−6 years was conducted in February 2021. Using a 40-point scale, a healthy diet score (HDS) was developed to evaluate children's dietary balance. The participants were divided into four groups based on their responses, and multiple regression analyses were performed with the HDS as the dependent variable. Maintaining regular mealtimes was associated with practices such as waking and going to bed earlier, less snacking, and eating breakfast every day. Even after adjusting for basic attributes, lifestyle habits, household circumstances, and other factors, regular mealtimes were still positively correlated with the HDS. These findings indicate that maintaining regular mealtimes is associated with higher HDS scores and better lifestyle habits. Furthermore, as the changed HDS was higher in the group whose mealtimes became regular during the pandemic, adopting regular mealtimes may lead to a more balanced diet.

Keywords: COVID-19; dietary balance; Japan; mealtime regularity; preschool children

1. Introduction

Since the novel coronavirus-19 (COVID-19) pandemic began in 2020, it has had a considerable impact on children's lives. In Japan, the first state of emergency was declared on 7 April 2020, which was accompanied by long-term closures of kindergartens, elementary schools, middle schools, and high schools. Many schools resumed in June but with special measures in place, such as distributed attendance and shortened classes. Since then, children have experienced a prolonged period of stress from changes, such as shorter summer vacations, fewer or cancelled events, and restricted club activities. As of June 2022, children still have to follow this "new normal" in various aspects of life.

It has been reported that worldwide, how and what children eat has changed during the pandemic. In a previous study in Italy among children aged 6−18 years with obesity, there was a significant increase in snack and soft drink consumption following the implementation of the COVID-19 lockdown [1]. In a survey of families with one or more children aged 18 months to 5 years in Canada, many respondents indicated that the COVID-19 pandemic had increased their consumption of snack foods [2]. Additionally, a study conducted by Androutsos et al. among children aged 2−18 years in Greece revealed that during the COVID-19 lockdowns, weight gain was associated with higher intakes of pasta, sweets, and snacks, as well as reduced physical activity [3]. In a survey of students in elementary school and older children in Japan, many indicated that both the opportunities for and the quantity of snacks had increased compared with before COVID-19 [4]. This shows how the spread of the novel coronavirus has had a considerable impact on children's lifestyle habits and diets.

However, in other studies only some children were found to have deteriorated lifestyle and eating habits during the pandemic. In the study by Adroutsos et al., no changes were observed with respect to aspects such as sleep and the frequency of eating unhealthy foods, such as fast foods, sweets, and snacks, in a majority of the children surveyed [3]. Furthermore, in a study investigating the factors associated with healthier or unhealthier eating habits in adults during the pandemic, 71.6% of the participants did not change their eating habits [5]. Among the factors positively associated with healthier dietary habits were annual household income, changes in household income, COVID-19 morbidity among friends, health literacy, and exercise frequency [5]. A change in caregivers owing to COVID-19 is thought to affect the eating habits of children living in the same household.

It is easier for children to maintain set wakeup times and mealtimes when the time at which they leave for and arrive home from school are constant. Therefore, the closure of educational institutions and child welfare facilities made it difficult to maintain set wakeup and bed times, which may have further affected children's eating habits. Following a survey of students in elementary schools and older children in Japan, it was reported that compared with before COVID-19, 30% of students in the fourth grade and above in elementary school had later or disturbed wakeup/bed times, and although indicated for less than 10% of the children, there was also an increase in irregular mealtimes and skipped meals [5]. In the 2015 National Nutrition Survey on Preschool Children in Japan, there a higher percentage of children indicated that they always eat breakfast among those with earlier wake-up and bed times [6]. It was further reported that later and more irregular meal and sleep times were associated with poorer physical and mental states in infants [7]. Previous studies among adults reported that mealtimes were associated with the diurnal rhythm of cardiac autonomic activity [8,9], highlighting the importance of maintaining regular mealtimes from early childhood. However, to the best of our knowledge, there have been few reports on whether regular mealtimes are associated with lifestyle habits and dietary balance in preschool children. In particular, it is unclear whether the dietary habits of children who maintained their mealtime regularity during the pandemic are more balanced than those who did not.

Therefore, with this study, we aimed to determine whether the pandemic affected mealtime regularity among preschool children and to compare children who maintained regular mealtimes during the pandemic with those who did not in order to determine the relationship with their lifestyle habits, including wakeup and bed times, as well as dietary balance. Furthermore, we aimed to investigate whether pandemic-related changes in mealtime regularity were related to changes in dietary balance, including chronological relationships.

2. Materials and Methods

2.1. Study Participants and Design

This online cross-sectional survey involving individuals registered with Cross Marketing Group (C company, Shinjuku, Tokyo), a company that provides meals to children aged 2–6 years, was conducted during the period of 24–25 February 2021. First, a screening survey was performed on all individuals registered with C company. Those who indicated that they had children aged 2–6 years and were the meal provider for the oldest child in this age range were eligible to participate in the study. Thereafter, the participants were sorted by region at ratios close to the distribution of households with children aged <6 years in the 2015 national census, and 2000 respondents were recruited until those figures were reached. All child-related questions pertained to the oldest child among those aged 2–6 years.

The questions included the respondent's age, residential prefecture, employment status, child's age, child's sex, birth order, primary daytime daycare location, and health status. The participants were also asked about current eating and lifestyle habits and whether these had changed due to the COVID-19 pandemic. These questions were designed from a maternal and child health perspective and were based on those used in the 2015 National Nutrition Survey on Preschool Children, Japan (Ministry of Health, Labor, and Welfare of Japan) [6].

2.2. Calculating the Healthy Diet Score (HDS)

A healthy diet score (HDS) was developed to comprehensively evaluate children's dietary balance. One goal of Japan's Fourth Basic Plan for the Promotion of Shokuiku is to increase the frequency of meals that include a combination of staple, main, and side dishes [10]. The Japanese Food Guide Spinning Top, which was created by the Ministry of Agriculture, Forestry, and Fisheries and the Ministry of Health, Labor, and Welfare of Japan, uses five categories to determine diet quality as follows: grain dishes, vegetable dishes, fish and meat dishes, fruit, and milk (dairy products) [11]. These five categories were used when categorizing meals in this study. In addition, unhealthy foods that should be avoided, such as fast foods and sweets, were evaluated as reverse-scored questions. Specifically, of the 13 food groups included in the food frequency survey, grains; fish; meat; eggs; soybeans/soy products; vegetables; fruit; milk/dairy products; and unsweetened beverages, such as tea were categorized as healthy, whereas sweetened beverages, sweets, instant noodles/cup ramen, and fast foods were categorized as unhealthy. The intake frequency of foods from the healthy food groups was scored as follows (Figure 1). Grains and vegetables: ≥2 times per day = 4 points, 1 time per day = 3 points, 4–6 times per week = 2 points, ≤3 times per week = 1 point, and <1 time per week = 0 points; and fish, meat, eggs, soybeans/soy products, fruit, milk/dairy products, and unsweetened beverages, such as tea: ≥1 time per day = 4 points, 4–6 times per week = 3 points, 1–3 times per week = 2 points, and <1 time per week = 1 point. The intake frequency of unhealthy foods, such as sweetened beverages, sweets, instant noodles/cup ramen, and fast foods, was reverse-scored as follows: ≥2 times per day = 0 points, 1 time per day = 1 point, 4–6 times per week = 2 points, 1–3 times per week = 3 points, and <1 time per week = 4 points. A mean value was calculated using the scores of all the main dishes (fish, meat, eggs, and soybeans/soy products). The scores for grain dishes, main dishes (mean); side dishes; fruit; dairy products; unsweetened beverages, such as tea; and unhealthy foods (reverse-scored) was totaled to obtain the HDS (0–40 points). The following secondary indicators were calculated based on intakes of: (1) grain dishes, main dishes (mean), and side dishes (0–12 points); (2) grain dishes, main dishes (mean), side dishes, fruit, and dairy products (0–20 points); (3) grain dishes, main dishes (using the total instead of the mean to consider the variety of main dishes), and side dishes (0–32 points); and (4) infrequently consuming unhealthy foods (0–16 points).

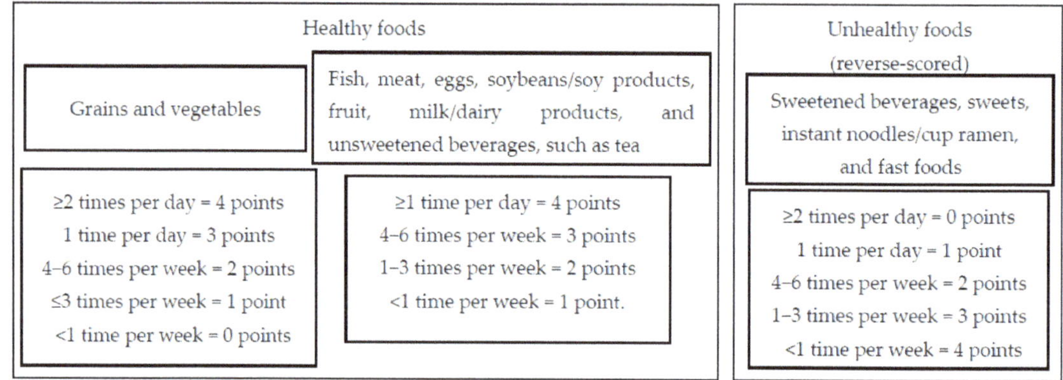

The scores for grain dishes, main dishes (a mean value of all the main dishes [fish, meat, eggs, soybeans/soy products]), side dishes (vegetables), fruit, milk/dairy products, unsweetened beverages such as tea, and unhealthy foods (reverse-scored) was totaled to obtain the HDS (0–40 points).

Figure 1. Calculating the healthy diet score (HDS).

To evaluate the validity of the HDS indicator developed for this study, the intake frequency of each food group was determined using the food diversity score (FDS) developed by Ishikawa et al., which is calculated using the same questions [12]. FDS is an 8-point scale that evaluates dietary diversity (grains, fish, meat, eggs, soybeans/soy products, vegetables, fruit, and milk/dairy products) based on the Food and Agriculture Organization's dietary assessment [13]. The correlation coefficient between the FDS and HDS was 0.574 ($p < 0.001$), indicating a significant positive correlation. A comparison of the FDS among the four groups based on mealtime regularity yielded similar results to comparisons using the HDS (highest score was in the group that originally had regular mealtimes that did not change). However, the differences between the four groups were larger using the HDS than the FDS. This indicates that the HDS is a useful indicator that considers the diversity of the food groups recommended in the Japanese diet (grain, main, and side dishes), as well as the infrequent consumption of unhealthy foods, such as sweets and instant noodles.

2.3. Changes in the HDS during COVID-19: Changed-Healthy Diet Score (C-HDS)

Here, we aimed to determine whether the quality of meals improved or deteriorated as mealtimes became more regular or irregular, respectively. The changed-healthy diet score (C-HDS) was calculated based on changes in the intake frequency of each food group compared with before COVID-19. The food categories used were the same as those for the HDS. Healthy foods included grains (rice or bread, for example); fish; meat; eggs; soybeans/soy products; vegetables; fruit; milk/dairy products; and unsweetened beverages, such as tea. Unhealthy foods included sweetened beverages, sweets, instant noodles/cup ramen, and fast foods. The change in intake frequency for a food group before and after COVID-19 was scored as follows: for increased intakes of a specific food group, healthy foods = +1 point, unhealthy foods = −1 point, and no change = 0 points (increased score: −4 to 9 points); and for decreased intakes of a specific food group, healthy foods = −1 point, unhealthy foods = +1 point, and no change = 0 points (decreased score: −9 to 4 points). These scores were totaled to determine the C-HDS. However, as 69.9% of the participants had a score of 0 points (no change), the scores were grouped into three categories for the regression analysis as follows: increased, no change, and decreased.

2.4. Ethical Considerations

After viewing the guidelines and principles of the Japan Marketing Research Association as summarized by company C and an explanation of the survey, informed consent to participate in the online survey was obtained. The survey explanation informed participants that the collected data would undergo statistical processing and be published in a manner whereby the individuals could not be identified and that the data would not be used for purposes other than the survey. Ethical approval was obtained from the Human Research Ethics Committee of Kagawa Nutrition University (approval number 317).

In addition, the survey data were used in a secondary analysis as part of an administrative promotion project under a Japanese Ministry of Health, Labor, and Welfare grant, "Research for Effective Development of Nutrition and Dietary Support for Healthy Development in Early Childhood" (grant number: 20DA2002).

2.5. Analysis

2.5.1. Data Set for Analysis

To prevent impersonation and fraudulent responses, 1 participant who did not answer almost any of the questions and 17 participants who did not provide answers about their relationship to the child, the child's sex, or birth order were excluded from the analysis. Furthermore, 2 participants who were grandparents, 29 participants who did not provide answers for questions addressing changes in meal regularity, and 101 participants who indicated that they did not want to provide answers about their dietary intake frequency at present or before/after COVID-19 were excluded from the analysis; a total of 1850 participants were included in the final analysis (Figure 2).

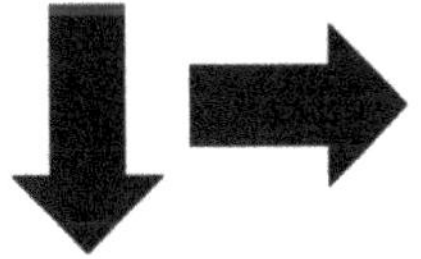

Figure 2. Participant flowchart.

Inconsistent answers for individual items were excluded from the final analysis. For example, answers that indicated that the respondents' children are now "rarely" involved in cooking but "considerably more" than before the COVID-19 ($n = 2$) were considered to be inconsistent answers. Regarding a child's height and weight, values above the 99th percentile or below the 1st percentile and calculated using the least squares mean method were considered input errors and excluded from the calculation of obesity (height, $n = 71$ participants; weight, $n = 33$ participants).

2.5.2. Statistical Analysis

The participants were divided into four groups based on their responses to questions addressing mealtime regularity after COVID-19 as follows: became regular (n = 125), originally regular and unchanged (n = 1514), became irregular (n = 63), and originally irregular and unchanged (n = 148). The distribution of independent variables was examined using the χ^2 test (residual analysis when a significant difference was noted) or one-way analysis of variance (multiple comparisons using the Games−Howell method when a significant difference was noted).

To determine the relationship between the current HDS and mealtime regularity, simple regression analyses were performed with the HDS as the dependent variable and the respondents' attributes, child's attributes, lifestyle habits, and household circumstances as independent variables. In model 1, multiple regression analysis was performed with the HDS as the dependent variable for each classification of the independent variables. In model 2, multiple regression analysis was performed with forced entry method of the items that exhibited significant associations with the HDS in model 1, the relationship of the respondent, mother's employment situation, child's sex, child's age, BMI percentile, and mealtime regularity. Mealtime regularity was converted into a binary dummy variable (1, originally regular; 0 otherwise). To account for the influence of missing values (BMI percentile (n = 155), others), the multiple regression analysis used for model 2 was performed with multiple imputation of missing values.

To examine the relationship between changes in the HDS (C-HDS) and in mealtime regularity after COVID-19, multiple logistic regression analysis was performed with the C-HDS as the dependent variable and the respondent's attributes, child's attributes, pandemic-related changes in the child's lifestyle and diet, and household circumstances as independent variables. A C-HDS score of ≤ 0 points was treated as a decrease, and a C-HDS score of ≥ 1 point was considered an increase. A C-HDS score of 0 points (no change) was used as the reference value in the multiple logistic regression analysis. A multiple logistic regression analysis was performed with forced entry of the following independent variables: changes in lifestyle habits and household circumstances that were significantly associated with the current HDS and mealtime regularity, respondent's relationship, mother's employment status, child's sex, child's age, and BMI percentile. Mealtime regularity was converted into a 3-value variable (1, became irregular; 2, became regular; 3, no change). Multiple logistic regression analysis was performed by multiple imputation of missing values. The Statistical Package for the Social Sciences, version 28 (IBM SPSS Inc., Chicago, IL, USA), was used for all statistical analyses, and statistical significance was set at $p < 0.05$.

3. Results

3.1. Comparison of Participant Characteristics according to Changes in Mealtime Regularity Compared with before COVID-19

Table 1 shows the participant characteristics according to changes in mealtime regularity compared with before COVID-19. In 41.6% of participants, the relationship to the child was through the father and in 58.4%, through the mother.

Table 2 shows the lifestyle habits and household circumstances according to the change in mealtime regularity compared with before COVID-19. There were significantly more participants in the originally regular group who woke up before 8 a.m. and who went to bed before 10 p.m. on weekdays and holidays. Screen time on weekdays was <2 h in 79.3% of participants, and this percentage was significantly higher in the originally regular group. Screen time on holidays was ≥ 2 h in 35.7% of participants, and this percentage was significantly higher in the originally irregular group. Significantly fewer participants in the 'became irregular' group defected almost every day. Regarding dietary status, significantly more participants in the originally regular group indicated that they snacked 0−1 time per day and ate breakfast every day.

Table 1. Participant characteristics according to changes in mealtime regularity compared with before COVID-19.

		Became Regular (*n* = 125)		Originally Regular, No Change (*n* = 1514)		Became Irregular (*n* = 63)		Originally Irregular, No Change (*n* = 148)	
Respondent's relationship to the child	Father	56	(44.8)	618	(40.8)	22	(34.9)	74	(50.0)
	Mother	69	(55.2)	896	(59.2)	41	(65.1)	74	(50.0)
Employment status of the child's mother	Employed	78	(62.4)	827	(54.6)	38	(60.3)	74	(50.0)
	Other	47	(37.6)	687	(45.4)	25	(39.7)	74	(50.0)
Child's sex	Boy	69	(55.2)	752	(49.7)	30	(47.6)	68	(45.9)
	Girl	56	(44.8)	762	(50.3)	33	(52.4)	80	(54.1)
Child's age (years)		3.2	±1.4	3.4	±1.4	3.2	±1.4	2.9	±1.4
Height (cm)		105.6	±12.1	105.9	±11.4	105.6	±12.1	102.9	±11.5
Weight (kg)		17.3	±4.6	17.3	±3.9	17.2	±4.1	16.5	±3.7
BMI percentile		46.7	±36.0	44.9	±34.2	39.7	±32.0	49.9	±33.2

Categorical variables are expressed as number of people (percentage), and continuous variables are expressed as mean ± standard deviation. BMI, body mass index.

Table 2. Lifestyle habits and household circumstances according to changes in mealtime regularity compared with before COVID-19.

		Became Regular (*n* = 125)		Originally Regular, No Change (*n* = 1514)		Became Irregular (*n* = 63)		Originally Irregular, No Change (*n* = 148)		*p* Value *
Wakeup time	Before 8 a.m. on weekdays and holidays	83	(66.4)	**1066**	**(70.4)**	<u>32</u>	(50.8)	<u>54</u>	(36.5)	<0.001
	Other	42	(33.6)	<u>448</u>	<u>(29.6)</u>	**31**	**(49.2)**	**94**	**(63.5)**	
Bed time	Before 10 p.m. on weekdays and holidays	91	(72.8)	**1156**	**(76.4)**	<u>34</u>	(54.0)	<u>53</u>	(35.8)	<0.001
	Other	34	(27.2)	<u>358</u>	(23.6)	**29**	**(46.0)**	**95**	**(64.2)**	
Physical activity time (weekdays)	<1 h (/day)	68	(54.4)	<u>708</u>	(47.0)	35	(55.6)	74	(51.0)	0.200
	≥1 h (/day)	57	(45.6)	**799**	**(53.0)**	28	(44.4)	71	(49.0)	
Physical activity time (holidays)	<1 h (/day)	69	(55.2)	743	(49.3)	35	(55.6)	74	(50.7)	0.485
	≥1 h (/day)	56	(44.8)	764	(50.7)	28	(44.4)	72	(49.3)	
Screen time (weekdays)	<2 h (/day)	97	(79.5)	**1219**	**(80.8)**	44	(71.0)	<u>98</u>	(66.7)	<0.001
	≥2 h (/day)	25	(20.5)	<u>289</u>	<u>(19.2)</u>	18	(29.0)	**49**	**(33.3)**	
Screen time (holidays)	<2 h (/day)	85	(69.7)	983	(65.2)	37	(59.7)	<u>76</u>	<u>(52.4)</u>	0.009
	≥2 h (/day)	37	(30.3)	525	(34.8)	25	(40.3)	**69**	**(47.6)**	
Frequency of defecation	Almost every day	94	(76.4)	1072	(70.9)	<u>34</u>	(54.0)	105	(71.4)	0.014
	Other	29	(23.6)	439	(29.1)	**29**	**(46.0)**	42	(28.6)	
Snack frequency	0−1 time (/day)	79	(64.8)	**1056**	**(70.1)**	<u>31</u>	(50.0)	<u>84</u>	(57.9)	<0.001
	2 times (/day)	37	(30.3)	390	(25.9)	**25**	**(40.3)**	40	(27.6)	
	≥3 times (/day)	6	(4.9)	<u>60</u>	<u>(4.0)</u>	6	(9.7)	**21**	**(14.5)**	
Breakfast frequency	Eat every day	112	(89.6)	**1442**	**(95.2)**	<u>47</u>	(74.6)	<u>116</u>	(78.4)	<0.001
	Other	13	(10.4)	<u>72</u>	<u>(4.8)</u>	**16**	**(25.4)**	**32**	**(21.6)**	
Eat together (breakfast)	Eat with an adult	106	(86.2)	**1249**	**(82.7)**	51	(81.0)	<u>97</u>	(67.4)	<0.001
	Other	17	(13.8)	<u>262</u>	<u>(17.3)</u>	12	(19.0)	**47**	**(32.6)**	
Eat together (dinner)	Eat with an adult	122	(97.6)	1457	(96.4)	61	(96.8)	<u>135</u>	(91.8)	0.041
	Other	3	(2.4)	55	(3.6)	2	(3.2)	**12**	**(8.2)**	
How often the respondent cooks	1−2 times per week	36	(28.8)	469	(31.2)	20	(31.7)	55	(37.9)	0.474
	3−6 times per week	27	(21.6)	330	(22.0)	17	(27.0)	34	(23.4)	
	Every day	62	(49.6)	702	(46.8)	26	(41.3)	56	(38.6)	

Table 2. *Cont.*

		Became Regular (*n* = 125)		Originally Regular, No Change (*n* = 1514)		Became Irregular (*n* = 63)		Originally Irregular, No Change (*n* = 148)		*p* Value *
Frequency with which the child is involved in cooking	Rarely	41	(32.8)	672	(44.4)	17	(27.4)	82	(55.4)	<0.001
	≥1 time per week	**84**	**(67.2)**	840	(55.6)	**45**	**(72.6)**	66	(44.6)	
Someone in the family works remotely		**51**	**(40.8)**	390	(25.8)	**25**	**(39.7)**	28	(18.9)	<0.001
Financial security	Yes	54	(43.5)	567	(37.5)	24	(38.1)	38	(25.7)	0.003
	Neither yes nor no	42	(33.9)	459	(30.4)	19	(30.2)	42	(28.4)	
	No	28	(22.6)	484	(32.1)	20	(31.7)	**68**	**(45.9)**	
Has free time	Yes	40	(32.3)	452	(29.9)	16	(25.4)	29	(19.6)	0.039
	Neither yes nor no	44	(35.5)	479	(31.7)	15	(23.8)	52	(35.1)	
	No	40	(32.3)	580	(38.4)	32	(50.8)	67	(45.3)	

The continuous variables presented in the table are expressed as mean ± standard deviation. * Chi-squared test (residual analysis performed when significant differences were noted; underlined numbers indicate adjusted standardized residual <−1.96; bold numbers indicate adjusted standardized residual >1.96).

Overall, 81.6% and 96.1% of the respondents indicated that their child ate breakfast and dinner with an adult, respectively. Significantly more children ate breakfast with an adult in the originally regular group, and significantly fewer ate dinner with an adult in the originally irregular group. Regarding the frequency of cooking, there were no significant differences between the groups. Significantly more participants in the originally irregular and unchanged groups indicated that they lacked financial security, and significantly more participants in the 'became irregular' group indicated that they had little free time.

3.2. Comparison of the Healthy Diet Score (HDS) according to Changes in Mealtime Regularity Compared with before COVID-19

Table 3 shows the HDS according to changes in mealtime regularity. The current HDS (0−40 points) of the originally regular and unchanged group was 31.6 ± 4.0 points, which was significantly higher than in the other groups. Intakes of grain dishes, main dishes (mean), and side dishes (0−12 points); grain dishes, main dishes (mean), side dishes, fruit, and dairy products (0−20 points); and grain dishes, main dishes (total), side dishes, fruit, and dairy products (0−32 points) were significantly higher in the 'became regular' group and originally regular and unchanged groups compared with the originally irregular and unchanged group. Significantly more participants infrequently consumed unhealthy food (0−16 points) in the originally regular group than in the 'became regular', originally irregular, and unchanged groups.

Regarding changes in food group intakes before and after COVID-19, the increased score (−4 to 9 points) was significantly higher in the 'became regular' group than in the originally regular and unchanged group, as well as the originally irregular and unchanged group. The decreased score (−9 to 4 points) was significantly lower in the 'became irregular' group than in the originally regular and unchanged group, as well as the originally irregular and unchanged group. The mean C-HDS (−13 to +13) in the 'became regular' group was 0.8 ± 2.3 points, which was significantly higher than in the other groups.

3.3. Multiple Regression Analyses with the Healthy Diet Score (HDS) as the Dependent Variable

Table 4 shows the results of the multiple regression analyses of the factors associated with the HDS. The HDS was significantly higher in the originally regular group and, even when adjusted for the other factors, was significantly associated with the investigated basic attributes of wakeup time (weekdays and holidays), bed time (weekdays and holidays), physical activity time (holidays), defecation frequency, snack frequency, frequency of breakfast for the child, eating together (breakfast and dinner), cooking frequency, and financial resources ($\beta = 0.131$, $p < 0.001$).

Table 3. Comparison of healthy diet score (HDS) according to changes in mealtime regularity compared with before COVID-19.

	Became Regular (n = 125)			Originally Regular, No Change (n = 1516)			Became Irregular (n = 63)			Originally Irregular, No Change (n = 148)			p Value
Current intake of the different food groups													
Healthy Diet Score * (0−40 points)	29.7	±4.7	a	31.6	±4.0	abc	29.6	±4.8	b	28.3	±4.6	c	<0.001
Grain dishes, main dishes (mean), and side dishes (0−12 points)	9.1	±2.4	a	9.3	±1.8	b	8.6	±2.5		8.2	±2.2	ab	<0.001
Grain dishes, main dishes (mean), side dishes, fruit, and dairy products (0−20 points)	15.3	±3.8	a	15.7	±3.0	b	14.8	±3.9		13.9	±3.3	ab	<0.001
Grain dishes, main dishes (total), side dishes, fruit, and dairy products ** (0−32 points)	23.8	±6.0	a	23.8	±4.6	b	22.8	±5.7		21.1	±5.1	ab	<0.001
Infrequently consuming unhealthy foods (0−16 points) [†]	11.0	±3.8	a	12.2	±2.7	ab	11.4	±3.3		11.0	±2.9	b	<0.001
Changes compared with before the pandemic [#]													
Changed Healthy Diet Score [#] (−13~+13)	0.8	±2.3	abc	0.1	±1.3	a	−0.1	±1.6	b	0.0	±1.0	c	<0.001
Increased score (−4~9 points)	0.9	±2.0	ab	0.1	±1.1	a	0.3	±1.5		0.0	±0.9	b	<0.001
Decreased score (−9~4 points)	−0.1	±1.0		0.0	±0.5	a	−0.4	±0.9	ab	0.0	±0.5	b	<0.001

The continuous variables presented in the table are expressed as mean ± standard deviation. One-way analysis of variance (multiple comparisons using the Games–Howell method; [a-c], performed for significant differences). HDS, healthy diet score; C-HDS, changed healthy diet score. * The intake frequency of each food group was determined and used to calculate the total intake frequency of grain dishes; main dishes (mean); side dishes; fruit; dairy products; unsweetened foods, such as tea; and unhealthy foods (reverse-scored question) as follows: grains and vegetables: ≥2 times per day = 4 points, 1 time per day = 3 points, 4−6 times per week = 2 points, ≤3 times per week = 1 point, <1 time per week = 0 points; fish, meat, eggs, soybeans/soy products, fruit, milk/dairy products, and unsweetened beverages such as tea: ≥1 time per day = 4 points, 4−6 times per week = 3 points, 1−3 times per week = 2 points, <1 time per week = 1 point. The intake frequency of unhealthy foods, such as sweetened beverages, sweets, instant noodles/cup ramen, and fast foods, was reverse-scored as follows: ≥2 times per day = 0 points, 1 time per day = 1 point, 4−6 times per week = 2 points, 1−3 times per week = 3 points, and <1 time per week = 4 points. ** Indicator considering the diversity of main dishes by using the total of the main dishes (fish, meat, eggs, and soybeans/soy products). [†] Healthy foods were defined as grains; fish; meat; eggs; soybeans/soy products; vegetables; fruit; milk/dairy products; and unsweetened beverages, such as tea. Unhealthy foods were defined as sweetened beverages, sweets, instant noodles/cup ramen, and fast foods. [#] The change in intake frequency for a food group before and after COVID-19 was scored as follows: for increased intakes of a specific food group, healthy foods = +1 point, unhealthy foods = −1 point, and no change = 0 points (increased score −4 to 9 points); and for decreased intakes of a specific food group, healthy foods = −1 point, unhealthy foods = +1 point, and no change = 0 points (decreased score: −9 to 4 points). These were totaled to determine the C-HDS.

3.4. Multiple Logistic Regression Analysis with the Change in Healthy Diet Score (C-HDS) as the Dependent Variable

Table 5 shows the results of the multiple logistic regression analysis using the C-HDS as the dependent variable. A significant increase in C-HDS was noted in the 'became regular' group (odds ratio: 2.21, 95% confidence interval: 1.35−3.61). A decreasing trend was observed in the C-HDS in the 'became irregular' group.

Table 4. Multiple regression analyses with the healthy diet score (HDS) as the dependent variable.

Item Category		Simple Regression		Model 1		Model 2		
		Standardized Coefficient β	p Value	Standardized Coefficient β	p Value	Standardized Coefficient	p Value	
Basic attributes	Relationship to child	0.163	<0.001	0.156	<0.001	0.110	<0.001	#
	Employment status of child's mother	−0.061	0.009	−0.038	0.119	−0.006	0.788	
	Child's sex	0.018	0.448	0.026	0.280	0.040	0.070	
	Child's age	−0.043	0.065	−0.019	0.425	−0.075	0.001	#
	BMI percentile	−0.051	0.037	−0.036	0.137	−0.016	0.462	
Lifestyle habits	Wakeup time (weekdays and holidays)	−0.143	<0.001	−0.052	0.028	−0.054	0.025	#
	Bed time (weekdays and holidays)	−0.174	<0.001	−0.079	0.001	−0.093	<0.001	#
	Physical activity time (weekdays)	0.079	0.008	0.033	0.196			
	Physical activity time (holidays)	0.079	0.001	0.045	0.075	0.077	<0.001	#
	Screen time (weekdays)	−0.101	<0.001	−0.009	0.757			
	Screen time (holidays)	−0.089	0.001	−0.027	0.326			
	Frequency of defecation	0.136	<0.001	0.101	<0.001	0.107	<0.001	#
	Snack frequency	−0.291	<0.001	−0.233	<0.001	0.236	<0.001	#
	Breakfast frequency	0.249	<0.001	0.141	<0.001	0.137	<0.001	#
	Originally regular meal times	0.229	<0.001	0.139	<0.001	0.131	<0.001	#

Table 4. *Cont.*

Item Category		Simple Regression		Model 1		Model 2	
		Standardized Coefficient β	*p* Value	Standardized Coefficient β	*p* Value	Standardized Coefficient	*p* Value
Household circumstances	Eat together (breakfast)	0.070	0.003	0.059	0.013	0.022	0.326
	Eat together (dinner)	0.062	0.008	0.051	0.031	0.034	0.134
	How often the respondent cooks						
	Every day	0.192	<0.001	0.187	<0.001	0.092	0.007 [#]
	3–6 times a week	−0.099	<0.001	−0.007	0.795	−0.013	0.650
	Rarely	-			-		-
	Frequency with which the child is involved in cooking	0.034	0.149	0.017	0.468		
	Someone in the family works remotely	−0.024	0.305	−0.022	0.336		
	Financial security						
	Yes	0.067	0.004	0.028	0.319	0.021	0.423
	No	−0.062	0.008	−0.061	0.034	−0.021	0.419
	Neither yes nor no	-			-		-
	Has free time						
	Yes	0.033	0.151	0.035	0.202		
	No	0.011	0.647	0.052	0.064		
	Neither yes nor no	-			-		-

Model 1, multiple regression analysis with forced entry of all items for each category. Model 2, multiple regression analysis with forced entry of the basic attributes and items that exhibited significant associations with the HDS in model 1. [#] $p < 0.05$ in the multiple regression analysis using data after multiple imputation of missing values. BMI, body mass index. Input variables: respondent's relationship to the child (0, father; 1, mother); mother's employment status (1, working; 0, other); child's sex (0, boy; 1, girl); child's age (continuous variable); BMI percentile (0, <75th percentile; 1, ≥75th percentile); wakeup time (0, weekdays and holidays before 8 a.m.; 1, otherwise); bed time (0, weekdays and holidays before 10 p.m.; 1, otherwise); physical activity time (weekdays and holidays) (0, <1 h; 1, ≥1 h); screen time (weekdays and holidays) (0, <2 h; 1, ≥2 h); defecation frequency (1, daily; 0, other); snack frequency (1, 0–1 time/day; 2, 2 times/day; 3, ≥3 times/day); frequency of breakfast (1, daily; 0, other); eating together (1, eat with an adult; 0, other); how often the respondent cooks ("rarely" input as reference value); frequency with which the child is involved in cooking (0, rarely; 1, ≥1 day/week); someone in household works remotely (1, yes; 0, no); and financial security/has free time ("not a lot" input as reference value).

Table 5. Multiple logistic regression analysis with the change in healthy diet score (C-HDS) as the dependent variable.

		n (%)		Increase [#]			*p* Value	Decrease [#]			*p* Value
				OR (95% CI)				OR (95% CI)			
Meal regularity	**Became regular**	125	(6.7)	2.21	(1.35	— 3.61)	0.002	1.50	(0.80	— 2.80)	0.205
	Became irregular	63	(3.4)	1.23	(0.60	— 2.53)	0.564	1.85	(0.92	— 3.72)	0.083
	No change	1664	(89.8)		1.00				1.00		

Multiple logistic regression analysis using data after multiple imputation of missing values ($n = 1825$). [#] Dependent variable reference: no change (C-HDS = 0 points). BMI, body mass index; CI, confidence interval; OR, odds ratio. Input variables: respondent's relationship to the child (0, father; 1, mother); mother's employment status (1, working; 0, other); child's sex (1, boy; 2, girl); child's age (continuous variable); BMI percentile (0, <75th percentile; 1, ≥75th percentile); change in wakeup and bed time (became regular, became irregular, or unchanged); change in the frequency and time of physical activity (increased, decreased, or unchanged); snack frequency (increased, decreased, or unchanged); breakfast frequency (increased, decreased, or unchanged); and how often the respondent cooks (increased, decreased, or unchanged).

4. Discussion

In addition to investigating how children's mealtime regularity relates to their current lifestyle habits and dietary status, we also investigated whether pandemic-related changes in mealtime regularity were associated with changes in dietary balance, including chronological relationships. Here, the results showed that maintaining regular mealtimes was associated with practices such as waking and going to bed earlier, less snacking, and eating breakfast every day. Furthermore, even after adjusting for basic attributes, lifestyle habits, household circumstances, and other factors, regular mealtimes were still significantly positively correlated with the HDS, whereas a change in regular mealtimes was significantly associated with an increased C-HDS.

Maintaining regular mealtimes has previously been shown to be associated with positive lifestyle habits. A cluster analysis using meal and sleep time data in preschool children after forming clusters, such as a group in which both meal and bed times were extremely late, a group with late and irregular times, and a group that woke early and regularly, showed that meal and sleep times tended to be linked [7]. In a previous study, missing breakfast was associated with negative lifestyle habits in primary school children, such as increased intake of soft drinks, high screen time, and low physical activity levels [14]. The "Health guide starting with diet—Raising children who enjoy eating" of the Japanese Ministry of Health, Labor, and Welfare describes the importance of "feeling hungry and having an appetite, then experiencing the pleasure of properly satisfying this" as a target for healthy child development through food and states that it is ideal for children to have a dietary rhythm [15]. Here, our findings were consistent with previous research, in that more participants in the group that originally had regular mealtimes indicated that they eat breakfast every day and wake up and go to bed early. In addition, even after adjusting for lifestyle habits and other factors, originally having regular mealtimes was positively associated with the HDS.

Having regular mealtimes means avoiding missing meals and maintaining constant intervals between meals. Moreover, in a study of children aged 7−18 years, missing breakfast, lunch, or dinner was associated with low vegetable and fruit intakes [16]. High intake of junk food have also been reported to increase the risk of missing meals [17]. Furthermore, a study of children aged 12−16 years reported that an irregular diet and daily junk food consumption were both associated with reduced mental and physical health [18]. It is thought that keeping a set interval between meals generates proper feelings of hunger and appetite, which leads to a balanced diet. In addition, it has previously been reported that children who frequently played outside after school and who ate "supplementary foods" instead of "non-essential foods" for snacks consumed more vegetables at dinner [19]. Therefore, it was presumed that feeling hungry before meals is associated with the quantity of vegetables consumed at meals. An irregular diet in preschool children has also been significantly associated with risk factors for lifestyle-related diseases, such as overweight and obesity [20]. Here, more regular mealtimes were associated with higher C-HDS scores, indicating that mealtime regularity is also chronologically associated with dietary balance. Although the results of this study could not demonstrate causal relationships between meal regularity and dietary balance, they do highlight the importance of regular mealtimes.

In this study, we developed the HDS as an indicator for evaluation of preschool children's dietary balance. Other indicators of dietary balance include the Health Eating Index (HEI) (2015), which uses the U.S. dietary reference intakes [21] and is calculated as a total score of nine recommended food groups and four items for which intake should be restricted. The HEI has been shown to be associated with child socioeconomic indicators [22] and with the frequency of fast-food consumption and eating dinner together as a family among preschool children [23]. Another such indicator is the diet diversity score (DDS), which evaluates the diversity of meals using various methods. Most use the 24 h memory method, meal records, or food frequency surveys to calculate the intakes of different food groups. The DDS has been reported to be associated with child anthropometric indices [24]. Qualitative dietary indicators evaluate how frequently the different food groups are consumed. Dietary diversity indicators have been reported to be related to BMI in preschool children [25] and parents who consider the content of their children's meals and snacks and aim to have them eat regular meals [12]. The 2015 National Nutrition Survey on Preschool Children in Japan included simple questions about the frequency of consuming foods from different food groups. Considering the burden these surveys place on respondents, using a similar simple indicator of dietary quality is preferable.

Japan's Fourth Basic Plan for the Promotion of Shokuiku set a target for increasing the proportion of the population that eats meals combining staple, main, and side dishes at least twice per day almost daily to more than 50.0%. In addition, the Japanese Food Guide Spinning Top [11] recommends evaluating dietary quality by considering five categories (grain dishes, main dishes, side dishes, fruit, and dairy products). The Mediterranean Diet Quality Index for Children and Adolescents (KIDMED) allocates +1 point for 12 recommended items, such as fruit and vegetables, and −1 point for four items that should be avoided, such as fast foods and sweets. Indicators such as the KIDMED, which also considers unhealthy foods, can also be useful [26]. The usefulness of the HDS and its relationship with other indicators requires further research.

The present study is significant because few studies have examined whether mealtime regularity is related to lifestyle habits and dietary balance in a large group of preschool children, including changes before and after COVID-19. However, this study is subject to some limitations. First, changes in mealtime regularity were categorized into four groups based on responses to a questionnaire. However, a definition of regular was not provided and was left open to interpretation by the respondents. Second, the HDS, which was developed for this study, was calculated according to the intake frequency of 13 food groups, and therefore may be less valid than indicators calculated based on the quantities consumed. Third, because this was an online study of people registered with a survey company, it may have been biased toward people who regularly use the Internet. Fourth, the survey required respondents to recall the situation before the pandemic and to evaluate changes in lifestyle habits and mealtime regularity now compared with before COVID-19. Thus, there is a possibility of recall bias. Fifth, apart from the impact of COVID-19, during the period of a year in early childhood, the content of a child's diet may also change, owing to growth. Furthermore, a second state of emergency was in effect during February 2021 when this study was being conducted. However, this only applied to certain regions and likely did not cover all the areas where the participants lived. Therefore, owing to differences in the COVID-19 situation throughout the country, the participants may have experienced different restrictions in their daily lives. Furthermore, there may have been individual differences in terms of awareness of and actions taken for infection control.

5. Conclusions

Our investigation of how mealtime regularity relates to lifestyle and dietary habits in children indicates that those who originally had regular mealtimes had higher HDS scores and better lifestyle habits, such as waking and going to bed earlier, less snacking, and eating breakfast every day. Furthermore, as C-HDS scores were higher in the group whose mealtimes became regular during the pandemic, adopting regular mealtimes may lead to a more balanced diet. In order to improve children's dietary balance, it is important to provide nutrition education that not only focuses on the composition of meals but also on mealtime regularity.

Author Contributions: Conceptualization, Y.T., Y.U., K.S. and K.E., methodology, Y.T., Y.U., K.S., M.S., H.F., Y.A., M.H. and K.E.; formal analysis, Y.T., Y.U. and S.S.; writing—original draft preparation, Y.T.; writing—review and editing, Y.T., Y.U., K.S., S.S. and K.E.; supervision, K.E.; project administration, K.E.; funding acquisition, K.E. All authors have read and agreed to the published version of the manuscript.

Funding: This study was a secondary analysis of survey data from the "Research for Effective Development of a Food, Nutrition and Dietary Support Guide for Healthy Development in Early Childhood," which was funded by the Ministry of Health, Labor and Welfare of Japan's Administrative Promotion Research Project (grant number: 20DA2002).

Institutional Review Board Statement: The study was conducted in accordance with the Declaration of Helsinki and approved by The Human Research Ethics Committee of Kagawa Nutrition University (approval number 317; 20 January 2021).

Informed Consent Statement: Informed consent was obtained from all participants involved in the study.

Data Availability Statement: The data sets used and/or analyzed during the current study are available from the corresponding author on reasonable request.

Conflicts of Interest: The authors declare no conflict of interest. The funders had no role in the design of the study; in the collection, analyses, or interpretation of data; in the writing of the manuscript, or in the decision to publish the results.

References

1. Pietrobelli, A.; Pecoraro, L.; Ferruzzi, A.; Heo, M.; Faith, M.; Zoller, T.; Antoniazzi, F.; Piacentini, G.; Fearnbach, S.N.; Heymsfield, S.B. Effects of COVID-19 Lockdown on Lifestyle Behaviors in Children with Obesity Living in Verona, Italy: A Longitudinal Study. *Obesity* **2020**, *28*, 1382–1385. [CrossRef] [PubMed]
2. Carroll, N.; Sadowski, A.; Laila, A.; Hruska, V.; Nixon, M.; Ma, D.W.L.; Haines, J.; On Behalf of The Guelph Family Health Study. The Impact of COVID-19 on Health Behavior, Stress, Financial and Food Security among Middle to High Income Canadian Families with Young Children. *Nutrients* **2020**, *12*, 2352. [CrossRef] [PubMed]
3. Androutsos, O.; Perperidi, M.; Georgiou, C.; Chouliaras, G. Lifestyle Changes and Determinants of Children's and Adolescents' Body Weight Increase during the First COVID-19 Lockdown in Greece: The COV-EAT Study. *Nutrients* **2021**, *13*, 930. [CrossRef] [PubMed]
4. National Center for Child Health and Development. Corona × Children's Questionnaire 1st Survey Report. Available online: https://www.ncchd.go.jp/center/activity/covid19_kodomo/report/CxC1_finalrepo_20210306revised.pdf (accessed on 24 January 2022). (In Japanese)
5. Shimpo, M.; Akamatsu, R.; Kojima, Y.; Yokoyama, T.; Okuhara, T.; Chiba, T. Factors Associated with Dietary Change since the Outbreak of COVID-19 in Japan. *Nutrients* **2021**, *13*, 2039. [CrossRef]
6. Ministry of Health, Labour, and Welfare of Japan. National Nutrition Survey on Preschool Children. 2015. Available online: https://www.mhlw.go.jp/file/06-Seisakujouhou-11900000-Koyoukintoujidoukateikyoku/0000134460.pdf (accessed on 8 February 2022). (In Japanese)
7. Fukuda, K.; Hasegawa, T.; Kawahashi, I.; Imada, S. Preschool children's eating and sleeping habits: Late rising and brunch on weekends is related to several physical and mental symptoms. *Sleep Med.* **2019**, *61*, 73–81. [CrossRef]
8. Yoshizaki, T.; Tada, Y.; Hida, A.; Sunami, A.; Yokoyama, Y.; Togo, F.; Kawano, Y. Influence of dietary behavior on the circadian rhythm of the autonomic nervous system as assessed by heart rate variability. *Physiol. Behav.* **2013**, *118*, 122–128. [CrossRef]
9. Yoshizaki, T.; Tada, Y.; Hida, A.; Sunami, A.; Yokoyama, Y.; Yasuda, J.; Nakai, A.; Togo, F.; Kawano, Y. Effects of feeding schedule changes on the circadian phase of the cardiac autonomic nervous system and serum lipid levels. *Eur. J. Appl. Physiol.* **2013**, *113*, 2603–2611. [CrossRef]
10. Ministry of Agriculture, Forestry, and Fisheries. The Fourth Basic Plan for the Promotion of Shokuiku. 2021. Available online: https://www.maff.go.jp/j/syokuiku/attach/pdf/kannrennhou-30.pdf (accessed on 8 February 2022).
11. Yoshiike, N.; Hayashi, F.; Takemi, Y.; Mizoguchi, K.; Seino, F. A new food guide in Japan: The Japanese food guide Spinning Top. *Nutr. Rev.* **2007**, *65*, 149–154. [CrossRef]
12. Ishikawa, M.; Eto, K.; Haraikawa, M.; Yoshiike, N.; Yokoyama, T. Relationship between parents' dietary care and food diversity among preschool children in Japan. *Public Health Nutr.* **2021**, *25*, 1–12. [CrossRef]
13. Food and Agriculture Organization of the United Nations (FAO). Dietary Assessment: A Resource Guide to Method Selection and Application in Low Resource Settings. Rome. 2018. Available online: https://www.fao.org/3/i9940en/I9940EN.pdf (accessed on 8 February 2022).
14. Kesztyüs, D.; Traub, M.; Lauer, R.; Kesztyüs, T.; Steinacker, J.M. Skipping breakfast is detrimental for primary school children: Cross-sectional analysis of determinants for targeted prevention. *BMC Public Health* **2017**, *17*, 258. [CrossRef]
15. Ministry of Health, Labour, and Welfare of Japan. Report of the Study Group on the Healthy Development of Children through Diet (From the Perspective of So-Called "Shokuiku"). Available online: https://www.mhlw.go.jp/shingi/2004/02/dl/s0219-4a.pdf (accessed on 8 February 2022). (In Japanese)
16. Pourrostami, K.; Heshmat, R.; Hemati, Z.; Heidari-Beni, M.; Qorbani, M.; Motlagh, M.E.; Raeisi, A.; Shafiee, G.; Ziaodini, H.; Beshtar, S.; et al. Association of fruit and vegetable intake with meal skipping in children and adolescents: The CASPIAN-V study. *Eat. Weight Disord.* **2020**, *25*, 903–910. [CrossRef] [PubMed]
17. Kelishadi, R.; Mozafarian, N.; Qorbani, M.; Motlagh, M.E.; Safiri, S.; Ardalan, G.; Keikhah, M.; Rezaei, F.; Heshmat, R. Is snack consumption associated with meal skipping in children and adolescents? The CASPIAN-IV study. *Eat. Weight Disord.* **2017**, *22*, 321–328. [CrossRef] [PubMed]
18. Zahra, J.; Ford, T.; Jodrell, D. Cross-sectional survey of daily junk food consumption, irregular eating, mental and physical health and parenting style of British secondary school children. *Child Care Health Dev.* **2014**, *40*, 481–491. [CrossRef]
19. Tada, Y.; Furukawa, C.; Miura, R.; Matsunaga, H.; Furusho, T.; Goto, S.; Usui, J.; Hashimoto, M.; Hida, A.; Kawano, Y. Relationship between Evening Meal Vegetable Consumption, Snacks, and Outdoor Play among Early Elementary School Children. *J. Jpn. Soc. Shokuiku* **2017**, *11*, 13–23. (In Japanese) [CrossRef]
20. Kostecka, M. The influence of preschool children's diets on the risk of lifestyle diseases. A pilot study. *Rocz. Panstw. Zakl. Hig.* **2018**, *69*, 139–145. [PubMed]
21. Krebs-Smith, S.M.; Pannucci, T.E.; Subar, A.F.; Kirkpatrick, S.I.; Lerman, J.L.; Tooze, J.A.; Wilson, M.M.; Reedy, J. Update of the Healthy Eating Index: HEI-2015. *J. Acad. Nutr. Diet.* **2018**, *118*, 1591–1602. [CrossRef]
22. Jun, S.; Zeh, M.J.; Eicher-Miller, H.A.; Bailey, R.L. Children's Dietary Quality and Micronutrient Adequacy by Food Security in the Household and among Household Children. *Nutrients* **2019**, *11*, 965. [CrossRef]
23. Carroll, J.E.; Price, G.; Longacre, M.R.; Hendricks, K.M.; Langeloh, G.; Beach, P.; Dalton, M.A.; Emond, J.A. Associations between advertisement-supported media exposure and dietary quality among preschool-age children. *Appetite* **2021**, *166*, 105465. [CrossRef]

24. Heidari-Beni, M.; Riahi, R.; Massoudi, S.; Qorbani, M.; Kelishadi, R. Association between dietary diversity score and anthropometric indices among children and adolescents: The weight disorders survey in the CASPIAN-IV study. *J. Sci. Food Agric.* **2021**, *101*, 5075–5081. [CrossRef]
25. Fernandez, C.; Kasper, N.M.; Miller, A.L.; Lumeng, J.C.; Peterson, K.E. Association of Dietary Variety and Diversity with Body Mass Index in US Preschool Children. *Pediatrics* **2016**, *137*, e20152307. [CrossRef]
26. Archero, F.; Ricotti, R.; Solito, A.; Carrera, D.; Civello, F.; Di Bella, R.; Bellone, S.; Prodam, F. Adherence to the Mediterranean diet among school children and adolescents living in Northern Italy and unhealthy food behaviors associated to overweight. *Nutrients* **2018**, *10*, 1322. [CrossRef] [PubMed]

Article

The Impact of Digital Screen Time on Dietary Habits and Physical Activity in Children and Adolescents

Agata Rocka [1], Faustyna Jasielska [1], Dominika Madras [1], Paulina Krawiec [2,*] and Elżbieta Pac-Kożuchowska [2]

1   Students' Scientific Group, Department of Pediatrics and Gastroenterology, Medical University of Lublin, Racławickie 1, 20-059 Lublin, Poland; agatarocka2@gmail.com (A.R.); faustyna.piedel@gmail.com (F.J.); dkmadras@gmail.com (D.M.)

2   Department of Pediatrics and Gastroenterology, Medical University of Lublin, Racławickie 1, 20-059 Lublin, Poland; elzbietapackozuchowska@umlub.pl

*   Correspondence: paulinakrawiec@umlub.pl; Tel.: +48-8171-85420; Fax: +48-81-7431-353

Abstract: Background: Over the last few decades, the time children spend using electronic devices has increased significantly. The aim of the study was to evaluate the impact of screen time on dietary behaviors and physical activity in children and adolescents. Methods: An online survey was conducted among parents of preschool and school-aged children during the COVID-19 lockdown in Poland. There were 3127 surveys used in the analysis. Results: Survey responses referred to 1662 (53%) boys and 1465 (47%) girls, with a mean age of 12.1 ± 3.4 years. During a routine weekday, most children (71%) spent >4 h on educational activities using electronic devices, and 43% of children spent 1–2 h using devices for recreational purposes. The majority of children (89%) were exposed to screens during meals, and ate snacks between main meals (77%). There was an association between screen time and the exposure to screens during meals, and between screen time and time spent performing physical activity. Conclusions: This study revealed that the majority of children were exposed to screens during meals, which is a risk factor of obesity. The promotion of the judicious use of digital devices and healthy dietary habits associated with the use of screens may be an important component of obesity prevention strategies.

Keywords: lifestyle; electronic devices; diet; obesity

Citation: Rocka, A.; Jasielska, F.; Madras, D.; Krawiec, P.; Pac-Kożuchowska, E. The Impact of Digital Screen Time on Dietary Habits and Physical Activity in Children and Adolescents. *Nutrients* **2022**, *14*, 2985. https://doi.org/10.3390/nu14142985

Academic Editor: Zhiyong Zou

Received: 16 June 2022
Accepted: 18 July 2022
Published: 21 July 2022

Publisher's Note: MDPI stays neutral with regard to jurisdictional claims in published maps and institutional affiliations.

Copyright: © 2022 by the authors. Licensee MDPI, Basel, Switzerland. This article is an open access article distributed under the terms and conditions of the Creative Commons Attribution (CC BY) license (https://creativecommons.org/licenses/by/4.0/).

1. Introduction

Although high-quality screen activities meet some education and entertainment needs, exposure to digital screens may be detrimental for children's physical health, cognitive skills, and psychosocial development [1].

Therefore, screen time in childhood and adolescence should be managed, weighing risks versus benefits. The American Academy of Pediatrics (AAP) recommends that children younger than 18–24 months of age should not ever use screen media, while older children should not use these media for more than a 1 h daily [2]. According to the World Health Organization (WHO), infants within the first year of life should not be exposed to digital screens. Screen time for children aged from 2 to 5 years old should not exceed 1 h per day [3]. Similar recommendations were released by the Canadian Pediatric Society, stating that children younger than 2 years old should not use screens, and children aged 2 to 5 years old should have limited screen time to less than 1 h per day [1]. The guidelines released by the Australian Department of Health for children older than 5 years old recommend that recreational screen time should not exceed 2 h per day [4]. The German recommendations advise reducing screen time to a minimum. Infants and toddlers should not use electronic devices; preschool children should have limited screen time to 30 min daily; primary school children should be limited to 60 min daily; and adolescents should be limited to 120 min daily [5]. On the other hand, the Royal College of Pediatrics and Child Health (RCPCH) indicates that due to the lack of solid scientific evidence, it is not possible to formulate a universal cut-off for children's screen time. Thus, the RCPCH recommends that families should negotiate screen time while considering individual child's needs, the quality of digital activities, and the extent to which using electronic devices displace physical and social activities and sleep [6].

However, a recent systematic review which performed a meta-analysis of 63 studies revealed that only 24.7% of children younger than 2 years of age and 35.6% of children aged between 2 and 5 years old met screen time guidelines [7]. Differences were noticed in children's adherence to appropriate screen time depending on the country [8].

It appears that the reasonable management of children's screen time is now more challenging than ever. Due to the coronavirus disease 2019 (COVID-19) outbreak, social distancing restrictions have been introduced, which obliged children and adolescents to engage with remote learning, limited opportunities for physical activity, and exacerbated unhealthy eating habits [9–11]. Thus, it is particularly important to collect contemporary data regarding screen time and to find effective strategies to promote the judicious use of digital devices and implement healthy lifestyle habits from early childhood. The aim of this study was to evaluate the relationships between digital screen time and eating habits and physical activity in preschool and school-aged children during the COVID-19 lockdown in Poland.

2. Materials and Methods

A cross-sectional study using an anonymous online survey was conducted among parents of Polish preschool and school-aged children from 11 March 2021 to 20 March 2021. We requested all Polish Educational Offices to send the letter of invitation to participate in our study to kindergartens and schools, which spread this invitation among parents of pupils. Some of the Polish Educational Offices promoted information about this study on their websites. The invitation to the study included a link to the online questionnaire.

A self-administrated caregiver-reported questionnaire was set up via Google Forms. The survey consisted of 34 closed and 11 open-ended questions on children's demographic characteristics, screen time, dietary habits, physical activity and sleep in the recent three months. The authors' questionnaire was available in Polish.

The final analysis included complete responses given by parents of children up to 18 years old who provided informed online consent prior to the survey. The exclusion criteria were as follows: responses given by parents of children older than 18 years old, incomplete data in the questionnaire, and lack of informed consent.

A total of 4437 responses were received, out of which 3127 met the inclusion criteria for the final analysis. There were 1310 polls excluded from the study due to a lack of informed consent for the study (n = 975), responses given by parents of children older than 18 years old (n = 121), and incomplete data (n = 214).

The statistical analysis was carried out using the Statistica version 13 program (StatSoft, Kraków, Poland). For nominal variables, to test differences between groups or relationships between parameters, the chi-squared test of homogeneity or independence was used. The statistically significant results were $p < 0.05$.

This study obtained approval of the Bioethics Committee of the Medical University of Lublin (KE-0254/15/2021). All participants provided online informed consent prior to survey initiation.

3. Results

Survey responses referred to 1662 (53%) boys and 1465 (47%) girls with age ranging from 1 to 18 years old. The mean age of the children was 12.10 ± 3.40 years old; the median was 12 years old. The majority of children lived in rural areas (n = 1603; 51.26%) and did not suffer from any underlying medical condition (n = 2489; 79.59%). The most common medical conditions in the study group were refractive errors (n = 245; 7.83%) and asthma (n = 90; 2.88%). General characteristics of the study group are summarized in Table 1.

Table 1. General characteristics of the study group.

Subject	Number of Subjects (n)	Percentage of Subject (%)
Sex		
Girls	1465	47.00
Boys	1662	53.00
Age groups [years]		
<3	20	0.64
3–6	146	4.67
7–11	1156	36.97
12–18	1805	57.72
Place of residence		
Village	1603	51.26
City under 20,000 residents	414	13.24
City with 20,000 to 100,000 residents	793	25.36
City over 100,000 residents	317	10.14
Medical history		
Children with chronic disease	638	20.40
Children without chronic disease	2489	79.60

The vast majority of parents declared that children used their own electronic devices (n = 2690; 86.00%). About one-third of respondents declared that they did not restrict access to electronic devices for their children (n = 1019; 32.00%).

During a regular weekday, the majority of children (n = 2220; 71%) spent more than 4 h on educational activities using electronics. During the weekend, 10.39% (325) of children spent more than 4 h on educational activities using electronic devices. Figures 1 and 2 show the time spent in front of screens on educational activities and entertainment during a regular weekday and weekend day. We also found an association between age and screen time, presented in Table 2, i.e., the older the child was, the longer screen exposure was reported. Answers of parents who denied that their children were using digital devices for educational or recreational purposes were not analyzed in these particular areas. There were 478 (15.29%) parents who declared that their children did not spend any time on educational activity using electronic devices during regular weekdays, and 747 (23.89%) during weekend days. There were 153 (4.92%) parents who declared that their children did not use electronic devices for recreational activities during regular weekdays, and 176 (5.63%) during weekend days.

Parents declared that children ate from 1 to 8 main meals per day (mean 4.43), with median 4.5. Moreover, 1788 (57%) of respondents declared that their children were eating meals at consistent times. The vast majority of children were exposed to digital screens during meals (n = 2782; 88.74%). Detailed analysis revealed that 345 (11.03%) children ate meals while using electronic devices every day, 775 (24.78%) did several times during a week, 1038 (33.20%) did several times during a month, and 617 (19.73%) rarely did. Another finding was an association between screen time and the exposure to screens during eating meals, as presented in Table 3. The longer screen time for educational and recreational activities was associated with more frequent exposure to screens during meals.

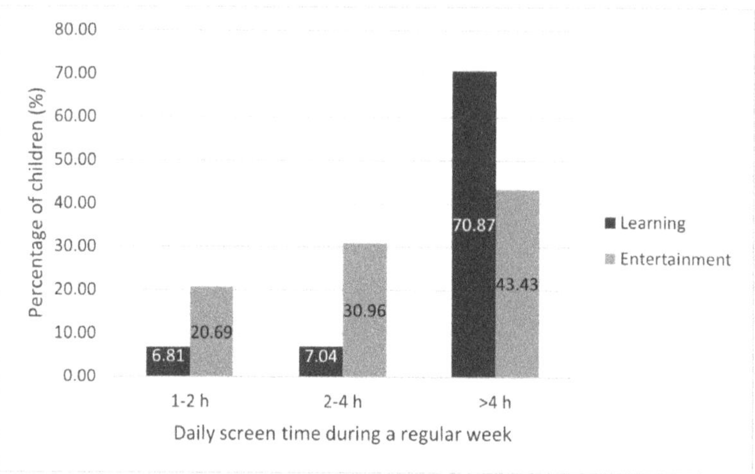

Figure 1. Screen time exposure during a regular weekday. Answers of parents who denied that their children were using digital devices for educational (15.20%) or recreational (4.92%) purposes are not presented.

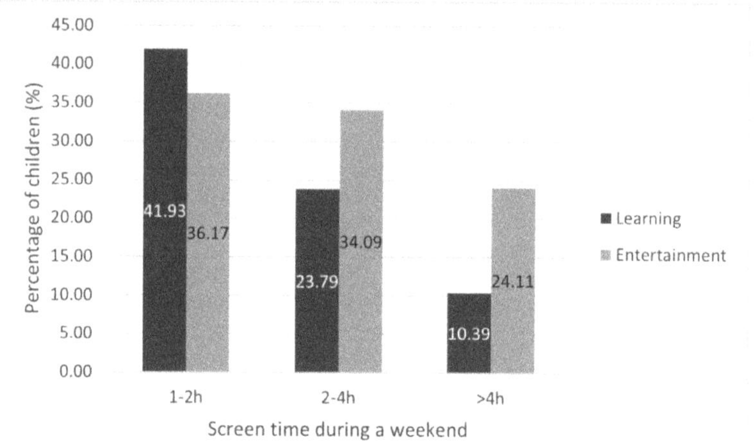

Figure 2. Screen time exposure during a weekend day. Answers of parents who denied that their children were using digital devices for educational (23.89%) or recreational (5.56%) purposes are not presented.

Table 2. Screen time in relation to the age of children.

Age group	Screen Time on Educational Activities During a Weekday		
	1–2 h n (%)	2–4 h n (%)	>4 h n (%)
<3 years old	0 (0.00%)	0 (0.00%)	1 (0.05%)
3–6 years old	11 (5.16%)	3 (1.36%)	0 (0.00%)
7–11 years old	190 (89.20%)	149 (67.73%)	520 (23.47%)
12–18 years old	12 (5.63%)	68 (30.91%)	1695 (76.49%)
Statistical analysis	$\text{Chi}^2 = 644.83; p < 0.001$		

Table 2. *Cont.*

| Age group | Screen time on educational activities during the weekend day | | |
	1–2 h n (%)	2–4 h n (%)	>4 h n (%)
<3 years old	1 (0.08%)	0 (0.00%)	0 (0.00%)
3–6 years old	4 (0.31%)	3 (0.40%)	0 (0.00%)
7–11 years old	494 (37.68%)	152 (20.43%)	35 (10.77%)
12–18 years old	812 (61.94%)	589 (79.17%)	290 (89.23%)
Statistical analysis	$Chi^2 = 130.67; p < 0.001$		

| Age group | Screen time on recreational activities during the weekday | | |
	1–2 h n (%)	2–4 h n (%)	>4 h n (%)
<3 years old	10 (0.74%)	2 (0.21%)	1 (0.15%)
3–6 years old	88 (6.48%)	25 (2.58%)	2 (0.31%)
7–11 years old	627 (46.17%)	338 (34.92%)	115 (17.77%)
12–18 years old	633 (46.61%)	603 (62.29%)	529 (81.76%)
Statistical analysis	$Chi^2 = 245.74; p < 0.001$		

| Age group | Screen time on recreational activities during the weekend day | | |
	1–2 h n (%)	2–4 h n (%)	>4 h n (%)
<3 years old	10 (0.88%)	2 (0.19%)	1 (0.13%)
3–6 years old	56 (4.95%)	64 (6.00%)	4 (0.53%)
7–11 years old	506 (44.74%)	416 (39.02%)	168 (22.28%)
12–18 years old	559 (49.43%)	584 (54.78%)	581 (77.06%)
Statistical analysis	$Chi^2 = 168.79; p < 0.001$		

Table 3. Association between the exposure to screen during meals and screen time.

| Exposure to Screens During Meals | Screen Time on Educational Activities During a Weekday | | |
	1–2 h n (%)	2–4 h n (%)	>4 h n (%)
Every day	14 (6.57%)	17 (7.73%)	273 (12.31%)
Often (several times a week)	38 (17.84%)	33 (15.00%)	616 (27.79%)
Sometimes (several times a month)	70 (32.86%)	77 (35.00%)	734 (33.11%)
Rarely (less than once a month)	52 (24.41%)	59 (26.82%)	396 (17.86%)
Never	39 (18.31%)	34 (15.45%)	198 (8.93%)
Statistical analysis	$Chi^2 = 62.09; p < 0.001$		

| Exposure to screens during meals | Screen time on educational activities during a weekend day | | |
	1–2 h n (%)	2–4 h n (%)	>4 h n (%)
Every day	136 (10.37%)	61 (8.20%)	81 (24.92%)
Often (several times a week)	307 (23.40%)	217 (29.17%)	99 (30.46%)
Sometimes (several times a month)	444 (33.84%)	282 (37.90%)	79 (24.31%)
Rarely (less than once a month)	284 (21.42%)	124 (16.67%)	47 (14.46%)
Never	144 (10.98%)	60 (8.06%)	19 (5.85%)
Statistical analysis	$Chi^2 = 98.51; p < 0.001$		

Table 3. *Cont.*

Exposure to screens during meals	Screen time on recreational activities during a weekday		
	1–2 h n (%)	2–4 h n (%)	>4 h n (%)
Every day	79 (5.82%)	95 (9.80%)	156 (24.11%)
Often (several times a week)	239 (17.60%)	291 (30.03%)	235 (36.32%)
Sometimes (several times a month)	499 (36.75%)	360 (37.15%)	147 (22.72%)
Rarely (less than once a month)	339 (24.96%)	151 (15.58%)	78 (12.06%)
Never	202 (14.87%)	72 (7.43%)	31 (4.79%)
Statistical analysis	$Chi^2 = 341.02; p < 0.001$		

Exposure to screens during meals	Screen time on recreational activities during a weekend day		
	1–2 h n (%)	2–4 h n (%)	>4 h n (%)
Every day	67 (5.92%)	75 (7.03%)	184 (24.40%)
Often (several times a week)	191 (16.89%)	283 (26.52%)	276 (36.60%)
Sometimes (several times a month)	377 (33.33%)	433 (40.58%)	193 (25.60%)
Rarely (less than once a month)	315 (27.85%)	182 (17.06%)	76 (10.08%)
Never	181 (16.00%)	94 (8.81%)	25 (3.32%)
Statistical analysis	$Chi^2 = 416.79; p < 0.001$		

There were 2410 (77%) children who reported eating snacks between main meals. The most common types of snacks were fruits (n = 2170; 68.40%), cookies (n = 1247; 39.88%), and crisps (n = 1100; 35.18%). Notably, one respondent might have given more than one answer. The most common types of snacks eaten while using electronic devices are summarized in Table 4.

Table 4. Types of snacks eaten while using electronic devices.

Type of Snacks	Number of Responses * (n)	The Percentage Contribution of Snacks (%)
Fruit	2170	69.40
Cookies	1247	39.88
Crisps	1100	35.18
Salty sticks	907	29.01
Vegetables	887	28.37
Chocolate	764	24.43
Bars	563	18.00
Delicacies	437	13.98
Salty peanuts	312	9.98
Sandwiches	188	6.01
Cereals with Milk	52	1.66
Popcorn	51	1.63
Yogurts	27	0.86
Other snacks	187	5.98

* One respondent could have given more than one answer.

Daily beverage consumption during screen time was reported in 309 (9.88%) children. In 919 (29.40%) cases, parents reported that children consumed drinks while using devices several times per week, and in 1505 (48.13%) cases, several times per month. The majority of children and adolescents drank beverages while using electronic devices (n = 2733; 87.41%). The most common types of drinks chosen when using electronic devices were water (n = 2480; 79.31%), fruit juices (n = 1499; 47.94%), and tea (n = 1666; 53.28%). Types of drinks consumed while using electronic devices are presented in Table 5. One respondent may have given more than one answer.

Table 5. Types of beverages while using electronic devices.

Type of Beverage	Number of Responses * (n)	The Percentage Contribution of Beverages (%)
Water	2480	79.31
Tea	1666	53.28
Fruit juices	1499	47.94
Carbonated drinks	558	18.80
Coffee	206	6.59
Energy drinks	109	3.49
Milk	47	1.50
Cocoa	60	1.92
Other drinks	133	4.25

* One respondent could have given more than one answer.

Parents reported that the mean time their children spent performing daily physical activity during a regular weekday was 2.08 ± 1.8 h, with a median of 1.5 h. During a weekend day, children spent 2.19 ± 1.7 h performing physical activity, with a median of 2 h.

There was a statistically significant relationship between screen time and the time spent on physical activity (Table 6). Although the majority of children spend between 1 and 4 h daily on physical activity, children with longer screen time tend to be less physically active. Even though the survey was taken during the COVID-19 pandemic, 55.5% of respondents declared that their child was attending physical education classes. However, one needs to remember that during lockdown, physical education classes in some schools were conducted via online platforms.

Table 6. Association between daily physical activity and screen time.

Daily Time on Physical Activity	Screen Time on Educational Activities During a Weekday		
	1–2 h n (%)	2–4 h n (%)	>4 h n (%)
<1 h	5 (2.43%)	18 (8.78%)	331 (15.64%)
1–2 h	135 (65.53%)	124 (60.49%)	1234 (58.32%)
2–4 h	35 (16.99%)	36 (17.56%)	353 (16.68%)
4–6 h	22 (10.68%)	18 (8.78%)	132 (6.24%)
>6 h	9 (4.37%)	9 (4.39%)	66 (3.12%)
Statistical analysis	$Chi^2 = 37.94$; $p < 0.001$		
Daily time on physical activity	Screen time on educational activities during a weekend day		
	1–2 h n (%)	2–4 h n (%)	>4 h n (%)
<1 h	159 (12.67%)	103 (14.63%)	61 (19.87%)
1–2 h	753 (60.00%)	409 (58.10%)	177 (57.65%)
2–4 h	221 (17.61%)	128 (18.18%)	36 (11.73%)
4–6 h	76 (6.06%)	47 (6.68%)	20 (6.51%)
>6 h	46 (3.67%)	17 (2.41%)	13 (4.23%)
Statistical analysis	$Chi^2 = 18.57$; $p = 0.02$		

Table 6. *Cont.*

Daily time on physical activity	Screen time on recreational activities during a weekday		
	1–2 h n (%)	2–4 h n (%)	>4 h n (%)
<1 h	120 (9.40%)	107 (11.59%)	136 (21.97%)
1–2 h	793 (62.15%)	555 (60.13%)	337 (54.44%)
2–4 h	232 (18.18%)	170 (18.42%)	83 (13.41%)
4–6 h	84 (6.58%)	72 (7.80%)	40 (6.46%)
>6 h	47 (3.68%)	19 (2.06%)	23 (3.72%)
Statistical analysis	$Chi^2 = 77.01; p < 0.001$		

Daily time on physical activity	Screen time on recreational activities during a weekend day		
	1–2 h n (%)	2–4 h n (%)	>4 h n (%)
<1 h	102 (9.44%)	98 (9.81%)	159 (22.08%)
1–2 h	652 (60.31%)	628 (62.86%)	397 (55.14%)
2–4 h	208 (19.24%)	175 (17.52%)	101 (14.03%)
4–6 h	82 (7.59%)	68 (6.81%)	44 (6.11%)
>6 h	37 (3.42%)	30 (3.00%)	19 (2.64%)
Statistical analysis	$Chi^2 = 78.25; p < 0.001$		

Interestingly, 2728 (87.00%) of parents believed there is an age limit until which child should not use electronic devices. Parents reported that the average age for adolescents making independent decisions about spending time in front of screens was 15 years.

4. Discussion

The dynamic developments in technology over last decades resulted in a significant increase in time spent using electronic devices by children [12]. Moreover, the COVID-19 pandemic exacerbated that trend [13]. A recent systematic review which performed a meta-analysis revealed that during the COVID-19 pandemic, there was a significant elevation of total screen time in 67% of children, and of recreational screen time in 60% of children [13]. The total screen time increased by 0.5 h per day (95% CI 0.3–0.9) in children younger than 5 years old, 0.9 h per day (95% CI 0.3–1.5) in adolescents aged 11 to 17 years old, and 1.4 h per day (95% CI 1.1–1.7) in children aged 6 to 10 years old was reported [13]. There was also an elevation of leisure screen time of 0.48 h daily (95%CI 0.29–0.67) in adolescents, 0.61 h daily (95% CI 0.4–0.82) in young children, and 1.04 h daily (95% CI 0.77–1.3) in primary aged children [13]. Moreover, there is also evidence for the decline in physical activity in children during the global pandemic [14]. Thus, it appears that due to the lockdowns, school closures and remote learning during the pandemic aggravated sedentary behaviors even in the youngest children, while physical activity has decreased [13–16].

To the best of our knowledge, this is one of the first studies among parents of children in all age groups which has evaluated the relationship between children's digital screen time and lifestyle behaviors during the COVID-19 pandemic in Poland. In the present study, the vast majority of children used their own electronic device. During a regular weekday, 70% children spent more than 4 h daily on educational activities using electronics, and more than 40% spent more than 4 h daily on entertainment using electronic devices. During weekends, electronic devices were used for more than 4 h daily on educational activities by 10% of children, and on entertainment by almost 25% of children. In the study group, unhealthy eating habits were revealed, i.e., the vast majority of children were exposed to digital screens during meals and ate snacks between main meals. There was an association between screen time and the exposure to screens during eating meals. The median time spent performing physical activity was 1.5 h during weekdays and 2 h during weekend days. There was also a significant relationship between daily screen time and physical activity.

In the current study, parents of preschool and school-aged children were included. Not only did the majority of children spend more than 4 h daily using electronic devices for educational activities, but a large proportion of them spent more than 2 h on entertainment-based screen time per day. Results from this study are consistent with previous findings. In a group of 166 children, Ferreira et al. showed that approximately 85% of children under 2 years of age and 80% of infants were exposed to electronic devices. Moreover, 79% of them spent up to 1 h per day in front of screens [17]. Kaur et al. also found that approximately 60% of children aged 2–5 years old had excessive screen time and did not follow AAP recommendations [18]. Thus, it now appears almost implausible to comply with recommendations on screen time for children [7,10,19,20].

Recent meta-analyses confirmed that a greater quantity of screen time is associated with an increased risk of obesity [21–23]. This phenomenon may be explained by the displacement of time for physical activity by screen-based sedentary behaviors [21,22,24]. Moreover, poor dietary habits, such as a higher intake of energy, more frequent consumption of fast food and sweets, and lower intake of fiber, vegetables, fruit, and fish were reported more often in children overexposed to digital devices [25,26]. One of the biggest contemporary health concerns is overweight and obesity in the pediatric population, which result from a lack of physical activity and unhealthy eating habits [27]. Notably, recent research has shown that the risk of obesity is higher in children who spend more time engaging in screen-based sedentary behaviors than in non-screen-based sedentary activities [28]. Sedentary digital media use in preadolescence was associated with an increased risk of overweight three years later [29]. Alturki et al. posited that having a smartphone was statistically significantly higher in a group of obese children than in children with normal weight [30]. Moreover, it has been revealed that during the COVID-19 pandemic, there was a trend in weight gaining, resulting in overweight and obesity in children and adolescents, which mostly affected those who were aged between 2 and 6 years [31].

Moreover, excessive exposure to digital screens negatively affects the duration and quality of sleep in childhood [21,22,32]. The link between screen time and poorer psychomotor and cognitive development has been also revealed [21,33]. On the other hand, although excessive screen time was associated with poorer language skills, the high-quality use of media was beneficial for child language [34]. There are some data also suggesting an association between screen time and the mental health of children and adolescents [21,22].

According to WHO guidelines, the median time spent performing physical activity in children aged 5 to 17 years should be 60 min per day [35]. In the present study, the median time for physical activity met WHO guidelines. Possible explanations of this finding may be the fact that the majority of children attended physical education classes. However, one needs to remember that during the lockdown, some of these school classes were conducted via online platforms, which raises concerns about the quality of children's physical activity in these times.

A significant association was found between reduced physical activity and a long time spent watching TV, as well as a long time spent in front of any screen [10,36]. Decreased physical activity in school is associated with longer screen time. Moreover, lower physical activity can be associated with spending more time in front of a screen at night [37]. Recent studies have indicated that there is a correlation between screen addiction behaviors and decreased physical activity in young adolescents [10]. It is important to replace screen-based sedentary behaviors with physical activity [3]. There are data which show decreases in physical activity in children during the COVID-19 pandemic [16,29,38]. Comparable findings were observed in Polish children in studies by Brzęk et al. and Łuszczki et al. [11,39]. It should be highlighted that only 30.77% of children met the WHO physical activity guidelines before the COVID-19 pandemic [39].

In the current study, there was an association between screen time and physical activity. Similar results were obtained in other studies. Mineshita et al. found that children with a longer usage of electronic devices were more likely to spend less time on physical activities [40]. The use of television screens for more than 2 h may be related with a reduction in physical fitness among juveniles as well [41]. On the other hand, however, Dahlgren et al. did not find any correlation between screen time and physical activity in a group of adolescents [42].

It is recommended that children should not use a digital screen during meals [2]. The results of the current study revealed that most of the children were exposed to screens during meals. This is a significant factor which may trigger obesity and eating disorders. Using digital technologies that divert attention away from meal negatively affects eating behavior and results in an excessive energy intake [43]. Moreover, children using digital devices during meals tend to consume more junk food [44]. Yong et al. found that during a week-long study, 85.3% of participants used a smartphone at least once during a meal, with an average frequency of one in three meals where phones were used [45].

In the current study, there was an association between screen time and the frequency of exposure to screens when eating meals. Kristo et al. also found a significant correlation between children's eating habits and the duration of tablet or smartphones use, but not with computer use [46]. Result of the study by Robinson et al. confirmed the influence of electronic devices on eating habits. It has been found that the more media devices children use during mealtime, the poorer the quality of the food [47]. Falbe et al. noted that longer screen time was associated with a lower consumption of vegetables and fruit, and a greater consumption of sweets, fast food, sweetened drinks, and salty snacks [48]. The results of the ALADINO study revealed that excessive screen time is correlated with a higher frequency of eating energy-dense and micronutrient-poor food, and a lower intake of fruit and vegetables [49]. According to Watts et al., time spent in front of a screen by adolescents was conducive to less healthy food choices [50]. Kelishadi et al. found that children who spent time in front of a screen for more than 4 h per day were more vulnerable to eating snacks compared with people with a daily screen time of less than 4 h [51]. Interestingly, in the present study, the majority of children ate snacks between main meals, most commonly fruit.

The vast majority of children from the study group drank beverages while using digital devices, mostly water, tea, and juices. Börnhorst et al. showed that one additional hour of screen time increased the consumption of sugary soft drinks, diet soft drinks, and flavored milk [52]. The cross-sectional Spanish study revealed that screen time exposure for longer than 1 h can increase the frequency of sweets, fast food, and soft drink intake [53].

Most parents declared that their child used their own electronic device. Moreover, 87% of parents believed there was an age limit until which child should not use electronic devices. Parents reported that the average age at which children should make independent decisions about spending time in front of screens is 15 years old. British recommendations also highlight the need to set appropriate amounts of time for the independent use of electrical devices by children. It is important that the limits are set by parents and children together, and should be based on the child's age. When limits are not respected, parents should clearly define the consequences. Moreover, the content of screen time also matters [54]. The question therefore arises as to what the best way is to educate parents on risks associated with prolonged screen time to enable them making rational decisions in this matter.

Our study has some limitations, including the limited response rate. This may be explained by the fact that the invitation to our study was sent indirectly to parents by Educational Offices. There was also a significant proportion of partial surveys which could not be used in the final analysis. Another limitation may be the fact that the study presents self-reported data by parents of children. Moreover, caution should be exercised when attempting to generalize the findings of the current study because it relates to an extreme event: the global COVID-19 pandemic. However, despite these limitations, this study identified important behaviors associated with screen time among children, which should be corrected through evidence-based interventions and prevention strategies.

5. Conclusions

In conclusion, screen time affects the lifestyle behaviors of children and adolescents. This study revealed that the vast majority of children were exposed to screens during meals. This behavior may increase risk of obesity due to higher energy intakes and the more frequent consumption of junk food. Starting in early childhood to promote the judicious use of digital devices and to build healthy dietary habits associated with the use of screens may be an important component in obesity prevention strategies: not only during the COVID-19 pandemic, but also in the future. Digital screen time may negatively affect physical activity. There is a need to develop effective strategies to limit excessive screen time and to promote healthy eating habits and physical activity in children. Understanding the mechanisms behind the impact of screens on children's physical activity and eating habits may help to create targeted parental control plans, which could result in long-term effects such as reducing the risk of obesity in the future.

Author Contributions: Conceptualization, P.K. and E.P.-K.; methodology, P.K. and E.P.-K.; software, A.R., F.J., D.M., P.K. and E.P.-K.; validation, A.R., F.J., D.M., P.K. and E.P.-K.; formal analysis, P.K. and E.P.-K.; investigation, A.R., F.J., D.M. and P.K.; resources, A.R., F.J., D.M., P.K. and E.P.-K.; data curation, A.R., F.J., D.M. and P.K.; writing—original draft preparation, A.R., F.J., D.M. and P.K.; writing—review and editing, P.K. and E.P.-K.; visualization, A.R., F.J., D.M. and P.K.; supervision, P.K. and E.P.-K.; project administration, P.K. and E.P.-K.; funding acquisition, E.P.-K. All authors have read and agreed to the published version of the manuscript.

Funding: The APC was funded by the Medical University of Lublin (DS406 to Elżbieta Pac-Kożuchowska).

Institutional Review Board Statement: The study was approved by the Bioethical Committee of the Medical University of Lublin (KE-0254/15/2021).

Informed Consent Statement: All participants provided online informed consent prior to survey initiation.

Data Availability Statement: The datasets analyzed during the current study are available from the corresponding author on reasonable request.

Conflicts of Interest: The authors declare no conflict of interest.

References

1. Canadian Paediatric Society, Digital Health Task Force, Ottawa, Ontario. Screen time and young children: Promoting health and development in a digital world. *Paediatr. Child Health* **2017**, *22*, 461–477. [CrossRef] [PubMed]
2. Council on Communication and Media; Hill, D.; Ameenuddin, N.; Reid Chassiakos, Y.L.; Cross, C.; Hutchinson, J.; Levine, A.; Boyd, R.; Mendelson, R.; Moreno, M.; et al. Media and Young Minds. *Pediatrics* **2016**, *138*, 20162591. [CrossRef]
3. WHO Guidelines on Physical Activity, Sedentary Behaviour and Sleep for Children under 5 Years of Age. Geneva: World Health Organization; 2019 (WHO/NMH/PND/19.2). Available online: https://apps.who.int/iris/bitstream/handle/10665/311664/9789241550536-eng.pdf?sequence=1&isAllowed=y (accessed on 12 June 2022).
4. Okely, A.D.; Ghersi, D.; Hesketh, K.D.; Santos, R.; Loughran, S.P.; Cliff, D.P.; Shilton, T.; Grant, D.; Jones, R.A.; Stanley, R.M.; et al. A collaborative approach to adopting/adapting guidelines—The Australian 24-Hour Movement Guidelines for the early years (Birth to 5 years): An integration of physical activity, sedentary behavior, and sleep. *BMC Public Health* **2017**, *17*, 869. [CrossRef] [PubMed]
5. Pfeifer, K.; Banzer, W.; Ferrari, N.; Füzéki, E.; Geidl, W.; Graf, C.; Hartung, V.; Klamroth, S.; Völker, K.; Vogt, L. Recommendations for physical activity. In *National Recommendations for Physical Activity and Physical Activity Promotion*; Rütten, A., Pfeifer, K., Eds.; FAU University Press: Erlangen, Germany, 2016; p. 28.
6. The Health Impacts of Screen Time: A Guide for Clinicians and Parents. The Royal College of Pediatrics and Child Health (RCPCH). Available online: https://www.rcpch.ac.uk/sites/default/files/2018-12/rcpch_screen_time_guide_-_final.pdf (accessed on 5 July 2022).
7. McArthur, B.A.; Volkova, V.; Tomopoulos, S.; Madigan, S. Global Prevalence of Meeting Screen Time Guidelines Among Children 5 Years and Younger: A Systematic Review and Meta-analysis. *JAMA Pediatrics* **2022**, *176*, 373–383. [CrossRef]
8. Whiting, S.; Buoncristiano, M.; Gelius, P.; Abu-Omar, K.; Pattison, M.; Hyska, J.; Duleva, V.; Musić Milanović, S.; Zamrazilová, H.; Hejgaard, T.; et al. Physical Activity, Screen Time, and Sleep Duration of Children Aged 6–9 Years in 25 Countries: An Analysis within the WHO European Childhood Obesity Surveillance Initiative (COSI) 2015–2017. *Obes. Facts* **2021**, *14*, 32–44. [CrossRef]

9. Carroll, N.; Sadowski, A.; Laila, A.; Hruska, V.; Nixon, M.; Ma, D.W.L.; Haines, J.; On Behalf of the Guelph Family Health Study. The Impact of COVID-19 on Health Behavior, Stress, Financial and Food Security among Middle to High Income Canadian Families with Young Children. *Nutrients* **2020**, *12*, 2352. [CrossRef]

10. Moitra, P.; Madan, J. Impact of screen time during COVID-19 on eating habits, physical activity, sleep, and depression symptoms: A cross-sectional study in Indian adolescents. *PLoS ONE* **2022**, *17*, 0264951. [CrossRef]

11. Łuszczki, E.; Bartosiewicz, A.; Pezdan-Śliż, I.; Kuchciak, M.; Jagielski, P.; Oleksy, Ł.; Stolarczyk, A.; Dereń, K. Children's Eating Habits, Physical Activity, Sleep, and Media Usage before and during COVID-19 Pandemic in Poland. *Nutrients* **2021**, *13*, 2447. [CrossRef]

12. Goode, J.A.; Fomby, P.; Mollborn, S.; Limburg, A. Children's Technology Time in Two US Cohorts. Children's Technology Time in Two US Cohorts. *Child Indic. Res.* **2020**, *13*, 1–26. [CrossRef]

13. Trott, M.; Driscoll, R.; Irlado, E.; Pardhan, S. Changes and correlates of screen time in adults and children during the COVID-19 pandemic: A systematic review and meta-analysis. *EClinicalMedicine* **2022**, *48*, 101452. [CrossRef]

14. Wunsch, K.; Kienberger, K.; Niessner, C. Changes in Physical Activity Patterns Due to the COVID-19 Pandemic: A Systematic Review and Meta-Analysis. *Int. J. Environ. Res. Public Health* **2022**, *19*, 2250. [CrossRef] [PubMed]

15. Musa, S.; Elyaman, R.; Dergaa, I. COVID-19 and screen-based sedentary behaviour: Systematic review of digital screen time and metabolic syndrome in adolescents. *PLoS ONE* **2022**, *17*, e0265560. [CrossRef] [PubMed]

16. Ferrante, G.; Mollicone, D.; Cazzato, S.; Lombardi, E.; Pifferi, M.; Turchetta, A.; Tancredi, G.; La Grutta, S. COVID-19 Pandemic and Reduced Physical Activity: Is There an Impact on Healthy and Asthmatic Children? *Front. Pediatrics* **2021**, *9*, 695703. [CrossRef] [PubMed]

17. Ferreira, J.; Prucha, B.; Pinto, O.; Souto, R.; Lima, R.P.; Morna, C. Screen Time Use in Children Less Than Five Years Old. *Nascer e Crescer—Birth Growth Med. J.* **2020**, *29*, 188–195. [CrossRef]

18. Kaur, N.; Gupta, M.; Malhi, P.; Grover, S. Prevalence of Screen Time Among Children Aged 2 to 5 Years in Chandigarh, a North Indian Union Territory. *J. Dev. Behav. Pediatrics* **2022**, *43*, 29–38. [CrossRef]

19. Xiang, M.; Zhang, Z.; Kuwahara, K. Impact of COVID-19 pandemic on children and adolescents' lifestyle behavior larger than expected. *Prog. Cardiovasc. Dis.* **2020**, *63*, 531–532. [CrossRef]

20. Guo, Y.F.; Liao, M.Q.; Cai, W.L.; Yu, X.X.; Li, S.N.; Ke, X.Y.; Tan, S.X.; Luo, Z.Y.; Cui, Y.F.; Wang, Q.; et al. Physical activity, screen exposure and sleep among students during the pandemic of COVID-19. *Sci. Rep.* **2021**, *11*, 8529. [CrossRef]

21. Stiglic, N.; Viner, R.M. Effects of screentime on the health and well-being of children and adolescents: A systematic review of reviews. *BMJ Open* **2019**, *9*, e023191. [CrossRef]

22. Li, C.; Cheng, G.; Sha, T.; Cheng, W.; Yan, Y. The Relationships between Screen Use and Health Indicators among Infants, Toddlers, and Preschoolers: A Meta-Analysis and Systematic Review. *Int. J. Environ. Res. Public Health* **2020**, *17*, 7324. [CrossRef]

23. Fang, K.; Mu, M.; Liu, K.; He, Y. Screen time and childhood overweight/obesity: A systematic review and meta-analysis. *Child. Care Health Dev.* **2019**, *45*, 744–753. [CrossRef]

24. De Araújo, L.G.M.; Turi, B.C.; Locci, B.; Mesquita, C.A.A.; Fonsati, N.B.; Monteiro, H.L. Patterns of Physical Activity and Screen Time Among Brazilian Children. *J. Phys. Act. Health* **2018**, *15*, 457–461. [CrossRef] [PubMed]

25. Tambalis, K.D.; Panagiotakos, D.B.; Psarra, G.; Sidossis, L.S. Screen time and its effect on dietary habits and lifestyle among schoolchildren. *Cent. Eur. J. Public Health* **2020**, *28*, 260–266. [CrossRef] [PubMed]

26. Shang, L.; Wang, J.; O'Loughlin, J.; Tremblay, A.; Mathieu, M.È.; Henderson, M.; Gray-Donald, K. Screen time is associated with dietary intake in overweight Canadian children. *Prev. Med. Rep.* **2015**, *2*, 265–269. [CrossRef] [PubMed]

27. Nittari, G.; Scuri, S.; Petrelli, F.; Pirillo, I.; di Luca, N.M.; Grappasonni, I. Fighting obesity in children from European World Health Organization member states. Epidemiological data, medical-social aspects, and prevention programs. *Clin. Ter.* **2019**, *170*, e223–e230. [CrossRef] [PubMed]

28. Hu, R.; Zheng, H.; Lu, C. The Association Between Sedentary Screen Time, Non-screen-based Sedentary Time, and Overweight in Chinese Preschool Children: A Cross-Sectional Study. *Front. Pediatrics* **2021**, *9*, 767608. [CrossRef]

29. Engberg, E.; Leppänen, M.H.; Sarkkola, C.; Viljakainen, H. Physical Activity Among Preadolescents Modifies the Long-Term Association Between Sedentary Time Spent Using Digital Media and the Increased Risk of Being Overweight. *J. Phys. Act. Health* **2021**, *18*, 1105–1112. [CrossRef]

30. Alturki, H.A.; Brookes, D.S.; Davies, P.S. Does spending more time on electronic screen devices determine the weight outcomes in obese and normal weight Saudi Arabian children? *Saudi Med. J.* **2020**, *41*, 79–87. [CrossRef]

31. Shalitin, S.; Phillip, M.; Yackobovitch-Gavan, M. Changes in body mass index in children and adolescents in Israel during the COVID-19 pandemic. *Int. J. Obes.* **2022**, *46*, 1160–1167. [CrossRef]

32. Janssen, X.; Martin, A.; Hughes, A.R.; Hill, C.M.; Kotronoulas, G.; Hesketh, K.R. Associations of screen time, sedentary time and physical activity with sleep in under 5s: A systematic review and meta-analysis. *Sleep Med. Rev.* **2020**, *49*, 101226. [CrossRef]

33. Madigan, S.; Browne, D.; Racine, N.; Mori, C.; Tough, S. Association Between Screen Time and Children's Performance on a Developmental Screening Test. *JAMA Pediatrics* **2019**, *173*, 244–250. [CrossRef]

34. Madigan, S.; McArthur, B.A.; Anhorn, C.; Eirich, R.; Christakis, D.A. Associations Between Screen Use and Child Language Skills: A Systematic Review and Meta-analysis. *JAMA Pediatrics* **2020**, *174*, 665–675. [CrossRef] [PubMed]

35. Chaput, J.-P.; Willumsen, J.; Bull, F.; Chou, R.; Ekelund, U.; Firth, J.; Jago, R.; Ortega, F.B.; Katzmarzyk, P.T. 2020 WHO guidelines on physical activity and sedentary behaviour for children and adolescents aged 5–17 years: Summary of the evidence. *Int. J. Behav. Nutr. Phys. Act.* **2020**, *17*, 141. [CrossRef] [PubMed]

36. Mozafarian, N.; Motlagh, M.E.; Heshmat, R.; Karimi, S.; Mansourian, M.; Mohebpour, F.; Qorbani, M.; Kelishadi, R. Factors Associated with Screen Time in Iranian Children and Adolescents: The CASPIAN-IV Study. *Int. J. Prev. Med.* **2017**, *8*, 31. [CrossRef] [PubMed]

37. Lyngdoh, M.; Akoijam, B.S.; Agui, R.S.; Sonarjit Singh, K. Diet, Physical Activity, and Screen Time among School Students in Manipur. *Indian J. Community Med.* **2019**, *44*, 134–137. [CrossRef]

38. Dallolio, L.; Marini, S.; Masini, A.; Toselli, S.; Stagni, R.; Bisi, M.C.; Gori, D.; Tessari, A.; Sansavini, A.; Lanari, M.; et al. The impact of COVID-19 on physical activity behaviour in Italian primary school children: A comparison before and during pandemic considering gender differences. *BMC Public Health* **2022**, *22*, 52. [CrossRef]

39. Brzęk, A.; Strauss, M.; Sanchis-Gomar, F.; Leischik, R. Physical Activity, Screen Time, Sedentary and Sleeping Habits of Polish Preschoolers during the COVID-19 Pandemic and WHO's Recommendations: An Observational Cohort Study. *Int. J. Environ. Res. Public Health* **2021**, *18*, 11173. [CrossRef]

40. Mineshita, Y.; Kim, H.K.; Chijiki, H.; Nanba, T.; Shinto, T.; Furuhashi, S.; Oneda, S.; Kuwahara, M.; Suwama, A.; Shibata, S. Screen time duration and timing: Effects on obesity, physical activity, dry eyes, and learning ability in elementary school children. *BMC Public Health* **2021**, *21*, 422. [CrossRef]

41. Greier, K.; Drenowatz, C.; Ruedl, G.; Riechelmann, H. Association between daily TV time and physical fitness in 6- to 14-year-old Austrian youth. *Transl. Pediatrics* **2019**, *8*, 371–377. [CrossRef]

42. Dahlgren, A.; Sjöblom, L.; Eke, H.; Bonn, S.E.; Trolle Lagerros, Y. Screen time and physical activity in children and adolescents aged 10–15 years. *PLoS ONE* **2021**, *16*, e0254255. [CrossRef]

43. La Marra, M.; Caviglia, G.; Perrella, R. Using Smartphones When Eating Increases Caloric Intake in Young People: An Overview of the Literature. *Front. Psychol.* **2020**, *11*, 587886. [CrossRef]

44. Jusienė, R.; Urbonas, V.; Laurinaitytė, I.; Rakickienė, L.; Breidokienė, R.; Kuzminskaitė, M.; Praninskienė, R. Screen Use During Meals Among Young Children: Exploration of Associated Variables. *Medicina* **2019**, *55*, 688. [CrossRef]

45. Yong, J.Y.Y.; Tong, E.M.W.; Liu, J.C.J. Meal-Time Smartphone Use in an Obesogenic Environment: Two Longitudinal Observational Studies. *JMIR mHealth uHealth* **2021**, *9*, 22929. [CrossRef]

46. Kristo, A.S.; Çinar, N.; Kucuknil, S.L.; Sikalidis, A.K. Technological Devices and Their Effect on Preschool Children's Eating Habits in Communities of Mixed Socioeconomic Status in Istanbul; A Pilot Cross-Sectional Study. *Behav. Sci.* **2021**, *11*, 157. [CrossRef] [PubMed]

47. Robinson, C.A.; Domoff, S.E.; Kasper, N.; Peterson, K.E.; Miller, A.L. The healthfulness of children's meals when multiple media and devices are present. *Appetite* **2022**, *169*, 105800. [CrossRef] [PubMed]

48. Falbe, J.; Willett, W.C.; Rosner, B.; Gortmaker, S.L.; Sonneville, K.R.; Field, A.E. Longitudinal relations of television, electronic games, and digital versatile discs with changes in diet in adolescents. *Am. J. Clin. Nutr.* **2014**, *100*, 1173–1181. [CrossRef] [PubMed]

49. Pérez-Farinós, N.; Villar-Villalba, C.; López Sobaler, A.M.; Dal Re Saavedra, M.Á.; Aparicio, A.; Santos Sanz, S.; Robledo de Dios, T.; Castrodeza-Sanz, J.J.; Ortega Anta, R.M. The relationship between hours of sleep, screen time and frequency of food and drink consumption in Spain in the 2011 and 2013 ALADINO: A cross-sectional study. *BMC Public Health* **2017**, *17*, 33. [CrossRef]

50. Watts, A.W.; Lovato, C.Y.; Barr, S.I.; Hanning, R.M.; Mâsse, L.C. Experiences of overweight/obese adolescents in navigating their home food environment. *Public Health Nutr.* **2015**, *18*, 3278–3286. [CrossRef]

51. Kelishadi, R.; Mozafarian, N.; Qorbani, M.; Maracy, M.R.; Motlagh, M.E.; Safiri, S.; Ardalan, G.; Asayesh, H.; Rezaei, F.; Heshmat, R. Association between screen time and snack consumption in children and adolescents: The CASPIAN-IV study. *J. Pediatrics Endocrinol. Metab.* **2017**, *30*, 211–219. [CrossRef]

52. Börnhorst, C.; Wijnhoven, T.M.; Kunešová, M.; Yngve, A.; Rito, A.I.; Lissner, L.; Duleva, V.; Petrauskiene, A.; Breda, J. WHO European Childhood Obesity Surveillance Initiative: Associations between sleep duration, screen time and food consumption frequencies. *BMC Public Health* **2015**, *15*, 442. [CrossRef]

53. Cartanyà-Hueso, À.; González-Marrón, A.; Lidón-Moyano, C.; Garcia-Palomo, E.; Martín-Sánchez, J.C.; Martínez-Sánchez, J.M. Association between Leisure Screen Time and Junk Food Intake in a Nationwide Representative Sample of Spanish Children (1–14 Years): A Cross-Sectional Study. *Healthcare* **2021**, *9*, 228. [CrossRef]

54. Torjesen, I. Parents should decide when children's screen time is too high, says first UK guidance. *BMJ* **2019**, *364*, l60. [CrossRef] [PubMed]

Article

Association of Household Type and Fast-Food Consumption in Korean Adolescents

Hwa Sook Kwon [1,†], Soo Hyun Kang [2,3,†], Yu Shin Park [2,3], Jung Gu Kang [4] and Eun Cheol Park [3,5,*]

1 Department of Administration, CHA Ilsan Medical Center, Seoul 10414, Korea; flower21_@naver.com
2 Department of Public Health, Graduate School, Yonsei University, Seoul 03722, Korea; kshyun@yuhs.ac (S.H.K.); dbtls0459@yuhs.ac (Y.S.P.)
3 Institute of Health Services Research, Yonsei University, Seoul 03722, Korea
4 CHA Ilsan Medical Center, Seoul 10414, Korea; kangski1958@gmail.com
5 Department of Preventive Medicine, College of Medicine, Yonsei University, Seoul 03722, Korea
* Correspondence: ecpark@yuhs.ac
† These authors contributed equally to this work.

Abstract: Background: Due to changing household types and weakening of family functions, children have fewer opportunities to develop healthy lifestyle patterns from contact with family members compared to the past. In this paper, we evaluate the association between household type and adolescents' fast-food consumption, focusing on whether they were living with their parents or not, and determine their reasons for not living with their parents. Methods: This cross-sectional study analyzed data from the Korea Youth Risk Behavior web-based survey between 2017 and 2020. The subjects were students in grades 7–12. The outcome variable was a frequency of fast-food intake of ≥ 5 times per week. The main independent variable was the type of household: (1) living with both parents; (2) living with a single parent (one of father, mother, stepfather, stepmother); (3) not living together, but having parents; and (4) having no parents. Results: Participants without parents were more likely to eat fast food frequently than those living with both parents. Among boys, not having parents and living in a dorm or boarding house or living with other family members or relatives were significantly associated with frequent fast-food intake; among girls, not having parents and living in a dorm or boarding house were significantly associated with frequent fast-food intake. Conclusion: Adolescents having no parents have a higher risk of frequent fast-food intake than those living with both parents. Further studies are needed to address household types in greater detail.

Keywords: household; fast food; adolescents; dietary behavior; family

Citation: Kwon, H.S.; Kang, S.H.; Park, Y.S.; Kang, J.G.; Park, E.C. Association of Household Type and Fast-Food Consumption in Korean Adolescents. *Nutrients* **2022**, *14*, 3024. https://doi.org/10.3390/nu14153024

Academic Editor: Zhiyong Zou

Received: 17 June 2022
Accepted: 20 July 2022
Published: 23 July 2022

Publisher's Note: MDPI stays neutral with regard to jurisdictional claims in published maps and institutional affiliations.

Copyright: © 2022 by the authors. Licensee MDPI, Basel, Switzerland. This article is an open access article distributed under the terms and conditions of the Creative Commons Attribution (CC BY) license (https://creativecommons.org/licenses/by/4.0/).

1. Introduction

Consumption of fast food, such as instant noodles, hamburgers, pizza, fried foods, and sugar-sweetened beverages, is reported to be higher in adolescents than in individuals of other age groups [1]. Fast food is directly related to total energy intake and is reported as a factor that degrades the quality of a diet [2–4]. As fast foods are commonly consumed with soda, the risk of being overweight, risk of obesity, and the risk of consuming an unbalanced diet may increase due to soda consumption [5,6]. Korean adolescents often eat out because they spend a lot of time outside their homes for after-school academies [7]. As the food industry grows and food choices available at convenience stores have become extensive, children are being increasingly exposed to substantial amounts of fast food. Similar to adolescents in other countries, the increasing dietary consumption of fast food (which is convenient, low cost, and easily available) among adolescents in Korea has raised concerns regarding nutritional imbalance and hindered growth and development [8]. The "Let's Move" program was launched in the US to fight childhood obesity and improve school lunches in February 2010, and stated that "as fast food became our everyday meal, children have been suffering from obesity and diabetes, which leads to several adult diseases" [1].

Adolescents' dietary behaviors are established through a complex interaction process which involves both internal factors (such as food preference, availability, and appearance) and external factors (such as the influence of parents, peers, and income level) [9]. If their dietary behavior is inappropriate, the growth, development, and nutritional health of adolescents, as well as their life-long health, may be permanently impaired [10,11]. Inappropriate dietary patterns formed in childhood tend to persist into adulthood, which has a considerable effect on adults' health and well-being [12–14]. Therefore, those family members who share the largest proportion of childhood have an immense influence on establishing adolescents' dietary patterns.

Changes in household types are a global trend; the number of households composed of couples and unmarried children in Korea has continually declined from 58.0% in 1990 to 29.1% in 2020, whereas the number of households comprising single parents has increased from 8.7% in 1990 to 9.7% in 2020 [15,16]. In recent years, the proportion of households composed of various configurations, such as grandparents-raising-grandchildren households (0.6%), no-children households (16.9%), and multicultural households (1.8%), has been increasing [16]. Accordingly, the functions of the family have been downscaled. Due to weakening of family functions and kinship, opportunities for children's lifestyle patterns to develop through observation, imitation, and learning from contact with family members have been reduced [17].

Therefore, this study aimed to examine the association between household type and adolescent fast-food consumption, focusing on whether they were living with their parents as well as their reasons for not living with parents.

2. Materials and Methods

2.1. Data and Study Population

Data for this cross-sectional study were obtained from the Korea Youth Risk Behavior Web-Based Survey (KYRBWS), which was conducted annually from 2017 to 2020 by the Korea Centers for Disease Control and Prevention (KCDC). The requirement for ethics approval from the KYRBWS was waived by the KCDC's institutional review board in accordance with the Bioethics and Safety Act of 2015 [18,19]. All data used in this study are publicly available on the KYRBWS website [20]. The KYRBWS complied with the Declaration of Helsinki [21], and all individuals who participated in the KYRBWS provided informed consent.

The KYRBWS used an anonymous and self-administered structured questionnaire with a complex research design that included multistage sampling, stratification, and clustering. The KYRBWS used an online survey system that did not allow respondents to proceed to the next section of the questionnaire unless all questions in the current section were answered. Responses that had logical errors or were outliers (e.g., the respondent incorrectly stated that he or she was younger than 12 years, and thus could not have been included in the KYRBWS because of its grade-level criteria) were processed as missing values. The questionnaire comprised approximately 120 items across 15 categories, including demographic characteristics and health-associated behaviors. The validity and reliability of this survey have been confirmed previously [22,23]. Students in grades 7 to 12 were the target population.

2.2. Variables

The outcome variable was the frequency of fast-food intake per week. All participants answered the question, "How often have you eaten fast food in the last seven days?" A frequency of never to four times a week was classified as "not eating often," whereas more than five to six times a week was classified as "eating frequently".

The main independent variable was household type. The type of cohabitation with parents was first evaluated by choosing all the current family members, then by additionally asking whether they lived with their father (including stepfather) and whether their father and mother (including stepmother) lived together. Family types were classified into four types according to whether they lived with their parents or not: (1) living with both parents; (2) living with a single parent (one of father, mother, stepfather, stepmother); (3) not living together but having parents; and (4) having no parents. For the secondary analyses, the two groups not living with parents were divided into three subgroups based on whom they lived with and where they lived: (1) living with other family members or relatives; (2) living in a dorm or boarding house; and (3) living in an orphanage.

The covariates included the survey year (2017–2020), school grade (7–12), self-reported economic status (low, middle, or high), residential areas (capital area, urban, or rural), smoking status (yes or no), alcohol use (yes or no), body mass index (under-weight, normal, or obese), physical activity (yes or no), and depressive symptoms (yes or no). Physical activity was defined using the following question: in the last seven days, on how many days has your heart rate increased more than usual or have you been out of breath due to physical activity (regardless of type) for a total of 60 min or more per day? Depressive symptoms were assessed using the following question: during the last 12 months, have you felt sadness or despair that interrupted your daily life for at least two weeks?

2.3. Statistical Analysis

All analyses were conducted separately by sex to account for sex-specific differences in dietary behavior. Differences in the frequency and proportion of the categorical variables were evaluated using the χ2 test. Multiple logistic regression analysis was performed to examine the association between household type and adolescents' fast-food consumption patterns, with adjustment for covariates to calculate the adjusted odds ratios (OR) and 95% confidential intervals (CI). All statistical analyses were performed using SAS software version 9.4 (SAS Institute), and used weighted logistic regression to account for the population's representative, clustered, and stratified sampling design. The results were considered statistically significant if the *p*-value was <0.05.

3. Results

A total of 245,839 students participated in the KYRBWS from 2017 to 2020. After excluding participants with missing values (*n* = 59,562), the final study population was 186,277 students (56,767 students in 2017, 55,124 in 2018, 34,990 in 2019, and 39,396 in 2020).

This study included 186,227 adolescents (89,280 boys and 96,997 girls) (Table 1). Among them, 3992 boys (4.5%) and 3505 girls (3.4%) reported consuming fast food five times or more in the last seven days. Among boys, 74,992 (85.6%) lived with both parents, 12,016 (13.7%) lived with a single parent, 1477 (1.7%) had parents but were not cohabitating, and 795 (0.9%) did not have parents. Among girls, 81,917 (85.3%) lived with both parents, 13,004 (13.5%) lived with a single parent, 1554 (1.6%) had parents but were not cohabitating, and 522 (0.5%) did not have parents. Among boys, the frequency of fast-food consumption was higher among those who did not have parents (9.1%) than among those living with both parents (4.3%), living with a single parent (5.0%), and living apart from their parents (5.1%). Among girls, the rate of fast-food consumption was higher among those who did not have parents (7.3%) than among those living with both parents (3.5%), living with a single parent (4.1%), and living apart from their parents (4.5%). Generally, both boys and girls had significantly different characteristics. However, there were no statistically significant regional differences in either boys or girls, or in physical activity among boys.

Participants who had no parents were more likely to eat fast food frequently than those who were living with both parents (boys: OR, 2.00, 95% CI, 1.52–2.62; girls: OR, 1.58, 95% CI, 1.09–2.30) (Table 2). The probability of frequent fast-food intake was increased among those who were living with a single parent (boys: OR, 1.08, 95% CI, 0.98–1.19; girls: OR, 1.03, 95% CI, 0.93–1.15) or living apart from their parents (boys: OR, 1.04, 95% CI, 0.80–1.35; girls: OR, 1.18, 95% CI, 0.89–1.62), although it was not statistically significant.

Table 1. General characteristics of the study population.

Variables	Frequent Fast-Food Intake												
	Boys							Girls					
	Total		Yes		No			Total		Yes		No	
	N	%	N	%	N	%	p-Value	N	%	N	%	N	%	p-Value
Total 186,277	89,280	100.0	3992	4.5	85,288	95.5		96,997	100.0	3505	3.6	93,492	96.4	
Type of Household													
Living with both parents	74,992	85.6	3243	4.3	71,749	95.7	<0.0001	81,917	85.3	2865	3.5	79,052	96.5	<0.0001
Living with single parent	12,016	13.7	602	5.0	11,414	95.0		13,004	13.5	532	4.1	12,472	95.9	
Living apart from parents	1477	1.7	75	5.1	1402	94.9		1554	1.6	70	4.5	1484	95.5	
No parents	795	0.9	72	9.1	723	90.9		522	0.5	38	7.3	484	92.7	
Grade													
Middle school 1st	15,932	18.2	605	3.8	15,327	96.2	<0.0001	17,049	17.8	499	2.9	16,550	97.1	<0.0001
Middle school 2nd	15,251	17.4	664	4.4	14,587	95.6		16,743	17.4	595	3.6	16,148	96.4	
Middle school 3rd	14,906	17.0	726	4.9	14,180	95.1		16,558	17.2	633	3.8	15,925	96.2	
High school 1st	14,233	16.2	667	4.7	13,566	95.3		15,441	16.1	537	3.5	14,904	96.5	
High school 2nd	14,617	16.7	639	4.4	13,978	95.6		15,648	16.3	619	4.0	15,029	96.0	
High school 3rd	14,341	16.4	691	4.8	13,650	95.2		15,558	16.2	622	4.0	14,936	96.0	
Income													
High	37,724	43.1	1742	4.6	35,982	95.4	<0.0001	35,883	37.4	1263	3.5	34,620	96.5	<0.0001
Middle	40,058	45.7	1658	4.1	38,400	95.9		48,021	50.0	1639	3.4	46,382	96.6	
Low	11,498	13.1	592	5.1	10,906	94.9		13,093	13.6	603	4.6	12,490	95.4	
Region													
Capital area	45,228	51.6	2009	4.4	43,219	95.6	0.4119	49,825	51.9	1790	3.6	48,035	96.4	0.2392
City area	38,908	44.4	1769	4.5	37,139	95.5		41,253	43.0	1522	3.7	39,731	96.3	
Rural	5144	5.9	214	4.2	4930	95.8		5919	6.2	193	3.3	5726	96.7	
Smoking													
Yes	7415	8.5	600	8.1	6815	91.9	<0.0001	3025	3.1	254	8.4	2771	91.6	<0.0001
No	81,865	93.4	3392	4.1	78,473	95.9		93,972	97.9	3,251	3.5	90,721	96.5	
Alcohol use													
Yes	13,821	15.8	866	6.3	12,955	93.7	<0.0001	11,832	12.3	672	5.7	11,160	94.3	<0.0001
No	75,459	86.1	3126	4.1	72,333	95.9		85,165	88.7	2833	3.3	82,332	96.7	
BMI													
Obese	17,695	20.2	854	4.8	16,841	95.2	0.0016	23,272	24.2	979	4.2	22,293	95.8	<0.0001
Normal	52,852	60.3	2378	4.5	50,474	95.5		64,531	67.2	2207	3.4	62,324	96.6	
Underweight	18,733	21.4	760	4.1	17,973	95.9		9194	9.6	319	3.5	8875	96.5	

Table 1. *Cont.*

Variables		Frequent Fast-Food Intake													
		Boys							Girls						
		Total		Yes		No			Total		Yes		No		
		N	%	N	%	N	%	*p*-Value	N	%	N	%	N	%	*p*-Value
Total 186,277		89,280	100.0	3992	4.5	85,288	95.5		96,997	100.0	3505	3.6	93,492	96.4	
Physical activity															
	Yes	77,940	89.0	3468	4.4	74,472	95.6	0.4237	68,451	71.3	2414	3.5	66,037	96.5	0.026
	No	11,340	12.9	524	4.6	10,816	95.4		28,546	29.7	1091	3.8	27,455	96.2	
Depressive symptom															
	Yes	17,488	19.6	1121	6.4	16,367	93.6	<0.0001	30,751	31.6	1614	5.2	29,137	94.8	<0.0001
	No	71,792	81.9	2871	4.0	68,921	96.0		66,246	69.0	1891	2.9	64,355	97.1	
Year															
	2017	27,916	31.9	1113	4.0	26,803	96.0	<0.0001	28,851	30.0	890	3.1	27,961	96.9	<0.0001
	2018	27,130	31.0	1089	4.0	26,041	96.0		27,994	29.2	1012	3.6	26,982	96.4	
	2019	15,840	18.1	847	5.3	14,993	94.7		19,150	19.9	794	4.1	18,356	95.9	
	2020	18,394	21.0	943	5.1	17,451	94.9		21,002	21.9	809	3.9	20,193	96.1	

Table 2. Results of factors associated with fast food intake.

Variables		Frequent Fast-Food Intake			
		Boys		**Girls**	
		OR	**95% CI**	**OR**	**95% CI**
Type of household					
	Living with both parents	1.00		1.00	
	Living with single parent	1.08	(0.98–1.19)	1.03	(0.93–1.15)
	Living apart from parents	1.04	(0.80–1.35)	1.18	(0.89–1.62)
	No parents	2.00	(1.52–2.62)	1.58	(1.09–2.30)
Grade					
Middle school	1st	0.89	(0.78–1.01)	0.83	(0.72–0.95)
	2nd	1.00	(0.87–1.14)	0.96	(0.84–1.10)
	3rd	1.10	(0.97–1.25)	1.04	(0.91–1.18)
High school	1st	1.06	(0.94–1.20)	0.92	(0.81–1.05)
	2nd	0.96	(0.85–1.08)	1.07	(0.95–1.22)
	3rd	1.00		1.00	
Income					
	High	1.00		1.00	
	Middle	0.90	(0.90–0.97)	0.96	(0.88–1.05)
	Low	1.01	(1.01–1.13)	1.16	(1.03–1.30)
Region					
	Capital area	1.00		1.00	
	City area	1.03	(0.95–1.11)	0.99	(0.92–1.08)
	Rural	1.00	(0.85–1.19)	0.86	(0.72–1.02)
Smoking					
	Yes	1.69	(1.52–1.89)	1.61	(1.36–1.90)
	No	1.00		1.00	
Alcohol use					
	Yes	1.24	(1.12–1.37)	1.37	(1.22–1.53)
	No	1.00		1.00	
BMI					
	Obese	1.14	(1.05–1.25)	1.32	(1.21–1.44)
	Normal	1.00		1.00	
	Underweight	0.89	(0.81–0.97)	0.99	(0.87–1.13)
Physical activity					
	Yes	1.00		1.00	
	No	1.03	(0.93–1.15)	1.07	(0.98–1.16)
Depressive symptom					
	Yes	1.49	(1.37–1.61)	1.74	(1.61–1.88)
	No	1.00		1.00	
Year					
	2017	1.00		1.00	
	2018	1.04	(0.94–1.15)	1.17	(1.05–1.30)
	2019	1.36	(1.22–1.51)	1.31	(1.17–1.47)
	2020	1.37	(1.23–1.52)	1.26	(1.12–1.41)

In subgroup analyses stratified by covariates, boys without parents had a strong association with frequent fast-food intake when they lived in a low- (OR, 2.60, 95% CI, 1.67–4.07) or high -income household (OR, 1.81, 95% CI, 1.16–2.83) in 2017 (OR, 2.35, 95% CI, 1.50–3.70) and in 2018 (OR, 2.26, 95% CI, 1.47–3.47) (Table 3). Boys living with a single parent in a high-income household had an increased risk of frequent fast-food intake (OR, 1.22, 95% CI, 1.02–1.46). Among girls without parents, there was a strong association with frequent fast-food intake when they had a low (OR, 2.13, 95% CI, 1.24–3.64) or high household income (OR, 2.30, 95% CI, 1.11–4.80).

Table 4 shows the secondary analyses regarding whom the participants were living with and where they were living for participants not living with their parents. Boys who did not have parents and were living in a dorm or boarding house (OR, 4.58, 95% CI, 2.27–9.23) or living with other family members or relatives (OR, 1.85, 95% CI, 1.36–2.53) demonstrated a significant association with frequent fast-food intake compared to those living with both parents. For girls, those who did not have parents and were living in a dorm or boarding house (OR, 4.40, 95% CI, 1.75–11.08) had a significant association with frequent fast-food intake compared to those living with both parents. Generally, participants living in other types of households had an increased risk of frequent fast-food intake compared to those living with both parents, although the statistical significance was marginal. Boys who lived apart from their parents with other family members or relatives and girls who lived apart from their parents in an orphanage had a decreased risk of frequent fast-food intake, although the statistical significance was again marginal.

Table 3. The results of subgroup analysis stratified by independent variables.

	Frequent Fast Food Intake						
Variables	Living with Both Parents	Living with Single Parent		Living Apart from Parents		No-Parents	
	OR	OR	95% CI	OR	95% CI	OR	95% CI
Boys							
Income							
High	1.00	1.22	(1.02–1.46)	1.10	(0.65–1.87)	1.81	(1.16–2.83)
Middle	1.00	1.01	(0.87–1.18)	0.92	(0.61–1.38)	1.64	(0.96–2.78)
Low	1.00	1.04	(0.85–1.27)	1.18	(0.75–1.85)	2.60	(1.67–4.07)
Year							
2017	1.00	1.07	(0.89–1.27)	1.29	(0.81–2.06)	2.35	(1.50–3.70)
2018	1.00	0.93	(0.75–1.15)	0.74	(0.42–1.30)	2.26	(1.47–3.47)
2019	1.00	1.18	(0.95–1.47)	1.19	(0.74–1.92)	1.89	(0.93–3.83)
2020	1.00	1.20	(1.00–1.45)	0.90	(0.51–1.59)	0.71	(0.25–2.04)
Girls							
Income							
High	1.00	1.22	(0.99–1.51)	1.61	(0.94–2.74)	2.30	(1.11–4.80)
Middle	1.00	0.97	(0.83–1.13)	0.99	(0.64–1.52)	0.62	(0.31–1.23)
Low	1.00	0.99	(0.82–1.21)	1.13	(0.69–1.84)	2.13	(1.24–3.64)
Year							
2017	1.00	0.90	(0.72–1.12)	1.07	(0.62–1.84)	1.21	(0.53–2.77)
2018	1.00	0.90	(0.74–1.10)	0.94	(0.56–1.59)	1.93	(0.98–3.82)
2019	1.00	1.23	(0.99–1.51)	1.07	(0.56–2.05)	1.54	(0.85–2.79)
2020	1.00	1.18	(0.95–1.46)	1.90	(1.11–3.25)	1.52	(0.63–3.67)

Table 4. Results of factors associated with fast food intake.

	Frequent Fast Food Intake			
Variables	Boys		Girls	
	OR	95% CI	OR	95% CI
Type of household				
Living with both parents	1.00		1.00	
Living with single parent	1.08	(0.98–1.19)	1.03	(0.93–1.15)
Living apart from parents				
Living with family or relative	0.93	(0.66–1.30)	1.23	(0.88–1.71)
Living in dormitory or boarding house	1.15	(0.74–1.78)	1.15	(0.70–1.89)
Living in orphanage	2.07	(0.84–5.12)	0.69	(0.27–1.73)
No parents				
Living with family or relative	1.85	(1.36–2.53)	1.17	(0.73–1.88)
Living in dormitory or boarding house	4.58	(2.27–9.23)	4.40	(1.75–11.08)
Living in orphanage	1.20	(0.49–2.99)	2.17	(0.94–4.99)

Subsequently, sensitivity analyses were conducted to confirm the robustness of the results regardless of the changing definition of frequent fast-food intake (Figure 1). In analyses with various cut-offs for defining the frequency, participants who lived with a single parent, apart from their parents, or who had no parents had an increased risk of frequent fast-food intake, except in the analysis of ever versus never fast-food intake. Although the statistical significance was marginal in a few analyses, participants who did not have parents had the highest risk of frequent fast-food intake in most analyses.

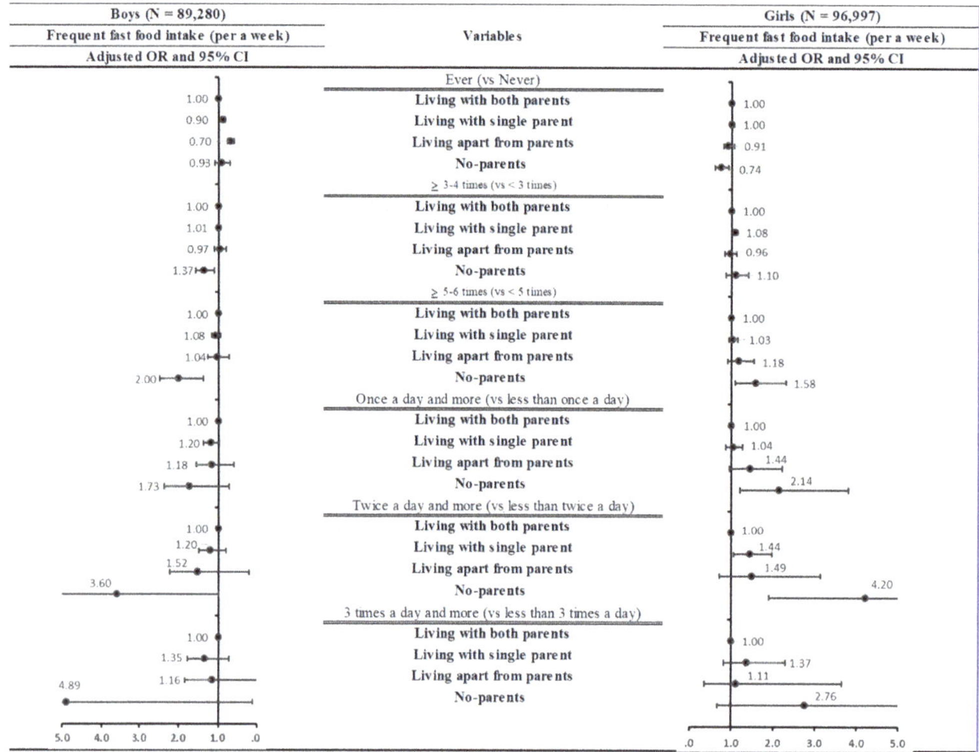

Figure 1. Results of sensitivity analysis by defining fast food intake. All covariates were adjusted. OR: Odds ratio; 95% CI: 95% Confidential Intervals.

4. Discussion

After adjusting for demographic, socio-economic, and health behavior factors, adolescents who did not have parents had a higher risk of frequent fast-food intake than those who lived with both parents. Among adolescents without parents, boys living in a dorm or boarding house or living with other family members or relatives as well as girls living in a dorm or boarding house had a notable association with frequent fast-food intake. Furthermore, regardless of the various definitions of frequency of fast-food intake, adolescents without parents were the most likely to consume fast-food.

In line with our study findings, the findings of previous studies have shown that parents play an instrumental role in establishing adolescents' dietary behavior. Family, especially parents, are the first subjects whose behavior is mirrored after birth by their children. Moreover, they are the individuals who the children imitate before they are old enough to attend a day care center. Children learn mainly through observation and imitating what they observe. Therefore, parents' words and deeds have a deep and subtle impact on children, and have a long-term impact on their adult life [17]. In a functioning family [24] or when children are involved in family meal preparation [25], children's dietary behaviors are more likely to be healthier. Both parents and grandparents play an influential role in the development of healthy dietary behavior.

In an Australian study, grandparents helping with childcare were found to play a role in preventing obesity-related behavior in young children by restricting access to certain foods, allowing grandchildren a high degree of input and control when planning mealtimes and food choices, and providing more encouragement of a balanced dietary intake compared to parents [26]. Grandparents can play the role of surrogate parents on behalf of children's real parents due to changes in household types, such as double income families, single-parent families, grandparents-raising-grandchildren families, and sibling-breadwinner families. Because of data limitations, we could not identify more detailed household types. However, we assume that certain factors, such as help from other family members or surrogate guardians on behalf of the parents, might influence the results in the analysis of household types. Further studies are needed to investigate household types in greater detail.

A previous study has reported that single-parent families are associated with a higher probability of an unbalanced diet than both-parent families. Due to the absence of dietary management by parents, children's meals are more likely to be replaced by fast food. In particular, in low-income households, irregular food intake such as binge eating and excessive intake of low-nutritional and high-calorie foods such as fast food and snacks were shown to be prevalent due to neglect by parents or guardians [27]. The higher risk of frequent fast-food intake among adolescents without parents found in our study might be explained by the absence of dietary management and neglect by parents or guardians.

Considering that adolescents in low-income households, especially adolescents without parents, are more likely to eat fast food, we speculated that adolescents in orphanages would be more likely to eat fast food than those living with family or living in a dorm due to their limited financial capacity to choose healthy foods. However, adolescents without parents and living in a dorm or boarding house were strongly associated with frequent fast-food intake. Care providers' supervision of eating behavior might be an explanation. In residential care facilities for children, regulations and supervision by the staff have an impact on the children's dietary patterns [28,29]. On the other hand, living in a dorm surrounded by peers might be associated with an increase in fast-food intake due to inexpensive price, better accessibility, and less supervision by guardians [30].

In Korea, a "children's meal card" (called the "Kkum-namu card") has been provided to children from low- income, single-parent, or near-poor families beginning in 2009, allowing them to have meals outside of school whenever have not eaten school meals [31]. It has been pointed out that the minimum amount of one meal with the "Children's meal card" is USD 3; this amount was very low for an appropriate meal, even in 2017. It was not enough to have a meal at restaurants, and was only enough for instant noodle cups or snacks in convenient stores. Hence, the minimum amount was increased to approximately 4 dollars in 2018 and then to 5, which was graded by age, in 2020. However, boys who were living with a single parent and girls who were living apart from parents in 2020 continued to have a higher risk of frequent fast-food intake in our study. Furthermore, to the best of our knowledge there is no evidence to show that the revised minimum amount of the "children's meal card" is sufficient to reduce these children's food insecurity. Therefore, further studies that examine the effect of the "children' meal card" on children's healthy and balanced diets are needed. Furthermore, more practical revisions or initiatives should be implemented to guarantee a well-balanced diet that reflects real-world costs.

This study has limitations. We could not identify the direction of the associations or establish causality, although appropriate methods were used to mitigate these limitations. Furthermore, the KYRBWS data were collected anonymously online and self-reported, which may have resulted in misclassification or biased results. Because the KYRBWS assessed cohabitation only with parents, not with grandparents or siblings, we could not evaluate whether the type of main guardian (i.e., grandparents, brother or sister, or other) was associated with fast-food intake among adolescents not living with their parents. In addition, certain associations had wide 95% CI or marginal statistical significance due to the small sample size, especially among those not living with parents.

Despite these limitations, this study identified the effect of both the household type and of the individuals adolescents were living with and where they were living on the frequency of fast-food intake. Furthermore, the multiyear national survey data (2017–2020) were obtained from approximately 186,000 adolescents based on random cluster sampling, ensuring that the results were sufficiently representative of Korean adolescents [32].

In this cross-sectional study, adolescents who did not have parents had a higher risk of frequent fast-food intake than those who lived with both parents. Among adolescents without parents, boys living in a dorm or boarding house or living with other family members or relatives and girls living in a dorm or boarding house had a notable association with frequent fast-food intake. In analysis with various definitions of frequency of fast-food intake, adolescents without parents were the most likely to consume fast food. These findings suggest that the type of household and cohabitation are associated with the frequency of fast food intake among adolescents. This research can provide evidence related to healthy dietary patterns among Korean adolescents. Consideration should be paid to the person(s) with whom they were living with and where they were living when conducting future research, as well as when revising public policies regarding well-balanced diets for children.

Author Contributions: Conceptualization, H.S.K., S.H.K. and J.G.K.; Formal Analysis: H.S.K., Y.S.P. and S.H.K.; Writing—Original Draft Preparation: H.S.K. and S.H.K.; Writing—Review and Editing: E.C.P. and S.H.K.; Supervision: E.C.P. All authors have read and agreed to the published version of the manuscript.

Funding: This research received no external funding.

Institutional Review Board Statement: The ethics committee of Yonsei University approved the use of these data. The study was exempted from the need for written informed consent by the Institutional Review Board (4-2021-1437).

Informed Consent Statement: The study was exempted from the need for written informed consent by the Institutional Review Board.

Data Availability Statement: The reports and microdata of KYRBS are released annually in December each year. The KYRBS website [http://yhs.cdc.go.kr, accessed on 1 June 2022] contains microdata files, reports and publications for the survey.

Acknowledgments: We would like to thank the KCDC for providing the raw data.

Conflicts of Interest: The authors declare no potential conflicts of interest with respect to the research, authorship, and/or publication of this article.

References

1. Jo, Y.W. *Analysis on Consumption of Fast-Foods and Sugar Sweetened Beverages, and Health Condition and Behavior*; Korea Centers for Disease Control and Prevention: Seoul, Korea, 2011.
2. Bowman, S.A.; Gortmaker, S.L.; Ebbeling, C.B.; Pereira, M.A.; Ludwig, D.S. Effects of Fast-Food Consumption on Energy Intake and Diet Quality Among Children in a National Household Survey. *Pediatrics* **2004**, *113*, 112–118. [CrossRef] [PubMed]
3. Bowman, S.A.; Vinyard, B.T. Fast Food Consumption of U.S. Adults: Impact on Energy and Nutrient Intakes and Overweight Status. *J. Am. Coll. Nutr.* **2004**, *23*, 163–168. [CrossRef]
4. French, S.A.; Story, M.; Neumark-Sztainer, D.; Fulkerson, J.A.; Hannan, P. Fast food restaurant use among adolescents: Associations with nutrient intake, food choices and behavioral and psychosocial variables. *Int. J. Obes.* **2001**, *25*, 1823–1833. [CrossRef] [PubMed]
5. Vartanian, L.R.; Schwartz, M.B.; Brownell, K.D. Effects of Soft Drink Consumption on Nutrition and Health: A Systematic Review and Meta-Analysis. *Am. J. Public Health* **2007**, *97*, 667–675. [CrossRef] [PubMed]
6. Giammattei, J.; Blix, G.; Marshak, H.H.; Wollitzer, A.O.; Pettitt, D.J. Television Watching and Soft Drink Consumption: Associations With Obesity in 11- to 13-Year-Old Schoolchildren. *Arch. Pediatrics Adolesc. Med.* **2003**, *157*, 882–886. [CrossRef]
7. Kowen, M.-R.; Jang, M.; Lee, N.-K. Transforming in awareness of relationship problems due to excessive private education in Korea. *Int. J. Qual. Stud. Health Well-Being* **2019**, *14*, 1586624. [CrossRef]
8. You, J.S.; Kim, S.M.; Chang, K.J. Nutritional Knowledge and Dietary Behavior of the 6th Grade Elementary School Students in Daejeon Area by Gender and Skipping Breakfast. *Korean J. Nutr.* **2009**, *42*, 256–267. [CrossRef]

9.	Hong, S.; Bae, H.C.; Kim, H.S.; Park, E.-C. Variation in meal-skipping rates of Korean adolescents according to socio-economic status: Results of the Korea Youth Risk Behavior Web-based Survey. *J. Prev. Med. Public Health* **2014**, *47*, 158–168. [CrossRef]

10.	Spear, B.A. Adolescent Growth and Development. *J. Am. Diet. Assoc.* **2002**, *102*, S23–S29. [CrossRef]

11.	Rees, J.M.; Neumark-Sztainer, D.; Kohn, M.; Jacobson, M. Improving the nutritional health of adolescents-position statement-society for adolescent medicine. *J. Adolesc. Health* **1999**, *24*, 461–462. [CrossRef]

12.	Kemm, J.R. Eating Patterns in Childhood and Adult Health. *Nutr. Health* **1987**, *4*, 205–215. [CrossRef] [PubMed]

13.	Nicklas, T.A.; Farris, R.P.; Smoak, C.G.; Frank, G.C.; Srinivasan, S.R.; Webber, L.S.; Berenson, G.S. Dietary factors relate to cardiovascular risk factors in early life. *Bogalusa Heart Study. Arterioscler. Off. J. Am. Heart Assoc. Inc.* **1988**, *8*, 193–199.

14.	Rosenheck, R. Fast food consumption and increased caloric intake: A systematic review of a trajectory towards weight gain and obesity risk. *Obes. Rev.* **2008**, *9*, 535–547. [CrossRef] [PubMed]

15.	Baek, Y.J. Family Structure and Diet of Children and Adolescents: Analysis of Data from the Fifth Korea National Health and Nutrition Examination Survey(KNHANES V) 2010–2011. Master's Thesis, Seoul National University, Seoul, Korea, 2014.

16.	Korea National Statistical Office. *Accual Statistic Information Report on Population and Housing Census of Korea*; Statistics Korea: Sejong, Korea, 2020.

17.	Qin, X.; Zhang, W.; Hua, L.; Zhu, Z.; Ye, T. Influence of family dietary environment on children's dietary behaviours. *Chin. Nurs. Res.* **2014**, *2014*, 4228–4230.

18.	Korea Centers for Disease Control and Prevention. *The Report of the Korea Youth Risk Behavior Web-Based Survey in 2017*; Korea Centers for Disease Control and Prevention: Cheongju, Korea, 2017.

19.	Jeong, W.; Kim, Y.K.; Lee, H.J.; Jang, J.; Kim, S.; Park, E.-C.; Jang, S.-I. Association of Bedtime with both Suicidal Ideation and Suicide Planning among Korean Adolescents. *Int. J. Environ. Res. Public Health* **2019**, *16*, 3817. [CrossRef] [PubMed]

20.	Korea Centers for Disease Control and Prevention. *The Statistics of 16th Korea Youth Risk Behavior Web-Based Survey*; Korea Centers for Disease Control and Prevention: Cheongju, Korea, 2021.

21.	Association, W.M. World Medical Association Declaration of Helsinki: Ethical Principles for Medical Research Involving Human Subjects. *JAMA* **2013**, *310*, 2191–2194.

22.	Bae, J.-S.; Joung, H.-J.; Kim, J.-Y.; Kwon, K.-N.; Kim, Y.-T.; Park, S.-W. Test-retest reliability of a questionnaire for the Korea Youth Risk Behavior Web-based Survey. *J. Prev. Med. Public Health* **2010**, *43*, 403–410. [CrossRef]

23.	Bae, J.-S.; Joung, H.-J.; Kim, J.-Y.; Kwon, K.-N.; Kim, Y.-J.; Park, S.-W. Validity of self-reported height, weight, and body mass index of the Korea Youth Risk Behavior Web-based Survey questionnaire. *J. Prev. Med. Public Health* **2010**, *43*, 396–402. [CrossRef]

24.	Li Wen, M.; Simpson, J.M.; Baur, L.A.; Rissel, C.; Flood, V.M. Family Functioning and Obesity Risk Behaviors: Implications for Early Obesity Intervention. *Obesity* **2011**, *19*, 1252–1258. [CrossRef]

25.	Berge, J.M.; MacLehose, R.F.; Larson, N.; Laska, M.; Neumark-Sztainer, D. Family Food Preparation and Its Effects on Adolescent Dietary Quality and Eating Patterns. *J. Adolesc. Health* **2016**, *59*, 530–536. [CrossRef]

26.	Bell, L.K.; Perry, R.A.; Prichard, I. Exploring Grandparents' Roles in Young Children's Lifestyle Behaviors and the Prevention of Childhood Obesity: An Australian Perspective. *J. Nutr. Educ. Behav.* **2018**, *50*, 516–521. [CrossRef] [PubMed]

27.	Lee, S.J.; Ryu, H.K. Analysis of the Dietary Life of Adolescents by Household Types in Korea using the Korea National Health and Nutrition Examination Survey. *Korean J. Community Living Sci.* **2021**, *32*, 285–304. [CrossRef]

28.	Markert, J.; Herke, M.; Bartels, A.; Gosse, K.; Roick, J.; Herz-Jakoby, A.; Täubig, V.; Schröer, W.; Richter, M. Food practices and nutrition of children and adolescents in residential care: A scoping review. *Appetite* **2021**, *167*, 105640. [CrossRef] [PubMed]

29.	He, M.; Tucker, P.; Gilliland, J.; Irwin, J.D.; Larsen, K.; Hess, P. The Influence of Local Food Environments on Adolescents' Food Purchasing Behaviors. *Int. J. Environ. Res. Public Health* **2012**, *9*, 1458–1471. [CrossRef] [PubMed]

30.	Nelson, M.C.; Gordon-Larsen, P.; North, K.E.; Adair, L.S. Body Mass Index Gain, Fast Food, and Physical Activity: Effects of Shared Environments over Time. *Obesity* **2006**, *14*, 701–709. [CrossRef] [PubMed]

31.	Jeon, H.J. A proposal for improvement of the business supporting poorly-fed children using mobile payment and information delivery platform. In Proceedings of the Korea Information Processing Society Conference, Seoul, Korea, 2020; pp. 37–40. [CrossRef]

32.	Kim, S.H.; Jeong, S.H.; Park, E.-C.; Jang, S.-I. Association of Cigarette Type Initially Smoked With Suicidal Behaviors Among Adolescents in Korea From 2015 to 2018. *JAMA Netw. Open* **2021**, *4*, e218803. [CrossRef] [PubMed]

nutrients

Article

The Effects of Increasing Fruit and Vegetable Intake in Children with Asthma on the Modulation of Innate Immune Responses

Banafsheh Hosseini [1,†], Bronwyn S. Berthon [1,†], Megan E. Jensen [2], Rebecca F. McLoughlin [1], Peter A. B. Wark [1,3], Kristy Nichol [1], Evan J. Williams [1], Katherine J. Baines [1], Adam Collison [1,2], Malcolm R. Starkey [1,2,4], Joerg Mattes [2,3] and Lisa G. Wood [1,2,*]

1   Priority Research Centre for Healthy Lungs, Hunter Medical Research Institute, University of Newcastle, Newcastle, NSW 2305, Australia; b.hosseini.bh@gmail.com (B.H.); bronwyn.berthon@newcastle.edu.au (B.S.B.); bec.mcloughlin@newcastle.edu.au (R.F.M.); peter.wark@health.nsw.gov.au (P.A.B.W.); kristy.nichol@newcastle.edu.au (K.N.); evan.j.williams@newcastle.edu.au (E.J.W.); katherine.baines@newcastle.edu.au (K.J.B.); adam.collison@newcastle.edu.au (A.C.); malcolm.starkey@newcastle.edu.au (M.R.S.)
2   Priority Research Centre Grow Up Well, Hunter Medical Research Institute, University of Newcastle, Newcastle, NSW 2305, Australia; megan.jensen@newcastle.edu.au (M.E.J.); joerg.mattes@newcastle.edu.au (J.M.)
3   Department of Respiratory and Sleep Medicine, John Hunter Hospital, Newcastle, NSW 2305, Australia
4   Department of Immunology and Pathology, Central Clinical School, Sub-Faculty of Translational Medicine and Public Health, Monash University, Melbourne, VIC 3004, Australia
*   Correspondence: lisa.wood@newcastle.edu.au
†   These authors contributed equally to this work.

Citation: Hosseini, B.; Berthon, B.S.; Jensen, M.E.; McLoughlin, R.F.; Wark, P.A.B.; Nichol, K.; Williams, E.J.; Baines, K.J.; Collison, A.; Starkey, M.R.; et al. The Effects of Increasing Fruit and Vegetable Intake in Children with Asthma on the Modulation of Innate Immune Responses. *Nutrients* 2022, 14, 3087. https://doi.org/10.3390/nu14153087

Academic Editor: Zhiyong Zou

Received: 20 June 2022
Accepted: 25 July 2022
Published: 27 July 2022

Publisher's Note: MDPI stays neutral with regard to jurisdictional claims in published maps and institutional affiliations.

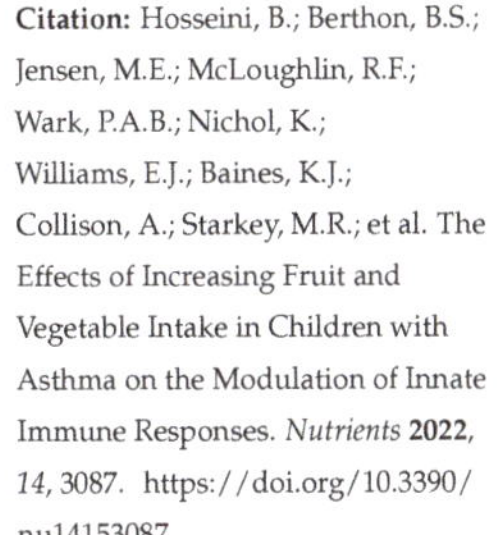

Copyright: © 2022 by the authors. Licensee MDPI, Basel, Switzerland. This article is an open access article distributed under the terms and conditions of the Creative Commons Attribution (CC BY) license (https://creativecommons.org/licenses/by/4.0/).

Abstract: Children with asthma are at risk of acute exacerbations triggered mainly by viral infections. A diet high in fruit and vegetables (F&V), a rich source of carotenoids, may improve innate immune responses in children with asthma. Children with asthma (3–11 years) with a history of exacerbations and low F&V intake ($\leq$3 serves/d) were randomly assigned to a high F&V diet or control (usual diet) for 6 months. Outcomes included respiratory-related adverse events and in-vitro cytokine production in peripheral blood mononuclear cells (PBMCs), treated with rhinovirus-1B (RV1B), house dust mite (HDM) and lipopolysaccharide (LPS). During the trial, there were fewer subjects with $\geq$2 asthma exacerbations in the high F&V diet group (n = 22) compared to the control group (n = 25) (63.6% vs. 88.0%, $p = 0.049$). Duration and severity of exacerbations were similar between groups. LPS-induced interferon (IFN)-γ and IFN-λ production showed a small but significant increase in the high F&V group after 3 months compared to baseline ($p < 0.05$). Additionally, RV1B-induced IFN-λ production in PBMCs was positively associated with the change in plasma lycopene at 6 months ($r_s = 0.35$, $p = 0.015$). A high F&V diet reduced asthma-related illness and modulated in vitro PBMC cytokine production in young children with asthma. Improving diet quality by increasing F&V intake could be an effective non-pharmacological strategy for preventing asthma-related illness by enhancing children's innate immune responses.

Keywords: childhood asthma; innate immunity; fruit and vegetables; carotenoids

1. Introduction

Asthma is the most prevalent chronic childhood disease, ranking among the top 20 conditions worldwide for disability-adjusted life years in children [1]. Children with asthma frequently experience exacerbations [2]. Acute asthma exacerbations are the primary cause of urgent healthcare visits, hospitalisations, mortality, and incur significant treatment costs [2]. Respiratory virus infections are the most common cause of asthma exacerbations in children [3], with rhinoviruses (RVs) being the most common cause of virus-associated exacerbations in children over the age of 3 years [4]. RV infection provokes asthma symptoms and can also impair lung function in patients with asthma [5]. Other environmental factors such as bacterial infection and allergen exposure play an essential role in asthma exacerbations. Several studies have shown a consistent relationship between levels of household lipopolysaccharide (LPS) and house dust mite (HDM) exposure [6] with asthma exacerbations [7–9].

Asthma is usually managed using inhaled corticosteroids (ICS); however, viral-induced exacerbations can occur despite ICS treatment in children with asthma [10,11]. One proposed non-pharmacological, adjuvant therapeutic approach to prevent and manage asthma is increasing the dietary intake of fruit and vegetables, a strategy that has been reported by the European Academy of Allergy and Clinical Immunology (EAACI) to decrease asthma incidence, particularly in children [12]. Fruit and vegetables are high in antioxidants and anti-inflammatory phytochemicals, including carotenoids (e.g., lutein, lycopene, β-cryptoxanthin, α-carotene and β-carotene) and other biologically active substances [13]. Dietary antioxidants can scavenge reactive oxygen species (ROS) [14], which are increased in the airways of patients with asthma [15], and thus, inhibit nuclear factor-kappa B (NFκB)-mediated inflammation [14]. However, this capacity is reduced following a low F&V diet [16].

Many epidemiological studies have shown that total fruit and vegetable (F&V) intake is inversely associated with the risk of asthma [17,18] and wheezing [19–21] and is positively associated with lung function [22,23]. Indeed, our previous meta-analyses showed inverse associations between fruit consumption and risk of prevalent wheeze and asthma severity [24]. Similarly, vegetable intake was inversely associated with the risk of asthma [24]. Extending these observations, in our previous randomised control trial (RCT) conducted in 137 adult patients with asthma, we demonstrated that those assigned to a low versus a high F&V diet (<3 vs. ≥7 serves of F&V per day) for 14 weeks had a 2.26-fold increased risk of an asthma exacerbation [25].

This study reports secondary findings from an RCT that delivered a 6-month dietary intervention to increase F&V intake in children with asthma [26]. We have previously demonstrated that high F&V diet for 6 months did not prevent exacerbations, though it was associated with improvements in lung function [26]. In this study, we aimed to extend our previous findings, hypothesising that a high F&V diet would protect children with asthma from exacerbations via enhancing innate immune responses.

2. Materials and Methods

2.1. Study Design and Participants

The complete study design and methodology have been described previously [26]. Participants were recruited via attendance to the emergency department or admission to the John Hunter Children's Hospital, Newcastle, Australia, and Maitland Hospital, Newcastle, Australia, following an asthma exacerbation, from September 2015 to July 2018.

Children aged 3–11 years were eligible if they had a physician diagnosis of asthma; recent exacerbation/s (≥1 exacerbation in the preceding 6 months or ≥2 in the past 12 months); stable asthma at the initial clinic visit (defined as no change in asthma medications, unscheduled medical visit for asthma, use of OCS or antibiotics in preceding 4 weeks); consuming ≤3 serves of F&V per day (assessed over the past week); willingness and ability to attend clinic appointments; desire to comply with proposed dietary changes; and agreement to collect blood samples for research purposes at clinic visits. Exclusion criteria included other respiratory conditions, diagnosed intestinal disorders, or consumption of nutritional supplements (in the previous 4 weeks).

Subjects randomly assigned to the high F&V diet were encouraged to meet age-appropriate Australian Dietary Guideline (ADG) recommendations for F&V serves/day [27]. The participants in the control group, blinded to the study hypothesis, continued their usual diet (≤3 serves/d F&V).

Food hampers were provided according to group allocation: the high F&V group received fortnightly food hampers, which included a selection of fresh and frozen fruit and vegetables, while the control group received monthly food baskets, which included a selection of carbohydrate-based foods, low in soluble fibre and antioxidants (bread, rice, pasta, cereal). The hampers were tailored to the subject's preferences, within the requirements of the dietary intervention.

Adherence was assessed by 24-h food recalls [27] collected at each clinic visit and during fortnightly telephone calls.

Children were screened to ensure they did not have any symptoms of cold/flu at the time of sample collection. The study was approved by the Hunter New England Ethics Committee (15/06/17/4.03) and registered with the University of Newcastle Human Research Ethics Committee. Written informed parental consent, and child assent (where applicable) was obtained before participation in the study. The children and their families were given an information brochure on general healthy eating for their age and a personalised consultation with a dietitian at the end of the trial.

2.2. Outcomes

The primary outcomes of this trial have been reported previously [26]. The secondary outcomes of this study included frequency of respiratory-related adverse events, severity and duration of asthma exacerbations, viral detection in nasal swabs collected during asthma-related events, and in-vitro cytokine production in PBMCs treated with different stimuli.

2.3. Clinical Assessment

Detailed methods of clinical assessments are reported elsewhere [26]. Briefly, at baseline, 3 months (±2 weeks) and 6 months (±2 weeks), all participants fulfilling the inclusion criteria attended the clinic at the Hunter Medical Research Institute, Newcastle, Australia, for clinical assessment and blood collection following a 12-h overnight fast.

2.4. Asthma Exacerbations and Upper Respiratory Tract Infection

Parent-reported asthma-related illness was categorised into three groups, based on the child's symptoms and signs: (1) exacerbation: a parent-reported asthma exacerbation alone without the presence of upper respiratory tract infection (URTI) symptoms; (2) URTI: a parent-reported URTI alone (runny/congested nose, sore throat, earache, sneezing, with or without fever), with no change in asthma symptoms; and (3) exacerbation with URTI: a parent-reported increase in cough, wheeze or shortness of breath with suspected URTI.

At the time of a suspected URTI or asthma exacerbation, the parent/guardian was asked to complete the validated Asthma flare-up diary for young children (ADYC) questionnaire to document the presence, duration and severity of an asthma exacerbation [28]. Parents were also instructed to collect a nasal swab at the onset of an asthma exacerbation or a suspected URTI (no later than day one or two of symptoms). For additional details, please see this article's Supplementary Materials (Supplementary File S1).

2.5. Laboratory Methods
2.5.1. Carotenoid Analysis

High-Performance Liquid Chromatography (HPLC) was used to measure plasma carotenoid concentrations using Agilent 1200 Series HPLC with Chemstations software (Agilent Corporation, Waldbronn, Germany) as described previously [29,30].

2.5.2. Identification of Respiratory Viruses in Nasal Samples

Viral RNA was extracted from 140 µL of each nasal sample using the QIAamp Viral RNA Mini Kit (Qiagen, Australia), following the manufacturer's instructions [31]. Samples were then stored at −80 °C until the qPCR assays were performed. For additional details, please see this article's Supplementary Materials (Supplementary File S1).

2.5.3. PBMC Isolation and Culture

PBMCs were isolated from whole blood by density gradient method [32] using Ficoll-PaqueTMPLUS (GE Healthcare, Sydney, Australia) and cultured with and without RV1B,

LPS or HDM for 48 h. For additional details, please see this article's Supplementary Materials (Supplementary File S1).

2.5.4. Cytokine Assays

Cell culture supernatant concentrations of IFN-γ, IL-1β and IL-6, were analysed using a bead-based multiplex assay (BD Bioscience, Sydney, Australia), and IFN-λ concentrations were measured using high-sensitivity commercial ELISA assays (R&D Systems, Sydney, Australia) as per the manufacturer's recommendations. For additional details, please see this article's Supplementary Materials (Supplementary File S1).

2.6. Statistical Analysis

To evaluate the true, undiluted, effect of the 6-month F&V intervention, a complete case per-protocol analysis (PPA) was completed, excluding intervention group subjects who consumed below ADG F&V serves in $\geq$four 24-h food recalls (n = 2). Data were reported as mean $\pm$ standard deviation or median [interquartile range]. Significant differences between groups were determined using an independent *t*-test (parametric data), the Mann–Whitney U-test, or the Wilcoxon signed-rank test as appropriate (non-parametric data). Group differences in change from baseline during the intervention were analysed using linear mixed models (LMMs) with group (intervention or control) and time (treated as categorical with levels at baseline (0 months), 3 and 6 months). LMMs were adjusted for the number of hospitalisations in the previous 12 months, and random effects were specified for time. LMMs use all data available at each time point; therefore, missing data imputation was not undertaken. Group differences in ADYC scores from baseline during the intervention were analysed using LMM adjusted for maintenance ICS use, accounting for multiple events per child. Pearson's chi-squared and Fisher's exact tests (two-tailed) were used to compare virus detection rates in nasal samples from both groups. Linear correlation between variables was measured using the Pearson product-moment correlation coefficient (r) or Spearman rank correlation (r_s), as per distribution. Statistical analyses were performed using STATA 15 (StataCorp, College Station, TX, USA). Significance was accepted when $p < 0.05$.

3. Results

3.1. Subjects' Characteristics

Sixty-seven children (median age of 5 (3, 7) years) with stable asthma were randomised to the high F&V diet (n = 33) or control group (n = 34). Of these, 16 subjects were either lost to follow-up or withdrew, with two excluded due to non-adherence to the dietary intervention. Forty-seven participants (intervention n = 22, control n = 25) were included in the per-protocol analysis (Figure S1).

Compared to the control group, significantly more individuals in the intervention group were exposed to tobacco smoke in-utero ($p = 0.015$), and they had $\geq$1 hospital admission for asthma in the previous 12 months ($p = 0.027$); all other baseline characteristics did not differ between groups (Table 1). Medication use in the last 12 months was also similar between groups (Table 2).

Table 1. Baseline demographics and clinical characteristics.

Baseline Characteristic	Intervention n = 22	Control n = 25	p
Age (years), median (range)	5 (3–10)	5 (3–11)	0.827
Age 3–6 years, n (%)	17 (77)	18 (72)	0.747
Age 7–11 years, n (%)	5 (23)	7 (28)	
Sex (Male: Female)	15:7	19:6	0.550
Race: White, n (%)	17 (77.3%)	20 (80.0%)	0.303
Height (cm), mean ± SD	117 ± 14	116 ± 16	0.988
Weight (kg), median (IQR)	21.6 (16.9, 25.1)	21.4 (16.9, 26.6)	0.664
BMI z-score, mean ± SD	0.1 ± 1.3	0.1 ± 1.4	0.922
BMI percentile, mean ± SD	49.4 ± 32.8	54.7 ± 32.4	0.623
Risk factors, n (%)			
Current food allergy	6 (27)	5 (20)	0.557
History of Eczema [#]	16 (73)	12 (48)	0.085
History of Hay fever [^]	16 (73)	12 (48)	0.085
Asthma in first degree relative [*]	15 (68)	16 (64)	0.592
Maternal Asthma	7 (32)	8 (32)	0.923
Paternal Asthma	6 (27)	10 (40)	0.418
Family history of Eczema	11 (52)	17 (68)	0.280
Family history of Hay fever	18 (86)	21 (84)	0.872
In-utero tobacco exposure [†]	5 (23)	0	**0.015**
Passive smoke exposure at home	3 (14)	1 (4)	0.318
Morbidity in previous 12 months			
ED visits for asthma, mean ± SD	1.7 ± 1.0	1.6 ± 0.9	0.672
≥1 hospital admission, n (%)	16 (73)	10 (40)	**0.027**
Hospitalisations, median (IQR)	1 (0,1)	0 (0, 1)	0.142

BMI z-scores and percentiles were calculated with reference to the Centre for Disease Control and Prevention 2000 Growth Charts. [#] Based on parental response to "Has your child ever had eczema?" [^] Based on parental response to "Has your child ever had a problem with sneezing, or a runny or blocked nose when he/she DID NOT have a cold or the flu?". [*] Based on parental response to "Has anyone in the child's immediate family ever had asthma? (Including mother, father or direct siblings)". [†] Based on parental response to "Does anyone living in the child's immediate home (where the child spends more than half of his/her time) smoke, even if he/she smokes outside?" Difference between groups analysed by the Wilcoxon Rank sum test (non-parametric data), two-sample *t*-test (parametric data) or Pearson's Chi-squared test/Fisher's exact test (testing equality of proportions) where appropriate. $p < 0.05$ considered statistically significant. Values in bold indicate statistically significant results. Abbreviations: IgE, immunoglobulin E; IQR, interquartile range; BMI, body mass index; ED, emergency department.

Table 2. Medication at baseline and in the previous 12 months.

Type of Medication	Intervention n = 22	Control n = 25	p-Value
Short courses of OCS, median (IQR)	3 (1, 4)	2 (1, 4)	0.974
≥2 Short courses of OCS, n (%)	17 (68)	13 (59)	0.526
ICS or ICS/LABA ever, n (%)	16 (73)	14 (56)	0.234
ICS intermittent, n (%) [*]	1 (5)	5 (20)	0.194
ICS maintenance, n (%) [^]	13 (59)	8 (32)	0.062
ICS/LABA maintenance, n (%) [^]	2 (9)	1 (4)	0.593
ICS dose, beclomethasone equiv., median (IQR)	400 (200, 400)	400 (200, 500)	0.573
Montelukast, n (%)	4 (18)	5 (20)	1.000
SABA only, n (%)	4 (18)	9 (36)	0.207
Intranasal CS, n (%)	3 (14)	3 (12)	1.000

[^] Reported to have been taken for most of the previous 12 months. Difference between groups analysed by the Wilcoxon Rank sum test (non-parametric data) or Pearson's Chi-squared test/Fisher's exact test (testing equality of proportions) where appropriate. $p < 0.05$ considered statistically significant. Abbreviations: IQR, interquartile range; OCS, oral corticosteroids; ICS, inhaled corticosteroids; LABA, long-acting β2-agonist; SABA; short-acting β2-agonist; CS, corticosteroid. [*] Reported to have been taken intermittently or on an as-needed basis.

3.2. Changes in Fruit and Vegetable Intake and Plasma Carotenoids

Efficacy of intervention has been reported previously [26]. F&V consumption over the trial duration was significantly higher in the intervention group than in the control group ($p < 0.001$), with no significant between-group difference detected at baseline. Similarly, after 3 and 6 months, there was a significant change in plasma carotenoid levels (as an objective biomarker of F&V intake) between groups, with no significant between-group difference observed at baseline [26].

3.3. Frequency and Severity of Asthma-Related Events

The high F&V diet group had significantly fewer subjects with 2 or more asthma exacerbations and URTI during the 6-month intervention (88.0% versus 63.6%, $p = 0.049$; Table 3). The frequency of reported exacerbations, URTI, and exacerbations with URTI was similar between the two groups ($p > 0.05$).

Table 3. Frequency of asthma exacerbations and upper respiratory tract infections reported by parent during the 6-month intervention study.

Event Type		Intervention (n = 22)		Control (n = 25)		*p*-Value
		n	%	n	%	
All Events (1+2+3)	None	2	9.0%	0	0%	0.214
	1 or more	20	91.0%	25	100%	0.214
	2 or more	14	63.6%	22	88%	**0.049**
Exacerbations (1)	None	12	54.6%	12	48%	0.654
	1 or more	10	45.4%	13	52%	0.654
	2 or more	5	22.7%	3	12%	0.446
URTI (2)	None	14	63.6%	13	52%	0.421
	1 or more	8	36.4%	12	48%	0.421
	2 or more	3	13.6%	6	24%	0.470
Exacerbations with URTI (3)	None	5	22.7%	4	16%	0.715
	1 or more	17	77.3%	21	84%	0.715
	2 or more	10	45.5%	15	60%	0.319

Exacerbation defined as increase in cough, wheeze or shortness of breath. Upper respiratory tract infection defined as a cold (runny/congested nose, sore throat, earache, sneezing, with or without fever). Exacerbations (1): Number of subjects who had asthma exacerbation alone, without any sign of respiratory infection. Upper respiratory tract infection (2): Number of subjects who had URTI alone with no change in asthma symptoms. Exacerbations with URTI (3): Number of subjects who had asthma exacerbation with suspected URTI symptoms. Difference between groups analysed using Pearson's Chi-squared test/Fisher's exact test (expected cell values < 5) where appropriate. Values in bold indicate statistically significant results.

A total of 135 ADYC diaries were completed and returned (intervention n = 20, 73 events; control n = 22, 62 events) with 71 valid ADYC exacerbation events included in analysis (intervention n = 18, 34 events; control n = 18, 37 events). No significant group difference was observed in the severity, duration, or use of β2-agonists or OCS for exacerbations recorded by parents using the ADYC (Table 4).

Table 4. Severity and duration of Asthma exacerbation (with and without upper respiratory tract infections) reported by parent during the 6-month intervention using the Asthma flare-up diary for young children (ADYC) [a].

Efficacy Outcomes	Intervention (n = 18)		Control (n = 18)		*p*-Value
	Events (Number)	Median (IQRI)	Events (Number)	Median (IQR)	
Asthma symptoms severity/event					
ADYC cumulative daily scores [b]	34	9.9 (5.2, 11.8)	37	9.8 (6.9, 12.8)	0.687
Exacerbation duration					
ADYC duration/event (days) [c]	34	4.0 (3.0, 7.0)	37	4.0 (3.0, 5.0)	0.522
β2-agonist use					
ADYC cumulative number of puffs/event [d]	33	42.0 (30.0, 92.0)	32	48.5 (27.0, 72.5)	0.526
OCS courses/child	8	1.0 (1.0, 2.0)	14	1.5 (1.0, 2.0)	0.462

[a] Values for each group are reported as median (interquartile range (IQR)), *p*-values for difference between groups in symptom severity, duration and B[2]-agonist use calculated using linear mixed model adjusting for maintenance ICS use, accounting for repeated measures; *p*-values for difference between groups in number of OCS courses calculated using Wilcoxon rank sum tests. The number of events for which valid and complete ADYCs were completed is indicated each outcome. Subjects who did not return valid ADYCs were excluded from this analysis. [b] Measured on the 17-item ADYC [28], on a scale of 1 (best) to 7 (worst), completed daily for the duration of exacerbations. The asthma symptoms severity is the sum of the daily ADYC scores per exacerbation event. The ADYC items included cough (n = 2), wheezing (n = 2), dyspnea (n = 4), night awakenings (n = 1), general wellbeing (n = 5), and child's response to salbutamol inhalations (n = 3). [c] Duration from the first day with two or more asthma symptoms to the last day with one or more asthma symptoms (cough, wheezing, and/or dyspnea) as indicated by ADYC. Up to one day without asthma symptoms could be included in an exacerbation. [d] Cumulative number of inhalations during asthma exacerbations, as indicated on the ADYCs.

3.4. Viral Detection in Nasal Swabs Collected during Asthma-Related Events

In total, 45 respiratory-related adverse events in the intervention group and 56 events in the control group were assessed by qPCR for common respiratory viruses. PCR analysis was positive for one or more respiratory viruses in 16 (35.6%) of the total events in the intervention group and 21 (37.5%) of the total events in control subjects. Overall, there was no difference between groups in virus detection rates for asthma exacerbations or URTIs. Rhinovirus was the most common respiratory virus detected during an event in the intervention (22.2%) and control (28.6%) groups. The second most prevalent virus during an event was Influenza B (6.6%) in the intervention group and Influenza A for the control group (9.0%). The results of the qPCR tests for each event are reported in (Table 5).

Table 5. qPCR for common respiratory viruses in asthma exacerbations and upper respiratory tract infections in children with asthma during 6-month dietary intervention study.

Event Type		Intervention	Control	*p*-Value
Total Events Tested for Virus (1 + 2 + 3)		45	56	
PCR virus-positive events (n)		16 (35.6%)	21 (37.5%)	0.840
	Rhinovirus	10 (22.2%) *	16 (28.6%) *	
	Coronavirus	1 (2.2%)	0	
Number of events positive for each respiratory virus	RSV [a] A	0	1 (1.8%) *	0.528
	RSV B	1 (2.2%)	3 (5.3%) *	
	Influenza A	2 (4.4%) *	5 (9.0%) *	
	Influenza B	3 (6.6%)	1 (1.8%) *	

Table 5. *Cont.*

Event Type		Intervention	Control	*p*-Value
Exacerbations (1) tested for virus		5	6	
PCR virus-positive events (n)		2 (40.0%)	2 (33.3%)	1.000
	Rhinovirus	1 (20.0%)	1 (16.7%)	
	Coronavirus	0	0	
Number of events positive for each	RSV [a] A	0	1 (16.7%)	
respiratory virus	RSV B	0	0	1.000
	Influenza A	0	0	
	Influenza B	1 (20.0%)	0	
URTI (2) tested for virus		11	11	
PCR virus-positive events (n)		4 (36.4%)	4 (36.4%)	1.000
	Rhinovirus	2 (18.2%) *	4 (36.4%) *	
	Coronavirus	1 (9.1%)	0	
Number of events positive for each	RSV A	0	0	
respiratory virus	RSV B	0	0	0.610
	Influenza A	2 (18.2%) *	1 (9.1%) *	
	Influenza B	0	1 (9.1%) *	
Exacerbation with URTI (3) tested for virus		29	39	
PCR virus-positive events (n)		10 (34.5%)	15 (38.5%)	0.597
	Rhinovirus	7 (24.1%)	11 (28.2%) *	
	Coronavirus	0	0	
Number of events positive for each	RSV A	0	0	
respiratory virus	RSV B	1 (3.4%)	3 (7.7%) *	0.120
	Influenza A	0	4 (10.2%) *	
	Influenza B	2 (6.8%)	0	

Data are presented as n (%). Exacerbations defined as increase in cough, wheeze or shortness of breath. Upper respiratory tract infection defined as a cold (runny/congested nose, sore throat, earache, sneezing, with or without fever). Exacerbations (1): Number of times that parent reported asthma exacerbation alone, without any sign of respiratory infection. Upper respiratory tract infection (2): Number of times that parent reported URTI alone with no change in asthma symptoms. Exacerbations with URTI (3): Number of times that parent reported asthma exacerbation with suspected URTI symptoms. [a] Respiratory syncytial virus; * Multiple-virus positive event(s). Difference between groups analysed using Pearson's Chi-squared test/Fisher's exact test (expected cell values < 5) where appropriate.

3.5. In Vitro PBMC Cytokine Production

No significant between-group difference was observed after stimulation of PBMCs at baseline (Table 6). However, after 3 months, LPS-induced IFN-γ and IFN-λ production showed a small but significant increase in the intervention group compared with baseline. No significant within or between-group changes were observed in LPS-induced IL-1β and IL-6 production in PBMCs during the 6-month intervention. There were no significant within or between-group changes in RV1B-induced IFN-γ, IFN-λ, IL-1β, and IL-6 production in PBMCs during the trial. Similarly, no significant within or between-group changes were observed in HDM-induced IFN-γ, IFN-λ, IL-1β, and IL-6 production in PBMCs over the duration of the study.

Overall, our model revealed that IL-1β and IL-6 production in PBMCs were more sensitive to LPS stimulation than HDM and RV1-B. In contrast, LPS and HDM stimulations were weak inducers of IFN-γ and IFN-λ compared to RV1B. This trend was similar in both groups at all time points studied.

Table 6. Cytokine secretion of PBMCs from children with asthma in response to Rhinovirus-1B, House Dust Mite and Lipopolysaccharide.

Variable	Baseline [a]	3 Months [a]	Adjusted Change [b] Coeff. [95% CI]	*p*-Value	6 Months [a]	Adjusted Change [b] Coeff. [95% CI]	*p*-Value
			Levels of cytokines in the supernatants of PBMCs stimulated with Rhinovirus-1B				
			IFN-γ (pg/mL)				
Intervention	19.62 (9.37, 32.39)	20.21 (7.92, 46.16)	4.72 [−11.08, 20.53]	0.558	16.64 (7.87, 43.66)	3.84 [−16.65, 24.34]	0.713
Control	14.08 (7.38, 46.20)	18.38 (11.38, 37.51)			15.29 (7.20, 32.26)		
			IFN-λ (pg/mL)				
Intervention	3.77 (1.66, 4.74)	2.57 (1.76, 5.08)	−1.58 [−28.06, 24.35]	0.890	2.14 (1.18, 6.42)	15.34 [−10.68, 41.38]	0.248
Control	6.68 (4.08, 39.29)	5.72 (1.43, 33.27)			5.5 (1.77, 15.85)		
			IL-1β (pg/mL)				
Intervention	22.47 (3.78, 71.17)	41.92 (5.83, 77.87)	−60.53 [−167.30, 46.24]	0.267	17.73 (4.41, 59.99)	−25.40 [−130.92, 80.11]	0.637
Control	21.25 (7.44, 92.67)	20.64 (7.93, 83.64)			9.36 (4.05, 42.13)		
			IL-6 (ng/mL)				
Intervention	1.55 (0.13, 8.91)	3.24 (0.68, 12.62)	−2.01 [−14.91, 10.88]	0.760	1.45 (0.29, 6.38)	−1.56 [−14.32, 11.19]	0.810
Control	1.26 (0.38, 9.38)	1.19 (0.48, 10.32)			0.59 (0.21, 1.74)		
			Levels of cytokines in the supernatants of PBMCs stimulated with House Dust Mite				
			IFN-γ (pg/mL)				
Intervention	1.02 ± 0.37	1.25 ± 0.39	0.23 [−0.14, 0.60]	0.221	1.03 ± 0.48	0.00 [−0.36, 0.36]	0.999
Control	0.98 ± 0.56	0.98 ± 0.48			0.90 ± 0.49		
			IFN-λ (pg/mL)				
Intervention	2.18 (0.82, 3.29)	2.91 (1.57, 5.08)	−1.89 [−11.50, 7.71]	0.699	1.67 (0.94, 5.75)	0.49 [−11.81, 12.80]	0.937
Control	2.17 (1.26, 5.94)	2.88 (1.11, 7.89)			2.29 (0.92, 6.40)		
			IL-1β (pg/mL)				
Intervention	13.89 (7.29, 42.49)	23.65 (7.58, 57.85)	13.81 [−88.27, −115.91]	0.791	31.36 [5.80, 65.04)	5.54 [−95.45, 106.54]	0.914
Control	14.71 (10.49, 54.65)	22.38 (9.04, 42.47)			24.91 [9.26, 78.49)		
			IL-6 (ng/mL)				
Intervention	1.87 (0.74, 7.42)	4.66 (0.75, 15.19)	1.15 [−13.90, 16.20]	0.881	2.20 (0.85, 15.12)	0.96 [−13.91, 15.83]	0.899
Control	5.56 (1.46, 12.78)	4.52 (1.04, 13.82)			10.11 (1.49, 20.10)		
			Levels of cytokines in the supernatants of PBMCs stimulated with Lipopolysaccharide				
			IFN-γ (pg/mL)				
Intervention	1.48 (1.07, 4.92)	**3.59 (1.37, 12.65)** [c]	−4.21 [−16.23, 7.81]	0.492	2.66 (1.35, 8.41)	−5.71 [−17.75, 6.32]	0.352
Control	2.37 (1.08, 13.65)	3.53 (1.56, 8.82)			5.52 (0.73, 12.74)		
			IFN-λ (pg/mL)				
Intervention	1.50 (0.6, 4.11)	**1.97 (0.64, 5.9)** [c]	0.61 [−6.89, 8.12]	0.872	1.86 (0.4, 4.53)	4.47 [−4.71, 13.66]	0.340
Control	4.22 (0.93, 10.12)	0.92 (0.2, 1.99)			2.06 (0.63, 6.79)		
			IL-1β (pg/mL)				
Intervention	1926.13 (1213.19, 4066.73)	4646.56 (1773.93, 6439.98)	997.36 [−1521.98, 3516.70]	0.438	1792.86 (1236.9, 4634.93)	−201.07 [−2663.57, 2261.42]	0.873
Control	1928.85 (1099.88, 8469.68)	1896.76 (1623.12, 9045.34)			2000.26 (968.92, 8832.58)		
			IL-6 (ng/mL)				
Intervention	70.04 (46.64, 103.49)	76.63 (38.87, 150.94)	20.14 [−18.31, 58.59]	0.305	78.34 (52.04, 114.49)	−2.33 [−40.56, 35.88]	0.905
Control	75.83 (48.56, 130.21)	65.34 (46.23, 94.99)			88.99 (71.16, 151.03)		

Intervention (n = 22), Control (n = 24). All variables adjusted for media. [a] Unadjusted mean ± SD or median (IQR) are presented for baseline, 3 month and 6 month measures. [b] Adjusted change means and 95% confidence intervals [CI] are the differences in change from baseline in intervention group compared to control group by linear mixed model adjusted for random effects and hospital admissions 12 months prior to recruitment. [c] $p < 0.05$ difference within group compared to baseline or 3 month. Values in bold indicate statistically significant results.

3.6. Associations

The associations between PBMC cytokine responses and plasma carotenoid levels at 6 months showed a positive correlation between RV1B-induced IFN-λ production and change in plasma lycopene (r_s = 0.35, *p* = 0.015). RV1B-induced IL-6 production was inversely associated with plasma lutein levels (r_s = −0.29, *p* = 0.045). Furthermore, HMD-induced IL-6 production was inversely correlated with change in plasma lutein (r_s = −0.28, *p* = 0.045). Similarly, inverse associations were identified between LPS-induced IL-1β production and change in plasma α-carotene (r_s = −0.32, *p* = 0.025), and β-carotene (r_s = −0.33, *p* = 0.020), as well as total carotenoids (r_s = −0.34, *p* = 0.016).

4. Discussion

This paper investigates the adverse respiratory events and innate immune responses of children with asthma following a 6-month high F&V dietary intervention. The intervention modified asthma-related illness, as the high F&V diet group had significantly fewer subjects with ≥2 asthma exacerbations/respiratory infections than the control group during the trial. Plasma carotenoids increased in the high F&V group, which correlated with self-reported F&V intake. LPS-induced IFN-γ and IFN-λ production was significantly increased in the intervention group after 3 months, compared to baseline; however, at the end of the 6-month trial, no significant within or between-group changes were observed in cytokine production from PBMCs stimulated with RV1B and HDM.

Previously our research group observed that when adults with asthma were assigned to a high F&V diet ($\geq$7 serves/day) and compared with those assigned to a low F&V diet (<3 serves/day), the low F&V diet group were 2.26 times more likely to exacerbate [25]. In the present study, we have also observed an effect of a high F&V diet on respiratory events, as the number of subjects who experienced at least two episodes of asthma exacerbation and URTI during the trial was significantly lower (63.6%) in the high F&V diet group compared with the control group (88%). Overall, the number of reported asthma-related events was lower in the intervention compared to the control group; however, this trend did not reach statistical significance. Further, there were no significant group differences in the severity and duration of asthma exacerbations recorded by parents using the ADYC. In this study, self-completion ADYC diaries were used to capture information about the onset and severity of exacerbation. Reliance on self-completed journals may not be suitable for evaluating exacerbation severity in clinical trials as they are not always completed correctly or returned. In our study, eleven (23.4%) parents did not return valid diaries because of the absence of episodes and poor compliance. Future RCTs should consider evaluating whether other methods, such as an online version and electronic reminders, would improve compliance. Additionally, parental perception of symptom severity can affect the scores [28]. Evidence shows that parents can have misperceptions about asthma and its management, as they tend to underestimate the severity of their child's asthma and overestimate asthma control [33,34].

There was no significant difference in virus detection in our study, with a virus detection rate of just 35.6% in the intervention group and 37.5% in control subjects. Previous studies reported higher rates of respiratory virus detection, ranging from 40% to 60% after asthma exacerbation in children [11,35]. In the present study, we focused on testing for a single virus of interest for each assay, whereas the studies with higher detection rates [11,35] used a gene expression panel that tests for multiple respiratory pathogens simultaneously. Moreover, in the study by Khetsuriani et al. [35], viruses were detected in 60% of patients using a combination of throat swabs and nasal samples. Therefore, it can be suggested that the difference in viral detection rates may be due to the sample collection method [35] and the differences in the use of PCR assays. Nonetheless, RV was the most prevalent respiratory virus detected in samples from the intervention (22.2%) and control (28.6%) groups. This is consistent with previous studies that reported that RVs are the most common precipitants of virus-associated exacerbations in children over the age of 3 years [4,11,36].

Cytokine responses in PBMCs stimulated with RV1B, LPS and HDM were also measured, with increased IFN-γ and IFN-λ production in LPS-exposed PBMCs from intervention group subjects at 3 months compared to baseline. IFNs can interact with specific cellular receptors and induce innate and adaptive immune responses, which are crucial for mediating host defences against viral and bacterial infections [37]. IFN-γ is one of the critical mediators of LPS-induced immune responses [38]. IFN-γ intensifies antimicrobial immune responses via inducing macrophage functions such as phagocytosis, respiratory burst activity, antigen presentation, and cytokine secretion. The functional significance of IFN-γ in antimicrobial defence is indicated by the increased susceptibilities of IFN-γ$^{-/-}$ and IFN-γR$^{-/-}$ mice to a wide range of infections [38]. We have previously shown that PBMCs from children with asthma had deficient IFN-γ production in response to both RV1B and LPS infection compared with healthy controls [39]. Other studies have also revealed that children with asthma, compared with age-matched controls, had defective or impaired IFN responses to respiratory viruses [40,41]. Therefore, an increase in IFN production in children with asthma could be expected to protect them from inflammatory injury.

To our knowledge, this is the first study to examine the effects of a high F&V diet on innate immune responses of PBMCs stimulated with RV1B, LPS and HDM in children with asthma. Two intervention studies [42,43] have examined the effect of withdrawal of antioxidant-rich foods (in particular F&V) from the diet in adults with asthma. Antioxidant withdrawal resulted in increased airway neutrophils [43] and upregulation of inflammatory and immune response genes in sputum cells, including the innate immune receptors TLR2, IL1R2, CD93, ANTXR2, the innate immune signaling molecules IRAK2, IRAK3, MAP3K8 and neutrophil proteases MMP25 and CPD [42]. Other research has mostly focused on the effects of isolated nutrients on immune responses in asthma. For example, in a murine model of asthma, administration of lycopene significantly increased IFN-γ expression [44]. Saedisomeolia et al. [45] showed that pre-incubation of airway epithelial cells with lycopene reduced the release of IL-6 following exposure to LPS. Similarly, in a mouse model of allergic airway inflammation [46], lycopene supplementation decreased allergen-induced release of the T helper 2–associated cytokines IL-4 and IL-5. In another study [47], pre-treatment of PBMCs with lutein was shown to suppress LPS- induced mRNA expression of IL-6 and IL-1β. Therefore, combined with the evidence for a whole-food intervention, it can be concluded that dietary antioxidants induce protective innate immune responses in asthma.

Correlation analysis of PBMC cytokine responses to RV1B, HDM, and LPS stimulation and plasma carotenoid levels showed that changes in total and individual carotenoids (such as lycopene, lutein, α- and β-carotene) were associated with higher IFN-λ production, while they were inversely associated with IL-6 and IL-1β production. Previous research has explored the association between dietary carotenoid intake and innate immune mediators. No association was found between intake of β-carotene and in vitro production of LPS-induced cytokines, including IL-1β, IL-6, IL-8, IL-10 and tumour necrosis factor (TNF)-α [48]. One explanation for this discrepancy may relate to the methods used for assessing carotenoid status. In the study by Vivek [48], dietary intake of carotenoids was evaluated using a food frequency questionnaire, whereas, in our study, we used an objective approach of directly measuring plasma carotenoid levels, thus providing a more accurate indication of in vivo carotenoid concentrations, which is a strength of this study. It would be reasonable to suggest that the protective effects of plasma carotenoid levels on innate immune responses observed in our study could be due to the consumption of a high F&V diet. Further research is needed to investigate and confirm this.

Findings from the Australian National Health Survey indicated that only 5.1% of all children met the recommended daily intake for F&V [49]. These trends are of public health concern. While cost may be a significant barrier to fruit and vegetable consumption, other factors such as taste, preference, and culture can also significantly impact F&V purchase and consumption [50]. There is growing evidence to indicate that participation in incentive programs such as community kitchens or school gardens can help promote consumption of F&V [51,52]. In our study, the intervention more than doubled F&V intake in children with asthma. Parental and children's knowledge of the protective effect of the F&V consumption on asthma is of great importance and needs to be included in the management of childhood asthma.

The current study has some limitations. Blinding of the treatment allocation was not feasible for this study [53]. However, this is typical of many dietary intervention studies. The participants were blinded to the study hypothesis to reduce bias, and a range of objective outcomes was included in the assessments. Furthermore, the attrition rate (26.8%) was high, which may be related to the parent burden, including time and commitment associated with study visit attendance, completing study questionnaires and participation in phone consultations and/or child discomfort during blood collection. However, this was expected and is comparable to other intervention studies in children with asthma [54,55]. Finally, due to insufficient PBMC samples, we could not determine whether cytokines responses would differ at 24 h or 72 h. However, a deficient interferon response to RV in asthmatic children was also reported in a previous study that harvested supernatants at 72 h [56].

5. Conclusions

In summary, we showed that increasing F&V intake decreased asthma-related illness. This study establishes the relevance of a dietary intervention strategy—increased F&V intake—for improving innate immune responses and reducing exacerbation rates in children with asthma. This is an appealing strategy with no side effects, which is likely to be widely accepted and adopted by children and their caregivers.

Supplementary Materials: The following supporting information can be downloaded at: https://www.mdpi.com/article/10.3390/nu14153087/s1, Figure S1: The Consolidated Standards of Reporting Trials diagram for the per-protocol analysis of the randomized controlled trial investigating the effect of a high fruit and vegetable intervention in children with asthma. Supplementary File S1: Methods [27,31,57–60].

Author Contributions: L.G.W., K.J.B., M.E.J., P.A.B.W., A.C., J.M. and M.R.S. designed and planned the study, supervised by L.G.W., M.E.J. and P.A.B.W.; B.S.B., M.E.J., B.H., R.F.M. and E.J.W. participated in recruitment and data collection. K.N. was involved in developing the laboratory methods for PBMC isolation and culture. B.H., R.F.M., K.N. and E.J.W. performed, analysed and interpreted laboratory analysis of blood samples. B.H. and B.S.B. performed the statistical analysis. B.H. wrote the article with input from all authors. All authors have read and agreed to the published version of the manuscript.

Funding: This work was supported by the Hunter Medical Research Institute Gastronomic Lunch and the University of Newcastle Priority Research Centre GrowUpWell. The funding bodies played no role in the design, the collection, analysis and interpretation of data, the writing of the report, or the decision to submit the article for publication.

Institutional Review Board Statement: This trial was reviewed and approved by the Hunter New England Health Human Research Ethics Committee (approval number: 15/06/17/4.03). Written, informed parental consent and written child assent (if applicable) were obtained for all participants. The trial was prospectively registered with the Australian New Zealand Clinical Trials Registry (ACTRN12615000851561).

Informed Consent Statement: Written informed parental consent, and child assent (where applicable) was obtained from all subjects involved in the study.

Data Availability Statement: The datasets used and analysed during the current study are available from the corresponding author on reasonable request.

Conflicts of Interest: The authors have no conflict of interest to declare.

References

1. Van Aalderen, W.M. Childhood asthma: Diagnosis and treatment. *Scientifica* **2012**, *2012*, 674204. [PubMed]
2. Contoli, M.; Message, S.D.; Laza-Stanca, V.; Edwards, M.R.; Wark, P.A.; Bartlett, N.W.; Kebadze, T.; Mallia, P.; Stanciu, L.A.; Parker, H.L.; et al. Role of deficient type III interferon-lambda production in asthma exacerbations. *Nat. Med.* **2006**, *12*, 1023–1026. [CrossRef] [PubMed]
3. Xepapadaki, P.; Papadopoulos, N.G. Childhood asthma and infection: Virus-induced exacerbations as determinants and modifiers. *Eur. Respir. J.* **2010**, *36*, 438–445. [CrossRef] [PubMed]
4. Steinke, J.W.; Borish, L. Immune Responses in Rhinovirus-Induced Asthma Exacerbations. *Curr. Allergy Asthma. Rep.* **2016**, *16*, 78. [CrossRef] [PubMed]
5. Beale, J.; Jayaraman, A.; Jackson, D.J.; Macintyre, J.D.; Edwards, M.R.; Walton, R.P.; Zhu, J.; Man, Y.; Betty, C.; Edwards, S.M.; et al. Rhinovirus-induced IL-25 in asthma exacerbation drives type 2 immunity and allergic pulmonary inflammation. *Sci. Transl. Med.* **2014**, *6*, 256ra134. [CrossRef] [PubMed]
6. Kanchongkittiphon, W.; Mendell, M.J.; Gaffin, J.M.; Wang, G.; Phipatanakul, W. Indoor environmental exposures and exacerbation of asthma: An update to the 2000 review by the Institute of Medicine. *Environ. Health Perspect.* **2015**, *123*, 6–20. [CrossRef] [PubMed]
7. Kumari, A.; Dash, D.; Singh, R. Lipopolysaccharide (LPS) exposure differently affects allergic asthma exacerbations and its amelioration by intranasal curcumin in mice. *Cytokine* **2015**, *76*, 334–342. [CrossRef] [PubMed]
8. Rabinovitch, N.; Liu, A.H.; Zhang, L.; Rodes, C.E.; Foarde, K.; Dutton, S.J.; Murphy, J.R.; Gelfand, E.W. Importance of the personal endotoxin cloud in school-age children with asthma. *J. Allergy Clin. Immunol.* **2005**, *116*, 1053–1057. [CrossRef] [PubMed]

9. Thorne, P.S.; Kulhankova, K.; Yin, M.; Cohn, R.; Arbes, S.J., Jr.; Zeldin, D.C. Endotoxin exposure is a risk factor for asthma: The national survey of endotoxin in United States housing. *Am. J. Respir. Crit. Care Med.* **2005**, *172*, 1371–1377. [CrossRef] [PubMed]

10. Global Initiative for Asthma. *Global Strategy for the Asthma Manangement and Prevention*; National Institutes of Health, National Heart, Lung, and Blood Institute: Bethesda, MD, USA, 2018.

11. Leung, T.F.; Chan, P.K.; Wong, G.W.; Fok, T.F.; Ng, P.C. Respiratory viruses and atypical bacteria triggering severe asthma exacerbation in children. *Hong Kong Med. J. Xianggang Yi Xue Za Zhi* **2013**, *19* (Suppl. S4), 11–14. [PubMed]

12. Garcia-Larsen, V.; Del Giacco, S.R.; Moreira, A.; Bonini, M.; Charles, D.; Reeves, T.; Carlsen, K.-H.; Haahtela, T.; Bonini, S.; Fonseca, J.; et al. Asthma and dietary intake: An overview of systematic reviews. *Allergy* **2016**, *71*, 433–442. [CrossRef] [PubMed]

13. Miller, H.E.; Rigelhof, F.; Marquart, L.; Prakash, A.; Kanter, M. Antioxidant Content of Whole Grain Breakfast Cereals, Fruits and Vegetables. *J. Am. Coll. Nutr.* **2000**, *19* (Suppl. S3), 312S–319S. [CrossRef]

14. Guilleminault, L.; Williams, E.J.; Scott, H.A.; Berthon, B.S.; Jensen, M.; Wood, L.G. Diet and Asthma: Is It Time to Adapt Our Message? *Nutrients* **2017**, *9*, 1227. [CrossRef] [PubMed]

15. Sahiner, U.M.; Birben, E.; Erzurum, S.; Sackesen, C.; Kalayci, O. Oxidative Stress in Asthma. *World Allergy Organ. J.* **2011**, *4*, 151–158. [CrossRef] [PubMed]

16. Wood, L.G.; Gibson, P.G. Dietary factors lead to innate immune activation in asthma. *Pharmacol. Ther.* **2009**, *123*, 37–53. [CrossRef]

17. La Vecchia, C.; Decarli, A.; Pagano, R. Vegetable consumption and risk of chronic disease. *Epidemiology* **1998**, *9*, 208–210. [CrossRef] [PubMed]

18. Troisi, R.J.; Willett, W.C.; Weiss, S.T.; Trichopoulis, D.; Rosner, B.; Speizer, F.E. A prospective study of diet and adult-onset asthma. *Am. J. Respir. Crit. Care Med.* **1995**, *151*, 1401–1408. [CrossRef]

19. Butland, B.K.; Strachan, D.P.; Anderson, H.R. Fresh fruit intake and asthma symptoms in young British adults: Confounding or effect modification by smoking? *Eur. Resp. J.* **1999**, *13*, 744–750. [CrossRef]

20. Ellwood, P.; Asher, M.I.; Bjorksten, B.; Burr, M.; Pearce, N.; Robertson, C.F. Diet and asthma, allergic rhinoconjunctivitis and atopic eczema symptom prevalence: An ecological analysis of the International Study of Asthma and Allergies in Childhood (ISAAC) data. *Eur. Respir. J.* **2001**, *17*, 436–443. [CrossRef]

21. Hijazi, N.; Abalkhail, B.; Seaton, A. Diet and childhood asthma in a society in transition: A study in urban and rural Saudi Arabia. *Thorax* **2000**, *55*, 775–779. [CrossRef]

22. Cook, D.G.; Carey, I.M.; Whincup, P.H.; Papacosta, O.; Chirico, S.; Bruckdorfer, K.R.; Walker, M. Effect of fresh fruit consumption on lung function and wheeze in children. *Thorax* **1997**, *52*, 628–633. [CrossRef] [PubMed]

23. Carey, I.; Strachan, D.; Cook, D. Effects of changes in fresh fruit consumption on ventilatory function in healthy British adults. *Am. J. Respir. Crit. Care Med.* **1998**, *158*, 728–733. [CrossRef] [PubMed]

24. Hosseini, B.; Berthon, B.S.; Wark, P.; Wood, L.G. Effects of Fruit and Vegetable Consumption on Risk of Asthma, Wheezing and Immune Responses: A Systematic Review and Meta-Analysis. *Nutrients* **2017**, *9*, 341. [CrossRef] [PubMed]

25. Wood, L.G.; Garg, M.L.; Smart, J.M.; Scott, H.A.; Barker, D.; Gibson, P.G. Manipulating antioxidant intake in asthma: A randomized controlled trial. *Am. J. Clin. Nutr.* **2012**, *96*, 534–543. [CrossRef]

26. Berthon, B.S.; McLoughlin, R.F.; Jensen, M.E.; Hosseini, B.; Williams, E.J.; Baines, K.J.; Taylor, S.L.; Rogers, G.B.; Ivey, K.L.; Morten, M. The effects of increasing fruit and vegetable intake in children with asthma: A randomized controlled trial. *Clin. Exp. Allergy* **2021**, *51*, 1144–1156. [CrossRef]

27. National Health and Medical Research Council. *Australian Dietary Guidelines*; National Health and Medical Research Council: Canberra, Australia, 2013.

28. Ducharme, F.M.; Jensen, M.E.; Mendelson, M.J.; Parkin, P.C.; Desplats, E.; Zhang, X.; Platt, R. Asthma Flare-up Diary for Young Children to monitor the severity of exacerbations. *J. Allergy Clin. Immunol.* **2016**, *137*, 744–749.e6. [CrossRef]

29. Barua, A.B.; Kostic, D.; Olson, J.A. New simplified procedures for the extraction and simultaneous high-performance liquid chromatographic analysis of retinol, tocopherols and carotenoids in human serum. *J. Chromatogr. B Biomed. Sci. Appl.* **1993**, *617*, 257–264. [CrossRef]

30. Wood, L.G.; Garg, M.L.; Blake, R.J.; Garcia-Caraballo, S.; Gibson, P.G. Airway and circulating levels of carotenoids in asthma and healthy controls. *J. Am. Coll. Nutr.* **2005**, *24*, 448–455. [CrossRef]

31. QIAGEN Viral RNA Mini Kit Handbook. Available online: https://www.qiagen.com/au/resources/resourcedetail?id=c80685c0-4103-49ea-aa72-8989420e3018&lang=en (accessed on 1 June 2021).

32. Gunawardhana, L.P.; Gibson, P.G.; Simpson, J.L.; Powell, H.; Baines, K.J. Activity and expression of histone acetylases and deacetylases in inflammatory phenotypes of asthma. Clinical and experi-mental allergy. *J. Br. Soc. Allergy Clin. Immunol.* **2014**, *44*, 47–57. [CrossRef]

33. Carroll, W.D.; Wildhaber, J.; Brand, P.L. Parent misperception of control in childhood/adolescent asthma: The Room to Breathe survey. *Eur. Respir. J.* **2012**, *39*, 90–96. [CrossRef]

34. Davis, K.J.; Disantostefano, R.; Peden, D.B. Is Johnny wheezing? Parent-child agreement in the Childhood Asthma in America survey. *Pediatric Allergy Immunol.* **2011**, *22 Pt 1*, 31–35. [CrossRef]

35. Khetsuriani, N.; Kazerouni, N.N.; Erdman, D.D.; Lu, X.; Redd, S.C.; Anderson, L.J.; Gerald Teague, W. Prevalence of viral respiratory tract infections in children with asthma. *J. Allergy Clin. Immunol.* **2007**, *119*, 314–321. [CrossRef]

36. Bizzintino, J.; Lee, W.M.; Laing, I.A.; Vang, F.; Pappas, T.; Zhang, G.; Martin, A.C.; Khoo, S.-K.; Cox, D.; Geelhoed, G.C.; et al. Association between human rhinovirus C and severity of acute asthma in children. *Eur. Respir. J.* **2011**, *37*, 1037–1042. [CrossRef] [PubMed]

37. Vlachiotis, S.; Andreakos, E. Lambda interferons in immunity and autoimmunity. *J. Autoimmun.* **2019**, *104*, 102319. [CrossRef] [PubMed]

38. Varma, T.K.; Lin, C.Y.; Toliver-Kinsky, T.E.; Sherwood, E.R. Endotoxin-induced gamma interferon production: Contributing cell types and key regulatory factors. *Clin. Diagn. Lab. Immunol.* **2002**, *9*, 530–543. [CrossRef] [PubMed]

39. Hosseini, B.; Berthon, B.S.; Starkey, M.R.; Collison, A.; McLoughlin, R.F.; Williams, E.J.; Nichol, K.; Wark, P.A.; Jensen, M.E.; Sena, C.R.D.; et al. Children With Asthma Have Impaired Innate Immunity and Increased Numbers of Type 2 Innate Lymphoid Cells Compared With Healthy Controls. *Front. Immunol.* **2021**, *12*, 664668. [CrossRef]

40. Durrani, S.R.; Montville, D.J.; Pratt, A.S.; Sahu, S.; DeVries, M.K.; Rajamanickam, V.; Gangnon, R.E.; Gill, M.A.; Gern, J.E.; Lemanske, R.F.; et al. Innate immune responses to rhinovirus are reduced by the high-affinity IgE receptor in allergic asthmatic children. *J. Allergy Clin. Immunol.* **2012**, *130*, 489–495. [CrossRef]

41. Johnston, S.L. Innate immunity in the pathogenesis of virus-induced asthma exacerbations. *Proc. Am. Thorac. Soc.* **2007**, *4*, 267–270. [CrossRef] [PubMed]

42. Baines, K.J.; Wood, L.G.; Gibson, P.G. The nutrigenomics of asthma: Molecular mechanisms of airway neutrophilia following dietary antioxidant withdrawal. *OMICS* **2009**, *13*, 355–365. [CrossRef]

43. Wood, L.G.; Garg, M.L.; Powell, H.; Gibson, P.G. Lycopene-rich treatments modify noneosinophilic airway inflammation in asthma: Proof of concept. *Free Radic. Res.* **2008**, *42*, 94–102. [CrossRef]

44. Lee, C.M.; Chang, J.H.; Moon, D.O.; Choi, Y.H.; Choi, I.W.; Park, Y.M.; Kim, G.Y. Lycopene suppresses ovalbumin-induced airway inflammation in a murine model of asthma. *Biochem. Biophys. Res. Commun.* **2008**, *374*, 248–252. [CrossRef]

45. Saedisomeolia, A.; Wood, L.G.; Garg, M.L.; Gibson, P.G.; Wark, P.A. Lycopene enrichment of cultured airway epithelial cells decreases the inflammation induced by rhinovirus infection and lipopolysaccharide. *J. Nutr. Biochem.* **2009**, *20*, 577–585. [CrossRef]

46. Hazlewood, L.C.; Wood, L.G.; Hansbro, P.M.; Foster, P.S. Dietary lycopene supplementation suppresses Th2 responses and lung eosinophilia in a mouse model of allergic asthma. *J Nutr. Biochem.* **2011**, *22*, 95–100. [CrossRef]

47. Chung, R.W.S.; Leanderson, P.; Lundberg, A.K.; Jonasson, L. Lutein exerts anti-inflammatory effects in patients with coronary artery disease. *Atherosclerosis* **2017**, *262*, 87–93. [CrossRef]

48. Vivek Iyer, B.L.H. *Does Typical Dietary Intake of Vitamin E, Vitamin C, or β-Carotene Correlate with In Vitro Production of Pro-Inflammatory Cytokines in Aged Adults?* University of Florida: Gainesville, FL, USA, 2012.

49. Mihrshahi, S.; Myton, R.; Partridge, S.R.; Esdaile, E.; Hardy, L.L.; Gale, J. Sustained low consumption of fruit and vegetables in Australian children: Findings from the Australian National Health Surveys. *Health Promot. J. Aust.* **2019**, *30*, 83–87. [CrossRef]

50. Ard, J.D.; Fitzpatrick, S.; Desmond, R.A.; Sutton, B.S.; Pisu, M.; Allison, D.B.; Franklin, F.; Baskin, M.L. The impact of cost on the availability of fruits and vegetables in the homes of schoolchildren in Birmingham, Alabama. *Am. J. Public Health* **2007**, *97*, 367–372. [CrossRef] [PubMed]

51. Iacovou, M.; Pattieson, D.C.; Truby, H.; Palermo, C. Social health and nutrition impacts of community kitchens: A systematic review. *Public Health Nutr.* **2013**, *16*, 535–543. [CrossRef] [PubMed]

52. McCormack, L.A.; Laska, M.N.; Larson, N.I.; Story, M. Review of the nutritional implications of farmers' markets and community gardens: A call for evaluation and research efforts. *J. Am. Diet Assoc.* **2010**, *110*, 399–408. [CrossRef]

53. Hébert, J.R.; Frongillo, E.A.; Adams, S.A.; Turner-McGrievy, G.M.; Hurley, T.G.; Miller, D.R.; Ockene, I.S. Perspective: Randomized Controlled Trials Are Not a Panacea for Diet-Related Research. *Adv. Nutr.* **2016**, *7*, 423–432. [CrossRef]

54. Evans, R., 3rd; Gergen, P.J.; Mitchell, H.; Kattan, M.; Kercsmar, C.; Crain, E.; Anderson, J.; Eggleston, P.; Malveaux, F.J.; Wedner, H.J. A randomized clinical trial to reduce asthma morbidity among inner-city children: Results of the National Cooperative Inner-City Asthma Study. *J. Pediatr.* **1999**, *135*, 332–338. [CrossRef]

55. Garaiova, I.; Muchova, J.; Nagyova, Z.; Wang, D.; Li, J.V.; Orszaghova, Z.; Michael, D.R.; Plummer, S.F.; Ďuračková, Z. Probiotics and vitamin C for the prevention of respiratory tract infections in children attending preschool: A randomised controlled pilot study. *Eur. J. Clin. Nutr.* **2015**, *69*, 373–379. [CrossRef]

56. Iikura, K.; Katsunuma, T.; Saika, S.; Saito, S.; Ichinohe, S.; Ida, H.; Saito, H.; Matsumoto, K. Peripheral blood mononuclear cells from patients with bronchial asthma show impaired innate immune responses to rhinovirus in vitro. *Int. Arch. Allergy Immunol.* **2011**, *155* (Suppl. S1), 27–33. [CrossRef] [PubMed]

57. Ducharme, F.M.; Mendelson, M.J.; Klassen, T.P. The Preschool Asthma Diary (PAD): A validated instrument for day-to-day monitoring of asthma severity in young children. *Pediatric Acad. Soc.* **2007**, 6310. Available online: http://www.abstracts2view.com/pasall (accessed on 1 June 2021).

58. Mishra, A.; Brown, A.L.; Yao, X.; Yang, S.; Park, S.J.; Liu, C.; Chung, J.H. Dendritic cells induce Th2-mediated airway inflammatory responses to house dust mite via DNA-dependent protein kinase. *Nat. Commun.* **2015**, *6*, 6224. [CrossRef] [PubMed]

59. Plevin, R.E.; Knoll, M.; McKay, M.; Arbabi, S.; Cuschieri, J. The role of lipopolysaccharide structure in monocyte activation and cytokine secretion. *Shock* **2016**, *45*, 22. [CrossRef]

60. Forbes, R.L.; Gibson, P.G.; Murphy, V.E.; Wark, P.A. Impaired type I and III interferon response to rhinovirus infection during pregnancy and asthma. *Thorax* **2012**, *67*, 209–214. [CrossRef]

Article

The Combined Effect of Birth Weight and Lifestyle on Clustered Cardio-Metabolic Risk Factors in Children and Adolescents: A National School-Based Cross-Sectional Survey

Di Shi [1,2], Jiajia Dang [1,2], Ning Ma [1,2], Yunfei Liu [1,2], Panliang Zhong [1,2], Shan Cai [1,2], Yinghua Ma [1,2], Zhiyong Zou [1,2], Yanhui Dong [1,2,*], Yi Song [1,2,*] and Jun Ma [1,2]

[1] Institute of Child and Adolescent Health, School of Public Health, Peking University, Beijing 100191, China; 2111210123@bjmu.edu.cn (D.S.); dangjj@bjmu.edu.cn (J.D.); mnmaning@bjmu.edu.cn (N.M.); liuyunfei_pku@163.com (Y.L.); 1610306219@pku.edu.cn (P.Z.); 1710306207@pku.edu.cn (S.C.); yinghuama@bjmu.edu.cn (Y.M.); harveyzou2002@bjmu.edu.cn (Z.Z.); majunt@bjmu.edu.cn (J.M.)

[2] National Health Commission Key Laboratory of Reproductive Health, Peking University, Beijing 100191, China

[*] Correspondence: dongyanhui@bjmu.edu.cn (Y.D.); songyi@bjmu.edu.cn (Y.S.); Tel.: +86-10-82801524 (Y.D.); +86-10-82801624 (Y.S.)

Abstract: Background: Due to the adverse effects of cardio-metabolic risk factors (CMRFs) in children and adolescents on their current and later life health, and the growing evidence that birth weight and lifestyle have on CMRFs, we aimed to estimate the combined effect of birth weight and lifestyle on clustered CMRFs in children and adolescents. Methods: We enrolled 11,509 participants aged 7–18 years old in a national school-based cross-sectional study in seven provinces in China in 2013. Information on CMRFs was collected through anthropometric measurements and blood sample testing. Information on birth weight, lifestyle and other basic information were investigated through children and adolescents' as well as parents' questionnaires. The generalized linear mixed model was applied to estimate the odd ratio (OR) and 95% confidence interval (95% CI) for the associations between CMRFs, clustered CMRFs and birth weight, lifestyle, and the combinations of birth weight and lifestyle. Results: Overall, the prevalence of clustered CMRFs was 3.6% in children and adolescents aged 7–18 years, higher in boys (4.4%) than girls (2.9%). The combination of LBW/ideal lifestyle (OR = 2.00, 95% CI: 1.07–3.72) was associated with higher risk of clustered CMRFs, as well as in adolescents aged 13–18 years and in boys. The combination of HBW/poor lifestyle (OR = 1.74, 95% CI: 1.13–2.68) was related to elevated risk of clustered CMRFs, especially in children aged 7–12 years. Conclusions: CMRFs in Chinese children and adolescents is concerning, ideal lifestyle could weaken the association of birth weight with clustered CMRFs, especially in younger age, indicating that programs to prevent abnormal birth weight or poor lifestyle or both among children and adolescents may reduce CMRFs in China.

Keywords: children; adolescents; cardio-metabolic risk factors; birth weight; lifestyle

Citation: Shi, D.; Dang, J.; Ma, N.; Liu, Y.; Zhong, P.; Cai, S.; Ma, Y.; Zou, Z.; Dong, Y.; Song, Y.; et al. The Combined Effect of Birth Weight and Lifestyle on Clustered Cardio-Metabolic Risk Factors in Children and Adolescents: A National School-Based Cross-Sectional Survey. *Nutrients* **2022**, *14*, 3131. https://doi.org/10.3390/nu14153131

Academic Editor: Iole Tomassini Barbarossa

Received: 7 July 2022
Accepted: 27 July 2022
Published: 29 July 2022

Publisher's Note: MDPI stays neutral with regard to jurisdictional claims in published maps and institutional affiliations.

Copyright: © 2022 by the authors. Licensee MDPI, Basel, Switzerland. This article is an open access article distributed under the terms and conditions of the Creative Commons Attribution (CC BY) license (https://creativecommons.org/licenses/by/4.0/).

1. Introduction

Cardiovascular diseases have remained a major health burden in adults [1], and most could stem from childhood and adolescence. Specifically, risk factors and behaviors that accelerated the process of cardiovascular diseases begin in childhood and even in the perinatal period [2]. Considering that some children and adolescents have already had identifiable cardio-metabolic risk factors (CMRFs) including hypertension, impaired fasting glucose, dyslipidemia, abdominal obesity at a very early age [3–6], plus clustered CMRFs, more than three CMRF items, may strongly predict the occurrence of cardiovascular diseases in adulthood [7–9]. Identifying it in vulnerable groups and risk factors that could alter CMRFs status might help reduce the prevalence of cardiovascular disease in the future.

There is evidence that low birth weight (LBW) children are more likely to develop cardio-metabolic diseases such as obesity, hypertension, and insulin resistance [10–15] while other studies have only found an association between LBW and hypertension [16,17]. Furthermore, studies have shown that high birth weight (HBW) mainly caused by maternal gestational diabetes may result in increased risk of hypertension, abdominal obesity, and insulin resistance during puberty [18–21] while other studies have not observed such an association [22,23]. The potential impact of HBW on CMRFs in children and adolescents remains insufficient and controversial. In addition, the age and sex differences between birth weight and CMRFs are unclear [24–28]. Therefore, the association between birth weight and clustered CMRFs and how the association varies across age span and different sexes is still unclear.

It is known that poor lifestyle, such as lack of physical activity [29], unhealthy eating habits [8], screen time [30], and sleep deprivation [31], are closely associated with cardio-metabolic risk factors. The high prevalence of abnormal birth weight occurring in China, which was 5.2% of LBW, 7.4% of HBW in 2013 [32], is coupled with the low prevalence of children and adolescents meeting recommended levels of healthy lifestyles with the economic and dietary pattern transitions [33]. Considering that both birth weight and lifestyle have independent effects on CMRFs, and that birth weight is unlikely to affect the lifestyle of children and adolescents [34,35], to what extent the combinations of birth weight and lifestyle act on the occurrence of CMRFs is scarce, especially in China. We hypothesized that birth weight and lifestyle might influence CMRFs, whereby the effects would be cumulative, ideal lifestyle might weaken the association of birth weight with clustered CMRFs, and such an association might vary between age and sex groups. A national school-based healthy lifestyles investigation conducted in seven provinces provided us an opportunity to assess the effects [36]. In this large representative sample of Chinese children and adolescents, we examined overall CMRFs, and aimed to examine the strength of birth weight and lifestyle on clustered CMRFs alone and as combined effects and the age and sex differences in association.

2. Materials and Methods

2.1. Study Design and Participants

The data was extracted from the baseline of a nationwide school-based multi-centered study of 7–18 years children and adolescents which was performed in seven provinces (Tianjin, Shanghai, Chongqing, Liaoning, Hunan, Ningxia, and Guangdong) in China in 2013. The multistage cluster random sampling method was used in this study and more detailed information is published elsewhere [36]. Briefly, three to four districts from each province were randomly selected at first. Second, 12–16 schools were chosen from each district. Third, two to three classes per grade of each school were selected randomly, after excluding children and adolescents with serious organic diseases or those who refused to sign informed consent. Children and adolescents in those classes were invited for blood sample collection. In total, 15,732 participants aged between 7–18 years with blood sample information, while cases with missing data on anthropometric measurement ($n = 664$), missing data on birth weight ($n = 2782$), missing all information about lifestyle factors ($n = 777$) were excluded. A total of 11,509 participants aged 7–18 years were included in the final analysis (Figure 1). The primary sample and the final sample showed no statistical difference in age, residence, and parental education level. However, the proportion of boys in the final sample was less than the primary sample (Table S1). This study has been approved by the Ethical Committee of Peking University (No. IRB0000105213034). All participants and their parents provided signed informed consent.

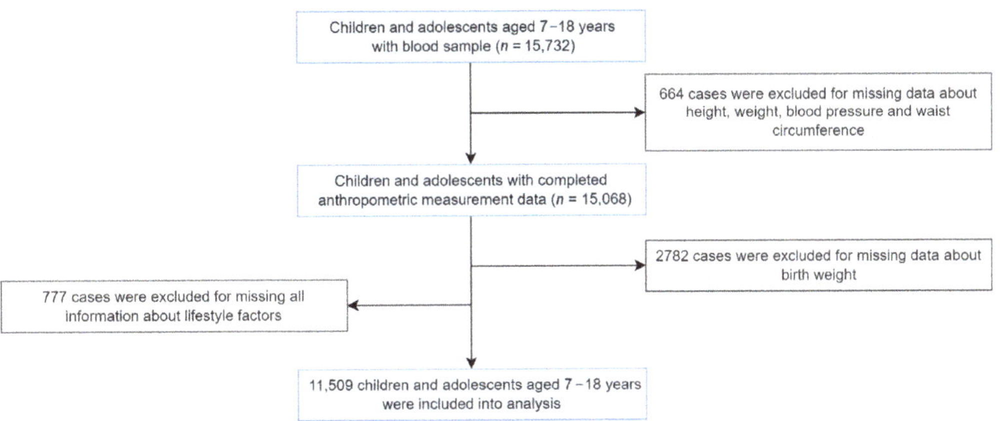

Figure 1. Flow diagram of participants' inclusion.

2.2. Study Measurements

2.2.1. Questionnaire Survey

Self-filled questionnaires for children and adolescents were used to obtain basic characteristics of sex, age, residence of children and adolescents, and their lifestyle, such as dietary consumption behaviors (fruits, vegetables, meat, and sugar-sweetened beverages), physical activity level and time, screen time and sleep duration. We collected the frequency and amount of each food consumed in the past week to calculate the dietary behavior of the children and adolescents. Parents self-administered questionnaires were used to obtain birth weight information for children and adolescents. Parents were asked to fill out the questionnaire based on the child's birth certificate or clinical record. If the documents were not available, they were required recall the information about the child's early life accurately. Parents were also requested to report the information on single-child status, delivery mode, delivery time, breastfeeding status of the children and adolescents, and parental education level and health behaviors. The questionnaires used in this study were all tested for logic and completeness. An expert panel then reviewed the questionnaires and considered them feasible and acceptable for children and adolescents and their parents.

2.2.2. Anthropometric Measurement and Blood Sample Detection

All children and adolescents received a standardization physical examination [36]. We used a portable stadiometer to measure the height of the participants standing upright without shoes, accurate to 0.1 cm. The participants wore lightweight underwear and weight was measured using a lever scale to the nearest 0.1 kg. Waist circumference was measured 1 cm above the navel, to 0.1 cm. The values of height, weight and waist circumference were the average value of two consecutive measurements.

We used a mercury sphygmomanometer, a stethoscope and an appropriate cuff to measure blood pressure. Children and adolescents were required to quietly rest for 5 min before starting the first measurement. A second measurement was performed after a 5-min interval. We calculated the average systolic and diastolic blood pressure as a result the final measurement of blood pressure. Venous blood was collected by a professional nurse after a 12-h fasting period. Serum glucose, total cholesterol (TC), triglyceride (TG), low-density lipoprotein cholesterol (LDL-C), and high-density lipoprotein cholesterol (HDL-C) were measured using an automatic biochemical analyzer by a qualified biological testing company.

Strict quality control has been carried out in the measurement of this study. All the measuring instruments have been corrected before use, and the measuring personnel have been trained in the standard process. Five percent of the subjects were retested daily, and failure to meet the criteria was considered a failure of the day's results and required full retesting.

2.3. Definition and Classification

2.3.1. Birth Weight

Normal birth weight (NBW) was considered to be children and adolescents with a birth weight from 2500 g to 3999 g. LBW was considered as birth weight less than 2499 g, HBW was considered as birth weight greater than 4000 g [37].

2.3.2. Lifestyle

Dietary consumption, physical activity, screen time and sleep duration were used to construct lifestyle scores. Healthy dietary consumption was defined as daily intake of fruits of $\geq$3 servings (one serving is about 100 g), vegetables of $\geq$4 servings (one serving is about 100 g), meat products of 2–3 servings (one serving is about 50 g), and weekly intake of sugar-sweetened beverage of <1 serving (one serving is about 250 mL) [38]. Adequate physical activity was defined as at least one hour of moderate to high intensity physical activity per day. Screen time was defined as less than two hours per day. Healthy sleep duration was defined as sleeping for more than 9 h per day [39]. Participants were grouped in the ideal lifestyle group if they had at least two or more healthy lifestyle factors, or into the poor lifestyle group if they had only one health lifestyle factor or fewer.

2.3.3. Combination of Birth Weight and Lifestyle

According to the combination of birth weight and lifestyle, the participants were divided into six groups: NBW/ideal lifestyle, NBW/poor lifestyle, LBW/ideal lifestyle, LBW/poor lifestyle, HBW/ideal lifestyle, HBW/poor lifestyle.

2.3.4. Clustered CMRFs

Hypertension was defined as blood pressure $\geq$ 95th percentile of age-, sex-, and height-specific references recommended by American Academy of Pediatrics in 2017 [40]. Impaired fasting glucose was defined as fasting glucose $\geq$ 5.6 mmol/L [41]. Dyslipidemia was defined according to the National Heart, Lung, and Blood Institute guidelines for cardiovascular health and risk reduction in children and adolescents: elevated TC was defined as $\geq$5.2 mmol/L, elevated LDL-C was defined as $\geq$3.4 mmol/L, abnormal HDL-C defined as $\leq$1.0 mmol/L, and elevated TG defined as $\geq$1.1 mmol/L for children below 9 years and $\geq$1.5 mmol/L for those aged 10 years or older [2]. Participants who met any one of the above criteria were divided into dyslipidemia. Abdominal obesity was defined as waist circumference $\geq$ 90th percentile of age–sex specific references recommended by National Health Commission of the People's Republic of China in 2018 [42]. Clustered CMRFs was defined as three or more factors being satisfied among all four items of hypertension, impaired fasting glucose, dyslipidemia, and abdominal obesity [43].

2.3.5. Confounding Variables

In this study, children and adolescents were divided into two age groups: 7–12 years and 13–18 years. Sex, residence, delivery mode, single-child status, parental smoking, and parental education were considered as confounding variables. Parental smoking was defined as either of parents smoked in the past week. Parental education was divided into junior high school and below and senior high school and above, with the highest education level of parents (either mother or father) being considered [44]. Missing values of the covariates (<8.6%) were imputed by multiple imputation. Sensitivity analysis results indicated that the direction and value of the findings before and after imputation were essentially unchanged, suggesting that the imputation variables would not affect the main results.

2.4. Statistics Analysis

Categorical variables were shown as number (percentage). Chi-square tests and Fisher's exact test were used to compare the differences of categorical data between sex groups and the differences of CMRFs and clustered CMRFs in different groups. ANOVA and Dunnett-*t* test were used to compare the differences between combinations of birth weight and lifestyles. A generalized linear mixed model was used to estimate the odds ratio (OR) and 95% confidence interval (CI) for association between CMRFs, clustered CMRFs and birth weight, lifestyle, and their combinations, using the school's name as the random-effect term to control the cluster effect of schools in different groups. Significance level was accepted at two-tailed $p < 0.05$. All data were analyzed by SPSS 26.0 and R 4.1.1.

3. Results

3.1. The Characteristics of Participants

The basic demographic characteristics of the total 11,509 participants from different sex groups were shown in Table 1. Children and adolescents with LBW accounted for 3.8% and HBW accounted for 9.0%, while poor lifestyle accounted for 46.2%. Girls tended to have unhealthier lifestyle ($p = 0.005$). Among the six combinations, the proportion of the number from high to low was NBW/ideal lifestyle (47.2%), NBW/poor lifestyle (40.1%), HBW/ideal lifestyle (4.7%), HBW/poor lifestyle (4.3%), LBW/ideal lifestyle (2.0%), LBW/poor lifestyle (1.8%). The prevalence of clustered CMRFs in children and adolescents aged 7–18 years was 3.6%, while 4.4% in boys and 2.9% in girls, the sex difference was statistically significant ($p < 0.001$).

Table 1. Demographic characteristics of eligible children and adolescents and their parents stratified by sex.

Variables		Total	Boys	Girls	χ^2/F	p
Age					8.908	0.003
	7–12 years	7090	3609 (63.0)	3481 (60.3)		
	13–18 years	4419	2123 (37.0)	2296 (39.7)		
Residence					0.936	0.333
	Rural	4523	2278 (39.7)	2245 (38.9)		
	Urban	6986	3454 (60.3)	3532 (61.1)		
Single-child status					82.888	<0.001
	Yes	7771	4099 (71.5)	3672 (63.6)		
	No	3738	1633 (28.5)	2105 (36.4)		
Parental education level					3.841	0.050
	Junior high school and below	4499	2292 (40.0)	2207 (38.2)		
	Senior high school and above	7010	3440 (60.0)	3570 (61.8)		
Birth weight					58.484	<0.001
	NBW	10,044	4913 (85.7)	5131 (88.8)		
	LBW	432	191 (3.3)	241 (4.2)		
	HBW	1033	628 (11.0)	405 (7.0)		
Fruits intake					1.465	0.226
	Inadequate	10,482	5202 (90.8)	5280 (91.4)		
	Adequate	1027	530 (9.2)	497 (8.6)		
Vegetables intake					7.865	0.005
	Inadequate	10,533	5204 (90.8)	5329 (92.2)		
	Adequate	976	528 (9.2)	448 (7.8)		
Meat intake					113.261	<0.001
	Improper	9543	4538 (79.2)	5005 (86.6)		
	proper	1966	1194 (20.8)	772 (13.4)		

Table 1. *Cont.*

Variables	Total	Boys	Girls	χ^2/F	p
Sugar-sweetened beverage				168.492	<0.001
Excessive	1316	877 (15.3)	439 (7.6)		
Proper	10,193	4855 (84.7)	5338 (92.4)		
Dietary consumption				30.410	<0.001
Unhealthy	8667	4189 (73.1)	4478 (77.5)		
Healthy	2842	1543 (26.9)	1299 (22.5)		
Physical activity				33.352	<0.001
Inadequate	6031	2849 (49.7)	3182 (55.1)		
Adequate	5478	2883 (50.3)	2595 (44.9)		
Screen time				70.682	<0.001
Excessive	3212	1802 (31.4)	1410 (24.4)		
Proper	8297	3930 (68.6)	4367 (75.6)		
Sleep duration				1.951	0.163
Inadequate	9196	4550 (79.4)	4646 (80.4)		
Adequate	2313	1182 (20.6)	1131 (19.6)		
Lifestyle				7.888	0.005
Poor lifestyle	5317	2573 (44.9)	2744 (47.5)		
Ideal lifestyle	6192	3159 (55.1)	3033 (52.5)		
Blood pressure				55.597	<0.001
Normal blood pressure	9478	4568 (79.7)	4910 (85.0)		
Hypertension	2031	1164 (20.3)	867 (15.0)		
Fasting glucoses				39.274	<0.001
Normal fasting glucoses	11,285	5574 (97.2)	5711 (98.9)		
Impaired fasting glucose	224	158 (2.8)	66 (1.1)		
Blood lipids				8.605	0.003
Normal blood lipids	8222	4166 (72.7)	4056 (70.2)		
Dyslipidemia	3287	1566 (27.3)	1721 (29.8)		
Waist circumference				4.408	0.036
Normal waist circumference	8870	4465 (77.9)	4405 (76.3)		
Abdominal obesity	2639	1267 (22.1)	1372 (23.7)		
CMRFs scores					<0.001
0	5725	2868 (50.0)	2857 (49.4)		
1	3815	1825 (31.8)	1990 (34.5)		
2	1553	791 (13.8)	762 (13.2)		
3	406	244 (4.3)	162 (2.8)		
4	10	4 (0.1)	6 (0.1)		
Combinations of birth weight and lifestyle [a]				13.794	<0.001
NBW/Ideal lifestyle	5430	2720 (47.5)	2710 (46.9)		Ref
NBW/Poor lifestyle	4614	2193 (38.3)	2421 (41.9)		0.050
LBW/Ideal lifestyle	225	109 (1.9)	116 (2.0)		0.992
LBW/Poor lifestyle	207	82 (1.4)	125 (2.2)		0.015
HBW/Ideal lifestyle	537	330 (5.8)	207 (3.6)		<0.001
HBW/Poor lifestyle	496	298 (5.2)	198 (3.4)		<0.001
Total	11,509	5732	5777		

Note: Chi square test was used to compare the two categorical variables. [a] ANOVA and Dunnett-*t* test were used to compare the differences between subgroups and NBW/Ideal lifestyle was used as reference group. Abbreviations: NBW, normal birth weight; LBW, low birth weight, HBW, high birth weight, CMRFs, cardio-metabolic risk factors, Ref, the reference group of Dunnett-*t* test.

3.2. CMRFs and Clustered CMRFs in Different Groups

As shown in Table 2, CMRFs and clustered CMRFs were differed in age and sex groups ($p < 0.05$). The proportion of hypertension, dyslipidemia and abdominal obesity differed significantly between urban and rural children and adolescents ($p < 0.001$). Birth weight and the combinations of birth weight and lifestyle were associated with dyslipidemia and abdominal obesity ($p < 0.05$). Lifestyle was associated with hypertension and dyslipidemia ($p < 0.05$). Higher risk of clustered CMRFs was related to meat intake, sugar-sweetened beverages, and sleep duration ($p < 0.05$).

Table 2. Prevalence of CMRFs and clustered CMRFs in different groups.

Variables	Total	Hypertension		Impaired Fasting Glucose		Dyslipidemia		Abdominal Obesity		Clustered CMRFs	
		n (%)	*p*	*n* (%)	*p*	*n* (%)	*p*	*n* (%)	*p*	*n* (%)	*p*
Sex			<0.001		<0.001		0.003		0.036		<0.001
Boys	5732	1164 (20.3)		158 (2.8)		1566 (27.3)		1267 (22.1)		248 (4.3)	
Girls	5777	867 (15.0)		66 (1.1)		1721 (29.8)		1372 (23.7)		168 (2.9)	
Age			0.020		<0.001		0.027		0.011		0.022
7–12 years	7090	1205 (17.0)		105 (1.5)		2077 (29.3)		1570 (22.1)		234 (3.3)	
13–18 years	4419	826 (18.7)		119 (2.7)		1210 (27.4)		1069 (24.2)		182 (4.1)	
Residence			<0.001		0.787		<0.001		<0.001		0.331
Rural	4523	963 (21.3)		90 (2.0)		1096 (24.2)		930 (20.6)		173 (3.8)	
Urban	6986	1068 (15.3)		134 (1.9)		2191 (31.4)		1709 (24.5)		243 (3.5)	
Birth weight			0.406		0.297		0.034		<0.001		0.528
NBW	10,044	1756 (17.5)		203 (2.0)		2877 (28.6)		2233 (22.2)		356 (3.5)	
LBW	432	85 (19.7)		7 (1.6)		141 (32.6)		89 (20.6)		19 (4.4)	
HBW	1033	190 (18.4)		14 (1.4)		269 (26.0)		317 (30.7)		41 (4.0)	
Fruits intake			0.428		0.479		0.79		0.293		0.614
Inadequate	10,482	1859 (17.7)		207 (2.0)		2990 (28.5)		2390 (22.8)		376 (3.6)	
Adequate	1027	172 (16.7)		17 (1.7)		297 (28.9)		249 (24.2)		40 (3.9)	
Vegetables intake			0.212		0.029		0.154		0.226		0.897
Inadequate	10,533	1873 (17.8)		214 (2.0)		2989 (28.4)		2400 (22.8)		380 (3.6)	
Adequate	976	158 (16.2)		10 (1.0)		298 (30.5)		239 (24.5)		36 (3.7)	
Meat intake			<0.001		0.397		0.002		0.732		0.017
Improper	9543	1738 (18.2)		181 (1.9)		2781 (29.1)		2194 (23.0)		363 (3.8)	
proper	1966	293 (14.9)		43 (2.2)		506 (25.7)		445 (22.6)		53 (2.7)	
Sugar-sweetened beverage			0.149		0.001		0.117		0.035		0.006
Excessive	1316	251 (19.1)		42 (3.2)		400 (30.4)		332 (25.2)		65 (4.9)	
Proper	10,193	1780 (17.5)		182 (1.8)		2887 (28.3)		2307 (22.6)		351 (3.4)	
Dietary consumption			<0.001		0.193		0.749		0.824		0.312
Unhealthy	8667	1603 (18.5)		177 (2.0)		2482 (28.6)		1983 (22.9)		322 (3.7)	
Healthy	2842	428 (15.1)		47 (1.7)		805 (28.3)		656 (23.1)		94 (3.3)	
Physical activity			0.166		0.724		0.066		0.531		0.424
Inadequate	6031	1036 (17.2)		120 (2.0)		1678 (27.8)		1397 (23.2)		210 (3.5)	
Adequate	5478	995 (18.2)		104 (1.9)		1609 (29.4)		1242 (22.7)		206 (3.8)	
Screen time			0.248		0.326		<0.001		0.415		0.585
Excessive	3212	588 (18.3)		56 (1.7)		845 (26.3)		753 (23.4)		121 (3.8)	
Proper	8297	1443 (17.4)		168 (2.0)		2442 (29.4)		1886 (22.7)		295 (3.6)	
Sleep duration			0.010		0.004		0.075		0.003		0.028
Inadequate	9196	1665 (18.1)		196 (2.1)		2661 (28.9)		2163 (23.5)		350 (3.8)	
Adequate	2313	366 (15.8)		28 (1.2)		626 (27.1)		476 (20.6)		66 (2.9)	
Lifestyle			0.004		0.263		0.002		0.226		0.545
Poor lifestyle	5317	978 (18.4)		113 (2.1)		1474 (27.7)		1258 (23.7)		200 (3.8)	
Ideal lifestyle	6192	1053 (17.0)		111 (1.8)		1813 (29.3)		1381 (22.3)		216 (3.5)	
Combinations of birth weight and lifestyle [a]			0.277		0.414		0.046		<0.001		0.077
NBW/Ideal lifestyle	5430	921 (17.0)	Ref	102 (1.9)	Ref	1589 (29.3)	Ref	1183 (21.8)	Ref	189 (3.5)	Ref
NBW/Poor lifestyle	4614	835 (18.1)	0.514	101 (2.2)	0.776	1288 (27.9)	0.512	1050 (22.8)	0.754	167 (3.6)	0.998
LBW/Ideal lifestyle	225	41 (18.2)	0.992	2 (0.9)	0.817	80 (35.6)	0.185	43 (19.1)	0.879	13 (5.8)	0.302
LBW/Poor lifestyle	207	44 (21.3)	0.441	5 (2.4)	0.987	61 (29.5)	1.000	46 (22.2)	1.000	6 (2.9)	0.995
HBW/Ideal lifestyle	537	91 (16.9)	1.000	7 (1.3)	0.886	144 (26.8)	0.724	155 (28.9)	0.001	14 (2.6)	0.828
HBW/Poor lifestyle	496	99 (20.0)	0.383	7 (1.4)	0.956	125 (25.2)	0.244	162 (32.7)	<0.001	27 (5.4)	0.117
Total	11,509	2031 (17.6)		224 (1.9)		3287 (28.6)		2639 (22.9)		416 (3.6)	

Note: [a] ANOVA and Dunnett-*t* test were used to compare the differences between subgroups and NBW/Ideal lifestyle was used as reference group. Abbreviations: NBW, normal birth weight; LBW, low birth weight, HBW, high birth weight, CMRFs, cardio-metabolic risk factors. Ref, the reference group of Dunnett-*t* test.

3.3. Multivariate Associations between CMRFs, Clustered CMRFs and Its Risk Factors

The associations between CMRFs, clustered CMRFs and birth weight in mixed effect models are shown in Table 3. After adjusting for confounding factors, LBW was related to elevated risk of hypertension in children and adolescents aged 13–18 years (OR = 1.81, 95% CI: 1.19–2.75) and boys (OR = 1.54, 95% CI: 1.06–2.25). HBW was associated with higher risk of abdominal obesity (OR = 1.66, 95% CI: 1.43–1.92) and was associated with clustered CMRFs in children and adolescents aged 7–12 years (OR = 1.70, 95% CI: 1.12–2.58) and girls (OR = 1.83, 95% CI: 1.19–3.08) (Figures S1–S3). Table 4 shows the associations between CMRFs, clustered CMRFs and lifestyle. A higher risk of hypertension was found in the inadequate sleep duration group in total model (OR = 1.18, 95% CI: 1.02–1.36), children aged 7–12 years (OR = 1.25, 95% CI: 1.07–1.45) and girls (OR = 1.28, 95% CI: 1.03–1.59). Abdominal obesity was associated with excessive screen time (OR = 1.17, 95% CI: 1.06–1.30) and inadequate sleep duration (OR = 1.14, 95% CI: 1.01–1.29) in children and adolescents aged 7–18 years. Poor lifestyle was related to abdominal obesity in children aged 7–12 years (OR = 1.15, 95% CI: 1.02–1.29). Unhealthy dietary consumption was associated with a higher risk of clustered CMRFs in boys (OR = 1.40, 95% CI: 1.01–1.94) and lower risk of clustered CMRFs in girls (OR = 0.68, 95% CI: 0.47–0.97) (Figures S4–S6).

Table 3. Associations between CMRFs, clustered CMRFs and birth weight stratified by age and sex groups.

	Birth Weight	Hypertension		Impaired Fasting Glucose		Dyslipidemia		Abdominal Obesity		Clustered CMRFs	
		OR (95% CI)	*p*	OR (95% CI)	*p*	OR (95% CI)	*p*	OR (95% CI)	*p*	OR (95% CI)	*p*
Total [a]	NBW	Ref		Ref		Ref		Ref		Ref	
	LBW	1.21 (0.92–1.60)	0.176	0.99 (0.46–2.17)	0.985	1.13 (0.88–1.44)	0.348	0.98 (0.76–1.27)	0.897	1.43 (0.87–2.36)	0.155
	HBW	1.02 (0.85–1.22)	0.864	0.68 (0.39–1.19)	0.171	0.99 (0.84–1.17)	0.898	1.66 (1.43–1.92)	<0.001	1.18 (0.84–1.67)	0.333
7–12 years [b]	NBW	Ref		Ref		Ref		Ref		Ref	
	LBW	0.94 (0.64–1.37)	0.737	0.80 (0.25–2.59)	0.706	1.21 (0.89–1.66)	0.218	0.82 (0.58–1.17)	0.281	1.24 (0.62–2.49)	0.541
	HBW	1.08 (0.85–1.36)	0.537	0.60 (0.26–1.41)	0.244	1.06 (0.86–1.30)	0.585	1.74 (1.44–2.10)	<0.001	1.70 (1.12–2.58)	0.013
13–18 years [b]	NBW	Ref		Ref		Ref		Ref		Ref	
	LBW	1.81 (1.19–2.75)	0.006	1.27 (0.44–3.65)	0.652	0.96 (0.63–1.45)	0.843	1.25 (0.86–1.83)	0.245	1.82 (0.88–3.75)	0.106
	HBW	0.92 (0.69–1.23)	0.567	0.74 (0.35–1.57)	0.434	0.84 (0.63–1.12)	0.235	1.54 (1.21–1.96)	<0.001	0.65 (0.35–1.21)	0.176
Boys [c]	NBW	Ref		Ref		Ref		Ref		Ref	
	LBW	1.54 (1.06–2.25)	0.024	1.07 (0.42–2.73)	0.883	1.18 (0.82–1.71)	0.369	1.13 (0.77–1.65)	0.534	1.47 (0.75–2.89)	0.257
	HBW	0.98 (0.78–1.23)	0.851	0.81 (0.45–1.48)	0.496	0.98 (0.79–1.22)	0.852	1.48 (1.21–1.81)	<0.001	0.92 (0.58–1.45)	0.713
Girls [c]	NBW	Ref		Ref		Ref		Ref		Ref	
	LBW	0.96 (0.63–1.46)	0.833	0.95 (0.45–2.01)	0.892	1.07 (0.76–1.49)	0.700	0.87 (0.62–1.23)	0.442	1.44 (0.68–3.03)	0.342
	HBW	1.02 (0.75–1.38)	0.894	0.85 (0.47–1.54)	0.591	0.99 (0.77–1.28)	0.943	1.99 (1.59–2.48)	<0.001	1.83 (1.19–3.08)	0.022

Note: Model [a] was adjusted for age, sex, residence, delivery mode, single-child status, parental smoking, parental education, and school effect. Model [b] was adjusted for sex, residence, delivery mode, single-child status, parental smoking, parental education, and school effect. Model [c] was adjusted for age, residence, delivery mode, single-child status, parental smoking, parental education, and school effect. NBW was used as reference group. NBW, normal birth weight; LBW, low birth weight; HBW, high birth weight; CMRFs, cardio-metabolic risk factors; OR, odds ratio; CI, confidence interval, Ref, the reference group.

As shown in Table 5, a higher risk of hypertension was found in LBW/ideal lifestyle group compared with NBW/ideal lifestyle in adolescents aged 13–18 years (OR = 2.43, 95% CI: 1.15–5.17) and boys (OR = 2.45, 95% CI: 1.25–4.79). Boys with NBW/poor lifestyle had a higher risk of impaired fasting glucose (OR = 1.65, 95% CI: 1.02–2.67). Abdominal obesity was significant higher in HBW/ideal lifestyle (OR = 1.54, 95% CI: 1.26–1.90) and HBW/poor lifestyle (OR = 1.89, 95% CI: 1.53–2.33) groups in children and adolescents aged 7–18 years. The effect value of the NBW/ideal lifestyle and NBW/poor lifestyle groups in abdominal obesity were statistically different from that of the HBW/poor lifestyle group ($p < 0.05$). In the age group of 7–12 years, abdominal obesity in HBW/ideal lifestyle and HBW/poor lifestyle groups was significantly different ($p < 0.05$). In the age group of 13–18 years, dyslipidemia in the LBW/ideal lifestyle and LBW/poor lifestyle groups was significantly different ($p < 0.05$). Higher risk of clustered CMRFs was found in the HBW/poor lifestyle group in total model (OR = 1.74, 95% CI: 1.13–2.68) and children aged 7–12 years (OR = 2.37, 95% CI: 1.39–4.05). It was also found in the LBW/ideal lifestyle group in total model (OR = 2.00, 95% CI: 1.07–3.72), adolescents aged 13–18 years (OR = 3.57, 95% CI: 1.23–10.37) and boys (OR = 2.78, 95% CI: 1.02–7.62) (Figures S7–S9).

Table 4. Associations between CMRFs, clustered CMRFs and lifestyles stratified by age and sex group.

	Lifestyles	Hypertension		Impaired Fasting Glucose		Dyslipidemia		Abdominal Obesity		Clustered CMRFs	
		OR (95% CI)	p	OR (95% CI)	p	OR (95% CI)	p	OR (95% CI)	p	OR (95% CI)	p
Total [a]	Unhealthy dietary consumption	1.10 (0.97–1.25)	0.132	1.30 (0.93–1.82)	0.129	0.95 (0.85–1.06)	0.360	0.98 (0.88–1.09)	0.695	1.02 (0.80–1.29)	0.905
	Inadequate physical activity	1.01 (0.91–1.12)	0.874	1.06 (0.80–1.40)	0.683	1.00 (0.91–1.10)	0.953	0.99 (0.91–1.09)	0.882	1.01 (0.82–1.24)	0.940
	Excessive screen time	1.04 (0.92–1.17)	0.572	0.99 (0.72–1.36)	0.945	0.98 (0.88–1.09)	0.758	1.17 (1.06–1.30)	0.003	1.16 (0.92–1.45)	0.209
	Inadequate sleep duration	1.18 (1.02–1.36)	0.027	1.11 (0.72–1.71)	0.651	0.97 (0.86–1.10)	0.673	1.14 (1.01–1.29)	0.035	1.17 (0.87–1.56)	0.293
	Poor lifestyle	1.08 (0.97–1.20)	0.181	1.14 (0.86–1.50)	0.368	0.97 (0.88–1.06)	0.500	1.08 (0.99–1.19)	0.085	1.09 (0.89–1.33)	0.431
7–12 years [b]	Unhealthy dietary consumption	1.00 (0.72–1.37)	0.980	1.51 (0.90–2.54)	0.116	0.94 (0.83–1.08)	0.401	1.02 (0.89–1.17)	0.756	1.02 (0.80–1.29)	0.905
	Inadequate physical activity	1.00 (0.88–1.15)	0.970	0.95 (0.63–1.41)	0.782	1.02 (0.91–1.15)	0.711	1.02 (0.90–1.14)	0.793	1.05 (0.80–1.38)	0.727
	Excessive screen time	1.01 (0.87–1.18)	0.859	1.08 (0.69–1.68)	0.747	0.98 (0.86–1.12)	0.798	1.17 (1.03–1.34)	0.018	1.07 (0.79–1.44)	0.682
	Inadequate sleep duration	1.25 (1.07–1.45)	0.005	1.27 (0.78–2.09)	0.335	0.99 (0.87–1.12)	0.824	1.19 (1.04–1.36)	0.010	1.24 (0.90–1.70)	0.181
	Poor lifestyle	1.12 (0.98–1.29)	0.098	1.05 (0.70–1.58)	0.807	0.95 (0.84–1.07)	0.353	1.15 (1.02–1.29)	0.023	1.03 (0.78–1.35)	0.852
13–18 years [b]	Unhealthy dietary consumption	1.07 (0.87–1.31)	0.542	1.15 (0.74–1.81)	0.532	0.96 (0.80–1.15)	0.675	0.97 (0.82–1.15)	0.743	1.10 (0.75–1.60)	0.636
	Inadequate physical activity	1.04 (0.88–1.24)	0.639	1.18 (0.79–1.76)	0.417	0.97 (0.83–1.13)	0.665	0.96 (0.83–1.11)	0.600	0.98 (0.71–1.34)	0.879
	Excessive screen time	1.03 (0.85–1.25)	0.764	0.83 (0.51–1.33)	0.429	0.97 (0.81–1.16)	0.708	1.16 (0.99–1.37)	0.075	1.18 (0.83–1.67)	0.364
	Inadequate sleep duration	1.01 (0.68–1.49)	0.973	0.75 (0.31–1.83)	0.532	0.95 (0.65–1.38)	0.776	0.96 (0.68–1.37)	0.822	1.10 (0.54–2.27)	0.792
	Poor lifestyle	1.03 (0.87–1.22)	0.721	1.19 (0.81–1.75)	0.384	1.01 (0.86–1.18)	0.937	1.02 (0.88–1.18)	0.784	1.17 (0.85–1.60)	0.338
Boys [c]	Unhealthy dietary consumption	1.11 (0.94–1.31)	0.213	1.31 (0.88–1.96)	0.184	1.00 (0.86–1.16)	0.999	1.03 (0.89–1.20)	0.667	1.40 (1.01–1.94)	0.046
	Inadequate physical activity	1.00 (0.87–1.15)	0.988	1.07 (0.76–1.49)	0.704	0.95 (0.83–1.09)	0.422	0.96 (0.84–1.10)	0.551	1.12 (0.86–1.46)	0.411
	Excessive screen time	1.05 (0.90–1.23)	0.507	1.06 (0.73–1.53)	0.769	0.94 (0.81–1.09)	0.435	1.17 (1.01–1.35)	0.034	1.10 (0.83–1.48)	0.504
	Inadequate sleep duration	1.12 (0.93 1.36)	0.236	1.08 (0.65–1.79)	0.764	1.05 (0.88–1.25)	0.613	1.15 (0.97–1.37)	0.113	1.20 (0.81–1.76)	0.369
	Poor lifestyle	1.08 (0.93–1.24)	0.304	1.22 (0.87–1.70)	0.244	0.97 (0.85–1.11)	0.677	1.09 (0.95–1.24)	0.213	1.28 (0.98–1.67)	0.070
Girls [c]	Unhealthy dietary consumption	1.11 (0.91–1.35)	0.304	1.04 (0.74–1.47)	0.817	0.90 (0.77–1.05)	0.192	0.96 (0.82–1.11)	0.551	0.68 (0.47–0.97)	0.034
	Inadequate physical activity	1.02 (0.87–1.20)	0.803	0.99 (0.74–1.32)	0.929	1.04 (0.91–1.19)	0.558	1.02 (0.89–1.16)	0.801	0.85 (0.62–1.17)	0.327
	Excessive screen time	1.00 (0.83–1.21)	0.973	0.93 (0.66–1.31)	0.676	1.03 (0.88–1.20)	0.738	1.18 (1.02–1.36)	0.031	1.21 (0.84–1.74)	0.313
	Inadequate sleep duration	1.28 (1.03–1.59)	0.028	1.06 (0.71–1.57)	0.789	0.91 (0.77–1.08)	0.270	1.11 (0.93–1.32)	0.263	1.18 (0.76–1.82)	0.456
	Poor lifestyle	1.09 (0.93–1.28)	0.304	0.97 (0.73–1.30)	0.857	0.95 (0.84–1.09)	0.486	1.08 (0.95–1.23)	0.237	0.85 (0.61–1.17)	0.323

Note: Model [a] was adjusted for age, sex, residence, delivery mode, single-child status, parental smoking, parental education, the interaction of sex and birth weight and school effect. Model [b] was adjusted for sex, residence, delivery mode, single-child status, parental smoking, parental education, and school effect. Model [c] was adjusted for age, residence, delivery mode, single-child status, parental smoking, parental education, and school effect. Healthy dietary consumption, adequate physical activity, proper screen time and adequate sleep duration were used as reference groups. NBW, normal birth weight; LBW, low birth weight; HBW, high birth weight; CMRFs, cardio-metabolic risk factors; OR, odds ratio; CI, confidence interval.

Table 5. Associations between CMRFs, clustered CMRFs and combinations of birth weight and lifestyle in mixed effect models stratified by age and sex groups.

	Combinations	Hypertension		Impaired Fasting Glucose		Dyslipidemia		Abdominal Obesity		Clustered CMRFs	
		OR (95% CI)	*p*	OR (95% CI)	*p*	OR (95% CI)	*p*	OR (95% CI)	*p*	OR (95% CI)	*p*
Total [a]	NBW/Ideal lifestyle	Ref		Ref		Ref		Ref *		Ref	
	NBW/Poor lifestyle	1.06 (0.95–1.19)	0.302	1.12 (0.83–1.49)	0.459	0.98 (0.89–1.08)	0.683	1.06 (0.96–1.17) *	0.237	1.04 (0.83–1.29)	0.754
	LBW/Ideal lifestyle	1.25 (0.84–1.85)	0.269	0.56 (0.14–2.34)	0.428	1.35 (0.97–1.88)	0.074	0.94 (0.66–1.36)	0.759	2.00 (1.07–3.72)	0.029
	LBW/Poor lifestyle	1.25 (0.84–1.84)	0.274	1.59 (0.62–4.10)	0.333	0.89 (0.61–1.29)	0.540	1.08 (0.76–1.54)	0.669	0.95 (0.41–2.19)	0.896
	HBW/Ideal lifestyle	0.94 (0.73–1.22)	0.652	0.70 (0.32–1.55)	0.384	0.98 (0.78–1.22)	0.841	1.54 (1.26–1.90)	<0.001	0.76 (0.43–1.33)	0.327
	HBW/Poor lifestyle	1.16 (0.90–1.49)	0.249	0.72 (0.33–1.58)	0.411	0.99 (0.78–1.25)	0.898	1.89 (1.53–2.33)	<0.001	1.74 (1.13–2.68)	0.012
7–12 years [b]	NBW/Ideal lifestyle	Ref		Ref		Ref		Ref *		Ref	
	NBW/Poor lifestyle	1.11 (0.96–1.29)	0.159	1.12 (0.74–1.71)	0.592	0.92 (0.81–1.05)	0.222	1.10 (0.97–1.25) *	0.134	0.89 (0.66–1.21)	0.470
	LBW/Ideal lifestyle	0.83 (0.49–1.43)	0.504	1.03 (0.24–4.37)	0.974	1.21 (0.80–1.83)	0.367	0.74 (0.45–1.21)	0.228	1.02 (0.36–2.84)	0.977
	LBW/Poor lifestyle	1.16 (0.68–1.99)	0.580	0.61 (0.08–4.59)	0.632	1.14 (0.72–1.80)	0.583	1.01 (0.62–1.66)	0.963	1.38 (0.54–3.53)	0.506
	HBW/Ideal lifestyle	1.06 (0.78–1.44)	0.720	0.79 (0.28–2.24)	0.653	0.97 (0.74–1.27)	0.798	1.55 (1.20–2.00) *	0.001	1.07 (0.56–2.04)	0.838
	HBW/Poor lifestyle	1.21 (0.86–1.72)	0.273	0.46 (0.11–1.92)	0.284	1.11 (0.82–1.49)	0.508	2.20 (1.67–2.88)	<0.001	2.37 (1.39–4.05)	0.002
13–18 years [b]	NBW/Ideal lifestyle	Ref		Ref		Ref		Ref *		Ref	
	NBW/Poor lifestyle	1.10 (0.86–1.39)	0.458	1.26 (0.78–2.05)	0.343	1.07 (0.87–1.31)	0.550	0.90 (0.74–1.10) *	0.316	1.18 (0.77–1.81)	0.442
	LBW/Ideal lifestyle	2.43 (1.15–5.17)	0.021	1.00 (0.66–1.62)	0.981	1.69 (0.86–3.31) #	0.129	1.38 (0.72–2.65)	0.338	3.57 (1.23–10.37)	0.019
	LBW/Poor lifestyle	1.32 (0.58–2.99)	0.506	2.03 (0.44–9.43)	0.366	0.51 (0.21–1.24)	0.137	0.98 (0.49–1.96)	0.950	1.00 (0.62–1.48)	0.970
	HBW/Ideal lifestyle	0.64 (0.34–1.19)	0.157	1.00 (0.64–1.52)	0.970	1.04 (0.64–1.69)	0.879	1.76 (1.16–2.67)	0.008	0.36 (0.08–1.58)	0.177
	HBW/Poor lifestyle	1.32 (0.78–2.23)	0.296	1.58 (0.58–4.34)	0.372	0.64 (0.35–1.16)	0.139	1.70 (1.09–2.65)	0.019	1.06 (0.40–2.83)	0.907
Boys [c]	NBW/Ideal lifestyle	Ref		Ref		Ref		Ref *		Ref	
	NBW/Poor lifestyle	1.15 (0.92–1.43)	0.215	1.65 (1.02–2.67)	0.043	1.03 (0.85–1.24)	0.757	1.07 (0.90–1.28) *	0.451	1.33 (0.91–1.95)	0.135
	LBW/Ideal lifestyle	2.45 (1.25–4.79)	0.009	0.92 (0.12–7.10)	0.935	1.14 (0.61–2.12)	0.682	1.03 (0.55–1.95)	0.920	2.78 (1.02–7.62)	0.047
	LBW/Poor lifestyle	1.36 (0.64–2.91)	0.428	1.02 (0.13–7.99)	0.988	0.68 (0.31–1.50)	0.342	1.08 (0.54–2.17)	0.829	0.56 (0.08–4.23)	0.576
	HBW/Ideal lifestyle	0.93 (0.58–1.50)	0.775	0.31 (0.04–2.31)	0.251	1.31 (0.90–1.90)	0.155	1.56 (1.10–2.19)	0.012	0.76 (0.29–1.98)	0.574
	HBW/Poor lifestyle	1.03 (0.63–1.67)	0.911	2.03 (0.79–5.21)	0.139	0.88 (0.57–1.37)	0.569	2.49 (1.73–3.57)	<0.001	1.19 (0.51–2.74)	0.689
Girls [c]	NBW/Ideal lifestyle	Ref		Ref		Ref		Ref *		Ref	
	NBW/Poor lifestyle	1.01 (0.79–1.29)	0.928	1.00 (0.68–1.49)	0.991	0.98 (0.81–1.17)	0.788	0.96 (0.81–1.15) *	0.689	0.68 (0.41–1.13) *	0.135
	LBW/Ideal lifestyle	1.24 (0.59–2.62)	0.572	0.80 (0.20–3.30)	0.758	1.40 (0.80–2.46)	0.244	0.91 (0.50–1.68)	0.771	1.60 (0.46–5.55)	0.462
	LBW/Poor lifestyle	1.36 (0.58–3.20)	0.477	1.04 (0.27–4.08)	0.956	1.12 (0.59–2.15)	0.723	0.90 (0.47–1.73)	0.753	1.19 (0.26–5.41)	0.818
	HBW/Ideal lifestyle	0.86 (0.48–1.54)	0.600	0.91 (0.33–2.47)	0.848	1.06 (0.68–1.65)	0.789	2.00 (1.37–2.93)	<0.001	1.38 (0.52–3.68)	0.520
	HBW/Poor lifestyle	1.53 (0.86–2.72)	0.151	0.80 (0.24–2.70)	0.715	1.02 (0.61–1.70)	0.950	2.08 (1.34–3.20)	0.001	1.92 (0.72–5.10)	0.191

Note: Model [a] was adjusted for age, sex, residence, delivery mode, single-child status, parental smoking, parental education, and school effect. Model [b] was adjusted for sex, residence, delivery mode, single-child status, parental smoking, parental education, and school effect. Model [c] was adjusted for age, residence, delivery mode, single-child status, parental smoking, parental education, and school effect. NBW/Ideal lifestyle was used as reference group. NBW, normal birth weight; LBW, low birth weight; HBW, high birth weight; CMRFs, cardio-metabolic risk factors; OR, odds ratio; CI, confidence interval, Ref, the reference group. #, the odds ratio difference from the group of LBW/Poor lifestyle was statistically significant, LBW/Poor lifestyle was used as reference group. * The odds ratio difference from the group of HBW/Poor lifestyle was statistically significant, HBW/Poor lifestyle was used as reference group.

4. Discussion

To the best of our knowledge, this was the first nationwide study addressing the effects of the combination of birth weight and lifestyle on clustered CMRFs and age and sex differences among children and adolescents in China. Our findings showed that the combined effect of LBW and ideal lifestyle was associated with clustered CMRFs, especially for those adolescents aged 13–18 years and boys. In addition, the significant combined effect of HBW and poor lifestyle on the clustered CMRFs in children and adolescents was found, especially for those children aged 7–12 years.

Previous studies have shown the association between CMRFs and birth weight in children and adolescents [10–15,45,46]. However, few studies focused on clustered CMRFs, possibly due to its low prevalence among children and adolescents. However, our study found that HBW had significant effects on clustered CMRFs in children aged 7–12 years and girls. In addition, the disease risk of the HBW group was higher than the corresponding NBW group regardless of lifestyle in children and adolescents aged 7–18 years, suggesting that HBW might be a health risk population that needs to be concentrated on, especially in younger children and girls. Currently it is believed that the cause of HBW is genetic factors [45] and uterine environmental factors. Maternal intermittent hyperglycemia leads to fetal hyperglycemia and fetal release of insulin, insulin-like growth factor and growth hormone which cause an increase in fetal fat deposition and weight gain [47–49]. Maternal abnormal lipid levels might also affect the placental nutrition supply, which prompts fetal overgrowth in early life [50]. Studies have shown that 16% of HBW children develop hypoglycemia within 24 h of life due to a placental glucose supply interruption [51,52]. All these factors might affect adipose tissue and pancreatic cell development, leading to increased BMI and impaired glucose metabolism, thus increasing the risk of obesity and metabolic complications in later life [53–55]. Furthermore, many studies support the evidence that LBW has an effect on hypertension in children and adolescents [16,17,56], which are consistent with our findings. For this phenomenon, the most reasonable hypothesis based on trial and clinical evidence is "congenital nephron reduction" [57,58], which means impaired intrauterine development can lead to impaired kidney development. A decrease in the number of nephrons could lead to compensatory hypertrophy, associated intraglomerular hypertension could then lead to glomerulosclerosis and hypertension within years [59,60]. In addition, HBW was confirmed to be associated with a higher risk of abdominal obesity in children and adolescents, which was consistent with the results of multiple other studies [61,62]. Early identification of abnormal birth weight children at health risk and the feasible prevention of CMRFs through lifestyle intervention are of great public health significance for the prevention of cardiovascular disease.

This study showed that the combination of LBW/ideal lifestyle was associated with elevated risk of clustered CMRFs. Due to the limitation of LBW sample size, this association was not found in LBW/poor lifestyle. However, it could be seen that the association between LBW and clustered CMRFs was mainly due to its association with hypertension, and studies have shown that the effect of LBW on hypertension can be weakened by a healthy lifestyle [22]. Therefore, maintaining a healthy lifestyle among low-birth-weight children and adolescents might also help to reduce their adverse health effects in current and later life. We also found that children and adolescents in the group of HBW/poor lifestyle had a greater risk than the group of HBW/ideal lifestyle, indicating the need to maintain a healthy lifestyle in children and adolescents with early-life adverse events to avoid cardiovascular metabolism-related diseases. Since the self-reported lifestyle of the latest week could not represent the health effect of long-term lifestyle in children and adolescents and other possible confounding variables, our study only found the association between poor lifestyle and abdominal obesity in children aged 7–12 years, the effect of sleep duration on abdominal obesity and hypertension and the role of screen time on abdominal obesity. Previous longitudinal studies have shown that healthy lifestyle was associated with CMRFs in children and adolescents. Accumulating evidence showed physical activity could affect C-reactive protein and had beneficial effects on the CMRFs in children [63].

In addition, a high-fat diet or deficiencies of specific nutrients could increase the risk of cardiovascular disease [64]. Furthermore, research has shown that screen time and poor sleep were important factors related to cardio-metabolic risk indicators in children [30,65]. However, considering that it is difficult for children and adolescents to achieve a completely healthy lifestyle, our research showed that as long as there were more than two of those healthy lifestyle factors, the risk of clustered CMRFs could be reduced, although significant results were only found in subgroups. This might be due to the small sample size of the subgroups. But prior studies have shown that ideal lifestyle had a positive effect on obesity in children and adolescents with abnormal birth weight [66,67], which could be referred by CMRFs [46].

We also found age and sex differences in clustered CMRFs. The prevalence of clustered CMRFs was higher in adolescents aged 13–18 years than those in children aged 7–12 years, higher in boys than in girls, which was consistent with previous findings [38,68–70]. However, a previous longitudinal study showed that birth weight had adverse effects on cardiometabolic metabolism in childhood, but further intensified in adolescence [55], which was not consistent with the conclusions of this study. We found the adverse effect of HBW on clustered CMRFs in the 7–12-year-old group and LBW on hypertension in the 13–18-year-old group, but we did not observe this association in other groups, which indicates that the earlier intervention is conducted, the bigger benefit obtained. Clustered CMRFs in adolescents might mean the track from childhood and the long-term accumulation of adverse outcomes of CMRFs, and simple lifestyle intervention might not work enough, and multiple and fortified measures are needed [27]. In addition, we observed an interesting phenomenon that unhealthy dietary consumption may lead to increased risk of clustered CMRFs in boys but may lead to decreased risk of clustered CMRFs in girls. This may be related to the different weight distribution between sex, with the proportion of overweight and obesity in boys is higher than that in girls, and the proportion of thinness in girls is higher than that in boys [71].

The sample size of this study was obtained from seven provinces in China and could be used as a representative national data. Furthermore, this study was the first one to date addressing the combined effects of birth weight and lifestyle on clustered CMRFs in children and adolescents, providing a strong rationale for promoting cardiovascular and metabolic health in children and adolescents. Moreover, the outcomes measured were based on blood samples, which were more accurate and objective. However, this study still has some limitations. Firstly, our data came from a cross-sectional survey, preventing us from making causal inferences. Secondly, the birth weight and lifestyle factors were collected according to questionnaires and recall bias might exist. However, as important information from a child's early life, parents were unlikely to forget a child's birth weight information, and children and adolescents surveyed could recall the lifestyle of the latest week clearly. Thirdly, residual confounding factors existed given that this study did not investigate maternal pre-pregnancy weight, pubertal development in children and adolescents, and other potentially unknown confounding information, which might have some impact on the result. Finally, according to the MCAR test, there were more boys with missing data, which might lead to the underestimation of the results due to higher clustered CMRFs prevalence in boys than girls.

5. Conclusions

In summary, we found that the combination of abnormal birth weight and poor lifestyle had a greater impact on clustered CMRFs in children and adolescents. This suggested that avoiding high birth weight through prenatal health education on nutritious diet and exercise guidelines and maintaining a healthy lifestyle during childhood and adolescents or both, such as sufficient physical activity, sleep duration, favorable diet consumption and proper screen time, might reduce the prevalence of CMRFs, especially in younger children.

Supplementary Materials: The following supporting information can be downloaded at: https://www.mdpi.com/article/10.3390/nu14153131/s1, Table S1: Missing completely at random (MCAR) test of the sample. Figure S1: Associations between hypertension, impaired fasting glucose and birth weight stratified by age and sex groups; Figure S2: Associations between dyslipidemia, abdominal obesity and birth weight stratified by age and sex groups; Figure S3: Associations between clustered CMRFs and birth weight stratified by age and sex groups; Figure S4: Associations between hypertension, impaired fasting glucose and lifestyle stratified by age and sex groups; Figure S5: Associations between dyslipidemia, abdominal obesity and lifestyle stratified by age and sex groups; Figure S6: Associations between clustered CMRFs and lifestyle stratified by age and sex groups; Figure S7: Associations between hypertension, impaired fasting glucose and combinations of birth weight and lifestyle stratified by age and sex groups; Figure S8: Associations between dyslipidemia, abdominal obesity and combinations of birth weight and lifestyle stratified by age and sex groups; Figure S9: Associations between clustered CMRFs and combinations of birth weight and lifestyle stratified by age and sex groups.

Author Contributions: Conceptualization, D.S. and Y.S.; data curation, J.D., N.M., Y.L., P.Z. and S.C.; formal analysis, D.S.; funding acquisition, Y.S.; investigation, Z.Z., Y.M. and J.M.; methodology, D.S., Y.D. and Y.S.; project administration, Y.D., Z.Z., Y.M., Y.S. and J.M.; resources, Y.D., Z.Z., Y.M., Y.S. and J.M.; software, D.S., J.D. and S.C.; supervision, Y.D., Y.S. and J.M.; validation, Y.D., Z.Z. and Y.S.; visualization, N.M., Y.L. and P.Z.; writing—original draft, D.S. and J.D.; writing—review and editing, N.M., Y.L., P.Z., S.C., Z.Z., Y.M., J.M., Y.D. and Y.S. All authors have read and agreed to the published version of the manuscript.

Funding: The present study was supported by the grants from Capital's Funds for Health Improvement and Research (2022-1G-4251 to Y.S.), the Natural Science Foundation of Beijing (Grant No. 7222247 to Y.S.; NO. 7222244 to Y.D.) and the National Natural Science Foundation (82103865 to Y.D.).

Institutional Review Board Statement: This study was conducted according to the guidelines laid down in the Declaration of Helsinki, and all procedures involving students were approved by the Ethics Committee of Peking University (NO. IRB0000105213034).

Informed Consent Statement: All participants and their parents voluntarily signed informed consent forms.

Data Availability Statement: The data supporting the conclusions of this article can be made available from the corresponding author upon request.

Acknowledgments: The authors gratefully acknowledge all the schoolteachers and doctors for their assistance in data collection and all students and their guardians for participating in the study.

Conflicts of Interest: The authors declare no conflict of interest.

References

1. Wang, Z.; Du, A.; Liu, H.; Wang, Z.; Hu, J. Systematic Analysis of the Global, Regional and National Burden of Cardiovascular from 1990 to 2017. *J. Epidemiol. Glob. Health* **2022**, *12*, 92–103. [CrossRef] [PubMed]
2. Lung Expert Panel on Integrated Guidelines for Cardiovascular Health and Risk Reduction in Children and Adolescents; National Heart, and Blood Institute. Expert panel on integrated guidelines for cardiovascular health and risk reduction in children and adolescents: Summary report. *Pediatrics* **2011**, *128* (Suppl. 125), S213–S256. [CrossRef]
3. Jackson, S.L.; Zhang, Z.; Wiltz, J.L.; Loustalot, F.; Ritchey, M.D.; Goodman, A.B.; Yang, Q. Hypertension among Youths—United States, 2001–2016. *MMWR Morb. Mortal. Wkly Rep.* **2018**, *67*, 758–762. [CrossRef] [PubMed]
4. Kit, B.K.; Kuklina, E.; Carroll, M.D.; Ostchega, Y.; Freedman, D.S.; Ogden, C.L. Prevalence of and Trends in Dyslipidemia and Blood Pressure among US Children and Adolescents, 1999–2012. *JAMA Pediatr.* **2015**, *169*, 272–279. [CrossRef] [PubMed]
5. May, A.L.; Kuklina, E.V.; Yoon, P.W. Prevalence of Cardiovascular Disease Risk Factors among US Adolescents, 1999–2008. *Pediatrics* **2012**, *129*, 1035–1041. [CrossRef]
6. Perak, A.M.; Ning, H.; Kit, B.K.; de Ferranti, S.D.; Van Horn, L.V.; Wilkins, J.T.; Lloyd-Jones, D.M. Trends in Levels of Lipids and Apolipoprotein B in US Youths Aged 6 to 19 Years, 1999–2016. *JAMA* **2019**, *321*, 1895–1905. [CrossRef]
7. Berenson, G.S.; Srinivasan, S.R.; Bao, W.; Newman, W.P., 3rd; Tracy, R.E.; Wattigney, W.A. Association between multiple cardiovascular risk factors and atherosclerosis in children and young adults. The Bogalusa Heart Study. *N. Engl. J. Med.* **1998**, *338*, 1650–1656. [CrossRef]

8. Juonala, M.; Viikari, J.S.; Kähönen, M.; Taittonen, L.; Laitinen, T.; Hutri-Kähönen, N.; Lehtimäki, T.; Jula, A.; Pietikäinen, M.; Jokinen, E.; et al. Life-time risk factors and progression of carotid atherosclerosis in young adults: The Cardiovascular Risk in Young Finns study. *Eur. Heart J.* **2010**, *31*, 1745–1751. [CrossRef]
9. Li, S.; Chen, W.; Srinivasan, S.R.; Bond, M.G.; Tang, R.; Urbina, E.M.; Berenson, G.S. Childhood cardiovascular risk factors and carotid vascular changes in adulthood: The Bogalusa Heart Study. *JAMA* **2003**, *290*, 2271–2276. [CrossRef]
10. Phillips, D.I.W.; Barker, D.J.P.; Hales, C.N.; Hirst, S.; Osmond, C. Thinness at birth and insulin resistance in adult life. *Diabetologia* **1994**, *37*, 150–154. [CrossRef]
11. Davies, A.A.; Smith, G.D.; May, M.T.; Ben-Shlomo, Y. Association between Birth Weight and Blood Pressure Is Robust, Amplifies with Age, and May Be Underestimated. *Hypertension* **2006**, *48*, 431–436. [CrossRef]
12. Barker, D.J.P.; Hales, C.N.; Fall, C.H.D.; Osmond, C.; Phipps, K.; Clark, P.M.S. Type 2 (non-insulin-dependent) diabetes mellitus, hypertension and hyperlipidaemia (syndrome X): Relation to reduced fetal growth. *Diabetologia* **1993**, *36*, 62–67. [CrossRef]
13. Stettler, N.; Stallings, V.A.; Troxel, A.B.; Zhao, J.; Schinnar, R.; Nelson, S.E.; Ziegler, E.E.; Strom, B.L. Weight gain in the first week of life and overweight in adulthood: A cohort study of European American subjects fed infant formula. *Circulation* **2005**, *111*, 1897–1903. [CrossRef]
14. Vohr, B.R.; Allan, W.; Katz, K.H.; Schneider, K.C.; Ment, L.R. Early predictors of hypertension in prematurely born adolescents. *Acta Paediatr.* **2010**, *99*, 1812–1818. [CrossRef]
15. Singhal, A.; Lucas, A. Early origins of cardiovascular disease: Is there a unifying hypothesis? *Lancet* **2004**, *363*, 1642–1645. [CrossRef]
16. Fonseca, M.J.; Severo, M.; A Lawlor, D.; Barros, H.; Santos, A. Direct and BMI-mediated effect of birthweight on childhood cardio-metabolic health—A birth cohort study. *Int. J. Obes.* **2019**, *43*, 1923–1931. [CrossRef]
17. Hovi, P.; Vohr, B.; Ment, L.R.; Doyle, L.W.; McGarvey, L.; Morrison, K.M.; Evensen, K.A.; van der Pal, S.; Grunau, R.E.; Brubakk, A.M.; et al. Blood Pressure in Young Adults Born at Very Low Birth Weight: Adults Born Preterm International Collaboration. *Hypertension* **2016**, *68*, 880–887. [CrossRef]
18. Sørensen, H.T.; Sabroe, S.; Rothman, K.J.; Gillman, M.; Fischer, P.; Sørensen, T.I. Relation between weight and length at birth and body mass index in young adulthood: Cohort study. *BMJ* **1997**, *315*, 1137. [CrossRef]
19. Seidman, D.S.; Laor, A.; Gale, R.; Stevenson, D.K.; Danon, Y.L. A longitudinal study of birth weight and being overweight in late adolescence. *Am. J. Dis. Child.* **1991**, *145*, 782–785. [CrossRef]
20. Darendeliler, F.; Poyrazoglu, S.; Sancakli, O.; Bas, F.; Gokcay, G.; Aki, S.; Eskiyurt, N. Adiponectin is an indicator of insulin resistance in non-obese prepubertal children born large for gestational age (LGA) and is affected by birth weight. *Clin. Endocrinol.* **2009**, *70*, 710–716. [CrossRef]
21. Giapros, V.; Evagelidou, E.; Challa, A.; Kiortsis, D.; Drougia, A.; Andronikou, S. Serum adiponectin and leptin levels and insulin resistance in children born large for gestational age are affected by the degree of overweight. *Clin. Endocrinol.* **2007**, *66*, 353–359.
22. Dong, B.; Dong, Y.-H.; Yang, Z.-G.; Wang, X.-J.; Zou, Z.-Y.; Wang, Z.; Ma, J. Healthy Body Weight may Modify Effect of Abnormal Birth Weight on Metabolic Syndrome in Adolescents. *Obesity* **2019**, *27*, 462–469. [CrossRef]
23. Fan, J.; Shi, X.; Jia, X.; Wang, Y.; Zhao, Y.; Bao, J.; Zhang, H.; Yang, Y. Birth weight, childhood obesity and risk of hypertension: A Mendelian randomization study. *J. Hypertens.* **2021**, *39*, 1876–1883. [CrossRef]
24. Hirschler, V.; Edit, S.; Miorin, C.; Guntsche, Z.; Maldonado, N.; Garcia, C.; Lapertosa, S.; Gonzalez, C.D.; Calzia, V.; Di Firma, R.; et al. Association between High Birth Weight and Later Central Obesity in 9-Year-Old Schoolchildren. *Metab. Syndr. Relat. Disord.* **2021**, *19*, 213–217. [CrossRef]
25. Kuciene, R.; Dulskiene, V.; Medzioniene, J. Associations between high birth weight, being large for gestational age, and high blood pressure among adolescents: A cross-sectional study. *Eur. J. Nutr.* **2018**, *57*, 373–381. [CrossRef]
26. Zhang, Y.; Li, H.; Liu, S.-J.; Fu, G.-J.; Zhao, Y.; Xie, Y.J.; Zhang, Y.; Wang, Y.-X. The associations of high birth weight with blood pressure and hypertension in later life: A systematic review and meta-analysis. *Hypertens. Res.* **2013**, *36*, 725–735. [CrossRef]
27. Pocobelli, G.; Dublin, S.; Enquobahrie, D.A.; Mueller, B.A. Birth Weight and Birth Weight for Gestational Age in Relation to Risk of Hospitalization with Primary Hypertension in Children and Young Adults. *Matern. Child Health J.* **2016**, *20*, 1415–1423. [CrossRef]
28. Luetic, G.G.; Menichini, M.L.; Deri, N.; Steinberg, J.; Carrá, A.; Cristiano, E.; Patrucco, L.; Curbelo, M.C.; Rojas, J.I. High birth weight and risk of multiple sclerosis: A multicentre study in Argentina. *Mult. Scler. Relat. Disord.* **2021**, *47*, 102628. [CrossRef]
29. Ekelund, U.; Luan, J.; Sherar, L.B.; Esliger, D.W.; Griew, P.; Cooper, A. Moderate to vigorous physical activity and sedentary time and cardiometabolic risk factors in children and adolescents. *JAMA* **2012**, *307*, 704–712. [CrossRef]
30. Danielsen, Y.; Júlíusson, P.; Nordhus, I.; Kleiven, M.; Meltzer, H.; Olsson, S.; Pallesen, S. The relationship between life-style and cardio-metabolic risk indicators in children: The importance of screen time. *Acta Paediatr.* **2010**, *100*, 253–259. [CrossRef]
31. Owens, J.A.; Weiss, M.R. Insufficient sleep in adolescents: Causes and consequences. *Minerva Pediatr.* **2017**, *69*, 326–336. [CrossRef]
32. Shen, L.; Wang, J.; Duan, Y.; Yang, Z. Prevalence of low birth weight and macrosomia estimates based on heaping adjustment method in China. *Sci. Rep.* **2021**, *11*, 15016. [CrossRef] [PubMed]
33. Chen, P.; Wang, D.; Shen, H.; Yu, L.; Gao, Q.; Mao, L.; Jiang, F.; Luo, Y.; Xie, M.; Zhang, Y.; et al. Physical activity and health in Chinese children and adolescents: Expert consensus statement (2020). *Br. J. Sports Med.* **2020**, *54*, 1321–1331. [CrossRef] [PubMed]
34. Hildebrand, M.; Øglund, G.P.; Wells, J.; Ekelund, U. Prenatal, birth and early life predictors of sedentary behavior in young people: A systematic review. *Int. J. Behav. Nutr. Phys. Act.* **2016**, *13*, 63. [CrossRef] [PubMed]

35. Ridgway, C.; Brage, S.; Sharp, S.J.; Corder, K.; Westgate, K.; van Sluijs, E.; Goodyer, I.M.; Hallal, P.C.; Anderssen, S.A.; Sardinha, L.; et al. Does Birth Weight Influence Physical Activity in Youth? A Combined Analysis of Four Studies Using Objectively Measured Physical Activity. *PLoS ONE* **2011**, *6*, e16125. [CrossRef] [PubMed]

36. Chen, Y.; Ma, L.; Ma, Y.; Wang, H.; Luo, J.; Zhang, X.; Luo, C.; Wang, H.; Zhao, H.; Pan, D.; et al. A national school-based health lifestyles interventions among Chinese children and adolescents against obesity: Rationale, design and methodology of a randomized controlled trial in China. *BMC Public Health* **2015**, *15*, 210. [CrossRef]

37. Barth, J.R.; William, H.; Jackson, R. Macrosomia: ACOG Practice Bulletin, Number 216. *Obstet. Gynecol.* **2020**, *135*, e118–e135.

38. Dang, J.; Chen, T.; Ma, N.; Liu, Y.; Zhong, P.; Shi, D.; Dong, Y.; Zou, Z.; Ma, Y.; Song, Y.; et al. Associations between Breastfeeding Duration and Obesity Phenotypes and the Offsetting Effect of a Healthy Lifestyle. *Nutrients* **2022**, *14*, 1999. [CrossRef]

39. Hirshkowitz, M.; Whiton, K.; Albert, S.M.; Alessi, C.; Bruni, O.; DonCarlos, L.; Hazen, N.; Herman, J.; Adams Hillard, P.J.; Katz, E.S.; et al. National Sleep Foundation's updated sleep duration recommendations: Final report. *Sleep Health.* **2015**, *1*, 233–243. [CrossRef]

40. Flynn, J.T.; Kaelber, D.C.; Baker-Smith, C.M.; Blowey, D.; Carroll, A.E.; Daniels, S.R.; de Ferranti, S.D.; Dionne, J.M.; Falkner, B.; Flinn, S.K.; et al. Clinical Practice Guideline for Screening and Management of High Blood Pressure in Children and Adolescents. *Pediatrics* **2017**, *140*, e20171904. [CrossRef]

41. Expert Committee on the Diagnosis and Classification of Diabetes Mellitus. Report of the expert committee on the diagnosis and classification of diabetes mellitus. *Diabetes Care* **2003**, *26* (Suppl. 21), S5–S20. [CrossRef]

42. National Health Commission of the People's Republic of China. Boundary Value of High Waist Circumference Screening for Children and Adolescents Aged 7–18 Years. Available online: http://www.nhc.gov.cn/wjw/pqt/201807/417de6982ab8493b9 1aba925b51a8a19.shtml (accessed on 13 June 2018).

43. Dou, Y.; Jiang, Y.; Yan, Y.; Chen, H.; Zhang, Y.; Chen, X.; Wang, Y.; Cheng, H.; Zhao, X.; Hou, D.; et al. Waist-to-height ratio as a screening tool for cardiometabolic risk in children and adolescents: A nationwide cross-sectional study in China. *BMJ Open* **2020**, *10*, e037040. [CrossRef]

44. Fernández-Alvira, J.M.; Mouratidou, T.; Bammann, K.; Hebestreit, A.; Barba, G.; Sieri, S.; Reisch, L.; Eiben, G.; Hadjigeorgiou, C.; Kovacs, E.; et al. Parental education and frequency of food consumption in European children: The IDEFICS study. *Public Health Nutr.* **2013**, *16*, 487–498. [CrossRef]

45. Renom Espineira, A.; Fernandes-Rosa, F.L.; Bueno, A.C.; de Souza, R.M.; Moreira, A.C.; de Castro, M.; Barbieri, M.A.; Bettiol, H.; Antonini, S.R. Postnatal growth and cardiometabolic profile in young adults born large for gestational age. *Clin. Endocrinol.* **2011**, *75*, 335–341. [CrossRef]

46. Araújo, J.; Severo, M.; Barros, H.; Mishra, G.D.; Guimarães, J.T.; Ramos, E. Developmental trajectories of adiposity from birth until early adulthood and association with cardiometabolic risk factors. *Int. J. Obes.* **2015**, *39*, 1443–1449. [CrossRef]

47. Ornoy, A. Growth and neurodevelopmental outcome of children born to mothers with pregestational and gestational diabetes. *Pediatr Endocrinol. Rev.* **2005**, *3*, 104–113.

48. Ornoy, A.; Wolf, A.; Ratzon, N.; Greenbaum, C.; Dulitzky, M. Neurodevelopmental outcome at early school age of children born to mothers with gestational diabetes. *Arch. Dis. Child Fetal Neonatal Ed.* **1999**, *81*, F10–F14. [CrossRef]

49. Silverman, B.L.; Rizzo, T.A.; Cho, N.H.; Metzger, B.E. Long-term effects of the intrauterine environment. The Northwestern University Diabetes in Pregnancy Center. *Diabetes Care* **1998**, *21*, B142.

50. DeVader, S.R.; Neeley, H.L.; Myles, T.D.; Leet, T.L. Evaluation of gestational weight gain guidelines for women with normal prepregnancy body mass index. *Obstet. Gynecol.* **2007**, *110*, 745–751. [CrossRef]

51. Groenendaal, F.; Elferink-Stinkens, P.M.; Registry, T.N.P. Hypoglycaemia and seizures in large-for-gestational-age (LGA) full-term neonates. *Acta Paediatr.* **2007**, *95*, 874–876. [CrossRef]

52. Schaefer-Graf, U.M.; Rossi, R.; Bührer, C.; Siebert, G.; Kjos, S.L.; Dudenhausen, J.W.; Vetter, K. Rate and risk factors of hypoglycemia in large-for-gestational-age newborn infants of nondiabetic mothers. *Am. J. Obstet. Gynecol.* **2002**, *187*, 913–917. [CrossRef] [PubMed]

53. Touger, L.; Looker, H.C.; Krakoff, J.; Lindsay, R.S.; Cook, V.; Knowler, W.C. Early growth in offspring of diabetic mothers. *Diabetes Care* **2005**, *28*, 585–589. [CrossRef] [PubMed]

54. Clausen, T.D.; Mathiesen, E.R.; Hansen, T.; Pedersen, O.; Jensen, D.M.; Lauenborg, J.; Schmidt, L.; Damm, P. Overweight and the Metabolic Syndrome in Adult Offspring of Women with Diet-Treated Gestational Diabetes Mellitus or Type 1 Diabetes. *J. Clin. Endocrinol. Metab.* **2009**, *94*, 2464–2470. [CrossRef] [PubMed]

55. Chiavaroli, V.; Marcovecchio, M.L.; De Giorgis, T.; Diesse, L.; Chiarelli, F.; Mohn, A. Progression of Cardio-Metabolic Risk Factors in Subjects Born Small and Large for Gestational Age. *PLoS ONE* **2014**, *9*, e104278. [CrossRef]

56. Raaijmakers, A.; Zhang, Z.Y.; Claessens, J.; Cauwenberghs, N.; van Tienoven, T.P.; Wei, F.F.; Jacobs, L.; Levtchenko, E.; Pauwels, S.; Kuznetsova, T.; et al. Does Extremely Low Birth Weight Predispose to Low-Renin Hypertension? *Hypertension* **2017**, *69*, 443–449. [CrossRef]

57. Hoy, W.E.; Hughson, M.D.; Bertram, J.F.; Douglas-Denton, R.; Amann, K. Nephron number, hypertension, renal disease, and renal failure. *J. Am. Soc. Nephrol.* **2005**, *16*, 2557–2564. [CrossRef]

58. Samuel, T.; Hoy, W.E.; Douglas-Denton, R.; Hughson, M.D.; Bertram, J.F. Determinants of Glomerular Volume in Different Cortical Zones of the Human Kidney. *J. Am. Soc. Nephrol.* **2005**, *16*, 3102–3109. [CrossRef]

59. Lopes, A.A.S.; Port, F.K. The low birth weight hypothesis as a plausible explanation for the black/white differences in hypertension, non-insulin-dependent diabetes, and end-stage renal disease. *Am. J. Kidney Dis.* **1995**, *25*, 350–356. [CrossRef]

60. Eriksson, J.; Forsén, T.; Tuomilehto, J.; Osmond, C.; Barker, D. Fetal and Childhood Growth and Hypertension in Adult Life. *Hypertension* **2000**, *36*, 790–794. [CrossRef]
61. Gomes, K.B.A.; Leal, V.S.; Oliveira, J.S.; Pereira, C.G.D.S.; Gonçalves, F.C.L.D.S.P.; Andrade, I.S.; Eickmann, S.H.; Lira, P.I.C.; Lima, M.C. Birth weight and overweight in adolescents: The erica project in the city of recife, pernambuco. *Rev. Paul. Pediatr.* **2021**, *39*, e2019380. [CrossRef]
62. Skilton, M.R.; Siitonen, N.; Würtz, P.; Viikari, J.S.; Juonala, M.; Seppälä, I.; Laitinen, T.; Lehtimäki, T.; Taittonen, L.; Kähönen, M.; et al. High birth weight is associated with obesity and increased carotid wall thickness in young adults: The cardiovascular risk in young Finns study. *Arterioscler. Thromb Vasc. Biol.* **2014**, *34*, 1064–1068. [CrossRef]
63. Andersen, L.B.; Riddoch, C.; Kriemler, S.; Hills, A.P. Physical activity and cardiovascular risk factors in children. *Br. J. Sports Med.* **2011**, *45*, 871–876. [CrossRef]
64. Kalea, A.Z.; Drosatos, K.; Buxton, J.L. Nutriepigenetics and cardiovascular disease. *Curr. Opin. Clin. Nutr. Metab. Care* **2018**, *21*, 252–259. [CrossRef]
65. Lissak, G. Adverse physiological and psychological effects of screen time on children and adolescents: Literature review and case study. *Environ. Res.* **2018**, *164*, 149–157. [CrossRef]
66. Qiao, Y.; Zhang, T.; Liu, H.; Katzmarzyk, P.T.; Chaput, J.-P.; Fogelholm, M.; Johnson, W.D.; Kuriyan, R.; Kurpad, A.; Lambert, E.; et al. Joint association of birth weight and physical activity/sedentary behavior with obesity in children ages 9–11 years from 12 countries. *Obesity* **2017**, *25*, 1091–1097. [CrossRef]
67. Deng, J.R.; Tan, W.Q.; Yang, S.Y.; Ao, L.P.; Liang, J.P.; Li, L.X.; Gao, Y.H.; Yang, Y.; Liu, L. High birth weight and its interaction with physical activity influence the risk of obesity in early school-aged children. *World J. Pediatr.* **2020**, *16*, 385–392. [CrossRef]
68. Baker, J.L.; Olsen, L.W.; Sørensen, T.I. Childhood body mass index and the risk of coronary heart disease in adulthood. *Ugeskr Laeger* **2008**, *170*, 2434–2437. [CrossRef]
69. Skinner, A.C.; Perrin, E.M.; Moss, L.A.; Skelton, J.A. Cardiometabolic Risks and Severity of Obesity in Children and Young Adults. *N. Engl. J. Med.* **2015**, *373*, 1307–1317. [CrossRef]
70. Liu, C.; Wu, S.; Pan, X. Clustering of cardio-metabolic risk factors and pre-diabetes among U.S. adolescents. *Sci. Rep.* **2021**, *11*, 5015. [CrossRef]
71. Wang, L.; Zhou, B.; Zhao, Z.; Yang, L.; Zhang, M.; Jiang, Y.; Li, Y.; Zhou, M.; Wang, L.; Huang, Z.; et al. Body-mass index and obesity in urban and rural China: Findings from consecutive nationally representative surveys during 2004–18. *Lancet* **2021**, *398*, 53–63. [CrossRef]

 *nutrients*

Article

Determinants of Longitudinal Changes in Cardiometabolic Risk in Adolescents with Overweight/Obesity: The EVASYON Study

Miguel Martín-Matillas [1], Dinalrilan Rocha-Silva [2], Abel Plaza-Florido [1,*], Manuel Delgado-Fernández [3], Amelia Marti [4,5,6], Pilar De Miguel-Etayo [7,8], Luis A. Moreno [7,8], Ascensión Marcos [9] and Cristina Campoy [10,11,12,13,*] on behalf of the EVASYON Study Group

[1] PROFITH "PROmoting FITness and Health through Physical Activity" Research Group, Sport and Health University Research Institute (iMUDS), Department of Physical Education and Sports, Faculty of Sport Sciences, University of Granada, 18071 Granada, Spain

[2] Department of Physical Education and Sports, Faculty of Sport Sciences, University of Granada, 18071 Granada, Spain

[3] PA-HELP "Physical Activity for Health Promotion, CTS-1018" Research Group, Sport and Health University Research Institute (iMUDS), Department of Physical Education and Sports, Faculty of Sport Sciences, University of Granada, 18071 Granada, Spain

[4] Department of Nutrition, Food Sciences and Physiology, University of Navarra, 31008 Pamplona, Spain

[5] Navarra Institute for Health Research (IdiSNA), 31008 Pamplona, Spain

[6] Physiopathology of Obesity and Nutrition Networking Biomedical Research Centre (CIBERobn CB12/03/30002), Spanish National Institute of Health Carlos III, 28029 Madrid, Spain

[7] Physiopathology of Obesity and Nutrition Networking Biomedical Research Centre (CIBERobn CB15/00043), Spanish National Institute of Health Carlos III, 28029 Madrid, Spain

[8] Growth, Exercise, Nutrition and Development (GENUD) Research Group, Faculty de Ciencias de la Salud, University de Zaragoza, Institute Agroalimentario de Aragón (IA2) and Institute de Investigación Sanitaria de Aragón (IIS Aragón), 50009 Zaragoza, Spain

[9] Department of Metabolism and Nutrition, Institute of Food Science, Technology and Nutrition (ICTAN), Spanish National Research Council (CSIC), 28040 Madrid, Spain

[10] Department of Pediatrics, School of Medicine, University of Granada, 18016 Granada, Spain

[11] EURISTIKOS Excellence Centre for Pediatric Research, Biomedical Research Centre, University of Granada, 18071 Granada, Spain

[12] Spanish Network of Biomedical Research in Epidemiology and Public Health (CIBERESP), Granada's Node, Institute of Health Carlos III, 28029 Madrid, Spain

[13] Granada's Biosanitary Institute (Ibs-Granada), 18012 Granada, Spain

* Correspondence: abeladrian@ugr.es (A.P.-F.); ccampoy@ugr.es (C.C.)

Citation: Martín-Matillas, M.; Rocha-Silva, D.; Plaza-Florido, A.; Delgado-Fernández, M.; Marti, A.; De Miguel-Etayo, P.; Moreno, L.A.; Marcos, A.; Campoy, C. Determinants of Longitudinal Changes in Cardiometabolic Risk in Adolescents with Overweight/Obesity: The EVASYON Study. *Nutrients* 2022, 14, 3241. https://doi.org/10.3390/nu14153241

Academic Editor: Zhiyong Zou

Received: 14 July 2022
Accepted: 5 August 2022
Published: 8 August 2022

Publisher's Note: MDPI stays neutral with regard to jurisdictional claims in published maps and institutional affiliations.

Copyright: © 2022 by the authors. Licensee MDPI, Basel, Switzerland. This article is an open access article distributed under the terms and conditions of the Creative Commons Attribution (CC BY) license (https://creativecommons.org/licenses/by/4.0/).

Abstract: We investigated which determinants (socioeconomic, early life factors, body composition changes, fitness changes and/or physical activity changes) best predicted longitudinal outcomes in cardiometabolic risk profile (Z-score change) in adolescents with OW/OB who underwent a 13-month multidisciplinary lifestyle intervention. A total of 165 adolescents (13–16 y; 46% boys) from the EVASYON study were included. Socioeconomic variables and early life factors were obtained from the medical records. Body composition was assessed using anthropometry. Fitness and physical activity were measured with field-based tests and questionnaires. Cardiometabolic risk factors (fasting glucose, HDL cholesterol, triglycerides, blood pressure and waist circumference) were derived from standard methods in the hospital. Body weight changes, sex and mother's education were selected in the stepwise process as the most important determinants of changes in cardiometabolic risk profile ($R^2 = 0.26$, $p = 0.002$; $R^2 = 0.14$, $p = 0.013$; and $R^2 = 0.14$, $p = 0.017$, respectively). Both boys and girls showed a lower cardiometabolic risk score with the reduction in body weight ($r = 0.535$, $p = 0.009$ and $r = 0.506$, $p = 0.005$, respectively). There was no interaction between sex and body weight change ($p = 0.614$). In conclusion, the simple measure of changes in body weight should be considered to track changes in cardiometabolic risk profile in adolescents with OW/OB.

Keywords: adolescents; parents' educational level; parents' occupational level; birthweight; breastfeeding; adiposity; cardiorespiratory fitness; cholesterol; blood pressure; lifestyle intervention

1. Introduction

Overweight/obesity (OW/OB) is an important public health concern. The worldwide prevalence of OB nearly tripled between 1975 and 2016. Over 340 million children and adolescents aged 5–19 were OW or OB in 2016. Common health consequences related to OW/OB are cardiovascular diseases (the leading cause of death in 2012) and other co-morbidities, such as diabetes, musculoskeletal disorders and some types of cancer [1]. Some estimations show that more than 90 million children and adolescents might have OW or OB in 2025 [2,3]. Importantly, OW/OB is positively associated with cardiometabolic risk factors (e.g., metabolic syndrome (MetS) markers) [4–7] and a higher risk of developing cardiovascular disease later in life [8]. In this regard, socioeconomic status and several health-related indicators, such as early life factors, body composition, fitness and physical activity (PA), have been related to cardiometabolic risk profile (using MetS markers) in children and adolescents with OW/OB [9–12].

Lifestyle interventions lasting from 3 to 12 months (i.e., mainly based on nutritional education and increase in PA levels) have been proposed to improve the cardiometabolic risk profile in children and adolescents with OW/OB [13–15]. In this regard, the EVASYON study implemented a 13-month multidisciplinary lifestyle intervention in adolescents with OW/OB aged 13 to 16 years, recruited from five hospitals—pediatric units—in five Spanish cities [16]. Previous studies derived from the EVASYON study reported a reduction in body weight, fat mass and individual cardiometabolic risk factors (e.g., blood pressure) after 13 months of a lifestyle intervention [17–19]. However, to the best of our knowledge, the best predictor (e.g., socioeconomic status, early life factors, changes in body composition, fitness and PA) of longitudinal changes in the cardiometabolic risk score profile (using MetS markers), induced by a long-term multidisciplinary lifestyle intervention in adolescents with OW/OB, is poorly understood.

Therefore, in this study, we aimed to identify the socioeconomic, early life factors, body composition, fitness and PA determinants of longitudinal changes in the cardiometabolic risk score after 13 months of multidisciplinary lifestyle intervention in adolescents with OW/OB.

2. Materials and Methods

2.1. Participants and Study Design

A total of 165 adolescents (76 boys and 89 girls) were included in the data analyses. The current study falls under the umbrella of the EVASYON study, which is a multidisciplinary program implemented in adolescents with OW/OB aged 13 to 16 years, recruited from five hospitals—pediatric units—in five Spanish cities (Granada, Madrid, Pamplona, Santander and Zaragoza). The EVASYON study was conducted by different teams of multidisciplinary professionals from each hospital with small groups, from 9 to 11 adolescents, over 13 months, including twenty visits within two specific phases: (1) intensive intervention phase (9 weekly visits) and (2) extensive intervention phase (11 monthly visits). Detailed description of the EVASYON study protocol was described elsewhere [16]. Briefly, the intervention included a personalized balanced diet, a PA program and psychological support within the family. Adolescents were instructed in several motivational strategies, life and time management strategies, including PA recommendations or sleep time, nutritional advice and family involvement. All measurements were performed at baseline (0 months), at the end of the intensive intervention (2 months) and at the end of the treatment program (13 months).

Before starting the EVASYON intervention program, all candidates were screened. The inclusion criteria were: (i) 13–16 years old, (ii) OW or OB in agreement with the International Obesity Task Force age- and sex-specific body mass index (BMI, kg/m^2) [20], (iii) Spanish or educated in Spain and (iv) no other diagnosed disease. The exclusion criterion was undergoing any pharmacological treatment.

The project followed the ethical standards recognized by the Declaration of Helsinki (reviewed in Hong Kong in September 1989 and in Edinburgh in 2000) and the EEC Good

Clinical Practice recommendations (document 111/3976/88, July 1990) and current Spanish legislation regulating clinical research in humans (Royal Decree 561/1993 on clinical trials). The study was approved by the Ethics Committee of each hospital that participated in the EVASYON study and by the Ethics Committee of the Spanish Council for Scientific Research (CSIC). The study was explained to all the participants before starting, and the volunteers, parents or tutors signed an informed consent form. Their participation was completely voluntary; they did not receive any rewards.

2.2. Socioeconomic Variables

Parental education (father and mother) and parental occupational levels (father and mother) were extracted from the medical reports that were completed through parents' interview. Parental education refers to the academic level indicating primary school, high school or university studies. Both parents were asked to answer a question concerning their current occupation according to the recommendations of the Spanish Society of Epidemiology [21]. Three categories of parental occupational levels were established: unskilled worker/unemployed, skilled worker and managerial, referred to as low, medium and high occupational level, respectively.

2.3. Early Life Factors

Birthweight (kg), size at birth (cm) and breastfeeding (none, less than 3 months, 4–6 months, 7–9 months, more than 9 months) were reported by mothers through medical interviews.

2.4. Body Composition Parameters

The method used for anthropometric measurements was the standardized protocol of the AVENA study [22]. The same trained researchers always obtained the measurements. Each measurement was performed three times non-consecutively, i.e., a complete set of measurements was performed and then repeated twice more. Weight and height were obtained by standardized procedures. Body mass index (BMI) was calculated as body weight (kg) divided by squared height (m^2), classifying the adolescents as overweight or obese [23]. The classification of obesity type according to age and sex (i.e., Overweight, Obesity type I, Obesity type II and Obesity type III) was calculated with a patented Excel-based tool [24], using the cut-off points defined by Bervoets et al. [25]. Peak height velocity (PHV) was calculated from sex, age and height using Moore et al.'s equations [26].

Skinfold thicknesses were measured to the nearest 0.1 mm on the left side of the body using a skinfold caliper (Holtain Caliper; Holtain Ltd., Wales, UK) at the following sites: (1) triceps, halfway between the acromion process and the olecranon process; (2) biceps, at the same level as the triceps skinfold, directly above the center of the cubital fossa; (3) subscapular, about 20 mm below the tip of the scapula, at an angle of 45 degrees to the lateral side of the body; (4) suprailiac, about 20 mm above the iliac crest and 20 mm toward the medial line; (5) thigh, in the midline of the anterior aspect of the thigh, midway between the inguinal crease and the proximal border of the patella; (6) calf, at the level of maximum calf circumference, on the medial aspect of the calf. The sum of 4 skinfolds included the biceps, triceps, subscapular and suprailiac, and the sum of 6 skinfolds included all the skinfolds measured [22].

Fat mass (FM) in kg and fat free mass (FFM) in kg were calculated from the adolescent's total body weight and the percentage of total body fat that was obtained using the equations reported by Slaughter et al. for children and adolescents [27].

2.5. Fitness and Physical Activity

2.5.1. Cardiorespiratory Fitness

Cardiorespiratory fitness (CRF) was assessed using the progressive 20 m shuttle run test published by Léger and Lambert in 1982 and revised in 1988 [28], where participants run as long as possible back and forth across a 20 m space at a specified music protocol that becomes 0.5 km/h faster each minute or stage. The last 0.5 stage completed is the individual score, and VO$_2$max was estimated with the Léger equation (i.e., CRF relative to body weight). In addition, CRF relative to FFM was calculated from individual absolute oxygen consumption divided by FFM.

2.5.2. Muscular Strength

A handgrip strength test was performed with the adolescents in a standard bipedal position and with the arm in complete extension without touching any part of the body with the hand dynamometer (TKK 5101; Takei, Tokyo, Japan). The dynamometer was adjusted for sex and hand size for each adolescent. The test was performed twice using both hands alternatively, allowing short rest between the measures. The final score is the average of the two hands, taking the maximum score from each attempt [29]. The handgrip strength test provides information about the maximal isometric strength that can be generated mainly by the hand and the arm. Relative handgrip strength (kg/kg) was calculated by dividing the original handgrip strength by the adolescents' total body weight. Standing long jump (cm) was performed in the standing position; the adolescent jumps as far as possible, trying to land with both feet together. The score is the maximum distance between the last heel mark and the take-off line of two attempts. The standing broad jump assesses lower-limb explosive strength.

2.5.3. Physical Activity

PA levels were obtained from the Physical Activity Questionnaire for Adolescents (PAQ-A) [30]. The adolescents were asked about their frequency of PA during the week and weekend days and to compute the score ranging from 1 to 5 points, 1 being the lowest level and 5 the highest level of PA.

2.6. Cardiometabolic Risk Factors
2.6.1. Fasting Glucose, Triglycerides and High-Density Lipoprotein

Fasting blood samples were taken in the morning (after a 10 h overnight fast); the adolescents went to the hospital for blood sampling between at 8 and 10 a.m. Blood samples were collected by puncture of the cubital vein (21.5 mL). For biochemical analyses, serum was separated from plasma by centrifugation, divided into aliquots and frozen and stored at −80 °C until analysis. Fasting glucose, triglycerides and high-density lipoproteins cholesterol (HDLc) were analyzed by the biochemical auto analyzer Olympus model AU2700 (Olympus, Melville, NY, USA). All serum analyses were performed at the end of the study in order to have the three samples from each subject analyzed in the same run to avoid systematic errors. The coefficients of variance of the lipid variables were 3% for triglycerides and 2% for HDLc [31].

2.6.2. Blood Pressure

Resting blood pressure was obtained using the left arm and measured using a validated digital automatic blood pressure monitor (OMRON M6, OMRON HEALTH CARE Co., Ltd., Kyoto, Japan) according to the International Protocol of the European Society of Hypertension [32]. For this study, the mean arterial blood pressure (MAP) in mmHg was used: MAP = Diastolic Blood Pressure + [0.333 × (Systolic Blood Pressure − Diastolic Blood Pressure)], according to the National High Blood Pressure Education Program Working Group on High Blood Pressure in Children and Adolescents [33].

2.6.3. Waist Circumference

The waist circumference (WC) was measured with an inelastic tape to the nearest millimeter, with the subject upright; the measuring tape was applied horizontally midway between the lowest rib margin and the iliac crest, at the end of a gentle exhalation [22].

2.6.4. Cardiometabolic Risk Score Using MetS

The cardiometabolic risk Z-score was calculated using the standardized Z-scores for each cardiometabolic risk factor at different time points, as follows: (Z-score fasting glucose + Z-score triglycerides + Z-score reverted HDLc + Z-score MAP + Z-score WC)/5. Higher scores indicate greater cardiometabolic risk [34].

2.7. Statistical Analysis

The participants' descriptive data are presented as means $\pm$ standard deviation or frequency and percentages for continuous and categorical variables, respectively. Student's *t*-tests and chi-squared tests were performed to investigate baseline differences for continuous and categorical variables, respectively, between boys and girls. Stepwise linear regression was performed to test the strongest potential determinants (i.e., socioeconomic, early life factors, body composition changes, fitness changes and PA changes) of the changes in the cardiometabolic risk score (Z-score change). The first model included changes in the above-mentioned variables in the intensive intervention phase. In contrast, the second model included changes in variables in the extensive intervention phase. All analyses were conducted using the Statistical Package for Social Sciences (SPSS, v. 21.0, IBM SPSS Statistics, IBM Corporation, Armonk, NY, USA). The significance level was set at 0.05. Figures were depicted using the GraphPad Prism software for Windows (version 5.0.0, GraphPad Software, San Diego, CA, USA).

3. Results

The descriptive characteristics of the participants at baseline are shown in Table 1. The study sample presented a chronological age of 14.48 $\pm$ 1.23 years and a biological maturation age (PHV) of 1.57 $\pm$ 1.21 years, with 54% girls. The adolescents' prevalence according to parental education was 37% for primary school for both parents, 37% for high school for fathers and 39% for mothers, and for university studies, 26% and 24% for fathers and mothers, respectively. With respect to the occupational level, fathers and mothers showed higher percentage in the lower professional level (>40%) than in the medium (>35%) or higher level (>16%). At baseline, obesity type I was the prevalent degree of obesity (44%) in the study sample, followed by overweight and obesity type II (25% and 21%, respectively), and finally, obesity type III (10%). Boys presented higher cardiorespiratory and muscular fitness, body weight, birth weight and adiposity than girls ($p < 0.05$). Girls reported higher PHV and fat free mass percentage than boys ($p < 0.05$).

Table 1. Descriptive characteristics of the participants at baseline.

Variables	All Samples		Boys		Girls		*p*-Value
Socioeconomic Variables							
Sex [n (%)]	165	(100%)	76	(46%)	89	(54%)	NA
Age (years)	14.48 $\pm$	1.23	14.40 $\pm$	1.19	14.55 $\pm$	1.25	0.42
PHV (years)	1.57 $\pm$	1.21	0.79 $\pm$	0.99	2.22 $\pm$	0.97	**<0.01**
Father's education							0.64
Primary School [n (%)]	56	(37%)	27	(38%)	29	(36%)	
High School [n (%)]	55	(37%)	23	(33%)	32	(40%)	
University [n (%)]	39	(26%)	20	(28%)	19	(24%)	
Mother's education							0.93
Primary School [n (%)]	54	(37%)	25	(35%)	29	(38%)	
High School [n (%)]	57	(39%)	18	(39%)	29	(38%)	
University [n (%)]	36	(24%)	28	(25%)	18	(24%)	
Father's occupational level							0.55
Low [n (%)]	61	(41%)	27	(40%)	34	(42%)	
Medium [n (%)]	51	(35%)	21	(31%)	30	(37%)	
High [n (%)]	36	(24%)	19	(28%)	17	(21%)	

Table 1. *Cont.*

Variables	All Samples			Boys			Girls			*p*-Value
Socioeconomic Variables										
Mother's occupational level										0.72
Low [n (%)]	76		(49%)	33		(46%)	43		(52%)	
Medium [n (%)]	54		(35%)	26		(36%)	28		(34%)	
High [n (%)]	25		(16%)	13		(18%)	12		(14%)	
Early life factors										
Birth weight (kg)	3.35	±	0.54	3.49	±	0.61	3.22	±	0.45	**<0.01**
Size at birth (cm)	50.65	±	2.65	50.78	±	2.57	50.52	±	2.74	0.57
Breastfeeding										0.71
None [n (%)]	20		(14%)	12		(17%)	8		(10%)	
<3 months [n (%)]	69		(47%)	31		(46%)	38		(48%)	
4–6 months [n (%)]	36		(24%)	15		(22%)	21		(26%)	
7–9 months [n (%)]	12		(8%)	6		(9%)	6		(8%)	
>9 months [n (%)]	10		(7%)	4		(6%)	6		(8%)	
Body composition parameters										
Weight (kg)	86.23	±	17.12	90.36	±	17.16	82.77	±	16.40	**<0.01**
Height (cm)	164.38	±	7.95	167.90	±	7.87	161.44	±	6.78	**<0.01**
BMI (kg/m^2)	31.76	±	5.07	31.87	±	4.73	31.66	±	5.37	0.80
Obesity types										0.43
Overweight [n (%)]	39		(25%)	14		(20%)	25		(30%)	
Obesity type I [n (%)]	68		(44%)	33		(47%)	35		(41%)	
Obesity type II [n (%)]	33		(21%)	18		(25%)	15		(18%)	
Obesity type III [n (%)]	15		(10%)	6		(8%)	9		(11%)	
Sum of 4 skinfolds (mm)	112.11	±	19.89	112.28	±	18.68	111.97	±	20.96	0.92
Sum of 6 skinfolds (mm)	186.44	±	25.30	186.21	±	24.54	186.63	±	26.05	0.91
Fat mass (kg)	28.24	±	7.97	33.36	±	8.28	23.92	±	4.32	**<0.01**
Fat mass (%)	32.57	±	5.19	36.62	±	3.94	29.15	±	3.32	**<0.01**
Fat free mass (kg)	58.05	±	12.03	57.01	±	10.01	58.93	±	13.50	0.32
Fat free mass (%)	67.43	±	5.18	63.38	±	3.94	70.84	±	3.31	**<0.01**
Fitness and physical activity										
CRF (Stages)	2.90	±	1.45	3.36	±	1.54	2.49	±	1.24	**<0.01**
CRF relative to BW (ml/kg/min)	35.61	±	4.17	36.95	±	4.02	34.40	±	3.95	**<0.01**
CRF relative to FFM (ml/kg/min)	53.18	±	7.85	58.27	±	5.62	48.62	±	6.68	**<0.01**
Handgrip strength (kg)	28.82	±	7.46	32.09	±	8.40	26.00	±	5.11	**<0.01**
Relative handgrip strength (kg/kg)	0.34	±	0.07	0.36	±	0.07	0.32	±	0.07	**<0.01**
Standing long jump (cm)	124.31	±	25.16	132.89	±	25.25	117.01	±	22.80	**<0.01**
Physical Activity levels (PAQ-A)	1.62	±	1.06	1.72	±	1.16	1.53	±	0.97	0.28
Cardiometabolic risk factors										
Fasting Glucose (mg/dL)	84.15	±	8.37	83.97	±	8.20	84.30	±	8.57	0.81
Triglycerides (mg/dL)	89.01	±	41.74	88.27	±	39.06	89.74	±	44.55	0.84
HDL cholesterol (mg/dL)	46.04	±	10.77	44.86	±	10.20	47.22	±	11.27	0.21
Mean arterial blood pressure (mmHg)	85.01	±	10.33	85.45	±	9.42	84.62	±	11.12	0.65
Waist circumference (cm)	98.71	±	12.79	104.00	±	11.02	94.25	±	12.53	**0.01**

Table 1 legend: NA, not applicable; PHV, peak height velocity; BMI, body mass index; Sum of 4 skinfolds (Biceps, Triceps, Subscapular, Suprailiac); Sum of 6 skinfolds (Biceps, Triceps, Subscapular, Suprailiac, Thigh, Calf); CRF, cardiorespiratory fitness; BW, body weight; FFM, fat free mass; PAQ-A, physical activity questionnaire for adolescents; HDL, high-density lipoprotein. *p*-values show differences between boys and girls. Bold numbers indicate *p*-value < 0.05.

Table 2 presents stepwise linear regressions between the potential determinants (i.e., socioeconomic, early life factors, body composition changes and fitness and PA changes) and the changes in the cardiometabolic risk score (Z-score change) in the intensive and the extensive intervention phases. None of the potential determinants was selected in the stepwise model in the intensive intervention phase. Meanwhile, body weight changes (post–pre intervention values), sex and mother's education were entered in the stepwise model as significant determinants in the extensive intervention phase ($R^2 = 0.26$, $p = 0.002$; $R^2 = 0.14$, $p = 0.013$; and $R^2 = 0.14$, $p = 0.017$, respectively). Body weight changes explained 26% of the variations in the cardiometabolic risk score (Z-score change), while the inclusion of sex and mother's education in the model showed an additional 28% (14% for sex and 14% for mother's education, respectively) of the explained variations in the cardiometabolic risk score (Z-score change).

Table 2. Stepwise linear regressions of potential determinants of longitudinal changes in the cardiometabolic risk score (Z-score change) in adolescents with overweight/obesity.

Intervention Phase	Significant Determinants ($p < 0.05$)	β	R2 of Change	*p*-Value
Intensive phase	Irrelevant determinants were included	NA	NA	NA
Extensive phase	Weight (Δ)	0.534	0.261	0.002
	Sex (1 = boys; 2 = girls)	−0.409	0.144	0.013
	Mother's education (1 = Primary School; 2 = High School; 3 = University)	−0.388	0.141	0.017

Table 2 legend: The potential determinants of longitudinal changes in the cardiometabolic risk, presented in Table 2, were included as predictors in the stepwise models. The cardiometabolic risk score (fasting glucose, triglycerides, HDL cholesterol, blood pressure and waist circumference) differences (i.e., Δ changes; post–pre) were included as the dependent variables.

Figure 1 presents the association according to sex between changes in body weight and changes in the cardiometabolic risk score (Z-score change) at the end of the extensive intervention phase. Both boys and girls showed a lower cardiometabolic risk score with the reduction in body weight (r = 0.535, $p = 0.009$ and r = 0.506, $p = 0.005$, respectively). There was no interaction between sex and body weight change ($p = 0.614$). Girls presented lower cardiometabolic risk scores at any level of body weight change compared to boys.

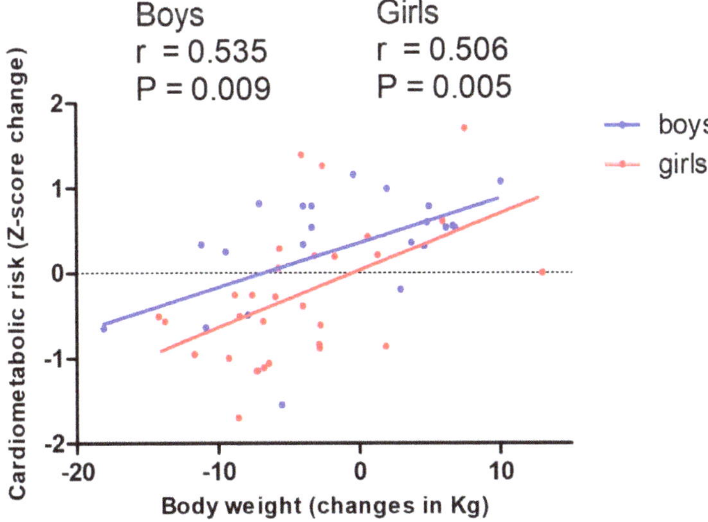

Figure 1. Scatter plot showing the association between changes in body weight and the cardiometabolic risk score (Z-score change) stratified by sex at the end of the extensive intervention phase.

4. Discussion

The EVASYON study is a multi-center project that aims to investigate the effect of a 13-month multidisciplinary lifestyle intervention on different health-related outcomes (e.g., body composition, cardiometabolic risk factors, among others) in adolescents with OW/OB aged 13 to 16 years [16]. In this regard, different studies from the EVASYON project reported an improvement in body weight, fat mass and cardiometabolic risk factors after 13 months of a lifestyle intervention in adolescents with OW/OB [17–19]. However, the determination of the best predictor from a broad set of health-related outcomes (socioeconomic, early life factors, body composition changes, fitness changes and PA changes) of longitudinal changes in the cardiometabolic risk profile (Z-score change) in adolescents with OW/OB remains, to the best of the authors' knowledge, unexplored. In this regard, the main findings of the present study were: (1) none of the potential determinants (i.e., socioeconomic, early life factors, body composition changes, fitness changes and PA changes) of the changes in the cardiometabolic risk score (Z-score change) were relevant in the intensive phase of the intervention; (2) instead, body weight changes, sex and mother's education predicted longitudinal changes in the cardiometabolic risk score (Z-score change) in adolescents with OW/OB in the extensive phase of the intervention. Reductions in body weight were associated with a lower cardiometabolic risk in boys and girls. Moreover, for the same reductions in body weight, the cardiometabolic risk was lower in girls compared with boys.

Low socioeconomic status is associated with a higher cardiometabolic risk and development of OW/OB during the early stages of life (i.e., from 0 to 15 years old) [35], and it seems to predict a higher risk of MetS in adulthood [36]. Although the associations between socioeconomic status and cardiometabolic risk are clearly detected in adults [37], the study results in adolescents are contradictory. Puolakka et al. [36] found that family socioeconomic status was associated with MetS and, in line with our results, Loucks et al. [38] found no relationship. Different methodological aspects between studies could partially explain these discrepancies (e.g., demographic heterogeneity in the study populations and sample size, SES measured by annual income [36] or the family income plus years of parental education [38]).

Additionally, parental education and occupation (low education level and unemployment, respectively) are related to the development of OW/OB and an adverse cardiometabolic risk profile in the pediatric population [9,35]. In this regard, Iguacel et al. [9] reported that mother's education was strongly associated with cardiometabolic risk in adolescents, independently of lifestyle. Additionally, the mother's education may impact children's adiposity due to the mother's influence on the diet during the early stage of life in kids and the duration of breastfeeding [39,40]. Similarly, we observed that mother's higher education was inversely associated with cardiometabolic risk (i.e., a higher level of education was related to a lower cardiometabolic risk) in adolescents with OW/OB. However, in our study, reductions in body weight were a stronger determinant of longitudinal changes in cardiometabolic risk than socioeconomic status variables.

Early life factors are also associated with adiposity and cardiometabolic health in adulthood [41–47]. Higher birth weight was associated with early development of insulin resistance and MetS [44] and increased risk of obesity [45]. Additionally, a lower birth weight was related to later risk of cardiovascular disease, obesity and diabetes [43]. In conclusion, low and high birth weights impact subsequent cardiometabolic health [41]. For size at birth, especially lower size was associated with subsequent cardiometabolic health [42]. In our study, weight and size at birth were not associated with cardiometabolic risk factors. Finally, there is also a relationship between breastfeeding and later diseases; a short period or absence of breastfeeding seems to be a risk factor for MetS [43,46]. Conversely, and along the same lines as Yakubov et al. [47], we found no association between cardiometabolic risk and any of the breastfeeding periods.

Interestingly, body weight reduction rather than changes in body composition indices (e.g., adiposity), fitness or PA was the strongest predictor of changes in the cardiometabolic risk profile (Z-score change) in adolescents with OW/OB. Of note, Ortega et al. reported that body mass index was a better predictor of cardiovascular disease mortality than body composition indices measured using hydrostatic weighing in a cohort of 60.335 participants followed up for a mean follow-up period of 15.2 years [48]. Similarly, we observed that the inexpensive and simple measure of body weight reflects the changes in the cardiometabolic risk profile (Z-score change) induced by a 13-month multidisciplinary lifestyle intervention better than any body composition indices. However, contrary to Ortega et al. [48], we did not use gold standard methods to determine body composition indices. Importantly, we believe that our findings are interesting for implementing a 13-month multidisciplinary lifestyle intervention in adolescents with OW/OB. A 30-year cohort study showed that the negative impact of higher body weight on cardiometabolic risk originates early in life [49].

Fitness and PA are inversely associated with cardiometabolic risk in youth [50,51]. Likewise, fitness may attenuate the adverse association between high body weight/adiposity and cardiometabolic risk in children and adolescents [34,52]. However, in the current study, changes in fitness and PA were not predictors of changes in cardiometabolic risk (Z-score change) in adolescents with OW/OB. In this regard, gold standard methods, such as indirect calorimetry and accelerometer, were not used in our study to assess fitness and PA, respectively. On the other hand, sex was the second determinant (the first determinant in the stepwise regression model was the reduction in body weight) of the changes in the cardiometabolic risk profile (Z-score change) in adolescents with OW/OB. We did not detect an interaction effect between changes in body weight and sex. Otherwise, for the same reductions in body weight, girls showed a lower cardiometabolic risk profile (Z-score) than boys. Interestingly, Barstad et al. examined sex differences in cardiometabolic risk factors in adolescents with obesity [53]. Specifically, girls presented lower levels of TG, blood pressure and higher HDL than boys with obesity, although PA levels were similar [48,53].

Our study presents some limitations that should be acknowledged. First, body composition indices were determined using skinfolds, and other methods, such as dual energy X-ray absorptiometry and air displacement plethysmography, may be more accurate. Second, PA and fitness were not measured using gold standard methods, such as accelerometry and indirect calorimetry. Third, PA questionnaires had to be applied in different seasons, and PA behavior could be affected by the seasonal effect. Fourth, our study lacks a control group to assess longitudinal changes induced by the lifestyle intervention in adolescents with OW/OB. Despite these limitations, our study tested several potential determinants (socioeconomic status, early life factors, body composition changes, fitness changes and PA changes) of changes in the cardiometabolic risk profile (Z-score) after a 13 month multidisciplinary lifestyle intervention in a cohort of 165 adolescents with OW/OB.

5. Conclusions

In the context of a 13-month multidisciplinary lifestyle intervention, changes in body weight rather than socioeconomic status, early life factors, body composition changes, fitness changes and PA changes were better predictors of changes in the cardiometabolic risk profile (Z-score change) in adolescents with OW/OB. In addition, the cardiometabolic risk profile was more favorable in girls than in boys for the identical decrease in body weight. In summary, the simple and inexpensive measure of body weight should be considered to track changes in the cardiometabolic risk profile in adolescents with OW/OB.

Author Contributions: Conceptualization, M.M.-M., D.R.-S., A.P.-F. and M.D.-F.; methodology and formal analysis; M.M.-M. and A.P.-F.; investigation, M.M.-M., D.R.-S., M.D.-F., A.M. (Amelia Marti), P.D.M.-E., L.A.M., A.M. (Ascensión Marcos) and C.C.; data curation, M.M.-M., D.R.-S., A.P.-F. and P.D.M.-E.; writing—original draft preparation, M.M.-M., D.R.-S. and A.P.-F.; writing—review and editing, all; project administration, M.D.-F., A.M. (Amelia Marti), L.A.M., A.M. (Ascensión Marcos) and C.C.; funding acquisition, A.M. (Amelia Marti), L.A.M., A.M. (Ascensión Marcos) and C.C. All authors have read and agreed to the published version of the manuscript.

Funding: The study was supported by the Ministry of Health and Consumption via the Carlos III Institute of Health (FIS Grant PI051579). The EVASYON study (PI052369) received the award for the best applied research project in 2009 from AESAN (Spanish Agency for Food Safety and Nutrition) from the Spanish Ministry of Health and Consumer Affairs. This study is part of Dinalrilan Rocha-Silva's PhD thesis from the Biomedicine Doctorate Program of the University of Granada.

Institutional Review Board Statement: The project followed the ethical standards recognized by the Declaration of Helsinki (reviewed in Hong Kong in September 1989 and in Edinburgh in 2000) and the EEC Good Clinical Practice recommendations (document 111/3976/88, July 1990) and the Spanish legislation regulating clinical research in humans (Royal Decree 561/1993 on clinical trials). The study was approved by the Ethics Committee of each hospital that participated in this project and by the Ethics Committee of the Spanish Council for Scientific Research (CSIC).

Informed Consent Statement: The study was explained to all the participants before starting, and the volunteers, parents or tutors signed an informed consent form. Their participation was completely voluntary; they did not receive any rewards.

Data Availability Statement: The data presented in this study are available on request from the corresponding author. The data are not publicly available due to restrictions concerning privacy and ethical reasons.

Acknowledgments: We gratefully acknowledge the enthusiasm and commitment of all the participating adolescents and their families.

Conflicts of Interest: The authors declare no conflict of interest.

References

1. WHO World Health Organization. Obesity and Overweight. 2021. Available online: https://www.who.int/news-room/fact-sheets/detail/obesity-and-overweight (accessed on 28 July 2022).
2. Garrido-Miguel, M.; Cavero-Redondo, I.; Álvarez-Bueno, C.; Rodríguez-Artalejo, F.; Moreno, L.A.; Ruiz, J.R.; Ahrens, W.; Martínez-Vizcaíno, V. Prevalence and Trends of Overweight and Obesity in European Children From 1999 to 2016: A Systematic Review and Meta-Analysis. *JAMA Pediatr.* **2019**, *173*, e192430. [CrossRef] [PubMed]
3. Lobstein, T.; Jackson-Leach, R. Planning for the Worst: Estimates of Obesity and Comorbidities in School-Age Children in 2025. *Pediatr. Obes.* **2016**, *11*, 321–325. [CrossRef] [PubMed]
4. Skinner, A.C.; Perrin, E.M.; Moss, L.A.; Skelton, J.A. Cardiometabolic Risks and Severity of Obesity in Children and Young Adults. *N. Engl. J. Med.* **2015**, *373*, 1307–1317. [CrossRef] [PubMed]
5. Ortega, F.B.; Lavie, C.J.; Blair, S.N. Obesity and Cardiovascular Disease. *Circ. Res.* **2016**, *118*, 1752–1770. [CrossRef] [PubMed]
6. Guerendiain, M.; Mayneris-Perxachs, J.; Montes, R.; López-Belmonte, G.; Martín-Matillas, M.; Castellote, A.I.A.I.; Martín-Bautista, E.; Martí, A.; Martínez, J.A.A.; Moreno, L.; et al. Relation between Plasma Antioxidant Vitamin Levels, Adiposity and Cardio-Metabolic Profile in Adolescents: Effects of a Multidisciplinary Obesity Programme. *Clin. Nutr.* **2017**, *36*, 209–217. [CrossRef]
7. Guerendiain, M.; Montes, R.; López-Belmonte, G.; Martín-Matillas, M.; Castellote, A.I.A.I.; Martín-Bautista, E.; Martí, A.; Martínez, J.A.A.; Moreno, L.; Garagorri, J.M.J.M.; et al. Changes in Plasma Fatty Acid Composition Are Associated with Improvements in Obesity and Related Metabolic Disorders: A Therapeutic Approach to Overweight Adolescents. *Clin. Nutr.* **2018**, *37*, 149–156. [CrossRef] [PubMed]
8. Falkner, B.; Gidding, S. Life-Course Implications of Pediatric Risk Factors for Cardiovascular Disease. *Can. J. Cardiol.* **2021**, *37*, 766–775. [CrossRef]
9. Iguacel, I.; Börnhorst, C.; Michels, N.; Breidenassel, C.; Dallongeville, J.; González-Gross, M.; Gottrand, F.; Kafatos, A.; Karaglani, E.; Kersting, M.; et al. Socioeconomically Disadvantaged Groups and Metabolic Syndrome in European Adolescents: The HELENA Study. *J. Adolesc. Health* **2021**, *68*, 146–154. [CrossRef]
10. Wang, J.; Perona, J.S.; Schmidt-RioValle, J.; Chen, Y.; Jing, J.; González-Jiménez, E. Metabolic Syndrome and Its Associated Early-Life Factors among Chinese and Spanish Adolescents: A Pilot Study. *Nutrients* **2019**, *11*, 1568. [CrossRef]
11. Ortega, F.B.; Ruiz, J.R.; Castillo, M.J.; Sjöström, M. Physical Fitness in Childhood and Adolescence: A Powerful Marker of Health. *Int. J. Obes.* **2008**, *32*, 1–11. [CrossRef]
12. Rizzo, N.S.; Ruiz, J.R.; Hurtig-Wennlöf, A.; Ortega, F.B.; Sjöström, M. Relationship of Physical Activity, Fitness, and Fatness with Clustered Metabolic Risk in Children and Adolescents: The European Youth Heart Study. *J. Pediatr.* **2007**, *150*, 388–394. [CrossRef] [PubMed]
13. Brand, C.; Martins, C.M.D.L.; Lemes, V.B.; Pessoa, M.L.F.; Dias, A.F.; Cadore, E.L.; Mota, J.; Gaya, A.C.A.; Gaya, A.R. Effects and Prevalence of Responders after a Multicomponent Intervention on Cardiometabolic Risk Factors in Children and Adolescents with Overweight/Obesity: Action for Health Study. *J. Sports Sci.* **2020**, *38*, 682–691. [CrossRef] [PubMed]

14. Roche, J.; Corgosinho, F.C.; Isacco, L.; Scheuermaier, K.; Pereira, B.; Gillet, V.; Moreira, G.A.; Pradella-Hallinan, M.; Tufik, S.; de Mello, M.T.; et al. A Multidisciplinary Weight Loss Intervention in Obese Adolescents with and without Sleep-Disordered Breathing Improves Cardiometabolic Health, Whether SDB Was Normalized or not. *Sleep Med.* **2020**, *75*, 225–235. [CrossRef] [PubMed]

15. Silveira, D.S.; Lemos, L.F.G.B.F.; Tassitano, R.M.; Cattuzzo, M.T.; Feitoza, A.H.P.; Aires, L.M.S.M.C.; Silva Mota, J.A.P.; Martins, C.M.D.L. Effect of a Pilot Multi-Component Intervention on Motor Performance and Metabolic Risks in Overweight/Obese Youth. *J. Sports Sci.* **2018**, *36*, 2317–2326. [CrossRef] [PubMed]

16. Martinez-Gomez, D.; Gomez-Martinez, S.; Puertollano, M.A.; Nova, E.; Wärnberg, J.; Veiga, O.L.; Martí, A.; Campoy, C.; Garagorri, J.M.; Azcona, C.; et al. Design and Evaluation of a Treatment Programme for Spanish Adolescents with Overweight and Obesity. the EVASYON Study. *BMC Public Health* **2009**, *9*, 414. [CrossRef] [PubMed]

17. De Miguel-Etayo, P.; Moreno, L.A.; Santabárbara, J.; Bueno, G.; Martín-Matillas, M.; Zapatera, B.; Julián, C.A.-S.; Martí, A.; Campoy, C.; Marcos, A.; et al. Body Composition Changes during a Multidisciplinary Treatment Programme in Overweight Adolescents: EVASYON Study. *Nutr. Hosp.* **2015**, *32*, 2525–2534. [CrossRef] [PubMed]

18. Romeo, J.; Martinez-Gomez, D.; Diaz, L.E.; Gómez-Martinez, S.; Marti, A.; Martin-Matillas, M.; Puertollano, M.A.; Veiga, O.L.; Martinez, J.A.; Wärnberg, J.; et al. Changes in Cardiometabolic Risk Factors, Appetite-Controlling Hormones and Cytokines after a Treatment Program in Overweight Adolescents: Preliminary Findings from the EVASYON Study. *Pediatr. Diabetes* **2011**, *12*, 372–380. [CrossRef]

19. De Miguel-Etayo, P.; Moreno, L.A.; Santabarbara, J.; Martín-Matillas, M.; Piqueras, M.J.; Rocha-Silva, D.; Marti del Moral, A.; Campoy, C.; Marcos, A.; Garagorri, J.M. Anthropometric Indices to Assess Body-Fat Changes during a Multidisciplinary Obesity Treatment in Adolescents: EVASYON Study. *Clin. Nutr.* **2015**, *34*, 523–528. [CrossRef]

20. Cole, T.J.; Bellizzi, M.C.; Flegal, K.M.; Dietz, W.H. Establishing a Standard Definition for Child Overweight and Obesity Worldwide: International Survey. *BMJ* **2000**, *320*, 1240–1243. [CrossRef]

21. Álvarez-Dardet, C.; Alonso, J.; Domingo, A.; Regidor, E. *La Medición de La Clase Social En Ciencias de La Salud. Informe de Un Grupo de Trabajo e La Sociedad Española de Epidemiología [The Measurement of Social Class in Health Sciences. Report of a Working Group of the Spanish Society of Epidemiology]*; SG Editores: Barcelona, Spain, 1995; ISBN 84-87621-35-X.

22. Moreno, L.A.; Joyanes, M.; Mesana, M.I.; González-Gross, M.; Gil, C.M.; Sarría, A.; Gutierrez, A.; Garaulet, M.; Perez-Prieto, R.; Bueno, M.; et al. Harmonization of Anthropometric Measurements for a Multicenter Nutrition Survey in Spanish Adolescents. *Nutrition* **2003**, *19*, 481–486. [CrossRef]

23. Cole, T.J.; Lobstein, T. Extended International (IOTF) Body Mass Index Cut-Offs for Thinness, Overweight and Obesity. *Pediatr. Obes.* **2012**, *7*, 284–294. [CrossRef] [PubMed]

24. Sanchez-Delgado, G.; Cadenas-Sanchez, C.; Martinez-Tellez, B.; Mora-Gonzalez, J.R.; Ruiz, J.R.; Ortega, F.B.P. An Excel-Based Classifier of BMI into Sex- and Age-Specific Weight Status in Youths Aged 2–18 Years According to the World Obesity Federation (Formerly IOTF) Cut-Points ©. 2020. OTRI-UGR-IPR-868. Available online: https://digibug.ugr.es/handle/10481/61590 (accessed on 19 May 2022).

25. Bervoets, L.; Massa, G. Defining Morbid Obesity in Children Based on BMI 40 at Age 18 Using the Extended International (IOTF) Cut-Offs. *Pediatr. Obes.* **2014**, *9*, e94–e98. [CrossRef]

26. Moore, S.A.; McKay, H.A.; Macdonald, H.; Nettlefold, L.; Baxter-Jones, A.D.G.; Cameron, N.; Brasher, P.M.A. Enhancing a Somatic Maturity Prediction Model. *Med. Sci. Sports Exerc.* **2015**, *47*, 1755–1764. [CrossRef] [PubMed]

27. Slaughter, M.H.; Lohman, T.G.; Boileau, R.A.; Horswill, C.A.; Stillman, R.J.; Van Loan, M.D.; Bemben, D.A. Skinfold Equations for Estimation of Body Fatness in Children and Youth. *Hum. Biol.* **1988**, *60*, 709–723. [CrossRef]

28. Léger, L.A.; Mercier, D.; Gadoury, C.; Lambert, J. The Multistage 20 Metre Shuttle Run Test for Aerobic Fitness. *J. Sports Sci.* **1988**, *6*, 93–101. [CrossRef] [PubMed]

29. Ruiz, J.R.; España-Romero, V.; Ortega, F.B.; Sjöström, M.; Castillo, M.J.; Gutierrez, A. Hand Span Influences Optimal Grip Span in Male and Female Teenagers. *J. Hand Surg. Am.* **2006**, *31*, 1367–1372. [CrossRef]

30. Martínez-Gómez, D.; Martínez-De-Haro, V.; Pozo, T.; Welk, G.J.; Villagra, A.; Calle, M.E.; Marcos, A.; Veiga, O.L. Reliability and Validity of the PAQ-A Questionnaire to Assess Physical Activity in Spanish Adolescents. *Rev. Esp. Salud Publica* **2009**, *83*, 427–439. [CrossRef] [PubMed]

31. Wallach, J.B. *Interpretation of Diagnostic Tests*; Lippincott Williams & Wilkins: Philadelphia, PA, USA, 2007.

32. Topouchian, J.A.; El Assaad, M.A.; Orobinskaia, L.V.; El Feghali, R.N.; Asmar, R.G. Validation of Two Automatic Devices for Self-Measurement of Blood Pressure According to the International Protocol of the European Society of Hypertension: The Omron M6 (HEM-7001-E) and the Omron R7 (HEM 637-IT). *Blood Press. Monit.* **2006**, *11*, 165–171. [CrossRef] [PubMed]

33. National High Blood Pressure Education Program Working Group on High Blood Pressure in Children and Adolescents. American Academy of Pediatrics The Fourth Report on the Diagnosis, Evaluation, and Treatment of High Blood Pressure in Children and Adolescents. *Pediatrics* **2004**, *114*, 555–576. [CrossRef]

34. Nyström, C.D.; Henriksson, P.; Martínez-Vizcaíno, V.; Medrano, M.; Cadenas-Sanchez, C.; Arias-Palencia, N.M.; Löf, M.; Ruiz, J.R.; Labayen, I.; Sánchez-López, M.; et al. Does Cardiorespiratory Fitness Attenuate the Adverse Effects of Severe/Morbid Obesity on Cardiometabolic Risk and Insulin Resistance in Children? A Pooled Analysis. *Diabetes Care* **2017**, *40*, 1580–1587. [CrossRef]

35. Wu, S.; Ding, Y.; Wu, F.; Li, R.; Hu, Y.; Hou, J.; Mao, P. Socio-Economic Position as an Intervention against Overweight and Obesity in Children: A Systematic Review and Meta-Analysis. *Sci. Rep.* **2015**, *5*, 11354. [CrossRef]

36. Puolakka, E.; Pahkala, K.; Laitinen, T.T.; Magnussen, C.G.; Hutri-Kähönen, N.; Tossavainen, P.; Jokinen, E.; Sabin, M.A.; Laitinen, T.; Elovainio, M.; et al. Childhood Socioeconomic Status in Predicting Metabolic Syndrome and Glucose Abnormalities in Adulthood: The Cardiovascular Risk in Young Finns Study. *Diabetes Care* **2016**, *39*, 2311–2317. [CrossRef] [PubMed]
37. Loucks, E.B.; Rehkopf, D.H.; Thurston, R.C.; Kawachi, I. Socioeconomic Disparities in Metabolic Syndrome Differ by Gender: Evidence from NHANES III. *Ann. Epidemiol.* **2007**, *17*, 19–26. [CrossRef] [PubMed]
38. Loucks, E.B.; Magnusson, K.T.; Cook, S.; Rehkopf, D.H.; Ford, E.S.; Berkman, L.F. Socioeconomic Position and the Metabolic Syndrome in Early, Middle, and Late Life: Evidence from NHANES 1999–2002. *Ann. Epidemiol.* **2007**, *17*, 782–790. [CrossRef] [PubMed]
39. Oude Groeniger, J.; De Koster, W.; Van Der Waal, J. Time-Varying Effects of Screen Media Exposure in the Relationship between Socioeconomic Background and Childhood Obesity. *Epidemiology* **2020**, *31*, 578–586. [CrossRef]
40. Camier, A.; Cissé, A.H.; Lioret, S.; Bernard, J.Y.; Charles, M.A.; Heude, B.; de Lauzon-Guillain, B. Infant Feeding Practices Associated with Adiposity Peak and Rebound in the EDEN Mother-Child Cohort. *Int. J. Obes.* **2022**, *46*, 809–816. [CrossRef]
41. Nordman, H.; Jääskeläinen, J.; Voutilainen, R. Birth Size as a Determinant of Cardiometabolic Risk Factors in Children. *Horm. Res. Paediatr.* **2020**, *93*, 144–153. [CrossRef]
42. Finken, M.J.J.; Van Der Steen, M.; Smeets, C.C.J.; Walenkamp, M.J.E.; De Bruin, C.; Hokken-Koelega, A.C.S.; Wit, J.M. Children Born Small for Gestational Age: Differential Diagnosis, Molecular Genetic Evaluation, and Implications. *Endocr. Rev.* **2018**, *39*, 851–894. [CrossRef]
43. Córdoba-Rodríguez, D.P.; Rodriguez, G.; Moreno, L.A. Predicting of Excess Body Fat in Children. *Curr. Opin. Clin. Nutr. Metab. Care* **2022**, *25*, 304–310. [CrossRef]
44. Agius, R.; Savona-Ventura, C.; Vassallo, J. Transgenerational Metabolic Determinants of Fetal Birth Weight. *Exp. Clin. Endocrinol. Diabetes* **2013**, *121*, 431–435. [CrossRef]
45. Yu, Z.B.; Han, S.P.; Zhu, G.Z.; Zhu, C.; Wang, X.J.; Cao, X.G.; Guo, X.R. Birth Weight and Subsequent Risk of Obesity: A Systematic Review and Meta-Analysis. *Obes. Rev.* **2011**, *12*, 525–542. [CrossRef] [PubMed]
46. González-Jiménez, E.; Montero-Alonso, M.A.; Schmidt-RioValle, J.; García-García, C.J.; Padez, C. Metabolic Syndrome in Spanish Adolescents and Its Association with Birth Weight, Breastfeeding Duration, Maternal Smoking, and Maternal Obesity: A Cross-Sectional Study. *Eur. J. Nutr.* **2015**, *54*, 589–597. [CrossRef] [PubMed]
47. Yakubov, R.; Nadir, E.; Stein, R.; Klein-Kremer, A. The Duration of Breastfeeding and Its Association with Metabolic Syndrome among Obese Children. *Sci. World J.* **2015**, *2015*, 1–4. [CrossRef] [PubMed]
48. Ortega, F.B.; Sui, X.; Lavie, C.J.; Blair, S.N. Body Mass Index, the Most Widely Used But Also Widely Criticized Index. *Mayo Clin. Proc.* **2016**, *91*, 443–455. [CrossRef]
49. Liao, Y.Y.; Chu, C.; Wang, Y.; Zheng, W.L.; Ma, Q.; Hu, J.W.; Yan, Y.; Wang, K.K.; Yuan, Y.; Chen, C.; et al. Long-Term Burden of Higher Body Mass Index from Childhood on Adult Cardiometabolic Biomarkers: A 30-Year Cohort Study. *Nutr. Metab. Cardiovasc. Dis.* **2021**, *31*, 439–447. [CrossRef]
50. Ortega, F.B.; Cadenas-Sanchez, C.; Lee, D.; Ruiz, J.R.; Blair, S.N.; Sui, X. Fitness and Fatness as Health Markers through the Lifespan: An Overview of Current Knowledge. *Prog. Prev. Med.* **2018**, *3*, e0013. [CrossRef]
51. Husøy, A.; Dalene, K.E.; Steene-Johannessen, J.; Anderssen, S.A.; Ekelund, U.; Tarp, J. Effect Modification by Cardiorespiratory Fitness on the Association between Physical Activity and Cardiometabolic Health in Youth: A Systematic Review. *J. Sports Sci.* **2021**, *39*, 845–853. [CrossRef]
52. Ortega, F.B.; Ruiz, J.R.; Labayen, I.; Lavie, C.J.; Blair, S.N. The Fat but Fit Paradox: What We Know and Don't Know about It. *Br. J. Sports Med.* **2018**, *52*, 151–153. [CrossRef]
53. Barstad, L.H.; Júlíusson, P.B.; Johnson, L.K.; Hertel, J.K.; Lekhal, S.; Hjelmesæth, J. Gender-Related Differences in Cardiometabolic Risk Factors and Lifestyle Behaviors in Treatment-Seeking Adolescents with Severe Obesity. *BMC Pediatr.* **2018**, *18*, 61. [CrossRef]

Article

Good Adherence to the Mediterranean Diet Lowered Risk of Renal Glomerular Impairment in Children: A Longitudinal Study

Menglong Li [1], Huidi Xiao [1], Wen Shu [1], Nubiya Amaerjiang [1], Jiawulan Zunong [1], Dayong Huang [2,*] and Yifei Hu [1,*]

[1] Department of Child, Adolescent Health and Maternal Care, School of Public Health, Capital Medical University, Beijing 100069, China
[2] Department of Hematology, Beijing Friendship Hospital, Capital Medical University, Beijing 100050, China
* Correspondence: hdayong@yahoo.com (D.H.); huyifei@yahoo.com (Y.H.); Tel.: +86-10-83911747 (Y.H.)

Citation: Li, M.; Xiao, H.; Shu, W.; Amaerjiang, N.; Zunong, J.; Huang, D.; Hu, Y. Good Adherence to the Mediterranean Diet Lowered Risk of Renal Glomerular Impairment in Children: A Longitudinal Study. *Nutrients* **2022**, *14*, 3343. https://doi.org/10.3390/nu14163343

Academic Editor: Margaret Allman-Farinelli

Received: 9 July 2022
Accepted: 11 August 2022
Published: 15 August 2022

Publisher's Note: MDPI stays neutral with regard to jurisdictional claims in published maps and institutional affiliations.

Copyright: © 2022 by the authors. Licensee MDPI, Basel, Switzerland. This article is an open access article distributed under the terms and conditions of the Creative Commons Attribution (CC BY) license (https://creativecommons.org/licenses/by/4.0/).

Abstract: Healthy diet patterns have a positive effect on chronic non-communicable diseases in the pediatric population, but the evidence is limited on the association between kidney impairment and adherence to a Mediterranean diet. We aim to determine the associations between Mediterranean diet adherence and longitudinal tubular and glomerular impairment in children. Based on four waves of urine assays conducted from October 2018 to November 2019, we assayed urinary β_2-microglobulin (β_2-MG) and microalbumin (MA) excretion to determine transient renal tubular and glomerular impairment during the follow-up of the child cohort (PROC) study in Beijing, China. We assessed Mediterranean diet adherence using the 16-item Mediterranean Diet Quality Index in children and adolescents (KIDMED) among 1914 primary school children. Poor, intermediate, and good adherence rates for the Mediterranean diet were 9.0% (KIDMED index 0–3), 54.4% (KIDMED index 4–7) and 36.5% (KIDMED index 8–12), respectively. A short sleep duration was more prevalent in children with lower Mediterranean diet adherence, with no significant differences presenting in the other demographic and lifestyle covariates. The results of linear mixed-effects models showed that a higher urinary MA excretion was inversely associated with a higher KIDMED score (β = −0.216, 95%CI: −0.358, −0.074, p = 0.003), after adjusting for sex, age, BMI z-score, SBP z-score, screen time, sleep duration and physical activity. Furthermore, in generalized linear mixed-effects models, consistent results found that transient renal glomerular impairment was less likely to develop in children with intermediate Mediterranean diet adherence (aOR = 0.68, 95%CI: 0.47, 0.99, p = 0.044) and in children with good Mediterranean diet adherence (aOR = 0.60, 95%CI: 0.40, 0.90, p = 0.014), taking poor Mediterranean diet adherence as a reference. We visualized the longitudinal associations between each item of the KIDMED test and kidney impairment via a forest plot and identified the main protective eating behaviors. Children who adhere well to the Mediterranean diet have a lower risk of transient glomerular impairment, underscoring the necessity of the early childhood development of healthy eating patterns to protect kidney health.

Keywords: Mediterranean diet; KIDMED test; kidney impairment; children; China

1. Introduction

Dietary patterns highlight the habitual consumption of foods, which are associated with the four leading causes of death— coronary heart disease, stroke, some cancers, and type II diabetes—in developed countries and are dramatically increasing in developing countries as well [1,2]. As a modifiable factor, a diet pattern can be categorized as "healthy" or "unhealthy" [1,3,4]. The Mediterranean diet is one of the healthy diet patterns based on a variety of whole foods, including fruits and vegetables, whole grains, legumes, nuts and seeds, and fishes, compared with highly processed foods [3]. Good adherence to the Mediterranean diet has a positive role in protecting people from chronic non-communicable diseases (NCDs) [5–7].

Early development and adherence to a healthy eating pattern is critical for children's growth and kidney health [8–11]. Western-style diets including fast food or fried high-energy foods have increased dramatically over the past few decades, gaining popularity among the younger generation in China [4]. It leads to a gradual decrease in the adoption of the traditional Chinese dietary pattern, which is similar in composition to the Mediterranean diet [12]. Diet measurement is often challenging, especially in children, given dietary patterns and food diversity and complexity, and the Mediterranean Diet Quality Index in children and adolescents (KIDMED) [13] provides an optimal and standardized tool to assess a healthier food mix. The Mediterranean diet has been strongly advocated in recent years to address the upsurging early onset of NCDs. However, the KIDMED index has not been well-practiced in Chinese children.

Substantial evidence from studies in the pediatric population has shown that adherence to the Mediterranean diet as assessed by the KIDMED index is associated with a reduced risk of obesity [14], metabolic syndrome [15,16], diabetes [17,18] and early cardiovascular diseases [19], while daily lifestyles may have a synergistic effect [20,21]. Paucity of evidence has shown the association between Mediterranean diet adherence and kidney impairment. We found only one cross-sectional study examining the association of Mediterranean diet adherence with the albuminuria level in Greek adolescents [11]. Albuminuria, which is assessed using elevated microalbumin (MA) excretion, reflects early renal glomerular impairment [22], and an elevated β_2-microglobulin (β_2-MG) reflects early renal tubular impairment [23]. In this study, we aim to determine the longitudinal associations between transient kidney impairment and Mediterranean diet adherence in children using the KIDMED index.

2. Materials and Methods

2.1. Study Design and Participants

Based on the child cohort (PROC) study, we enrolled 1914 children aged 6 to 8 years attending primary school in Beijing, China (detailed elsewhere [24]). We conducted 4 waves of urine assays from October 2018 to November 2019 (detailed elsewhere [25,26]). In brief, the first wave of urine assays was performed from October to November 2018 and waves 2–4 were performed within one week in November 2019 (see details elsewhere [27]). All participants provided at least one urine sample, for a sample size of 6968 visits across 4 waves.

2.2. Urine Collection and Measurements

The urine collection and measurement procedures are detailed elsewhere [25,26]. Briefly, we performed a baseline fasting urine assay for wave 1, a 24 h urine assay (Sunday through Monday morning) for wave 2, a Wednesday fasting urine assay for wave 3 and a Friday fasting urine assay for wave 4. Urinary β_2-MG and MA excretion were measured in waves 1–4. An elevated β_2-MG > 0.2 mg/L was used to define renal tubular impairment [28] and an elevated MA $\geq$ 20 mg/L was used to define renal glomerular impairment [29].

2.3. Assessment of Adherence to the Mediterranean Diet

The 16-item Mediterranean Diet Quality Index in children and adolescents (KIDMED) [13] was used to assess Mediterranean diet adherence among children aged 6–8 years at baseline. A total of 12 items with positive implications for the Mediterranean diet were designated with a +1 value, while 4 items with a negative meaning were designated with a −1 value. The summed index ranged from 0 to 12 points and can be classified into 3 grades: (1) 0–3, low Mediterranean diet adherence; (2) 4–7, intermediate Mediterranean diet adherence (needs to be improved); (3) 8–12, optimal/good Mediterranean diet adherence.

2.4. Data Collection of Covariates

Anthropometric measurements including weight, height, body mass index (BMI), systolic blood pressure (SBP) and diastolic blood pressure (DBP) were analyzed in this study. Height z-scores, weight z-scores and BMI z-scores were calculated using 2007 WHO criteria, and age- and sex-specific SBP z-scores were determined. Lifestyle covariates included sleep duration, screen time and physical activity by parental self-administrated questionnaires (Children's Sleep Habits Questionnaire [CSHQ] [30,31]; the Chinese version of the Children's Leisure Activities Study Survey [CLASS-C] [25,32]). A sleep duration of <10 h/d was used to define short sleep. Computer/cell-phone screen time of ≥2 h/d was used to define long screen time. Physical activity of <1 h/d was used to define insufficient physical activity [25].

2.5. Statistical Analysis

The main outcomes were the repeated measurements of urinary β_2-MG and MA excretion and the categorical transient renal tubular and glomerular impairment. Descriptive statistics are presented according to the KIDMED index (Mediterranean diet adherence). Variables with missing values were handled using multiple imputations, and a total of 50 complete datasets were generated for the final analysis. Sex or lifestyle covariates are presented as counts and percentages. Z-scores of height, weight and BMI, SBP and DBP are described as mean ± standard deviation (SD), and urinary β_2-MG and MA excretion are described as median and interquartile range (IQR). The χ^2 test, analysis of variance and Kruskal–Wallis test were performed to compare the differences between the three groups of poor, intermediate, and good adherence to the Mediterranean diet. We generated linear mixed-effects models to determine the associations between renal impairment indicators and KIDMED scores with estimated coefficients and 95% confidence intervals (95%CI). The longitudinal associations of kidney impairment and KIDMED index were determined using generalized linear mixed-effects models with crude and adjusted odds ratios (cOR and aOR) and 95%CIs. Multivariable models were adjusted for sex, age, BMI z-score, SBP z-score, screen time, sleep duration and physical activity level, while the weekday of the urine assay was included as a random effect. The results in Table 1 are based on the first imputed dataset, while Tables 2 and 3 and Figure 1 contain valid statistical inferences of the parameters based on 50 datasets using PROC MIANALYZE, and Table 4 is also based on the first imputed dataset. Statistical significance was determined using a two-tailed p value of 0.05. All analyses were performed using Statistical Analysis System V.9.4 (SAS Institute Inc., Cary, NC, USA).

Table 1. Participants' characteristics categorized by KIDMED index of children aged 6–9 in Beijing, China (N = 1914).

Characteristics	KIDMED Index (Mediterranean Diet Adherence)			p
	0–3 (Low *n* = 173)	4–7 (Intermediate *n* = 1042)	8–12 (Good *n* = 699)	
At baseline				
Boy (*n* (%))[1]	90 (52.0)	506 (48.6)	360 (51.5)	0.41
Age (year) [2]	6.6 ± 0.3	6.6 ± 0.3	6.6 ± 0.3	0.64
Height z-score [2]	0.73 ± 0.94	0.62 ± 0.96	0.71 ± 0.95	0.12
Weight z-score [2]	0.78 ± 1.42	0.67 ± 1.43	0.74 ± 1.38	0.41
Body mass index z-score [2]	0.45 ± 1.58	0.37 ± 1.54	0.41 ± 1.52	0.76
Systolic blood pressure (mmHg) [2]	100 ± 9	101 ± 9	101 ± 8	0.27
Diastolic blood pressure (mmHg) [2]	56 ± 6	56 ± 6	56 ± 6	0.60
Short sleep (<10 h/d) [1]	149 (86.1)	792 (76.0)	500 (71.5)	<0.001
Long screen time (≥2 h/d) [1]	10 (5.8)	55 (5.3)	30 (4.3)	0.57
Insufficient physical activity (<1 h/d) [1]	136 (78.6)	789 (75.7)	526 (75.3)	0.65

Table 1. *Cont.*

Characteristics	KIDMED Index (Mediterranean Diet Adherence)			p
	0–3 (Low n = 173)	4–7 (Intermediate n = 1042)	8–12 (Good n = 699)	
Urinary β_2-MG excretion (mg/L) [3]				
Wave 1	0.09 (0.04–0.15)	0.08 (0.04–0.13)	0.08 (0.04–0.12)	0.34
Wave 2	0.15 (0.12–0.18)	0.14 (0.12–0.18)	0.14 (0.12–0.18)	0.15
Wave 3	0.16 (0.13–0.21)	0.17 (0.13–0.21)	0.16 (0.13–0.20)	0.017
Wave 4	0.17 (0.13–0.22)	0.16 (0.13–0.21)	0.16 (0.13–0.21)	0.42
Urinary MA excretion (mg/L) [3]				
Wave 1	9.67 (6.57–13.14)	9.19 (6.54–12.78)	8.76 (6.30–11.80)	0.027
Wave 2	7.15 (6.20–9.20)	6.70 (6.10–8.50)	6.60 (6.00–8.40)	0.046
Wave 3	7.30 (6.30–9.20)	7.10 (6.30–9.50)	7.00 (6.20–8.80)	0.10
Wave 4	6.90 (6.00–10.50)	6.80 (6.00–9.00)	6.65 (6.00–8.70)	0.089

KIDMED: the 16-item Mediterranean Diet Quality Index in children and adolescents; β_2-MG: β_2-microglobulin; MA: microalbumin.[1] Comparison by KIDMED group using χ^2 test. [2] Means and standard deviations (SDs) compared between KIDMED groups using analysis of variance. [3] Medians and interquartile ranges (IQRs) compared between KIDMED groups using Kruskal–Wallis test.

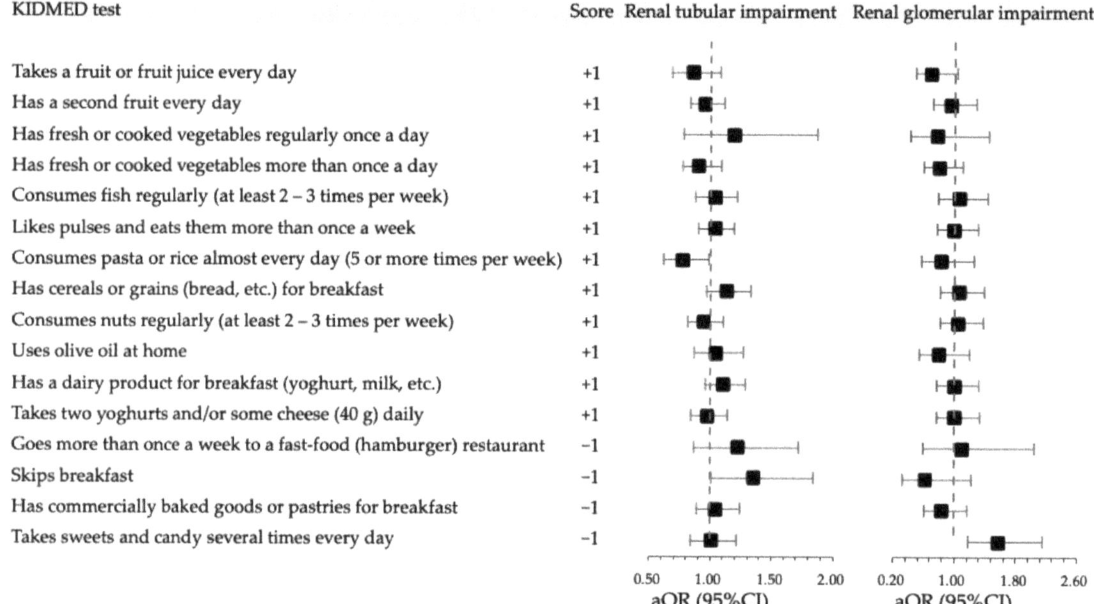

Figure 1. Forest plot for longitudinal associations of 16 items in KIDMED test with transient kidney impairment. All 16 items of the KIDMED test were included in the multivariable models, adjusting for sex, age, body mass index z-score, systolic blood pressure z-score, screen time, sleep duration and physical activity level, while the weekday of the urine assay was included as a random effect.

Table 2. Linear associations of indicators of kidney impairment with KIDMED scores among children aged 6–9 in Beijing, China using linear mixed-effects models.

Dependent Variable	Independent Variable	Model 1		Model 2		Model 3	
		Estimate (95%CI)	p	Estimate (95%CI)	p	Estimate (95%CI)	p
Urinary β_2-MG excretion	KIDMED score	−0.001 (−0.003, 0.001)	0.20	−0.001 (−0.002, 0.001)	0.28	−0.001 (−0.003, 0)	0.13
Urinary MA excretion	KIDMED score	−0.218 (−0.359, −0.076)	0.003	−0.206 (−0.347, −0.065)	0.004	−0.216 (−0.358, −0.074)	0.003

Model 1: unadjusted; model 2: adjusted for sex, age, body mass index z-score; model 3: based on model 2, further adjusted for systolic blood pressure z-score, screen time, sleep duration and physical activity level. (KIDMED: 16-item Mediterranean Diet Quality Index in children and adolescents; β_2-MG: β_2-microglobulin; MA: microalbumin; CI: confidence interval. All models included one random effect: the weekday of the urine assay).

Table 3. Longitudinal associations of kidney impairment with KIDMED index among children aged 6–9 in Beijing, China using generalized linear mixed-effects models.

Dependent Variable	Independent Variable	Model 1		Model 2		Model 3	
		cOR (95%CI)	p	aOR (95%CI)	p	aOR (95%CI)	p
Renal tubular impairment	KIDMED index 0–3	1		1		1	
	KIDMED index 4–7	0.90 (0.72, 1.12)	0.33	0.90 (0.72, 1.12)	0.35	0.88 (0.70, 1.10)	0.25
	KIDMED index 8–12	0.86 (0.69, 1.09)	0.22	0.87 (0.69, 1.10)	0.26	0.84 (0.67, 1.07)	0.16
Renal glomerular impairment	KIDMED index 0–3	1		1		1	
	KIDMED index 4–7	0.71 (0.50, 1.03)	0.068	0.70 (0.48, 1.00)	0.051	0.68 (0.47, 0.99)	0.044
	KIDMED index 8–12	0.61 (0.41, 0.91)	0.016	0.62 (0.41, 0.92)	0.018	0.60 (0.40, 0.90)	0.014

Model 1: unadjusted; model 2: adjusted for sex, age, body mass index z-score; model 3: based on model 2, further adjusting for+ systolic blood pressure z-score, screen time, sleep duration and physical activity level. (KIDMED: 16-item Mediterranean Diet Quality Index in children and adolescents; cOR: crude odds ratio; aOR: adjusted odds ratio; CI: confidence interval. All models included one random effect: the weekday of the urine assay).

Table 4. Frequency of item-specific KIDMED test with kidney impairment (N = 6968).

KIDMED Test Item	Score	Renal Tubular Impairment *n* (%)	Renal Glomerular Impairment *n* (%)
Takes a fruit or fruit juice every day			
Yes	+1	1190 (19.1)	294 (4.7)
No	0	161 (21.8)	49 (6.6)
Has a second fruit every day			
Yes	+1	752 (18.7)	187 (4.6)
No	0	599 (20.4)	156 (5.3)
Has fresh or cooked vegetables regularly once a day			
Yes	+1	1312 (19.4)	327 (4.8)
No	0	39 (18.0)	16 (7.4)
Has fresh or cooked vegetables more than once a day			
Yes	+1	1044 (19.0)	244 (4.4)
No	0	307 (20.8)	99 (6.7)
Consumes fish regularly (at least 2–3 times per week)			
Yes	+1	409 (19.5)	99 (4.7)
No	0	942 (19.4)	244 (5.0)
Likes pulses and eats them more than once a week			
Yes	+1	813 (19.4)	200 (4.8)
No	0	538 (19.4)	143 (5.2)
Consumes pasta or rice almost every day (5 or more times per week)			
Yes	+1	1210 (18.9)	308 (4.8)
No	0	141 (24.5)	35 (6.1)
Has cereals or grains (bread, etc.) for breakfast			
Yes	+1	851 (20.1)	208 (4.9)
No	0	500 (18.3)	135 (4.9)
Consumes nuts regularly (at least 2–3 times per week)			
Yes	+1	516 (19.4)	128 (4.8)
No	0	835 (19.4)	215 (5.0)
Uses olive oil at home			
Yes	+1	196 (19.0)	47 (4.6)
No	0	1155 (19.5)	296 (5.0)
Has a dairy product for breakfast (yoghurt, milk, etc.)			
Yes	+1	872 (19.8)	202 (4.6)
No	0	479 (18.7)	141 (5.5)
Takes two yoghurts and/or some cheese (40 g) daily			
Yes	+1	464 (19.4)	104 (4.4)
No	0	887 (19.4)	239 (5.2)
Goes more than once a week to a fast-food (hamburger) restaurant			
Yes	−1	54 (23.0)	14 (6.0)
No	0	1297 (19.3)	329 (4.9)
Skips breakfast			
Yes	−1	78 (22.8)	12 (3.5)
No	0	1273 (19.2)	331 (5.0)
Has commercially baked goods or pastries for breakfast			
Yes	−1	322 (19.7)	76 (4.6)
No	0	1029 (19.3)	267 (5.0)
Takes sweets and candy several times every day			
Yes	−1	199 (19.7)	76 (7.5)
No	0	1152 (19.3)	267 (4.5)

KIDMED: the 16-item Mediterranean Diet Quality Index in children and adolescents.

3. Results

3.1. Sociodemographic Characteristics

This study enrolled 1914 children of 6.6 ± 0.3 years old with 7 ± 2 scores of KIDMED test at baseline. Poor, intermediate, and good adherence rates for the Mediterranean diet were 9.0% (KIDMED index 0–3), 54.4% (KIDMED index 4–7) and 36.5% (KIDMED index 8–12), respectively. A short sleep duration was more prevalent in children with lower

Mediterranean diet adherence, while there were no significant differences in sex, age, height z-score, weight z-score, BMI z-score, SBP, DBP, long screen time or insufficient physical activity between the three groups. We observed significant differences between the three groups in wave 3 of the urinary β_2-MG excretion, and in waves 1–2 of the urinary MA excretion between the three groups (Table 1). The prevalence of transient renal tubular impairment (elevated β_2-MG > 0.2 mg/L) was 8.8% (168 of 1914), 15.9% (277 of 1746), 25.7% (425 of 1651) and 29.0% (481 of 1657) from wave 1 to 4, while that of transient renal glomerular impairment (elevated MA $\geq$ 20 mg/L) was 5.6% (107 of 1914), 5.5% (96 of 1746), 4.4% (72 of 1651) and 4.1% (68 of 1657), respectively.

3.2. Linear Associations of Kidney Impairment Indicators with KIDMED Scores

Linear mixed-effects models were generated to determine associations between the indicators of kidney impairment and the KIDMED scores. The results in the unadjusted model 1 and the adjusted model 2 (adjusting for sex, age, and BMI z-score) showed that the urinary MA excretion was associated with the KIDMED score, whereas the urinary β_2-MG excretion was not significantly associated with the KIDMED score. The adjusted model 3 showed that a higher urinary MA excretion was inversely associated with a higher KIDMED score (β = -0.216, 95%CI: -0.358, -0.074, p = 0.003), adjusting for sex, age, BMI z-score, SBP z-score, screen time, sleep duration and physical activity (Table 2).

3.3. Longitudinal Associations of Kidney Impairment with KIDMED Index

The results of the generalized linear mixed-effects models were consistent with the linear models. As the KIDMED index increased (with better Mediterranean diet adherence), the children were less likely to develop renal glomerular impairment. For renal tubular impairment, we did not observe a significant longitudinal association with the KIDMED index. In the unadjusted model 1 and the adjusted model 2 (adjusting for sex, age and BMI z-score), taking the KIDMED index 0–3 as a reference, the effect of the KIDMED index 4–7 on renal glomerular impairment showed marginal significance, while the effect of the KIDMED index 8–12 showed consistent significance. The results from model 3 showed the likelihood of renal glomerular impairment among children with intermediate Mediterranean diet adherence (aOR = 0.68, 95%CI: 0.47, 0.99, p = 0.044) and those with good Mediterranean diet adherence (aOR = 0.60, 95%CI: 0.40, 0.90, p = 0.014), adjusting for sex, age, BMI z-score, SBP z-score, screen time, sleep duration, and physical activity (Table 3).

3.4. Longitudinal Associations of Kidney Impairment with KIDMED Test

We further presented the frequency of children with kidney impairment in terms of each item of the KIDMED test (Table 4) and the association of kidney impairment with each item of KIDMED test (Figure 1). Adjusting for sex, age, BMI z-score, SBP z-score, screen time, sleep duration, and physical activity, renal tubular impairment was less likely to occur among those that consumed pasta or rice almost every day (aOR = 0.77, 95%CI: 0.61, 0.98, p = 0.033) and more likely to occur among those that skipped breakfast (aOR = 1.35, 95%CI: 1.00, 1.83, p = 0.050). Children who took sweets and candy several times every day had a greater risk to develop renal glomerular impairment (aOR = 1.59, 95%CI: 1.18, 2.15, p = 0.003) (Figure 1).

4. Discussion

The study used longitudinal data from the general pediatric population in China with a considerably large sample size to determine the associations between Mediterranean diet adherence (assessed using the KIDMED index) and transient kidney impairment. We found that children with a higher KIDMED index were less likely to develop renal glomerular impairment. Good adherence to the Mediterranean diet is associated with a 40% lower risk of having renal glomerular impairment, after adjusting for demographic and lifestyle covariates, including sex, age, BMI z-score, SBP z-score, screen time, sleep duration and physical activity. This finding underscores a major need for early development and adherence to a healthy eating pattern to enhance kidney health in children.

Poor, intermediate, and good adherence rates for the Mediterranean diet were 9.0% (KIDMED index 0–3), 54.4% (KIDMED index 4–7) and 36.5% (KIDMED index 8–12), respectively, while the proportions in the KIDMED-developed population of Spanish children aged 2–14 years were 2.9%, 48.6% and 48.5% [13]. The rate differences suggest slightly lower Mediterranean diet adherence compared with Spanish children. Studies on the assessment of Mediterranean diet adherence have mainly been conducted among Mediterranean European countries such as Spain, Greece, Chile, Portugal, Lebanon and Italy with mean KIDMED scores ranging from 5 to 8 [8], which is similar to our mean KIDMED scores of 7. These results demonstrate the good applicability of the KIDMED in the Chinese pediatric population. Moreover, we observed that a short sleep duration was more prevalent in the context of lower Mediterranean diet adherence. A cross-sectional study among 503 university students reported that good Mediterranean diet adherence was associated with overall good sleep quality and sleep composition, which was associated with the intake of foods rich in melatonin, tryptophan and phytonutrients [33].

For the linear association of the indicators of kidney impairment and the Mediterranean diet adherence, we observed a consistent longitudinal negative association between the KIDMED score and urinary MA excretion in the unadjusted analyses, with respect to the adjusted sex, age, BMI z-score, SBP z-score, screen time, sleep duration and physical activity. This result is consistent with a cross-sectional study of Greek adolescents that reported an inverse correlation between the KIDMED score and albumin to creatinine ratio (ACR, r = −0.111, p = 0.041) [11]. The kidney contributes to many physiological and biological mechanisms in the body [34], while microalbuminuria (renal glomerular impairment) is associated with systemic inflammation of the body and endothelial dysfunction [35]. Early damage to the glomerular filtration barrier results in impaired size and charge selectivity, leading to increased MA excretion [36]. One pathophysiological explanation for this negative correlation is that the Mediterranean diet has the endothelial protective and anti-atherosclerotic properties [11,37]. No significant longitudinal associations between the KIDMED score and urinary β_2-MG excretion were observed. The difference in the molecular mass of β_2-MG (11.8 kDa) and MA (67 kDa) partly explains the differences associated with the Mediterranean diet, i.e., the excretion of MA requires more extensive impairment [38]. Furthermore, studies have shown that increased glomerular albumin leakage stimulates proinflammatory and profibrotic signals that directly lead to tubulointerstitial impairment [36]. The exposure of renal tubules to albumin triggers inflammatory responses and toxic effects, leading to interstitial impairment, fibrosis and dysfunction, ultimately resulting in irreversible kidney impairment [36,39]. Therefore, Mediterranean diet adherence may play a broader and more important role in preventing kidney damage.

We further assessed the longitudinal association between transient kidney impairment and Mediterranean diet adherence and validated the stability of the inverse association between renal glomerular impairment and Mediterranean diet adherence. We found that children with moderate or good Mediterranean diet adherence had a 32% and 40% lower risk of glomerular impairment, respectively, compared with children with low Mediterranean diet adherence, adjusting for sex, age, BMI z-score, SBP z-score, screen time, sleep duration and physical activity level. This result provides quantitative evidence for evaluating the direct effect of the Mediterranean diet on glomerular impairment in children. We found a similar non-significant negative association between renal tubular impairment and Mediterranean diet adherence. A further breakdown of the 16-item KIDMED test analysis revealed that eating pasta or rice almost every day was associated with a 23% lower risk of renal tubular impairment. China is a country with a traditional grain-based diet, and most children eat grains more than once a day [4,12], which may be related to the protection of the renal tubular system. We also identified two unhealthy eating habits as leading causes of transient kidney impairment, namely, skipping breakfast increases the risk of tubular impairment by 35% and eating sweets and candies several times a day increases glomerular impairment by 59%. While we did not expect children in Beijing or in China to have absolute adherence to the Mediterranean diet, the KIDMED test involving many healthy compositions are similar to the traditional Chinese dietary pattern or other healthy diet patterns. In this perspective, using KIDMED advocates a healthy dietary pattern rather than promoting a Mediterranean diet per se. The early development of healthy eating patterns and the correction of unhealthy eating habits will help protect kidney health in the general pediatric population with respect to healthy lifestyles.

The major strength of this study was that we used the quantified tool of the KIDMED index to assess Mediterranean diet adherence, thus ensuring the stability of the results. However, our findings are based on the original version of the KIDMED questionnaire [13], and there was a revision in 2019 to emphasize the benefits of consuming whole fruit rather than fruit juice and whole grains [40]. We used appropriate statistical methods such as imputation methods [41] for missing data, linear mixed-effects models for linear associations and generalized linear mixed-effects models for longitudinal associations, to maximize longitudinal information and avoid reporting bias. In addition, we selected and adjusted for key covariates such as sex, age, BMI, SBP and lifestyle factors to assess the effect of the Mediterranean diet on transient kidney impairment. This study was limited by not considering the potential interaction of the two outcomes and other factors of renal function. Urine β_2-MG and MA excretion were tested by different machines across the study waves, but longitudinal data minimized these effects.

5. Conclusions

We present the longitudinal associations of transient kidney impairment with Mediterranean diet adherence in the general Chinese pediatric population. We found that children who adhered to a Mediterranean diet were associated with a reduced risk of transient renal glomerular impairment. Our findings underscore the necessity of the early development of healthy diet patterns and the correction of unhealthy eating habits in children to protect kidney health in the context of more relevant healthy lifestyles.

Author Contributions: Y.H. and D.H. conceptualized and designed the study. M.L., W.S., H.X., N.A. and J.Z. carried out the survey; M.L. performed statistical analysis of the data; H.X. checked data analysis process; M.L. drafted the manuscript. D.H. and Y.H. edited, helped interpret and revise the manuscript. All authors have read and agreed to the published version of the manuscript.

Funding: This research was funded by the National Natural Science Foundation of China (Y.H., grant No. 82073574), the Beijing Natural Science Foundation (Y.H., grant No. 7202009), and the Capital's Funds for Health Improvement and Research (Y.H., Grant No. 2022-1G-4262).

Institutional Review Board Statement: The protocol of the child cohort designed to study sensitization, puberty, obesity and cardiovascular risk (PROC) was reviewed and approved by the Ethics Committee of Capital Medical University (No. 2018SY82) that applied the guidance of the Declaration of Helsinki, later amendments, and comparable international ethical standards.

Informed Consent Statement: Informed consent was obtained from all participants' parents.

Data Availability Statement: The data that support the findings of this study are not publicly available but are available from the corresponding author on reasonable request.

Acknowledgments: We gratefully acknowledge the teachers of six primary school of Shunyi District, Beijing and the staff of the Shunyi District Center for Disease Control and Prevention and the Shunyi District Education Commission for their support and assistance to the field work. We appreciate all the constructive comments and suggestions from the reviewers.

Conflicts of Interest: The authors declare no conflict of interest related to this work. The funders had no role in the study design; in the collection, analyses, or interpretation of data; in the writing of the manuscript, or in the decision to publish the results.

References

1. Gomez-Delgado, F.; Katsiki, N.; Lopez-Miranda, J.; Perez-Martinez, P. Dietary habits, lipoprotein metabolism and cardiovascular disease: From individual foods to dietary patterns. *Crit. Rev. Food Sci. Nutr.* **2021**, *61*, 1651–1669. [CrossRef] [PubMed]
2. Zeraattalab-Motlagh, S.; Jayedi, A.; Shab-Bidar, S. Mediterranean dietary pattern and the risk of type 2 diabetes: A systematic review and dose-response meta-analysis of prospective cohort studies. *Eur. J. Nutr.* **2022**, *61*, 1735–1748. [CrossRef] [PubMed]
3. De Cosmi, V.; Mazzocchi, A.; Milani, G.P.; Agostoni, C. Dietary Patterns vs. Dietary Recommendations. *Front. Nutr.* **2022**, *9*, 883806. [CrossRef]
4. Chen, M.; Xiong, J.; Zhao, L.; Cheng, G.; Zhang, Q.; Ding, G. Epidemiological research progress on the relationship between children's dietary patterns and health. *Chin. J. Prev. Med.* **2022**, *56*, 139–145.
5. Guasch-Ferré, M.; Willett, W.C. The Mediterranean diet and health: A comprehensive overview. *J. Intern. Med.* **2021**, *290*, 549–566. [CrossRef]
6. Di Renzo, L.; Gualtieri, P.; Lorenzo, A.D. Diet, Nutrition and Chronic Degenerative Diseases. *Nutrients* **2021**, *13*, 1372. [CrossRef]
7. Bach, K.E.; Kelly, J.T.; Palmer, S.C.; Khalesi, S.; Strippoli, G.F.M.; Campbell, K.L. Healthy Dietary Patterns and Incidence of CKD: A Meta-Analysis of Cohort Studies. *Clin. J. Am. Soc. Nephrol.* **2019**, *14*, 1441–1449. [CrossRef]
8. Romero-Robles, M.A.; Ccami-Bernal, F.; Ortiz-Benique, Z.N.; Pinto-Ruiz, D.F.; Benites-Zapata, V.A.; Casas Patiño, D. Adherence to Mediterranean diet associated with health-related quality of life in children and adolescents: A systematic review. *BMC Nutr.* **2022**, *8*, 57. [CrossRef]
9. Martíncrespo-Blanco, M.C.; Varillas-Delgado, D.; Blanco-Abril, S.; Cid-Exposito, M.G.; Robledo-Martín, J. Effectiveness of an Intervention Programme on Adherence to the Mediterranean Diet in a Preschool Child: A Randomised Controlled Trial. *Nutrients* **2022**, *14*, 1536. [CrossRef]
10. Archero, F.; Ricotti, R.; Solito, A.; Carrera, D.; Civello, F.; Di Bella, R.; Bellone, S.; Prodam, F. Adherence to the Mediterranean Diet among School Children and Adolescents Living in Northern Italy and Unhealthy Food Behaviors Associated to Overweight. *Nutrients* **2018**, *10*, 1322. [CrossRef]
11. Mazaraki, A.; Tsioufis, C.; Dimitriadis, K.; Tsiachris, D.; Stefanadi, E.; Zampelas, A.; Richter, D.; Mariolis, A.; Panagiotakos, D.; Tousoulis, D.; et al. Adherence to the Mediterranean diet and albuminuria levels in Greek adolescents: Data from the Leontio Lyceum ALbuminuria (3L study). *Eur. J. Clin. Nutr.* **2011**, *65*, 219–225. [CrossRef]
12. Zhen, S.; Ma, Y.; Zhao, Z.; Yang, X.; Wen, D. Dietary pattern is associated with obesity in Chinese children and adolescents: Data from China Health and Nutrition Survey (CHNS). *Nutr. J.* **2018**, *17*, 68. [CrossRef]
13. Serra-Majem, L.; Ribas, L.; Ngo, J.; Ortega, R.M.; García, A.; Pérez-Rodrigo, C.; Aranceta, J. Food, youth and the Mediterranean diet in Spain. Development of KIDMED, Mediterranean Diet Quality Index in children and adolescents. *Public Health Nutr.* **2004**, *7*, 931–935. [CrossRef]
14. Galan-Lopez, P.; Domínguez, R.; Pihu, M.; Gísladóttir, T.; Sánchez-Oliver, A.J.; Ries, F. Evaluation of Physical Fitness, Body Composition, and Adherence to Mediterranean Diet in Adolescents from Estonia: The AdolesHealth Study. *Int. J. Environ. Res. Public Health* **2019**, *16*, 4479. [CrossRef]
15. Kim, R.J.; Lopez, R.; Snair, M.; Tang, A. Mediterranean diet adherence and metabolic syndrome in US adolescents. *Int. J. Food Sci. Nutr.* **2021**, *72*, 537–547. [CrossRef]
16. Martino, F.; Puddu, P.E.; Lamacchia, F.; Colantoni, C.; Zanoni, C.; Barillà, F.; Martino, E.; Angelico, F. Mediterranean diet and physical activity impact on metabolic syndrome among children and adolescents from Southern Italy: Contribution from the Calabrian Sierras Community Study (CSCS). *Int. J. Cardiol.* **2016**, *225*, 284–288. [CrossRef]
17. Dominguez-Riscart, J.; Buero-Fernandez, N.; Garcia-Zarzuela, A.; Morales-Perez, C.; Garcia-Ojanguren, A.; Lechuga-Sancho, A.M. Adherence to Mediterranean Diet Is Associated With Better Glycemic Control in Children With Type 1 Diabetes: A Cross-Sectional Study. *Front. Nutr.* **2022**, *9*, 813989. [CrossRef]
18. Della Corte, C.; Mosca, A.; Vania, A.; Alterio, A.; Iasevoli, S.; Nobili, V. Good adherence to the Mediterranean diet reduces the risk for NASH and diabetes in pediatric patients with obesity: The results of an Italian Study. *Nutrition* **2017**, *39-40*, 8–14. [CrossRef]
19. Agostinis-Sobrinho, C.; Santos, R.; Rosário, R.; Moreira, C.; Lopes, L.; Mota, J.; Martinkenas, A.; García-Hermoso, A.; Correa-Bautista, J.E.; Ramírez-Vélez, R. Optimal Adherence to a Mediterranean Diet May Not Overcome the Deleterious Effects of Low Physical Fitness on Cardiovascular Disease Risk in Adolescents: A Cross-Sectional Pooled Analysis. *Nutrients* **2018**, *10*, 815. [CrossRef]
20. Wärnberg, J.; Pérez-Farinós, N.; Benavente-Marín, J.C.; Gómez, S.F.; Labayen, I.; Zapico, A.G.; Gusi, N.; Aznar, S.; Alcaraz, P.E.; González-Valeiro, M.; et al. Screen Time and Parents' Education Level Are Associated with Poor Adherence to the Mediterranean Diet in Spanish Children and Adolescents: The PASOS Study. *J. Clin. Med.* **2021**, *10*, 795. [CrossRef]
21. Lampropoulou, M.; Chaini, M.; Rigopoulos, N.; Evangeliou, A.; Papadopoulou-Legbelou, K.; Koutelidakis, A.E. Association Between Serum Lipid Levels in Greek Children with Dyslipidemia and Mediterranean Diet Adherence, Dietary Habits, Lifestyle and Family Socioeconomic Factors. *Nutrients* **2020**, *12*, 1600. [CrossRef] [PubMed]
22. Wu, H.Y.; Huang, J.W.; Peng, Y.S.; Hung, K.Y.; Wu, K.D.; Lai, M.S.; Chien, K.L. Microalbuminuria screening for detecting chronic kidney disease in the general population: A systematic review. *Ren. Fail.* **2013**, *35*, 607–614. [CrossRef] [PubMed]
23. Kadam, P.; Yachha, M.; Srivastava, G.; Pillai, A.; Pandita, A. Urinary beta-2 microglobulin as an early predictive biomarker of acute kidney injury in neonates with perinatal asphyxia. *Eur. J. Pediatrics* **2022**, *181*, 281–286. [CrossRef]

24. Li, M.; Shu, W.; Zunong, J.; Amaerjiang, N.; Xiao, H.; Li, D.; Vermund, S.H.; Hu, Y. Predictors of non-alcoholic fatty liver disease in children. *Pediatric Res.* **2021**, 1–9. [CrossRef]
25. Li, M.; Shu, W.; Amaerjiang, N.; Xiao, H.; Zunong, J.; Vermund, S.H.; Huang, D.; Hu, Y. Interaction of Hydration Status and Physical Activity Level on Early Renal Damage in Children: A Longitudinal Study. *Front. Nutr.* **2022**, *9*, 910291. [CrossRef]
26. Li, M.; Amaerjiang, N.; Li, Z.; Xiao, H.; Zunong, J.; Gao, L.; Vermund, S.H.; Hu, Y. Insufficient Fruit and Vegetable Intake and Low Potassium Intake Aggravate Early Renal Damage in Children: A Longitudinal Study. *Nutrients* **2022**, *14*, 1228. [CrossRef]
27. Amaerjiang, N.; Li, M.; Xiao, H.; Zunong, J.; Li, Z.; Huang, D.; Vermund, S.H.; Pérez-Escamilla, R.; Jiang, X.; Hu, Y. Dehydration Status Aggravates Early Renal Impairment in Children: A Longitudinal Study. *Nutrients* **2022**, *14*, 335. [CrossRef]
28. Barton, K.T.; Kakajiwala, A.; Dietzen, D.J.; Goss, C.W.; Gu, H.; Dharnidharka, V.R. Using the newer Kidney Disease: Improving Global Outcomes criteria, beta-2-microglobulin levels associate with severity of acute kidney injury. *Clin. Kidney J.* **2018**, *11*, 797–802. [CrossRef]
29. Shatat, I.F.; Qanungo, S.; Hudson, S.; Laken, M.A.; Hailpern, S.M. Changes in Urine Microalbumin-to-Creatinine Ratio in Children with Sickle Cell Disease over Time. *Front. Pediatrics* **2016**, *4*, 106. [CrossRef]
30. Owens, J.A.; Spirito, A.; McGuinn, M. The Children's Sleep Habits Questionnaire (CSHQ): Psychometric properties of a survey instrument for school-aged children. *Sleep* **2000**, *23*, 1043–1051. [CrossRef]
31. Amaerjiang, N.; Xiao, H.; Zunong, J.; Shu, W.; Li, M.; Pérez-Escamilla, R.; Hu, Y. Sleep disturbances in children newly enrolled in elementary school are associated with parenting stress in China. *Sleep Med.* **2021**, *88*, 247–255. [CrossRef]
32. Li, H.; Chen, P.; Zhuang, J. Revision and reliability validity assessment of children's leisure activities study survey. *Chin. J. Sch. Health* **2011**, *32*, 268–270.
33. Naja, F.; Hasan, H.; Khadem, S.H.; Buanq, M.A.; Al-Mulla, H.K.; Aljassmi, A.K.; Faris, M.E. Adherence to the Mediterranean Diet and Its Association With Sleep Quality and Chronotype Among Youth: A Cross-Sectional Study. *Front. Nutr.* **2021**, *8*, 805955. [CrossRef]
34. Naber, T.; Purohit, S. Chronic Kidney Disease: Role of Diet for a Reduction in the Severity of the Disease. *Nutrients* **2021**, *13*, 3277. [CrossRef]
35. Swärd, P.; Tofik, R.; Bakoush, O.; Torffvit, O.; Nilsson, P.M.; Christensson, A. Patterns of urinary albumin and IgM associate with markers of vascular ageing in young to middle-aged individuals in the Malmö offspring study. *BMC Cardiovasc. Disord.* **2020**, *20*, 358. [CrossRef]
36. Lambers Heerspink, H.J.; Gansevoort, R.T. Albuminuria Is an Appropriate Therapeutic Target in Patients with CKD: The Pro View. *Clin. J. Am. Soc. Nephrol.* **2015**, *10*, 1079–1088. [CrossRef]
37. Leone, A.; Fernández-Montero, A.; de la Fuente-Arrillaga, C.; Martínez-González, M.; Bertoli, S.; Battezzati, A.; Bes-Rastrollo, M. Adherence to the Mediterranean Dietary Pattern and Incidence of Nephrolithiasis in the Seguimiento Universidad de Navarra Follow-up (SUN) Cohort. *Am. J. Kidney Dis. Off. J. Natl. Kidney Found.* **2017**, *70*, 778–786. [CrossRef]
38. Djukanović, L.; Djordjević, V.; Ležaić, V.; Cukuranović, R.; Marić, I.; Bukvić, D.; Marinković, J.; Cukuranović, J.; Rajić, M.; Stefanović, V. Urinary protein patterns in patients with Balkan endemic nephropathy. *Int. Urol. Nephrol.* **2013**, *45*, 1661–1669. [CrossRef]
39. Chen, S.; Chen, J.; Li, S.; Guo, F.; Li, A.; Wu, H.; Chen, J.; Pan, Q.; Liao, S.; Liu, H.F.; et al. High-Fat Diet-Induced Renal Proximal Tubular Inflammatory Injury: Emerging Risk Factor of Chronic Kidney Disease. *Front. Physiol.* **2021**, *12*, 786599. [CrossRef]
40. Altavilla, C.; Caballero-Pérez, P. An update of the KIDMED questionnaire, a Mediterranean Diet Quality Index in children and adolescents. *Public Health Nutr.* **2019**, *22*, 2543–2547. [CrossRef]
41. Blazek, K.; van Zwieten, A.; Saglimbene, V.; Teixeira-Pinto, A. A practical guide to multiple imputation of missing data in nephrology. *Kidney Int.* **2021**, *99*, 68–74. [CrossRef]

Article

Validation of the Diet Quality Questionnaire in Chinese Children and Adolescents and Relationship with Pediatric Overweight and Obesity

Huan Wang [1,2], Anna W. Herforth [3,4], Bo Xi [2] and Zhiyong Zou [1,*]

1   National Health Commission Key Laboratory of Reproductive Health, Institute of Child and Adolescent Health, School of Public Health, Peking University, Beijing 100191, China
2   Department of Epidemiology, School of Public Health, Cheeloo College of Medicine, Shandong University, Jinan 250012, China
3   Department of Global Health and Population, Harvard T.H. Chan School of Public Health, Boston, MA 02115, USA
4   Division of Human Nutrition and Health, Wageningen University and Research, P.O. Box 230, 6700 AE Wageningen, The Netherlands
*   Correspondence: harveyzou2002@bjmu.edu.cn

Abstract: The low-burden Diet Quality Questionnaire (DQQ) has been developed to rapidly assess diet quality globally. Poor diet is often correlated with body size, and certain dietary risk factors can result in overweight and obesity. We aimed to examine the extent to which the DQQ captured food group consumption among children and adolescents in China, and to understand the association of several new indicators of diet quality scores derived from the DQQ with overweight and obesity, using the 2011 wave of the China Health and Nutrition Survey. The DQQ questions are constructed using sentinel foods—that is, food items that are intended to capture a large proportion of the population consuming the food groups. The overall Global Dietary Recommendations (GDR) score, GDR-Healthy score, and GDR-Limit score are novel indicators of diet quality that reflect dietary risk factors for non-communicable diseases derived from the DQQ questions. Multivariable logistic regression analysis was used to examine the associations of the GDR scores with overweight and obesity in the sample. The DQQ questions captured over 95% of children who consumed the food groups. Additionally, we found that the GDR-Limit score was positively associated with general obesity (odds ratio (OR) = 1.43, 95% confidence interval (CI): 1.17–1.74) and abdominal obesity (OR = 1.22, 95% CI: 1.05–1.43), whereas the overall GDR score was negatively related to general obesity (OR = 0.85, 95% CI: 0.74–0.97). The low-burden DQQ could be a valid tool to assess diet quality for the Chinese pediatric population aged 7–18 years. Poor diet quality, as determined by the GDR-Limit score, is associated with the increased risk of obesity in Chinese children and adolescents.

Keywords: Diet Quality Questionnaire; global dietary recommendations; overweight; obesity; children and adolescents

Citation: Wang, H.; Herforth, A.W.; Xi, B.; Zou, Z. Validation of the Diet Quality Questionnaire in Chinese Children and Adolescents and Relationship with Pediatric Overweight and Obesity. *Nutrients* 2022, 14, 3551. https://doi.org/10.3390/nu14173551

Academic Editor: Peter Pribis

Received: 4 May 2022
Accepted: 27 May 2022
Published: 29 August 2022

Publisher's Note: MDPI stays neutral with regard to jurisdictional claims in published maps and institutional affiliations.

Copyright: © 2022 by the authors. Licensee MDPI, Basel, Switzerland. This article is an open access article distributed under the terms and conditions of the Creative Commons Attribution (CC BY) license (https://creativecommons.org/licenses/by/4.0/).

1. Introduction

Diets have shifted dramatically and rapidly in China in recent decades with socio-economic development and urbanization [1,2]. Between 1991 and 2009, higher daily fat intake, lower daily protein intake, and an increasing percentage of energy from fat mainly characterized the diets of Chinese children and adolescents aged 7–17 years [3], which subsequently contributed to the prevalence of pediatric overweight and obesity [3]. More recently, a nationwide study of more than one million Chinese school-aged children and adolescents showed that the mean prevalence of overweight and obesity increased markedly from 5.3% in 1995 to 20.5% in 2014 with economic development, highlighting an additional focus on healthy diets and physical activity [4]. Childhood obesity could not only track into adulthood [5] but also correlate with non-communicable diseases (NCDs) and mortality in early adulthood [6], making pediatric obesity a pressing concern in the prevention of obesity and obesity-related adverse outcomes later in life.

Although increasing work has been conducted on diet quality and obesity, diet quality scores do not always correlate with obesity strongly or in expected directions. The Healthy Eating Index (HEI) was reported to be negatively associated with obesity, whereas diversity-based indices were positively associated with obesity in adults [7]. A systematic review and meta-analysis found no significant associations of dietary diversity score with overweight, obesity, or abdominal obesity in either adults or children [8]. Another systemic review provided convincing evidence on the null association between the HEI and body mass index (BMI) in adults and children [9]. These inconsistent findings might be attributed to the considerable heterogeneity of diet assessment or diverse populations with different demographics and socio-economic contexts. For example, diet quality was mainly assessed from a single 24 h recall, multiple 24 h recalls, or a food frequency questionnaire, and many diet quality indicators were country-specific [7–10], which complicates diet assessment and limits widespread use. Moreover, some diet quality scores, such as diet diversity scores, are designed to measure specific aspects of diet quality, such as nutrient adequacy, rather than total diet quality or risk factors for obesity or NCDs [11]. When selecting and applying indicators, it is important to use them for the purpose intended. Few indicators have been developed that specifically relate to diet-related NCDs.

Measuring diet quality in the pediatric population has additional challenges and is less well studied. Many diet quality indices in the pediatric population have been developed and modified, yet few indices were validated [12,13]. To better monitor healthy diets globally, simple and feasible approaches are highly required to be uniform and standardized, which promotes comparability over time and across countries [14].

Recently, a low-burden Diet Quality Questionnaire (DQQ) entirely based on 29 food groups has been developed to collect dietary data, which takes only five minutes to administer using "yes/no" questions about foods or drinks and is easily understood by respondents [11,15,16]. The novel DQQ is designed to rapidly assess diet quality in populations and has been adapted for more than 100 countries. The DQQ has been developed for the general population but so far has been validated and implemented only for adults (aged 15 years and older). The DQQ questions are constructed using sentinel foods—that is, food items which are intended to capture a large proportion of the population consuming the food groups. The DQQ sentinel food items have not yet been validated for use in populations younger than 15 years of age.

The DQQ data can be used to construct several diet quality indicators at the population level, such as the Minimum Diet Diversity for Women, Food Group Diversity Score, and Global Dietary Recommendations (GDR) scores [15]. The GDR scores are constructed to reflect dietary risk factors for NCDs [15]. Thus, these scores may be plausibly associated with obesity, because dietary risks for NCDs and for obesity are similar. In particular, we would expect that the GDR-Limit score may be associated with obesity, because it captures the consumption of food groups that are generally energy-dense and high in fat and/or sugar.

The present study has two aims, using data from the 2011 wave of the China Health and Nutrition Survey (CHNS). Firstly, we aim to validate the DQQ sentinel foods for Chinese children and adolescents aged 7–18 years. Secondly, we examine the associations of the GDR scores with overweight and obesity. The motivation of these analyses is to understand whether the DQQ could be applied for pediatric use in China, and to understand the relationship of GDR scores (indicators derived from the DQQ) with obesity in children and adolescents.

2. Materials and Methods

2.1. Study Population

Data were obtained from the 2011 wave of the CHNS, which aimed to understand the interplay of socio-economic transition and nutrition and health-related outcomes in

China [17]. The CHNS is an international collaborative project between the Carolina Population Center at the University of North Carolina at Chapel Hill and the National Institute for Nutrition and Health at the Chinese Center for Disease Control and Prevention. This study was approved by corresponding institutional review committees and all participants provided written informed consent for inclusion before they participated in the survey [18]. Further information on the survey design and the publicly available datasets can be found in the cohort profile [17] and at the CHNS website [19].

There were 15,725 participants in the 2011 wave of the CHNS, and we included all children and adolescents (hereafter referred to as "children") aged 7–18 years with complete records on diet and anthropometrics (*n* = 1506). After further exclusion of those with implausible dietary intakes (carbohydrate intake $\geq$1500 g/day, calcium intake $\geq$3000 mg/day, or sodium intake $\geq$30 g/day), a total of 1501 children were included in this cross-sectional study (Figure 1).

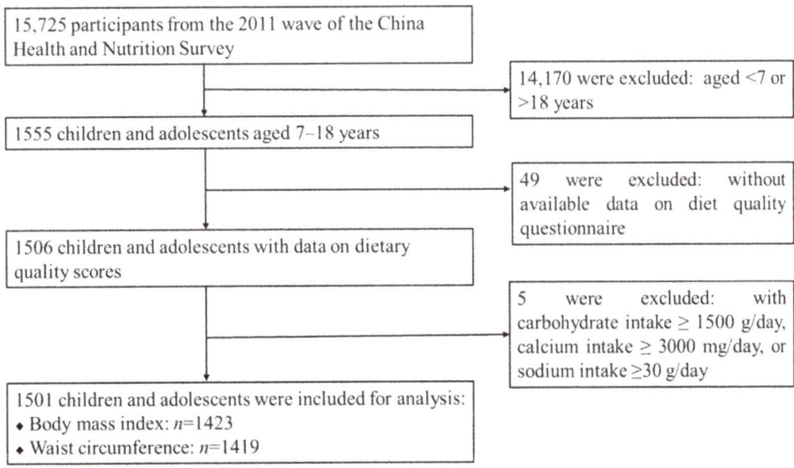

Figure 1. Flow chart of the inclusion/exclusion of participants.

2.2. Dietary Data Collection

The quantitative dietary data were collected using 24 h dietary recall by trained investigators for three consecutive days that were randomly allocated from Monday to Sunday. For children younger than 12 years, someone who prepared the food for the household was asked to recall the children's dietary intakes. More details on the dietary interview have been described elsewhere [1]. Nutrient intakes were calculated mainly using the 2009 Chinese food composition database [20], complemented by the 2018/2019 Chinese food composition databases [21,22].

2.3. Dietary Assessment

Food intake of the first day from the consecutive three 24 h dietary recalls was coded into 29 food groups following the DQQ tool. The DQQ has been adapted to represent foods in the Chinese context that could reliably capture the food group consumption for the Chinese population, and the identification of sentinel food items for China has been described elsewhere [16]. The China DQQ and further information are available at the Global Diet Quality Project website [15].

The GDR-Healthy score, GDR-Limit score, and overall GDR score were constructed from the dietary intake data: (1) GDR-Healthy score: reflecting five global recommendations on health-protective foods for healthy diets (fruits and vegetables, beans and other legumes, nuts and seeds, whole grains, and dietary fiber); (2) GDR-Limit score: reflecting six global recommendations on dietary components to limit (total fat, saturated fat, dietary sodium, free sugars, processed meat, and unprocessed red meat); (3) overall GDR score: subtracts the GDR-Limit score from the GDR-Healthy score, and reflects all 11 recommendations. The GDR score and its subcomponents were validated against quantitative intakes aligned with each of the recommendations. Specific food groups included for the GDR-Healthy score, GDR-Limit score, and overall GDR score are presented in Table S1. The GDR-Healthy and GDR-Limit scores ranged from 0 to 9 points and the overall GDR score ranged from −9 to 9 points. A lower overall GDR score, lower GDR-Healthy score, and higher GDR-Limit score indicate poorer diet quality [11].

2.4. Physical Examination

Height, weight, and waist circumference (WC) were measured by trained field investigators following standardized procedures, as recommended by the World Health Organization (WHO) [23]. Height (accurate to 0.1 cm) and weight (accurate to 0.1 kg) were measured using calibrated weighing and height scales when participants stood straight in light clothes and without shoes. BMI was calculated as weight (kg) divided by height squared (m^2). WC (accurate to 0.1 cm) was measured at the midpoint between the lowest rib and the iliac crest using non-elastic tapes in the standing position after normal expiration.

2.5. Definitions of Overweight and Obesity

Overweight and general obesity were defined using the sex- and age-specific BMI cut-offs for screening among children and adolescents aged 2–18 years, released by the National Health and Family Planning Commission of China in 2018 [24]. Abdominal obesity was defined as WC values ≥ the sex- and age-specific 90th percentiles for Chinese children [25]. Additionally, according to the WHO BMI for age z-scores [26] and the international WC cut-offs [27], overweight, general obesity, and abdominal obesity were re-defined to test the robustness of the results.

2.6. Statistical Analysis

The sentinel food analysis was conducted by ranking all foods in each food group in descending order according to the cumulative frequency of food consumption. Additionally, the percentages of the consumption of sentinel food items compared with all food items in the respective 29 food groups were calculated.

Continuous variables were expressed as means ± standard deviations (age, weight, height, BMI, and WC) or medians ± interquartile ranges (dietary intakes and urbanicity index), and categorical variables were shown as numbers (percentages). The independent *t*-test, Wilcoxon rank test, and Chi-square test were used to compare differences in characteristics between boys and girls.

The Cochran–Armitage test was used to examine trends in the prevalence of overweight and obesity across diet quality scores. The multivariable logistic regression analyses were used to evaluate the associations of the GDR scores with overweight and obesity; adjusted odds ratios (OR) and 95% confidence intervals (CI) were estimated after adjusting for sex, age, residence, and urbanicity index. Stratified analyses by sex (boys vs. girls), age (7–12 vs. 13–18 years), and residence (rural vs. urban) were conducted. All analyses were performed using SAS 9.4 (SAS Institute, Cary, NC, USA). Two-sided *p* values < 0.05 were considered statistically significant.

3. Results

3.1. Participant Characteristics

A total of 1501 children from the 2011 wave of the CHNS with a mean age of 11.7 years were included in this study, 59.9% of whom lived in rural areas. Approximately one tenth of children were overweight (11.0%) or obese (10.0%) and approximately one fifth had abdominal obesity (20.1%). Boys (51.7% of the sample) had significantly higher levels of weight, height, BMI, and WC than girls; boys also consumed more energy, carbohydrates, protein, and fat per day than girls. Boys were more likely to be overweight or obese than girls, whereas there was no significant difference in the prevalence of abdominal obesity between boys and girls (Table 1).

Table 1. Characteristics of participants by sex.

Characteristics	Total (n = 1501)	Boys (n = 776)	Girls (n = 725)	p Value [b]
Age, years	11.72 ± 3.30	11.75 ± 3.31	11.69 ± 3.28	0.725
Weight, kg	41.56 ± 19.76	43.13 ± 16.38	39.89 ± 22.69	0.002
Height, cm	147.13 ± 17.29	149.24 ± 18.59	144.90 ± 15.50	<0.001
BMI, kg/m^2	18.42 ± 3.93	18.66 ± 3.88	18.17 ± 3.96	0.020
WC, cm	65.06 ± 14.01	66.21 ± 12.64	63.85 ± 15.23	0.002
Energy, kcal/day [a]	1525.20 ± 858.45	1657.62 ± 926.57	1412.82 ± 781.48	<0.001
Carbohydrate, g/day [a]	214.04 ± 138.98	234.58 ± 142.68	193.15 ± 122.90	<0.001
Protein, g/day [a]	53.73 ± 34.41	59.02 ± 35.15	49.63 ± 31.45	<0.001
Fat, g/day [a]	49.65 ± 45.55	52.92 ± 46.20	45.32 ± 44.05	<0.001
Urbanicity index [a]	73.84 ± 35.76	71.29 ± 35.79	76.30 ± 36.06	0.366
Residence, n (%)				0.935
Rural	899 (59.89)	464 (59.79)	435 (60.00)	
Urban	602 (40.11)	312 (40.21)	290 (40.00)	
BMI categories, n (%)				0.001
Non-overweight/obesity	1125 (79.06)	552 (75.41)	573 (82.92)	
Overweight	156 (10.96)	100 (13.66)	56 (8.10)	
General obesity	142 (9.98)	80 (10.93)	62 (8.97)	
WC categories, n (%)				0.651
Non-obesity	1134 (79.92)	586 (80.38)	548 (79.42)	
Abdominal obesity	285 (20.08)	143 (19.62)	142 (20.58)	

BMI, body mass index; WC, waist circumference. Continuous variables are expressed as means ± standard deviations, and categorical variables as numbers (percentages). [a] Continuous variables are expressed as medians ± interquartile ranges. [b] Differences in characteristics between boys and girls were tested by independent t-test, Wilcoxon rank test, or Chi-square test.

3.2. DQQ Sentinel Food Validation for Children and Adolescents

In almost every food group, the sentinel food items captured over 95% of children aged 7–18 years who consumed the food groups, suggesting that the DQQ was a valid tool to collect the most common food consumption groups of the Chinese children (Figure 2). For example, people who consumed the sentinel food items (rice, noodles, steamed buns, and bread) accounted for 99.1% of those who consumed grains as a staple food (Table S2). For the vitamin A-rich fruits group, persimmon, mango, papaya, cantaloupe, and hawthorn captured 96.9% of children who consumed this food group; however, the sentinel food items only included the first four foods (capturing 93.8% of children). The specific sentinel food items for each food group are shown in Table S2.

3.3. Associations of Diet Quality Scores with Overweight and Obesity

The prevalence of overweight, general obesity, and abdominal obesity is presented in Figures 3 and 4. Overall, the prevalence of general obesity was higher as the GDR-Limit score increased but was lower as the overall GDR score increased (both p for trend < 0.05). The observed trend in the overall GDR score appears to be driven by the GDR-Limit score. The prevalence of abdominal obesity also gradually increased with an increment in the GDR-Limit score (p for trend < 0.05). Detailed numbers of overweight and obese children who had each GDR score are shown in Table S3.

After adjustment for sex, age, residence, and urbanicity index, the continuous GDR-Limit score was positively associated with general obesity (OR = 1.43, 95% CI: 1.17–1.74) and abdominal obesity (OR = 1.22, 95% CI: 1.05–1.43), whereas the continuous overall GDR score was negatively associated with general obesity (OR = 0.85, 95% CI: 0.74–0.97) (Table 2). As the diet quality scores were categorized according to their distribution, compared with children with a zero-point GDR-Limit score, those with a GDR-Limit score $\geq$2 point had increased odds of general obesity (OR = 2.66, 95% CI: 1.53–4.62) and abdominal obesity (OR = 1.60, 95% CI: 1.07–2.40) (Table 2).

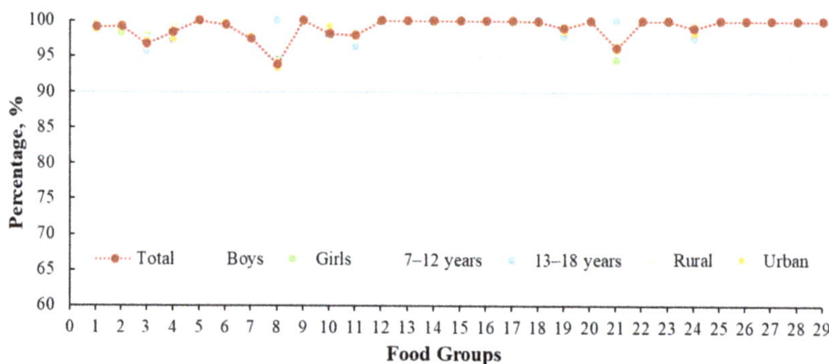

Figure 2. Percentage (%) of the consumption of sentinel food items compared with all food items in respective 29 food groups by sex, age, and residence. Note: 1: Staple foods made from grains; 2: Whole grains; 3: White root/tubers; 4: Legumes; 5: Vitamin A-rich orange vegetables; 6: Dark green leafy vegetables; 7: Other vegetables; 8: Vitamin A-rich fruits; 9: Citrus; 10: Other fruits; 11: Grain-baked sweets; 12: Other sweets; 13: Eggs; 14: Cheese; 15: Yogurt; 16: Processed meats; 17: Unprocessed red meat (ruminant); 18: Unprocessed red meat (nonruminant); 19: Poultry; 20: Fish and seafood; 21: Nuts and seeds; 22: Packaged ultra-processed salty snacks; 23: Instant noodles; 24: Deep fried foods; 25: Fluid milk; 26: Sweetened tea/coffee/milk drinks; 27: Fruit juice; 28: Sugar-sweetened beverages (sodas); 29: Fast food.

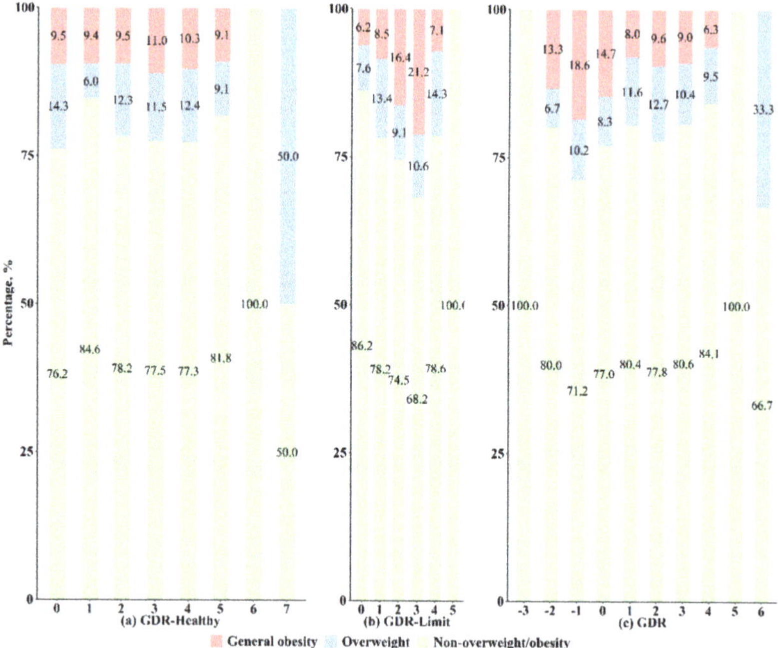

Figure 3. Prevalence of overweight and obesity by Global Dietary Recommendations (GDR) scores.

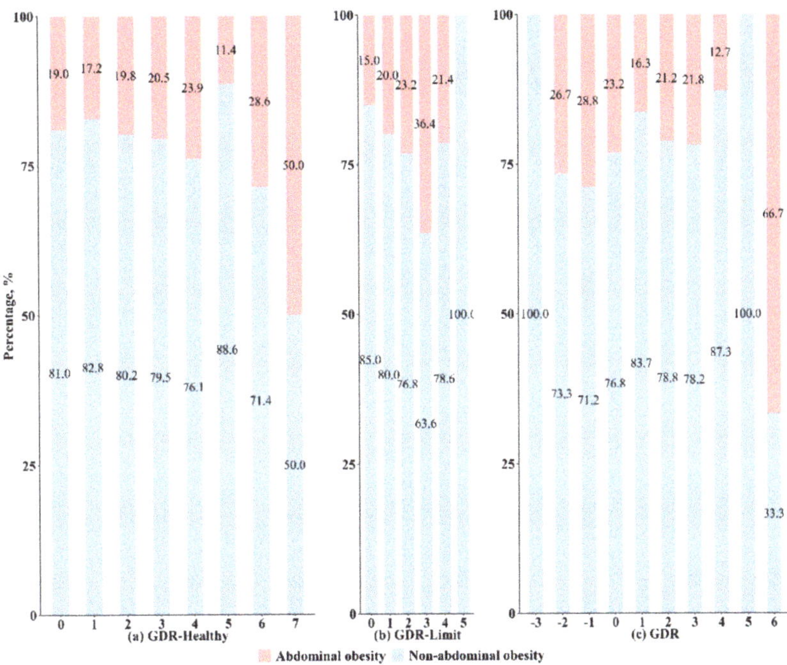

Figure 4. Prevalence of abdominal obesity by Global Dietary Recommendations (GDR) scores.

Table 2. Associations of Global Dietary Recommendations scores with overweight and obesity.

Scores	Overweight		General Obesity		Abdominal Obesity	
	OR (95% CI)	*p* Value	OR (95% CI)	*p* Value	OR (95% CI)	*p* Value
GDR-Healthy						
Continuous	1.07 (0.92–1.24)	0.393	0.98 (0.83–1.15)	0.812	1.03 (0.91–1.16)	0.644
Categories						
≤1	1.00 (Ref.)		1.00 (Ref.)		1.00 (Ref.)	
2	1.86 (1.05–3.29)	0.034	1.01 (0.59–1.73)	0.973	1.13 (0.76–1.69)	0.553
≥3	1.72 (0.99–2.99)	0.056	1.05 (0.63–1.73)	0.863	1.17 (0.80–1.71)	0.419
GDR-Limit						
Continuous	1.03 (0.84–1.27)	0.797	1.43 (1.17–1.74)	<0.001	1.22 (1.05–1.43)	0.012
Categories						
0	1.00 (Ref.)		1.00 (Ref.)		1.00 (Ref.)	
1	1.69 (1.05–2.70)	0.030	1.33 (0.78–2.26)	0.296	1.23 (0.86–1.77)	0.262
≥2	1.21 (0.69–2.11)	0.516	2.66 (1.53–4.62)	0.001	1.60 (1.07–2.40)	0.022
Overall GDR						
Continuous	1.04 (0.91–1.18)	0.572	0.85 (0.74–0.97)	0.016	0.94 (0.86–1.04)	0.244
Categories						
<0	1.00 (Ref.)		1.00 (Ref.)		1.00 (Ref.)	
0	0.87 (0.36–2.14)	0.769	0.89 (0.44–1.80)	0.750	0.81 (0.46–1.44)	0.480
≥1	1.17 (0.59–2.70)	0.554	0.53 (0.29–0.98)	0.043	0.63 (0.39–1.04)	0.070

CI, confidence interval; GDR, global dietary recommendations; OR, odds ratio; Ref, reference group. Logistic regression analyses were used to calculate the odds ratios and 95% confidence intervals with adjustment for sex, age, residence, and urbanicity index.

Stratified analyses by sex (boys vs. girls), age (7–12 vs. 13–18 years), and residence (rural vs. urban) are shown in Figure 5, and similar positive associations between the GDR-Limit score and obesity were found in the subgroups (Figure 5 and Table S4). In the sensitivity analysis, after redefining overweight and obesity, the GDR-Limit score was also positively associated with obesity (Table S5).

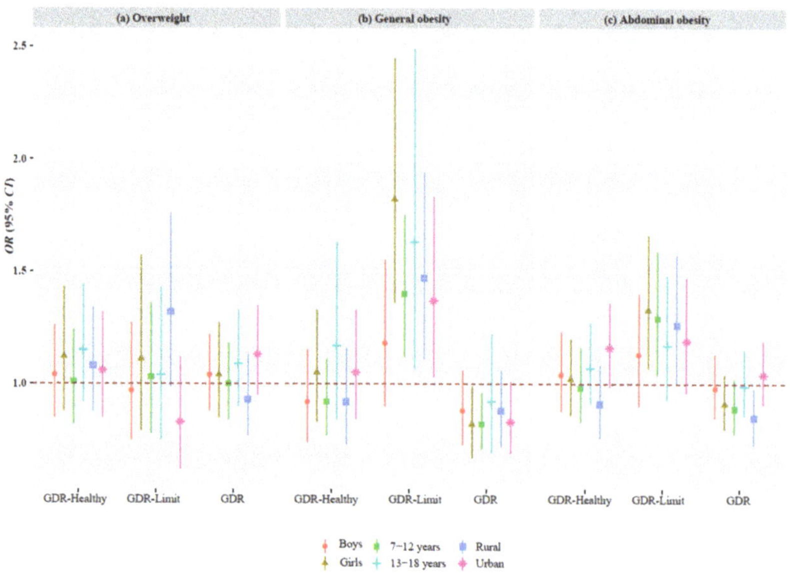

Figure 5. Associations of the Global Dietary Recommendations scores with overweight and obesity by sex, age, and residence. Note: CI, confidence interval; GDR, global dietary recommendations; OR, odds ratio. Logistic regression analyses were used to calculate the odds ratios and 95% confidence intervals with adjustment for sex, age, residence, and urbanicity index.

4. Discussion

In this national cross-sectional analysis of 1501 children and adolescents aged 7–18 years from the 2011 wave of the CHNS, we found that the sentinel foods in the DQQ captured over 95% of children who consumed the food groups, indicating that it is a valid tool for diet quality assessment in this age group. The GDR-Limit score (derived from the DQQ questions) was positively associated with general and abdominal obesity, whereas the overall GDR score was negatively associated with general obesity. The present study suggests that the DQQ tool and new indicators of diet quality are valid for Chinese children and adolescents, and the poor diet quality determined by the GDR-Limit score is associated with the increased risk of obesity.

Our finding that the sentinel food items captured over 95% of the food consumption of children suggests that the China-adapted DQQ tool has the potential to assess the diet quality in the Chinese pediatric population, aligned with global diet quality frameworks in the general population. Although there are several diet quality indicators for Chinese children and adolescents, such as the Chinese Children Dietary Index [28], Chinese Healthy Eating Index [29], and Chinese Healthy Eating Index for School-Age Children [30], these indices rely on 24 h diet recalls and food–nutrient conversion tables, which is time-consuming and poses a heavy burden for both investigators and interviewees. Additionally, these indices only reflect adherence to the Dietary Guidelines for Chinese residents, impeding the comparison of diet quality across countries. Given the increasing attention to adolescent nutrition globally [31,32], aligning with the global framework is useful. Although existing global platforms targeting adolescents are scarce, there are potential global indicators for individuals aged 2–19 years, including the GDR scores calculated easily from the DQQ [14]. Therefore, our study results support the application of the DQQ tool in China and accelerate the process of monitoring healthy diets for children and adolescents at national and global levels.

In our study, we used the GDR scores from the China-adapted DQQ to assess diet quality, which reflects adherence to the WHO Global Dietary Recommendations [11]. We observed higher odds of obesity with a higher GDR-Limit score, and lower odds with a higher overall GDR score. On the other hand, the GDR-Healthy score was not significantly associated with obesity; thus, our findings support the importance of reducing intakes of unhealthy foods as the most important factor for reducing the risk of obesity. The overall GDR score was aligned with the 11 Global Dietary Recommendations on fruits and vegetables, beans and other legumes, nuts and seeds, whole grains, dietary fiber, total fat, saturated fat, dietary sodium, free sugar, processed meat, and unprocessed red meat [11]. Importantly, WHO proposes these recommendations generally based on evidence related to diet-related NCD risks [33–35]; therefore, the GDR scores obtained from the DQQ are promising diet quality indicators related to the risk of obesity and other diet-related NCDs in the Chinese population.

This is the first study to validate the DQQ sentinel food items in Chinese children and adolescents. It is also the first study to assess the association between the GDR scores and overweight and obesity in a pediatric population. Several limitations of this study should be noted. First, we only obtained data from the 2011 wave of the CHNS; it is possible that the relationship between diet quality and obesity has changed in the last 10 years. Second, although age, sex, residence, and a comprehensive urbanicity index (proxy of modernization and urbanization) were accounted for in our analyses, residual confounding factors cannot be ruled out, such as physical activity, sedentary behavior, and pubertal development status. Third, although this study finds that the DQQ sentinel foods are valid for children and adolescents, further research is warranted to validate the application of the DQQ in terms of the ability of children and adolescents to reliably report their diet. In general, dietary recall and reporting are challenging for the pediatric age range [36].

5. Conclusions

The DQQ sentinel food items could be applied for use in populations aged 7–18 years. The GDR-Limit score is associated with the increased risk of obesity, and the low-burden DQQ could be a valid tool to assess diet quality for Chinese children and adolescents.

Supplementary Materials: The following are available online at https://www.mdpi.com/article/10.3390/nu14173551/s1, Table S1: Global Dietary Recommendations scores constructed from the Diet Quality Questionnaire; Table S2: Percentage (%) of the consumption of sentinel food items compared with all food items in respective 29 food groups by sex, age, and residence; Table S3: Number and percentage (%) of the overweight and obesity by the Global Dietary Recommendations (GDR) scores; Table S4: Subgroup analysis of associations between the Global Dietary Recommendations scores and overweight and obesity by sex, age, and residence; Table S5: Sensitivity analysis of associations between the Global Dietary Recommendations scores and overweight and obesity.

Author Contributions: Conceptualization, A.W.H., Z.Z., B.X. and H.W.; methodology, Z.Z., B.X. and H.W.; software, H.W.; validation, Z.Z.; formal analysis, H.W.; investigation, Z.Z.; resources, Z.Z.; data curation, Z.Z.; writing—original draft preparation, Z.Z. and H.W.; writing—review and editing, H.W., A.W.H., B.X. and Z.Z.; supervision, Z.Z. and A.W.H.; project administration, Z.Z.; funding acquisition, Z.Z. and A.W.H.; All authors have read and agreed to the published version of the manuscript.

Funding: The study was funded by the National Natural Science Foundation of China (82073573 to Z.Z.) and the EU and BMZ through GIZ (Knowledge for Nutrition, K4N, to A.W.H.). The funders had no role in the study design, data collection and analysis, decision to publish, or manuscript preparation.

Institutional Review Board Statement: The study was approved by the Institutional Review Committee of the University of North Carolina at Chapel Hill, and the National Institute for Nutrition and Health, Chinese Center for Disease Control and Prevention.

Informed Consent Statement: Informed consent was voluntarily provided by all participants and their legal guardians.

Data Availability Statement: The dataset in the present study was open-access and can be freely obtained from the CHNS website: https://www.cpc.unc.edu/projects/china/data/datasets/data_downloads/longitudinal (accessed on 20 November 2021).

Acknowledgments: This study used data from the China Health and Nutrition Survey (CHNS). We gratefully acknowledge the National Institute of Nutrition and Food Safety, China Centre for Disease Control and Prevention; the Carolina Population Centre, University of North Carolina at Chapel Hill; the National Institutes of Health (NIH; R01-HD30880, DK056350, and R01-HD38700); and the Fogarty International Centre, NIH, for their financial contributions towards the CHNS data collection and analysis files since 1989. We also thank all the participants and the staff of the CHNS involved in this study.

Conflicts of Interest: The authors declare no conflict of interest. The funders had no role in the design of the study, data collection, analyses, or interpretation of data, in the writing of the manuscript, or in the decision to publish the results.

References

1. Zhai, F.Y.; Du, S.F.; Wang, Z.H.; Zhang, J.G.; Du, W.W.; Popkin, B.M. Dynamics of the Chinese diet and the role of urbanicity, 1991–2011. *Obes. Rev.* **2014**, *15* (Suppl. 1), 16–26. [CrossRef]
2. Popkin, B.M. Synthesis and implications: China's nutrition transition in the context of changes across other low- and middle-income countries. *Obes. Rev.* **2014**, *15* (Suppl. 1), 60–67. [CrossRef]
3. Zong, X.N.; Li, H. Physical growth of children and adolescents in China over the past 35 years. *Bull. World Health Organ.* **2014**, *92*, 555–564. [CrossRef]
4. Dong, Y.H.; Jan, C.; Ma, Y.H.; Dong, B.; Zou, Z.Y.; Yang, Y.D.; Xu, R.B.; Song, Y.; Ma, J.; Sawyer, S.M.; et al. Economic development and the nutritional status of Chinese school-aged children and adolescents from 1995 to 2014: An analysis of five successive national surveys. *Lancet Diabetes Endocrinol.* **2019**, *7*, 288–299. [CrossRef]
5. Simmonds, M.; Llewellyn, A.; Owen, C.G.; Woolacott, N. Predicting adult obesity from childhood obesity: A systematic review and meta-analysis. *Obes. Rev.* **2016**, *17*, 95–107. [CrossRef] [PubMed]
6. Horesh, A.; Tsur, A.M.; Bardugo, A.; Twig, G. Adolescent and childhood obesity and excess morbidity and mortality in young adulthood—A systematic review. *Curr. Obes. Rep.* **2021**, *10*, 301–310. [CrossRef] [PubMed]
7. Asghari, G.; Mirmiran, P.; Yuzbashian, E.; Azizi, F. A systematic review of diet quality indices in relation to obesity. *Br. J. Nutr.* **2017**, *117*, 1055–1065. [CrossRef]
8. Qorbani, M.; Mahdavi-Gorabi, A.; Khatibi, N.; Ejtahed, H.S.; Khazdouz, M.; Djalalinia, S.; Sahebkar, A.; Esmaeili-Abdar, M.; Hasani, M. Dietary diversity score and cardio-metabolic risk factors: An updated systematic review and meta-analysis. *Eat. Weight Disord.* **2022**, *27*, 85–100. [CrossRef]
9. Miller, V.; Webb, P.; Micha, R.; Mozaffarian, D. Defining diet quality: A synthesis of dietary quality metrics and their validity for the double burden of malnutrition. *Lancet Planet. Health* **2020**, *4*, e352–e370. [CrossRef]
10. Verger, E.O.; Le Port, A.; Borderon, A.; Bourbon, G.; Moursi, M.; Savy, M.; Mariotti, F.; Martin-Prevel, Y. Dietary diversity indicators and their associations with dietary adequacy and health outcomes: A systematic scoping review. *Adv. Nutr.* **2021**, *12*, 1659–1672. [CrossRef]
11. Herforth, A.W.; Wiesmann, D.; Martínez-Steele, E.; Andrade, G.; Monteiro, C.A. Introducing a suite of low-burden diet quality indicators that reflect healthy diet patterns at population level. *Curr. Dev. Nutr.* **2020**, *4*, nzaa168. [CrossRef]
12. Hernández-Ruiz, Á.; Díaz-Jereda, L.A.; Madrigal, C.; Soto-Méndez, M.J.; Kuijsten, A.; Gil, Á. Methodological Aspects of Diet Quality Indicators in Childhood: A Mapping Review. *Adv. Nutr.* **2021**, *12*, 2435–2494. [CrossRef]
13. Dalwood, P.; Marshall, S.; Burrows, T.L.; McIntosh, A.; Collins, C.E. Diet quality indices and their associations with health-related outcomes in children and adolescents: An updated systematic review. *Nutr. J.* **2020**, *19*, 118. [CrossRef]
14. World Health Organization. Proceedings of the Report of the Technical Consultation on Measuring Healthy Diets: Concepts, Methods and Metrics, Virtual Meeting, 18–20 May 2021; World Health Organization: Geneva, Switzerland, 2021. Licence: CC BY-NC-SA 3.0 IGO.
15. Global Diet Quality Project. Available online: https://www.globaldietquality.org/about (accessed on 8 January 2022).
16. Ma, S.; Herforth, A.W.; Vogliano, C.; Zou, Z. Most Commonly-Consumed Food Items by Food Group, and by Province, in China: Implications for Diet Quality Monitoring. *Nutrients* **2022**, *14*, 1754. [CrossRef]
17. Popkin, B.M.; Du, S.F.; Zhai, F.Y.; Zhang, B. Cohort Profile: The China Health and Nutrition Survey—Monitoring and understanding socio-economic and health change in China, 1989–2011. *Int. J. Epidemiol.* **2010**, *39*, 1435–1440. [CrossRef]
18. Zhang, B.; Zhai, F.Y.; Du, S.F.; Popkin, B.M. The China Health and Nutrition Survey, 1989–2011. *Obes. Rev.* **2014**, *15* (Suppl. 1), 2–7. [CrossRef]
19. China Health and Nutrition Survey. Available online: https://www.cpc.unc.edu/projects/china (accessed on 8 January 2022).

20. Yang, Y.X.; Wang, G.Y.; Pan, X.C. *China Food Composition*, 2nd ed.; Peking University Medical Press: Beijing, China, 2009.

21. Yang, Y.X. *China Food Composition*, 6th ed.; Book 2; Peking University Medical Press: Beijing, China, 2019.

22. Yang, Y.X. *China Food Composition*, 6th ed.; Book 1; Peking University Medical Press: Beijing, China, 2018.

23. World Health Organization. *Physical Status: The Use and Interpretation of Anthropometry*; Report of a WHO Expert Committee; World Health Organization Technical Report Series; World Health Organization: Geneva, Switzerland, 1995; Volume 854, pp. 1–452.

24. *WS/T 586-2018*; Screening for Overweight and Obesity among School-Age Children and Adolescents. China Standard Press: Beijing, China, 2018; National Health and Famliy Plannig Commission of the People's Republic of China.

25. Ma, G.S.; Ji, C.Y.; Ma, J.; Mi, J.; Sung, R.Y.; Xiong, F.; Yan, W.L.; Hu, X.Q.; Li, Y.P.; Du, S.M.; et al. Waist circumference reference values for screening cardiovascular risk factors in Chinese children and adolescents aged 7–18 years. *Chin. J. Epidemiol.* **2010**, *31*, 609–615. [CrossRef]

26. World Health Organization. Growth Reference 5–19 Years. Available online: www.who.int/growthref/who2007_bmi_for_age/en/ (accessed on 8 January 2022).

27. Xi, B.; Zong, X.N.; Kelishadi, R.; Litwin, M.; Hong, Y.M.; Poh, B.K.; Steffen, L.M.; Galcheva, S.V.; Herter-Aeberli, I.; Nawarycz, T.; et al. International waist circumference percentile cutoffs for central obesity in children and adolescents aged 6 to 18 years. *J. Clin. Endocrinol. Metab.* **2020**, *105*, e1569–e1583. [CrossRef]

28. Cheng, G.; Duan, R.; Kranz, S.; Libuda, L.; Zhang, L. Development of a Dietary Index to Assess Overall Diet Quality for Chinese School-Aged Children: The Chinese Children Dietary Index. *J. Acad. Nutr. Diet.* **2016**, *116*, 608–617. [CrossRef]

29. Yuan, Y.Q.; Li, F.; Dong, R.H.; Chen, J.S.; He, G.S.; Li, S.G.; Chen, B. The Development of a Chinese Healthy Eating Index and Its Application in the General Population. *Nutrients* **2017**, *9*, 977. [CrossRef]

30. Wu, H.; Yuan, Y.Q.; Wang, Y.C.; Zhou, X.F.; Liu, S.J.; Cai, M.Q.; He, G.S.; Li, S.G.; Zang, J.J.; Chen, B. The development of a Chinese Healthy Eating Index for School-age Children and its Application in children from China Health and Nutrition Survey. *Int. J. Food Sci. Nutr.* **2021**, *72*, 280–291. [CrossRef]

31. Patton, G.C.; Neufeld, L.M.; Dogra, S.; Frongillo, E.A.; Hargreaves, D.; He, S.; Mates, E.; Menon, P.; Naguib, M.; Norris, S.A. Nourishing our future: The Lancet Series on adolescent nutrition. *Lancet* **2022**, *399*, 123–125. [CrossRef]

32. Al-Jawaldeh, A.; Bekele, H.; de Silva, A.; Gomes, F.; Untoro, J.; Wickramasinghe, K.; Williams, J.; Branca, F. A new global policy framework for adolescent nutrition? *Lancet* **2022**, *399*, 125–127. [CrossRef]

33. World Cancer Research Fund/American Institute for Cancer Research. Continuous Update PROJECT Expert Report 2018. Recommendations and Public Health and Policy Implications. Available online: https://www.wcrf.org/wp-content/uploads/2021/01/Recommendations.pdf (accessed on 8 January 2022).

34. World Health Organization. Healthy Diet. Available online: https://www.who.int/news-room/fact-sheets/detail/healthy-diet (accessed on 8 January 2022).

35. World Health Organization. *Diet, Nutrition and the Prevention of Chronic Diseases*; Report of a Joint WHO/FAO Expert Consultation; WHO Technical Report Series 916; World Health Organization: Geneva, Switzerland, 2003.

36. Livingstone, M.B.; Robson, P.J.; Wallace, J.M. Issues in dietary intake assessment of children and adolescents. *Br. J. Nutr.* **2004**, *92*, S213–S222. [CrossRef]

Article

A Mediterranean-Diet-Based Nutritional Intervention for Children with Prediabetes in a Rural Town: A Pilot Randomized Controlled Trial

Isabel María Blancas-Sánchez [1,2], María Del Rosal Jurado [3], Pilar Aparicio-Martínez [3,4,*], Gracia Quintana Navarro [2], Manuel Vaquero-Abellan [3,4,*], Rafael A. Castro Jiménez [4] and Francisco Javier Fonseca Pozo [2,4,5]

[1] Emergency Department, Reina Sofia's University Hospital, Andalusian Health Care System, 14004 Cordoba, Spain
[2] Grupo Investigación GC09 Nutrigenomics, Metabolic Syndrome, Instituto Maimónides de Investigación Biomédica de Córdoba (IMIBIC), Reina Sofia's University Hospital, 14004 Cordoba, Spain
[3] Departamento de Enfermería, Farmacología y Fisioterapia, Campus de Menéndez Pidal, Universidad de Córdoba, 14071 Córdoba, Spain
[4] Grupo Investigación GC12 Clinical and Epidemiological Research in Primary Care, Instituto Maimónides de Investigación Biomédica de Córdoba (IMIBIC), Reina Sofia's University Hospital, 14071 Cordoba, Spain
[5] Distrito Sanitario Córdoba Guadalquivir, Andalusian Health Care System, 14004 Cordoba, Spain
* Correspondence: n32apmap@uco.es (P.A.-M.); en1vaabm@uco.es (M.V.-A.); Tel.: +34-957-218-573 (P.A.-M.)

Citation: Blancas-Sánchez, I.M.; Del Rosal Jurado, M.; Aparicio-Martínez, P.; Quintana Navarro, G.; Vaquero-Abellan, M.; Castro Jiménez, R.A.; Fonseca Pozo, F.J. A Mediterranean-Diet-Based Nutritional Intervention for Children with Prediabetes in a Rural Town: A Pilot Randomized Controlled Trial. *Nutrients* **2022**, *14*, 3614. https://doi.org/10.3390/nu14173614

Academic Editors: Zhiyong Zou and Kalliopi Karatzi

Received: 25 July 2022
Accepted: 30 August 2022
Published: 1 September 2022

Publisher's Note: MDPI stays neutral with regard to jurisdictional claims in published maps and institutional affiliations.

Copyright: © 2022 by the authors. Licensee MDPI, Basel, Switzerland. This article is an open access article distributed under the terms and conditions of the Creative Commons Attribution (CC BY) license (https://creativecommons.org/licenses/by/4.0/).

Abstract: Prediabetes is a pathological condition in which the blood glucose concentration is higher than normal concentrations but lower than those considered necessary for a type 2 diabetes mellitus diagnosis. Various authors have indicated that the Mediterranean Diet is one of the dietary patterns with the most healthy outcomes, reducing high levels of HbA1c, triglycerides, BMI, and other anthropometric parameters. The main objective of this study was to determine the efficacy of the nutritional intervention for children with prediabetes, including the effectiveness of this nutritional education regarding anthropometric parameters. A randomized pilot trial with two groups, an experimental group (EG) and a control group (CG), using intervention in dietary habits with nutritional reinforcement was carried out on 29 children with prediabetes from a rural area. The nutritional intervention was analyzed through astrophotometric and glycemic measurements and validated surveys. Results: The results indicated improvement in eating habits, adherence to the Mediterranean diet, anthropometric measurements, mainly body mass index and perimeters, and analytical parameters, with a significant decrease in glycated hemoglobin in the EG compared to the CG ($p < 0.001$). Although the results showed that both groups' anthropometric parameters improved, a more significant decrease was observed in the experimental group compared to the control.

Keywords: child-nutrition disorders; prediabetic state; nutrition therapy; diet; Mediterranean; rural health

1. Introduction

Nutrition can be defined as incorporating nutrients for adequate development and growth, nourishment, and metabolic balance [1]. During the last decades, there has been a switch to a higher intake of carbohydrates (45%), the majority being sucrose (27%) and lactose (27%), which has caused increases in body fat [2,3]. A healthy diet and suitable nutrition have been linked to gestational, perinatal, pediatric, and adult health, decreasing the probability of multiple acute and chronic diseases [4,5].

High body mass index (BMI), associated with overweight and metabolic disorders, is a risk factor for severe health problems during adulthood [6,7], such as cardiovascular diseases (CVDs) [8]. In 2009, the World Health Organization (WHO) indicated that one in ten children of school age (5–17 years) was overweight or obese [3,9]; the current rate is one out of five children [8,10,11]. It is currently estimated that in European countries, up to 20% of school-age children are overweight, and 5% are obese [12]. These rates and their rapid increase have been linked to an unbalanced diet and a sedentary lifestyle [13–15].

Children with high BMI have a greater risk of suffering from metabolic syndrome, a clinical entity that can damage the vascular endothelium, favoring the appearance of atherosclerosis and the subsequent development of CVDs [12,16–18]. Furthermore, this syndrome is a risk for future problems related to high glycated hemoglobin (HbA1c) values, increased insulin resistance, and changes in visceral fat [19,20]. High and borderline HbA1c levels, which can be used to define prediabetes, are under study in pediatric populations due to their predictive power [21,22].

Specifically, a concerning disorder increasing in adolescents and young adults is prediabetes [23]. Prediabetes is a metabolic disorder characterized by blood-glucose concentrations higher than those considered normal but not high enough to diagnose diabetes mellitus 2 (MD2) [24]. An American cross-sectional analysis from 2005 to 2016 indicated that the prevalence of prediabetes among adolescents was 18.0% [23], linked to higher BMI, low-density lipoprotein cholesterol levels, systolic blood pressure, and central adiposity and lower insulin sensitivity. By contrast, the prevalence of children with prediabetes was lower (around 2.51% globally, with a 95% confidence interval (IC) of 1.61–3.41%), and lower still in European countries (4.52%, IC 0.53–9.56%) and rural areas (2.47, 95% IC 0.93–4.02). However, this prevalence increased globally by 1.73% from 2011 to 2018 compared to the first decade of this century [25].

In children, prediabetes has been linked to age, gender, high BMI, family history of diabetes, low physical activity, living in an urban area, dietary habits, lifestyle, and sociodemographic and economic factors [25–29]. A recent study carried out in New Zealand indicated that the prediabetes rate (16%, with 6% in European children) was associated with ethnicity, anthropometric measures, and physical activity levels [30]. These results support the notion that anthropometric measures are critical in prediabetes and lifestyle [25–30]. Nutrition is understudied as a factor compared with other indices, such as BMI [27].

Various authors have indicated that the Mediterranean diet is one of the dietary patterns with the healthiest outcomes, reducing high levels of HbA1c, triglycerides, high BMI, and other anthropometric parameters [6,31–34]. The study Prevention with the Mediterranean Diet (PREDIMED) [6,31] concluded that up to 40% of the children studied reduced the appearance of MD2, without associated weight loss, after an intervention based on the Mediterranean diet. Other authors observed that a nutritional intervention based on adherence to the Mediterranean diet resulted in a significant reduction in weight in the experimental group (EG) compared to the control group (CG) [35]. Another study focused on determining the effectiveness of the vegetarian versus the Mediterranean diet highlighted that both diets effectively reduced body weight, body mass index, fat mass, and lipid profile [34]. Moreover, the Mediterranean diet has reduced the risk for various metabolic and endocrine diseases, such as diabetes or metabolic syndrome, preventing or delaying the progression to MD2 [35,36]. Additionally, compared to the healthy diet, based on the American Diabetes Association's recommendation, the Mediterranean diet reduced HbA1c and triglyceridemia [37].

In Spain, a nutritional intervention based on the Mediterranean diet was shown to decrease BMI and reduce or maintain blood glucose control (reducing HbA1c 0.3–2.0% at three to six months in young adults) [36], thereby improving metabolic levels [32,38]. Nonetheless, the adherence to diets among people that suffer from high HbA1c levels is low [36]; consequently, it is relevant to improve its adherence, mainly via nutritional education, which is the most robust scientifically supported intervention [36]. However, few studies have addressed the efficacy of these interventions in prediabetes in areas with reduced access to health services or smaller populations, such as rural areas [39], and in children [25,32]. The only reported clinical trial focused on a healthy-lifestyle intervention in Spanish children, combining lifestyle education and a psycho-educational program, is the PREvention of DIabetes in KIDs (PREDIKID) study [40]. Nonetheless, this trial is based on the efficacy of exercise combined with lifestyle education and does not determine the effectiveness of the Mediterranean diet in reducing the HbA1c [40].

Therefore, the objectives of this study were as follows: to determine the efficacy of the nutritional intervention on children with prediabetes; to evaluate the improvement of an individualized and directed nutritional intervention compared to a standardized diet (an adapted Mediterranean diet vs. a healthy standardized diet according to the American Diabetes Association's recommendation); and to determine the effectiveness of nutritional education regarding anthropometric parameters and adherence to a Mediterranean diet in children with prediabetes.

2. Materials and Methods

2.1. Study Design

This research focused on a pre-and post-nutritional, randomized, single-blind intervention on parallel groups of schoolchildren, EG and CG, with prediabetes in a rural area, with an average income per household of around EUR 19 thousand.

The study design was divided into two phases, pre-and post-intervention, comparing a nutritional intervention based on the adapted catalog with recommendations of healthcare professionals with an adapted Mediterranean diet (EG) and a standardized healthy diet based on the current nutritional advice from healthcare centers (CG). In each phase, a follow-up was carried out with five visits (32 days apart), from the initial visit (v0) in weeks one and two, in which the study variables were gathered (anthropometric values and blood sample), to a final visit (v4), when the study variables were collected after the nutritional intervention. The study was approved by the Regional Ethical Committee, registered on the regional database (ID:2353) and at Clinical Trials (NCT05424107) (https://clinicaltrials.gov/ct2/show/NCT05424107?type=Intr&cond=PreDiabetes&cntry=ES&draw=2&rank=5, accessed on 21 June 2022).

2.2. Study Population, Sampling, and Data Collection

The study's target population was a sample of primary- and secondary-school children presenting altered HbA1c levels [24]. The inclusion criteria were the presentation of HbA1c levels between 5.7 and 6.4% and informed and signed consent from the minor and the minor's parents or guardians. The exclusion criteria were the presence of HbA1c levels below 5.7% or above 6.4%, pathological metabolic disease, such as diabetes or metabolic syndrome, and a lack of signed informed consent from the minor and their legal guardian.

The sampling was based on estimating high BMI in rural areas of Spain (12.6%), a confidence interval of 95%, and a precision of 4.5. The minimum sample was set as 233 children based on BMI, estimating 2.47 with a 95% IC 0.93–1.02 of the sample with prediabetes in rural areas [25].

The participants were recruited through non-probabilistic convenience sampling between March 2018 and February 2019. For the recruitment of the participants, informative sessions were held in the educational centers for legal guardians, parents, and participants. In these sessions, the study objective and planning were explained, along with the voluntary nature of their participation, and the questions or doubts raised by the tutors, parents, and pediatric patients were answered. After these information sessions, the last session was held before data collection. The interested parties signed the informed consent, which was ratified by the tutors, patients, and researchers who were present.

Most of the participants (238 out of 254 (93.7%)) agreed to undergo blood tests measuring different analytical data, including HbA1c. Out of this sample, 30 students (8.46% of the total sample) met the inclusion criteria for this intervention, which matched the prevalence of children with prediabetes in European countries (4.52%, IC 0.53–9.56%) [25]. The 29 children included in the study were assigned in a simple random way to each of the two groups: 14 EG and 15 CG. This randomization was done by an outsider researcher via an online program based on their identification number (https://www.random.org/lists/ accessed on 1 January 2019). Such program allowed to the randomly assigned each child to permuted group. One child was not included because the guardians' decision followed the CONSORT flow diagram [41] (Figure 1).

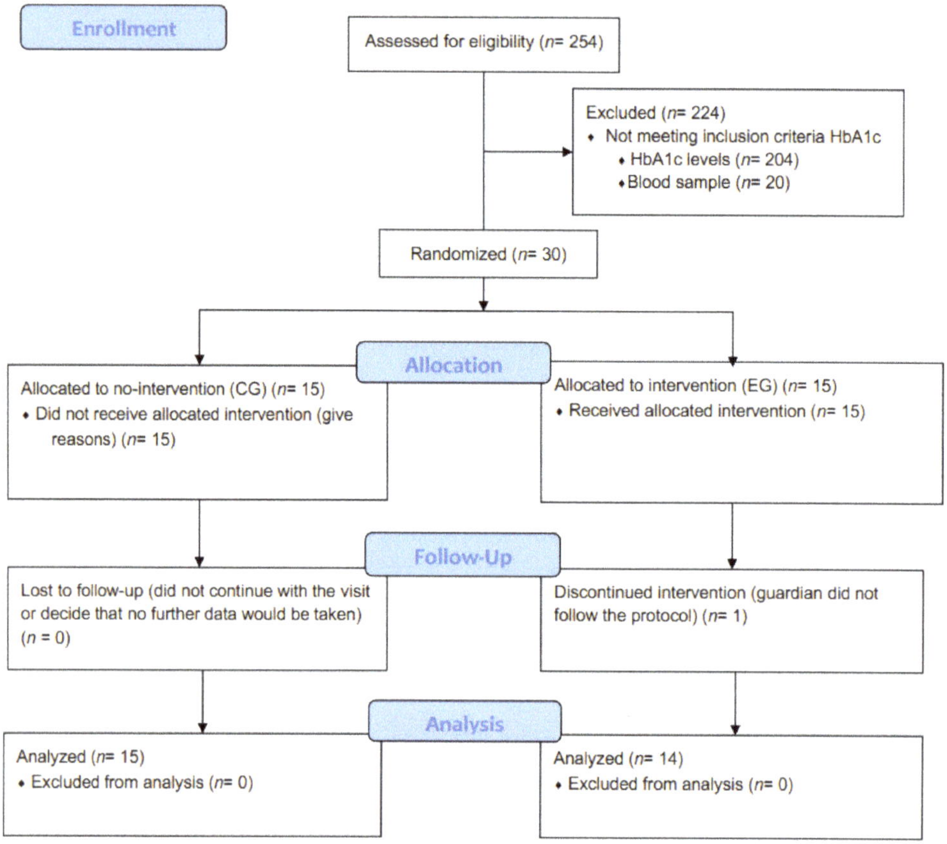

Enrollment

Assessed for eligibility (*n*= 254)

Excluded (*n*= 224)
- Not meeting inclusion criteria HbA1c
  - HbA1c levels (*n*= 204)
  - Blood sample (*n*= 20)

Randomized (*n*= 30)

Allocation

Allocated to no-intervention (CG) (*n*= 15)
- Did not receive allocated intervention (give reasons) (*n*= 15)

Allocated to intervention (EG) (*n*= 15)
- Received allocated intervention (*n*= 15)

Follow-Up

Lost to follow-up (did not continue with the visit or decide that no further data would be taken) (*n* = 0)

Discontinued intervention (guardian did not follow the protocol) (*n*= 1)

Analysis

Analyzed (*n*= 15)
- Excluded from analysis (*n*= 0)

Analyzed (*n*= 14)
- Excluded from analysis (*n*= 0)

Figure 1. Flow diagram of the selection and randomization of the sample following the CONSORT guidelines.

2.3. Nutritional Intervention and Evaluation

The nutritional intervention and the outcome data (anthropometric parameters, glycemic values, and questionnaire on adherence to the Mediterranean diet and food-consumption frequency) are described according to four visits distributed over six weeks. The structure of the nutritional intervention and planification implemented was as follows (Figure 2).

Visit 0 (weeks one and two): The initial one-hour visit was conducted individually on all study subjects (29 children). The anthropometric parameters were taken on these visits. The first assessment of the children's eating habits and adherence to a Mediterranean diet was carried out. In addition, a socioeconomic evaluation was carried out on all the parents through an original and specific questionnaire consisting of nine questions. During the same visit, the EG received targeted nutritional education, with specific recommendations on the Mediterranean diet. On the other hand, the CG only received an information sheet with standardized recommendations for healthy eating, called the Decalogue of Healthy Eating.

Visits 1 (week six), 2 (week eleven), and 3 (week fifteen): These three visits took 30 min to complete. They included an evaluation of dietary adherence, the resolution of doubts about diet and food, and nutritional reinforcement for the children and families in the

EG. Visit 4 (week twenty): The final evaluation was conducted for both groups' children. The Mediterranean diet adherence questionnaires were filled out, and the anthropometric variables and blood samples were retaken.

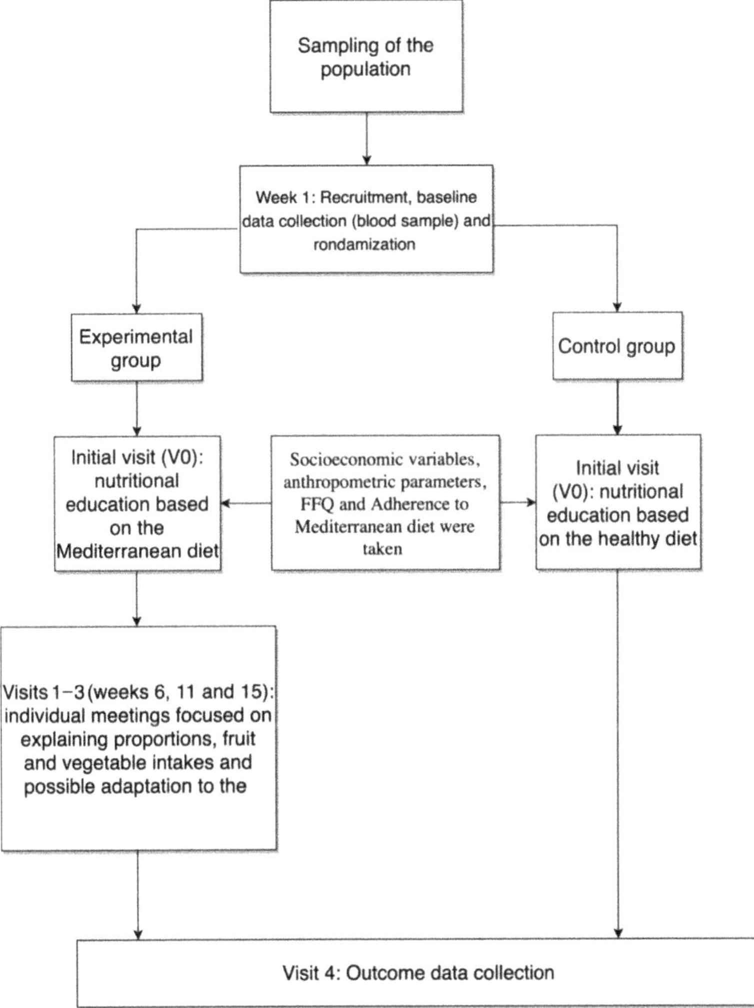

Figure 2. Flow chart of the nutritional intervention, visit follow-ups, and outcome data collection.

2.4. Parameters and Measuring Instruments

The parameters measured were HbA1c levels, height (m^2), weight (kg), waist circumference (cm), hip circumference (cm), arm circumference (cm), BMI (kg/m^2), validated nutritional assessment questionnaire focused on adherence to the KIDMED, Food Frequency Questionnaire (FFQ), and the decalogue of healthy eating. All these measurements were taken up to six months from the V0.

The anthropometric evaluation was carried out on all the study subjects, measuring the following parameters: height, weight, BMI, waist circumference, hip circumference, arm circumference, body fat percentage, and fat-free-mass percentage. The measurements obtained using a tape measure and Omron BF511 impedanciometry were grouped according to the recommendations of the World Health Organization.

Validated nutritional assessment questionnaires that focused on the frequency of food consumption, quality questionnaires, and adherence to the Mediterranean diet, such as the KIDMED, were applied. In addition, the decalogue of healthy eating was used to determine adherence to the change in diet.

The Food Frequency Questionnaire (FFQ) includes 137 foods classified into 14 groups (dairy products, eggs, meat or meat products, fish or shellfish, vegetables, potatoes, fruits, nuts, legumes, cereals, olive oil, pastries, cakes or sweets, and alcoholic beverages). Frequencies were recorded using a 9-category Likert-type scale (from "rarely" to "6 or more times per day"). In addition, the rate of consumption of common foods in this population group, such as juices (nectars, concentrates, etc.), sweets or sugary breakfast cereals (with chocolate, honey, etc.), was collected, along with the habitual culinary techniques used in the children's houses. Energy and nutrient intakes were calculated using Spanish food composition tables.

The quality index of the KIDMED aimed at children and adolescents comprises 16 questions related to adherence to the Mediterranean diet and seven items complementary to the test. All questions were answered positively or negatively. Of the 16 main questions, question numbers 6, 11, 14, and 16, in their affirmative answers, had a negative meaning; therefore, they were worth (-1). On the other hand, the remaining questions whose affirmative answers represented a positive value concerning the Mediterranean diet were scored with $(+1)$. Negative responses received no score (0). According to the test, the results were grouped into different levels of adherence to the Mediterranean diet: low (score 0 to 3), medium (score 4–7), and high (8 to 12).

Lastly, the decalogue of healthy eating, developed for the internal control of adherence to the prescribed dietary recommendations, allowed the dietitians and the children/parents to determine which nutritional aspects were quickest and simplest to improve. On the test, the results were grouped into different levels of adherence to the diet: low (score 0 to 3), medium (score 4–7) and high (8 to 10).

Furthermore, the educational material designed and prepared specifically for the intervention group included printed documents and PowerPoint presentations for the group visits, focused on food cards and hidden nutrient techniques, healthy menus, diabetes, and the decalogue of the Mediterranean diet.

2.5. Statistical Analysis

The descriptive analysis of the categorical variables was expressed as a percentage, and the quantitative variables were expressed through measures of central tendency (mean $\pm$ standard deviation and 95% confidence interval). The normalization test (Shapiro–Wilk) indicated normalization inside the sample ($p = 0.24$), although there were differences for the total sample ($p < 0.05$). Differences in continuous variables between groups (EG and CG groups) were analyzed by Student's test or Mann–Whitney Student when adequate. Paired measurements, and categorical variables were analyzed using the chi-squared test. All statistical analyses were performed using the statistical package SPSS (version 28.0) for Windows.

2.6. Ethical Aspects

This research followed the Helsinki Code and the Principles of Biomedicine and obtained the approval of the Ethical Committee in the Regional Committee of the provincial Hospital (Number 2353).

3. Results

Twenty-nine children with prediabetes participated in the study, with a mean age of ten years, fifteen of whom were male (52% of the sample). The boys showed higher values than the girls in terms of weight (53.31 ± 11.8 males vs. 31.77 ± 3.57 females), height (149.4 ± 9.51 males vs. 134.4 ± 5.87 females), waist circumference (82.03 ± 8.67 vs. 60.82 ± 5.61), arm circumference (24.47 ± 3.06 vs. 18.57 ± 1.74), and BMI (23.6 ± 4.49

vs. 17.54 ± 2.09 ($p < 0.01$). Despite the initial differences regarding gender, there were no significant differences between girls and boys regarding the Hb1Ac ($p = 0.085$). Regarding the family characteristics, most of the guardians or parents were married (89.7%), with a primary education level (48.3%), a housewife (37.9%), with an 8-day shift (47.4%), with a period of unemployment between 6 months and one year (41.7%), with two dependents (62.1%) and received more than EUR 900 per month (62.1%).

Both groups (CG and EG) had no significant differences for the variables studied during the initial visit (V0) (Table 1). In this sense, the results of the glycemic profiling showed how both groups had similar values at the first visit ($5.86 \pm 0.13\%$ in EG vs. $5.91 \pm 0.18\%$ CG; $p = 0.47$ for HbA1c and 12.07 ± 12.2 mIU/L in EG vs. 11.7 ± 7.2 mIU/L in CG; $p = 0.72$ for insulin) (Table 1).

Table 1. The study included anthropometric characteristics of the 29 children with prediabetes according to the randomization group (initial visit (V0)).

Variables	Control Group (CG) ($n = 15$)	Experimental Group (EG) ($n = 14$)	Differences between GC vs. GE (p-Value)
Gender (boys)	7 (46.7%)	8 (57.1%)	0.6
Weight (kg)	41.12 ± 12.04	44.18 ± 16	0.36
Height (m)	140.95 ± 9.29	143.2 ± 27.4	0.25
BMI [1] (kg/m^2)	20.2 ± 4.1	20.9 ± 4.3	0.42
Waist circumference (cm)	68.9 ± 12.8	74.8 ± 15.3	0.4
Arm circumference (cm)	21.3 ± 3.2	21.9 ± 4.7	0.13
Hip circumference (cm)	88.9 ± 20.3	85.8 ± 12.5	0.51
BF% [1]	29.9 ± 2.2	31.5 ± 3.9	0.58
FF% [1]	24.3 ± 7.9	25.3 ± 7.9	0.41
Hb1Ac	5.86 ± 0.2	5.91 ± 0.1	0.47
Insulin level	11.72 ± 7.16	12 ± 10.01	0.68

[1] BMI: Body mass index; BF%: Body fat percentage; FF%: Fat-free percentage

The results of the intervention (Table 2) showed that both groups experienced significant decreases in waist, arm, and hip circumferences ($p < 0.05$). The results indicated no significant differences between the GE and GC regarding the decrease in the perimeters. Nonetheless, the children from the EG showed a more statistically significant decrease in waist circumference (-3.19 cm; $p = 0.001$ EG vs. -2.32 cm; $p = 0.02$ CG) and hip circumference (-2.51 cm; $p = 0.012$ EG vs. -2.36 cm; $p = 0.018$ CG). Furthermore, the fat-free mass percentage increased after the intervention, and this increase was greater in the CG compared to the EG (2.98%; $p = 0.003$ in CG vs. 2.47%; $p = 0.016$ in EG). The results also showed that there was no significant difference after the nutritional intervention regarding BMI and body fat percentage ($p > 0.05$) (Table 2).

Table 2. According to the randomization group, values after the nutritional intervention of 29 children with prediabetes.

Variables	Control Group (CG) ($n = 15$)	Differences between V0 and V4 (p-Value) [a]	Experimental Group (EG) ($n = 14$)	Differences between V0 and V4 (p-Value)	Differences GC vs. GE (p-Value) [b]
Weight (kg)	40.27 ± 11.64	-0.85 (0.23)	42.97 ± 15.87	-1.83 (0.13)	0.46
Height (m)	142.45 ± 10.17	1.5 (0.85)	144.73 ± 13.39	1.5 (0.75)	0.91
BMI [1] (kg/m^2)	19.58 ± 4.07	-0.62 (0.19)	20.61 ± 4.4	-0.69 (0.13)	0.62
Waist circumference (cm)	66.48 ± 12.74	-2.32 (0.02)	71.61 ± 13.65	-3.19 (0.001)	0.58
Arm circumference (cm)	18.94 ± 3.06	-2.36 (0.018)	19.39 ± 4.7	-2.51 (0.012)	0.45
Hip circumference (cm)	88.77 ± 23.32	-0.13 ($p < 0.001$)	85.42 ± 18.25	-0.38 (<0.001)	0.17
BF% [1]	29.59 ± 3.11	-0.31 (0.39)	31 ± 3.71	-0.5 (0.22)	0.26
FF% [1]	27.28 ± 8.9	2.98 (0.003)	27.77 ± 10.87	2.47 (0.016)	0.82

Table 2. *Cont.*

Variables	Control Group (CG) (n = 15)	Differences between V0 and V4 (p-Value) [a]	Experimental Group (EG) (n = 14)	Differences between V0 and V4 (p-Value)	Differences GC vs. GE (p-Value) [b]
Hb1Ac	5.56 ± 0.2	−0.3 (<0.001)	5.51 ± 0.1	−0.27 (0.001)	0.51
Insulin (mIU/L)	10.8 ± 5.89	−0.8 (0.41)	7.3 ± 4.59	−4.8 (0.006)	0.046

[a] The Student's t for related samples and the Wilcoxon test were applied according to the normalization of the samples. [b] The Student's t for independent samples and Mann–Whitney U test were applied to the normalization of the samples. [1] BMI: body mass index; BF%: body fat percentage; FF%: fat-free percentage

Another parameter that showed improvement after the intervention was HbA1c and insulin levels (Table 2). In both groups, HbA1c values decreased significantly between the first and last visit (p = 0.001 in CE and p < 0.001 in CG), but insulin had a significant change only in the intervention group (7.3 mIU/L; p = 0.006 in EG vs. 10.8 mIU/L; p = 0.41 in CG), indicating significant differences between EG and CG for insulin level (p = 0.046) (Figure 3). In addition, after the intervention, the frequency of HbA1c over normal values decreased significantly (58.85% in EG and 46.6% in CG) (Figure 4).

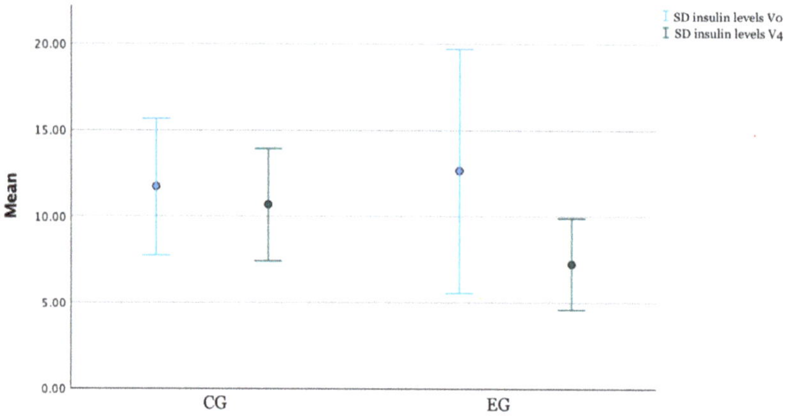

Figure 3. Changes in insulin levels after the interventions in each EG and CG (V4).

According to the KIDMED test, showed that 44.8% of the sample (n = 29) had low adherence to the Mediterranean Diet, 48.3% showed medium adherence, and only 6.9% showed good compliance with the healthy dietary model. The results of the KIDMED were 3.2 ± 2.1 (95% CI 2.01–4.39) for the CG and 4.57 ± 2.47 (CI 95% 3.14–5.99) for the EG, with no significant differences between groups at the beginning (p = 0.69). After the intervention, the results of the KIDMED test improved in both groups, significantly improving the EG and compared to CG (2.5; p = 0.012 EG vs. 1.89, p = 0.056 CG) (Figure 5). In this sense, after the intervention, the EG population went from having low adherence (42.9%) to acceptable adherence (only 14.3% showed low adherence).

Before the intervention, the mean energy intake of the 29 children with prediabetes was 2883 ± 430 kcal/day. The primary energy contribution in the diet was made by carbohydrates (50%, completes 27%, and simple 23%), followed by fats (35%, monounsaturated 17%, saturated 12%, and polyunsaturated 6%), and proteins (15%); the intake of carbohydrates and fats in the EG decreased after the intervention (p < 0.01). The FFQ indicated a significant decrease in food intake, mainly in terms of carbohydrates and fat, in both the EG and the CG (p < 0.01). Furthermore, the results after the intervention regarding the FFQ score showed that the children from the EG experienced a higher reduction in food intake (47.14; IC 95% 21.9–72.33 points in the FFQ; p < 0.001 for the children in the EG vs. 23.3; IC 95% 8.9–37.43 points in the FFQ; p = 0.002 children in CG).

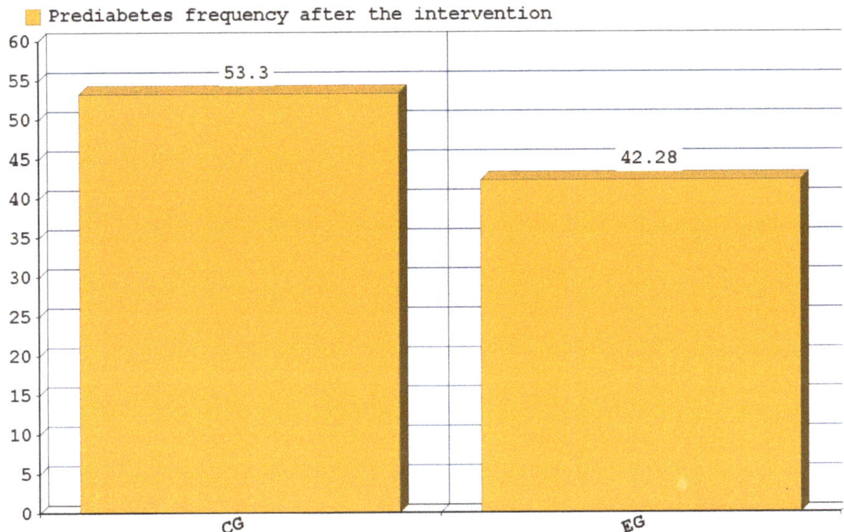

Figure 4. Prediabetes frequency after the intervention (V4) in the control and experimental group.

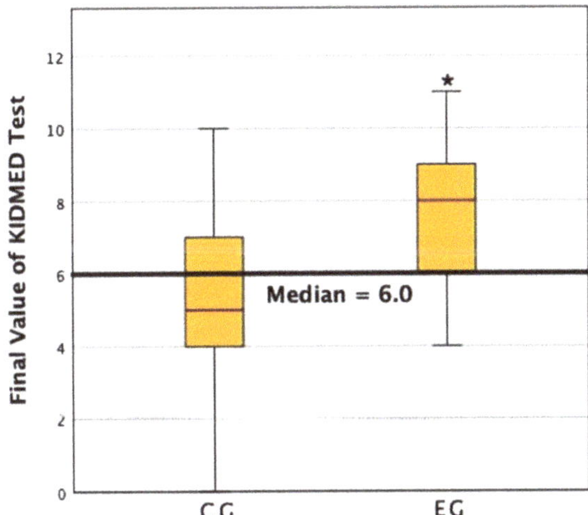

Figure 5. Changes in KIDMED between groups from the first visit to the last. An asterisk denotes a statistically significant difference between the interventions.

According to the adherence to the decalogue of healthy eating, variation in percentage concerning items included in the decalogue of the healthy eating questionnaire was shown (Table 3). The differences between visits in the consumption of vegetables and whole meal bread increased, and a decrease in the consumption of carbonated beverages was noted for both groups ($p < 0.001$) (Table 3). Although there were improvements in both groups (Table 3), the results for all the items indicated a significant improvement in the EG (0.8; $p = 0.021$) compared to the GC (0.4; $p = 0.25$). The EG children decreased their weekly consumption of red and processed meats by 70%; half of them did not consume carbonated and sugary drinks, industrial pastries, snacks, or sweets (Table 3). Finally, water became the main drink for 100% of the children ($p < 0.05$), indicating no significant differences between groups ($p = 0.89$) (Table 3).

Table 3. Percentage of compliance with each item included in the decalogue of healthy eating after visit 4.

Items	CG		Differences between Visits	GE		Differences between Visits	Differences between GC and GE per Items
	Initial Visit	Last Visit		Initial Visit	Last Visit		
Consume ≥ 2 servings of vegetables per day, one of them raw	8.9%	67.7%	<0.001	7.1%	80%	<0.001	0.41
Use 3–4 tablespoons of extra virgin olive oil raw and for cooking	68%	80%	0.25	78.6%	100%	0.014	0.02
Eat ≥ 3 whole servings of fruit per day	4.3%	21.12%	0.02	7.1%	30%	0.016	0.158
Consume natural juices or smoothies	6.7%	33.3%	0.02	0%	80%	<0.001	<0.001
Preferably consume more whole grains (bread, pasta, and rice) than refined ones	0%	57.85	<0.001	0%	70%	<0.001	0.16
Eat ≥ 3 servings of legumes per week	45%	68%	0.04	50%	90%	0.004	0.014
Eat ≥ 3–4 handfuls per week of natural nuts	16.3%	36.7%	0.016	14.3%	50%	<0.001	0.045
Consume ≥ 3 servings of oily and/or white fish per week	48%	80%	<0.001	42.9%	80%	<0.001	0.45
Preferably consume lean meats without skin (chicken, turkey, rabbit, and pork tenderloin)	65%	73.3%	0.16	64.3%	100%	<0.001	0.012
Do not consume red and/or processed meats weekly (ribs, chops, hamburger, commercial sausages, etc.)	23.3%	66.7%	0.002	0%	70%	<0.001	0.16
Do not consume carbonated and/or sugary drinks weekly	13.3%	43.3%	<0.001	7.1%	50%	<0.001	0.84
Do not consume industrial pastries. treats and snacks weekly	0%	43.3%	<0.001	0%	50%	<0.001	0.049
Consume quality and minimally processed dairy products (natural yogurt, fresh cheese, etc.)	0%	13.3%	0.04	0%	20%	0.02	0.05
Water is the main daily drink	60%	100%	0.004	71.4%	100%	0.01	0.89

4. Discussion

This pilot study analyzed the efficacy of a nutritional intervention based on the Mediterranean diet in comparison with a healthy standardized diet to reduce HbA1c and insulin levels and, therefore, reduce the prevalence of prediabetes in a rural pediatric population. Furthermore, this study analyzed the impact of nutritional interventions guided by health care providers to promote healthy habits and control risk factors, such as high BMI.

The initial results of the data indicated that the frequency of prediabetes in a rural town seemed to be in line with the data from European countries and lower-income areas [25]. These initial data indicate that the prevalence of prediabetes also increased and was more common in boys [25–30].

The results for both groups indicated that the BMI did not significantly decrease in the Mediterranean or healthy diet, which matched the results of several studies regarding the possibility of no significant decrease in weight or BMI [6,31–34]. However, the waist, arm, and hip diameters significantly improved in both groups; the decrease was more notable in the experimental group than in the control. These results coincide with those of Albert Pérez et al. (2018), which indicated that nutritional interventions based on diet or exercise decrease anthropometric values and metabolic parameters [42]. Additionally, both groups experienced an increase in fat-free percentage, with the increase higher in the CG, which was in line with a recent meta-analysis regarding prediabetes that indicated the positive impact of the Mediterranean diet and of healthy diets following the American Diabetes Association's recommendations [37].

Moreover, the changes in the HbA1c in both groups highlighted how the Mediterranean diet and healthy diet, monitored by health care professionals, are effective interventions to control and reduce the risk of MD2. These results also matched the findings of the last executive summary of the Spanish Atherosclerosis Society [36], which highlighted that the Mediterranean diet and healthy diets based on the American Diabetes Association's recommendations (mainly the Dietary Approaches to Stop Hypertension, also known as DASH) reduce HbA1c levels and risk factors in adults. The only parameter that differed between the groups was insulin levels, which were related to insulin resistance. This seems to match the results of a previous analysis that indicated how the Mediterranean diet results in a modification in compound-specific gene expression via the involvement of the nuclear high-mobility group A1 (HMGA1) protein [43,44]. These results imply that the Mediterranean diet seems to provoke modifications in metabolic complications in humans, children, or adults, reducing pro- and anti-inflammatory adipocytokines and insulin resistance [44].

Another exciting result was the improvement in adherence to the Mediterranean diet and the modifications in food consumption. Both groups experienced a reduction in food intake, with a corresponding reduction in energy, possibly associated with the lessening of the anthropometric parameters and improvements in Hb1Ac. However, the children from the EG experienced a significantly decreased frequency of low adherence to the diet and increased consumption of healthy foods, with a simultaneous increase in Mediterranean diet adherence. Moreover, after the intervention, the five most frequently consumed foods among our study population were carbonated and sugary drinks, skim milk products, whole-milk products, vegetables, and fruits. The pre-intervention data coincided with a previous analysis in Croatia, which reflected how children (8 and 9 years old) have low adherence to the Mediterranean diet, with their consumption of processed and sugary foods at levels higher than those recommended, which can lead to insulin resistance and other health issues [25,36,45]. However, the post-intervention food preferences in this study were opposed to those presented in the ANIBES Study [46], in which the five most frequently consumed foods were bread, pastries, meats, olive oil, and sausages, suggesting that other factors, such as the role of health care professionals, intervene in food-intake changes. In this sense, the study by Costarelli et al. reflected the efficiency of nutritional interventions in schools, although adherence to the Mediterranean diet is related to the profiles of those who implement these interventions [47]. When teachers oversee interventions based on improving adherence to the Mediterranean diet, the results of the control and experimental groups do not differ significantly.

Despite the differences regarding the children's preferences, these previous findings [25,36,45,48] highlight that the low adherence to this diet is one of the factors that are increasing the prevalence of diabetes and other issues [25–30], reflecting how a nutritional intervention, such as that presented in the current study, with follow-ups by health professionals, can improve adherence, decrease inadequate food intake, and improve glycemic values.

Limitations, Implications for the Field, and Future Research

As with any research, this study has limitations. The major limitation was the sample size of 29 individuals, 15 in the CG and 14 in the EG, resulting from the study's voluntariness and the established inclusion criteria. This limitation implies that the findings may be exaggerated. It is, therefore, relevant to conduct this type of analysis with a larger sample. Another limitation was related to the coexistence of other health programs carried out in schools; furthermore, sociodemographic and individual conditions need to be considered since income or parental status may have modified the food preferences and the effectiveness of the results. In this sense, since the groups were located in a rural area of Southern Spain, the results could be limited in their generalizability to other children with prediabetes, even with similar-sized samples.

Despite these limitations, the presented results showed how the eating habits of children with prediabetes improve with a dietary intervention based on nutritional education that allows them to discern unhealthy consumption. The eating habits improved exponentially in both groups, particularly through the adaptation to and inclusion of the Mediterranean diet, indicating that the role of healthcare specialists in any nutritional intervention results in improved healthy eating [32]. Given the results, it would be interesting to establish a targeted dietary intervention with consecutive personalized visits by specialized health personnel at the primary care level and even in schools.

Additionally, since there is currently a lack of studies aimed at the pediatric population [42], this study shows that nutritional intervention effectively reduces glycemic parameters and, therefore, the prevalence of prediabetes in the child population and values related to insulin resistance decrease accordingly. These findings suggest that if an adequate strategy is carried out in dietary re-education from an early age through educational programs, such as the Food, Nutrition, and Gastronomy Program for Early Childhood Education (PANGEI) [47], the establishment of healthy habits from early infancy improves quality of life and seems to ensure proper development [32,49]. Moreover, recent studies show how nutritional interventions are effective in this population, maintaining these dietary changes over time (minimum 36 months), but few include biochemical parameters aimed at prediabetes [42].

For this reason, future research will be focused on a nutritional intervention carried out by health care professionals (such as nurses) with school-based follow-ups, monitored by a nutritionist with a larger sample of children and during a minimum period of 36 months.

5. Conclusions

This pre-and post-intervention research aimed at a rural pediatric population with levels close to diabetes showed how the Mediterranean diet effectively reduces biochemical and anthropometric parameters. This efficacy was measured through a nutritional evaluation using validated questionnaires, visits to the children to improve their eating habits, and their adherence to the Mediterranean diet.

Author Contributions: Conceptualization, I.M.B.-S., F.J.F.P. and M.V.-A.; methodology, I.M.B.-S., F.J.F.P. and M.V.-A.; software, G.Q.N., I.M.B.-S. and P.A.-M.; validation, I.M.B.-S., F.J.F.P. and P.A.-M.; formal analysis, I.M.B.-S., M.D.R.J. and P.A.-M.; investigation, I.M.B.-S. and M.D.R.J.; resources, I.M.B.-S., R.A.C.J. and F.J.F.P.; data curation, I.M.B.-S., M.D.R.J. and P.A.-M.; writing—original draft preparation, I.M.B.-S., M.D.R.J., P.A.-M., F.J.F.P. and M.V.-A.; writing—review and editing, G.Q.N., I.M.B.-S., R.A.C.J., P.A.-M., F.J.F.P. and M.V.-A.; visualization, I.M.B.-S. and P.A.-M.; supervision, F.J.F.P. and M.V.-A.; project administration, F.J.F.P. and R.A.C.J.; funding acquisition, F.J.F.P. and R.A.C.J. All authors have read and agreed to the published version of the manuscript.

Funding: The APC was funded by the Maimonides Institute of Biomedical Research of Cordoba (IMIBIC), Cordoba, Spain, through the Foundation for Biomedical Research of Cordoba (FIBICO).

Institutional Review Board Statement: The study was conducted following the Declaration of Helsinki and approved by the Institutional Review Board from Reina's Sofia Hospital (15 October 2013, number 2353).

Informed Consent Statement: Informed consent was obtained from all subjects involved in the study, their parents or legal guardians before the intervention.

Data Availability Statement: The data is available via contacting the corresponding author.

Acknowledgments: We would like to thank the parents and children who decided to participate in this research.

Conflicts of Interest: The authors declare no conflict of interest.

References

1. Stephenson, J.; Heslehurst, N.; Hall, J.; Schoenaker, D.A.J.M.; Hutchinson, J.; Cade, J.E.; Poston, L.; Barrett, G.; Crozier, S.R.; Barker, M.; et al. Before the Beginning: Nutrition and Lifestyle in the Preconception Period and Its Importance for Future Health. *Lancet* **2018**, *391*, 1830–1841. [CrossRef]
2. Mozaffarian, D. Dietary and Policy Priorities for Cardiovascular Disease, Diabetes, and Obesity: A Comprehensive Review. *Circulation* **2016**, *133*, 187–225. [CrossRef] [PubMed]
3. Valerio, G.; Baroni, M.G.; Brufani, C.; Forziato, C.; Grugni, G.; Licenziati, M.R.; Maffeis, C.; Miraglia Del Giudice, E.; Morandi, A.; Pacifico, L.; et al. Childhood Obesity Classification Systems and Cardiometabolic Risk Factors: A Comparison of the Italian, World Health Organization and International Obesity Task Force References. *Ital. J. Pediatr.* **2017**, *43*, 19. [CrossRef] [PubMed]

4. Atzendorf, J.; Apfelbacher, C.; Gomes de Matos, E.; Kraus, L.; Piontek, D. Patterns of Multiple Lifestyle Risk Factors and Their Link to Mental Health in the German Adult Population: A Cross-Sectional Study. *BMJ Open* **2018**, *8*, e022184. [CrossRef] [PubMed]
5. Morales-Suárez-Varela, M.; Rubio-López, N.; Ruso, C.; Llopis-Gonzalez, A.; Ruiz-Rojo, E.; Redondo, M.; Pico, Y. Anthropometric Status and Nutritional Intake in Children (6–9 Years) in Valencia (Spain): The ANIVA Study. *Int. J. Environ. Res. Public Health* **2015**, *12*, 16082–16095. [CrossRef]
6. Brambilla, P.; Pozzobon, G.; Pietrobelli, A. Physical Activity as the Main Therapeutic Tool for Metabolic Syndrome in Childhood. *Int. J. Obes.* **2011**, *35*, 16–28. [CrossRef]
7. Kosti, R.I.; Panagiotakos, D.B. The Epidemic of Obesity in Children and Adolescents in the World. *Cent. Eur. J. Public Health* **2006**, *14*, 151–159. [CrossRef]
8. Aranceta-Bartrina, J.; Gianzo-Citores, M.; Pérez-Rodrigo, C. Prevalence of Overweight, Obesity and Abdominal Obesity in the Spanish Population Aged 3 to 24 Years. The ENPE Study. *Rev. Esp. Cardiol.* **2020**, *73*, 290–299. [CrossRef]
9. World Health Organization. *Taking Action on Childhood Obesity Report*; World Health Organization: Geneva, Switzerland, 2018.
10. Echeverría, G.; McGee, E.; Urquiaga, I.; Jiménez, P.; D'Acuña, S.; Villarroel, L.; Velasco, N.; Leighton, F.; Rigotti, A. Inverse Associations between a Locally Validated Mediterranean Diet Index, Overweight/Obesity, and Metabolic Syndrome in Chilean Adults. *Nutrients* **2017**, *9*, 862. [CrossRef]
11. Peng, W.; Goldsmith, R.; Berry, E.M. Demographic and Lifestyle Factors Associated with Adherence to the Mediterranean Diet in Relation to Overweight/Obesity among Israeli Adolescents: Findings from the Mabat Israeli National Youth Health and Nutrition Survey. *Public Health Nutr.* **2017**, *20*, 883–892. [CrossRef]
12. Gracia-Arnaiz, M. Taking Measures in Times of Crisis: The Political Economy of Obesity Prevention in Spain. *Food Policy* **2017**, *68*, 65–76. [CrossRef]
13. Aguilar-Morales, I.; Colin-Ramirez, E.; Rivera-Mancía, S.; Vallejo, M.; Vázquez-Antona, C. Performance of Waist-to-Height Ratio, Waist Circumference, and Body Mass Index in Discriminating Cardio-Metabolic Risk Factors in a Sample of School-Aged Mexican Children. *Nutrients* **2018**, *10*, 1850. [CrossRef]
14. Ahmed, A.M.; Ragab, S.H.; Ismail, N.A.; Awad, M.A.; Kandil, M.E. Echocardiographic Assessment of Epicardial Adipose Tissue in Obese Children and Its Relation to Clinical Parameters of Metabolic Syndrome. *J. Saudi Heart Assoc.* **2013**, *25*, 108. [CrossRef]
15. Castillo, E.H.; Borges, G.; Talavera, J.O.; Orozco, R.; Vargas-Alemán, C.; Huitrón-Bravo, G.; Diaz-Montiel, J.C.; Castañón, S.; Salmerón, J. Body Mass Index and the Prevalence of Metabolic Syndrome among Children and Adolescents in Two Mexican Populations. *J. Adolesc. Health* **2007**, *40*, 521–526. [CrossRef] [PubMed]
16. Hahad, O.; Prochaska, J.H.; Daiber, A.; Münzel, T. Environmental Noise-Induced Effects on Stress Hormones, Oxidative Stress, and Vascular Dysfunction: Key Factors in the Relationship between Cerebrocardiovascular and Psychological Disorders. *Oxid. Med. Cell. Longev.* **2019**, *2019*, 4623109. [CrossRef]
17. Markopoulou, P.; Papanikolaou, E.; Analytis, A.; Zoumakis, E.; Siahanidou, T. Preterm Birth as a Risk Factor for Metabolic Syndrome and Cardiovascular Disease in Adult Life: A Systematic Review and Meta-Analysis. *J. Pediatr.* **2019**, *210*, 69–80.e5. [CrossRef] [PubMed]
18. Huang, Y.; Pomeranz, J.; Wilde, P. Adoption and Design of Emerging Dietary Policies to Improve Cardiometabolic Health in the US. *Curr. Atheroscler. Rep.* **2018**, *20*, 25. [CrossRef]
19. Alp, H.; Eklioğlu, B.S.; Atabek, M.E.; Karaarslan, S.; Baysal, T.; Altın, H.; Karataş, Z.; Şap, F. Evaluation of Epicardial Adipose Tissue, Carotid Intima-Media Thickness and Ventricular Functions in Obese Children and Adolescents. *J. Pediatr. Endocrinol. Metabol.* **2014**, *27*, 827–835. [CrossRef] [PubMed]
20. Zhang, Y.; Wang, S. Profiles of Body Mass Index and Blood Pressure among Young Adults Categorised by Waist-to-Height Ratio Cut-Offs in Shandong, China. *Ann. Hum. Biol.* **2019**, *46*, 409–414. [CrossRef]
21. Eklioğlu, B.S.; Atabek, M.E.; Akyürek, N.; Alp, H. Prediabetes and Cardiovascular Parameters in Obese Children and Adolescents. *J. Clin. Res. Pediatr. Endocrinol.* **2016**, *8*, 80–85. [CrossRef]
22. Ighbariya, A.; Weiss, R. Insulin Resistance, Prediabetes, Metabolic Syndrome: What Should Every Pediatrician Know? *J. Clin. Res. Pediatri. Endocrinol.* **2018**, *9* (Suppl. S2), 49–57. [CrossRef] [PubMed]
23. Andes, L.J.; Cheng, Y.J.; Rolka, D.B.; Gregg, E.W.; Imperatore, G. Prevalence of Prediabetes Among Adolescents and Young Adults in the United States, 2005–2016. *JAMA Pediatr.* **2020**, *174*, e194498. [CrossRef] [PubMed]
24. Cosson, E.; Hamo-Tchatchouang, E.; Banu, I.; Nguyen, M.-T.; Chiheb, S.; Ba, H.; Valensi, P. A Large Proportion of Prediabetes and Diabetes Goes Undiagnosed When Only Fasting Plasma Glucose and/or HbA1c Are Measured in Overweight or Obese Patients. *Diabetes Metab.* **2010**, *36*, 312–318. [CrossRef] [PubMed]
25. Han, C.; Song, Q.; Ren, Y.; Chen, X.; Jiang, X.; Hu, D. Global Prevalence of Prediabetes in Children and Adolescents: A Systematic Review and Meta-Analysis. *J. Diabetes* **2022**, *14*, 434–441. [CrossRef] [PubMed]
26. Al Amiri, E.; Abdullatif, M.; Abdulle, A.; Al Bitar, N.; Afandi, E.Z.; Parish, M.; Darwiche, G. The Prevalence, Risk Factors, and Screening Measure for Prediabetes and Diabetes among Emirati Overweight/Obese Children and Adolescents. *BMC Public Health* **2015**, *15*, 1298. [CrossRef]
27. Pandey, U.; Midha, T.; Rao, Y.K.; Katiyar, P.; Wal, P.; Kaur, S.; Martolia, D.S. Anthropometric Indicators as Predictor of Prediabetes in Indian Adolescents. *Indian Heart J.* **2017**, *69*, 474–479. [CrossRef]
28. Pandey, M.; Chalk, W. Lockdown: 'Noisy Neighbours Are Ruining My Life'. *BBC News*, 12 May 2020; p. 3.

29. Cabana, F.; Jasmi, R.; Maguire, R. Great Ape Nutrition: Low-Sugar and High-Fibre Diets Can Lead to Increased Natural Behaviours, Decreased Regurgitation and Reingestion, and Reversal of Prediabetes. *Int. Zoo Yearb.* **2018**, *52*, 48–61. [CrossRef]
30. Mazahery, H.; Gammon, C.; Lawgun, D.; Conlon, C.; Beck, K.; Von Hurst, P. Pre-Diabetes Prevalence and Associated Factors in New Zealand School Children: A Cross-Sectional Study. *N. Z. Med. J.* **2021**, *134*, 76–90.
31. Estruch, R.; Ros, E.; Salas-Salvadó, J.; Covas, M.-I.; Corella, D.; Arós, F.; Gómez-Gracia, E.; Ruiz-Gutiérrez, V.; Fiol, M.; Lapetra, J.; et al. Primary Prevention of Cardiovascular Disease with a Mediterranean Diet. *N. Engl. J. Med.* **2013**, *368*, 1279–1290. [CrossRef]
32. Crespo-Blanco, M.C.; Varillas-Delgado, D.; Blanco-Abril, S.; Cid-Exposito, M.G.; Robledo-Martín, J. Effectiveness of an Intervention Programme on Adherence to the Mediterranean Diet in a Preschool Child: A Randomised Controlled Trial. *Nutrients* **2022**, *14*, 1536. [CrossRef]
33. Finicelli, M.; Di Salle, A.; Galderisi, U.; Peluso, G. The Mediterranean Diet: An Update of the Clinical Trials. *Nutrients* **2022**, *14*, 2956. [CrossRef] [PubMed]
34. Sofi, F.; Dinu, M.; Pagliai, G.; Cesari, F.; Gori, A.M.; Sereni, A.; Becatti, M.; Fiorillo, C.; Marcucci, R.; Casini, A. Low-Calorie Vegetarian versus Mediterranean Diets for Reducing Body Weight and Improving Cardiovascular Risk Profile. *Circulation* **2018**, *137*, 1103–1113. [CrossRef] [PubMed]
35. Salas-Salvadó, J.; Díaz-López, A.; Ruiz-Canela, M.; Basora, J.; Fitó, M.; Corella, D.; Serra-Majem, L.; Wärnberg, J.; Romaguera, D.; Estruch, R.; et al. Effect of a Lifestyle Intervention Program with Energy-Restricted Mediterranean Diet and Exercise on Weight Loss and Cardiovascular Risk Factors: One-Year Results of the PREDIMED-Plus Trial. *Diabetes Care* **2019**, *42*, 777–788. [CrossRef] [PubMed]
36. Pascual Fuster, V.; Pérez Pérez, A.; Carretero Gómez, J.; Caixàs Pedragós, A.; Gómez-Huelgas, R.; Pérez-Martínez, P. Executive Summary: Updates to the Dietary Treatment of Prediabetes and Type 2 Diabetes Mellitus. *Endocrinol. Diabetes Nutr.* **2021**, *68*, 277–287. [CrossRef]
37. van Zuuren, E.J.; Fedorowicz, Z.; Kuijpers, T.; Pijl, H. Effects of Low-Carbohydrate- Compared with Low-Fat-Diet Interventions on Metabolic Control in People with Type 2 Diabetes: A Systematic Review including GRADE Assessments. *Am. J. Clin. Nutr.* **2018**, *108*, 300–331. [CrossRef]
38. Rojo-Martínez, G.; Valdés, S.; Soriguer, F.; Vendrell, J.; Urrutia, I.; Pérez, V.; Ortega, E.; Ocón, P.; Montanya, E.; Menéndez, E.; et al. Incidence of Diabetes Mellitus in Spain as Results of the Nation-Wide Cohort Diabetes Study. *Sci. Rep.* **2020**, *10*, 2765. [CrossRef]
39. Azzi, J.L.; Azzi, S.; Lavigne-Robichaud, M.; Vermeer, A.; Barresi, T.; Blaine, S.; Giroux, I. Participant Evaluation of a Prediabetes Intervention Program Designed for Rural Adults. *Can. J. Diet. Pract. Res.* **2020**, *81*, 80–85. [CrossRef]
40. Arenaza, L.; Medrano, M.; Amasene, M.; Rodríguez-Vigil, B.; Díez, I.; Graña, M.; Tobalina, I.; Maiz, E.; Arteche, E.; Larrarte, E.; et al. Prevention of Diabetes in Overweight/Obese Children through a Family Based Intervention Program including Supervised Exercise (PREDIKID Project): Study Protocol for a Randomized Controlled Trial. *Trials* **2017**, *18*, 372. [CrossRef]
41. Schulz, K.F.; Altman, D.G.; Moher, D.; the CONSORT Group. CONSORT 2010 Statement: Updated Guidelines for Reporting Parallel Group Randomised Trials. *BMC Med.* **2010**, *8*, 18. [CrossRef]
42. Albert Pérez, E.; Mateu Olivares, V.; Martínez-Espinosa, R.; Molina Vila, M.; Reig García-Galbis, M. New Insights about How to Make an Intervention in Children and Adolescents with Metabolic Syndrome: Diet, Exercise vs. Changes in Body Composition. A Systematic Review of RCT. *Nutrients* **2018**, *10*, 878. [CrossRef]
43. Dominguez-Riscart, J.; Buero-Fernandez, N.; Garcia-Zarzuela, A.; Morales-Perez, C.; Garcia-Ojanguren, A.; Lechuga-Sancho, A.M. Adherence to Mediterranean Diet Is Associated with Better Glycemic Control in Children with Type 1 Diabetes: A Cross-Sectional Study. *Front. Nutr.* **2022**, *9*, 813989. [CrossRef] [PubMed]
44. Mirabelli, M.; Chiefari, E.; Arcidiacono, B.; Corigliano, D.M.; Brunetti, F.S.; Maggisano, V.; Russo, D.; Foti, D.P.; Brunetti, A. Mediterranean Diet Nutrients to Turn the Tide against Insulin Resistance and Related Diseases. *Nutrients* **2020**, *12*, 1066. [CrossRef] [PubMed]
45. Bučan Nenadić, D.; Kolak, E.; Selak, M.; Smoljo, M.; Radić, J.; Vučković, M.; Dropuljić, B.; Pijerov, T.; Babić Cikoš, D. Anthropometric Parameters and Mediterranean Diet Adherence in Preschool Children in Split-Dalmatia County, Croatia—Are They Related? *Nutrients* **2021**, *13*, 4252. [CrossRef] [PubMed]
46. Ruiz, E.; Ávila, J.M.; Castillo, A.; Valero, T.; Del Pozo, S.; Rodriguez, P.; Bartrina, J.A.; Gil, Á.; González-Gross, M.; Ortega, R.M.; et al. The ANIBES Study on Energy Balance in Spain: Design, Protocol and Methodology. *Nutrients* **2015**, *7*, 970. [CrossRef]
47. Ministerio de Sanidad, Servicios Sociales e Igualdad; Agencia Española de Consumo, Seguridad Alimentaria y Nutrición (AECOSAN). *Programa de Alimentación, Nutrición y Gastronomía Para Educación Infantil (PANGEI). El Gusto Es Mío*; Secretaría General Técnica, Centro de Publicaciones, Ministerio de Educación, Cultura y Deporte: Madrid, Spain, 2016; ISBN 030-16-552-0.
48. Costarelli, V.; Koretsi, E.; Georgitsogianni, E. Health-Related Quality of Life of Greek Adolescents: The Role of the Mediterranean Diet. *Qual. Life Res.* **2013**, *22*, 951–956. [CrossRef]
49. Costarelli, V.; Michou, M.; Svoronou, E.; Koutava, N.; Symvoulidou, M.; Anastasiou, K.; Abeliotis, K. 'Healthy Children, Healthy Planet': A Pilot School-Based Educational Intervention. *Health Educ. J.* **2022**, *81*, 183–195. [CrossRef]

 *nutrients*

Article

Determinants of Complementary Feeding Indicators: A Secondary Analysis of Thailand Multiple Indicators Cluster Survey 2019

Abhirat Supthanasup [1], Nisachol Cetthakrikul [2,*], Matthew Kelly [3], Haribondhu Sarma [3] and Cathy Banwell [3]

[1] School of Human Ecology, Sukhothai Thammathirat Open University, Nonthaburi 11120, Thailand
[2] International Health Policy Program, Ministry of Public Health, Nonthaburi 11000, Thailand
[3] National Centre for Epidemiology and Population Health, Australian National University, Canberra, ACT 2601, Australia
* Correspondence: nisachol@ihpp.thaigov.net

Abstract: Child complementary feeding (CF) practices meet dietary recommendations more often among educated, high-income groups. Much of the evidence for this association addresses inadequate CF for addressing child undernutrition. However, in many countries, including Thailand, child malnutrition assessments must now address under- and over-nutrition. More comprehensive data is needed to understand this complex situation. This study uses data from the Thailand Multiple Indicators Survey 2019, to identify the determinants of CF practices among 6–23-month children (n = 4125) using the newly developed WHO indicators. Logistic regression analysis was used to measure associations between sociodemographic factors and CF practices. In a fully adjusted model, child age, primary caregivers' education, and household incomes were statistically associated with (in)appropriate CF practices. Older children aged 9–23 months, not only have better minimum dietary diversity (MDD), minimum acceptable diet (MAD), and egg and/or flesh food consumption (EFF), but also tend to consume more unhealthy foods. The proportion of inappropriate CF practices was higher among children living with caregivers other than their mothers. While maternal education and household income were positively associated with MDD and MAD, children of mothers from middle-class households consumed more sweetened beverages. Therefore, nutrition programs addressing different feeding problems should be developed specifically for different primary caregiver and demographic groups.

Keywords: complementary feeding; determinants; infant feeding; Thailand; young child

Citation: Supthanasup, A.; Cetthakrikul, N.; Kelly, M.; Sarma, H.; Banwell, C. Determinants of Complementary Feeding Indicators: A Secondary Analysis of Thailand Multiple Indicators Cluster Survey 2019. *Nutrients* **2022**, *14*, 4370. https://doi.org/10.3390/nu14204370

Academic Editor: Zhiyong Zou

Received: 4 September 2022
Accepted: 13 October 2022
Published: 18 October 2022

Publisher's Note: MDPI stays neutral with regard to jurisdictional claims in published maps and institutional affiliations.

Copyright: © 2022 by the authors. Licensee MDPI, Basel, Switzerland. This article is an open access article distributed under the terms and conditions of the Creative Commons Attribution (CC BY) license (https://creativecommons.org/licenses/by/4.0/).

1. Introduction

Child malnutrition remains an alarming global issue, despite the remarkable progress in improving social and economic development [1]. Malnutrition, which is defined as "deficiencies, excesses, or imbalances in a person's intake of energy and/or nutrients", addresses two broad groups of conditions- undernutrition and overnutrition [2]. International organisations estimate that global stunting and wasting prevalence among children under five declined from 33.1% and 7.5% in 2000 to 22.0% and 6.7% in 2020, respectively [3]. However, the prevalence of overweight among children under five increased from 5.4% in 2000 to 5.7% in 2020 [3]. The prevalence of the two malnourished states displays significant variation across contexts and settings.

Child undernutrition has the greatest impact in low-and middle-income countries (LMICs). An analysis using national survey data from 50 LMICs showed a moderate decline in child stunting prevalence from 40.1% and 23.3% in 2000 to 31.9% and 17.5% in 2015, respectively [4]. Moreover, undernutrition has the greatest impact on Asian and African children. Approximately, 79 million Asian children younger than five years are affected by stunting and 31.9 million by wasting [3]. At the same time, the levels and trends of child overnutrition have been growing in many regions, especially Southeast Asia. In this region, the prevalence increased from only 3.7% in 2000 to 7.5% in 2020 [3]. Moreover, the impact of child overnutrition is greatest in upper-middle-income (8.8%) and high-income (7.8%) countries when compared with low-income (3.7%) and lower-middle-income (4.0%) countries [3].

In Thailand, a Multiple Indicator Cluster Survey (MICS) revealed that stunted, underweight, and wasted prevalence among young children had slightly improved between 2012 and 2019. The stunting rate decreased from 16.3% in 2012 to 10.5% in 2016 before slightly increasing to 13% in 2019 [5–7]. The trend in child underweight is similar to those in stunting, where the rate decreased from 9.2% in 2012 to 6.7% in 2016 and then rose to 8% in 2019. However, MICS shows a slight drop in levels of child wasting from 6.7% in 2012 to 5.4% in 2016, and then an increase to 8% in 2019. As well, like other upper-middle-income countries, Thailand is facing increasing levels of child overnutrition. Overweight prevalence among Thai children under five was under 2% in 1986 [8], it then steadily increased to 9% in the next 33 years [7]. Therefore, Thailand is now encountering a double burden of child malnutrition.

Inappropriate complementary feeding is a cause of malnutrition, both under and over nutrition, in children. Children being fed poor quality semi-solid food and/or with low meal frequency were associated with undernutrition, which includes three groups of conditions: stunting, wasting, and underweight [9,10]. Meanwhile, the early introduction of complementary feeding has been associated with the risk of early cessation of breastfeeding and the risk of overweight and obesity [11–13]. Caregivers, most often mothers, take responsibility for young child feeding, with their practices influencing child nutritional status. Child feeding practices are also influenced by multidimensional determinants, including personal, household and community factors. Previous studies tend to focus on determining the socioeconomic determinants of child feeding practices, especially in LMICs. Children that are older, and from better maternal education backgrounds, higher household wealth status and higher access to antenatal care are associated with better complementary feeding practices [14–17]. However, the determinants of child feeding practices vary in different contexts and settings.

In 2008, the WHO introduced indicators for assessing infant and young child feeding practices. These indicators were designed to assess trends and diversity in young child diets, identify children at risk of malnutrition and monitor progress in risk reduction, and evaluate interventions [18]. As the food landscape has changed, the WHO proposed a new set of indicators in 2021, covering dietary diversity, food groups, and unhealthy food and beverage consumption [19]. The summary of changes between the 2008 and 2021 set of indicators is described in Table S1.

In Thailand, the determinants of child health and nutrition have undergone fundamental changes as the country has experienced rapid socio-demographic and economic change over the past decades. Thailand became an upper-middle-income country in 2011, and is experiencing the dual burdens of over and under nutrition. Consequently, gaining a better understanding of the current relationship between socioeconomic and demographic status and child feeding practices is needed. Therefore, the aim of this study is to use the newly developed WHO infant feeding indicators to assess patterns of complementary feeding among Thai children 6–23 months of age, the distribution of these patterns, and to identify the potential risk factors associated with inappropriate complementary feeding practices in Thailand.

2. Materials and Methods

This study is a cross-sectional quantitative analysis of secondary data. We analysed data for 4125 children aged 6–23 months obtained from the nationally representative Multiple Indicators Cluster Survey (MICS) 2019 to assess the association between complementary feeding patterns and child, caregivers, and other characteristics.

2.1. Data Sources and Variables

Multiple Indicators Cluster Survey (MICS) is an international cross-sectional household survey program developed and introduced by the United Nations Children's Fund (UNICEF) to support countries to collect useful data about women and children. Countries can employ MICS to measure indicators or monitor the progress of Sustainable Development Goals. Furthermore, findings from MICS are able to be empirical evidence for developing policies to improve children's diets [7].

This study analysed the Multiple Indicators Cluster Survey 2019 (MICS6), conducted by the National Statistical Office (NSO) in Thailand, between May and November 2019. MICS6 used face-to-face interviews by the trained field staff of NSO. NSO collected data from 17 of Thailand's 77 provinces in five regions (Bangkok, Central, North, South, and Northeast). Field workers entered all data into tablets directly during the interviews. Each interview took around one hour. If the sampled respondent was not home or physically present during the first visit, NSO staff revisited at least three times [7].

Our study only used data from interviews with mothers or caregivers of children aged 6–23 months old. The main independent variables were child's factors (gender and age), maternal/ caregivers' factors (age, education, and nationality), and household and community level factors (household wealth index, residential areas, and geographical regions). The WHO complementary feeding indicators were dependent variables.

2.2. Complementary Feeding Indicators of WHO

We applied the new and updated infant and young child feeding indicators of the WHO [19]. Table 1 summarised the new WHO complementary feeding indicators.

Table 1. WHO complementary feeding indicators *.

Indicator	Short Name	Age Group	Definition
Introduction of solid, semi-solid or soft foods 6–8 months	ISSF	Infants 6–8 months of age	Percentage of infants 6–8 months of age who consumed solid, semi-solid or soft foods during the previous day.
Minimum dietary diversity 6–23 months	MDD	Children 6–23 months of age	Percentage of children 6–23 months of age who consumed foods and beverages from at least five out of eight defined food groups during the previous day.
Minimum meal frequency 6–23 months	MMF	Children 6–23 months of age	Percentage of children 6–23 months of age who consumed solid, semi-solid or soft foods (but also including milk feeds for non-breastfed children) the minimum number of times or more during the previous day. The minimum number of times is defined as: • two feedings of solid, semi-solid or soft foods for breastfed infants aged 6–8 months; • three feedings of solid, semi-solid or soft foods for breastfed children aged 9–23 months; • four feedings of solid, semi-solid or soft foods or milk feeds for non-breastfed children aged 6–23 months whereby at least one of the four feeds must be a solid, semi-solid or soft feed.

Table 1. *Cont.*

Indicator	Short Name	Age Group	Definition
Minimum milk feeding frequency for non-breastfed children 6–23 months	MMFF	Children 6–23 months of age	Percentage of non-breastfed children 6–23 months of age who consumed at least two milk feeds during the previous day.
Minimum acceptable diet 6–23 months	MAD	Children 6–23 months of age	Percentage of children 6–23 months of age who consumed a minimum acceptable diet during the previous day. The minimum acceptable diet is defined as: • for breastfed children: receiving at least the minimum dietary diversity and minimum meal frequency for their age during the previous day; • for non-breastfed children: receiving at least the minimum dietary diversity and minimum meal frequency for their age during the previous day as well as at least two milk feeds.
Egg and/or flesh food consumption 6–23 months	EFF	Children 6–23 months of age	Percentage of children 6–23 months of age who consumed egg and/or flesh food during the previous day.
Sweet beverage consumption 6–23 months	SwB	Children 6–23 months of age	Percentage of children 6–23 months of age who consumed a sweet beverage during the previous day.
Unhealthy food consumption 6–23 months	UFC	Children 6–23 months of age	Percentage of children 6–23 months of age who consumed selected sentinel unhealthy foods (such as fried foods, confections) during the previous day.
Zero vegetable or fruit consumption 6–23 months	ZVF	Children 6–23 months of age	Percentage of children 6–23 months of age who did not consume any vegetables or fruits during the previous day.

* World Health Organization. (2021). Indicators for assessing infant and young child feeding practices: definitions and measurement methods.

2.3. Data Analysis

STATA software version 17 was used for all calculations (serial license number: 401709350741). There were three steps to the analysis. First, was a descriptive analysis to explain the characteristics of the sample and frequency of the complementary feeding indicators. Second, univariate logistic regression analysis was used to calculate the odds ratio (OR) to explore the association between complementary feeding indicators and each independent variable. Third, multivariate logistic regression was applied to find an adjusted odds ratio (AOR) to examine the association between the dependent variable and adjusted each independent variables, mutually adjusted for all other independent variables. We included all independent variables in the multivariate logistic regression analysis, even though they did not show statistically significant association in our analysis, because previous papers had suggested there may be an association. Statistical significance was measured at the 95% confidence level (p-value < 0.05).

3. Results

3.1. Characteristics of the Sample

The number of sampled children aged between 6–23 months was 4125. Male and female children were nearly equally represented in the sample, at 51.49% and 48.51% respectively. Most children were in the age range of 12–23 months (69.8%). The proportion of children cared for by people other than their mothers increased by the child's age. Overall, around 30% of children were currently breastfed (reported receiving any breastmilk in the past 24 h). Moreover, around 38% of children with mothers as primary caregivers were currently being breastfed, while only 3% of those being cared for by others reported receiving breastmilk in the past 24 h. In terms of characteristics of mothers and other caregivers, about three-fourths of mothers were 20–35 years old. Approximately 96% of other caregivers were aged above 35. Mainly, mothers of these children hold a high school degree and above (86.57%), while the majority of other caregivers had only attained a primary level of education (75.40%). The majority of children, especially those being cared for by others, were in rural areas in the northeast of Thailand (Table 2).

Table 2. Percentage of individual-, caregiver-, household-, and community-level characteristics of children aged 6–23 months, Thailand 2019 (*n* = 4125).

| Characteristics | All (*n* = 4125) | Children Provide Care by | | Characteristics | All (*n* = 4125) | Children Provide Care by | |
		Mother (*n* = 3152)	Other (*n* = 973)			Mother (*n* = 3152)	Other (*n* = 973)
Child characteristics				Household characteristics			
Male	51.49	51.33	52.00	Household wealth index			
Currently breastfed *	29.86	37.74	2.91	Poorest	23.22	20.69	31.45
Age (months)				Second	22.52	20.69	28.47
6–8	13.16	14.34	9.35	Middle	21.87	22.27	20.55
9–11	17.04	17.7	14.90	Fourth	18.30	19.67	13.87
12–17	32.15	32.07	32.37	Richest	14.08	16.69	5.65
18–23	37.65	35.88	43.37	Community level characteristics			
Maternal/caregiver characteristics				Residence			
Age (years)				Urban	32.70	35.50	23.64
15–19	26.81	20.86		Geographical region			
20–35	66.61	72.02	4.46	Bangkok and central	30.64	32.87	23.43
>35	6.58	7.12	95.54	North	14.06	14.82	11.60
Education				Northeast	32.99	26.30	54.68
Kindergarten and primary	18.13	13.43	75.40	South	22.30	26.02	10.28
High school and above	81.87	86.57	24.60				
Primary language							
Thai	72.41	86.64	26.31				

* reported receiving any breastmilk in the past 24 h.

3.2. Complementary Feeding Indicators

Table 3 presents the complementary feeding indicators of children being cared for by mothers and others across child age groups. Most children in the 6–8 months group had received solid, semi-solid, or soft foods (91.56% and 93.41% in groups being cared for by mothers and others, respectively). When comparing the proportions of other indicators between primary caregivers' groups, there were significant differences in all indicators. Overall, less than half of 6–8 month old children meet minimum dietary diversity standards (MDD). There were higher proportions of children being cared for by mothers who achieved MDD than those with others (45.35% vs. 32.97%, $p < 0.001$ in the 6–8 months group; 65.95% vs. 61.38%, $p = 0.009$; 75.96% vs. 67.62%, $p < 0.001$ in the 12–17 months group; 74.09% vs. 67.54%, $p < 0.001$ in 18–23 months group). There were higher proportions of children being cared for by mothers achieving minimum acceptable diet standards (MAD) than those with others (42.19% vs. 27.27%, $p < 0.001$ in the 6–8 months group; 64.31% vs. 58.45%, $p = 0.001$ in the 9–11 months group; 70.39% vs.65.10%, $p = 0.002$ in the 12–17 months group; 70.37% vs. 63.68%, $p < 0.001$ in the 18–23 months group). Overall, most children aged 6–23 months had eggs or flesh meats and vegetables or fruit in the past 24 h, but the rates were lower for infants aged 6–8 months. The consumption rate of eggs or flesh meats and vegetables or fruit were lower in those children being cared for by others when compared to mothers. Furthermore, the consumption rate of sweet beverages and unhealthy foods increased as children aged. These consumption rates were higher in children being cared for by others when compared to mothers.

Table 3. Percentage of children who met the complementary feeding indicators according to their age range and primary caregivers (*n* = 4125).

Indicators	Child Age	All (*n* = 4125)	Primary Caregiver		*p*-Value *
			Mother (*n* = 3152)	Other (*n* = 973)	
Introduction of solid, semi-solid or soft foods 6–8 months (ISSSF)	6–8 months	92.00%	91.56%	93.41%	0.061
Minimum dietary diversity (MDD)	6–8 months	42.42%	45.35%	32.97%	<0.001
	9–11 months	64.87%	65.95%	61.38%	0.009
	12–17 months	73.99%	75.96%	67.62%	<0.001
	18–23 months	72.53%	74.09%	67.54%	<0.001
Minimum meal frequency (MMF)	6–8 months	88.80%	86.48%	96.34%	<0.001
	9–11 months	96.10%	96.32%	95.42%	0.184
	12–17 months	97.28%	96.57%	99.65%	<0.001
	18–23 months	95.64%	95.61%	95.68%	0.934
Minimum milk feeding frequency for non-breastfed children (MMFF)	6–8 months	95.64%	96.61%	93.98%	<0.001
	9–11 months	97.96%	99.29%	93.65%	<0.001
	12–17 months	97.18%	97.09%	97.55%	0.456
	18–23 months	93.50%	93.48%	93.57%	0.947
Minimum acceptable diet (MAD)	6–8 months	38.67%	42.19%	27.27%	<0.001
	9–11 months	62.93%	64.31%	58.45%	0.001
	12–17 months	69.14%	70.39%	65.10%	0.002
	18–23 months	68.80%	70.37%	63.68%	<0.001
Egg and/or flesh food consumption (EFF)	6–8 months	70.21%	72.54%	62.64%	<0.001
	9–11 months	88.85%	89.89%	85.52%	<0.001
	12–17 months	94.64%	95.15%	93.02%	0.010
	18–23 months	96.41%	96.61%	95.79%	0.230
Sweet Beverage consumption (SwB)	6–8 months	67.98%	62.17%	86.81%	<0.001
	9–11 months	72.65%	68.28%	86.82%	<0.001
	12–17 months	74.01%	69.90%	87.30%	<0.001
	18–23 months	70.81%	70.60%	71.53%	0.572
Unhealthy food consumption (UFC)	6–8 months	52.72%	50.00%	61.54%	<0.001
	9–11 months	65.09%	62.66%	72.92%	<0.001
	12–17 months	72.24%	71.61%	74.29%	0.100
	18–23 months	78.04%	77.23%	80.66%	0.023
Zero vegetable or fruit consumption (ZVF)	6–8 months	35.18%	33.48%	40.66%	<0.001
	9–11 months	16.90%	14.18%	25.69%	<0.001
	12–17 months	13.79%	12.67%	17.46%	<0.001
	18–23 months	12.65%	11.43%	16.58%	<0.001

* Chi-square test.

3.3. Determinants of Complementary Feeding Indicators

Findings of univariate logistic regression revealed that either living with mothers or not, the age of children had an association with almost all complementary feeding indicators, except, MMFF. For children who lived with mothers, sociodemographic factors (region or wealth) related to MDD, MMFF, MAD, EFF, SWB, UFC, and ZVF, while sociodemographic factors of children who lived with others associated with MDD, MAD, EFF, SWB, and ZVF. Education levels of primary caregivers had an association with some indicators in the 'living with mothers' group namely MDD, MMF, MAD, EFF, SWB, and ZVF, but the variable did not associate with the 'living with others' group. (Supplementary Tables S2 and S3).

Table 4 presents the results of the multivariate logistic regression analysis in which all the main variables were included. The child's age had strong associations with complementary feeding indicators. Older children tend to have better MDD, MMF, MAD, and EFF. After adjusting for covariates, children aged 12–17 months had strong associations with MDD in the group being cared for by mothers (adjusted odd ratios (AOR) = 4.35 [95% CI: 3.38–5.60]), and the group being cared for by others (4.75 [1.66–13.62]). There were associations between child age and EFF in children being cared for by both mothers and others; compared with those in youngest age group, those in aged 9–11 months (3.35 [2.32–4.85]) and (6.90 [1.62–29.31), 12–17 months (7.18 [4.89–10.54) and (18.95 [4.28–83.95]), and 18–23 months (9.54 [6.22–14.61]) and (84.02 [8.64–817.36]). Children of mothers with higher education and from wealthier households had better appropriate complementary feeding (MDD and MAD). Children of mothers with higher education backgrounds were more likely to achieve MDD (1.41 [1.10–1.81]) and MAD (1.41 [1.10–1.81]). Moreover, higher maternal education increased chances of achieving EFF. There was no association between other caregivers' education and household wealth index, and EFF. However, better household incomes were associated with around three-fold increased change of achieving MAD in groups of children being cared for by others.

Conversely, older children in groups being cared for by mothers had strong associations with consuming unhealthy foods. Compared with children aged 6–8 months, those in the aged 9–11 months (1.87 [1.15–3.04]), 12–17 months (2.72 [1.75–4.23]), and 18–23 months (3.44 [2.20–5.37) had increased chance of unhealthy food consumption. Higher household income was a protective factor against feeding children unhealthy foods among those lived with mothers. Moreover, children of mothers from middle-class households and living in Bangkok and Central Thailand tend to consume sweet beverages. However, current breastfeeding was protective against sweet beverage consumption. Older children of mothers with better household income tend to consume more fruits and vegetables when compared with the youngest and lowest income groups. Further details are given in Table 5.

Table 4. Factors associated with complementary feeding indicators, ISSF, MDD, MMF, MMFF, MAD, EFF, the multivariate logistic regression analysis.

	ISSSF [1]		MDD [2]		MMF [3]		MMFF [4]		MAD [5]		EFF [6]	
	AOR (95%CI)											
	Primary Caregiver		**Primary Caregiver**		**Primary Caregiver**		**Primary Caregiver**		**Primary Caregiver**		**Primary Caregiver**	
	Mother	**Other**	**Mother**	**Other**	**Mother**	**Other**	**Mother**	**Other**	**Mother**	**Other**	**Mother**	**Other**
Child gender (Reference category male)												
Female	0.67 (0.33–1.37)	5.95 (0.35–101.68)	1.04 (0.89–1.23)	0.81 (0.45–1.46)	0.85 (0.60–1.21)	0.62 (0.14–2.85)	0.86 (0.54–1.39)	0.62 (0.16–2.46)	1.04 (0.8–1.23)	1.10 (0.21–5.62)	0.87 (0.65–1.15)	0.76 (0.25–2.35)
Child age (months) (Reference category 6–8 months)												
9–11	-	-	2.35 ** (1.80–3.07)	3.01 (0.95–9.50)	3.54 ** (2.06–6.07)	0.22 (0.01–6.20)	4.54 (0.90–22.86)	1.87 (0.10–35.98)	2.40 ** (1.84–3.13)	3.28 * (1.00–10.76)	3.35 ** (2.32–4.85)	6.90 * (1.62–29.31)
12–17	-	-	4.35 ** (3.38–5.60)	4.75 ** (1.66–13.62)	3.03 ** (1.86–4.93)	N/A	1.43 (0.54–3.76)	2.72 (0.21–34.72)	3.52 ** (2.74–4.51)	5.61 ** (1.90–16.59)	7.18 ** (4.89–10.54)	18.95 ** (4.28–83.95)
18–23	-	-	3.83 ** (2.99–4.92)	3.25 * (1.16–9.10)	1.53 (0.98–2.39)	0.11 (0.00–2.47)	0.52 (0.22–1.24)	0.38 (0.04–3.40)	3.25 ** (2.56–4.20)	3.59 * (1.24–10.39)	9.54 ** (6.22–14.61)	84.02 ** (8.64–817.36)
Mother's/caregiver's age (years) (Reference category > 35 years)												
15–19	1.70 (0.18–16.02)	N/A	1.05 (0.73–1.51)	1.65 (0.34–7.94)	1.02 (0.47–2.22)	0.19 (0.01–3.81)	0.55 (0.19–1.62)	0.07 * (0.01–0.75)	1.01 (0.71–1.45)	0.91 (0.19–4.33)	1.03 (0.55–1.96)	0.67 (0.06–7.02)
20–35	0.50 (0.18–1.38)	N/A	0.87 (0.71–1.08)		0.91 (0.59–1.40)		0.54 (0.26–1.12)		0.85 (0.69–1.05)		0.83 (0.57–1.21)	
Mother's/caregiver's language (Reference category Thai)												
Non-Thai	2.65 (0.54–13.01)	1.20 (0.05–29.39)	1.24 (0.92–1.67)	1.40 (0.28–7.07)	0.79 (0.88–2.34)	1.71 (0.02–128.34)	0.67 (0.31–1.44)	0.34 (0.02–6.58)	1.16 (0.87–1.54)	1.10 (0.21–5.62)	2.87 ** (1.54–5.35)	0.18 (0.01–2.19)
Mother's/caregiver's education (Reference category Kindergarten and Primary)												
≥High school	1.58 (0.57–4.37)	N/A	1.41 * (1.10–1.81)	0.93 (0.45–1.91)	1.44 (0.88–2.34)	0.63 (0.10–4.00)	1.30 (0.65–2.62)	1.29 (0.21–7.88)	1.41 * (1.10–1.81))	1.09 (0.53–2.25)	1.76 * (1.18–2.62)	1.58 (0.39–6.44)
Household wealth index (Reference category poorest)												
Second	2.48 (0.83–7.44)	4.63 (0.16–132.80)	1.42 * (1.11–1.84)	2.74 * (1.29–5.79)	1.05 (0.61–1.80)	6.94 (0.68–71.02)	1.36 (0.67–2.76)	4.03 (0.77–20.98)	1.40 * (1.09–1.80)	2.95 * (1.39–6.27)	1.25 (0.81–1.94)	0.72 (0.19–2.72)
Middle	3.17 (0.97–10.33)	5.26 (0.22–123.14)	1.41 * (1.09–1.83)	3.08 * (1.29–7.40)	0.97 (0.55–1.71)	2.35 (0.27–20.34)	1.22 (0.61–2.45)	3.96 (0.41–38.61)	1.36 * (1.05–1.75)	3.32 * (1.38–8.03)	1.16 (0.74–1.80)	1.32 (0.28–6.31)

Table 4. *Cont.*

	ISSSF [1]		MDD [2]		MMF [3]		MMFF [4]		MAD [5]		EFF [6]	
					AOR (95%CI)							
	Primary Caregiver		Primary Caregiver		Primary Caregiver		Primary Caregiver		Primary Caregiver		Primary Caregiver	
	Mother	Other	Mother	Other	Mother	Other	Mother	Other	Mother	Other	Mother	Other
Fourth	1.09 (0.43–2.78)	N/A	1.62 ** (1.24–2.13)	0.06 (0.96–7.30)	0.55 * (0.31–0.96)	2.36 (0.17–33.21)	1.85 (0.85–4.01)	7.18 (0.61–84.86)	1.47 * (1.12–1.92)	2.93 * (1.06–8.12)	1.25 (0.78–1.99)	6.22 (0.46–83.43)
Richest	0.82 (0.32–2.10)	0.03 (0.00–1.24)	2.20 ** (1.63–2.97)	N/A	0.78 (0.41–1.48)	N/A	3.47 * (1.28–9.35)	N/A	2.09 ** (1.56–2.80)	N/A	1.60 (0.96–2.69)	N/A
Residence (Reference category urban)												
Rural	0.94 (0.43–2.05)	35.18 * (1.51–822.10)	1.06 (0.88–1.26)	0.89 (0.42–1.88)	1.09 (0.75–1.58)	0.34 (0.02–4.83)	0.94 (0.55–1.61)	0.82 (0.14–4.64)	1.08 (0.91–1.29)	0.81 (0.38–1.73)	1.08 (0.80–1.46)	0.84 (0.19–3.79)
Geographical region (Reference category Bangkok and central)												
North	1.17 (0.38–3.63)	0.04 (0.00–1.74)	0.66 ** (0.51–0.86)	0.18 * (0.05–0.64)	1.24 (0.69–2.21))	3.04 (0.09–102.21)	0.63 (0.31–1.26)	0.47 (0.07–3.11)	0.61 ** (0.48–0.79)	0.21 * (0.06–0.74)	1.02 (0.67–1.55)	0.10 * (0.02–0.65)
Northeast	1.20 (0.46–3.13)	7.47 (0.22–251.49)	1.10 (0.88–1.36)	2.08 (0.99–4.34)	0.95 (0.60–1.49)	1.96 (0.33–11.79)	1.10 (0.57–2.11)	4.44 (0.84–23.38)	1.10 (0.88–1.36)	2.25 * (0.04–1.06)	2.20 ** (1.50–3.24)	0.82 (0.22–3.15)
South	1.00 (0.37—2.67)	0.24 (0.00–11.63)	1.27 * (1.00–1.61)	0.60 (0.41–4.60)	1.40 (0.84–2.32)	N/A	1.06 (0.53–2.11)	N/A	1.23 (0.98–1.55)	1.43 (0.43–4.84)	1.85 ** (1.24–2.76)	N/A
Currently breastfeeding *** (Reference category No)												
Yes	0.26 ** (0.11–0.63)	27.52 (0.09–8842.30)	1.35 ** (1.13–1.62))	2.14 (0.25–17.93)	0.10 ** (0.06–0.16)	0.01 * (0.00–0.40)	-	-	0.98 (0.82–1.17)	0.33 (0.02–5.64)	0.50 ** (0.37–0.67)	0.38 (0.03–5.43)

* $p < 0.05$; ** $p < 0.01$, N/A = not available; *** reported receiving any breastmilk in the past 24 h; [1] ISSF = Introduction of solid, semi-solid or soft foods 6–8 months; [2] MDD = Minimum dietary diversity 6–23 months; [3] MMF = Minimum meal frequency 6–23 months; [4] MMFF = Minimum milk feeding frequency for non-breastfed children 6–23 months; [5] MAD = Minimum acceptable diet 6–23 months; [6] EFF = Egg and/or flesh food consumption 6–23 months; AOR = Adjusted odds ratio; 95%CI = 95% confidence interval.

Table 5. Factors associated with complementary feeding indicators, SwB, UFC, ZVF, the multivariate logistic regression analysis.

	SwB [1]		UFC [2]		ZVF [3]	
	AOR (95%CI)					
	Primary Caregiver		**Primary Caregiver**		**Primary Caregiver**	
	Mother	**Other**	**Mother**	**Other**	**Mother**	**Other**
Child gender (Reference category male)						
Female	1.03 (0.87–1.23)	1.87 (0.81–4.32)	1.10 (0.88–1.38)	1.47 (0.57–3.83)	1.00 (0.82–1.24)	1.41 (0.70–2.83)
Child age (months) (Reference category 6–8 months)						
9–11	1.12 (0.84–1.50)	2.21 (0.35–13.74)	1.87 * (1.15–3.04)	0.47 (0.03–7.40)	0.34 ** (0.25–0.47)	0.80 (0.24–2.67)
12–17	0.97 (0.74–1.26)	2.75 (0.58–13.11)	2.72 ** (1.75–4.23)	0.40 (0.03–4.79)	0.29 ** (0.22–0.39)	0.43 (0.14–1.30)
18–23	0.78 (0.59–1.02)	0.38 (0.09–1.52)	3.44 ** (2.20–5.37)	2.07 (0.16–26.62)	0.27 ** (0.20–0.36)	0.50 (0.17–1.51)
Mother's/caregiver's age (years) (Reference category > 35 years)						
15–19	0.92 (0.63–1.34)	0.19 (0.03–1.06)	0.86 (0.53–1.40)	8.73 (0.48–157.89)	0.92 (0.59–1.43)	2.10 (0.37–11.90)
20–35	0.94 (0.76–1.17)		1.01 (0.76–1.36)		1.00 (0.77–1.30)	
Mother's/caregiver's language (Reference category Thai)						
Non-Thai	1.08 (0.80–1.45)	0.64 (0.08–5.10)	0.79 (0.54–1.16)	0.50 (0.06–4.41)	0.67 * (0.45–0.98)	0.52 (0.20–1.34)
Mother's/caregiver's education (Reference category Kindergarten and Primary)						
≥ High school	1.27 (0.98–1.65)	1.44 (0.54–3.81)	1.16 (0.82–1.62)	0.16 * (0.05–0.58)	0.73 * (0.54–0.98)	0.52 (0.20–1.34)
Household wealth index (Reference category poorest)						
Second	1.06 (0.82–1.38)	0.45 (0.14–1.46)	0.87 (0.60–1.28)	6.24 * (1.50–25.89)	0.77 (0.57–1.04)	0.72 (0.32–1.65)
Middle	1.29 (0.98–1.69)	0.27 * (0.08–0.91)	0.66 * (0.46–0.96)	2.81 (0.71–11.07)	0.67 * (0.49–0.92)	0.51 (0.19–1.41)
Fourth	1.69 ** (1.25–2.27)	0.56 (0.13–2.46)	0.54 ** (0.36–0.80)	0.83 (0.20–3.49)	0.54 ** (0.38–0.76)	0.18 (0.04–0.76)
Richest	1.18 (0.87–1.60)	3.11 (0.06–172.67)	0.48 ** (0.31–0.73)	3.68 (0.12–113.02)	0.45 ** (0.31–0.66)	N/A
Residence (Reference category urban)						
Rural	0.92 (0.76–1.10)	1.89 (0.74–4.79)	1.19 (0.92–1.52)	1.34 (0.37–4.83)	1.03 (0.82–1.29)	2.03 (0.76–5.42)
Geographical region (Reference category Bangkok and central)						
North	0.75 * (0.57–0.99)	1.89 (0.74–4.79)	0.73 (0.49–1.06)	0.42 (0.06–3.10)	1.10 (0.79–1.54)	2.74 (0.80–9.38)
Northeast	0.96 (0.76–1.21)	0.44 (0.10–1.92)	0.97 (0.69–1.34)	0.16 * (0.03–0.74)	1.11 (0.85–1.46)	0.51 (0.21–1.22)
South	1.00 (0.78–1.27)	1.63 (0.58–0.29)	0.43 ** (0.32–0.58)	0.21 (0.03–1.30)	0.89 (0.66–1.21)	0.43 (0.08–2.57)
Currently breastfeeding * (Reference category No)**						
Yes	0.16 ** (0.14–0.20)	0.02 ** (0.00–0.23)	0.08 (0.68–1.64)	4.68 (0.18–119.32)	1.04 (0.84–1.30)	0.46 (0.04–5.56)

* $p < 0.05$; ** $p < 0.01$; *** reported receiving any breastmilk in the past 24 h; [1] SwB = Sweet beverage consumption 6–23 months; [2] UFC = Unhealthy food consumption 6–23 months; [3] ZVF = Zero vegetable or fruit consumption 6–23 months; AOR = Adjusted odds ratio; 95%CI = 95% confidence interval.

4. Discussion

This analysis of a nationally representative survey of Thai children, mothers and other caregivers reveals mixed patterns of both appropriate and inappropriate child feeding practices with some indicators at concerning levels. Overall, the complementary feeding practices of other caregivers were less favourable than those of mothers. Although we cannot identify who the non-mother caregivers are due to the limitation of MICS6 data their age range points to older generations. A study in the UK compared dietary provision between parents and grandparents found that parents scored higher for promoting balance and variety [20]. The results of this study also highlight the need of expanding interventions promoting healthier complementary feeding practices that target non-mother caregivers, including grandparents.

In this study, there were several factors that were associated with WHO complementary feeding indicators. Child age was regularly associated with inappropriate child feeding practices. Fewer children in the 6–8 months range, especially those being cared for by others (non-mothers), met the MDD compared with older age groups. Also, this group of children had a significantly lower MAD and lower EFF, and higher ZVF. Importantly, achieving MDD was also associated with consuming at least one egg or flesh food and at least one fruit or vegetable. It could point to an inappropriate diet composition for younger children, which predominantly consists of starchy food with little or no eggs, flesh foods, fruits and vegetables. This type of diet is considered low in nutrient density and with poor mineral availability [21], increasing the risk of malnutrition. In Thailand, traditional weaning foods are rice and banana [22]. A study of modern feeding practices that analysed recipes shared on Thai online peer support groups revealed that animal-source foods were not commonly fed until children were aged eight months and older due to allergy concerns [23]. However, a recent literature review points to the role of delayed introduction of allergenic food in increasing the risk for allergy development [24]. Moreover, a recent systematic review revealed that introducing a variety of vegetables at the early stage of weaning promotes vegetable acceptance [25]. Therefore, programs to improve complementary feeding practices need to focus on encouraging parents to feed their children eggs, flesh foods, and vegetables from the beginning of weaning process.

While young Thai children still have challenges meeting dietary diversity needs, older children tend to consume more unhealthy foods. A systematic review of energy-dense, nutrient-poor foods highlighted the contribution that consumption of these types of food make to a substantial proportion of the energy intake among children younger than 23 months old in low- and middle-income countries [26]. While low dietary diversity is associated with child stunting in many settings [27,28], feeding young children with energy-dense foods is likely to increase the risk of overweight. Interestingly, early undernutrition followed by later overweight has promoted central adiposity and insulin resistance [29] which increase the risk of later developing non-communicable diseases [30].

It is often found that higher maternal education is associated with appropriate child feeding practices regarding to minimum dietary diversity, minimum meal frequencies, and minimum acceptable diet [15,16]. Our findings are in line with these previous studies in that children of mothers with higher educational level displayed better complementary feeding indicators (MDD and MAD). Also, a positive association was found between maternal education level and new feeding indicators, egg and/or flesh food consumption and vegetable and fruit consumption.

Findings from this study provide further support for associations between household income level and complementary feeding indicators on MDD and MAD. These results support the potential role of household incomes in providing children a wide variety of foods. For example, household income was associated with feeding diverse complementary foods in Nepal [31]. However, interestingly, feeding a variety of food to children might not be limited to healthy foods. In this study, children from middle-class families had higher sweet beverage consumption compared with the poorest and richest households. Geographical variation also plays a role in sweet beverage consumption.

Parents living in Bangkok and the central Thai region were more likely to provide their children with sugary beverages. There is limited evidence on associations between children's sweet beverages consumption and household income and geography. However, urbanisation and rising household incomes provide middle-class families easy access to a variety of foods, including sweet beverages in accessible settings. These findings show that parental feeding practices often do not comply with the current guidelines. Recently, organizations including UNICEF and the European Society for Paediatric Gastroenterology, Hepatology, and Nutrition (ESPGHAN) published complementary feeding guidelines which recommended no provision of sugar-sweetened beverages [32,33]. Early life exposure to sweet beverages increases risks of obesity, dental caries, liver fat and non-alcoholic fatty liver disease (NAFLD) in later life [34–36]. Thus, information or other interventions are needed to encourage middle-class parents to avoid giving sweet beverages to their children. Interestingly, our findings pointed out that children who received breastmilk had a lower sweet beverage consumption. These findings align with a previous Brazilian study showing that breastfed children were less likely to consume sugary foods later in life [37].

Strengths and Limitations

The strengths of this study include that it used generalizable national survey data. Therefore, sample groups represented the population in Thailand. Second, this study presents the association between determinants (individual and community), and complementary feeding indicators, of problems in child feeding in Thailand. Policymakers and relevant stakeholders can apply our findings to develop better complementary feeding policies and measures.

The limitations of this study are as follows; first, some populations who do not have a registered household number such as homeless people, illegal migrants, and people living in slum areas, were excluded from the survey because MICS6 selected households from the household registry from the Department of Provincial Administration. Second, MICS used a quantitative approach to collect data. Consequently, the perspectives on, and reasons for, complementary feeding practices, such as providing sweet beverages or unhealthy foods, were not explored. Further qualitative studies should explore these issues in more depth. Third, the study is cross-sectional, meaning we cannot ascertain causal relationships. Forth, all data on complementary feeding were collected using the 24-h recall which may not reflect children's diets over a longer period. There is a possibility that some respondents may provide answers that they think will be favoured by others. This presumes that they know what the preferred practice is. Last, the sample of children cared for by people who were not their mothers may be relatively small especially when they are separated into different sub-categories, such as child age, and caregiver age.

5. Conclusions

This study points out the sociodemographic and economic factors of inappropriate complementary feeding practices among Thai children aged 6–23 months according to the newly developed WHO IYCF indicators. Overall, our study demonstrated that the child's age and the characteristics of primary caregivers, including non-mother caregivers, were crucial determinants of complementary feeding practice. Although most children aged 6–8 months were introduced solid, semi-solid, or soft foods, more than half of them still have challenges meeting MDD and MAD. Moreover, egg/flesh food and fruit and vegetable consumption increased with child's age. Consequently, younger children, especially those cared for by non-mothers often miss essential dietary components. While older children are likely to achieve MDD, the higher percentage of them consumed SwB and UFC when compared to those 6–8 months of age. Moreover, dietary diversity was a positive attribute for children of middle-class mothers, this group also had high sweet beverage consumption, indicating diversity alone may not be only a positive factor for diets. Our findings can be applied to develop or update complementary feeding guidelines to educate mothers about appropriate feeding practices. Guidelines may need to specifically address non-mother caregivers. Furthermore, stakeholders can employ the results to tailor-made nutritional programs and interventions addressing inappropriate complementary feeding among different groups of children.

Supplementary Materials: The following supporting information can be downloaded at: https://www.mdpi.com/article/10.3390/nu14204370/s1, Table S1: Comparison indicators in 2008 and 2021 and their key revision; Table S2: Factors associated with complementary feeding indicators in children who lived with their mothers, the univariate logistic regression analysis; Table S3: Factors associated with complementary feeding indicators in children who lived with others, the univariate logistic regression analysis.

Author Contributions: A.S. and N.C. initially conceived the original study design, and C.B. and M.K. helped refine the study design. A.S. and N.C. analysed data. M.K., C.B. and H.S. reviewed and contributed feedback and critical comments on data analysis. A.S. and N.C. wrote the first draft of the manuscript. All authors have read and agreed to the published version of the manuscript.

Funding: The APC was funded by the Capacity Building on Health Policy and Systems Research program (HPSR Fellowship) under cooperation between the Bank for Agriculture and Agricultural Co-operatives (BAAC), National Health Security Office (NHSO), and International Health Policy Program Foundation (IHPF).

Institutional Review Board Statement: The NSO carried out the MICS study under the requirements of the Thai Statistical Act B.E. 2550 (2007). Before interviewing participants, NSO provided all respondents with information necessary to ascertain their willingness to participate in MICS6, such as confidentiality, the anonymity of their information, and their right to refuse to answer particular questions and to stop the interview. Afterward, participants gave their verbal consent (National Statistical Office, 2020). NSO permitted us to access and use MICS6 data for achieving the objective of this study. Ethical review and approval were therefore not required for this analysis. We were not provided with any identifying information on participants.

Informed Consent Statement: Not applicable.

Data Availability Statement: Not applicable.

Acknowledgments: The authors would like to thank the NSO for the value database. Also, we would like to acknowledge HPSR Fellowship program, International Health Policy Program Foundation for financial support.

Conflicts of Interest: The authors declare no conflict of interest.

References

1. Siddiqui, F.; Salam, R.A.; Lassi, Z.S.; Das, J.K. The Intertwined Relationship Between Malnutrition and Poverty. *Front. Public Health* **2020**, *8*, 453. [CrossRef]
2. World Health Organization. Malnutrition. 2020. Available online: https://www.who.int/news-room/questions-and-answers/item/malnutrition#:~[]:text=Malnutrition%20refers%20to%20deficiencies%2C%20excesses,of%20energy%20and%2For%20nutrients (accessed on 1 April 2022).
3. United Nations Children's Fund; World Health Organization; World Bank Group. *Levels and Trends in Child Malnutrition: Key Findings of the 2021 Edition of the Joint Child Malnutrition Estimates*; United Nations Children's Fund: New York, NY, USA, 2021.
4. Victora, C.G.; Christian, P.; Vidaletti, L.P.; Gatica-Domínguez, G.; Menon, P.; Black, E.R. Revisiting maternal and child undernutrition in low-income and middle-income countries: Variable progress towards an unfinished agenda. *Lancet* **2021**, *397*, 1388–1399. [CrossRef]
5. National Statistical Office of Thailand. Thailand Multiple Indicator Cluster Survey 2012, Exclusive Summary. 2012. Available online: https://mics-surveys-prod.s3.amazonaws.com/MICS4/East%20Asia%20and%20the%20Pacific/Thailand/2012-2013/Summary/Thailand%202012%20MICS%20Summary_English.pdf (accessed on 1 April 2022).
6. National Statistical Office of Thailand. *Thailand Multiple Indicator Cluster Survey 2015–2016, Final Report*; National Statistical Office of Thailand: Bangkok, Thailand, 2016.
7. National Statistical Office of Thailand. *Thailand Multiple Indicator Cluster Survey 2019, Survey Findings Report*; National Statistical Office of Thailand: Bangkok, Thailand, 2020.
8. Ministry of Public Health. *The Third National Food and Nutrition Survey in Thailand*; Nutrition division, Ministry of Public Health: Bangkok, Thailand, 1986. (In Thai)

9. Masuke, R.; Msuya, S.E.; Mahande, J.M.; Diarz, E.J.; Stray-Pedersen, B.; Jahanpour, O.; Mgongo, M. Effect of inappropriate complementary feeding practices on the nutritional status of children aged 6–24 months in urban Moshi, Northern Tanzania: Cohort study. *PLoS ONE* **2021**, *16*, e0250562. [CrossRef] [PubMed]
10. Zhou, H.; Wang, X.-L.; Ye, F.; Zeng, X.L.; Wang, Y. Relationship between child feeding practices and malnutrition in 7 remote and poor counties, PR China. *Asia Pac. J. Clin. Nutr.* **2012**, *21*, 234–240.
11. Laving, A.R.; Hussain, S.R.; Atieno, D.O. Overnutrition: Does Complementary Feeding Play a Role? *Ann. Nutr. Metab.* **2018**, *73*, 15–18. [CrossRef] [PubMed]
12. Schneider, B.C.; Gatica-Domínguez, G.; Assunção, M.C.F.; Matijasevich, A.; Barros, A.J.D.; Santos, I.S.; Silveira, M.F. Introduction to complementary feeding in the first year of life and risk of overweight at 24 months of age: Changes from 2004 to 2015 Pelotas (Brazil) Birth Cohorts. *Br. J. Nutr.* **2020**, *124*, 620–630. [CrossRef]
13. Wang, J.; Wu, Y.; Xiong, G.; Chao, T.; Jin, Q.; Liu, R.; Hao, L.; Wei, S.; Yang, N.; Yang, X. Introduction of complementary feeding before 4months of age increases the risk of childhood overweight or obesity: A meta-analysis of prospective cohort studies. *Nutr. Res.* **2016**, *36*, 759–770. [CrossRef]
14. Gewa, C.A.; Leslie, T.F. Distribution and determinants of young child feeding practices in the East African region: Demographic health survey data analysis from 2008–2011. *J. Health Popul. Nutr.* **2015**, *34*, 6. [CrossRef] [PubMed]
15. Kabir, I.; Khanam, M.; Agho, K.E.; Mihrshahi, S.; Dibley, M.J.; Roy, S.K. Determinants of inappropriate complementary feeding practices in infant and young children in Bangladesh: Secondary data analysis of Demographic Health Survey 2007. *Matern. Child Nutr.* **2011**, *8*, 11–27. [CrossRef] [PubMed]
16. Khanal, V.; Sauer, K.; Zhao, Y. Determinants of complementary feeding practices among Nepalese children aged 6–23 months: Findings from demographic and health survey 2011. *BMC Pediatr.* **2013**, *13*, 131. [CrossRef]
17. Mihrshahi, S.; Kabir, I.; Roy, S.K.; Agho, K.E.; Senarath, U.; Dibley, M.J.; for the South Asia Infant Feeding Research Network (SAIFRN). Determinants of Infant and Young Child Feeding Practices in Bangladesh: Secondary Data Analysis of Demographic and Health Survey 2004. *Food Nutr. Bull.* **2010**, *31*, 295–313. [CrossRef] [PubMed]
18. World Health Organization. *Indicators for Assessing Infant and Young Child Feeding Practices. Part 1: Definitions*; World Health Organization: Geneva, Switzerland, 2008.
19. World Health Organization. *Indicators for Assessing Infant and Young Child Feeding Practices. Definitions and Measurement Methods*; World Health Organization: Geneva, Switzerland, 2021.
20. Marr, C.; Breeze, P.; Caton, S.J. A comparison between parent and grandparent dietary provision, feeding styles and feeding practices when caring for preschool-aged children. *Appetite* **2021**, *168*, 105777. [CrossRef] [PubMed]
21. Michaelsen, K.F.; Hoppe, C.; Roos, N.; Kaestel, P.; Stougaard, M.; Lauritzen, L.; Mølgaard, C.; Girma, T.; Friis, H. Choice of Foods and Ingredients for Moderately Malnourished Children 6 Months to 5 Years of Age. *Food Nutr. Bull.* **2009**, *30*, S343–S404. [CrossRef]
22. Vong-Ek, P. *Patterns of Maternal Beliefs Affecting the Duration of Breastfeeding in the Central and the Northeast of Thailand (No. 152)*; Institute for Population and Social Research, Mahidol University: Nakhon Pathom, Thailand, 1990.
23. Supthanasup, A.; Banwell, C.; Kelly, M.; Yiengprugsawan, V. Recipe Components and Parents' Infant and Young Child Feeding Concerns: A Mixed-Methods Study of Recipe Posts Shared in Thai Facebook Groups for Parents. *Nutrients* **2021**, *13*, 1186. [CrossRef] [PubMed]
24. Larson, K.; McLaughlin, J.; Stonehouse, M.; Young, B.; Haglund, K. Introducing Allergenic Food into Infants' Diets: Systematic Review. *MCN Am. J. Matern./Child Nurs.* **2017**, *42*, 72–80. [CrossRef]
25. Barends, C.; Weenen, H.; Warren, J.; Hetherington, M.M.; de Graaf, C.; de Vries, J.H. A systematic review of practices to promote vegetable acceptance in the first three years of life. *Appetite* **2019**, *137*, 174–197. [CrossRef]
26. Pries, A.M.; Filteau, S.; Ferguson, E.L. Snack food and beverage consumption and young child nutrition in low- and middle-income countries: A systematic review. *Matern. Child Nutr.* **2019**, *15*, e12729. [CrossRef] [PubMed]
27. Darapheak, C.; Takano, T.; Kizuki, M.; Nakamura, K.; Seino, K. Consumption of animal source foods and dietary diversity reduce stunting in children in Cambodia. *Int. Arch. Med.* **2013**, *6*, 29. [CrossRef]
28. Rah, J.H.; Akhter, N.; Semba, R.D.; De Pee, S.; Bloem, M.W.; Campbell, A.A.; Moench-Pfanner, R.; Sun, K.; Badham, J.; Kraemer, K. Low dietary diversity is a predictor of child stunting in rural Bangladesh. *Eur. J. Clin. Nutr.* **2010**, *64*, 1393–1398. [CrossRef] [PubMed]
29. Wells, J.C.; Sawaya, A.L.; Wibaek, R.; Mwangome, M.; Poullas, M.S.; Yajnik, C.S.; Demaio, A. The double burden of malnutrition: Aetiological pathways and consequences for health. *Lancet* **2020**, *395*, 75–88. [CrossRef]
30. Leocádio, P.C.L.; Lopes, S.C.; Dias, R.P.; Alvarez-Leite, J.I.; Guerrant, R.L.; Malva, J.O.; Oriá, R.B. The Transition From Undernutrition to Overnutrition Under Adverse Environments and Poverty: The Risk for Chronic Diseases. *Front. Nutr.* **2021**, *8*, 676044. [CrossRef] [PubMed]
31. Subedi, N.; Paudel, S.; Rana, T.; Poudyal, A.K. Infant and young child feeding practices in Chepang communities. *J. Nepal Health Res. Counc.* **2012**, *10*, 141–146.
32. United Nations Children's Fund. *Improving Young Children's Diets during the Complementary Feeding Period. UNICEF Programming Guidance*; United Nations Children's Fund: New York, NY, USA, 2020.

33. Fewtrell, M.; Bronsky, J.; Campoy, C.; Domellöf, M.; Embleton, N.; Mis, N.F.; Hojsak, I.; Hulst, J.M.; Indrio, F.; Lapillonne, A.; et al. Complementary Feeding: A Position Paper by the European Society for Paediatric Gastroenterology, Hepatology, and Nutrition (ESPGHAN) Committee on Nutrition. *J. Pediatr. Gastroenterol. Nutr.* **2017**, *64*, 119–132. [CrossRef] [PubMed]
34. Bernabé, E.; Ballantyne, H.; Longbottom, C.; Pitts, N. Early Introduction of Sugar-Sweetened Beverages and Caries Trajectories from Age 12 to 48 Months. *J. Dent. Res.* **2020**, *99*, 898–906. [CrossRef] [PubMed]
35. Geurtsen, M.L.; Santos, S.; Gaillard, R.; Felix, J.F.; Jaddoe, V.W.V. Associations Between Intake of Sugar-Containing Beverages in Infancy with Liver Fat Accumulation at School Age. *Hepatology* **2020**, *73*, 560–570. [CrossRef]
36. Malik, V.S.; Pan, A.; Willett, W.C.; Hu, F.B. Sugar-sweetened beverages and weight gain in children and adults: A systematic review and meta-analysis. *Am. J. Clin. Nutr.* **2013**, *98*, 1084–1102. [CrossRef] [PubMed]
37. Bortolini, G.A.; Giugliani, E.R.J.; Gubert, M.B.; Santos, L.M.P. Breastfeeding is associated with children's dietary diversity in Brazil. *Ciênc. Saúde Coletiva* **2019**, *24*, 4345–4354. [CrossRef] [PubMed]

MDPI
St. Alban-Anlage 66
4052 Basel
Switzerland
www.mdpi.com

Nutrients Editorial Office
E-mail: nutrients@mdpi.com
www.mdpi.com/journal/nutrients

Disclaimer/Publisher's Note: The statements, opinions and data contained in all publications are solely those of the individual author(s) and contributor(s) and not of MDPI and/or the editor(s). MDPI and/or the editor(s) disclaim responsibility for any injury to people or property resulting from any ideas, methods, instructions or products referred to in the content.

www.ingramcontent.com/pod-product-compliance
Lightning Source LLC
LaVergne TN
LVHW071622170726
843515LV00009B/2454